THE OXFORD HANDBOOK OF

MEMBRANE COMPUTING

Editors

Gheorghe Păun (Bucharest, Romania)
Grzegorz Rozenberg (Leiden, The Netherlands)
Arto Salomaa (Turku, Finland)

Advisory Board

E. Csuhaj-Varjú (Budapest, Hungary)
R. Freund (Vienna, Austria)
M. Gheorghe (Sheffield, UK)
O.H. Ibarra (Santa Barbara, USA)
V. Manca (Verona, Italy)
G. Mauri (Milan, Italy)
M.J. Pérez-Jiménez (Seville, Spain)

THE OXFORD HANDBOOK OF

MEMBRANE COMPUTING

Edited by
GHEORGHE PĂUN,
GRZEGORZ ROZENBERG
and
ARTO SALOMAA

OXFORD
UNIVERSITY PRESS

Great Clarendon Street, Oxford OX2 6DP

Oxford University Press is a department of the University of Oxford.
It furthers the University's objective of excellence in research, scholarship,
and education by publishing worldwide in

Oxford New York

Auckland Cape Town Dar es Salaam Hong Kong Karachi
Kuala Lumpur Madrid Melbourne Mexico City Nairobi
New Delhi Shanghai Taipei Toronto

With offices in

Argentina Austria Brazil Chile Czech Republic France Greece
Guatemala Hungary Italy Japan Poland Portugal Singapore
South Korea Switzerland Thailand Turkey Ukraine Vietnam

Oxford is a registered trade mark of Oxford University Press
in the UK and in certain other countries

Published in the United States
by Oxford University Press Inc., New York

© Oxford University Press 2010

The moral rights of the authors have been asserted
Database right Oxford University Press (maker)

First published 2010

All rights reserved. No part of this publication may be reproduced,
stored in a retrieval system, or transmitted, in any form or by any means,
without the prior permission in writing of Oxford University Press,
or as expressly permitted by law, or under terms agreed with the appropriate
reprographics rights organization. Enquiries concerning reproduction
outside the scope of the above should be sent to the Rights Department,
Oxford University Press, at the address above

You must not circulate this book in any other binding or cover
and you must impose the same condition on any acquirer

British Library Cataloguing in Publication Data

Data available

Library of Congress Cataloging in Publication Data

Data available

Typeset by SPI Publisher Services, Pondicherry, India
Printed in Great Britain
on acid-free paper by
CPI Antony Rowe, Chippenham, Wiltshire

ISBN 978–0–19–955667–0

1 3 5 7 9 10 8 6 4 2

Preface

Membrane computing is a branch of natural computing aiming to abstract computing models, called membrane systems, from the structure and the functioning of the living cell as well as from the cooperation of cells in tissues, organs, and other populations of cells. This direction of research was initiated in 1998 and it has developed quickly into a vigorous scientific discipline. Already in 2003, Thompson Institute for Scientific Information, ISI, declared this area as a fast emerging research front in computer science. Indeed now, about ten years since the circulation of the paper that initiated membrane computing on the Internet (it was published in *The Journal of Computer and System Sciences* in 2000), the bibliography of this domain contains more than 1200 titles—among them about 25 PhD theses, about 30 collective volumes, and about 15 special issues of journals devoted to membrane computing. Many classes of P systems (as membrane systems are currently called) were introduced, with mathematical, computer science or biological motivation. A substantial broad body of theory has been created. Applications were reported especially in biology and bio-medicine, but also in unexpected directions, such as economics, approximate optimization, and computer graphics. A number of implementations are currently attempted.

Two specialized meetings are organized every year: the Workshop on Membrane Computing, started in 2000 in Curtea de Argeș, Romania (the tenth edition took place again in Curtea de Argeș, in August 2009), and the Brainstorming Week on Membrane Computing, organized in Seville, Spain, at the beginning of each February. Several conferences devoted to bio-inspired computing either have membrane computing in their scope or have satellite workshops devoted to membrane computing. Proceedings and/or special issues of journals are published as a result of these meetings. This huge amount of information accumulated so fast that now it cannot be covered in a monograph written by a single author or a small group of authors (as was still possible in 2002, when one of us wrote a monograph in membrane computing published by Springer-Verlag). Therefore we have decided to produce the *Handbook of Membrane Computing*.

The goal of the handbook is to provide an overview of the state-of-the-art (as of 2008) written by the leading researchers in this area. The handbook should serve as the main source of information on membrane computing on both the general/informal level and on the technical level. Through the references of individual chapters, it also provides a comprehensive guide to literature. Moreover, it is a source of interesting/challenging research problems and trends in membrane computing.

The handbook is organized in such a way that each chapter can be read independently of other chapters—each chapter has its own prerequisites and its own bibliography. To facilitate the reading of individual chapters, we have provided a chapter on basic notions, notations, and results of computability and formal language theory (Chapter 3)—these are used throughout the handbook. Also, especially for the newcomers to this research area, we provide an introduction to and an overview of membrane computing.

More specifically, the handbook consists of 20 chapters (one of them with sub-chapters), and three introductory/prerequisite chapters. Together they represent the main directions of research in membrane computing, covering theory, applications, and implementation/software issues. An additional bibliographical list, containing only books, collective volumes, PhD theses, and special issues of journals, closes the volume, providing valuable additional information about the literature.

The interested reader can find a complete bibliography of membrane computing (with a lot of additional information about this research area) at the web address http://ppage.psystems.eu (also accessible from http://psystems.disco.unimib.it, where the page was hosted for almost a decade).

Chapter Summaries

The goal of Chapter 1 (*An Introduction to and an Overview of Membrane Computing*) is to introduce basic concepts (together with the motivation behind them) and formalism of membrane computing as well as to provide an overview of main research directions. In this way, Chapter 1 is a sort of road map to both membrane computing and the organization of the present handbook. It is advisable to read this chapter before reading any other chapter (except for Chapters 2 and 3).

Chapter 2 (*Cell Biology for Membrane Computing*) provides some useful prerequisites for the comprehension of the cellular structures and the processes from which the models of membrane computing have been inspired. One starts by describing with some details the differences between the two distinct types of cells, prokaryotic and eukaryotic cells, with a particular emphasis on several aspects of the cellular membrane (components, transport and receptor proteins, electrochemical potential) and of intracellular processes (metabolism, gene expression, signal transduction). Then, one presents two supracellular levels of biological organization, that is, the structure of tissues in multicellular organisms and population-like behaviors within colonies of bacterial cells.

In turn, Chapter 3 (*Computability Elements for Membrane Computing*) introduces most of the computability elements (concepts, definitions, and results) necessary for the subsequent chapters of the book. One recalls basic facts from language and automata theory, as well as from computational complexity theory. The references given at the end of the chapter provide details and more specialized notions and results, less frequently used in membrane computing.

Chapter 4 (*Catalytic P Systems*) introduces and investigates one of the basic classes of P systems. One describes the computational power of the original model of P systems and of specific variants of these catalytic P systems, especially when only catalytic rules are used. There are presented several universality results, but also transition modes which only yield regular sets. Moreover, catalytic P systems are not only considered as generating devices for sets of (vectors of) natural numbers, yet also as acceptors and as devices to generate formal languages. Finally, variants of deterministic systems as well as more powerful variants of catalysts are considered.

Chapter 5 (*Communication P Systems*) considers various models of P systems that evolve only by communicating objects between regions—especially based on formal counterparts of the biological processes of *symport and antiport*. Their computational power is examined using different combinations of transition modes and halting conditions. The descriptional complexity of systems using symport and antiport rules is investigated and corresponding results with respect to the number of membranes, the number of used symbols, the size and the number of rules are presented.

The next chapter (*P Automata*) considers accepting P systems, in most cases with only communication rules; such devices combine characteristics of classical automata and distributed natural systems being in interaction with their environment. One presents the most important variants of P automata and their properties, with special emphasis put on their computational power.

Chapter 7 (*P Systems with String Objects*) presents P systems with objects described by strings, processed by rewriting rules (usual rules or with replication). The computing power of such systems is investigated.

The next chapter (*Splicing P Systems*) continues the study of P systems with string objects, giving an overview of P systems operating on strings by using the *splicing* operation (the abstraction of a biological process involving DNA molecules and enzymes). Various models of splicing (tissue) P systems are presented and the most important results concerning them are recalled. Universal splicing P systems are also described.

Chapter 9 (*Tissue and Population P Systems*) passes from cell-like P systems (with the structure of membranes described by a tree) to tissue-like P systems (with the membranes arranged in the nodes of an arbitrary graph). The main classes of tissue and population P systems working on symbol and string objects are presented and some generalizations and applications of them are briefly mentioned.

Chapter 10 (*Conformon P Systems*) deals with P systems whose objects are described both by their name and a numerical value (such a pair is called a *conformon*). Conformon P systems have been studied both as computational devices and as a platform to model biological processes. This chapter gives a complete overview of the results concerning their computational power and briefly discusses a way to model the dynamics of HIV infection in this framework.

Chapter 11 (*Active Membranes*) presents the class of P systems whose rules directly involve the membranes where they act, also making evolve the membrane structure during the computation. Such systems are called *with active membranes*. An important kind of rules are those which can enlarge the number of membranes, in special, rules for membrane division. Besides presenting various types of rules with active membranes, one briefly examines here the computing power of various classes of P systems using such rules. The use of the enhanced parallelism provided, e.g., by dividing membranes for solving computationally hard problems is considered in other chapters of the handbook.

An important/active research direction is addressed in Chapter 12 (*Complexity—Membrane Division, Membrane Creation*), namely computational complexity theory aspects for P systems. Complexity classes associated with different models of P systems are defined. Limitations on the efficiency of basic transition P systems lead to consider other features (rules for division and creation of membranes) able to produce an exponential amount of computational resources in polynomial time. A comprehensive overview of existing results is presented. In particular, frontiers between efficiency and non-efficiency in terms of (minimal) ingredients needed to (efficiently) solve computationally hard problems are analyzed. Many interesting characterizations of the $\mathbf{P} \neq \mathbf{NP}$ conjecture arise for these unconventional computing models.

Chapter 13 (*Spiking Neural P Systems*) describes a variant of P systems that models the way neurons interact to form a "computation". One recalls the basic definitions, several normal forms, and then one considers systems that operate in an "asynchronous" mode. Universality results, as well as closure and decidability properties are presented. Finally, one shows how spiking neural P systems can be used to solve computationally hard (**NP**-complete) decision problems, both numerical (such as SUBSET SUM) and non–numerical (SAT and 3-SAT).

In the previous chapters, only P systems with objects placed in membranes are considered. Chapter 14 (*P Systems with Objects on Membranes*) presents several classes of P systems where objects are placed either only on membranes or both on membranes and in the compartments delimited by them. Three main types of systems are considered: (i) those inspired by Cardelli's brane calculi (objects evolve together with membranes, by operations of the exo-, endo- phagocytosis type), (ii) systems where objects bound on membranes control the evolution of objects in the neighboring regions (they are called "Ruston models"), and (iii) systems where objects are placed both on membranes and in regions and they can change their place (called "Trento models").

A natural and fruitful link is explored in Chapter 15 (*Petri Nets and Membrane Computing*) between P systems and Petri nets. Petri nets are a well-established model of concurrent and distributed computation featuring a wealth of tools for the analysis and verification of their behavioral properties. Like P systems, Petri nets are in essence multiset rewriting systems. Using this key commonality, the authors describe a faithful translation from basic P systems to Petri nets, then one sketches the changes required to deal with promoters and inhibitors and with dynamically changing membrane structures. To capture the compartmentalization of P systems, the Petri net model is extended with localities and one shows how to adapt the notion of a Petri net process accordingly. This makes it possible to describe ongoing concurrent behavior of P systems in terms of causalities between the reactions that are taking place.

Chapter 16 (*Semantics of P Systems*) is devoted to the P systems semantics, particularly to their operational semantics. First one presents an abstract syntax of the P systems, and one defines a structural operational semantics of P systems by means of three sets of inference rules corresponding to maximally parallel rewriting, parallel communication, and parallel dissolving. Then one describes an implementation of the P systems based on the operational semantics, together with some results on its correctness. Using a representation given by register membranes, it is possible to describe the evolution involving rules with promoters and inhibitors. The evolution is expressed in terms of both dynamic and static allocation of resources to rules, and prove that these semantics are equivalent. Dynamic allocation allows translation of the maximally parallel application of membrane rules into sequential rewriting.

Chapter 17 (*Software for P Systems*) provides an overview of the many computer programs produced in the last years in order to simulate P systems. These software products have both a didactic purpose and are useful in the various applications of membrane computing. The chapter presents a brief state-of-the-art in this respect.

With Chapter 18 (*Probabilistic/Stochastic Models*) there starts a series of chapters devoted to applications of P systems. Here, probabilistic and stochastic P systems and modeling principles regarding main molecular interactions occurring in cellular biology are introduced. Methods supporting formal verification and analysis are presented and a number of case studies show the power and suitability of these models.

Chapter 19 (*Fundamentals of Metabolic P Systems*) deals with a special class of deterministic P systems, introduced for expressing biological metabolism. Their dynamics are computed by a special type of regulated multiset rewriting. In this chapter the basic results on metabolic P systems are indicated and a procedure is outlined for constructing in this framework models of biological processes from time series of observed dynamics.

The next chapter (*Metabolic P Dynamics*) continues and generalizes the investigations from the previous chapter. A general framework for analyzing dynamical problems is outlined, and some definitions of dynamical concepts relevant to biology are proposed in terms of quasi determinism and fluctuations. A dynamical perspective of P systems is elaborated in terms of metabolic P systems, by showing significant examples of oscillatory patterns. A discrete approach to anabolism and catabolism is outlined, which proves the versatility of metabolic P systems in accounting for energetic aspects of biochemical reactions.

An unexpected application of P systems is presented in Chapter 21 (*Membrane Algorithms*), in the form of a framework for devising distributed evolutionary algorithms involving various ingredients from membrane computing. What membrane algorithms inherit from P systems and what is newly brought to P systems by membrane algorithms are explained. Five applications of membrane algorithms, to the traveling salesman problem, to the job-shop scheduling problem, to the min storage problem, and two types of applications to the function optimization problem, are briefly described.

Chapter 22 (*Membrane Computing and Computer Science*) is organized around the main computer science application areas studied in the general framework of P systems: sorting, computer graphics, cryptography, and parallel architectures. The topics illustrate how the general theory of P systems and applied techniques meet together to give solutions to some important computer science problems.

Chapter 23 (*Other Developments*) presents a series of further research directions in membrane computing, such as P colonies (computing systems composed of as simple as possible agents, in the form of membranes with a limited contents and a small number of rules of a restricted type), the role of time in membrane computing (the power of P systems without a clock, of P systems with the result not depending on the time, and so on), membrane computing and self-assembly (for instance, computing by self-assembling certain structures from "bricks" of given types, in terms of tissue and population P systems of very restricted types), membrane computing and X-machines (one briefly describes how some classes of P systems can be transformed to communicating X-machines and vice-versa, the benefits gained out of these transformations, as well as the potential of combining the two models for solving complex problems or specifying dynamic systems, such as multi-agent systems), P systems with ingredients inspired from quantum computing, membrane computing and economics (many economic processes can be modeled and simulated by means of P systems; the possibilities and limits of this approach are discussed and a case study is presented), the connection between membrane computing and mobile ambients, further research topics.

The handbook ends with a selective bibliography of membrane computing, indicating the books, collective volumes, PhD theses, special issues of journals devoted to this research area.

*

The editors are deeply indebted to all contributors for their pleasant cooperation and their dedication to the timely production of the handbook.

We are also grateful to Oxford University Press, especially to Alison Jones, Dewi Jackson, Carol Bestley, Keith Mansfield, and Helen Eaton for efficient cooperation in producing the handbook.

Gheorghe Păun
Grzegorz Rozenberg
Arto Salomaa

September 2008

Contents

List of Contributors		xiv
1.	An Introduction to and an Overview of Membrane Computing Gh. Păun, G. Rozenberg	1
2.	Cell Biology for Membrane Computing D. Besozzi, I.I. Ardelean	28
3.	Computability Elements for Membrane Computing Gh. Păun, G. Rozenberg, A. Salomaa	58
4.	Catalytic P Systems R. Freund, O.H. Ibarra, A. Păun, P. Sosík, H.-C. Yen	83
5.	Communication P Systems R. Freund, A. Alhazov, Y. Rogozhin, S. Verlan	118
6.	P Automata E. Csuhaj-Varjú, M. Oswald, G. Vaszil	144
7.	P Systems with String Objects C. Ferretti, G. Mauri, C. Zandron	168
8.	Splicing P Systems S. Verlan, P. Frisco	198
9.	Tissue and Population P Systems F. Bernardini, M. Gheorghe	227
10.	Conformon P Systems P. Frisco	251
11.	Active Membranes Gh. Păun	282

12.	Complexity—Membrane Division, Membrane Creation M.J. Pérez-Jiménez, A. Riscos-Núñez, Á. Romero-Jiménez, D. Woods	302
13.	Spiking Neural P Systems O.H. Ibarra, A. Leporati, A. Păun, S. Woodworth	337
14.	P Systems with Objects on Membranes M. Cavaliere, S.N. Krishna, A. Păun, Gh. Păun	363
15.	Petri Nets and Membrane Computing J. Kleijn, M. Koutny	389
16.	Semantics of P Systems G. Ciobanu	413
17.	Software for P Systems D. Díaz-Pernil, C. Graciani, M.A. Gutiérrez-Naranjo, I. Pérez-Hurtado, M.J. Pérez-Jiménez	437
18.	Probabilistic/Stochastic Models P. Cazzaniga, M. Gheorghe, N. Krasnogor, G. Mauri, D. Pescini, F.J. Romero-Campero	455
19.	Fundamentals of Metabolic P Systems V. Manca	475
20.	Metabolic P Dynamics V. Manca	499
21.	Membrane Algorithms T.Y. Nishida, T. Shiotani, Y. Takahashi	529
22.	Membrane Computing and Computer Science R. Ceterchi, D. Sburlan	553
23.	Other Developments	584
23.1.	P Colonies A. Kelemenová	584
23.2.	Time in Membrane Computing M. Cavaliere, D. Sburlan	594

23.3.	Membrane Computing and Self-Assembly M. Gheorghe, N. Krasnogor	605
23.4.	Membrane Computing and X-Machines P. Kefalas, I. Stamatopoulou, M. Gheorghe, G. Eleftherakis	612
23.5.	Q-UREM P Systems A. Leporati	621
23.6.	Membrane Computing and Economics Gh. Păun, R.A. Păun	632
23.7.	Mobile Membranes and Mobile Ambients B. Aman, G. Ciobanu	645
23.8.	Other Topics Gh. Păun, G. Rozenberg	654

Selective Bibliography   664
Index of Notions   668

List of Contributors

Artiom Alhazov Academy of Sciences of Moldova, Institute of Mathematics and Computer Science, Academiei, MD-2028, Chişinău, Moldova, artiom@math.md

Bogdan Aman Romanian Academy, Institute of Computer Science, and "A.I.Cuza" University of Iaşi, Blvd. Carol 1 no.8, 700505 Iaşi, Romania, baman@iit.tuiasi.ro

Ioan I. Ardelean Institute of Biology of the Romanian Academy, Centre of Microbiology, Splaiul Independentei 296, PO Box 56–53, Bucharest, Romania, and Ovidius University, Bulevardul Mamaia 124, 900527 Constanta, Romania, Ioan.ardelean@ibiol.ro

Daniela Besozzi Università degli Studi di Milano, Dipartimento di Informatica e Comunicazione, Via Comelico 39, 20135 Milano, Italy, besozzi@dico.unimi.it

Francesco Bernadini Leiden Institute of Advanced Computer Science, Leiden University, Niels Bohrweg 1, 2333 CA Leiden, The Netherlands, bernardi@liacs.nl

Matteo Cavaliere The Microsoft Research–University of Trento, Centre for Computational and Systems Biology, Trento, Italy, cavaliere@cosbi.eu

Paolo Cazzaniga Dipartimento di Informatica, Sistemistica e Comunicazione, Università degli Studi di Milano–Bicocca, Viale Sarca 336, 20126 Milano, Italy, cazzaniga@disco.unimib.it

Rodica Ceterchi Faculty of Mathematics and Computer Science, University of Bucharest, Academiei 14, Bucharest, Romania, rceterchi@gmail.com

Gabriel Ciobanu Romanian Academy, Institute of Computer Science, and "A.I.Cuza" University of Iaşi, Blvd. Carol 1 no.8, 700505 Iaşi, Romania, Gabriel@info.uaic.ro; Gabriel@iit.tuiasi.ro

Erzsébet Csuhaj-Varjú Computer and Automation Research Institute, Hungarian Academy of Sciences, Kende u. 13–17, H-111 Budapest, Hungary, csuhaj@sztaki.hu

Daniel Díaz-Pernil Research Group on Natural Computing, Department of Computer Science and Artificial Intelligence, University of Sevilla, Avda. Reina Mercedes s/n, 41012 Sevilla, Spain, sbdani@us.es

George Eleftherakis Department of Computer Science, CITY College, 13 Tsimiski str, 54624 Thessalonika, Greece, eleftherakis@city.academic.gr

Claudio Ferretti Dipartimento di Informatica, Sistemistica e Comunicazione, Università degli Studi di Milano–Bicocca, Viale Sarca 336, I-20126 Milano, Italy, ferretti@disco.unimib.it

Rudolf Freund Institute of Computer Languages, Vienna University of Technology, Favoritenstrasse 9–11, A-1040 Vienna, Austria, rudi@emcc.at

Pierluigi Frisco School of Mathematical and Computer Sciences, Heriot-Watt University, Edinburgh, EH14 4AS, UK, pier@macs.hw.ac.uk

Marian Gheorghe Department of Computer Science, The University of Sheffield, Regent Court, Portobello Street, Sheffield S1 4DP, UK, m.gheorghe@dcs.shef.ac.uk

Carmen Graciani Research Group on Natural Computing, Department of Computer Science and Artificial Intelligence, University of Sevilla, Avda. Reina Mercedes s/n, 41012 Sevilla, Spain, cgdiaz@us.es

Miguel A. Gutiérrez-Naranjo Research Group on Natural Computing, Department of Computer Science and Artificial Intelligence, University of Sevilla, Avda. Reina Mercedes s/n, 41012 Sevilla, Spain, magutier@us.es

Oscar H. Ibarra Department of Computer Science, University of California, Santa Barbara, CA 93106, USA, ibarra@cs.uscb.edu

Petros Kefalas Department of Computer Science, CITY College, 13 Tsimiski str, 54624 Thessalonika, Greece, Kefalas@city.academic.gr

Alica Kelemenová Institute of Computer Science, Silesian University, Bezručovo nám 13, 746 01 Opava, Czech Republic, and Department of Computer Science, Catholic University, Nám. A. Hlinku 56, 034 01 Ružomberok, Slovakia, alica.kelemenova@fpf.slu.cz

Jetty Kleijn Leiden Institute of Advanced Computer Science, Leiden University, Niels Bohrweg 1, 2333 CA Leiden, The Netherlands, kleijn@iliacs.nl

Maciej Koutny School of Computing Science, Newcastle University, Newcastle-upon-Tyne, NE1 7RU, UK, Maciej.koutny@ncl.ac.uk

Natalio Krasnogor School of Nottingham Computer Science, The University of Nottingham, Jubilee Campus, Nottingham, NG8 1BB, UK, natalio.krasnogor@nottingham.ac.uk

Shankara Narayanan Krishna Department of Computer Science and Engineering, Indian Institute of Technology, Bombay, Poway, Mumbai 400 076, India, krishnas@cse.iitb.ac.in

Alberto Leporati Dipartimento di Informatica, Sistemistica e Comunicazione, Università degli Studi di Milano–Bicocca, Viale Sarca 336, 20126 Milano, Italy, leporati@disco.unimib.it

Vincenzo Manca Department of Computer Science, Verona University, Strada Le Grazie, 15–37134 Verona, Italy, vincenzo.manca@univr.it

Giancarlo Mauri Dipartimento di Informatica, Sistemistica e Comunicazione, Università degli Studi di Milano–Bicocca, Viale Sarca 336, I-20126 Milano, Italy, mauri@disco.unimib.it

Taishin Y. Nishida Faculty of Engineering, Toyama Prefectural University, Kosugi-machi, Toyama 939-0398, Japan, nishida@pu-toyama.ac.jp

Marion Oswald Faculty of Informatics, Vienna University of Technology, Favoritenstr. 9, A-1040 Vienna, Austria, marion@emcc.at

Andrei Păun National Institute of Research and Development for Biological Sciences, Bioinformatics Department, Splaiul Independentei 296, PO Box 56–53, Bucharest, Romania, and Universidad Politecnica de Madrid, Departamento de Inteligencia Artificial, Campus de Montegancedo S/N, Boadillo del Monte, 28660 Madrid, Spain, andreipaun@gmail.com

Gheorghe Păun Institute of Mathematics of the Romanian Academy, PO Box 1-764, 014700 Bucharest, Romania, and Research Group on Natural Computing, Department of Computer Science and Artificial Intelligence, University of Sevilla, Avda. Reina Mercedes s/n 41012, Sevilla, Spain, George.paun@imar.ro, gpaun@us.es

Radu A. Păun Department of Economics, University of Maryland at College Park, 3105 Tydings Hall, College Park, MD 20742, USA, and Competition Council of Romania Research Department, 1, Piata Presei Libere, Building D, 013701 Bucharest, Romania, paun@econ.umd.edu, radu.paun@consiliulconcurentei.ro

Ignacio Pérez-Hurtado Research Group on Natural Computing, Department of Computer Science and Artificial Intelligence, University of Sevilla, Avda. Reina Mercedes s/n, 41012 Sevilla, Spain, perezh@us.es

Mario J. Pérez-Jiménez Research Group on Natural Computing, Department of Computer Science and Artificial Intelligence, University of Sevilla, Avda. Reina Mercedes s/n, 41012 Sevilla, Spain, marper@us.es

Dario Pescini Dipartimento di Informatica, Sistemistica e Comunicazione, Università degli Studi di Milano–Bicocca, Viale Sarca 336/14, 20126 Milano, Italy, pescini@disco.unimib.it

Agustín Riscos-Núñez Research Group on Natural Computing, Department of Computer Science and Artificial Intelligence, University of Sevilla, Avda, Reina Mercedes s/n, 41012 Sevilla, Spain, ariscosn@us.es

Yurii Rogozhin Academy of Sciences of Moldova, Institute of Mathematics and Computer Science, Academiei, MD-2028, Chişinău, Moldova, and Rovira i Virgili

University, Research Group on Mathematical Linguistics, Pl. Imperial Tàrraco 1, 43005 Tarragona, Spain, rogozhin@math.md

Francisco J. Romero-Campero School of Nottingham Computer Science, The University of Nottingham, Jubilee Campus, Nottingham, NG8 1BB, UK, fxs@cs.nott.ac.uk

Álvaro Romero-Jiménez Research Group on Natural Computing, Department of Computer Science and Artificial Intelligence, University of Sevilla, Avda. Reina Mercedes s/n, 41012 Sevilla, Spain, alvaro.romero@cs.us.es

Grzegorz Rozenberg Leiden Institute of Advanced Computer Science, Leiden University, Niels Bohrweg 1, 2333 CA Leiden, The Netherlands, and Department of Computer Science, University of Colorado at Boulder, Boulder, CO 80309, USA, rozenber@liacs.nl

Arto Salomaa Turku Centre for Computer Science: TUCS, Lemminkäisenkatu 14A, 20520 Turku, Finland, asalomaa@utu.fi

Dragoş Sburlan Faculty of Mathematics and Informatics, Ovidius University, 124 Mamaia Bd, Constantza, Romania, dsburlan@univ-ovidius.ro

Tatsuya Shiotani Faculty of Engineering, Toyama Prefectural University, Kosugi-machi, Toyama 939-0398, Japan

Petr Sosik Universidad Politecnica de Madrid, Departamento de Inteligencia Artificial, Campus de Montegancedo S/N, Boadillo del Monte, 28660 Madrid, Spain and, Institute of Computer Science, Silesian University, 74601 Opava, Czech Republic, petr.sosik@fpf.slu.cz

Ioanna Stamatopoulou Department of Computer Science, CITY College, 13 Tsimiski str, 54624 Thessalonika, Greece, istamatopoulou@seerc.org

Yoshiyuki Takahashi Faculty of Engineering, Toyama Prefectural University, Kosugi-machi, Toyama 939-0398, Japan

György Vaszil Computer and Automation Research Institute, Hungarian Academy of Sciences, Kende u. 13–17, H-111 Budapest, Hungary, vaszil@sztaki.hu

Sergey Verlan Academy of Sciences of Moldova, Institute of Mathematics and Computer Science, Academiei, MD-2028, Chişinău, Moldova, and Université Paris 12, LACL, Department Informatique, 61 av. Général de Gaulle, 94010 Creteil, France, verlan@univ-paris12.fr

Damien Woods Research Group on Natural Computing, Department of Computer Science and Artificial Intelligence, University of Sevilla, Avda. Reina Mercedes s/n, 41012 Sevilla, Spain, dwoods@us.es

Sara Woodworth Department of Computer Science, University of California, Santa Barbara, CA 93106, USA, swood@cs.uscb.edu

Hsu-Chun Yen Dept of Electrical Engineering, National Taiwan University, Taipei, Taiwan 106, ROC, yen@cc.ee.ntu.edu.tw

Claudio Zandron Dipartimento di Informatica, Sistemistica e Comunicazione, Università degli Studi di Milano–Bicocca, Viale Sarca 336, I-20126 Milano, Italy, zandron@disco.unimib.it

CHAPTER 1

AN INTRODUCTION TO AND AN OVERVIEW OF MEMBRANE COMPUTING

GHEORGHE PĂUN
GRZEGORZ ROZENBERG

1.1 INTRODUCTION

THE research area of membrane computing originated as an attempt to formulate a model of computation motivated by the structure and functioning of a living cell—more specifically, by the role of membranes in compartmentalization of living cells into "protected reactors". Therefore, initial models were based on a cell-like (hence hierarchical) arrangement of *membranes* delimiting compartments where *multisets* of chemicals (called *objects*) evolve according to given *evolution rules*. These rules were either modeling chemical reactions and had the form of (multiset) rewriting rules, or they were inspired by other biological processes, such as passing objects through membranes (either in symport or antiport fashion), and had the form of communication rules. These initial models were then modified by incorporating various additional features motivated by considerations rooted in biology, mathematics, or computer science. The next important step in the development of research in membrane computing was to also consider other (nonhierarchical) arrangements of membranes. While hierarchical (cell-like) arrangements of membranes correspond to trees, tissue-like membrane systems consider arbitrary graphs as underlying structures, with membranes placed in the nodes while edges correspond to communication channels. The latest developments in this line of research are neural-like membrane systems which are motivated by spiking neural networks.

All classes of computing devices considered in membrane computing are now generically called *P systems*. Most of the initial research problems and results were related to the computing power of various types of P systems, and most classes of P systems turned out to be equivalent to Turing machines, hence computationally complete/universal. Then, research concerning complexity issues became popular, yielding results relating classic complexity classes to the computational efficiency of various classes of P systems. Research on applications of P systems started relatively late, but it is now a very active research direction. Application areas include biology and bio-medicine as well as optimization, economics, and computer science. Software implementation packages were developed and (attempts of) biochemical implementations were reported. An active and important line of research consists of establishing bridges with other well-known models of computations motivated by computer science, mathematics or biology. Interesting relationships were established with, among others, Petri nets, process algebra, dynamical systems, X-machines, brane calculi, and fuzzy sets based models.

The chapters of this handbook present central notions and results from all the main directions of research. The goal of this chapter is to introduce basic notions and formalisms of membrane computing. We avoid unnecessary formalizing but rather stress the intuition for and the motivation behind the development of basic concepts. This chapter also provides an overview of main research directions and in this way becomes a sort of road map to membrane computing and to the structure of the handbook.

Since the handbook covers in detail all the topics discussed in this chapter and the chapters of the handbook provide extensive references, we do not give references within this chapter. However, at the end of this chapter we provide a short list of basic readings and the address of the P systems website, which together cover both the origins of and the current research in membrane computing.

1.2 Strings and Multisets

The main data structure used in membrane computing is a multiset (sometimes also called a *bag*), which is essentially a set with multiplicities associated with its elements. In this section we discuss basic notions related to multisets and some of their relationships with strings.

Strings and languages are discussed in detail in Chapter 3; here we recall only some basic notions. Given an alphabet V (a finite and nonempty set of abstract symbols), a string over V is obtained by juxtaposing symbols of V. For instance, for the alphabet $V = \{a, b, c\}$, a, $abbac$, $cccc$ are strings over V (the last string is usually written as c^4). The empty string is denoted by λ. The length of a string x, denoted by $|x|$, is the number of occurrences of symbols it contains. Thus, $|a| = 1$, $|abbac| = 5$, and $|c^4| = 4$, while $|\lambda| = 0$. The number of occurrences of a symbol $a \in V$ in a string x over V is denoted by $|x|_a$. If an alphabet V is ordered, say $V = (a_1, a_2, \ldots, a_n)$, and x is a string over V, then $(|x|_{a_1}, |x|_{a_2}, \ldots, |x|_{a_n})$ is the *Parikh vector* of x, denoted by $\Psi_V(x)$. For example, for $V = (a, b, c)$, $\Psi_V(abbac) = (2, 2, 1)$ and $\Psi_V(c^4) = (0, 0, 4)$. The mapping Ψ_V is extended in the obvious way to sets of strings (languages), yielding then sets of Parikh vectors.

For a set U, a *multiset over* U is a mapping $M : U \longrightarrow \mathbf{N}$, where $\mathbf{N}$ is the set of nonnegative integers. For $a \in U$, $M(a)$ is the *multiplicity of a in M*, and the *support of M*, denoted by $supp(M)$, is the set $\{a \in U \mid M(a) \neq 0\}$. If $supp(M) = \emptyset$, then M is the *empty* multiset, denoted by $\emptyset$. In this setup the multiplicity of each object is finite; clearly, to allow also for infinite multiplicities one modifies the definition of M to $M : U \longrightarrow \mathbf{N} \cup \{\infty\}$.

If the set U is finite, $U = \{a_1, \ldots, a_n\}$, then the multiset M can be explicitly given in the form $\{(a_1, M(a_1)), \ldots, (a_n, M(a_n))\}$, thus specifying for each element of U its multiplicity in M. In membrane computing, the usual way to represent a multiset $M = \{(a_1, M(a_1)), \ldots, (a_n, M(a_n))\}$ over a finite set $U = \{a_1, \ldots, a_n\}$ is by using strings in the obvious way: $w = a_1^{M(a_1)} a_2^{M(a_2)} \ldots a_n^{M(a_n)}$ and all permutations of w represent M; the empty multiset is represented by λ. Hence, if U is ordered, $U = (a_1, \ldots, a_n)$, then the Parikh vector of w, $\Psi_U(w)$, is exactly the vector $(M(a_1), \ldots, M(a_n))$ of multiplicities. For instance, if $U = (a, b, c)$, then the multiset $\{(a, 2), (b, 1), (c, 3)\}$ is represented by the string $a^2 b c^3$ (and by all permutations of it). This compact representation is so frequent in membrane computing, that we customarily say "the multiset w" instead of "the multiset represented by string w".

It is worthwhile pointing out here the differences (and the similarities) between three data structures used in membrane computing: sets (no multiplicity, no order), multisets (multiplicity but no order), and strings (multiplicity and linear order).

We recall now a number of basic relations between and operations on multisets (all multisets are over a common set U).

Let $M_1, M_2 : U \longrightarrow \mathbf{N}$ be multisets. We say that M_1 is *included* in M_2, and write $M_1 \subseteq M_2$, if $M_1(a) \leq M_2(a)$, for all $a \in U$. The inclusion is *strict*, and we write $M_1 \subset M_2$, if $M_1 \subseteq M_2$ and $M_1 \neq M_2$. We will use the same relations $\subseteq, \subset$ also for strings representing multisets, e.g. we write $w \subseteq z$ when $M_w \subseteq M_z$, for multisets M_w and M_z represented by w and z, respectively.

The *union* of M_1 and M_2 is the multiset $M_1 \cup M_2 : U \longrightarrow \mathbf{N}$ defined by $(M_1 \cup M_2)(a) = M_1(a) + M_2(a)$, for all $a \in U$. In terms of strings, this corresponds to concatenation: if w_1, w_2 represent the multisets M_1, M_2, respectively, then $w_1 w_2$ (and each permutation of it) represents the multiset $M_1 \cup M_2$. The *intersection* of M_1 and M_2 is the multiset $M_1 \cap M_2 : U \longrightarrow \mathbf{N}$ defined by $(M_1 \cap M_2)(a) = \min\{M_1(a), M_2(a)\}$, for all $a \in U$. The *difference* $M_1 - M_2$ is defined here only when M_2 is included in M_1 and it is the multiset $M_1 - M_2 : U \longrightarrow \mathbf{N}$ given by $(M_1 - M_2)(a) = M_1(a) - M_2(a)$, for all $a \in U$. If $n \in \mathbf{N}$ and $M : U \longrightarrow \mathbf{N}$ is a multiset, then the *product* of M by n (the *amplification* by n) is the multiset $n \otimes M : U \longrightarrow \mathbf{N}$ defined by $(n \otimes M)(a) = n \cdot M(a)$, for all $a \in U$. (When M is represented by a string w, then we also write $n \otimes w$ to denote a string representation of $n \otimes M$.)

The *cardinality* (sometimes called the *weight*) of a multiset $M : U \longrightarrow \mathbf{N}$ is defined by $card(M) = \sum_{a \in U} M(a)$; if M is represented by a string w, then the cardinality of M equals $|w|$.

1.3 Membrane Structure

The central notion of membrane computing is the notion of *membrane* understood as a three-dimensional vesicle, which geometrically can be considered as a ball in the Euclidean space. Thus a membrane delimits a space, separating in this way "inside" from "outside". The (inside) space delimited by a membrane serves as a "protected reactor"—a space where specific reactions take place, using molecules in this space.

Inspired by the structure of the living cell, in our basic model we will consider hierarchical arrangements of finite number of membranes, with membranes nested within other membranes (hence the balls in the space included within other balls). In this chapter, unless indicated otherwise, we will deal with such cell-like membrane structures (referred to simply as *membrane structures*).

We will introduce now, rather informally, a number of notions which are useful in discussing membrane structures.

Membrane m' is *included* in membrane m if the space delimited by m' is strictly included in the space delimited by m. If, moreover, there is no membrane m'' such that m'' is included in m and m' is included in m'', then m' and m are *neighbors*, with m called the *upper neighbor of m'*, and m' called a *lower neighbor of m*. A membrane m with no lower neighbors is called *elementary*, and a membrane m with no upper neighbor is called a *skin membrane*. Motivated by the structure of a living cell, we require that in the cell-like membrane structures we consider in this chapter there is always a unique skin membrane. Then the space "outside" the skin membrane is called the *environment*.

For the sake of reference we use labels to identify membranes (where labels come from a given alphabet of labels). In the models we consider in this chapter the labeling of membranes is injective, i.e. different membranes get different labels. The number of membranes is called the *degree* of a membrane structure.

A *region* is either a space delimited by an elementary membrane or a space delimited by a non-elementary membrane and all of its lower neighbors. Thus the environment is not a region.

Since each region is uniquely determined by its (upper) delimiting membrane, we carry over to regions the terminology introduced for membranes. Thus we have neighboring regions, and for two neighboring regions we can talk about the upper and the lower neighbor. We use membrane labels to label regions in an obvious way: regions inherit labels of their (upper) delimiting membranes. Also, by convention, the upper neighbor of the skin region is the environment.

Finally, we say that different membranes m and m' are *sibling* membranes if there is a membrane m'' which is the upper neighbor for both m and m'; we also say that regions associated with m and m' are siblings.

Graphically, we will represent membrane structures in the two-dimensional plane using Euler-Venn diagrams representing the inclusion relationships between the regions delimited by membranes. Figure 1.1, a sort of logo of membrane computing, is an example of such a representation. It also illustrates some of the basic terminology introduced above. Note, e.g. that region 8 is elementary and it is a lower neighbor of region 6, which is the upper neighbor for both region 8 and region 9, which therefore are siblings. Also, regions 2, 3, and 4 are siblings, and so are regions 5, 6, and 7.

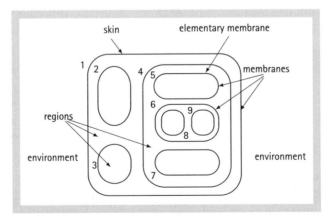

Fig. 1.1 A membrane structure.

We will need a linear representation (specification) of membrane structures, and will use a standard parenthesis notation to this aim, which we will briefly recall now.

For a given membrane structure S, let $l_1, \ldots, l_n$ be the labels of membranes in S, where l_1 is the label of the skin membrane. The membrane labeled by l_i, $1 \leq i \leq n$, is represented by an ordered pair of parentheses $[\ ]_{l_i}$, and a parentheses expression for μ is constructed as follows:

1. $[\ ]_{l_1}$ is a *correct form* for S.
2. If $z = z_1 [\ ]_{l_i} z_2$ is a correct form for S, for some $i \in \{1, 2, \ldots, n\}$, and membranes labeled by $k_1, \ldots, k_p \in \{2, \ldots, n\}$ are all lower neighbors of the membrane labeled by l_i, then for each string $x = [\ ]_{k_{j_1}} [\ ]_{k_{j_2}} \cdots [\ ]_{k_{j_p}}$, where $j_1, \ldots, j_p$ is a permutation of $1, 2, \ldots, p$, $z_1 [\ x\ ]_{l_i} z_2$ is also a correct form for S.
3. A *parentheses expression* for μ is a correct form for S which includes the right hand parenthesis $]_{l_i}$ for each $i \in \{1, 2, \ldots, n\}$.

We say that a parentheses expression for S is *over* the alphabet of labels $L = \{l_1, \ldots, l_n\}$.

Clearly, a parentheses expression μ for a membrane structure S contains all the information about the inclusion relationships between the regions delimited by membranes of M, and hence all the neighborhood information about M (regions, upper and lower neighbors, etc.). One can produce the Euler-Venn diagram of M from μ. As a matter of fact, parentheses expressions for membrane structures are a standard formal way to specify membrane structures in membrane computing literature.

Thus, e.g. for the membrane structure S from Fig. 1.1,

$$[\ [\ ]_2 [\ [\ ]_7 [\ ]_5 [\ [\ ]_8 [\ ]_9]_6]_4 [\ ]_3]_1$$

is a parentheses expression for S obtained through the following sequence of correct forms for S:

$$[\ ]_1,$$
$$[\ [\ ]_2 [\ ]_4 [\ ]_3]_1,$$
$$[\ [\ ]_2 [\ [\ ]_7 [\ ]_5 [\ ]_6]_4 [\ ]_3]_1,$$
$$[\ [\ ]_2 [\ [\ ]_7 [\ ]_5 [\ [\ ]_8 [\ ]_9]_6]_4 [\ ]_3]_1.$$

Note that, in general, the same parentheses expression can be obtained by a different sequence of correct forms (whenever a membrane structure includes two sibling membranes that are not elementary).

More importantly, in general there are (many) different parentheses expressions for a given membrane structure. This is the case whenever a membrane structure contains sibling membranes (because then in step (2) of the above procedure one can choose among various permutations of $1, 2, \ldots, p$).

Thus, for the membrane structure S from Fig. 1.1, also

$$[\,[\,]_2[\,[\,]_5[\,]_7[\,[\,]_8[\,]_9]_6]_4[\,]_3]_1,$$

as well as

$$[\,[\,]_3[\,]_2[\,[\,]_5[\,]_7[\,[\,]_8[\,]_9]_6]_4]_1,$$

among others, are parentheses expressions for S.

A mathematically natural way to formalize membrane structures is to use rooted unordered node-labeled trees. For example, the membrane structure from Fig. 1.1 corresponds to the tree given in Fig. 1.2. One can then also use standard ways of associating parentheses expressions with trees to get parentheses expressions for membrane structures.

We have decided to present membrane structures through Euler-Venn diagrams because this corresponds directly to the cell-like intuition of compartments, and some crucial notions such as regions come up more naturally in this way. Also, the terminology that we use (neighbors, etc.) is closer to this underlying intuition of spaces delimited by membranes—the standard terminology of mother and children nodes is more suitable when discussing intuition/issues such as inheritance of properties, hierarchical control, or generating/producing processes.

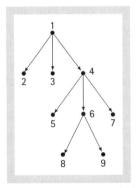

Fig. 1.2 A tree describing the membrane structure from Fig. 1.1.

1.4 Multiset Rewriting Rules

Multiset rewriting rules are an abstraction of biochemical reactions taking place in a cell, and are written in the form $u \to v$, where u and v are strings (representing multisets of objects from a given set O). In order to apply such a rule in a region (compartment) of a membrane structure, it is necessary that the multiset w of objects present in that compartment includes u, i.e. $u \subseteq w$. For instance, the simple use of the rule $a^2bc^3 \to bd^4$ transforms two copies of a, one of b, and three of c into four copies of d, while b is reproduced (it plays here the role of a catalyst). This rule can be applied to the multiset $a^3bc^6d^2$, but not to abc^6d^2, because $a^2bc^3 \subseteq abc^6d^2$ does not hold.

For a rule $u \to v$ we say that $|u|$ (the weight of the multiset on the left hand side) is the *weight* of the rule; a rule with the weight at least two is said to be *cooperative*. The rule from the previous example is of this type. A rule of weight 1 is called *non-cooperative*. *Catalytic rules* are cooperative rules of the form $ca \to cv$, where c, a are objects with $c \neq a$, v a multiset, and c does not appear in v (c only assists a in evolving into v, but c itself is not undergoing any transformation).

In P systems, multiset rewriting rules are located in regions, and they act on objects present in the region where the rules reside.

The communication between regions is provided by passing/exchanging objects between regions through membranes that separate them. Such a communication is a basic mechanism of cooperation between regions, and it is incorporated in rewriting rules by *target indications*. The rules are of the form $u \to v$, where u is a string over O, and v is a string over $O \times \{here, out, in\}$. Thus, each element of v is of the form (a, tar), where a is an object from O and tar is either *here* or *out* or *in*. The meaning of these target indications is as follows.

If $tar = here$, then a remains in the region where the rule resides.

If $tar = out$, then a is sent to the upper neighbor of the region where the rule resides (recall that the upper neighbor of the skin region is the environment).

If $tar = in$, then a is sent non-deterministically to one of the lower neighbors. Note that elementary regions do not have lower neighbors, and so a rule including $tar = in$ residing in an elementary region cannot be applied.

Although in the basic/generic version of a P system the evolution rules are multiset rewriting rules of the form discussed above, some other versions of P systems also include other sorts of target indications. Hence, e.g. one may have $tar = in_j$, where j is a membrane label. In this case a is sent to the lower neighbor labeled by j – recall that regions inherit labels of (upper) membranes delimiting them. Clearly, if there is no lower neighbor labeled by j, then a rule using this target indication is not applicable. Another example is the use of target indication of the form $tar = go$. In this case a is sent non-deterministically to one of the neighbors of the region where the rule resides.

In order to make the notation for rules more concise, one may group together objects with the same target indication, and also the target indication *here* may be omitted. For instance, the rule $a^2bc^3 \to (b, here)(a, here)(c, here)(a, here)(d, out)(a, out)(c, in)(a, in)$ may be written in a more compact way as $a^2bc^3 \to ba^2c(da, out)(ca, in)$.

The application of a single rule proceeds as follows. First of all, we assume the existence of a global clock that measures time by marking time units. When a rule $u \to v$ residing in region h is used, it acts only on objects placed in h (i.e. objects specified in u are available in h in required multiplicities). Then the following happens within one time unit.

1. Objects present in u in the specified multiplicities are removed from h.
2. Objects present in v in the specified multiplicities are produced.
3. The produced objects are placed either in h or in neighbors of h according to the target indications in v.

Hence, e.g. if h is not elementary, the rule $a^2bc^3 \to ba^2c(da, out)(ca, in)$ is applicable if h contains at least two objects a, one object b, and three objects c. The application of this rule:

1. removes from h two objects a, one object b, and three objects c;
2. produces one object b, four objects a, two objects c, and one object d;
3. the produced object b remains in h, two produced objects a remain in h, one produced object c remains in h, the produced object d and one produced object a are sent to the upper neighbor, one produced object a is sent to a (non-deterministically chosen) lower neighbor, and one produced object c is sent to a (non-deterministically chosen) lower neighbor.

Note that if h is the skin region, then the produced object d and one produced object a are sent to the environment. Since there are no evolution rules placed in the environment, the skin region never receives objects from the environment and so objects that leave the membrane structure are "lost" (can never return).

1.5 MAXIMAL PARALLELISM

In the previous section we discussed an application of a single rule in order to get a basic setup for considering an application of the set of rules residing in a region. Clearly, there are various possible regimes of applying the available sets of rules to multisets of objects present in regions at a given time unit. The originally proposed (and still mostly considered) mode of application is *maximal parallelism*—it was motivated by the parallel nature of biochemical reactions taking place in a living cell as well as by the mathematical convenience of this kind of parallelism.

The maximal parallelism means that rules should be used in parallel to the maximum degree possible. For example, assume that currently the multiset of objects present in region h is $w = a^3b^2c^2$ and the set of rules R residing in h is $R = \{r_1, \ldots, r_5\}$, with

$$r_1 : ab \to v_1, \; r_2 : c \to v_2, \; r_3 : bc \to v_3, \; r_4 : a^3c^2 \to v_4, \; r_5 : ad \to v_5,$$

for some $v_1, \ldots, v_5$.

Note that applying r_1 removes ("consumes") the multiset ab, while two parallel applications of r_1 remove the multiset a^2b^2. The remaining "still available" multiset ac^2 does not allow one more parallel application of r_1, but it allows for a parallel application of r_2, it even allows two parallel applications of r_2 which remove the multiset c^2. The remaining multiset a does not allow an application of any rule at all, and therefore we say that the parallel application of the *multiset of rules* $\{(r_1, 2), (r_2, 2)\}$ is maximally parallel. The reader can easily check that, e.g. the application of $\{(r_3, 2)\}$ is maximally parallel (because no rule can be applied to the remaining multiset of objects a^3), and the application of $\{(r_1, 1), (r_2, 1), (r_3, 1)\}$ is maximally parallel (because no rule can be applied to the remaining multiset of objects a^2). On the other hand, the application of $\{(r_1, 2), (r_2, 1)\}$ is not maximally parallel, because r_2 can be still applied to the remaining multiset of objects ac.

In other words, we assign non-deterministically objects to rules, until no further rule can be enabled. The obtained multiset of rules can then be applied in the maximally parallel sense.

Consider now the generic case of a multiset of objects w and a set of rules $R = \{r_i \mid 1 \leq i \leq k\}$, with $r_i : u_i \to v_i$ for each $i = 1, 2, \ldots, k$. Let us consider a multiset $M = \{(r_1, n_1), \ldots, (r_k, n_k)\}$ over R (hence with $n_i \geq 0, 1 \leq i \leq k$).

We say that the multiset of rules M is *applicable* to the multiset of objects w if

$$Z = \bigcup_{i=1}^{k} n_i \otimes u_i \subseteq w.$$

Then, M is maximal (among the multisets applicable to w) if for no $j \in \{1, 2, \ldots, k\}$ we have

$$u_j \cup Z \subseteq w.$$

Note that for a given multiset of objects w and a given multiset of rules R, there can be several applicable multisets of rules (any submultiset of an applicable multiset is applicable) and, similarly, several applicable multisets can be maximal (as it was the case in the above example). Denoting the set of all multisets of rules applicable to w by $Appl(R, w)$, we can define the *set of multisets of rules applicable to w in the maximally parallel mode* in a formal way as follows:

$$Appl_{max}(R, w) = \{R' \mid R' \in Appl(R, w) \text{ and there is}$$
$$\text{no } R'' \in Appl(R, w) \text{ with } R'' \supsetneq R'\}.$$

When defining transitions in a P system, any of these maximal multisets can be chosen in order to perform a maximally parallel transition step, hence the evolution of the system has branchings chosen in a non-deterministic manner.

1.6 Cell-like P Systems with Multiset Rewriting Rules

We are ready now to define P systems in a more formal fashion. In order to define a P system, we have to specify the alphabet of objects, with the subalphabet of catalysts, the alphabet of membrane labels, the membrane structure, the multisets of objects present in the compartments of the membrane structure (at the beginning of the computation), and the rules for evolving the objects—in the basic model the evolution rules are multiset rewriting rules.

Thus, a P system (with multiset rewriting rules) of degree $m \geq 1$ is a construct

$$\Pi = (O, H, \mu, w_1, \ldots, w_m, R_1, \ldots, R_m, i_0),$$

where:

1. O is the alphabet of objects;
2. H is the alphabet of membrane labels;
3. μ is a membrane structure of degree m;
4. $w_1, \ldots, w_m \in O^*$ are the multisets of objects associated with the m regions of μ;
5. $R_i, 1 \leq i \leq m$, are the finite sets of multiset rewriting rules associated with the m regions of μ;
6. $i_0 \in H \cup \{e\}$ specifies the input/output region of Π, where e is a reserved symbol not in H.

Note that in the above we assume that H is ordered (hence the regions are ordered), and so w_i and R_i are the multiset of objects and the set of rules associated with the ith region. As a matter of fact, a standard choice for the alphabet of labels is $H = \{1, 2, \ldots, m\}$—we will also do it in this chapter unless made clear otherwise. The standard way to specify a membrane structure is by giving a parentheses expression for it. Therefore, in this chapter we assume that μ is a parentheses expression over H. The intuition behind $w_1, \ldots, w_m$ is that these are multisets of objects present in regions 1 through m at the beginning of the computations (which we will define below). Finally, i_0 specifies a distinguished space where either one looks for an output of a computation (when a generative mode is used) or one places an input for a computation (when an accepting mode is used)—this is discussed in detail later on. Here $i_0 = e$ means that the environment is the distinguished space. In case when the standard alphabet of labels $H = \{1, 2, \ldots, m\}$ is used, 0 is used as the label for the environment ($e = 0$).

A configuration of Π is an m-tuple of multisets specifying the objects "currently" present in each of the m regions of the system. $C_0 = (w_1, \ldots, w_m)$ is the *initial configuration* of Π. Starting from the initial configuration and using rules from $R_1, \ldots, R_m$ in the maximally parallel fashion in each region, simultaneously in all regions (synchronized by a global clock marking time units), one gets a sequence of consecutive configurations. Each passage from a configuration C to a successor configuration C' is called a *transition* and denoted by $C \Longrightarrow C'$. Due to the non-determinism in applying the rewriting rules, for a given configuration C one can have several possible successors. A sequence of transitions $C_0 \Longrightarrow C_1 \Longrightarrow \ldots \Longrightarrow C_s$ is called a *computation* of Π. Such a computation is *halting* if no rule can be applied to the last configuration C_s.

Thus, the transitions take place according to a global clock which marks the time for all regions: in each time unit all regions evolve synchronously and in each region rules are used in the maximally parallel way.

We end this section with a simple example of a P system Π with multiset rewriting rules specified as follows:

$\Pi = (O, \mu, w_1, w_2, R_1, R_2, i_0)$, where

$O = \{a, b, c, d, e, f\}$,

$\mu = [\,[\ ]_2\,]_1$,

$w_1 = aa$,

$w_2 = d$,

$R_1 = \{r_1 : a \to (a, out), r_2 : a \to (c, in)\}$,

$R_2 = \{r_3 : d \to e, r_4 : d \to f, r_5 : ce \to cee, r_6 : cf \to c(b, out)\}$, and

$i_0 = 2$.

Note that since we have not specified H above, it is assumed that H is standard, i.e. $H = \{1, 2\}$.

The initial configuration is (aa, d). Then, non-deterministically, objects aa either go to the environment (the application of rule r_1), or get transformed into c and enter membrane 2 (the application of rule r_2). Also, in the first step of a computation, a choice is made in region 2, between rules r_3 and r_4, which transform the object d into either e or f. Altogether, we have six possibilities of choosing a maximal applicable multiset of rules in the first step: $\{(r_1, 2), (r_3, 1)\}$, $\{(r_1, 2), (r_4, 1)\}$, $\{(r_1, 1), (r_2, 1), (r_3, 1)\}$, $\{(r_1, 1), (r_2, 1), (r_4, 1)\}$, $\{(r_2, 2), (r_3, 1)\}$, $\{(r_2, 2), (r_4, 1)\}$. The choices $\{(r_1, 1), (r_2, 1), (r_3, 1)\}$ and $\{(r_2, 2), (r_3, 1)\}$ lead to infinite computations: having at least a copy of c and a copy of e in region 2, one can use forever rule r_5 (increasing each time the number of copies of e). All other computations halt (above the double arrow indicating a transition we give the used multiset of rules):

$$(aa, d) \xRightarrow{r_1 r_1 r_3} (\lambda, e),$$

$$(aa, d) \xRightarrow{r_1 r_1 r_4} (\lambda, f),$$

$$(aa, d) \xRightarrow{r_1 r_2 r_4} (\lambda, cf) \xRightarrow{r_6} (b, c),$$

$$(aa, d) \xRightarrow{r_2 r_2 r_4} (\lambda, ccf) \xRightarrow{r_6} (b, cc).$$

Consequently, Π halts with one of the multisets e, f, c, cc in the output region (region 2).

A detailed presentation of P systems with multiset rewriting rules is provided in Chapter 4.

1.7 COMMUNICATION RULES

Rewriting rules discussed above were associated with (located in) regions and were acting on multisets of objects residing in these regions. Another type of rule is associated with (located on) membranes, and they govern the movement/exchange of objects (located in neighboring regions) through these membranes. These types of rules are referred to as *communication rules*, because exchange of objects between neighboring regions is the way that regions "communicate" with each other. Symport and antiport rules are typical examples of communication rules (other types of communication rules are also discussed in the handbook)—they are motivated by the coupled cross-membrane transport of ions and molecules (see Chapter 2 for the biological background).

Symport rules are either of the form (u, in) or of the form (u, out), and antiport rules are of the form $(u, out; v, in)$, where u, v are strings over an alphabet O of objects. If these communication rules are associated with membrane h, then their application has the following effect:

- for (u, in), the multiset u of objects is moved from the upper neighbor of region h to region h;
- for (u, out), the multiset u of objects is moved from region h to the upper neighbor of region h;

- for $(u, out; v, in)$, simultaneously the multiset u of objects from region h is moved to the upper neighbor of region h, and the multiset v of objects from the upper neighbor of region h is moved to region h.

Obviously, for these rules to be applicable it is required that the multisets of objects u and v are present/available in the corresponding regions.

Also, we need to comment here that there is no problem with having symport rules (u, in) or antiport rules $(u, out; v, in)$ associated with the skin membrane, as the P systems using communication rules (symport/antiport rules) will have objects also present in the environment and will allow objects from the skin region to be moved to the environment and back.

To simplify the terminology, we will say that the multiset of objects *enters* (through) membrane h whenever *in* is used, and it *exits* (through) membrane h whenever *out* is used.

The length of u in a symport rule or the maximum of $|u|, |v|$ in an antiport rule as above is called the *weight* of the rule.

For example, the symport rule (ab, in) requires that objects a and b enter simultaneously the membrane (on which the rule is located), while (ab, out) requires that objects a and b exit simultaneously the membrane. An antiport rule $(a, out; bc, in)$ indicates the simultaneous exit of object a through the membrane and the entrance of objects bc through the membrane. All these rules are of weight 2. (Sometimes, one distinguishes the length of each string in an antiport rule, so that $(|u|, |v|)$ is said to be the *diameter* of the rule $(u, out; v, in)$; thus, the diameter of the rule $(a, out; bc, in)$ is $(1, 2)$.)

1.8 Cell-like P Systems with Communication Rules

A P system with communication rules is defined in a similar fashion to a P system with multiset rewriting rules, but with some essential differences. First of all, we consider now communication rules rather than rewriting rules—sets of such rules are associated with membranes rather than with regions. Moreover, since there are communication rules only, one needs a supply of objects that can be moved into the system. Otherwise, the multiset of objects present in the system will never change and consecutive configurations of a computation would present merely distributions of this fixed multiset through the regions of the system. Such a supply is ensured by specifying a subset E of the set of all objects O, and assuming that objects from E are available in the environment in arbitrary multiplicities (thus their multiset is $M : E \longrightarrow \{\infty\}$).

The basic model of a P system with communication rules evolves through applications of symport/antiport rules. It is formally defined as follows.

A P system with symport/antiport rules of degree $m \geq 1$ is a construct

$$\Pi = (O, H, E, \mu, w_1, \ldots, w_m, R_1, \ldots, R_m, i_0),$$

where $O, H, \mu, w_1, \ldots, w_m,$ and i_0 are as in the definition of a P system with multiset rewriting rules, $E \subseteq O$, and $R_1, \ldots, R_m$ are finite sets of symport and antiport rules associated with the m membranes of μ.

Note that objects can be sent to the environment (e.g. by symport rules (u, out) associated with the skin membrane), as well as objects can be brought from the environment into the skin region (e.g. by symport rules (u, in) associated with the skin membrane). In particular, objects from E can be brought into the system, and, irrespectively of how many objects are brought from the environment, arbitrarily many still remain there. Special attention has to be paid to symport rules associated with the skin membrane. For example, if a symport rule (c, in) with $c \in E$ is placed on the skin membrane, then its application (in the maximally parallel fashion) would lead to an infinite iteration of introducing object c into the skin region. For this reason, symport rules of the form (u, in) with $u \in E^+$ are not allowed. Instead, each such rule u must contain at least one occurrence of an object which is not in E. Such "accompanying" objects are available in the environment only if they were present in the system in its initial configuration and at some previous moment of time they were sent out to the environment by the system.

The configurations of a symport/antiport P system have now to specify both the multisets present in the compartments of μ and also the multiset of objects present in the environment different from objects in E. The transitions among configurations (based on the maximally parallel use of rules), computations, and halting computations are defined as for P systems with multiset rewriting rules.

It is important to notice that because rules associated with neighboring membranes compete for objects, the maximal parallelism can no longer be defined locally, within regions (as it was done for multiset rewriting rules)—it must be defined globally on the level of the whole system.

Consider now an example, illustrated in Fig. 1.3 (since the communication rules are associated with membranes, we write them next to respective membranes). This P system Π with symport/antiport rules is formally specified as follows:

$$\Pi = (\{a, b, c, d\}, \{d\}, [\ [\ ]_2\ ]_1, bcc, aabb, R_1, R_2, 1), \text{ where}$$

$$R_1 = \{r_1 : (b, out), r_2 : (bd^2, in)\}, \text{ and}$$

$$R_2 = \{r_3 : (a, out; bc, in), r_4 : (aa, out), r_5 : (bd, in)\}.$$

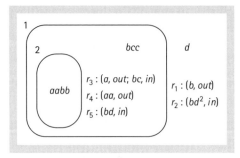

Fig. 1.3 A simple symport/antiport system with competing rules.

The reader can easily see that using r_3 does not allow the use of any other rule—therefore this is a maximally parallel choice. Similarly, using r_4 and r_1 simultaneously is maximally parallel. For the first choice, the computation stops after the first step. For the second choice, object b (brought into the environment by r_1) will bring into the skin region (by r_2) two copies of d—arbitrarily many copies of d are always available in the environment. The alternating use of rules r_1, r_2 can be iterated any number of times, and the only way to stop this process of bringing d into the skin region is to "hide" the "carrier" b in region 2 by using rule r_5.

Thus, a halting computation which starts with rules r_4, r_1 is of the following form:

$$(\lambda, bcc, aabb) \xRightarrow{r_4 r_1} (b, aacc, bb) \xRightarrow{r_2} (\lambda, aaccbdd, bb) \xRightarrow{r_1} (b, aaccdd, bb)$$
$$\xRightarrow{r_2} \cdots \xRightarrow{r_1} (b, aaccd^{2n}, bb) \xRightarrow{r_2} (\lambda, aaccbd^{2n+2}, bb)$$
$$\xRightarrow{r_5} (\lambda, aaccd^{2n+1}, b^3 d), \text{ for some } n \geq 0.$$

Now, in order to formally define the maximal parallelism for symport/antiport rules, we have to take into consideration all rules, associated with all membranes—sets of rules associated with different membranes are considered disjoint (this can be achieved by labeling the rules by distinct labels). Thus, we first define the multisets of rules *applicable* to the multisets of objects present in regions of the system (note that now we have to take into account the indications *in*, *out* when choosing the multisets of objects to be used), and then the set of applicable multisets of rules which are maximal in the sense of the inclusion can be defined analogously to the way it was done in Section 1.5.

Chapter 5 is devoted to P systems with communication rules.

1.9 MINIMAL PARALLELISM AND OTHER TRANSITION MODES

The maximal parallelism is the mostly used regime of applying the rules in a P system, but there are other possible modes of applications of rules. For instance, one can use any multiset of (jointly) applicable rules in a region (the *asynchronous mode*), or only one applicable rule in the whole system (the *sequential mode*), or a specified number of rules in the whole system or in each compartment (*bounded parallelism*), etc.

An interesting application mode is that of *minimal parallelism*, whose intuition is as follows: each region which *can* evolve (using at least one rule) *must* evolve (by means of at least one rule); in the case of symport/antiport rules, this principle applies to membranes rather than to regions.

Formally, let us consider a P system Π with multiset rewriting rules, with m regions, having associated the multisets w_i and the sets of rules R_i, $1 \leq i \leq m$. A multiset M of rules is said to be *minimally applicable* to $C = (w_1, \ldots, w_m)$ if (1) M is a multiset of rules applicable to C with the multiplicities of the rules being either 0 or 1, (2) for each $i = 1, 2, \ldots, m$ we have at most one rule $r_{ik} \in R_i$ with the multiplicity 1, and (3) for no $j = 1, 2, \ldots, m$ such that all rules from R_j have multiplicity 0 in M, we can change the multiplicity of any rule in R_j to 1 so that the obtained multiset is still applicable to C.

A transition in Π is done in the minimally parallel mode if it uses a minimally applicable multiset of rules or any supermultiset of it (note that an applicable supermultiset can be obtained only by possibly increasing the number of used rules in compartments where already a rule had the multiplicity 1, but if in a compartment we have the multiplicity zero for all rules, then, by condition (3) above, no rule can be applied). Thus, in comparison with the maximal parallelism, a considerable degree of non-determinism is possible (note that the maximal parallelism is one of the possible evolutions of the minimally parallel mode, as a maximally applicable multiset of rules is a supermultiset of a minimally applicable one).

In the case of P systems with symport/antiport rules, in order to define a minimally applicable multiset of rules we usually consider the sets R_i, of rules associated with membranes, and then choose from each of them at most one rule with multiplicity 1. It may happen that the rules associated with neighboring membranes compete for objects in "shared" regions. In such a case we choose non-deterministically at most one rule from each set R_i, assign the multiplicity 1 to them, and then check conditions (1), (2), and (3) given above. If these conditions are satisfied, then we have obtained a minimally applicable multiset of rules; this multiset or any supermultiset of it can be applied yielding a minimally parallel transition.

Minimal parallelism is discussed in several chapters of the handbook, e.g. in Chapters 4, 5, 11, and 12.

1.10 Results of Computations

Now that we have introduced the basic notions concerning the structure and functioning of P systems, we move to discuss two basic modes of using P systems as computing devices: the *generating* mode and the *recognizing* mode.

In the generating mode, a P system starts from the initial configuration and evolves/computes according to its rules and the specified mode of using them. During a specific computation, the system makes non-deterministic choices of rules to be applied. Among such computations some are defined to be *successful*, and successful computations yield results. The standard way to define successful computations is through *halting*. The result of a halting computation is the multiset of objects present in the region i_0 when the computation halts. This can be the number of objects present in region i_0 (the weight of the multiset), or, if we want to distinguish between objects, we can take as the result of the computation the vector giving the multiplicities of objects in region i_0 in a halting configuration. As mentioned already, the distinguished space i_0 can also be the environment ($i_0 = e$), and then the result consists of the objects present in the environment at the end of a computation. If this is the case and we deal with communication P systems, then we have to ignore the objects from E, whose multiplicity is infinite. This suggests the possibility of considering a predetermined subset $T \subseteq O$ of "terminal objects" (like in Chomsky grammars and Lindenmayer systems), and then to count only the objects in T.

The notion of a successful computation can be defined in many ways. For instance, instead of the global halting considered above (no region can evolve, i.e. no rule can be used in the whole system), also the *local halting* was considered in the literature: a computation stops when at least one region "dies" (reaches a configuration where no rule can be applied in that region). Yet another possibility is to use a specific signal to prompt the reading of the result: e.g. if a specific object enters a specific membrane, then we read the result (even if the computation can continue).

Then, there are various possible forms of results. Numbers or vectors of numbers were already mentioned. When we read the result in the environment, we can also associate a string with a computation: the objects which leave the system are arranged/sequenced in the order they are sent to the environment; if several objects leave the system simultaneously then all their permutations are allowed as a substring of the generated string. One can also associate a string with a computation by following the membrane *trace* of a specified object as it travels through the system: e.g. a symbol b_i is generated each time that the traveling object enters membrane i.

It is important to note that when a string is associated with a computation, then we get a qualitative difference between the internal data structure, a multiset, and the result, a string, which contains "positional information".

The generating mode of using P systems as computing devices corresponds to the use of grammars in the theory of computation. The use of a recognizing mode in P systems corresponds to the use of automata in the theory of computation. In this mode, a multiset w is introduced into region i_0 of system Π (in this case, i_0 must specify a region of the system, not the environment), and w is accepted/recognized if and only if the computation starting in such a configuration eventually stops (i.e. if the regions $1, 2, \ldots, m$ of the system contain initially the multisets $w_1, w_2, \ldots, w_m$, respectively, then the computation starts from the configuration $(w_1, \ldots, w_{i_0-1}, w_{i_0} \cup w, w_{i_0+1}, \ldots, w_m)$).

As with grammars and automata, a clear duality exists between the generating and the accepting modes. Note also the essential role of non-determinism in the generating mode: a deterministic system (never branching, always having only one choice of rules for defining the transitions) either halts and then generates only one result, or does not halt and generates "nothing". The things are different in the accepting case, when we can impose a deterministic behavior, i.e. to have at most one successor for all configurations that are reachable from the initial configuration of Π. Such a behavior is desirable in many cases, e.g. when the system is used in the accepting mode for solving a *decision problem* in an efficient way: the input multiset w is an encoding of an instance of a decision problem, and the system has to say, in a finite number of computation steps, whether this instance has the positive or the negative answer.

An interesting issue appears here. P systems are basically non-deterministic. However, when computing functions we need unique answers. An obvious way to ensure uniqueness is to require determinism, but there are also other ways. For example, one can require *confluence*: it does not matter how the system evolves providing that it "converges" to a unique configuration, and then continues deterministically—then we also get a unique result in this way. An even more "liberal" framework is provided by *weakly confluent* systems: it does not matter how the system evolves providing that all computations halt and all of them provide the same answer.

1.11 OTHER MAJOR CLASSES OF P SYSTEMS

Besides the two main classes of P systems, with multiset rewriting rules and communication rules, several other classes are considered in the literature. The motivation for considering various other models comes mostly from three (somewhat contradictory) directions: mathematics, computer science, and biology. Moreover, specific models may be driven either by theoretical issues or by applications. The handbook presents in detail many different classes of P systems. In this section we present a brief review of some of these different models.

1.11.1 P Systems with Active Membranes

A possible weakness of P systems considered in this chapter so far, especially from a biological point of view, is the fact that the membrane structure is static, it does not evolve during the computation. To solve this problem, one considers also rules for evolving membranes. Among the first membrane evolving rules were rules for *dissolving* membranes: a special symbol δ is considered, and an application of a multiset rewriting rule of the form $u \to v\delta$ has the same effect as an application of the rule $u \to v$, except that, after this rule is applied, the membrane delimiting the region (where the rule is applied) is dissolved. After the dissolution the objects from the region become objects of the upper neighbor of the dissolved region, while the rules of the dissolved region are removed (they were describing the biochemistry of a "protected reactor" which no longer exists). The skin membrane is never dissolved.

More intricate are the rules *with active membranes*, where the membranes themselves are involved in the rules, and hence they may evolve during computations. P systems using such rules are discussed in detail in Chapter 11. Here we mention only a couple of ideas concerning the evolution of membranes. Important types of rules are rules for *for dividing membranes*. For example, a rule $[\,a\,]_h \to [\,b\,]_h[\,c\,]_h$, where a, b, c are objects, and h is a label of a membrane, implies the division of membrane h, induced by object a: two copies of this membrane are produced, with all objects of region h replicated and object a replaced in the first copy by b and in the second copy by c.

Using such rules, the number of objects as well as the number of membranes can grow exponentially. This implies that different membranes carry the same label, but this causes no problems in evolving these membranes (and corresponding objects), because the labels of membranes identify the rules to be used.

This is also the case for *membrane creation* rules. They are global and their basic form is $a \to [\,b\,]_h$, where a, b are objects and h a label. Through an application of such a rule, a new elementary membrane is created from a, with label h and containing object b. Sometimes, such rules are localized, i.e. they are of the form $[\,a \to [\,b\,]_h\,]_g$; in this case the rule $a \to [\,b\,]_h$ is applied only to objects a from region g.

Several other types of rules for evolving membranes were considered, mainly inspired by biology (e.g. corresponding to the operations of exocytosis or endocytosis), but we do not consider them in this chapter. We conclude this section by mentioning that the enhanced parallelism provided by rules for increasing the number of membranes is a basic tool for devising efficient solutions to computationally hard problems in terms of P systems. Moreover, in systems with evolving membranes, the membrane structure (the tree describing it) can become the main data structure dealt within the computation and so the P systems become tree processing devices. These issues are discussed, e.g. in Chapters 11 and 12.

1.11.2 P Systems with String Objects

In the systems above, the objects were considered atomic, in the etymological sense, and were given by symbols from a given alphabet. However, it is natural to consider also P systems whose objects are structured. There are many natural motivations to consider such structured objects—they come from computer science, mathematics, and biology. For instance, many molecules such as proteins, DNA, RNA have string-like structures. Hence one can describe such objects by strings over a given alphabet. For P systems operating on string-like objects we have to specify the alphabet, the membrane structure, the strings present in the system at the beginning of the computation, the rules for processing the strings, and the way to define the result of a computation. The strings of a P system can be considered either as a set (languages) or as a multiset. Then, the rules must be string processing rules—one can consider, among others, parallel string rewriting such as in Lindenmayer systems, or sequential string rewriting such as in Chomsky grammars. Another fruitful idea is to use string evolving rules which are specific for DNA processing, such as splicing or insertion-deletion. To generate an exponential workspace (for "solving" computationally hard problems) one can use operations which replicate strings. P systems with string objects are discussed in detail in Chapters 7 and 8.

1.11.3 Tissue-like and Neural-like P Systems

Until now we have considered hierarchical arrangements of membranes, which correspond to cell-like membrane structures described by trees. Tissue-like membrane structures are described by graphs, where membranes, also called cells, are (placed in) nodes of a graph. Here, an edge between two nodes corresponds to a communication channel between membranes placed in these nodes. If a communication channel between two membranes exists, then they can communicate, e.g. by means of antiport-like rules. For instance, two membranes i, j connected by a communication channel can use rules of the form $(i, u/v, j)$, where u, v are multisets of objects. The application of such a rule moves objects specified by u from membrane i to membrane j, with objects specified by v moving in the opposite direction. If no communication channel (edge) exists between two membranes, then they can still communicate through the environment: one of them sends objects out to the environment (using rules of the form $(i, u/v, 0)$, with 0 labeling the environment), and the other one can take objects from the environment (by analogous rules). Also symport-like rules can be used.

Several variations of this basic mechanism are considered in the literature, especially concerning the structure/functioning of the communication channels. For instance, in so-called population P systems, such channels can be either created or destroyed during computations (hence the structure of edges between membranes is dynamical). This corresponds to having two states for each possible channel: open and closed. A direct generalization of this two state situation is to associate a finite number of states with each channel, thus controlling the communication in a more intricate way. The rules associated with a channel (i, j) can then be of the form $(s, u/v, s')$, where s, s' are states, and u, v are multisets of objects: if the channel is in state s, the multisets u, v can be exchanged between membranes, and the channel changes its state to s'.

Computations are defined in the customary way (usually, based on the maximally parallel way of applying the rules), and so are the results of computations. Also membrane division (cell multiplication) can easily be introduced into these models.

Neural-like P systems are closely related to tissue-like P systems. Typical differences with tissue-like P systems are as follows. Each cell (called a neuron) has now a state which controls the evolution of objects; the rules are not communication rules as above, but rather rewriting rules of the form $su \to s'v$, where s, s' are states, and $u \to v$ is a usual multiset rewriting rule, with target indications in v, which direct the movement of objects from one cell to another. One of the cells is designated as the output cell, which also sends objects to the environment, providing in this way an output. When moved between cells, objects can also be replicated, and then copies sent to all cells to which a channel ("synapse") is available from the cell where the rule is applied.

Still closer to the neural functioning are the so-called spiking neural P systems (in short, SN P systems), where only one object is considered, called the *spike* (electrical impulse). The spikes of a neuron are processed by rules of the form $E/a^c \to a; d$, where E is a regular expression over the alphabet $\{a\}$, and c and d are natural numbers. The idea is that if the neuron contains a number of spikes which is given (admitted) by the regular expression E, then c spikes are consumed, and one is produced. The produced spike is sent, after the delay of d time units, to all neurons connected by an outgoing synapse (edge) to the neuron where the rule is used. Delay d corresponds to the idea of the refractory period in neurology. Also, an output neuron is specified, and its spikes are sent to the environment and in this way produce a *spike train*, through which results of a computation can be defined in several ways.

It is important to notice that in SN P systems time is used as a support of information, in the sense that the time distance between consecutive spikes passing along an axon plays an important role in computations—in fact (through spike trains) time defines the results of computations.

The tissue and population P systems are discussed in Chapter 9, while Chapter 13 discusses SN P systems.

1.12 Overview of Directions of Research and (Types of) Results

The main directions of research in membrane computing concern computing power and efficiency, relationships to other models of computation, applications and implementations. We briefly review some of the research trends.

1.12.1 Computing Power and Efficiency

The initial goal of membrane computing was to define models of computation inspired by cell biology. Indeed a great variety of models were defined and their computing power investigated, mainly by relating them to standard models of computation. It turned out that most classes of P systems are Turing complete. Since the proofs of Turing completeness were always constructive, they provided a way to obtain universal P systems, i.e. P systems that can simulate any other P system when given a "code" of the system to be simulated. For this reason these results are often referred to in the literature as *universality* results rather than computational completeness results.

This universality holds for all sorts of P systems discussed in this chapter: cell-like, tissue-like, or neural-like, with symbol objects or string objects, working in the generative or in the accepting modes. Universality holds for these classes of P systems in their most general form, but also for their quite restricted forms, with restrictions on the number of membranes, form of the rules, etc. In most cases, trade-offs (retaining the computing power) were found between the number of membranes, the complexity of rules, and/or the number of objects used. Thus, on this very abstract level, one can conclude that "the cell is a powerful computer", both when it works alone and in tissue-like or neural-like network configurations.

The computational power (the "competence") is only one of the important questions to be dealt with when defining a new computing model. The other fundamental question concerns the computing *efficiency*, i.e. the resources used for solving problems, especially time and space. It turns out that P systems can be efficient problem solvers—this holds for P systems equipped with ways for producing an exponential workspace in linear time. The efficiency speed-up is obtained by trading space for time, with the space produced in a bioinspired way.

There are three basic ways to construct such an exponential space in cell-like P systems: membrane division, membrane creation (combined with the creation of exponentially many objects), and string replication. Such constructions are done through cell division in tissue-like systems and through object replication in neural-like systems. Also, the possibility to use a precomputed exponential workspace, which is unstructured and nonactive (e.g. with the regions containing no objects) was considered in spiking neural P systems, where the workspace cannot grow exponentially in linear time during the computation.

In all these cases, polynomial or pseudo-polynomial solutions to NP-complete problems were obtained. Roughly speaking, the framework for dealing with complexity matters is that of *accepting P systems with input*: a family of P systems of a given type is constructed based on a given problem, and then an instance of the problem is introduced as an input to such systems. Working in deterministic or confluent mode, in a given time one of the answers yes/no is obtained in the form of specific objects sent to the environment. The family of systems should be constructed in a uniform way (starting from the size of problem instances) by a Turing machine working in polynomial time. A less constrained framework is also considered, where a *semi-uniform* construction is allowed, carried out in polynomial time by a Turing machine, but starting from the instance to be solved.

This direction of research is currently very active. A whole variety of problems is considered here, and characterizations of classic complexity classes as well as of the **P≠NP** conjecture have been obtained in this framework. Details can be found in Chapter 12.

1.12.2 Applications and Implementations

The first, natural (considering the origin of membrane computing) domain of applications for P systems was biological processes. As a modeling framework for this domain, P systems can be seen as models approaching reality at the micro level, where "reactants" and "reactions" can be known and are treated individually, as opposed to the macro approach (e.g. by means of differential equations), which deals with populations of reactants which are large enough to be better approximated by infinite rather than by finite, discrete sets. There are several attractive features of P systems as models of biological processes: inherent compartmentalization, easy extensibility, direct intuitive appearance for biologists, easy programmability, etc. P systems are particularly suitable when one deals with a reduced number of objects or with slow reactions, which is the case for a considerable number of biological processes, especially those related to networks of pathway controls, genetic processes, protein interactions. Chapters 18, 19, 20 discuss those matters in detail.

Similar, from many points of view, to the biochemical reality is the economic reality (e.g. at the market level), where compartments can be defined as spaces where various "objects" (goods, parts of goods, money, working time, contracts, and so on and so forth) "react" according to well-specified rules. This direction of research needs to be developed, but the general idea seems to be well-timed now as multiagent computer based approaches (simulations) are more and more used in economics, while "exact" methods, e.g. of the kind provided by operational research, seem to be less applicable to non-trivial, complex economic phenomena.

The applications to computer science are quite diverse: computer graphics, sorting/ranking, cryptography, modelling/simulating circuits, parallel architectures, etc. They demonstrate and rely on the considerable expressive power of P systems and their versatility, but the results have not yet found practical use in computer science. Perhaps a promising exception here is the use of membrane computing ideas in evolutionary computing. The so called membrane algorithms (see Chapter 21) seem to be rather efficient and useful, in terms of the convergence speed, the quality of the provided solutions, and the average and the worst solutions.

Most of these applications are based on programs/software for simulating P systems. There are also software products developed for didactic reasons, distributed implementations, as well as hardware attempts to implement P systems. There is also a specialized programming language (*P-lingua*) under development, as well as a plan to implement a P system in a biolab. Various details concerning the above topics can be found in a number of chapters of this handbook as well as on the website of P systems.

1.13 CONCLUDING REMARKS

In this chapter we introduced some basic notions of membrane computing, discussed the motivation behind them, and gave an overview of research directions and types of results both in theory and applications. The handbook provides in-depth coverage of membrane computing. It is not meant to provide exhaustive coverage, which may be practically impossible by now as the bibliography of membrane computing now includes over 1200 positions, with about 35 collective volumes and special issues of journals.

We conclude this chapter with a sort of bird's eye view (non-exhaustive list) of topics, themes, and issues encountered in the current literature on membrane computing. Together with the contents of this chapter, it should provide the reader with an impression of the richness and the diversity of this field of research.

(1) Basic notions:
- Objects: symbols, strings of symbols, spikes, arrays, trees, numerical variables, conformons, other data structures, combinations.
- Data structures: multisets, sets (hence languages in the case of strings), fuzzy sets and fuzzy multisets.
- Location of objects: in compartments, on membranes, combined.
- Forms of rules: multiset or string rewriting, communication rules—e.g. symport/antiport, boundary rules, rules with active membranes, combined, array/tree processing, spike processing.
- Controls of application of rules: catalysts, priority, promoters, inhibitors, activators, sequencing, energy.
- Membrane configurations: cell-like (tree), tissue-like (arbitrary graph), with static or with dynamic communication channels.
- Type of membrane structure: static, dynamic, precomputed (arbitrarily large).
- Timing: synchronized, non-synchronized, local synchronization, time-free.
- Application of rules: maximal parallelism, minimal parallelism, bounded parallelism, sequential.
- Successful computations: global halting, local halting, specified events signaling the end of a computation, non-halting.
- Modes of using a system: generating, accepting, input-output function.
- Types of evolution: deterministic, non-deterministic, confluent, weakly confluent, probabilistic.
- Ways to define the output: internal, external, traces, tree of membrane structure, spike train, time elapsed between two events.
- Types of outputs: set of numbers, set of vectors of numbers, languages, set of arrays, yes/no answer, set of trees.

(2) Research lines:
- Research issues: computing power, computing efficiency, descriptional complexity, normal forms, hierarchies, implementations/simulations, semantics, model checking, verification, time evolution (dynamical systems approaches).
- Types of theoretical results: universality, collapsing hierarchies, infinite hierarchies, normal forms, polynomial solutions to **NP**-complete problems and even to **PSPACE**-complete problems (with time/space trade off), classifications, comparisons with Chomsky and Lindenmayer hierarchies, comparisons with classic complexity classes, new complexity classes.

- Applications: biology/biomedicine, population dynamics, ecosystems, economics, optimization, computer graphics, linguistics, computer science, cryptography.
- Links to other models: brane calculi, Petri nets, process algebra, X-machines, lambda-calculus, ambient calculus.
- Other (sometimes "exotic") issues: computing beyond Turing, reproduction, calculi of configurations, multiset theory.
- Major open research topics: borderline between universality and non-universality, between efficiency and non efficiency, comparison of complexity classes based on deterministic, confluent or weakly-confluent evolution, on uniform or semi-uniform constructions, using precomputed resources, "non-crisp" approaches (e.g. fuzzy and rough set approaches), hardware implementations, biolab implementations, applications in economics, programming languages for membrane computing.

Acknowledgements

We are indebted to R. Brijder, M. Main, and K. Salomaa for their comments on an earlier version of this chapter.

Basic Readings and the Website

[1] G. Ciobanu, Gh. Păun, M.J. Pérez-Jiménez, eds.: *Applications of Membrane Computing.* Springer, 2006.
[2] Gh. Păun: Computing with membranes. *J. Computer and System Sci.*, 61 (2000), 108–143. Paper circulated first as a Turku Center for Computer Science (TUCS) Report 208, November 1998 (www.tucs.fi).
[3] Gh. Păun: *Membrane Computing. An Introduction.* Springer, 2002.
[4] Gh. Păun, M.J. Pérez-Jiménez: Spiking neural P systems. An overview. In vol. *Advancing Artificial Intelligence through Biological Process Applications* (A.B. Porto, A. Pazos, W. Buno, eds.), Medical Information Science Reference, Hershey, 2008, 60–73.
[5] Gh. Păun, G. Rozenberg: A guide to membrane computing. *Theoretical Computer Sci.*, 287 (2002), 73–100.
[6] The Membrane Computing Website: http://ppage.psystems.eu

CHAPTER 2

CELL BIOLOGY FOR MEMBRANE COMPUTING

DANIELA BESOZZI

IOAN I. ARDELEAN

2.1 INTRODUCTION

NATURAL systems have frequently represented a source of inspiration in computer science for the development of algorithmic methodologies and formal theories, whose effectiveness and broadest range applicability can be validated by numberless examples, starting from the classical and well-grounded theories of artificial neural networks or genetic algorithms, to the more recent research areas of molecular computing, of ant colony and particle swarm optimization, to mention a few. Membrane computing can be classified into this family of computational models, as it arose as an abstraction of the compartmentalized structure of living cells, and the way biochemical substances are processed within (and moved between) membrane-bounded regions. Starting from Păun's seminal paper [26], several aspects and processes of cells have been considered for the definition of many different computing devices in membrane computing.

The aim of this chapter is to introduce some notions of cell biology that can be helpful to understand the cell-inspiring basis underlying membrane computing.

The biological concepts are here presented considering three levels of organization of the living world: we first describe subject topics at the *subcellular level* (that is, the cellular components and the structural organization of cells) and at the *single-cell level* (the cellular processes and cell's functioning), and then we consider the *supracellular levels* of organization, occurring within multicellular organisms and populations of bacterial cells.

The subcellular and the single-cell levels are described in Section 2.2, where we start by describing the two distinct types of cells that form all living organisms, namely prokaryotic and eukaryotic cells, and then focus on one of the most peculiar eukaryotic cells in evolved organisms, the neuron (Section 2.2.1). We also present several aspects of the cellular membranes, such as their principal components (e.g., phospholipids, transport, and receptor proteins), the transmembrane electrochemical potential, and the presence of other specialized structure. Then we deal with intracellular processes, such as the metabolic reactions related to the cell energetics, the control and regulation of gene expression, the transduction of external signals into the cell, and two ways that cells can adopt to communicate with, or react to, the external environment (Section 2.2.3).

The subcellular and the single-cell levels are described in Section 2.2, where some distinct morphological and functional structures existing in multicellular organisms. In particular, we provide a sketch of the four principal types of tissues in vertebrates (the connective, epithelial, muscular, and nervous tissues), and then describe the cell–cell adhesion structures that are needed for tissular cells to communicate and coordinate each other. In Section 2.4 we deal with two population-like behaviors that can take place within colonies of bacterial cells: we describe both an intercellular communication mechanism (quorum sensing), and a colony-based form of life (the biofilms) that several microorganisms are able to switch on in particular environmental conditions.

For a more detailed analysis of the topics considered in this chapter, and for further notions about the numerous cellular aspects that have not been included here, we refer the reader to some comprehensive monographs dealing with cells and cellular processes in general [1, 16, 8, 31, 24], or with microorganisms [18], animals [11] and plants [29], in particular. We emphasize that the description of the subjects treated here is consistent with the current biological knowledge that can be found in literature.

2.2 THE CELL

This section is devoted to the cell, the simplest unit of biological organization that constitutes all living organisms. We start by describing the two distinct types of cells, eukaryotic and prokaryotic, focusing in particular on neurons. Then we consider several aspects related to the cellular membranes, such as their composition, structure, main constituents, and functions. We conclude the section with an introduction to some processes occurring at the single-cell level, such as gene expression and metabolism, as well as to some communication pathways occurring between cells and the external environment.

2.2.1 Eukaryotic and Prokaryotic Cells

We outline here the differences and similarities between the two distinct types of cells existing in the biological world, and then give a more detailed description of one of the most peculiar animal cells, the neuron.

The structure and contents of cells. All living organisms, whether unicellular or multicellular, are composed of the same kind of molecules—possibly assembled in different ways—and share some similarity in cellular structure and organization.

Concerning composition, the most abundant biochemical compounds, which are usually imported into the cell from the extracellular environment, are small organic molecules (sugars, vitamins, etc.), inorganic ions (calcium, sodium, potassium, etc.) and, especially, water. In contrast, the polymeric macromolecules such as lipids, proteins, nucleic acids, and polysaccharides, are less plentiful and usually need to be synthesized inside the cell. Proteins, that carry out numerous functions in the cell, are the most diverse (in type and number) among the cellular macromolecules. Each protein can be constituted by one or more polypeptides (each one made up from a limited set of building blocks, called amino acids), which fold into a specific three dimensional structure that assures the proper functionality of proteins. The specific aminoacidic sequences that identify all protein polypeptides are encoded in genes, which are specific portions of the genetic makeup of cells. Genes are constituted by deoxyribonucleic acid (DNA), another polymeric macromolecule consisting of four distinct monomers called nucleotides. By means of genes, the cell directs the synthesis of proteins, controlling their amounts and the time they are to be made for the correct functioning of all cellular processes. The *central dogma of biology* explains how genes are expressed into proteins, passing through the synthesis of a third kind of cellular macromolecule, called the (messenger) ribonucleic acid (RNA) (see Section 2.2.3).

Concerning the structural organization, all biological systems consist of only two distinct types of cells: *eukaryotic* and *prokaryotic* cells (see left and right side in Fig. 2.1, respectively). The main ultrastructural difference between the two types of cells is the presence of a true nucleus in eukaryotic cells, surrounded by a membrane protecting the genetic material of the cell. On the contrary, in prokaryotic cells the genetic material is not surrounded by any membrane. Despite the differences concerning the internal architecture (see further on for more details), all prokaryotic

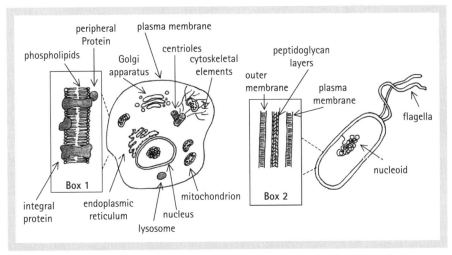

Fig. 2.1 Eukaryotic (animal) cell, left side, and prokaryotic (bacterial) cell, right side.

and eukaryotic cells are delimited by a membrane, called the *plasma membrane*, which functions as a selectively permeable barrier between the cytoplasm (the whole cell interior) and the environment around the cell. In particular, the plasma membrane governs a large number of functions: the bidirectional movement of chemicals between the cell and the external environment (e.g. the inflow of nutrients and the outflow of waste solutes), the maintenance of the membrane potential and of internal pH, the transduction of external signals into the cell, the antigenic response against foreign materials, the passage of metabolites between adjacent cells within a tissue, and so on. Nearly all these jobs are carried out by specific proteins (which will be described in Section 2.2.2) that are associated with the membrane.

In the last decades, the progress made through detailed analysis of the cells showed that there are prokaryotic cells which, from a biochemical point of view, are closer to some eukaryotic cells than to other prokaryotic cells. These properties are suggested to organize the living world into three major (evolutive) groups of organisms, called domains—*Bacteria*, *Archaea*, and *Eukarya*—which represents the dominant scientific view nowadays. The organisms in the domains *Bacteria* and *Archaea* are single-celled and correspond to the prokaryotic cell type (Fig. 2.1, right side). They are generally characterized by a very simple internal organization: a single plasma membrane-closed compartment where the chemical compounds are dispersed inside an unstructured aqueous phase, called *cytosol*. Inside the cytosol there can be some types of intracytoplasmic structures, which can differ from species to species. One structure that is found in all prokaryotic cells is the *nucleoid*: it corresponds to the prokaryotic chromosome (usually occurring in one copy), the aggregated mass of DNA that is not covered by a membrane. Other well defined intracytoplasmic structures are photosynthetic membranes existing in both phototrophic anoxygenic bacteria and oxygenic cyanobacteria. Bacteria also possess a *cell wall* (Fig. 2.1, Box 2), a external coat of variable thickness which contains layers of a polymer, called peptidoglycan; the cell wall covers the plasma membrane, protecting and giving shape to the cell. In Gram-positive bacteria (so called because they retain Gram's stain), the cell wall (80 nm in thickness) is composed mainly of many peptidoglycan layers, whereas in Gram-negative bacteria it is very thin and constituted by only few peptidoglycan layers covered by an ulterior external membrane. The space between this external membrane and the plasma membrane, called the *periplasm*, contains many types of proteins involved in active transport, protein secretion, and chemotaxis. Classically, the periplasmic space occurs only in Gram-negative bacteria, but nowadays there are claims that a periplasmic space, placed between the plasma membrane and the peptidoglycan layer, can be identified also in Gram-positive bacteria.

Eukaryotic cells, in contrast, are characterized by a more complex internal architecture, consisting of several cytoplasmic compartments as well as other structural elements (Fig. 2.1, left side). First of all, they possess a true *nucleus*, a membrane-enclosed structure that contains the genetic material of the cell, which is organized in several chromosomes whose number is, in almost all cases, characteristic for that living species. In humans, for example, each somatic cell (that is, each cell forming the body) contains 46 chromosomes. Eukaryotic cells constitute the organisms belonging to the third evolutive domain, *Eukarya*, which are usually multicellular, like animals, plants and fungi, though also unicellular eukaryotes exist (e.g. ciliates and yeasts). All eukaryotic cells are furnished with several internal membranes which bound specific compartments (or organelles) inside the cytoplasm; these are the regions where specific cellular functions are performed. For instance, the nucleus is the place where genes are expressed and DNA is replicated for cell division. The nucleus is one of the few internal compartments enclosed by two membranes; similarly, *mitochondria*, the numerous organelles where the cellular energetic metabolism takes place, possess an outer membrane and a (largely infolded) inner membrane. The rough and smooth *endoplasmic reticula* are the compartments specialized in the synthesis of proteins and lipids, while *Golgi apparatus*, a stack of membrane-bound vesicles, directs the freshly synthesized membrane constituents to their right location within the cell (e.g. new membrane proteins directed to the plasma membrane). *Peroxisomes* are the compartments where fatty acids and amino acids are degraded, while inside *lysosomes* (only found in animal cells) the digestion of foreign materials or waste cellular products takes place. Other cytoplasmic organelles exist only in specific types of cells, such as the chloroplasts and vacuoles of plant cells. *Chloroplasts* are the sites where photosynthesis occurs, they possess several internal membrane-bounded and flattened sacs (the thylakoids) that contain the light absorbtion

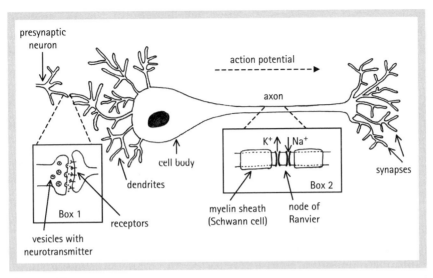

Fig. 2.2 The neuron.

green pigments (chlorophyll). The *vacuole* is a large organelle filled with water and soluble compounds (like sugars), that generates a hydrostatic pressure that is counterbalanced by the plant *cell wall*, a structure that surrounds the cell, giving support and rigidity to it.

Besides the compartmentalized internal structure, eukaryotic cells also contain a complex *cytoskeleton*, which is a highly dynamical network of fibrous elements (the so-called microtubules, actin filaments, and intermediate filaments) that allow the immobilization and the controlled movements of cellular structures and organelles, and at the same time assure structural stability and help the cell to maintain its shape. For this purpose, the work of cytoskeletal elements is usually combined with external mechanical pressure, thus providing the numerous forms that eukaryotic cells can assume in multicellular organisms. In fact, some cells are subject to form variation and need a very flexible boundary (e.g. migrating blood cells), others are characterized by a fixed geometry and assume their shape according to the neighboring cells (e.g. cells of epithelial tissues), while for some cells (e.g. neural cells) the form is an unchangeable cellular aspect that has to be maintained during the whole cell life.

The neuron. A typical mammalian *neuron* is characterized by four parts (the cell body, the axon, the synapses, and the dendrites) that accomplish very different functions in the cell (see Fig. 2.2).

Distinct types of neuronal cells can show different characteristics and morphologies concerning these four parts: for instance, efferent (or motor) neurons—that innervate, e.g. muscle cells—have a long single axon usually insulated by a sheath of myelin; afferent (or sensory) neurons—which transmit sensory inputs from receptor cells—have a two-branched axon that split immediately after the cell body; multipolar interneurons—which connect afferent and efferent neurons—have a highly branched dendritic domain and a single axon. The cell body is the domain that contains the nucleus and the other organelles, it is the cellular site where almost all proteins and membrane components of the neuron are synthesized. The axon is a long cellular extension that departs from the cell body, its function is to connect each neuron to a circuit of many other neurons through the axon terminals, called synapses. The axon is specialized for the rapid transmission of electric impulses (the so-called *action potential*—see also the description of transmembrane potential and transport proteins in Section 2.2.2) unidirectionally within the cell, from the cell body towards the synapses. This electric impulse originates from the numerous electric disturbances that are conveyed from the dendrites to the junction between the cell body and the axon, and there, if the signal is strong enough, the mechanism of impulse conduction is triggered on. In human neurons, especially in motor neurons that possess very long axons, the axon is coated by repeated portions of myelin sheaths (they are stacks of specialized plasma membranes of another type of neuronal cell, the Schwann cells, that wind around the axon), separated each other by unmyelinated areas, called nodes of Ranvier (see Fig. 2.2, Box 2). The function of the myelination is to achieve a 10 to 100-fold increase in the velocity of signal conduction. When the electric signal conducted through the axon reaches the synapses, it induces a rise in the cytosolic (local) concentration of calcium that, as a consequence, triggers the release of neurotransmitters from the excited neuron (see Fig. 2.2, Box 1). The neurotransmitter is a chemical solute contained inside small intracellular vesicles that fuse with the synaptic membrane (by a mechanism called exocytosis—Section 2.2.3). After this membrane fusion, the neurotransmitter diffuses into the narrow space between the excited neuron and the postsynaptic cells, and then binds to specific (excitatory or inhibitory) receptors placed on the plasma membrane of the postsynaptic cells. The binding triggers the opening of ion channels on the postsynaptic cell, thus causing a change in its electric potential. Therefore, the electric signal coming from the presynaptic neuron and spreading along its axon, is initially transformed into a *chemical* signal transmitted at the junction between connected cells, and then it turns back into an *electric* signal spread along the postsynaptic cell. If the postsynaptic cell is a neuron, the neurotransmitter is received at the dendrites, that are extensions of the cell body specialized for the conduction of the electric impulse towards the cell body. The dendrites are usually highly branched, especially in the neurons of the central nervous system (Section 2.3), thus allowing each neuron to receive chemical signals from a very large number of other presynaptic neurons. The postsynaptic cells can also be different from neurons, for instance, they can be muscle or gland cells, in which case the neurotransmitter can either stimulate contraction or induce hormone secretion, respectively. The retrograde signaling of chemicals, going from the postsynaptic neurons back to presynaptic cell, is another process that can take place in the brain, and it is supposed to play a role in the mechanisms of learning.

2.2.2 Cellular Membranes

In this section we focus on the constituents, the functions and some specialized structures of cellular membranes and, in particular, of the plasma membrane. We delineate with some detail the topic of membrane-associated proteins, and specify the numerous roles they partake in the settling of the transmembrane electrochemical potential, the transport of chemicals across the membrane, and the reception and transmission of signals.

The components of cellular membranes. The plasma membrane, as well as the membranes surrounding the internal organelles in eukaryotic cells, is constituted of several kinds of molecules. The most copious constituents of membranes are lipids, proteins, and carbohydrates, whose abundance varies greatly among the membranes surrounding distinct organelles, as well as among different types of cells. In particular, membrane lipids can vary in both quantity and type. The lipids most abundantly occurring in the plasma membranes are phospholipids and glicolipids for prokaryotic cells; phospholipids, glicolipids, sphyngolipids, and cholesterol for eukaryotic cells. The internal membranes of eukaryotic cells are predominantly composed of phospholipids.

There exist four major classes of phospholipids, differing with respect to their molecular composition. Most of them are structurally similar to a two-pronged fork, where two opposite ends can be distinguished (see Fig. 2.3, Box 1): a hydrophilic part constituted by polar—that is, electrically charged—chemical groups, usually called the "head", and a hydrophobic part constituted by apolar chemical groups, called the "tails" (the prongs of the fork). Molecules like phospholipids, characterized by both hydrophobic and hydrophilic parts, are called *amphipathic*. Thanks to this property, phospholipids auto-assemble in sheetlike bilayers, consisting of two external hydrophilic facets, where phospholipid heads can interact with the aqueous solvent, and an internal hydrophobic facet, where phospholipid tails establish interactions among each other. The cellular membranes consist of such closed bilayered-structures, of about 7 nm in thickness, which separate two aqueous compartments.

Membrane proteins. Membrane proteins are classifiable into many heterogeneous classes, according to the function they accomplish within the cell, the molecular weight, the tridimensional structure, etc. Some membrane proteins have enzymatic functions and perform a whole variety of tasks: for instance, they help in regulating the selective permeability of the membrane, the cell signaling and membrane trafficking [6]. Other proteins can be associated with carbohydrates: these proteins, called *glicoproteins*, carry out very important tasks such as hormone

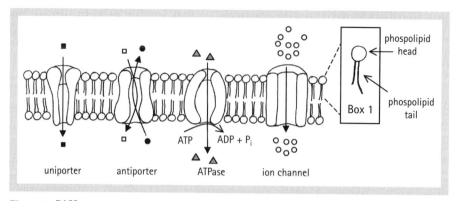

Fig. 2.3 Different transport proteins and solute movements in cellular membranes.

receptiveness, antigenic response, and cellular interactions. Membrane proteins can be divided into two major types, with respect to the way they are associated to the lipid bilayer (Fig. 2.1, Box 1). On the one hand, *peripheral* proteins can be found at both sides of the membrane, facing either the cytoplasm or the external environment. These proteins can be either partially dipped into the external hydrophilic facets of the bilayer, or weakly bound to the phospholipids heads, or else tethered to other membrane proteins; in no case, anyway, do they have direct interactions with the hydrophobic core of the bilayer [6]. On the other hand, *integral* (or transmembrane) proteins are constituted by one or more aminoacidic domains which span the whole depth of the bilayer, hence facing both sides of the membrane. Integral proteins usually have a very specific orientation with respect to the internal and the external sides of the membrane: this fact confers to the protein different physical and chemical properties, which provide an essential role for its function (e.g. for the inward or outward transport of metabolites, or for signal transduction). We will describe later on three principal functions of transmembrane proteins: the transport of ions and solutes, the histocompatibility, and the reception of extracellular signals.

The lipid and protein composition of a cellular membrane determines its fluidity and dynamical mobility. Indeed, biological membranes are not rigid structures: due to thermal motions, phospholipids can rotate around their longitudinal axis, and also diffuse laterally within the layer where they reside, thus continuously randomizing the position of the individual molecules. Phospholipids can also transversely diffuse (flip-flop) from one layer of the membrane to the other one: this process is not as common as lateral diffusion and has to be catalyzed by transmembrane enzymes called flippases.

The first successful conceptualization of the architecture of cellular membranes was proposed by Singer and Nicholson [33, 32], and it is known as the *fluid mosaic model*. This model was developed on the principle of Brownian motion, which explains the diffusion and collisions of molecules as the effect of agitation movements. According to this model, the cellular membrane is a bidimensional "sea of lipids" where lipids, as well as integral proteins (considered as monomeric and low abundant structures) can freely float undisturbed. Recently, several criticisms have been put forward against the fluid mosaic model, since new experiments evidenced that the lateral movements of both membrane lipids and proteins can be actually constrained by various conditions [15, 19]. Therefore, new models for the membrane organization have been proposed, suggesting that cellular membranes can be actually seen as "compartmentalized fluids" [10], where membrane proteins have differential interactions with other proteins and lipids, the membrane thickness is variable (due to the formation of microdomains of clustered molecules), and diffusion is not driven by simple Brownian motions (see also [4] for a formalization of these concepts).

The transmembrane electrochemical potential. The different concentrations of solutes on the internal (cytoplasmic) and external sides of the plasma membrane of most eukaryotic and prokaryotic cells, generate an *electrochemical potential* [24]. In general, if two aqueous compartments with distinct concentrations of *neutral* solutes are separated by a *permeable* membrane, then the solutes simply diffuse across the membrane from the compartment of higher concentration towards the other one, until the chemical equilibrium between the two compartments is achieved. In other words, the thermodynamic processes tend to push the *chemical* gradient (that is, the difference in solute concentrations) towards zero. Actually, cellular membranes face two aqueous solutions usually composed of ions, proteins, and other *charged* solutes; therefore, an *electrical* gradient across the membrane is also present. In this case, the spontaneous movement of such electrically charged solutes is governed by both the electrical potential and the chemical gradient: the flow of the charged solutes across the membrane is driven by a force that tends to nullify the sum of the two gradients, that is, to balance the transmembrane *electrochemical* potential.

However, the membrane bilayers are not permeable to all solutes. Small uncharged polar molecules and gases (e.g. oxygen, carbon dioxide) can undergo simple diffusion across cellular membranes down their concentration gradients (at no energy expense by the cell itself); other chemicals, such as water and urea, are only slightly diffusible and usually their passage is accelerated by transport proteins. In contrast, large uncharged polar molecules (glucose), ions (calcium, sodium, potassium, etc.), and charged polar molecules (amino acids, ATP, etc.) cannot freely cross cellular membranes. This selective permeability assures the maintenance of the proper transmembrane electrochemical potential value, that is critical for the physiological functioning of many cellular processes (e.g. the conduction of the electric impulse in neurons, the contraction phenomena in muscle cells, the uptake of amino acids or other molecules, etc.). The magnitude of the electrochemical potential across the plasma membrane of most animal cells is in between −30 mV and −70 mV, at rest—with the cytoplasmic compartment always negative with respect to the extracellular region—and generally it does not vary in time. In some specialized (electrically active) cells, the electrochemical potential undergoes sudden variations in time. In neurons, this phenomenon is termed the *action potential*, and consists of repeated cycles of rapid depolarizations (when the neuronal potential becomes less negative, passing from around −60 mV with inside negative, to the approximate value +50 mV, with inside positive) and rapid repolarizations (when the membrane potential returns to the resting value). This electrical activity, that moves in one direction along the axon at a speed of up to 100 m/s, is the result of the opening and closing of specific transport proteins (voltage-gated ion channels) which, by allowing the controlled movement of ions inwards and outwards, can influence the electrochemical potential along the axonal membrane.

Transport proteins. Given the selective permeability of cellular membranes, the difference in the electrical and chemical concentrations across the bilayer is maintained thanks to specific transmembrane proteins, called *transporters*, which allow the (electrogenic) passage of specific ions and solutes. Transporters can mediate the passage of solutes in two distinct ways. When the solutes move down their electrochemical gradient, or "downhill", the process is termed *passive transport*, and it happens with no energy expense by the cell itself (that is, the whole transport proceeds in an *exergonic* way); in this case, solutes are never accumulated above their equilibrium point. When the solutes move against their electrochemical gradient, or "uphill", the process is called *active transport*; in this case, instead, the solutes can be accumulated above their equilibrium concentration. Active transport requires the expense of additional energy, since the movement of solutes is not favorable from a thermodynamic point of view (that is, the process is *endergonic*). The cellular energy required to carry out the catalytic active transport can be taken from other exergonic reactions such as ATP hydrolysis, sunlight absorption, or the coupling with a downhill flow of other solutes.

Two major classes of transporters can be identified, *carriers* and *channels*, differing for their structural and functional properties, the stereospecificity (i.e. chemical affinity) for the molecular species—or a class of closely related solutes—that they transport, the rate of transmembrane movement, the saturability (that is, the existence of a solute concentration threshold above which the transporter will not function at a higher rate), etc. Carriers are usually saturable monomeric proteins, which bind the transported solutes with high specificity, and move them at low flow rates. Channels are usually oligomeric complexes (with helical transmembrane domains or barrel-shaped structures), which show less stereospecificity than carriers but larger transport rates. Within the broad classification of transporters into carriers and channels, several other families of related transport proteins can be distinguished [24]. In Fig. 2.3 we give a simple graphical representation of some carriers and channels hereafter described.

Among carriers, the family of *porters* can facilitate the downhill diffusion of a solute, or couple the uphill transport of one molecule with the simultaneous downhill transport of a different type of molecule. In the first case, the carriers are called *uniporters*, while in the latter case they are called *symporters*, if the transport of the two molecules occurs in the same direction across the membrane, or *antiporters*, in the case that the passage occurs in opposite directions. The transport of a molecule by means of a porter requires a conformational change in the protein structure, a process which slows down the transport rate (about 10^2–10^4 molecules per second). To carry out the passage across the membrane, a conformational change is necessary also for the carriers belonging to the different family of *P-type ATPases* (or ATP-powered pumps). In this case, the structural transformation of the transporter is driven by the protein phosphorylation, which takes place via the hydrolysis of ATP molecules (see also Section 2.2.3). The final effect is the capability to move ions or other small molecules against their electrochemical gradient in an active way. The transport rate of pumps is about 1–10^3 molecules per second. Well-known examples of P-type pumps are the sodium–potassium ATPase in the plasma membrane of animal cells, the calcium ATPase in the plasma membrane and in the sarcoplasmic and endoplasmic reticulum, and the proton-potassium ATPase in the epithelial cells of the stomach.

Other active-transport carriers are the *F-type* and *V-type* ATPases: they both catalyze the uphill passage of protons and share a similar structure, consisting of a transmembrane protein complex forming a pore for proton movement, and a peripheral part for ATP-binding and hydrolysis. F-type ATPases are needed for energy-conserving reactions in mitochondria and chloroplasts, while V-type ATPases are responsible for the acidification of intracellular compartments (lysosomes, endosomes, the Golgi apparatus) in many organisms.

Among channels, we can distinguish between the families of aquaporins and ion channels. *Aquaporins* allow a very rapid movement of water molecules in a continuous flux, at rates of about 10^9 molecules per second. For instance, the erythrocytes— the red blood cells responsible for oxygen transport through the bloodstream—are particularly rich in aquaporins in their plasma membrane, which permit these cells to rapidly swell or shrink (by an inward or outward passage of water molecules, respectively) according to the extracellular osmotic gradient. By working in coordination with pumps, *ion channels* control the membrane permeability to various ions (sodium, potassium, etc.) and hence the transmembrane electrochemical gradient. In contrast to carriers, the transport rate of ion channels can be up to 10^7–10^8 ions per second; in addition, they are not saturable (that is, the flow does not reach a maximum rate when the solute concentration is high). Ion channels can be open or closed, according to specific cellular events: *ligand-gated* channels open in response to the binding of extracellular or intracellular molecules (e.g. extracellular neurotransmitter at postsynaptic cells, intracellular second messengers, etc.), while *voltage-gated* channels respond to changes in membrane electrical potential. The channel opening is usually very fast, occurring in fraction of milliseconds, and the period of opening is also transient and short (around few milliseconds). These properties of ion channels make them play a central role in the propagation of the action potential in neurons, and in the triggering of contraction in muscle cells. In fact, in the axon, selective voltage-gated channels are—together with the sodium-potassium ATPase—the main responsible for the sequential and transient increase in the membrane permeability for sodium and potassium ions, as they cyclically open and close in response to an initial variation in the membrane potential. In particular, sodium channels are characterized by refractoriness, that is, they get inactivated for a short period of time immediately after their opening, and will not open again until the resting transmembrane potential is reestablished. This property assures that the action potential is spread unidirectionally downstream along the axon as, thanks to the refractory period, the electrochemical potential cannot be modified backwards. The axon channels are mainly concentrated at the nodes of Ranvier (the portions of the axon membrane that freely face the extracellular fluid), thus assuring that the action potential "jumps" from an unmyelinated area to the successive one, without undergoing any attenuation (see Fig. 2.2).

Finally, the family of *ABC (ATP-Binding Cassette) transporters* is composed of ATP-driven proteins that move several solutes (e.g. proteins, amino acids, lipids, metal ions, etc.) against their concentration gradient, either inwards or outwards the cell. Most ABC transporters act as pumps, and have high stereospecificity for the transported solutes, others function as (ATP hydrolysis-gated) ion channels. Two typical examples of proteins in this family are the multidrug transporter MDR1 [14], which is responsible for tumor resistance against drugs, and the flippases, which shift lipids from one layer of cellular membranes to the other one.

Cell surface receptors. In addition to the transmembrane proteins that are specific for the transport of diverse solutes across the lipid bilayer, the cell surface is equipped with other kinds of proteins that are needed to detect extracellular molecules, recognize non-self entities inside the organism, or provide the adhesion of the cell to other cells and to the extracellular substance. All these processes take place through the binding of an external *ligand* (the signaling molecule), or a molecule occurring on the surface of another cell in the case of cell adhesion mechanisms, to a complementary transmembrane protein called *receptor*. The binding induces a conformational change of the receptor, thus activating a proper cell response (see also the description of signal transduction pathways in Section 2.2.3). Besides the already cited ligand-gated ion channels, the signal transducing receptors can be classified into two broader classes, according to the related type of intracellular protein that "receives" the external signal. *G protein-coupled receptors* activate GTP-binding proteins (the so-called *G proteins*) that act as intracellular "molecular switches", since they can be either in an active or inactive form. *Protein kinase-linked receptors* activate cytosolic or membrane-associated kinases that act by phosphorylating other intracellular proteins. In both cases, through the activation of these effectors, the cell is able to induce a positive or negative regulation of a cascade of downstream enzymatic reactions, that might eventually control gene expression inside the nucleus.

Other cell surface proteins, specific for the detection of particles or infectious agents that are extraneous to the organism, can be found on the plasma membrane of the eukaryotic cells of the immune system (the white blood cells called *lymphocytes*) or of connective tissues (the *macrophages*). The recognition process takes place through the binding between an *antigen*, a substance that can elicit a response of the adaptive immune system, and an *antibody* (or immunoglobulin), a protein that is able to interact with a particular site (called epitope) of the antigen thus facilitating its destruction. The response of the immune system mediated by cell surface receptors involves a particular class of lymphocytes, called *T cells*. These cells are able to react to (fragments of) foreign antigens that are present on the plasma membrane of infected host cells. For instance, they are able to kill virus-infected host cells that display viral antigens on their surfaces. The antigen-activated receptors of T cells are deeply embedded within the lipid bilayer, so they can transduce the signal of non-self detection to the cell interior and therefore trigger the proliferation of immune cells. The attack of T cells against infected cells is also mediated by cell–cell adhesion molecules, that assure the tight binding between the cytotoxic immune cells and the target infected cell.

A different family of cell surface proteins, generally called *cell adhesion molecules* (CAM), are critical for the formation of tissues in multicellular organisms, and for connecting the cytoskeleton of one cell either to other adjacent cells or to the extracellular matrix. The main types of cell–cell junctional systems will be discussed in Section 2.3.

Specializations of the plasma membrane. The plasma membrane of differentiated cells in vertebrate organisms, as well as the external envelope of prokaryotic cells, can present numerous extraflexions, invaginations, or other peculiar structures, which serve the scope of enhancing the cell functionality. For instance, in the monolayered epithelium of the small intestine (see also Section 2.3), each cell is furnished with thousands of finger-like projections, called *microvilli*, which increase the surface of the plasma membrane. A direct consequence of the presence of microvilli is that the number of transport proteins occurring on the plasma membrane is very high, therefore enhancing the absorbing capacity (internalization of nutrients and other metabolites) of the epithelium. Other epithelial cells can present on their luminal surfaces hundreds of regularly lined-up, hairlike appendages called *cilia*, which are highly specialized structures constituted by a complex organization of cytoskeletal microtubules and other motor proteins. By moving in a whip-like coordinated (but not constantly synchronized) motion, cilia can drift fluids over the surface of a tissue. For instance, in mammals the epithelial cells lining the respiratory tract and the oviduct are furnished with a huge number of cilia (around $10^8/cm^2$ or more) whose functions are, respectively, to sweep away the mucus, particles, and bacteria from the lungs, and to drive eggs toward the uterus. In contrast, the unicellular protozoans called ciliates synchronously use cilia both to swim in the surrounding environment and to bring food inside.

An efficient motility machine of cells is the *flagellum*, a flexible extension found in some prokaryotic cells as well as in some eukaryotic cells. The movement of flagella can be undulating, oscillating, or coil-fashioned, and it enables the cell to swim through a fluid. For instance, the mammalian spermatozoon has a single flagellum, the unicellular green alga *Chlamydomonas reinhardtii* has two flagella, which are used to propel cells forward. In contrast to eukaryotic flagella, which are constituted by membrane projections and contain microtubule elements, bacterial flagella have a simpler ultrastructure. They are composed of repeated subunits of the protein flagellin, which form a rigid helical filament, driven by a complex rotary motor embedded in the bacterial cell wall (see also the chemotactic phenomena described in Section 2.2.3).

Other locomotion machines of bacterial cells are the *type-IV pili*. They are hairlike appendages that bacteria use to adhere to a solid substrate (an abiotic surface or other bacterial cells), and then move along it with a particular motion, called *twitching*, which is the effect of the continuous extension and retraction movements of pili. This motion is 100 times slower than the flagellar swimming of bacterial planktonic cells, and appears as small, intermittent jerks along the substrate. The bacterium *P. aeruginosa* uses twitching to swim along a surface [25, 34], looking for the most appropriate spot to attach and start the formation of structured and sessile bacterial communities (see also the description of biofilms in Section 2.4).

2.2.3 Intracellular Processes and Communication with the Environment

Cellular biochemical networks consist of several processing modules that allow a living cell to communicate with the extracellular environment—by receiving, integrating, and amplifying signals, and then activating an appropriate response—or to utilize externally provided metabolites to obtain the energy necessary for all cell activities. Ultimately, all these reactions are connected through complex interplay mechanisms to a very important intracellular process, that is, gene expression. In this section, after a basic description of metabolic processes, we provide a general image of the regulation of metabolism and of cellular activity, which are implemented through the control of both gene expression (at the level of transcription and translation) and enzyme activity (at the so-called post-translational level) in eukaryotes and prokaryotes. Finally, we focus on two distinct ways to communicate with the environment: endocytosis and exocytosis in eukaryotic cells, chemotaxis in bacterial cells.

Metabolism and cellular energetics. The term *metabolism* refers to the totality of cellular reactions, which (didactically) can be divided into two classes: *catabolic reactions* are those that degrade chemical compounds into simpler ones and release energy, while *anabolic reactions* are the chemical processes that use energy to synthesize the complex molecules that are ultimately required for the ultrastructural components of the cell (cellular membranes, proteins, nucleic acids, etc.). The energy that is necessary for cells to operate through these biochemical pathways is obtained from the outside world via some intermediate compounds: *autotrophic organisms*, such as plants and some bacteria, use radiant energy directly provided from sunlight, while *heterotrophic organisms*, such as animals, utilize the chemical energy contained inside nutrients. The principal form of energy generated and exploited inside all types of cells is held by the molecule *adenosine triphosphate*, or ATP. Through the *hydrolysis* of one molecule of ATP, that is, its transformation into a "discharged" molecule of ADP (adenosine diphosphate), the cell can gain 7.3 kcal/mol of free energy, which can be used to power other energetically unfavorable reactions. For this purpose, the cellular concentration of ATP has to be maintained at proper levels, and regenerated whenever it tends to decrease. The chemical process from which ATP is mainly produced is the degradation of fatty acids and sugars; from one molecule of glucose, for instance, eukaryotic and prokaryotic cells can generate 36 and 38 ATP molecules, respectively. The synthesis of ATP molecules requires a complex cascade of enzymatic reactions, which involve the *oxidation-reduction* of molecules, whereby either one proton or one electron is transferred from one chemical compound (the reducer, or donor) to another one (the oxidant, or acceptor). In the majority of cells, these enzymatic cascade occurs via *aerobic oxidation*, while in leaf cells and in some unicellular organisms (e.g. cyanobacteria) it takes place also via oxygenic *photosynthesis*. In the aerobic oxidation of glucose, for instance, the first cascade phase (termed glycolysis) does not require any molecular oxygen and occurs in the cytosol of both prokaryotic and eukaryotic cells. The second cascade phase, instead, requires molecular oxygen and takes place in the mitochondria of eukaryotes (through other phases, termed Krebs cycle and oxidative phosphorylation [24]), or at particular cytoplasmic regions of the plasma membrane of prokaryotes. In contrast, photosynthesis produces oxygen and carbohydrates: it occurs on the thylakoid membrane of chloroplasts, where light energy is both converted into ATP and stored into carbohydrates, while oxygen is released into the environment.

Though apparently very different, the energy conserving steps of aerobic oxidation and of photosynthesis are driven by the same kind of biochemical process, called *chemiosmosis*, and share the same kind of membrane protein (ATP-synthase) to regenerate ATP from the discharged ADP molecules. According to the chemiosmotic theory (see, e.g. [23]), the energy required to synthesize ATP comes from the electric potential and the proton concentration (pH) gradient across the (mitochondrial or chloroplast) membrane. This force, called *proton-motive*, is produced by the coupling of a step-by-step transport of electrons—from reducer to oxidant molecules, via a chain of electron-carrier protein complexes—with the pumping of protons across the (inner mitochondrial, or thylakoid, or bacterial) membrane against their concentration gradient. The subsequent downhill passage of protons then activates the synthesis of ATP, catalyzed by a member of the F-type membrane protein family.

Gene expression and regulation. The control of metabolism is one of the essential properties of all living organisms belonging to the three domains of life. For example, based on the control of its biochemical reactions, a bacterial cell synthesizes the amino acids it needs only in the case that they are not present in the external growing medium and cannot be taken up from the exterior of the cell. This way the cell more efficiently uses its resources of chemicals and energy, as the process of transmembrane transport is faster and less energy consuming than the intracellular synthesis of amino acids. In both eukaryotic and prokaryotic cells, the control of metabolism is performed at the level of gene expression, during transcription and translation, and with post-translational modifications. *Transcription* is the process by which a messenger ribonucleic acid (mRNA) is synthesized from a (complementary) template of DNA, which takes place after the binding between the DNA and an enzyme called *RNA polymerase*. Transcription can be facilitated or inhibited by the so-called transcription factors, a group of several distinct proteins that bind to specific portions of DNA and act as activators or repressors, thus controlling the transfer of the genetic information from genes to mRNA. *Translation* corresponds to the first phase of the synthesis of proteins (more precisely, polypeptides) from an mRNA template: it takes place by means of large molecular complexes, called *ribosomes*, through three consecutive steps termed initiation, elongation, and termination. In prokaryotic cells, where the DNA is not covered by a nuclear membrane, the process of translation can begin alongside with transcription: as soon as a part of the transcribed mRNA is accessible by the ribosomal units, the transformation of the ribonucleotide sequence into the corresponding amino acids chain can start. In contrast, in eukaryotic cells, translation occurs inside the cytoplasm, after the mRNA sequences have been transported outside the nucleus (which is the site of eukaryotic transcription). Before this (mature) cytoplasmic mRNA can be translated, it must undergo several post-transcriptional modifications, the most relevant being the *splicing* processing. This operates by removing certain stretches (called *introns*) of the nuclear pre-mRNA, and reassembling the remaining protein-coding segments (called *exons*); sometimes, this process gives rise to different functional mRNA sequences, thus allowing a single gene to code for multiple proteins. Finally, the term *post-translational modification* refers mainly to different types of alterations of proteins, occurring as one of the latest phases of their biosynthesis.

These modifications can involve the addition of chemical functional groups (e.g. phosphate, acetyl, methyl groups), the linking to other peptides or proteins (e.g. ubiquitination), and the change of protein's structure (e.g. by the formation of disulfide bridges), etc.

The controlling processes occurring at each of these levels interact with the other level both in time and in space. For example, post-translational mechanisms govern the cell's responses in the range of about 10^{-4}–10^2 seconds, while transcription and translation govern responses in the range of about 10^2–10^8 seconds [21], thus covering a wide range of time scales. The control occurring at the level of transcription is responsible for the fact that, in each cell, only some of the genes are active either in specific conditions, or in specific phases of the cell cycle. In the bacterium E. coli, for example, around 800 genes (up to a total of 4500 genes) are active in transcription in normal conditions. When something important changes in the growing medium (e.g. temperature, oxygen concentration, nutrient availability, etc.), then the transcription of some genes can be either stopped or started. In other words, the activity of genes can be modulated in time, by constitutively expressing some of them while partially or totally repressing others.

In bacteria the majority of the genes are found grouped together in operons. The *operon* is a cluster of genes that encode proteins involved in the same metabolic process. One of the best known operons is called *lac*. It consists of three structural genes coding for three proteins which enable the cell to utilize lactose as nutrient. During transcription, the genetic information contained in the operon is transcribed into a unique molecule of mRNA (termed polycistronic mRNA) which delivers the information for the three proteins coded by the genes within the operon. In the last decade, it became more evident that the response of a given bacterium to an environmental or internal change involves not only a single operon, but also the activity of many operons organized in networks (generically called *global regulatory systems*), which are so complex and diverse that a specialized nomenclature is used to describe them: regulon, modulon, and stimulon (see [28] for further details). The control of transcription in bacteria can occur either at the level of single operon or on whole networks of operons. For instance, the positive and negative regulation of transcription initiation of an individual operon occurs via several mechanisms, such as the modulation of promoter strength (that is, the ability to capture RNA polymerases, which can be significantly enhanced by activators), the intervention of initiation factors called alternative sigma factors (whose intracellular concentration changes during major events), the chemical modification of DNA in the promoter region or the degree of supercoiling of DNA, etc. On the other side, the control of transcription termination occurs when, for example, a specific protein (e.g. the factor Rho) blocks the activity of the RNA polymerase. In eukaryotic cells, genes are not organized into operons thus, contrary to bacteria, each molecule of mRNA contains the information for the synthesis of only one type of protein (in this case, the mRNA is termed monocistronic). The control of gene expression in eukaryotic cells involves both non specific proteins, which are part of the transcription initiation machinery common to all genes, and regulatory transcription factors, which are instead specific to one or more genes. All these coarse and fine control mechanisms modulate the synthesis of mRNA within a narrow range, thus adjusting the concentration of mRNA to the needs of the cell at that moment.

The control at the level of translation occurs, in bacteria, by repressor proteins, single-stranded RNA sequences that are complementary and bind to mRNA molecules (antisense RNA), or by metabolites binding to a specific segment of mRNA (riboswitch), thus blocking translation, while in eukariotes it occurs by phosphorylation of specific proteins, antisense RNA or via the control of mRNA stability, which is influenced by specific RNA-binding proteins. At the level of enzymatic activity, the post-translational control is a very fast process (taking place in the order of seconds) enabling the cell to adjust its metabolic pathways to unexpected changes in the surrounding environment. The main mechanisms of post-translational control common to the three domains of life involve several processes: the reversible binding of a small sized molecule (called allosteric effector) to an enzyme that changes the conformational state of the enzyme and therefore its activity; the covalent modifications of an enzyme that control its activity by adding to (or removing from) the protein small chemicals such as phosphate or methyl groups; the control of biosynthetic pathways by feedback inhibition; the degradation of proteins; the differential compartmentalization of enzymes and metabolites within separate sites, such as cellular structures or membrane protected spaces; and so on. For a thorough description of gene expression and regulation in prokaryotes and eukaryotes we refer to [31].

Signal transduction. Signal transduction is the process by which extracellular signals are converted into specific cellular responses through a cascade of reactions that are initially mediated by molecules synthesized and released by other cells. In multicellular organisms, for instance, signaling molecules can act on distant or close cells (via hormones or neurotransmitters, respectively), as well as on the cell itself (via growth factors), and are detected by specific cell surface receptors. In general, the interaction between the external signaling molecule and its specific receptor leads to the appropriate cell response, through the synthesis or the recruitment of intracellular signaling molecules called *second messengers* (such as cyclic AMP, inositol triphosphate, calcium, etc.). In turn, second messengers activate the cellular machineries that induce other metabolic processes (that eventually involve gene expression regulation through the activation of repressor or promoter proteins), until the transduction event terminates and the cell can return to the prestimulated state.

Different types of cells can possess distinct sets of membrane receptors, therefore they can differentially respond to the same type of ligand. However, it is a widely accepted opinion that cells are characterized by evolutionary highly conserved signal transduction pathways, that are all shared by distinct receptors and by common second messengers (see the notion of "bow-tie" organization in [17, 13]). These pathways, in addition, are usually activated by integrated mechanisms, so that multiple signals can be simultaneously received inside the cell, and give rise to a unified proper response that preserves the homeostasis of both the cell and the organism.

Endocytosis and exocytosis. Eukaryotic cells use several processes, generally known as *endocytosis*, to internalize external macromolecules or microorganisms. On the other side, there exists an apposite mechanism, called *exocytosis*, by which the cell can release waste or secretion products into the extracellular space. Though acting in opposite directions (inwards and outwards, respectively), both endocytosis and exocytosis directly involve the plasma membrane of the cell and small membrane-bounded regions called *vesicles*.

Exocytosis takes place as follows: the freshly synthesized proteins are carried inside small transport vesicles—just in time sorted from the rough endoplasmic reticulum, through the Golgi complex—which move towards the plasma membrane and eventually fuse with it. The proteins contained inside these vesicles are thus targeted to the cell surface and released outside the cell, a phenomenon that is especially evident in gland cells. A less usual form of exocytosis occurs directly from the plasma membrane without the intervention of intracellularly created vesicles; in this case, evaginations towards the extracellular space are formed and then they bud off, thus encapsulating small portions of the cytosol and reducing the plasma membrane total surface.

Endocytic mechanisms can be broadly classified into *phagocytosis*, *pinocytosis* and *receptor-mediated endocytosis*. Phagocytosis requires the protrusion of the plasma membrane into cup-shaped extensions, held by cytoskeletal (actin) elements, which envelope large external particles (e.g. bacteria) and direct them to the lysosomes for degradation or digestion. In macrophage cells, in particular, this process is usually driven by antibody-antigen recognizing proteins occurring on the surfaces of both the macrophage and the bacterium. In pinocytosis, or macropinocytosis, the cell internalizes external fluid and molecules in a (usually) non-specific manner: invaginations of small portions of the plasma membrane pinch off and form uncoated vesicles, that are then directed to specific compartments (the endosomes) for further processing. The endocytosis strictly mediated by receptors is a more selective pathway and takes place through a different mechanism. In this case, the macromolecules that have to be internalized (e.g. insulin, transferrin, cholesterol-rich low density lipoprotein, etc.), specifically bind to their transmembrane receptors that are placed in correspondence to plasma membrane pits, coated by specific classes of different proteins (clathrin, caveolin, etc.). When the binding between the receptors and the extracellular ligands takes place, these coated pits pinch off into coated vesicles, which are targeted to and fuse with the endosomes, where the internalized macromolecules undergo further processing, while coating proteins and receptors are recycled back to the plasma membrane. We refer to [20] for a detailed description of other coating protein-dependent and independent pathways of endocytosis.

Chemotaxis. Chemotaxis is a typical behavior of planktonic (free floating) prokaryotes—though it can be exhibited also by some eukaryotic cells—which is ruled by an efficient signal transduction pathway. Chemotaxis allows cells to perform directional and biased motions in their surrounding environment, in the presence of gradients of specific chemical ligands which act as attractants or as repellents. The chemotactic pathway has been well characterized at the molecular scale in bacteria. It consists of several transmembrane and cytoplasmic proteins which act, respectively, as signal sensors and response regulators [12]. These proteins govern the reversal switch of the flagellar motors (molecular complexes that interact with the filaments of the flagellum) through a cascade of post-translational modifications. The movements of swimming bacteria commonly consist of alternating switches between clockwise and counterclockwise rotations of the flagella. When the flagella rotate clockwise, they are unorganized and cause random changes in the swimming direction—this motion is termed *tumbling*. When the flagella are rotating counterclockwise, they work together in a coordinated way and produce a smooth swimming in a single direction—this motion is termed *running*. In the presence of a homogeneous environment, the frequency of the switches between running and tumbling allows the bacterium to perform random walks and a sampling of its surroundings. In the presence of a chemical ligand gradient acting either as attractant or repellent, the bacterium quickly induces a reduced frequency of the flagellar switch, which causes a consequent longer running and motion in a biased direction (that is, towards the attractant or away from the repellent concentration gradient). However, when the concentration of the ligand remains constant in time, then the frequency of the switching is reset from the pure running to the steady-state tumbling value. This bacterial behavior is a well known example of biological robustness and adaptation, two important properties which allow bacteria to respond to the environmental changes in an optimal way: by showing insensitiveness to the values of the ligand concentrations in a stable environment, bacteria are able to sense a broad range of ligand concentration gradients, thus always moving in the best direction for their feeding and survival [2, 36].

2.3 TISSUES

In "superior" multicellular organisms (those derived from a zygotic cell by repeated cellular divisions), a progressive organization of differentiated types of cells takes place: they assemble together in the formation of morphologically and functionally distinct complexes, called *tissues*. This process, called *histodifferentiation*, is due to the fact that, even if all cells of an organism share a common set of genetic instructions, they can be different from one another and accomplish various functions as they are able to express distinct parts of their genetic makeup. Tissues originate during the embryogenetic development: starting from three cell layers called *endoderm*, *ectoderm*, and *mesoderm* (they are invaginations of cells, named gastrula, that develop in an early embryonic form after the cleavage of a fertilized single cell egg), various groups of cells initiate to differentiate and rough out those organism parts that will then give rise to organs and apparatus. For instance, the gastrointestinal tract, respiratory tract, and endocrine organs originate from the endoderm, the nervous systems and skin originate from the ectoderm, while the muscles, bones, urinary and reproductive systems originate from the mesoderm.

In this section, we first focus on the four principal types of vertebrate tissues (the connective, epithelial, muscular, and nervous tissues), by providing a brief description of the structure and functions of each one. Then, by taking as prototypal example the epithelial tissue, we depict the cell–cell adhesion structures (metabolic and mechanical junctional systems) that are needed for cells to communicate and coordinate each other.

Tissues. The *connective tissues* have supporting, protective, and trophic functions for organs and for other tissues in the body. The bone, cartilage, and adipose tissues are all examples of connective tissue, each one consisting of different types of differentiated cells (the osteoblasts, chondrocytes, and adipocytes, respectively). Unlike other tissues, the connective tissue contains relatively fewer cells, and is characterized by an abundant *extracellular matrix*, an amorphous and complex network of macromolecules (collagen and elastic fibers, glicoproteins, etc.) within which a few cells are sparsely distributed. The components of the extracellular matrix are mostly secreted by cells called *fibroblasts*, while another type of connective cell, the *macrophages*, accomplish the function of removing foreign or defective material from the tissue through the processes of phagocytosis and pinocytosis (Section 2.2.3).

On the other side, the *epithelial tissue* represents an opposite example with respect to the inherent structural organization (see Fig. 2.4, left side). It is constituted by a spare extracellular matrix, within which copious cells are closely connected to one another by means of cell–cell adhesions (see the description of cell–cell junctional systems further on), and it is organized in monolayered or stratified sheets called epithelia. The principal functions of epithelial tissues are: the lining of internal cavities, such as lungs, intestine, stomach, and of the free surfaces of the body (i.e. the skin); the protection against mechanical, chemical or physical injuries; and the reception of various types of stimuli, etc. The epithelial cells can have distinct forms (i.e. flat or columnar) and functions, according to

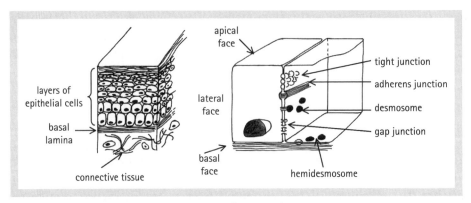

Fig. 2.4 A multilayered epithelial tissue (left side) and cell–cell junctional systems in monolayered epithelial cells (right side).

their position in the organism. For instance, the epithelial cells lining the intestine are rich in microvilli (Section 2.2.2) and are specialized in the absorption of the products of digestion (glucose, amino acids), which they transport from the organ lumen into the bloodstream. The parietal cells in the gastric lining, instead, are specialized in the acidification of the stomach lumen by the secretion of hydrochloric acid.

The *nervous tissue* has an extremely complex structure, highly specialized for the transmission of (electric) signals, the control and the coordination of the body. The nervous system in vertebrates is composed of the central nervous system (CNS, formed by the brain and the spinal cord, that directly receives the sensory inputs from eyes, ears, tongue and nose), and the peripheral sensory and motor nervous systems (PSN and PMN, respectively). The PSN forwards the somatic and visceral reception inputs to the CNS, while the PMN receives the signals from the CNS and transmits them to either voluntary or automatic muscles (like skeletal muscles and heart, respectively). Millions of cells constitute the nervous systems; the principal among these are the neurons (Section 2.2.1), whose functions are to integrate the information that is constantly occurring inside and outside the body (sensory neurons), and then convey it either to other nervous cells or to muscular, glandular, or epithelial cells, in order to trigger the appropriate body response (motor neurons). These functions are accomplished through the electric and the chemical signaling capabilities of neurons, which make them able to process and conduct the information within a single cell (via the action potential), and to transmit this information between connected cells (via neurotransmitters at synapses). For this purpose, for instance, brain neurons are structured in very complex, interconnected circuitries, which are far from being static and rigid organizations: neuronal plasticity during development, or due to experience adaptation, is generally accepted to play a major role in learning, memory, and cognition processes. Other cells that constitute the nervous system are the neuroglial cells, which fill out the space between neurons, and accomplish nutritive and modulating functions for neurons.

Finally, two types of *muscular tissue* exist, called smooth and striated, which are responsible for the force and movements of the body, thanks to the production of electric impulses that drive the contractile mechanisms of muscular excitable cells. The smooth muscle is usually found in the wall of tubular organs (e.g. the intestine and many blood vessels), and is constituted of long thin cells that are able to contract slowly and to maintain the contraction state for a long time. On the other side, cardiac and skeletal muscles, which form the striated muscular tissue, are composed of other types of cells that work in a very different mode. The cardiac muscle cells are branched and electrically connected each other by means of gap junctions (see below); this organization allows them to simultaneously contract in a rhythmic, automatic, and continuous way to assure the heart beating. Skeletal muscles, are formed of bundles of fibers—each one corresponding to a giant cell—whose (voluntary) contraction is mediated by the cells of the CNS that innervate the muscle.

As a concluding remark about tissues, we sketch out the tissular organization in plants, where fewer cell types than in animals exist, and three major kinds of tissues can be distinguished. The vegetal cells constituting the *ground tissue* are principally responsible for the structural prop and metabolism (especially for photosynthesis). The protection and the absorption of nutrients is carried out by the cells in the *dermal tissue*, while the transport of metabolites and water throughout the plant is assured by the cells of the *vascular tissue*. We refer the interested reader to [29] for a thorough description of vegetal cells and tissues.

Cell–cell junctional systems. Some cells possess a specific orientation within the tissue, called *polarity*, which characterizes two opposite sides of the cell with respect to the structural and the molecular points of view. In monolayered epithelial cells, for instance, the region of the cell facing the lumen is morphologically and functionally distinct from the opposite region; these regions are called, respectively, the *apical* and *basal* faces (see Fig. 2.4, right side). The other sides of the cell, that are bound to the adjacent cells within the same epithelial layer, are called the *lateral* faces. The plasma membrane of epithelial cells can present various specialized regions; in Section 2.2.2 we described the morphological structures (microvilli, cilia) occurring on the apical face. Here we focus on the specializations occurring on the basolateral face, which correspond to junctional systems that allow neighboring cells to work in a coordinated manner—by controlling the movement of ions and small molecules between them—and anchor the epithelial layer to the underneath extracellular matrix. Cell–cell junctions can be classified into two major functional groups: *adherent junctions* and *communicating junctions* (Fig. 2.4, right side). On the one hand, adherent junctions are needed to mechanically connect each other cells, and can be further distinguished into *occluding* and *anchoring* junctions. On the other hand, communicating junctions are transport systems that allow the passage of small molecules between adjacent cells.

The occluding junctions that are located immediately below the apical surface are called *tight junctions*, or "zonula occludens". Their function is to seal together the plasma membranes of neighboring cells, forming a sort of "belt" around the cells, and create an impermeable barrier against the diffusion of water-soluble substances between the opposite faces of the epithelium. For example, tight junctions preclude that the products of digestion within the intestine lumen can cross the interstices among epithelial cells, hence directly entering the bloodstream. Moreover, these barriers can also prevent the lateral diffusion of some membrane proteins and lipids between the apical and the basolateral faces, thus restricting specific proteins to well fixed regions of the plasma membrane. This morphological organization is fundamental, for instance, for the transepithelial movement of glucose and amino acids (carried out by the intestinal epithelial cells), which occurs thanks to the correct distribution of specific symporters placed at the apical face, and of ATPases and uniporters placed at the basolateral face. Anchoring junctions consist of transmembrane adhesion proteins, which form membrane-spanning structures connecting the cytoskeleton elements of adjacent cells. They mechanically attach each cell to its neighbors (at their lateral faces), and also to the underneath extracellular matrix (at the basal face). There exist different types of anchoring junctions: *adherens junctions*, or "zonula adherens", and *desmosomes*, or "macula adherens". They are, respectively, belt-like and button-like structures of contact between cells, which confer mechanical strength to the tissue. Similar in structure to spot desmosome, *hemidesmosomes* are placed on the basal face of the cell and bind the cells to the extracellular matrix.

Gap junctions are communicating complexes which allow the flow of small molecules (e.g. calcium, cyclic AMP, precursors of DNA and RNA) from the cytosol of one cell to the cytosol of a neighboring one. In this way, cells can share their metabolites, and therefore are coordinated by an electrochemical coupling. For instance, gap junctions are needed for the synchronized contractions of heart muscle cells, for gene regulation in differentiation and embryogenesis processes, etc. Like ion channels, gap junctions are not continuously open, but flip between the open and closed state in response to changes in the cell.

2.4 POPULATIONS OF BACTERIAL CELLS

The transition from unicellular to multicellular organisms occurred in the course of evolution, when two processes—the specialization of cell function and the subdivision of jobs among single cells—began to increase inside the colonies constituted by pooled unicellular organisms. In Section 2.3 we sketched how the process of histodifferentiation in multicellular organisms assures the existence of many types of cells and tissues, that perform very different functions. On the other hand, there also exist multicellular organisms constituted of (usually) unspecialized cells, which develop from the union of previously isolated cells (e.g. *Dictyostelium discoideum*), or form aggregated colonies of cells derived from the division of a unicellular organism (e.g. *Gonium*) (see, e.g. [11]).

In this section we consider two phenomena occurring within populations of bacterial cells, which usually keep their own individualities though they may behave in a coordinated fashion as a unique multicellular-like organism. Very often, this happens with the aim of reacting suitably to specific environmental conditions. For instance, in harsh environments the bacterium *M. xanthus* can develop multicellular fruiting bodies, which contain particularly resistant spores that assure the survival of the microorganism [25]. In the following, we focus on two population-like behaviors of bacteria, and describe an intercellular communication mechanism, termed *quorum sensing*, and a sessile and colony-based form of life, known as *biofilm*.

Quorum sensing mechanisms. Bacteria have developed cross-talking mechanisms to sense the presence of other bacterial cells in the surrounding environment, and to trigger coordinated and synchronized behaviors, therefore performing whole-population activities. This communication process, termed *quorum sensing* [35, 22], allows bacteria to respond to variations in cell density by inducing the regulation of gene expression and cell physiology. Quorum sensing mechanisms are able to control and enhance several processes, like host-attack (virulence of pathogenic bacteria) or symbiosis, antibiotic production, cell motility, biofilm formation, etc. The communication between bacterial cells is mediated by signaling molecules, called *autoinducers*, that each cell is able to synthesize, release outside the cell, and detect in its vicinity. If a threshold concentration of autoinducer is reached (that is, if a sufficient number of bacteria are close enough to each other), then these molecules diffuse back into the cells where they promote their own synthesis within the cell, by means of a feedback control process. So, the cellular and external concentrations of these molecules increase as a function of the cell density. In a cascade of events, this process can stimulate or inhibit the expression of other genes, thus regulating in a synchronized, multicellular-like way, the physiological activities of the bacterial community. Gram-positive and Gram-negative bacteria use distinct quorum sensing signaling molecules: in general, Gram-negative bacteria produce acylated homoserine lactones as autoinducers, while Gram-positive bacteria use oligopeptides. Moreover, recent experiments have shown that quorum sensing can occur within cells of the same species, as well as between different bacterial species. Anyway, each quorum sensing system is optimized for the survival of a particular species in the peculiar, natural habitat where it resides. The first quorum sensing system was described more than 30 years ago in *Vibrio fischeri*, a luminous marine bacterium that lives in symbiosis (mutually useful relationship) with the squids, within their light organ (see [35, 22] and references therein). Inside this organ, thanks to nutrient abundance and good proliferation conditions, *Vibrio fischeri* bacteria can grow to an optimal cell density and trigger the expression of the genes which codify for bioluminescent molecules. In turn, the squid uses the produced light to escape from predators as it is thought that it camouflages itself by projecting light downward from its light organ.

Biofilms. Many microbial species are able, in particular conditions, to evolve as surface-attached communities called *biofilms* (see, e.g. [9, 25, 34]). Biofilms are constituted by a huge number of cells, of a single or of several species, brought together inside a matrix of extracellular polymeric substances produced by the bacteria themselves. These microbial communities show genotypic and phenotypic characteristics well-differentiated from the planktonically growing cells of the same species, and they provide an advantage in terms of a better adaptation to the environment and an enhanced resistance against antimicrobial agents. Direct observations have shown that biofilms are largely found in nature; for instance, the Gram-negative bacterium *P. aeruginosa* is one of the most studied single-species forming biofilm, because of its strong pulmonary infectiveness and the capability to constitute antibiotic resistant communities on the surface of medical implants or industrial water systems.

The development of a biofilm initiates in response to environmental signals, like the availability of nutrients, and requires a sophisticated mechanism of intercellular and intracellular signaling to trigger on a series of coordinated behaviors that result in the establishment of the microcolony. Quorum sensing mechanisms are also supposed to play a role in biofilm development. The processes for the growth and maturation of a biofilm encompass several stages: initially, the free-living cells reversibly attach to the surface, after having scanned it by twitching (see Section 2.2.2) in search for the appropriate location. Then, the surface-adhered cells activate the production of a matrix of extracellular polymeric substances (polysaccharides, proteins, and nucleic acids), which constitutes the structural and protective support for the biofilm and provides a firm attachment to the surface. The biofilm can therefore mature into a highly structured architecture (for instance, *P. aeruginosa* biofilms are mushroom-shaped), where microcolonies of cells are interconnected by a net of fluid-filled channels, through which water can leak from the bulk phase and deliver nutrients to the cells deeply embedded in the biofilm. The last stage of the biofilm formation comprises the detachment of single (or groups of) cells from the biofilm, and it is supposed to occur whenever the environment where the biofilm is growing becomes nutritionally deprived: the starvation pushes cells to search for other sources of nutrients, thus returning to the planktonic mode of growth.

2.5 FINAL REMARKS

Several aspects of cell biology at the three organization levels considered in this chapter—single cell, tissues, and colonies of cells—have been exploited in membrane computing as a source of inspiration for new classes of P systems. A few remarkable examples, that will be treated in detail in other chapters of this handbook, are represented by *symport/antiport communication rules*, inspired by the functioning of membrane porters; *spiking neural P systems*, based on the connections among neurons and the firing of action potentials; *tissue P systems* and *population P systems*, based on the structure and the communication mechanisms occurring within biological tissues or populations of single cells; *splicing P systems*, which use string-objects and exploit evolution rules inspired by the splicing operations of genes; P systems with *active membranes*, where the division of a compartment and its contents into two (or more) sibling regions is performed, etc. Other examples of P systems as computing models can also be found in the subsequent chapters of the handbook.

On the other hand, P systems have been recently exploited as discrete and compartmentalized models for the description and analysis of various biological and biochemical systems. We refer to Chapters 18, 19, 20, and to [3, 5, 30, 27] for more examples and details about the investigation of biological systems by means of P systems.

REFERENCES

[1] B. ALBERTS, A. JOHNSON, J. LEWIS, M. RAFF, K. ROBERTS, P. WALTER: *Molecular Biology of the Cell*. 4th edition, Garland Science, New York, 2002.

[2] N. BARKAI, S. LEIBLER: Robustness in simple biochemical networks. *Nature*, 387 (1997), 913–917.

[3] F. BERNARDINI: *Membrane Systems for Molecular Computing and Biological Modelling*. PhD Thesis, University of Sheffield, Sheffield, UK, 2005.

[4] D. BESOZZI, G. ROZENBERG: Formalizing spherical membrane structures and membrane proteins populations. *Lecture Notes in Computer Sci.*, 4361 (2006), 18–41.

[5] L. BIANCO: *Membrane Models of Biological Systems*. PhD Thesis, University of Verona, Verona, Italy, 2007.

[6] W. CHO, R.V. STAHELIN: Membrane-protein interactions in cell signaling and membrane trafficking. *Annu. Rev. Biophys. Biomol. Struct.*, 34 (2005), 119–151.

[7] G. CIOBANU, GH. PĂUN, M.J. PÉREZ-JIMÉNEZ, eds.: *Applications of Membrane Computing*. Springer, 2006.

[8] G.M. COOPER: *The Cell: A Molecular Approach*. 2nd Edition, Sinauer Associates Inc., Sunderland, Massachusetts, 2000.

[9] D. DE BEER, P. STOODLEY: Microbial biofilms. *Prokaryotes*, 1 (2006), 904–937.

[10] D.M. ENGELMAN: Membranes are more mosaic than fluid. *Nature*, 438 (2005), 578–580.
[11] C.P. HICKMAN, L.S. ROBERTS, A. LARSON: *Integrated Principles of Zoology*. 11th edition, McGraw-Hill, 2000.
[12] M.S. JURICA, B.L. STODDARD: Mind your B's and R's: bacterial chemotaxis, signal transduction and protein recognition. *Current Biology*, 6 (1998), 809–813.
[13] H. KITANO: Biological robustness. *Nature Reviews Genetics*, 5 (2004), 826–837.
[14] H. KITANO: Cancer as a robust system: implications for anticancer therapy. *Nature Reviews Cancer*, 4 (2004), 227–235.
[15] A. KUSUMI, C. NAKADA, K. RITCHIE, K. MURASE, K. SUZUKI, H. MURAKOSHI, R.S. KASAI, J. KONDO, T. FUJIWARA: Paradigm shift of the plasma membrane concept from the two-dimensional continuum fluid to the partitioned fluid: high-speed single-molecule tracking of membrane molecules. *Annu. Rev. Biophys. Biomol. Struct.*, 34 (2005), 351–378.
[16] H. LODISH, A. BERK, S.L. ZIPURSKY, P. MATSUDAIRA, D. BALTIMORE, J.E. DARNELL: *Molecular Cell Biology*. 4th edition, W.H. Freeman and Co., New York, 2000.
[17] H.W. MA, A.P. ZENG: The connectivity structure, giant strong component and centrality of metabolic networks. *Bioinformatics*, 19 (2003), 1423–1430.
[18] M.M. MADIGAN, J.M. MARTINKO, J. PARKER: *BROCK Biology of Microorganisms*. 10th edition, Prentice Hall, 2003.
[19] D. MARGUET, P.F. LENNE, H. RIGNEAULT, H.T. HE: Dynamics in the plasma membrane: how to combine fluidity and order. *The EMBO Journal*, 25 (2006), 3446–3457.
[20] S. MAYOR, R.E. PAGANO: Pathways of clathrin-independent endocytosis. *Nature Reviews Molecular Cell Biology*, 8 (2007), 603–612.
[21] H.H. MCADAMS, A. ARKIN: Simulation of prokaryotic genetic circuits. *Annu. Rev. Biophys. Biomol. Struct.*, 27 (1998), 199–224.
[22] M.B. MILLER, B.L. BASSLER: Quorum sensing in Bacteria. *Annu. Rev. Microbiol.*, 55 (2001), 165–99.
[23] P. MITCHELL: Keilin's respiratory chain concept and its chemiosmotic consequences. *Science*, 206 (1979), 1148–1159.
[24] D.L. NELSON, M.M. COX: *Lehninger Principles of Biochemistry*. 4th edition, W.H. Freeman and Company, 2004.
[25] G. O'TOOLE, H.B. KAPLAN, R. KOLTER: Biofilm formation as microbial development. *Annu. Rev. Microbiol.*, 54 (2000), 49–79.
[26] GH. PĂUN: Computing with membranes. *J. Computer and System Sci.*, 61 (2000), 108–143.
[27] D. PESCINI: *Modelling, Analysis and Stochastic Simulations of Biological Systems*. PhD Thesis, University of Milano-Bicocca, Milano, Italy, 2008.
[28] L.M. PRESCOTT, J.P. HARLEY, D.A. KLEIN: *Microbiology*. 4th edition, McGraw-Hill, 1999.
[29] P.H. RAVEN, R.F. EVERT, S.E. EICHHORN: *Biology of Plants*. 7th edition, W.H. Freeman and Company, New York, 2005.
[30] F.J. ROMERO-CAMPERO: *P Systems, a Computational Modelling Framework for Systems Biology*. PhD Thesis, University of Sevilla, Spain, 2008.
[31] P.J. RUSSELL: *Genetics*. 5th edition, Benjamin-Cummings Publishing Company, 1997.
[32] S.J. SINGER: Some early history of membrane molecular biology. *Annual Review of Physiology*, 66 (2004), 1–27.

[33] S.J. SINGER, G.L. NICOLSON: The fluid mosaic model of the structure of cell membranes. *Science*, 175 (1972), 720–731.
[34] P. STOODLEY, K. SAUER, D.G. DAVIES, J.W. COSTERTON: Biofilms as complex differentiated communities. *Annu. Rev. Microbiol.*, 56 (2002), 187–209.
[35] C.M. WATERS, B.L. BASSLER: Quorum sensing: cell-to-cell communication in Bacteria. *Annu. Rev. Cell Dev. Biol.*, 21 (2005), 319–346.
[36] T. YI, Y. HUANG, M.I. SIMON, J. DOYLE: Robust perfect adaptation in bacterial chemotaxis through integral feedback control. *PNAS*, 97 (2000), 4649–4653.
[37] S.H. ZINDER, M. DWORKIN: Morphological and physiological diversity. *Prokaryotes*, 1 (2006), 185–220.

CHAPTER 3

COMPUTABILITY ELEMENTS FOR MEMBRANE COMPUTING

GHEORGHE PĂUN
GRZEGORZ ROZENBERG
ARTO SALOMAA

3.1 INTRODUCTION

MEMBRANE computing can be seen as a *commutative version of language and automata theory*, including language and automata theory (sometimes recently developed) research areas such as regulated rewriting, Lindenmayer systems, grammar systems, theoretical DNA computing, i.e. areas where parallelism and distribution play an important part. Indeed, the main data structure investigated in membrane computing is the *multiset*, a set of abstract objects (in general, symbols from a given alphabet) with multiplicities associated with its elements, and obviously a multiset can be interpreted as the commutative version of a string (the class of all strings equivalent modulo permutation). One of the basic operations on multisets, also corresponding to bio-chemical reactions taking place in a cell, is multiset rewriting, which is the direct counterpart of string rewriting. This directly leads to repeating in terms of multisets the whole program of language theory—implicitly, to developing a grammatical approach and an automata approach to multiset processing. That is why not only many notions were introduced in membrane computing which recall notions from formal language theory and from automata theory, but also results corresponding to those in these classic areas of theoretical computer science were looked for and proved in membrane computing, often also importing proof techniques and, of course, the well-established test bed classifications, such as Chomsky hierarchy, Lindenmayer families of languages, and Turing machines and their restrictions.

Thus, it is natural to recall here several notions and results from language and automata theory, in the classic form, dealing with strings and also in the multiset form.

The language and automata approach is fundamental in introducing computing devices and investigating their *computing power* (competence), but an important subsequent issue is that of *computing efficiency* (performance), and now comes into the stage the theory of computational complexity. We recall only a few basic elements concerning the time–space complexity classes; technical definitions, of complexity classes for specific classes of membrane systems (P systems) will be given in the chapter devoted to complexity studies in membrane computing.

Further details of formal language theory and automata theory, as well as of computational complexity, can be found in the many monographs and textbooks available. The chapter ends with some of the most comprehensive titles, but many others can be found in the literature. Further references can be found in the introductory chapter of the book [12].

In what follows, we do not give precise references (e.g. the original source or recent versions) for the mentioned notions and results, as most of these notions and results are classic and tracing their evolution would uselessly fragment and lengthen our presentation. Also, we do not give any proof hint, not even examples; this chapter chiefly intends to fix notations and terminology, not to be a comprehensive introduction to languages, grammars, automata, complexity.

3.2 BASIC ELEMENTS OF FORMAL LANGUAGE THEORY

3.2.1 Words and Languages

An *alphabet* is a finite non-empty set of abstract symbols. Any sequence of symbols from an alphabet V is called *string* over V. For an alphabet V we denote by V^* the set of all strings of symbols from V. The empty string is denoted by λ, and the set of non-empty strings over V, that is $V^* - \{\lambda\}$, is denoted by V^+. Each subset of V^* is called a *language* over V. A language which does not contain the empty string (hence is a subset of V^+) is said to be *λ-free*.

If $x = x_1 x_2 x_3$ for some $x_1, x_2, x_3 \in V^*$, then x_2 is called a *substring* of x; if, in the previous writing, $x_1 = \lambda$, then x_2 is a *prefix* of x, and, if $x_3 = \lambda$, then x_2 is a *suffix* of x. The sets of all prefixes, suffixes, and substrings of a string x are denoted by $Pref(x)$, $Suf(x)$, $Sub(x)$, respectively.

The number of occurrences in a string $x \in V^*$ of symbols from V is denoted by $|x|$ and this is the *length* of x. The number of occurrences of symbols $a \in U$ for some $U \subseteq V$ in $x \in V^*$ is denoted by $|x|_U$. Singleton sets $\{a\}$ are often identified with a. For instance, if U is a singleton, $U = \{a\}$, then we simply write $|x|_a$ instead of $|x|_{\{a\}}$. For a language $L \subseteq V^*$, the set $length(L) = \{|x| \mid x \in L\}$ is called the *length set* of L.

The *Parikh vector* associated with a string $x \in V^*$ with respect to the alphabet $V = \{a_1, \ldots, a_n\}$ (the ordering of the symbols is important here) is $\Psi_V(x) = (|x|_{a_1}, |x|_{a_2}, \ldots, |x|_{a_n})$. For a language $L \subseteq V^*$ we define $\Psi_V(L) = \{\Psi_V(x) \mid x \in L\}$; this is called the *Parikh image* of L.

Membrane computing mainly deals with multisets, not with strings, but these two notions are intimately related, as we have already seen in Chapter 1.

A set of languages is usually called a *family* of languages. For a family FL of languages, we denote by $NFL, PsFL$ the family of length sets and of Parikh images, respectively, of languages in FL.

A set Q of vectors in $\mathbf{N}^n$, for some $n \geq 1$, is called *linear* if there are the vectors $v_i \in \mathbf{N}^n$, $0 \leq i \leq m$, such that

$$Q = \{v_0 + \sum_{i=1}^{m} a_i v_i \mid a_1, \ldots, a_m \in \mathbf{N}\}.$$

A finite union of linear sets is said to be *semilinear*. For instance, the set $\{(n, n, 0) \mid n \geq 1\}$ is linear, $\{(n, n, 0) \mid n \geq 1\} \cup \{(1, 1, 1)\}$ is a semilinear non-linear set, but the next two sets are not semilinear: $\{(nm) \mid n, m \geq 2\}$, $\{(n, m) \mid n \geq 1, 1 \leq m \leq 2^n\}$.

A language $L \subseteq V^*$ is semilinear if $\Psi_V(L)$ is a semilinear set. The family of semilinear languages is denoted by $SLIN$.

Because the languages are sets, we can perform on them the usual Boolean operations of union, intersection, difference, complement. There also are specific operations—we mention here only a few of them.

The *concatenation* of two languages L_1, L_2 is $L_1 L_2 = \{xy \mid x \in L_1, y \in L_2\}$. Also the operations of taking the prefixes, substrings, or suffixes are extended in the natural way from strings to languages, and we denote by $Pref(L)$, $Sub(L)$, $Suf(L)$ the respective sets.

Then, we can define:

$$L^0 = \{\lambda\},$$

$$L^{i+1} = LL^i, \ i \geq 0,$$

$$L^* = \bigcup_{i=0}^{\infty} L^i \text{ (the } *\text{-Kleene closure)},$$

$$L^+ = \bigcup_{i=1}^{\infty} L^i \text{ (the } +\text{-Kleene closure)}.$$

For alphabets V, U, a mapping $h : V \longrightarrow U^*$, extended to $h : V^* \longrightarrow U^*$ by $h(x_1 x_2) = h(x_1)h(x_2)$, for $x_1, x_2 \in V^*$ (clearly, $h(\lambda) = \{\lambda\}$), is called a *morphism*. If $h(a) \neq \lambda$ for each $a \in V$, then h is a λ-*free* morphism. A morphism $h : V^* \longrightarrow U^*$ is called a *coding* if $h(a) \in U$ for each $a \in V$ and a *weak coding* if $h(a) \in U \cup \{\lambda\}$ for each $a \in V$. If $h : (V_1 \cup V_2)^* \longrightarrow V_1^*$ is the morphism defined by $h(a) = a$ for $a \in V_1$, and $h(a) = \lambda$ otherwise, then we say that h is a *projection* (associated with V_1) and we denote it by pr_{V_1}. For a morphism $h : V^* \longrightarrow U^*$, we define a mapping $h^{-1} : U^* \longrightarrow 2^{V^*}$ (and we call it an *inverse morphism*) by $h^{-1}(w) = \{x \in V^* \mid h(x) = w\}$, $w \in U^*$.

If $L \subseteq V^*$, $k \geq 1$, and $h : V^* \longrightarrow U^*$ is a morphism such that $h(x) \neq \lambda$ for each $x \in Sub(L)$, $|x| = k$, then we say that h is k-*restricted* on L.

Two further pairs of operations with languages are the *left/right quotient* and the *left/right derivative*. For instance, the left quotient of a language $L_1 \subseteq V^*$ with respect to $L_2 \subseteq V^*$ is

$$L_2 \backslash L_1 = \{w \in V^* \mid \text{there is } x \in L_2 \text{ such that } xw \in L_1\},$$

while the *left derivative* of a language $L \subseteq V^*$ with respect to a string $x \in V^*$ is

$$\partial_x^l(L) = \{w \in V^* \mid xw \in L\}.$$

The right quotient and the right derivative are defined in a symmetric manner.

A language that can be obtained from the letters of an alphabet V and λ by using finitely many times the operations of union, concatenation, and Kleene $*$ is called *regular*. This family, denoted by REG, has many other characterizations. One of them is in terms of *regular expressions*.

Given an alphabet V, (i) λ and each $a \in V$ are regular expressions (over V), (ii) if E_1, E_2 are regular expressions over V, then also $(E_1) \cup (E_2)$, $(E_1)(E_2)$, and $(E_1)^+$ are regular expressions over V, and (iii) nothing else is a regular expression over V. With each expression E we associate a language $L(E)$, defined in the following way: (i) $L(\lambda) = \{\lambda\}$ and $L(a) = \{a\}$, for all $a \in V$, (ii) $L((E_1) \cup (E_2)) = L(E_1) \cup L(E_2)$, $L((E_1)(E_2)) = L(E_1)L(E_2)$, and $L((E_1)^+) = L(E_1)^+$, for all regular expressions E_1, E_2 over V. Non-necessary parentheses are omitted when writing a regular expression, and $(E)^+ \cup \{\lambda\}$ is written in the form $(E)^*$.

A language $L \subseteq V^*$ is regular if and only if there is a regular expression E over V such that $L(E) = L$.

Operations with languages can take languages in a given family FL and produce as a result a language which belongs to the same family (and then we say that FL is *closed* under that operation), or not, and this is an important property of families of languages.

A family of languages is *non-trivial* if it contains at least one language different from $\emptyset$ and $\{\lambda\}$. From now on, all families of languages we consider are supposed to be non-trivial.

A family of languages is said to be a *trio* if it is closed under λ-free morphisms, inverse morphisms, and intersection with regular languages. A trio closed under union is called a *semi-AFL* (AFL is the abbreviation of "abstract family of languages"). A semi-AFL closed under concatenation and Kleene + is said to be an *AFL*. A trio/semi-AFL/AFL is said to be *full* if it is closed under arbitrary morphisms (and Kleene * in the case of AFLs). A family of languages closed under none of the six AFL operations is said to be an *anti-AFL*.

The six AFL operations are not independent. For instance, if FL is a family of λ-free languages which is closed under concatenation, λ-free morphisms, and inverse morphisms, then FL is closed under intersection with regular languages and union, hence FL is a semi-AFL.

3.2.2 Chomsky Hierarchy

A test bed for any new computing device (identifying languages or related sets) is the Chomsky hierarchy, a well-granulated and well-investigated hierarchy of languages, having both grammatical and automata characterizations. Generally speaking, a *grammar* is a (finite) device *generating* in a well specified sense the strings of a language, while an automaton is a (finite) device which *analyzes* the strings over a given alphabet, accepting those strings which belong to a specified language.

The Chomsky hierarchy is defined starting from Chomsky grammars, which are particular cases of *rewriting systems*, where the operation used in processing the strings is the rewriting (the replacement of a "short" substring of the processed string by another short substring). This idea is central also to membrane computing, where however one "rewrites" a "small" sub-multiset of a multiset by means of a rewriting-like rule.

A *Chomsky grammar* is a quadruple $G = (N, T, S, P)$, where N, T are disjoint alphabets (N is called the *non-terminal alphabet* and T is the *terminal alphabet*), $S \in N$ is the *axiom*, and P is a finite subset of $(N \cup T)^* N (N \cup T)^* \times (N \cup T)^*$; the elements (u, v) of P are written in the form $u \to v$ and are called *rules* or *production rules*.

For $x, y \in (N \cup T)^*$ we write

$$x \Longrightarrow_G y \quad \text{iff} \quad x = x_1 u x_2, \; y = x_1 v x_2,$$

for some $x_1, x_2 \in (N \cup T)^*$ and $u \to v \in P$.

We say that x *directly derives* y (with respect to G). When G is understood we write $\Longrightarrow$ instead of $\Longrightarrow_G$. The reflexive and transitive closure of the relation $\Longrightarrow_G$ is denoted by $\Longrightarrow_G^*$. Each string $w \in (N \cup T)^*$ such that $S \Longrightarrow_G^* w$ is called a *sentential form*.

The language generated by G, denoted by $L(G)$, is defined by

$$L(G) = \{x \in T^* \mid S \Longrightarrow^* x\}.$$

Two grammars G_1, G_2 are called *equivalent* if $L(G_1) - \{\lambda\} = L(G_2) - \{\lambda\}$ (the two languages coincide modulo the empty string).

According to the form of their rules, the Chomsky grammars are classified as follows. A grammar $G = (N, T, S, P)$ is called:

- *length-increasing* or *monotonous*, if for all $u \to v \in P$ we have $|u| \leq |v|$.
- *context-sensitive*, if each $u \to v \in P$ has $u = u_1 A u_2$, $v = u_1 x u_2$, for $u_1, u_2 \in (N \cup T)^*$, $A \in N$, and $x \in (N \cup T)^+$. (In length-increasing and context-sensitive grammars the production $S \to \lambda$ is allowed, provided that S does not appear in the right-hand sides of rules in P.)
- *context-free*, if for each production $u \to v \in P$ we have $u \in N$.
- *linear*, if each rule $u \to v \in P$ has $u \in N$ and $v \in T^* \cup T^*NT^*$.
- *right-linear*, if each rule $u \to v \in P$ has $u \in N$ and $v \in T^* \cup T^*N$.
- *left-linear*, if each rule $u \to v \in P$ has $u \in N$ and $v \in T^* \cup NT^*$.
- *regular*, if each rule $u \to v \in P$ has $u \in N$ and $v \in T \cup TN \cup \{\lambda\}$.

The arbitrary, length-increasing, context-free, and regular grammars are also said to be of *type 0*, *type 1*, *type 2*, and *type 3*, respectively.

The family of languages generated by length-increasing grammars is equal to the family of languages generated by context-sensitive grammars; the families of languages generated by right- or by left-linear grammars coincide and they are equal to the family of languages generated by regular grammars, as well as with the family of regular languages (as defined above).

We denote by *RE, CS, CF, LIN,* and *REG* the families of languages generated by arbitrary, context-sensitive, context-free, linear, and regular grammars, respectively (RE stands for *recursively enumerable*). By *FIN* we denote the family of finite languages.

The following strict inclusions hold:

$$FIN \subset REG \subset LIN \subset CF \subset CS \subset RE.$$

This is *the Chomsky hierarchy*, the constant reference for investigations related to the power of membrane systems. The important role of the Chomsky hierarchy is

Table 3.1 Closure properties of the families in the Chomsky hierarchy.

	RE	CS	CF	LIN	REG
Union	yes	yes	yes	yes	yes
Intersection	yes	yes	no	no	yes
Complement	no	yes	no	no	yes
Concatenation	yes	yes	yes	no	yes
Kleene*	yes	yes	yes	no	yes
Intersection with regular languages	yes	yes	yes	yes	yes
Morphisms	yes	no	yes	yes	yes
λ-free morphisms	yes	yes	yes	yes	yes
Inverse morphisms	yes	yes	yes	yes	yes
Left/right quotient	yes	no	no	no	yes
Left/right quotient with regular languages	yes	no	yes	yes	yes
Left/right derivative	yes	yes	yes	yes	yes

owing to several reasons: the family RE of languages generated by type-0 Chomsky grammars is exactly the family of languages which are recognized by Turing machines, and according to the Turing–Church thesis this is the maximal level of algorithmic computability; the Chomsky hierarchy is well structured, hence we have a detailed classification of computing machineries (in between the above mentioned classes of grammars there are other types of grammars which were considered in the literature); the families in Chomsky hierarchy are characterized by natural computing devices (e.g. automata).

Actually, in many cases, what is used in membrane computing are the length sets and Parikh images associated with languages in Chomsky hierarchy, and the previous hierarchy has the following counterparts:

$$NFIN \subset NREG = NLIN = NCF \subset NCS \subset NRE,$$

$$PsFIN \subset PsREG = PsLIN = PsSLIN = PsCF \subset PsCS \subset PsRE.$$

The closure properties of the families listed above are indicated in Table 3.1.

Therefore, RE, CF, REG are full AFLs, CS is an AFL (not full), and LIN is a full semi-AFL.

The difference between families CS and RE in terms of closure properties can be illustrated also with the following significant result.

For every language $L \subseteq T^*, L \in RE$, there are $L' \in CS$ and $c_1, c_2 \notin T$, such that $L' \subseteq L\{c_1\}\{c_2\}^*$, and for each $w \in L$ there is $i \geq 0$ such that $wc_1 c_2^i \in L'$.

Therefore, (i) each RE language is equal to a CS language modulo a tail of the form $c_1 c_2^i$ for some $i \geq 0$; (ii) each recursively enumerable language is the projection of a context-sensitive language, and (iii) for each RE language L there is a CS language L_1 and a regular language L_2 such that $L = L_1/L_2$.

These results prove that the families RE and CS are "almost equal", the difference between them lies in an arbitrarily long tail to be added to the strings of a language; this tail carries no information other than its length, hence from a syntactical point of view the two languages can be considered indistinguishable. In another formulation, in order to obtain the power of type-0 grammars (of Turing machines, hence computationally complete), it is sufficient to have (i) context-sensitivity and (ii) (enough) erasing.

When the erasing is limited we cannot go beyond context-sensitive languages. A formal counterpart of this assertion is the fact that *the family CS is closed under restricted morphisms*. Another important formulation of this intuition is the following useful result: the *workspace theorem*.

Let $G = (N, T, S, P)$ be a type-0 grammar and consider a derivation in G $\delta : S = w_0 \Longrightarrow w_1 \Longrightarrow \ldots \Longrightarrow w_m = y$, $y \in T^*$, $m \geq 1$. We define $WS(\delta, G) = \max_{0 \leq i \leq m} |w_i|$, and, for $y \in L(G)$, we put $WS(y, G) = \min\{WS(\delta, G) \mid \delta : S \Longrightarrow^* y\}$. We say that G has a (linearly) bounded workspace if there is a constant k such that, for each $y \in L(G) - \{\lambda\}$ we have $WS(y, G) \leq k|y|$.

Then, the workspace theorem says that *if G has a bounded workspace, then $L(G) \in CS$*.

3.2.3 Normal Forms, Necessary Conditions

We recall now a series of basic results in formal language theory, useful also for membrane computing, starting with normal forms for various classes of grammars in Chomsky hierarchy.

1. (Kuroda normal form) *For every type-0 grammar G, an equivalent grammar $G' = (N, T, S, P)$ can be effectively constructed, with the rules in P of the forms $A \to BC$, $A \to a$, $A \to \lambda$, $AB \to CD$, for $A, B, C, D \in N$ and $a \in T$.*

2. (Penttonen normal form) *For every type-0 grammar G, an equivalent grammar $G' = (N, T, S, P)$ can be effectively constructed, with the rules in P of the forms $A \to x$, $x \in (N \cup T)^*$, $|x| \leq 2$, and $AB \to AC$ with $A, B, C \in N$.*
   Similar results hold true for length-increasing grammars; in this case, rules of the form $A \to \lambda$ are no longer allowed, but only a completion rule $S \to \lambda$ if the generated language also contains the empty string (then S cannot appear in the right hand side of any rule).

3. (Geffert normal forms) (i) *Each recursively enumerable language can be generated by a grammar $G = (N, T, S, P)$ with $N = \{S, A, B, C\}$ and the rules in P of the forms $S \to uSv$, $S \to x$, with $u, v, x \in (T \cup \{A, B, C\})^*$, and only one non-context-free rule, $ABC \to \lambda$.*
   (ii) *Each recursively enumerable language can be generated by a grammar $G = (N, T, S, P)$ with $N = \{S, A, B, C, D\}$ and the rules in P of the forms*

$S \to uSv$, $S \to x$, with $u, v, x \in (T \cup \{A, B, C, D\})^*$, and only two non-context-free rules, $AB \to \lambda$, $CD \to \lambda$.

Let us pass now to conditions for a language to be context-free.

1. (Bar–Hillel/*uvwxy*/pumping lemma) *If $L \in CF$, $L \subseteq V^*$, then there are $p, q \in \mathbf{N}$ such that every $z \in L$ with $|z| > p$ can be written in the form $z = uvwxy$, with $u, v, w, x, y \in V^*$, $|vwx| \leq q$, $vx \neq \lambda$, and $uv^i wx^i y \in L$ for all $i \geq 0$.*
2. (Parikh theorem) *Every context-free language is semilinear.*
   As a consequence, every context-free language over a one-letter alphabet is regular.

A characterization of context-free languages is obtained as follows. The *Dyck language*, D_n, over the alphabet $T_n = \{a_1, a'_1, \ldots, a_n, a'_n\}$, $n \geq 1$, is the language generated by the context-free grammar

$$G = (\{S\}, T_n, S, \{S \to \lambda, S \to SS\} \cup \{S \to a_i S a'_i \mid 1 \leq i \leq n\}).$$

The pairs (a_i, a'_i), $1 \leq i \leq n$, can be interpreted as labeled left and right parentheses, and then D_n consists of all strings of correctly nested parentheses.

3. (Chomsky–Schützenberger theorem) *Every context-free language L can be written in the form $L = h(D_n \cap R)$, where h is a morphism, D_n, $n \geq 1$, is a Dyck language, and R is a regular language.*

3.2.4 Lindenmayer Systems

Another well-known framework for studying formal languages is constituted by Lindenmayer systems, which, like membrane computing, are biologically inspired and, also, parallel computing devices.

A 0L (0-interactions Lindenmayer) system is a construct $G = (V, w, P)$, where V is an alphabet, $w \in V^*$ (axiom), and P is a finite set of rules of the form $a \to v$ with $a \in V$, $v \in V^*$, such that for each $a \in V$ there is at least one rule $a \to v$ in P (we say that P is *complete*). For $w_1, w_2 \in V^*$ we write $w_1 \Longrightarrow w_2$ if $w_1 = a_1 \ldots a_n$, $w_2 = v_1 \ldots v_n$, for $a_i \to v_i \in P$, $1 \leq i \leq n$. The generated language is $L(G) = \{x \in V^* \mid w \Longrightarrow^* x\}$.

If for each rule $a \to v \in P$ we have $v \neq \lambda$, then we say that G is *propagating*; if for each $a \in V$ there is only one rule $a \to v$ in P, then G is *deterministic*. If we distinguish a subset T of V and we define the generated language as $L(G) = \{x \in T^* \mid w \Longrightarrow^* x\}$, then we say that G is *extended*.

The family of languages generated by 0L systems is denoted by $0L$; we add the letters P, D, E in front of $0L$ if propagating, deterministic, or extended 0L systems are used, respectively.

A *tabled* 0L system, abbreviated T0L, is a system $G = (V, w, P_1, \ldots, P_n)$, such that each triple (V, w, P_i), $1 \leq i \leq n$, is a 0L system; each P_i, $1 \leq i \leq n$, is called a *table*. The language generated by G is

$$L(G) = \{x \in V^* \mid w \Longrightarrow_{P_{j_1}} w_1 \Longrightarrow_{P_{j_2}} \ldots \Longrightarrow_{P_{j_m}} w_m = x,$$
$$m \geq 0, 1 \leq j_i \leq n, 1 \leq i \leq m\}.$$

(Each derivation step is performed by the rules of the same table.)

A T0L system is deterministic when each of its tables is deterministic. The propagating and the extended features are defined in the usual way.

The family of languages generated by T0L systems is denoted by $T0L$; the letters E and D are added when extended or deterministic systems are used.

We give here only a simple example, of the D0L system

$$G = (\{a, b\}, a, \{a \to ab, b \to a\}),$$

which generates a language $L(G)$ such that $NL(G) = \{1, 2, 3, 5, 8, 13, \ldots\}$, i.e. the Fibonacci sequence starting with 1 and 2.

In what concerns the relations between Lindenmayer families and Chomsky families (useful in the forthcoming chapters of the Handbook), we mention the following results: $D0L$ is incomparable with FIN, REG, LIN, CF, whereas $CF \subset E0L \subset ET0L \subset CS$; $ND0L - NCF \neq \emptyset$, $PsET0L \subset PsCS$. It is not known whether or not $PsE0L$ is strictly included in $PsET0L$.

An important property of ET0L systems is the following normal form: *each language $L \in ET0L$ can be generated by an ET0L system with only two tables*. It is also worth mentioning that $ET0L$ is a full AFL.

3.2.5 Automata and Transducers

As suggested above, automata are computing devices which start from the strings over a given alphabet and *analyze* them (we also say *recognize* or *accept*), thus identifying a language, the set of all accepted strings.

The five basic families of languages in the Chomsky hierarchy, REG, LIN, CF, CS, RE, are also characterized by recognizing automata, but we present here only two of these devices, those which, in some sense, define the two poles of computability: finite automata and Turing machines.

A (non-deterministic) *finite automaton* is a construct

$$M = (K, V, s_0, F, \delta),$$

where K and V are disjoint alphabets, $s_0 \in K$, $F \subseteq K$, and $\delta : K \times V \longrightarrow 2^K$; K is the set of states, V is the alphabet of the automaton, s_0 is the initial state, F is the set of final states, and δ is the transition mapping. If $card(\delta(s, a)) \leq 1$

for all $s \in K, a \in V$, then we say that the automaton is *deterministic*. A relation $\vdash$ is defined in the following way on the set $K \times V^*$: for $s, s' \in K, a \in V, x \in V^*$, we write $(s, ax) \vdash (s', x)$ if $s' \in \delta(s, a)$; by definition, $(s, \lambda) \vdash (s, \lambda)$. If $\vdash^*$ is the reflexive and transitive closure of the relation $\vdash$, then the language of the strings recognized by automaton M is defined by

$$L(M) = \{x \in V^* \mid (s_0, x) \vdash^* (s, \lambda), s \in F\}.$$

Both deterministic and non-deterministic finite automata characterize the same family of languages, namely REG. The power of finite automata is not increased if we also allow *λ-transitions*, that is, if δ is defined on $K \times (V \cup \{\lambda\})$ (the automaton can also change its state when reading no symbol on its tape), or when the input string is scanned in a two-way manner, going along it to right or to left, without changing its symbols.

An important related notion is that of a *sequential transducer* which is defined as a finite automaton with outputs associated with its moves: the transition mapping is then defined by $\delta : K \times V \longrightarrow 2^K \times V^*$, i.e. a string is also produced when reading a symbol in the input string.

A finite automaton can be imagined as having an input tape, with a string written on it (one symbol in each cell), which is scanned from left to right, starting in the initial state; if a final state is reached when exhausting the string, then the input string is accepted. In turn, a Turing machine can be imagined as a finite automaton with the possibility of scanning the tape both to left and right and with the possibility of rewriting the symbol from the scanned cell (including erasing symbols and writing a symbol in an empty cell).

Formally, a *Turing machine* is a construct

$$M = (K, V, T, B, s_0, F, \delta),$$

where K, V are disjoint alphabets (the set of states and the tape alphabet), $T \subseteq V$ (the input alphabet), $B \in V - T$ (the blank symbol), $s_0 \in K$ (the initial state), $F \subseteq K$ (the set of final states), and δ is a partial mapping from $K \times V$ to the power set of $K \times V \times \{L, R\}$ (the move mapping; if $(s', b, d) \in \delta(s, a)$, for $s, s' \in K, a, b \in V$, and $d \in \{L, R\}$, then the machine reads the symbol a in state s and passes to state s', replaces a with b, and moves the read-write head to the left when $d = L$ and to the right when $d = R$). If $card(\delta(s, a)) \leq 1$ for all $s \in K, a \in V$, then M is said to be *deterministic*.

An *instantaneous description* of a Turing machine as above is a string xsy, where $x \in V^*, y \in V^*(V - \{B\}) \cup \{\lambda\}$, and $s \in K$. In this way we identify the contents of the tape, the state, and the position of the read-write head: it scans the first symbol of y. Observe that the blank symbol may appear in x, y, but not in the last position of y; both x and y may be empty. We denote by ID_M the set of all instantaneous descriptions of M.

On the set ID_M one defines the *direct transition* relation $\vdash_M$ as follows:

$$xsay \vdash_M xbs'y \text{ iff } (s', b, R) \in \delta(s, a),$$
$$xs \vdash_M xbs' \text{ iff } (s', b, R) \in \delta(s, B),$$
$$xcsay \vdash_M xs'cby \text{ iff } (s', b, L) \in \delta(s, a),$$
$$xcs \vdash_M xs'cb \text{ iff } (s', b, L) \in \delta(s, B),$$

where $x, y \in V^*, a, b, c \in V, s, s' \in K$. (Note that the read-write head cannot go left from the first position.)

The language recognized by a Turing machine M is defined by

$$L(M) = \{w \in T^* \mid s_0 w \vdash_M^* xsy \text{ for some } s \in F, x, y \in V^*\}.$$

Another possibility is to define the language accepted by a Turing machine as consisting of the input strings $w \in T^*$ such that the machine, starting from the configuration $s_0 w$, reaches a configuration where no further move is possible (we say that the machine *halts*); in this case, the set F of final states is no longer necessary. The two modes of defining the language $L(M)$ are equivalent, the identified families of languages are the same, namely *RE*, and this is true both for deterministic and non-deterministic machines.

A very useful characterization of *NRE* (the family of sets of numbers which are Turing computable) is obtained by means of *register machines* (also called counter machines, program machines, etc.). Such a device is a construct $M = (m, H, l_0, l_h, I)$, where m is a natural number (the number of registers), H is a set of labels, $l_0, l_h \in H$ are distinguished labels (the initial and the halting one), and I is the set of *labeled program instructions*. The instructions can be of the following forms:

- $l_i : (ADD(r), l_j, l_k)$; add 1 to register r and continue with one of the instructions with labels l_j, l_k, non-deterministically chosen;
- $l_i : (SUB(r), l_j, l_k)$; if the contents of register r is greater than zero, subtract 1 from register r and go to label l_j, otherwise go to the instruction with label l_k;
- $l_h : \text{HALT}$.

Each label l_i is associated with only one instruction, while l_h is only associated with the halt instruction.

A register machine M generates a subset $N(M)$ of $\mathbf{N}$ in the following way: M starts with all registers empty, by executing the instruction with label l_0; one proceeds as indicated by the labels and as made possible by the contents of registers; if M reaches the halting instruction while containing the number n in the first register, then n is introduced in $N(M)$.

We can use a register machine also in the accepting mode: we start with number n in register 1 and all other registers empty, by executing the instruction with label l_0; if the computation halts, then number n is accepted. Accepting register machines also characterize NRE. In the accepting case, the machine can be *deterministic*, i.e. the ADD instructions $l_i : (\text{ADD}(r), l_j, l_k)$ can be considered as having $l_j = l_k$. In the generating mode, in the halting configuration we may assume that all registers different from register 1 are empty; in the accepting mode, all registers may be assumed empty in the halting configuration.

In a natural way, we can generate or accept vectors of numbers: instead of considering only the first register, we distinguish several registers (say, the first k) which are used for the result of a computation (in the generative case) or for introducing the vector to be analyzed (in the accepting case).

A register machine which does not have the possibility of checking whether a register is zero is called *partially blind*; in this case, the SUB instructions are of the form $l_i : (\text{SUB}(r), l_j)$. If register r is non-empty, then the subtraction takes place and one continues with instruction with label l_j, otherwise the machine stops, aborting the computation. These machines are strictly less powerful than Turing machines.

3.2.6 Regulated Rewriting

As we have seen before, Chomsky grammars of type 0 characterize Turing computability, which is not the case for more restricted classes of grammars, in particular, for context-free grammars. On the other hand, context-free rules are "nice", they rewrite only one symbol (hence the derivation in such a grammar can be described by a tree); the context-free grammars and languages also have many positive closure and decidability properties. That is why it is of interest to consider restrictions in the use of rules of a context-free grammar in order to increase their power, and at this time several dozens of such restrictions are known. The oldest one, the *matrix grammars*, were already introduced, in a particular form, in 1965, and turn out to be very useful for membrane computing.

A context-free *matrix* grammar (with appearance checking) is a construct $G = (N, T, S, M, F)$, where N, T are disjoint alphabets (of non-terminals and terminals, respectively), $S \in N$ (axiom), M is a finite set of *matrices*, that is, sequences of the form $(A_1 \to x_1, \ldots, A_n \to x_n)$, $n \geq 1$, of context-free rules over $N \cup T$ (with $A_i \in N$, $x_i \in (N \cup T)^*$, in all cases), and F is a set of occurrences of rules in M.

For $y, z \in (N \cup T)^*$ we write $y \Longrightarrow z$ if there is a matrix $(A_1 \to x_1, \ldots, A_n \to x_n)$ in M and the strings $w_i \in (N \cup T)^*$, $1 \leq i \leq n+1$, such that $y = w_1$, $z = w_{n+1}$, and, for each $1 \leq i \leq n$, either (1) $w_i = w'_i A_i w''_i$, $w_{i+1} = w'_i x_i w''_i$, for some $w'_i, w''_i \in (N \cup T)^*$, or (2) $w_i = w_{i+1}$, A_i does not appear in w_i, and the rule $A_i \to x_i$ appears in F. The reflexive and transitive closure of this relation is denoted by $\Longrightarrow^*$. Then, the generated language is

$$L(G) = \{w \in T^* \mid S \Longrightarrow^* w\}.$$

Note that the rules from F are skipped when they cannot be applied (that is why the matrix grammars of this general form are called *with appearance checking*). If the set F is empty, then the grammar is said to be without appearance checking (and $F = \emptyset$ is omitted when writing the grammar).

The family of languages generated by context-free matrix grammars with appearance checking is denoted by MAT_{ac}^λ (the superscript indicates that λ-rules are allowed); when using only λ-free rules, we denote the corresponding family by MAT_{ac}, and when only grammars without appearance checking are used, also the subscript ac is omitted.

The following results about matrix languages are known:

1. $CF \subset MAT \subseteq MAT^\lambda \subset RE$,
2. $MAT \subset CS$, $CS - MAT^\lambda \neq \emptyset$,
3. MAT contains non-semilinear languages,
4. Each language $L \in MAT^\lambda$, $L \subseteq a^*$, is regular (hence $D0L - MAT^\lambda \neq \emptyset$),
5. $MAT \subset MAT_{ac} \subset CS$,
6. $MAT^\lambda \subset MAT_{ac}^\lambda = RE$,

with consequences of the following forms in terms of length sets and Parikh images:

1. $NMAT = NREG = NSLIN$, but $PsMAT - PsCF \neq \emptyset$,
2. $Ps\,D0L - PsMAT^\lambda \neq \emptyset$,
3. $PsCS - PsMAT^\lambda \neq \emptyset$.

The questions of whether or not the inclusion $MAT \subseteq MAT^\lambda$ is proper and whether or not MAT^λ contains languages which are not context-sensitive are old *open problems* of this area.

When generating languages from a given family it is not necessary to use matrix grammars of the general form. For instance, the following useful result holds.

A matrix grammar $G = (N, T, S, M, F)$ is said to be in the *binary normal form* if $N = N_1 \cup N_2 \cup \{S, \#\}$, with these three sets mutually disjoint, and the matrices in M are in one of the following forms:

1. $(S \to XA)$, with $X \in N_1$, $A \in N_2$,
2. $(X \to Y, A \to x)$, with $X, Y \in N_1$, $A \in N_2$, $x \in (N_2 \cup T)^*$,
3. $(X \to Y, A \to \#)$, with $X, Y \in N_1$, $A \in N_2$,
4. $(X \to \lambda, A \to x)$, with $X \in N_1$, $A \in N_2$, and $x \in T^*$.

Moreover, there is only one matrix of type 1 and F consists exactly of all rules $A \to \#$ appearing in matrices of type 3; # is a trap-symbol, once introduced, it is never removed. A matrix of type 4 is used only once, in the last step of a derivation.

It is known that *for each matrix grammar there is an equivalent matrix grammar in the binary normal form* (hence each language in RE can be generated by a grammar in the binary normal form). This result can be strengthened in various ways.

For an arbitrary matrix grammar $G = (N, T, S, M, F)$, let us denote by $ac(G)$ the cardinality of the set $\{A \in N \mid A \to \alpha \in F\}$ and by $Var(G)$ the cardinality of N. Then, one can show (see [3]) that four non-terminals are sufficient in order to characterize RE by matrix grammars and out of them only three are used in appearance checking rules. Furthermore, if the total number of non-terminals is not restricted, then each recursively enumerable language can be generated by a matrix grammar G such that $ac(G) \leq 2$.

Consequently, to the properties of a grammar G in the binary normal form we can add the fact that $ac(G) \leq 2$. One says that this is *the strong binary normal form* for matrix grammars.

Another variant of the binary normal form is the following one. A matrix grammar with appearance checking $G = (N, T, S, M, F)$ is in the *Z-binary normal form* if $N = N_1 \cup N_2 \cup \{S, Z, \#\}$, with these three sets mutually disjoint, and the matrices in M are in one of the following forms:

1. $(S \to XA)$, with $X \in N_1$, $A \in N_2$,
2. $(X \to Y, A \to x)$, with $X, Y \in N_1$, $A \in N_2$, $x \in (N_2 \cup T)^*$, $|x| \leq 2$,
3. $(X \to Y, A \to \#)$, with $X \in N_1$, $Y \in N_1 \cup \{Z\}$, $A \in N_2$,
4. $(Z \to \lambda)$.

Moreover, there is only one matrix of type 1, F consists exactly of all rules $A \to \#$ appearing in matrices of type 3, and, if a sentential form generated by G contains the symbol Z, then it is of the form Zw, for some $w \in T^*$ or $w \in (T \cup N \cup \{\#\})^*$ with $|w|_\# \geq 1$ (that is, the appearance of Z makes sure that either all other symbols are terminal or at least one # appears in the string). As above, # is a trap-symbol, and the (unique) matrix of type 4 is used only once, in the last step of a derivation.

As above, *for each language $L \in RE$ there is a matrix grammar with appearance checking G in the Z-binary normal form such that $L = L(G)$.*

Note that in the case of the Z-binary normal form we do not have a bound on the number of non-terminals used in the appearance checking mode, but a special non-terminal symbol exists, Z, which appears if and only if the derivation is terminal (modulo the erasing of this special symbol) or the trap-symbol is present.

As we have mentioned, many other ways to control the derivation of context-free grammars were considered. We only mention here some of them, with only a few hints about their definition and functioning.

In a *programmed* grammar $G = (N, T, S, P)$, the rules are of the form $(b : A \to z, E, F)$, where b is a label, $A \to z$ is a context-free production, and E, F are two sets of labels of productions of G. (E is said to be the *success field*, and F is the *failure field* of the production.) When applying such a rule, if the context-free rule $A \to z$ can be executed, then it is applied and the next production to be applied is chosen from those with the label in E, otherwise, we choose a production labeled by some element of F, and try to apply it. If the failure field F of all rules is empty, then the grammar is said to be without appearance checking. The derivation starts from S, by using any rule which can be applied (hence irrespective of its label), and ends when a terminal string is obtained.

Four families of languages are obtained (using grammars with or without λ-rules, with or without appearance checking), which precisely characterize the four families of matrix languages considered above.

In a context-free *ordered* grammar $G = (N, T, S, P, >)$, we have a new ingredient, the partial order relation $>$ over P. A production p can be applied to a sentential form only if it can be applied as a context-free rule and there is no production $r \in P$ such that r is applicable and $r > p$ holds.

The family of languages generated by ordered grammars is denoted by ORD^λ, with the superscript being omitted when only λ-free rules are used. The following relations are known:

1. $ET0L \subset ORD \subset MAT_{ac}$,
2. $ORD^\lambda \subset RE$.

Another important class of regulated grammars is that of *random context grammars*, which are constructs $G = (N, T, S, P)$, where N, T, S are as above and P is a finite set of triples of the form $p = (A \rightarrow w; E, F)$, where $A \rightarrow w$ is a context-free production over $N \cup T$ and E, F are subsets of N. Then, p can be applied to a string $x \in (N \cup T)^*$ only if all symbols of E appear in x and no symbol of F appears in X. (E is said to be the set of *permitting* and F is said to be the set of *forbidding* context conditions of p.)

We denote by RC_{ac}^λ the family of languages generated by random context grammars with erasing rules; when all sets F are empty, then the subscript ac is removed, and when only λ-free rules are used we remove the superscript λ. We have:

1. $CF \subset RC \subseteq MAT \subset RC_{ac} = MAT_{ac}$,
2. $RC \subseteq RC^\lambda \subseteq MAT^\lambda \subset RC_{ac}^\lambda = RE$.

The checking elements of E, F can also be strings; then, for using a rule, all the strings from E should appear in the sentential form to be rewritten and no string from F should appear in this sentential form. Such grammars (called *semi-conditional*) generate all recursively enumerable or all context-sensitive languages, depending on whether λ-rules are used or not, respectively.

3.2.7 Grammar Systems

Traditionally, formal language theory investigates grammars working alone, but a natural and fruitful idea is to consider several grammars (grammar-like devices) cooperating in producing a single language, i.e. *grammar systems*. We briefly recall here the two main classes of grammar systems investigated so far, the sequential ones (known as *cooperating distributed grammar systems*) and the *parallel communicating grammar systems*.

A *cooperating distributed* (in short, CD) grammar system of degree $n \geq 1$, is a construct

$$\Gamma = (N, T, S, P_1, P_2, \ldots, P_n),$$

where N, T are disjoint alphabets (the nonterminal and the terminal alphabet, respectively), $S \in N$ (the axiom), and $P_1, \ldots, P_n$ are finite sets of context-free rules over $N \cup T$, called the *components* of Γ.

Several cooperation strategies were considered; we mention here only the five basic ones.

For two sentential forms w, w' over $N \cup T$, we write $w \Longrightarrow_i^* w'$, $w \Longrightarrow_i^{=k} w'$, $w \Longrightarrow_i^{\leq k} w'$, $w \Longrightarrow_i^{\geq k} w'$, $w \Longrightarrow_i^t w'$, for some $k \geq 1$, if w' can be obtained from w as follows: (1) by any number of derivation steps, (2) by k derivation steps, (3) by at most k derivation steps, (4) by at least k derivation steps, (5) by a sequence of derivation steps using rules from P_i which cannot be continued, respectively. The language generated by Γ in the mode $a \in \{*, t\} \cup \{\leq k, = k, \geq k \mid k \geq 1\}$, is defined by

$$L_a(\Gamma) = \{ x \in T^* \mid S \Longrightarrow_{j_1}^a w_1 \Longrightarrow_{j_2}^a \ldots \Longrightarrow_{j_m}^a w_m = x,$$

for some $m \geq 1, 1 \leq j_s \leq n, 1 \leq s \leq m\}$.

Informally stated, the components of the system work in turn, according to the cooperation protocol a, on a common sentential form. When a terminal string is obtained, this string is said to be generated by the system.

We denote by $CD_n(a)$ the family of languages generated by CD grammar systems of degree at most $n \geq 1$, in the mode a; when the degree is not bounded the subscript n is replaced by $*$.

Here are a few results about the generative power of CD grammar systems:

1. $CF = CD_*(*) = CD_1(a) = CD_*(= 1) = CD_*(\leq k) = CD_*(\geq 1)$, for all $k \geq 1$, $a \in \{t\} \cup \{= j, \geq j \mid j \geq 1\}$,
2. $CF = CD_1(t) = CD_2(t) \subset CD_3(t) = ET0L$,
3. $CD_*(a) \subseteq MAT^\lambda$, for all $a \in \{= k, \geq k \mid k \geq 1\}$.

A *parallel communicating* (PC, for short) *grammar system* of degree $n \geq 1$, is a construct

$$\Gamma = (N, T, K, (S_1, P_1), \ldots, (S_n, P_n)),$$

where N, T, K are pairwise disjoint alphabets, with $K = \{Q_1, \ldots, Q_n\}$, $S_i \in N$, and P_i are finite sets of rewriting rules over $N \cup T \cup K$, $1 \leq i \leq n$; the elements of N are *non-terminal symbols*, those of T are *terminals*; the elements of K are called *query symbols*; the pairs (S_i, P_i) are the *components* of the system. Note that the query symbols are associated in a one-to-one manner with the components. When discussing the type of the components in the Chomsky hierarchy, the query symbols are interpreted as non-terminals.

For $(x_1, \ldots, x_n), (y_1, \ldots, y_n)$, with $x_i, y_i \in (N \cup T \cup K)^*, 1 \leq i \leq n$ (we call such an n-tuple a *configuration*), and $x_1 \notin T^*$, we write $(x_1, \ldots, x_n) \Longrightarrow_r (y_1, \ldots, y_n)$ if one of the following two cases holds:

(i) $|x_i|_K = 0$ for all $1 \leq i \leq n$; then $x_i \Longrightarrow_{P_i} y_i$ or $x_i = y_i \in T^*, 1 \leq i \leq n$;
(ii) there is i, $1 \leq i \leq n$, such that $|x_i|_K > 0$; we write such a string x_i as

$$x_i = z_1 Q_{i_1} z_2 Q_{i_2} \ldots z_t Q_{i_t} z_{t+1},$$

for $t \geq 1$, $z_j \in (N \cup T)^*, 1 \leq j \leq t+1$; if $|x_{i_j}|_K = 0$ for all $1 \leq j \leq t$, then

$$y_i = z_1 x_{i_1} z_2 x_{i_2} \ldots z_t x_{i_t} z_{t+1},$$

[and $y_{i_j} = S_{i_j}, 1 \leq j \leq t$]; otherwise $y_i = x_i$. For all unspecified i we have $y_i = x_i$.

Point (i) defines a *rewriting* step (performed componentwise, synchronously, using one rule in all components whose current strings are not terminal); (ii) defines a *communication* step: the query symbols Q_{i_j} introduced in some x_i are replaced by the associated strings x_{i_j}, provided that these strings do not contain further query symbols. The communication has priority over rewriting (a rewriting step is allowed only when no query symbol appears in the current configuration). The work of the system is blocked in two cases: when circular queries appear, and when no query symbol is present but a rewriting step is not possible because a component cannot rewrite its sentential form, although it is a non-terminal string.

The relation $\Longrightarrow_r$ considered above is said to be performed in the *returning* mode: after communicating, a component resumes working from its axiom. If the text in brackets, [and $y_{i_j} = S_{i_j}, 1 \leq i \leq t$], is removed, then we obtain the *non-returning* mode of derivation: after communicating, a component continues the processing of the current string. We denote by $\Longrightarrow_{nr}$ the obtained relation.

The language generated by Γ is the set of terminal strings generated by its first component, when starting from $(S_1, \ldots, S_n)$, that is

$$L_f(\Gamma) = \{w \in T^* \mid (S_1, \ldots, S_n) \Longrightarrow_f^* (w, a_2, \ldots, a_n),$$

$$\text{for } a_i \in (N \cup T \cup K)^*, 2 \leq i \leq n\}, \ f \in \{r, nr\}.$$

(In the last configuration of a derivation, the strings in the components $2, \ldots, n$ are ignored; moreover, it is supposed that the work of Γ stops when a terminal string is obtained by the first component.)

A PC grammar system is said to be *centralized* if only the first component can introduce query symbols, and *non-centralized* otherwise.

We denote by $NCPC_n X$ the family of languages generated by PC grammar systems with at most $n \geq 1$ components, with rules of type $X \in \{REG, LIN, CF\}$, centralized (indicated by C) and non-returning (indicated by N); when non-centralized systems are used we remove the symbol C, and when returning systems are used we remove the symbol N; when no bound on the number of components is imposed we replace the subscript n by $*$.

Here are a few results about the generative capacity of these systems:

1. $Y_n REG - LIN \neq \emptyset$, $Y_n LIN - CF \neq \emptyset$, for all $n \geq 2$,
   $Y_n REG - CF \neq \emptyset$, for all $n \geq 3$, and for all $Y \in \{PC, CPC, NPC, NCPC\}$,
2. $Y_n REG - CF \neq \emptyset$, for all $n \geq 2$, and $Y \in \{NPC, NCPC\}$,
3. $CPC_2 REG \subset CF$, $PC_2 REG \subseteq CF$,
4. $CPC_* REG \subseteq MAT$,
5. $PC_* CF = NPC_* CF = RE$.

3.2.8 Splicing and Insertion-Deletion

Strings can be processed in various ways, among which rewriting is the most traditional. Many other operations were considered, with biological or linguistic inspiration. In general, they cannot be directly transferred to multisets, but in membrane computing one also investigates systems whose elements are described by strings. That is why we briefly introduce here two types of operations, both of them modeling operations taking place at the genome level (in both cases we also define language generating devices based on these operations).

The first is the splicing operation, considered in 1987 by T. Head, as a formal counterpart of the recombination of DNA molecules under the influence of restriction enzymes.

Consider an alphabet V and two symbols $\#, \$$ not in V. A *splicing rule* over V is a string $r = u_1 \# u_2 \$ u_3 \# u_4$, where $u_1, u_2, u_3, u_4 \in V^*$. For such a rule r and for $x, y, w, z \in V^*$ we define

$$(x, y) \vdash_r (w, z) \quad \text{iff} \quad x = x_1 u_1 u_2 x_2, \; y = y_1 u_3 u_4 y_2,$$
$$w = x_1 u_1 u_4 y_2, \; z = y_1 u_3 u_2 x_2,$$
$$\text{for some } x_1, x_2, y_1, y_2 \in V^*.$$

(One cuts the strings x, y in between u_1, u_2 and u_3, u_4, respectively, and one recombines the fragments obtained in this way.)

This operation can be extended to languages in the following natural way. A pair $\sigma = (V, R)$, where V is an alphabet and $R \subseteq V^* \# V^* \$ V^* \# V^*$ is a set of splicing rules, is called an *H scheme*. For an H scheme σ and a language $L \subseteq V^*$ we define

$$\sigma(L) = \{z \in V^* \mid (x, y) \vdash_r (w, z) \text{ or } (x, y) \vdash_r (z, w),$$

$$\text{for some } x, y \in L, \text{ and } r \in R\},$$

$$\sigma^0(L) = L,$$

$$\sigma^{i+1}(L) = \sigma^i(L) \cup \sigma(\sigma^i(L)), \text{ for } i \geq 0,$$

$$\sigma^*(L) = \bigcup_{i \geq 0} \sigma^i(L).$$

Then, an *extended H system* is a construct $\gamma = (V, T, A, R)$, where V is an alphabet, $T \subseteq V$ (terminal alphabet), $A \subseteq V^*$ (axioms), and $R \subseteq V^*\#V^*\$V^*\#V^*$ is a set of splicing rules over V. The language generated by γ is defined by

$$L(\gamma) = \sigma^*(A) \cap T^*,$$

where $\sigma = (V, R)$ is the H scheme associated with γ.

For two families FL_1, FL_2 of languages, we denote by $EH(FL_1, FL_2)$ the family of languages $L(\gamma)$ generated by H systems $\gamma = (V, T, A, R)$ with $A \in FL_1$ and $R \in FL_2$ (because the splicing rules are written as strings, it makes sense to speak about the type of the "language" R in a given hierarchy of languages).

The following results are basic in the splicing area:

1. $EH(REG, FIN) = REG$, $EH(CF, FIN) = CF$.
2. $EH(FIN, REG) = RE$.

When the use of splicing rules is restricted—and there are several ways to do this, in most cases extending to splicing systems restrictions from regulated rewriting area, such as context conditions (promoters, inhibitors, target languages), matrix, programmed, ordered systems, etc.—then characterizations of RE can be obtained even for H systems with a finite set of rules.

Insertion–deletion operations were investigated both in linguistics and language theory and, recently, in DNA computing.

An insertion rule is of the form (u, x, v), with the meaning that a string $w = w_1 uv w_2$ is transformed by such a rule to $w' = w_1 uxv w_2$ (the string x was inserted in the context (u, v)). A rule of the same form can be used in the deletion manner: from $z = z_1 uxv z_2$ we pass to $z' = z_1 uv z_2$ (the substring x was deleted from the context (u, v)). Then, an *insertion–deletion system* (in short, an *ins–del system*) can be defined as a construct $\gamma = (V, T, A, I, D)$, where V is an alphabet, $T \subseteq V$ (terminal alphabet), $A \subseteq V^*$ (a finite set of axioms), and I, D are finite sets of insertion and deletion rules, respectively. The generated language consists of all strings over T^* which can be obtained by starting from the strings from A and using iteratively rules from I and from D.

The complexity of an insertion–deletion system is described by several parameters, the most important ones being the length of inserted or deleted strings and the length of the contexts controlling the place where these operations are performed. Specifically, the family of languages generated by ins–del systems which insert strings of length at most m between strings of length at most n, and delete strings of length at most p from contexts composed of strings of length at most q is denoted by $INS_m^n DEL_p^q$. When any of the parameters m, n, p, q is not bounded, it is replaced with $*$.

We recall only a few results concerning the power of ins–del systems:

- $INS_2^2 DEL_0^0$ contains non-semilinear languages.
- $RE = INS_3^0 DEL_2^0 = INS_2^0 DEL_3^0 = INS_1^1 DEL_2^0 = INS_1^1 DEL_1^1 = INS_2^0 DEL_1^1$.
- $INS_*^1 DEL_0^0 \cup INS_2^0 DEL_2^0 \subseteq CF$.

3.3 ELEMENTS OF COMPLEXITY

The power of computing models, compared with the power of Turing machines and other classic models of computing, is the most fundamental issue to address when defining a new computing device, but similarly important (and relevant from a practical point of view) is the question concerning the *efficiency* of a model. How fast can a computing device solve problems from a given class? Is it possible to solve problems in a reasonable time and using reasonable computing resources? Dually: which problems can be solved in such circumstances? The problems which can be solved with given resources form a *complexity class*.

In order to define a complexity class we need to specify several parameters. First, we have to choose a given *model of computation*, a way to define "algorithms", and in the complexity theory the standard one is the Turing machine (the multi-tape version, with the input tape being a read-only tape, is the one customarily chosen, but this is not essential for our needs). Second, for defining a complexity class we need to fix a *mode of computation*. Turing machines can be deterministic or non-deterministic, hence we have at least these two modes. Third, we have to specify the *resource* we want to deal with; *time* is perhaps the most important, but the *space* used for computing is also important. One can imagine several others, related to the "effort" of computing, to the effort of programming the computing device, and—in the case of distributed computing devices—the cost of communication among parts, the cost of synchronization, etc.

Having specified these elements, a complexity class is defined with respect to functions $f : \mathbb{N} \longrightarrow \mathbb{N}$ (with certain minimal properties, for instance non-decreasing, computable, etc.) in the following way: the complexity class **md-res**$_M(f)$ consists of all problems which can be solved in mode *md* by devices D from class M such that for any input x, device D needs at most $f(|x|)$ units from the resource *res*. Note the important fact that the size of the input is described by its length (in a given representation). When the input is a number, n, the length of the binary representation of n is taken as the size of the input.

The usual problem in complexity is a decidability problem, a question with two possible answers, "yes" and "no". A deterministic algorithm (Turing machine) for solving such a problem Q will take an instance I_Q of the problem and after a while will provide one of the two possible outputs, "yes" or "no"; if the time needed to solve the problem is bounded by $f(|I_Q|)$, then we say that the problem belongs to **DTIME**(f).

Turing machines are the standard models of computing used in complexity theory not only because they are mathematically elegant and historically important, but also because they are robust from the complexity classes points of view. Roughly speaking, this means that the complexity classes remain the same if we define them by taking the class M of devices to be the class of deterministic Turing machines, and then changing the type of the Turing machine (with several tapes, for instance) or even the model of computation (but not the mode of computing, e.g. passing to non-deterministic computations).

The most important complexity classes are **DTIME**(f), **DSPACE**(f), which contain the problems which can be solved in a deterministic mode in a time, respectively, a space bounded by the function f, and **NTIME**(f), **NSPACE**(f), which contain the problems which can be solved in a non-deterministic mode in a time, respectively, a space bounded by the function f.

The difference between deterministic and non-deterministic algorithms (solutions to problems) is crucial. In the non-deterministic case, we say that a problem is in **NTIME**(f) if, for every input of size n there is at least one computation of the algorithm (of the non-deterministic Turing machine, for instance) which gives us the answer "yes" after $f(n)$ time units. Note the important detail that it is sufficient for at least one computation to exist which gives us the positive answer in time $f(n)$; it does not matter how many computations fail to give this answer. If no computation exists which gives the answer "yes", then we conclude that the problem has the answer "no". Therefore, the conclusion "no" is obtained at the limit, knowing that no computation provides the answer "yes". The difference from the deterministic case is obvious.

A very illustrative reformulation of the difference between deterministic and non-deterministic complexity classes is the following. In the deterministic case one has to explore all branches of the computability tree, giving the unique answer one can get after completing this exploration, "yes" or "no". In the non-deterministic case one has only to guess a successful root-to-leaf branch, hence to explore a single branch of the computability tree. In the first case the time of computing is proportional to the number of nodes in the whole computability tree; in the second case the time is proportional to the depth of the tree, the length of the longest branch.

Obviously, deterministic classes are included in the non-deterministic classes.

Now, the important step is to choose the mapping f. The most widely investigated complexity classes start from the constant, logarithmic, linear, polynomial, and exponential functions. In general, it is considered that the borderline between *tractable* and *intractable* is that between *polynomial* and *exponential*. That is why polynomial functions are very important—also mentioning that "constants and lower monomials do not count": a polynomial of degree k is identified in this framework with n^k. Then, the union of all classes **DTIME**(n^k), $k \geq 1$, is usually denoted by **P**, while the union of all classes **NTIME**(n^k), $k \geq 1$, is usually denoted by **NP**. These are the classes of problems which can be solved in deterministic and non-deterministic polynomial time respectively. The inclusion $\mathbf{P} \subseteq \mathbf{NP}$ is clear; whether or not this is an equality is probably the best known and the most important open problem in computer science today, the so-called $\mathbf{P} \stackrel{?}{=} \mathbf{NP}$ problem.

However, is not enough to know that a problem is not in the class **P**. Is the problem inherently difficult, hence in $\mathbf{NP} - \mathbf{P}$ (provided that, as generally expected, $\mathbf{P} \neq \mathbf{NP}$), or are we not yet able to find a deterministic polynomial time solution to it? Then, is the problem **NP**-*complete*? This last question is related to the following idea. The problems can be *reduced* from one problem to another. We say that problem B reduces to problem A if there is a transformation R which, for any input x of B, produces an input $R(x)$ to A such that the answer of A to the input $R(x)$ is the answer of B to the input x. Thus, if B reduces to A, in order to solve B it is enough to compute $R(x)$ and to solve A for the input $R(x)$. Of course, a reduction is acceptable if it is less complex than the problem B itself—with respect to classes **P** and **NP**, a reduction R is acceptable if it can be done in polynomial time. In such a case, when B reduces to A, we may say that A is at least as complex as B, hence it is important to know whether a problem A has the property that "many" (all) problems from a given class can be reduced to A. In such a case we may say that A is among the hardest problems of that class. For most natural complexity classes, **P** and **NP** included, there are problems such that *all* problems from that class can be reduced to them. Such problems are called *complete* for the respective classes. Of particular interest are the problems which are **NP**-complete. All problems for which we do not have (yet) a deterministic polynomial algorithm can be reduced to them, hence the **NP**-complete problems are "the most difficult from **NP**". In practical applications, this is considered a sufficient guarantee that such a problem is not tractable.

There are a lot of **NP**-complete problems known, many of them with a simple formulation, and many of them of a direct practical interest (in some sense, the converse is also true: most practical problems are intractable). Actually, knowing that a problem is in **P** does not automatically have a practical importance. In theory we only consider the degree of the polynomial function f, the coefficients "do not matter", which is not at all true for a real computation. Furthermore, the theory deals with the worst-case complexity, while for real problems the average-case complexity might be more relevant, etc. So, an important challenge is to define computing models—and then to implement them—able to overcome this difficulty. A classic hope in this respect is *parallel computing* (made possible, at least theoretically, at a massive level, by various areas of molecular computing).

In some sense, the parallelism tries to simulate the non-determinism: if we have enough processors, then we can explore all branches of a computation tree at the same time, and thus we could get the answer "yes" or "no"—hence the answer provided by a deterministic computation—after exhausting the tree, in a time equal to the depth of the tree, that is, in the time needed by a non-deterministic computation.

A widely used computing model for defining parallel computing complexity classes are the Boolean circuits. Roughly speaking, they are families of ascending nets with Boolean gates in the nodes, starting from given truth values and computing a complex logical function, in parallel; each circuit in the family starts from an input of a given size, hence such a family is of the form $\mathcal{C} = (C_1, C_2, \ldots)$, such that C_i has i inputs. In order to accept such a family as a good parallel "computer", we have first to ensure that the circuits C_i are constructed in a *uniform manner* (a Turing machine should exist, able to construct C_i starting from 1^i in polynomial time—starting from 1^i is meant to ensure that the circuits indeed form a family, they are related to the same problem, and the difference from C^i and C^j lies in the size of the input they can handle). The complexity classes are then defined by taking into account both the time necessary for computing a function with n inputs and the number of gates from C_n. The time is equal to the depth of the circuit. Thus, one compares the depth of circuits C_n with their number of nodes.

The functions that can be computed in a polylogarithmic time, $\log^k n$, by using a polynomial number of gates, n^k are considered practically tractable for parallel computing (by Boolean circuits). One says that the time is *parallel time*, while the number of gates is called the (amount of) *work* (power), which leads to the notation $\mathbf{NC} = \bigcup_k \mathbf{PT/WK}(\log^k n, n^k)$. This class has nice properties, similar to those of classes **P** and **NP**: it is stable under changing the computing model (for instance, from Boolean circuits to PRAMs—parallel random access machines), it is closed under reductions, but many important problems about it are still unsolved. Perhaps the most relevant is the question of whether or not the inclusion $\mathbf{NC} \subseteq \mathbf{P}$ is proper or not. (The inclusion is clear, as we deal with a polynomial number of processors working a polynomial number of steps.) Otherwise stated: are all sequentially polynomial problems efficiently parallelizable (solvable in polylogarithmic time on a polynomial number of processors), or are there inherently sequential problems? As in the case of the $\mathbf{P} \stackrel{?}{=} \mathbf{NP}$ problem, the conjecture widely accepted is that the answer is negative, and the most probable candidate problems from **P** which are not in **NC** are the P-complete problems. There are many such problems, so it seems that Boolean circuits, PRAMs, and other related parallel models of computing are not able to handle them in an efficient manner.

We conclude this short and informal excursion through complexity theory by advising the reader to consult one of the many monographs available in this area, and with the remark that in Chapter 12 complexity classes specific to membrane computing will be introduced.

References

[1] E. Csuhaj-Varju, J. Dassow, J. Kelemen, Gh. Păun: *Grammar Systems. A Grammatical Approach to Distribution and Cooperation.* Gordon and Breach, London, 1994.

[2] J. Dassow, Gh. Păun: *Regulated Rewriting in Formal Language Theory.* Springer, Berlin, 1989.

[3] R. Freund, Gh. Păun: On the number of non-terminals in graph-controlled, programmed, and matrix grammars. *Lecture Notes in Computer Sci.*, 2055 (2001), 214–225.

[4] M.R. Garey, D.J. Johnson: *Computers and Intractability. A Guide to the Theory of NP-Completeness.* W.H. Freeman, San Francisco, 1979.

[5] S. Ginsburg: *Algebraic and Automata-Theoretic Properties of Formal Languages.* North-Holland, Amsterdam, 1975.

[6] M. Harrison: *Introduction to Formal Language Theory.* Addison-Wesley, Reading, MA, 1978.

[7] G.T. Herman, G. Rozenberg: *Developmental Systems and Languages.* North-Holland, Amsterdam, 1975.

[8] J.E. Hopcroft, J.D. Ullman: *Introduction to Automata Theory, Languages and Computation.* Addison-Wesley, Reading, MA, 1979.

[9] M. Minsky: *Computation: Finite and Infinite Machines.* Prentice Hall, 1967.

[10] Ch.P. Papadimitriou: *Computational Complexity.* Addison-Wesley, Reading, MA, 1994.

[11] Gh. Păun: *Marcus Contextual Grammars.* Kluwer, Dordrecht, Boston, 1997.

[12] Gh. Păun: *Membrane Computing. An Introduction.* Springer, Berlin, 2002.

[13] Gh. Păun, G. Rozenberg, A. Salomaa: *DNA Computing. New Computing Paradigms.* Springer, Heidelberg, 1998.

[14] G. Rozenberg, A. Salomaa: *The Mathematical Theory of L Systems.* Academic Press, New York, 1980.

[15] G. Rozenberg, A. Salomaa, eds. *Handbook of Formal Languages.* 3 volumes, Springer, Berlin, 1997.

[16] A. Salomaa: *Formal Languages.* Academic Press, New York, 1973.

[17] D. Wood: *Theory of Computation.* Harper and Row, New York, 1987.

CHAPTER 4

CATALYTIC P SYSTEMS

RUDOLF FREUND

OSCAR H. IBARRA

ANDREI PĂUN

PETR SOSÍK

HSU-CHUN YEN

4.1 INTRODUCTION

THE catalytic P systems considered in this chapter go back to the original model of P systems introduced in [24]. Multisets of objects are placed in the regions of a hierarchic membrane structure and evolve by applying catalytic or non-cooperative rules in the maximally parallel way. We define catalytic P systems as devices for generating sets of (vectors of) natural numbers in the following way:

Definition 4.1 *A catalytic P system of degree $m \geq 1$ is a construct*

$$\Pi = (O, C, \mu, w_1, \ldots, w_m, R_1, \ldots, R_m, i_0) \text{ where:}$$

1. *O is the alphabet of objects;*
2. *$C \subseteq O$ is the alphabet of catalysts;*
3. *μ is a membrane structure of degree m with membranes labeled in a one-to-one manner with the natural numbers $1, 2, \ldots, m$;*

4. $w_1, \ldots, w_m \in O^*$ are the multisets of objects initially present in the m regions of μ;
5. $R_i, 1 \leq i \leq m$, are finite sets of evolution rules over O associated with the regions $1, 2, \ldots, m$ of μ; these evolution rules are of the forms $ca \to cv$ or $a \to v$, where c is a catalyst, a is an object from $O - C$, and v is a string from $((O - C) \times \{here, out, in\})^*$;
6. $i_0 \in \{0, 1, \ldots, m\}$ indicates the output region of Π.

The membrane structure and the multisets represented by w_i, $1 \leq i \leq m$, in Π constitute the *initial configuration* of the system. A transition between configurations means applying a maximal multiset of evolution rules in parallel. The application of a rule $u \to w$ in a region containing a multiset M results in subtracting from M the multiset identified by u, and then in adding the multiset identified by w to the respective region, whereby the objects eventually are transported through membranes due to the targets *in*, *out*, and *here* (we usually omit the target *here*). In a more general way, instead of *in* we may also use the targets in_j specifying the exact region j where the object should enter through an inner membrane. The P system Π is called *purely catalytic* if every evolution rule involves a catalyst.

The system continues taking parallel steps until there remain no applicable rules in any region of Π; then the system halts. The number or the Parikh vector of the multiset of objects from O contained in the output region i_0 at the moment when the system halts is considered to be the *result* of the underlying computation of Π; $i_0 = 0$ designates the environment outside the skin membrane to contain the results. By collecting the results of all computations possible in Π we get the set of natural numbers and vectors generated by Π, denoted by $N(\Pi)$ and $Ps(\Pi)$, respectively. The families of all sets of numbers or sets of vectors computed by [purely] catalytic P systems with at most m membranes and the set of catalysts containing at most k elements are denoted by $NOP_m([p]cat_k)$ and $PsOP_m([p]cat_k)$, respectively. (We use the bracket notation $[p]$ to specify the parameter p to be present or not at the same time.)

We may relax the previous definition by specifying a set of terminal objects Σ and only count the number of the symbols of $\Sigma \subseteq O$ to be found in the specified output membrane at the end of halting computations; in that way, we obtain *extended [purely] catalytic P systems* of the form

$$\Pi = (O, \Sigma, C, \mu, w_1, \ldots, w_m, R_1, \ldots, R_m, i_0)$$

and $N_\Sigma(\Pi)$ and $Ps_\Sigma(\Pi)$, respectively, as well as the corresponding families $XO_E P_m([p]cat_k)$ with $X \in \{N, Ps\}$. The most interesting case of extended [purely] catalytic P systems especially referred to in the rest of this chapter is when we take $\Sigma = O - C$, thus obtaining $N_{-C}(\Pi)$ and $Ps_{-C}(\Pi)$, respectively, as well as the corresponding families $XO_{-C} P_m([p]cat_k)$ with $X \in \{N, Ps\}$.

In addition to these generating P systems we may also consider *accepting P systems* where the multiset to be analyzed is put into region i_0 (then called *input membrane*) together with the corresponding initial multiset w_{i_0} and accepted by a halting computation (in which case extended systems are of no interest). The families of all sets (of vectors) of natural numbers accepted in that way by halting computations in [purely] catalytic P systems with at most m membranes and the set of catalysts containing at most k elements are denoted by $X_a OP_m([p]cat_k)$, $X \in \{N, Ps\}$. In the case of deterministic systems, i.e. where to each configuration reachable from the start configuration at most one applicable multiset of rules exists, we add the symbol D in front of O.

Finally, we may also consider functions $f : \mathbf{N}^\alpha \to \mathbf{N}^\beta$ to be computed by [purely] catalytic P systems with the input multiset put into region i_0 together with w_{i_0} and the result also obtained in region i_0 when the system halts; the corresponding families of functions computed by [purely] catalytic P systems with at most m membranes and the set of catalysts containing at most k elements are denoted by $F_{\alpha,\beta} Y P_m([p]cat_k)$, $Y \in \{O, O_E, O_{-C}\}$. In the case of deterministic systems, we add the symbol D in front of O. For every family defined for [purely] catalytic P systems, we replace any of the parameters m and k by $*$ if it is unbounded.

In the following section we will show that $NOP_1(cat_2) = NRE$ and $NOP_1(pcat_3) = NRE$, i.e. [purely] catalytic P systems are computationally complete. The proofs elaborated in [11] even show that any partial recursive function $f : \mathbf{N}^\alpha \to \mathbf{N}^\beta$ can be computed by deterministic (purely) catalytic P systems with only one membrane and with only $\alpha + 2$ ($\alpha + 3$) catalysts, i.e.

$$F_{\alpha,\beta} O_{-C} P_1(cat_{\alpha+2}) = F_{\alpha,\beta} O_{-C} P_1(pcat_{\alpha+3}) = F_{\alpha,\beta} RE$$

where $F_{\alpha,\beta} RE$ denotes the family of all partially recursive functions $f : \mathbf{N}^\alpha \to \mathbf{N}^\beta$. We then give some partial results for catalytic P systems with less than two catalysts.

Considering [purely] catalytic P systems as accepting mechanisms, the deterministic variants turn out to be less powerful than their non-deterministic counterparts, i.e. deterministic [purely] catalytic P systems are not universal, which implies

$$Ps_a DOP_*([p]cat_*) \subsetneq Ps_a OP_*([p]cat_*) = Ps RE.$$

Only when adding order relations on the rules, universality can be regained also for deterministic [purely] catalytic P systems.

Allowing to look ahead k steps in any computation before choosing the multiset of rules to be applied, yields the concept of k-determinism. Even the non-deterministic (purely) catalytic P systems then turn out to be at least 4-deterministic.

As special extensions of the concept of catalysts we consider bistable catalysts, which may change between two states when being used in a rule, as well as mobile catalysts, which may not only have the target *here*, but also the targets *in* and *out*.

We also investigate the influence of variants of transition modes and halting on the computational power of (purely) catalytic P systems. Finally we describe a way to define languages by using catalytic P systems.

4.2 Computational Completeness, Universality

The following observations concerning the computational power of register machines are essential for the main results established in this chapter:

Proposition 4.1 *For any partially recursive function $f : \mathbf{N}^\alpha \to \mathbf{N}^\beta$ there exists a deterministic $(\alpha + 2 + \beta)$-register machine M computing f in such a way that, when starting with $(n_1, \ldots, n_\alpha) \in \mathbf{N}^\alpha$ in registers 1 to α, M has computed $f(n_1, \ldots, n_\alpha) = (r_1, \ldots, r_\beta)$ if it halts in the final label h with registers $\alpha + 3$ to $\alpha + 2 + \beta$ containing r_1 to r_β, and all other registers being empty; if the final label cannot be reached, $f(n_1, \ldots, n_\alpha)$ remains undefined. The registers $\alpha + 3$ to $\alpha + 2 + \beta$ are never decremented.*

The following two corollaries are immediate consequences of the preceding proposition (by taking $\alpha = 0$ and $\beta = 0$, respectively):

Corollary 4.1 *For any recursively enumerable set $Q \subseteq \mathbf{N}^\beta$ of vectors of non-negative integers there exists a non-deterministic $(\beta + 2)$-register machine M generating Q in such a way that, when starting with all registers 1 to $\beta + 2$ being empty, M non-deterministically computes and halts with n_i in registers $i + 2$, $1 \leq i \leq \beta$, and registers 1 and 2 being empty if and only if $(n_1, \ldots, n_\beta) \in Q$. The registers 3 to $\beta + 2$ are never decremented.*

Corollary 4.2 *For any recursively enumerable set $Q \subseteq \mathbf{N}^\alpha$ of vectors of non-negative integers there exists a deterministic $(\alpha + 2)$-register machine M accepting Q in such a way that M halts with all registers being empty if and only if M starts with some $(n_1, \ldots, n_\alpha) \in Q$ in registers 1 to α and the registers $\alpha + 1$ to $\alpha + 2$ being empty.*

Based on Corollary 4.1 with $\beta = 1$, we now are able to establish our main result for the generative power of (purely) catalytic P systems:

Theorem 4.1 $NO_{-C}P_1(cat_2) = NO_{-C}P_1(pcat_3) = NRE$.

Proof. We only prove the inclusion $NRE \subseteq NOP_1(cat_2)$ by constructing a catalytic P system Π which simulates the non-deterministic register machine $M = (m, H, l_0, l_h, I)$ from Corollary 4.1 for $\beta = 1$, i.e. $m = 3$, and non-deterministically generates a representation of any number from the set $N(M)$ in NRE by the corresponding number of symbols o_3:

$\Pi = (O, \{c_1, c_2\}, [_1 \,]_1, w, R, 1)$,

$O = \{\#\} \cup \{c_1, c'_1, c''_1, c_2, c'_2, c''_2\} \cup \{o_k \mid 1 \le k \le 3\}$

$\cup \{p_j, \tilde{p}_j, p'_j, p''_j, \bar{p}_j, \bar{p}'_j, \bar{p}''_j, \hat{p}_j, \hat{p}'_j, \hat{p}''_j \mid j : (\text{SUB}(r), k, l) \in I\}$

$\cup \{p_j, \tilde{p}_j \mid j : (\text{ADD}(r), k, l) \in P\} \cup \{p_{l_h}, \tilde{p}_{l_h}\}$,

$R = \{x \to \# \mid x \in \{p_j, \tilde{p}_j, p'_j, p''_j, \bar{p}_j, \bar{p}''_j, \hat{p}_j, \hat{p}''_j$

$\mid j : (\text{SUB}(r), k, l) \in I\}\}$

$\cup \{x \to \# \mid x \in \{c'_1, c''_1, c'_2, c''_2\}\} \cup \{\# \to \#\}$

$\cup \{c_1 p_{l_h} \to c_1, c_2 \tilde{p}_{l_h} \to c_2\}$

$\cup \{c_1 \tilde{p}_j \to c_1 \mid j : (\text{ADD}(r), k, l) \in I, r \in \{1, 2, 3\}\}$

$\cup \{c_2 p_j \to c_2 p_k \tilde{p}_k o_r, c_2 p_j \to c_2 p_l \tilde{p}_l o_r$

$\mid j : (\text{ADD}(r), k, l) \in I, r \in \{1, 2, 3\}\}$

$\cup \{c_r p_j \to c_r \hat{p}_j \hat{p}'_j, c_r p_j \to c_r \bar{p}_j \bar{p}'_j \bar{p}''_j, c_r o_r \to c_r c'_r, c_r c'_r \to c_r c''_r,$

$c_{3-r} c''_r \to c_{3-r}, c_r \hat{p}'_j \to c_r \#, c_{3-r} \hat{p}'_j \to c_{3-r} \hat{p}''_j, c_r \hat{p}''_j \to c_r p_k \tilde{p}_k,$

$c_r \bar{p}_j \to c_r, c_{3-r} \bar{p}''_j \to c_{3-r} p''_j, c_{3-r} p''_j \to c_{3-r} p'_j,$

$c_r p'_j \to c_r p_l \tilde{p}_l \mid j : (\text{SUB}(r), k, l) \in I, r \in \{1, 2\}\}$

$\cup \{c_{3-r} y \to c_{3-r} \mid y \in \{\bar{p}_j, \hat{p}_j, \bar{p}'_j\}, j : (\text{SUB}(r), k, l) \in I, r \in \{1, 2\}\}$.

Every label i of H is represented by the pair of symbols $p_i \tilde{p}_i$ in order to keep both catalysts busy. Thus, an add-instruction $j : (\text{ADD}(r), k, l) \in I$ is accomplished by the parallel execution of the two rules $c_1 \tilde{p}_j \to c_1$ and $c_2 p_j \to c_2 p_i \tilde{p}_i o_r$ thereby non-deterministically choosing $i \in \{k, l\}$.

The simulation of a subtract-instruction $j : (\text{SUB}(r), k, l) \in I$ on register $r \in \{1, 2\}$, starts with non-deterministically guessing whether the contents of register r is not empty (we use $c_r p_j \to c_r \hat{p}_j \hat{p}'_j$) or empty (we use $c_r p_j \to c_r \bar{p}_j \bar{p}'_j \bar{p}''_j$); in parallel, in both cases $c_{3-r} \tilde{p}_j \to c_{3-r}$ is applied with the second catalyst. At the end of the simulation, we have the pair of label symbols $p_k \tilde{p}_k$ or the pair $p_l \tilde{p}_l$ and may continue with simulating the instruction labeled with k or l, respectively.

If we have chosen the case with register r not being empty, then we continue with applying $c_r o_r \to c_r c'_r$ and $c_{3-r} \hat{p}_j \to c_{3-r}$. In case of a wrong choice, i.e. if no symbol o_r is present, due to maximal parallelism the symbol $\hat{p}'_j$ has to be consumed by using $c_r \hat{p}'_j \to c_r \#$, thus introducing the trap symbol $\#$ which causes a non-terminating computation with the rule $\# \to \#$; having taken the correct choice, the symbol $\hat{p}'_j$ waits one step to be used in the rule $c_{3-r} \hat{p}'_j \to c_{3-r} \hat{p}''_j$ together with the rule $c_r c'_r \to c_r c''_r$. The simulation then finishes with applying the rules $c_r \hat{p}''_j \to c_r p_k \bar{p}_k$ and $c_{3-r} c''_r \to c_{3-r}$ in parallel.

If we have chosen the case with register r being empty, then we continue with applying $c_r \bar{p}_j \to c_r$ and $c_{3-r} \bar{p}''_j \to c_{3-r} p''_j$. When applying $c_{3-r} p''_j \to c_{3-r} p'_j$, the second catalyst should not have the chance to be used with the rule $c_r o_r \to c_r c'_r$, because then in the last step of the simulation the symbols c'_r and p'_j would compete for the catalyst c_r, but only one could be taken, whereas the other one would be enforced to be trapped by using $c'_r \to \#$ or $p'_j \to \#$. In contrast, the symbol $\bar{p}'_j$ was allowed to wait for being used in the last step of the computation in the rule $c_{3-r} \bar{p}'_j \to c_{3-r}$ together with the rule $c_r p'_j \to c_r p_l \bar{p}_l$.

The following table summarizes in which way a subtract-instruction $j : (\text{SUB}(r), k, l) \in P$ is simulated depending on the contents of register $r \in \{1, 2\}$:

simulation of the subtract-instruction $j : (\text{SUB}(r), k, l)$ if the contents of register r is	
not empty	empty
$c_r p_j \to c_r \hat{p}_j \hat{p}'_j$	$c_r p_j \to c_r \bar{p}_j \bar{p}'_j \bar{p}''_j$
$c_{3-r} \bar{p}_j \to c_{3-r}$	$c_{3-r} \bar{p}_j \to c_{3-r}$
$c_r o_r \to c_r c'_r$	$c_r \bar{p}_j \to c_r$
$c_{3-r} \hat{p}_j \to c_{3-r}$	$c_{3-r} \bar{p}''_j \to c_{3-r} p''_j$
$c_r c'_r \to c_r c''_r$	
$c_{3-r} \hat{p}'_j \to c_{3-r} \hat{p}''_j$	$c_{3-r} p''_j \to c_{3-r} p'_j$
$c_r \hat{p}''_j \to c_r p_k \bar{p}_k$	$c_r p'_j \to c_r p_l \bar{p}_l$
$c_{3-r} c''_r \to c_{3-r}$	$c_{3-r} \bar{p}'_j \to c_{3-r}$

We should like to mention that at any time c_r can be used in the rule $c_r o_r \to c_r c'_r$, but, carried out at the wrong time, the application of this rule will immediately cause the introduction of the trap symbol # and therefore lead to a non-halting computation. Moreover, as already argued above, making the wrong choice when simulating a subtract-instruction also leads to the (enforced) introduction of the trap symbol and therefore to a non-halting computation. Hence, only correct simulations of successful computations of M lead to a halting computation of the catalytic P system Π (which finishes with applying the rules $c_1 p_{l_h} \to c_1$ and $c_2 \bar{p}_{l_h} \to c_2$ in parallel, thus eliminating the pair of symbols $p_{l_h}, \bar{p}_{l_h}$ representing the final halt-instruction of M). According to this construction described above, we conclude $N_{-C}(\Pi) = N(M)$.

As every non-catalytic rule in R is of the form $x \to \#$, i.e. involves the trap symbol only used in non-halting computations, by introducing a third catalyst c_3 and replacing every non-catalytic rule in R of the form $x \to \#$ by the corresponding catalytic rule $c_3 x \to c_3 \#$, we immediately obtain a purely catalytic P system

Π' with $N_{-C}(\Pi') = N_{-C}(\Pi)$ and therefore $N_{-C}(\Pi') = N(M)$, which observation completes the proof.

The proof given above can easily be extended to vectors of natural numbers, because the crucial point was the simulation of the subtract-instructions; due to Corollary 4.1, we still need only two registers that eventually are decremented, hence, we get the following more general result:

Corollary 4.3 *For any $m \geq 1$ and $k \geq 2$,*

$$PsRE = PsO_{-C}P_m(cat_k) = PsO_{-C}P_m(pcat_{k+1}).$$

If we require the results to be taken as the whole contents of the output membrane, we obviously need a second membrane to receive these results, because due to the definition of catalysts, these symbols cannot be removed at the end of a computation. Yet let us take the construction of the catalytic P system Π from the proof of Theorem 4.1, add a second membrane inside the skin membrane and replace each rule $c_1 \tilde{p}_j \to c_1$ in R by $c_1 \tilde{p}_j \to c_1(o_3, in)$ for every j : $(ADD(3), k, l) \in I$ to obtain a set of rules R' from R. Then for the catalytic P system $\Pi' = (O, C, [_1[_2\ ]_2]_1, w, R', 2)$ we obviously get $N(\Pi') = N_{-C}(\Pi)$. Hence, we infer

$$NOP_2(cat_2) = NOP_2(pcat_3) = NRE.$$

As an immediate consequence, we obtain the following general result as a counterpart to Corollary 4.3:

Corollary 4.4 *For any $m \geq 2$ and $k \geq 2$,*

$$PsRE = PsOP_m(cat_k) = PsOP_m(pcat_{k+1}).$$

Based on Proposition 4.1, in [11] it was shown that any partially recursive function $f : \mathbf{N}^\alpha \to \mathbf{N}^\beta$ can be computed by a deterministic (purely) catalytic P system with only one membrane and with only $\alpha + 2$ (resp. $\alpha + 3$) catalysts:

Theorem 4.2 *For any $\alpha, \beta \geq 1$,*

$$F_{\alpha,\beta}RE = F_{\alpha,\beta}O_{-C}P_m(cat_{\alpha+2}) = F_{\alpha,\beta}O_{-C}P_1(pcat_{\alpha+3}).$$

Considering (purely) catalytic P systems as accepting mechanisms, the preceding results, based on Corollary 4.2, immediately yield the following characterization of $PsRE$ by accepting catalytic P systems and accepting purely catalytic P systems, the input vector $(x_1, \ldots, x_n)$ being encoded as $o_1^{x_1} \ldots o_n^{x_n}$:

Theorem 4.3 $Ps_aOP_1(cat_*) = Ps_aOP_1(pcat_*) = PsRE.$

In [20], several variants of small universal register machines are described, one of them using only eight registers and operations as above. Hence, we may construct a universal purely catalytic P system as follows: We take the universal register machine U_{32} from Korec with eight registers (see [20] for the details of this universal machine) and construct the corresponding purely catalytic P system Π_U with only one membrane and nine catalysts. Then, given another purely catalytic P system Π, we first have to construct the corresponding register machine M_Π computing $N_{-C}(\Pi)$—we have omitted the tedious construction in the proof of Theorem 4.1; then the encoding $code(M_\Pi)$ of M_Π needed as input for U_{32} has to be computed. Thus, one input for Π_U is $o_2^{code(M_\pi)}$ where 2 is assumed to be number of the register in U_{32} which takes the code of M_Π, and the other input is $o_1^{x_1}$ where x_1 is the number we want to check. By this construction, Π_U accepts a natural number x_1 with input $o_1^{x_1} o_2^{code(M_\pi)}$ if and only if Π accepts the input $o_1^{x_1}$. Hence, Π_U is a purely catalytic P system which is universal for accepting purely catalytic P systems. Similar constructions can be made for catalytic P systems thus yielding a catalytic P system with one membrane and eight catalysts which is universal for accepting catalytic P systems. In the same way, universal (purely) catalytic P systems accepting or generating vectors of natural numbers can be constructed.

4.3 Catalytic P Systems with One Catalyst

With respect to the number of membranes, the results established so far in this chapter obviously are already optimal. Considering the number of catalysts, the cases that remain to be investigated are catalytic P systems with no catalyst or with one catalyst as well as purely catalytic P systems with one or two catalysts. For the rest of this section, we shall only consider (purely) catalytic P systems with only one membrane. We first will show that P systems with no catalyst and purely catalytic P systems with one catalyst can only generate semilinear sets:

Theorem 4.4 $PsO_{-C} P_1(cat_0) = PsO_{-C} P_1(pcat_1) = PsREG.$

Proof. Let $\Pi = (O, \{c\}, [\,_1\,]_1, w, R, 1)$ be a purely catalytic P system with only one catalyst c in the skin membrane, and let Σ be the set of all objects $a \in O$ that do not appear on the left-hand side of a catalytic rule in R. Hence, a computation in Π halts if and only if it reaches a configuration $[z]_1$ such that $z = cv$ for some $v \in \Sigma^*$. We define a context-free grammar G simulating the computations in Π by sequential derivations in G as follows (S is a new symbol not contained in O):

$G = ((O - (\Sigma \cup \{c\})) \cup \{S\}, \Sigma, S, P),$

$P = \{S \to w\}$

$\cup \{X \to v \mid cX \to cv \in R, X \in O - (\Sigma \cup \{c\}), v \in (O - \{c\})^*\}.$

Hence, after the application of the start production $S \to w$, a derivation step in G using the production $X \to v$ directly corresponds with a transition step in Π using the catalytic rule $cX \to cv$. Thus, we conclude that $Ps(L(G)) = Ps_{-C}(\Pi)$.

On the other hand, given a context-free grammar $G = (N, T, S, P)$, we construct the corresponding purely catalytic P system

$$\Pi = (N \cup T \cup \{c\}, \{c\}, [_1\ ]_1, S, R, 1)$$
$$R = \{cX \to cv \mid X \to v \in P, X \in N, v \in (N \cup T)^*\}$$

with $Ps_{-C}(\Pi) = Ps(L(G))$. The observation that $PsCF=PsREG$ completes the proof for the case $pcat_1$.

The constructions for the case cat_0 are simple variants of the case $pcat_1$, hence, we omit the details.

Considering catalytic P systems with only one catalyst in one membrane, we conjecture that these systems are not universal. Yet so far only partial results into that direction could be obtained, see [10].

Let $\Pi = (O \cup \{c\}, \{c\}, [_1\ ]_1, w, R, 1)$ be a catalytic P systems with the single catalyst c in the skin membrane. In addition, let the set O be divided into three disjoint subsets O', O'', and O''' such that the rules in R obey to the following constraints:

1. for $a \in O'$, there exists no rule in R with a on the left-hand side;
2. for $a \in O''$, the rules in R are of the form $ca \to cv$ and $v \in (O' \cup O'')^*$;
3. for $a \in O'''$, the rules in R are of the form $a \to v$ with $v \in (O' \cup O''')^*$.

Catalytic P systems fulfiling the requirements stated above are called 1-*separated*, and the corresponding set of sets of natural numbers generated by such P systems is denoted by $NO_{-C}P_1(sepcat_1)$. The main idea of 1-separated catalytic P systems is that catalytic rules and non-catalytic rules are separated with respect to the symbols they generate and therefore cannot interfere, i.e. the objects generated by catalytic rules cannot be affected by non-catalytic rules and vice versa.

If we relax the third condition stated above for 1-separated catalytic P systems and allow the non-catalytic rules to be of the most general form $a \to v$ with $v \in O^*$, then these catalytic P systems are called *weakly* 1-*separated* and the corresponding set of sets of natural numbers generated by such P systems is denoted by $NO_{-C}P_1(wsepcat_1)$.

The only restriction remaining in the case of weakly 1-separated P systems in comparison with the non-restricted variant is that "catalytic objects" $a \in O''$ cannot generate "non-catalytic objects" $a \in O'''$, yet this feature already allows us to show that $NO_{-C}P_1(wsepcat_1) \subsetneq NRE$:

Theorem 4.5 $PsO_{-C}P_1(sepcat_1) = PsO_{-C}P_1(wsepcat_1) = PsREG$.

Proof. We only have to prove $\mathit{PsO_{-C}}\,P_1(\mathit{wsepcat}_1) \subseteq \mathit{PsREG}$. Let $\Pi = (O \cup \{c\}, \{c\}, [_1\ ]_1, w, R, 1)$ be a weakly 1-separated P system with only one catalyst c in the skin membrane and with the rules in R fulfilling the conditions as defined above. We construct the context-free grammar (with S being a new object not contained in O)

$$G = ((O - O') \cup \{S\}, O', S, P),$$
$$P = \{S \to w\}$$
$$\cup \{X \to v \mid cX \to cv \in R, X \in O'', v \in (O' \cup O'')^*\}$$
$$\cup \{X \to v \mid X \to v \in R, X \in O''', v \in O^*\}.$$

By the definition of weakly 1-separated P systems, the objects generated by catalytic rules can only be affected by catalytic rules; for the application of the catalytic rules the maximally parallel mode has no regulating effect, because all the possible evolutions of the catalytic objects from O'', even those generated at some time by the non-catalytic rules, can be simulated by sequentially applying the corresponding productions in the context-free grammar G. As we only consider halting computations in Π, the possible evolutions of the non-catalytic objects from O''' in Π can be simulated sequentially in G, too, because there is no conflicting interplay with the objects from O''. On the other hand, derivations in G only yield results that can also be obtained as the result of a halting computation in Π using the maximally parallel mode. In sum, we conclude $\mathit{Ps}(L(G)) = \mathit{Ps}_{-C}(\Pi)$.

4.4 Acceptance with Catalytic P Systems

Whereas in the generating case no additional ingredients are needed to obtain computational completeness for non-deterministic (purely) catalytic P systems, for deterministic (purely) catalytic P systems used as accepting mechanisms computational completeness can only be obtained when using such ingredients, e.g., priorities.

In the deterministic case, at each step of the computation from the initial configuration, the maximal multiset of rules that is applicable is unique. Thus, every deterministic (catalytic) P system has exactly one computation which either may be finite (i.e. halting) or infinite (non-halting). Given a (catalytic) P system

$$\Pi = (O, C, \mu, w_1, \ldots, w_m, R_1, \ldots, R_m, i_0),$$

we consider the initial configurations $C_0(n_1) = (w_1, \ldots, w_{i_0} o_1^{n_1}, \ldots, w_m)$ for every $n_1 \in \mathbf{N}$ and define the set of numbers accepted by Π as

$$N_a(\Pi) = \{n_1 \in \mathbf{N} \mid \Pi \text{ halts when starting from } C_0(n_1)\}.$$

Obviously, this definition can easily be extended to vectors of natural numbers $(n_1, \ldots, n_k)$ for arbitrary $k \in \mathbf{N}$, i.e. with $C_0((n_1, \ldots, n_k)) = (w_1, \ldots, w_{i_0} o_1^{n_1} \ldots o_k^{n_k}, \ldots, w_m)$ we define

$$Ps_a(\Pi) = \{(n_1, \ldots, n_k) \in \mathbf{N}^k \mid \Pi \text{ halts when starting from } C_0((n_1, \ldots, n_k))\}$$

where the objects o_i, $1 \leq i \leq k$, are the input symbols.

From the results proved so far in this chapter, we immediately infer the following results for the families of sets of (vectors of) natural numbers accepted by (purely) catalytic P systems:

Theorem 4.6 *For all $k \geq 2$ and $d \geq 1$, $N_a OP_d(cat_k) = N_a OP_d(pcat_{k+1}) = NRE$ and $Ps_a OP_d(cat_*) = Ps_a OP_d(pcat_*) = PsRE$.*

Yet, as we shall elaborate in the following sections, the corresponding deterministic variants of (purely) catalytic P systems are not as powerful as their non-deterministic counterparts, i.e. the corresponding families $N_a DOP_d([p]cat_*)$ and $Ps_a DOP_d([p]cat_*)$ only contain recursive sets.

4.4.1 Deterministic Purely Catalytic P Systems

We here consider deterministic purely catalytic P systems, i.e. all rules are of the form $ca \rightarrow cv$ with c being a catalyst. Moreover, our main result is first established for systems with only one membrane. For such a system Π and any configuration C_1 of Π, by $|C_1|_M$ we denote the number of symbols from $M \subseteq O$ in the skin membrane; by $\Psi(C_1)$ we denote the Parikh vector of symbols from $O - C$ in C_1 (with respect to a given ordering of the symbols in $O - C$). Moreover, we write $C_1 \Longrightarrow^{S_1 \ldots S_k} C_2$ to denote the reachability of the configuration C_2 from the configuration C_1 through applying the sequence $S_1 \ldots S_k$ of multisets of rules or $C_1 \Longrightarrow^+ C_2$ if the actual sequence is irrelevant. We write $C_1 \Longrightarrow^* C_2$ if $C_2 = C_1$ or $C_1 \Longrightarrow^+ C_2$.

The following lemma is essential for the proof of the main results established in the following:

Lemma 4.1 *(Dickson's lemma, see [8]) For any k-tuple $(n_1, \ldots, n_k)$ of natural numbers let $R(n_1, \ldots, n_k)$ be the set of k-tuples of numbers $(m_1, \ldots, m_k)$ with $m_i \geq n_i$ for all $1 \leq i \leq k$. Then any union of such sets $R(n_1, \ldots, n_k)$ is a finite union.*

Due to the nature of determinism as well as to the number of catalysts being bounded, an infinite computation of a deterministic purely catalytic P system is "periodic" in the sense as stated in the following theorem.

94   CATALYTIC P SYSTEMS

Theorem 4.7 *Given a deterministic purely catalytic P system*

$$\Pi = (O, C, [\,_1\,]_1, w_1, R_1),$$

with $C_0 = (w_1)$ denoting the initial configuration, the following three statements are equivalent:

(1) Π does not halt;
(2) there exist C_1 and C_1' such that $C_0 \Longrightarrow^* C_1 \Longrightarrow^+ C_1'$ and $\Psi(C_1) \leq \Psi(C_1')$;
(3) the computation of Π is of the form $C_0 \Longrightarrow^{T_1 \ldots T_r (S_1 \ldots S_k)^\omega}$, where $T_1, \ldots, T_r$, $S_1, \ldots, S_k$ are multisets of rules, i.e. following a finite prefix $T_1 \ldots T_r$, the computation is "periodic" with $S_1 \ldots S_k$ being repeated forever.

Proof. The following claims are essential for what follows (detailed proofs can be found in [18]):

1. **(Claim 1)** Given a computation $C_1 \Longrightarrow^{H_1} C_2 \Longrightarrow^{H_2} \ldots C_{i-1} \Longrightarrow^{H_{i-1}} C_i$ and a configuration C_1' with $\Psi(C_1) \leq \Psi(C_1')$, there exist multisets $H_1', \ldots, H_{i-1}'$ and configurations $C_2', \ldots, C_i'$ such that
   (i) $C_1' \Longrightarrow^{H_1'} C_2' \Longrightarrow^{H_2'} \ldots C_{i-1}' \Longrightarrow^{H_{i-1}'} C_i'$,
   (ii) $H_j \subseteq H_j'$, for all j with $1 \leq j \leq i-1$, and
   (iii) $\Psi(C_j) \leq \Psi(C_j')$, for all j with $1 \leq j \leq i$.

2. **(Claim 2)** If $C_1 \Longrightarrow^{H_1 \ldots H_k} C_2$, $\Psi(C_1) \leq \Psi(C_2)$, and $C_2 \Longrightarrow^{H_1 \ldots H_k} C_3$, then it must be the case that $C_1 \Longrightarrow^{H_1 \ldots H_k} C_2 \Longrightarrow^{H_1 \ldots H_k} C_3 \ldots \Longrightarrow^{H_1 \ldots H_k} C_i \Longrightarrow^{H_1 \ldots H_k} C_{i+1} \ldots$, with $\Psi(C_i) \leq \Psi(C_{i+1})$ for all $i \geq 1$, i.e., $H_1 \ldots H_k$ is repeated forever.

We first show (1) $\Longrightarrow$ (2). Assume that Π does not terminate. Let $C_0 \Longrightarrow C_1 \ldots \Longrightarrow C_l \ldots$ be the infinite computation in Π; for each $l \in \mathbb{N}$, now consider $R_l = \{(m_1, \ldots, m_d) \mid (m_1, \ldots, m_d) \geq \Psi(C_l)\}$ (where d denotes the number of symbols in $O - C$). According to Dickson's lemma, the union $\bigcup_{l \in \mathbb{N}} R_l$ is a finite union, which implies that there exist $i < j$ such that $\Psi(C_i) \leq \Psi(C_j)$; hence, (2) holds. Now we prove (2) $\Longrightarrow$ (3). Let $H_1 \ldots H_k$ be the sequence of rule sets such that $C_1 \Longrightarrow^{H_1 \ldots H_k} C_1'$ and $\Psi(C_1) \leq \Psi(C_1')$. According to Claim 1, we get rule sets $H_1^1, H_2^1, \ldots, H_k^1$ and a configuration C_2' such that $C_1 \Longrightarrow^{H_1 \ldots H_k} C_1' \Longrightarrow^{H_1^1 \ldots H_k^1} C_2'$ as well as $H_l \subseteq H_l^1$ for all $1 \leq l \leq k$ and $\Psi(C_1') \leq \Psi(C_2')$. Iteratively applying this construction, we get rule sets $H_1^t, H_2^t, \ldots, H_k^t$ and configurations C_{t+1}', $t \geq 1$, such that $C_1 \Longrightarrow^{H_1 \ldots H_k} C_1' \Longrightarrow^{H_1^1 \ldots H_k^1} C_2' \ldots \Longrightarrow^{H_1^t \ldots H_k^t} C_{t+1}'$ as well as, for each j with $1 \leq j \leq t$, $H_l^j \subseteq H_l^{j+1}$ for all $1 \leq l \leq k$ and $\Psi(C_j') \leq \Psi(C_{j+1}')$. Since the number of catalytic symbols (which bounds the degree of maximal parallelism) is a constant, for each l with $1 \leq l \leq k$ there must be a t_l such that $H_l^{t_l} = H_l^{t_l+m}$ for all $m \geq 1$. Now choose s to be the maximum among all these t_l, i.e., $H_l^s = H_l^{s+1}$ for all $1 \leq l \leq k$. Thus, we have $C_s' \Longrightarrow^{H_1^s \ldots H_k^s} C_{s+1}' \Longrightarrow^{H_1^{s+1} \ldots H_k^{s+1}} C_{s+2}'$. By letting $S_l = H_l^s$, $1 \leq l \leq k$, Claim 2 guarantees that $S_1 \ldots S_k$ repeats forever from C_s', i.e. (3) holds. (3) $\Longrightarrow$ (1) is obvious. This completes the proof of the theorem.

From (3) of Theorem 4.7, we immediately get the following characterization of the reachability sets of deterministic purely catalytic P systems:

Corollary 4.5 *The reachability set $\{\Psi(C_1) \mid C_0 \Longrightarrow^* C_1\}$ of a deterministic purely catalytic P system $\Pi = (O, C, [_1 \ ]_1, w_1, R_1)$ is semilinear, where $C_0 = (w_1)$ is the initial configuration.*

As an immediate consequence of the preceding results, we also infer that deterministic purely catalytic P systems with only one membrane are not computationally complete because their halting problem is decidable:

Corollary 4.6 *The problem of determining whether the deterministic purely catalytic P system $\Pi = (O, C, [_1 \ ]_1, w_1, R_1)$ halts is decidable.*

Proof. According to (2) of Theorem 4.7, the computation of Π does not halt if and only if there exist C_1 and C_1' such that $C_0 \Longrightarrow^* C_1 \Longrightarrow^+ C_1'$ and $\Psi(C_1) \leq \Psi(C_1')$; by simulating the computation of Π we either find that it halts or else we find the sequence $S_1 \ldots S_k$ being repeated forever according to the construction given in the proof of Theorem 4.7 establishing (3). Hence, the halting problem is decidable.

The following method of constructing a P system with only one membrane from a P system with an arbitrary membrane structure can be applied not only to (purely) catalytic P systems, but also to many other variants of P systems with a static membrane structure to be considered later on in this *Handbook*:

Lemma 4.2 *Let $\Pi = (O, C, H, \mu, w_1, \ldots, w_m, R_1, \ldots, R_m, i_0)$ be a catalytic P system with an arbitrary membrane structure μ. Then we can effectively construct a catalytic P system with only one membrane $\Pi' = (O', C, [_1 \ ]_1, w_1', R_1')$ such that the sets of (vectors of) natural numbers accepted by Π and Π' are equal, i.e., $N_a(\Pi) = N_a(\Pi')$ and $Ps_a(\Pi) = Ps_a(\Pi')$.*

Proof. The symbols $a \in O$ in membrane region i are represented by (a, i), i.e., we take $O' = \{(a, i) \mid a \in O, i \in H\}$. For any multiset w (and the corresponding string) appearing in region i in a configuration C_1 of Π, by (w, i) we denote the corresponding multiset (string) over O' representing w in the corresponding configuration C_1' of Π'. Thus, we take the axiom $w_1' = (w_1, 1) \ldots (w_m, m)$. The single set of rules R_1' is obtained from the rule sets $R_1, \ldots, R_m$ by replacing a rule $ca \to c(b_1, tar_1) \ldots, (b_l, tar_l)$ or $a \to (b_1, tar_1) \ldots, (b_l, tar_l)$ from R_i by the corresponding rule(s) $(c, i)(a, i) \to (c, i)(b_1, mem_1) \ldots, (b_l, mem_l)$ where mem_j denotes (one of) the membrane(s) indicated by tar_j, $1 \leq j \leq l$. Due to this construction, the transitions in Π are simulated by Π' in only one membrane, starting from the initial configurations $((w_1, 1) \ldots (w_{i_0} o_1^{n_1} \ldots o_k^{n_k}, i_0) \ldots (w_m, m))$; thus, we obtain $N_a(\Pi) = N_a(\Pi')$ and $Ps_a(\Pi) = Ps_a(\Pi')$.

Applying Lemma 4.2 to Corollary 4.6, we obtain the following result:

Theorem 4.8 *Deterministic purely catalytic P systems are not computationally complete.*

Proof. Based on Corollary 4.6, there exists a decision procedure Δ such that given a deterministic purely catalytic P system Π with one membrane and an input $(\Pi, o_1^{n_1} \ldots o_k^{n_k})$, Δ can tell whether Π halts on $w_1 o_1^{n_1} \ldots o_k^{n_k}$ or not. It is obvious that a decision procedure Δ_P can be constructed such that, given an arbitrary deterministic purely catalytic P system Π, Δ_P computes the corresponding one-membrane deterministic purely catalytic P system Π' from Π according to the construction given in the proof of Lemma 4.2 and then can tell whether Π' halts on $w'_1(o_1^{n_1}, i_0) \ldots (o_k^{n_k}, i_0)$ or not. Hence, all the sets of (vectors of) natural numbers in $N_a DOP_*(pcat_*)$ and $Ps_a DOP_*(pcat_*)$, respectively, are recursive, which observation completes the proof.

As non-deterministic purely catalytic P systems are computationally complete, we thus have proved that deterministic purely catalytic P systems are less powerful than non-deterministic purely catalytic P systems, i.e. $N_a DOP_*(pcat_*) \subsetneq N_a OP_*(pcat_*)$ and $Ps_a DOP_*(pcat_*) \subsetneq Ps_a OP_*(pcat_*)$.

4.4.2 Deterministic Catalytic P Systems

Now we consider the full class of deterministic catalytic P systems, where the rules are of the form $ca \to cw$ or $a \to w$. Intuitively, what makes the reachability set of a deterministic purely catalytic P systems "simpler" is that any infinite computation of such a system is *periodic* in the sense described in (3) of Theorem 4.7. Such a periodic behavior is partly due to the fact that the maximum degree of parallelism during the computation of a deterministic purely catalytic P systems is bounded by the number of catalytic symbols in the initial configuration. Note, however, that the degree of parallelism may become unbounded if the deterministic catalytic P system uses rules of type $a \to v$. In fact, the semilinearity result no longer holds for the full class of deterministic catalytic P systems as the following example indicates. It is interesting to note that the degree of parallelism in this example is unbounded.

Example 4.1 *Consider the deterministic catalytic P system*

$$\Pi = (\{a_1\}, \emptyset, [\,_1\,]_1, a_1, \{a_1 \to a_1 a_1\}),$$

with only one non-catalytic rule $a_1 \to a_1 a_1$. Obviously, from the initial configuration (a_1), in n steps we obtain the configuration $(a_1^{2^n})$. Hence, the set of natural numbers corresponding with the set of all reachable configurations is $\{2^n \mid n \geq 0\}$, which is clearly not semilinear.

Although the reachability set of a deterministic (not necessarily purely) catalytic P system is not semilinear in general, being deterministic does make the computational power of the model weaker than its non-deterministic counterpart. In what follows, we propose a graph-theoretic approach for reasoning about the behaviors of deterministic catalytic P systems.

Consider a deterministic catalytic P system $\Pi = (O, C, [_1\ ]_1, w_1, R_1)$, where $C = \{c_1, \ldots, c_k\}$ and $\Sigma = O - C = \{a_1, \ldots, a_m\}$ is the set of non-catalytic symbols. Let $C_0 = (w_1)$ be the initial configuration which contains (possibly multiple copies of) the catalysts c_i, $1 \leq i \leq k$, and a multiset over Σ. For $c_i, c_j \in C$ and $a \in \Sigma$, two rules r_1 and r_2 from R_1 are said to be *in conflict* if one of the following conditions holds:

- $r_1 : c_i a \to c_i w_1$, $r_2 : c_j a \to c_j w_2$, and $i \neq j$;
- $r_1 : c_i a \to c_i w_1$, $r_2 : c_i a \to c_i w_2$, and $w_1 \neq w_2$;
- $r_1 : c_i a \to c_i w_1$, $r_2 : a \to w_2$;
- $r_1 : a \to w_1$, $r_2 : a \to w_2$, and $w_1 \neq w_2$.

In each of the cases described above, the rules r_1 and r_2 compete for the same non-catalytic symbol a (a is said to be *involved* in two conflicting rules). At no time can a deterministic catalytic P system ever enable a rule that is in conflict with another rule. Thus, conflicting rules can be removed without affecting the computation of a deterministic catalytic P system, regardless of the initial configuration. Note that rules $c_1 a_1 \to c_1 w_1$ and $c_1 a_2 \to c_1 w_2$ are not in conflict if $a_1 \neq a_2$, yet in fact, Π being deterministic means that a reachable configuration must not contain both a_1 and a_2, i.e. in this case determinism may depend on the initial configuration.

Let $G(\Pi) = (\Sigma, E)$ be the directed labeled graph with E exactly containing the edges (a_i, r, a_j) such that r is a rule (of the form $ca_i \to cw$ or $a_i \to w$ with a_j in w) from R_1 not in conflict with another rule in R_1. The examination of $G(\Pi)$ reveals an important property: for each node a_i, the outgoing edges (if they exist) have the same label (this means that the applicable rule r associated with such a node a_i is unique. As we shall see later, such a "deterministic" nature of the graph $G(\Pi)$ offers an intuitive reason for the deterministic catalytic P systems not being computationally complete. To set the stage for this result, we require the following lemma (see [18]):

Lemma 4.3 *Consider the deterministic catalytic P system*

$$\Pi = (O, C, [_1\ ]_1, w_1, R_1)$$

where $C = \{c_1, \ldots, c_k\}$ and $\Sigma = O - C = \{a_1, \ldots, a_m\}$, and let $C_0 = (w_1)$ be the initial configuration. Then:

(1) *Π does not halt if and only if there is a loop reachable from some node a_{i_0} with $|w_1|_{a_{i_0}} > 0$ in $G(\Pi)$.*

(2) *Let w_1' be a multiset such that w_1' contains the same catalysts as w_1 as well as, for all j with $1 \leq j \leq m$, $|w_1'|_{a_j} \leq 1$ and $|w_1'|_{a_j} = 1$ if and only if $|w_1|_{a_j} \geq 1$.*

Let Π' be the deterministic catalytic P system obtained by replacing w_1 with w'_1 in Π. Then Π halts if and only if Π' halts.

(3) The problem of determining whether Π halts or not is decidable in polynomial time.

Deterministic catalytic P systems also have the following *monotonicity property*, whose proof is similar to that of Claim 1 in Theorem 4.7.

Lemma 4.4 *If a deterministic catalytic P system $\Pi = (O, C, [\ _1\ ]_1, w_1, R_1)$ does not halt with the initial configuration $C_1 = (w_1)$, then Π does not halt with any other initial configuration C_2 such that C_2 contains the same catalysts as C_1 and $\Psi(C_2) \geq \Psi(C_1)$.*

Hence, we obtain the following result:

Theorem 4.9 *For a deterministic catalytic P system Π and a fixed string w, the set $L = \{o_1^{n_1} \ldots o_k^{n_k} \mid \Pi \text{ halts on } wo_1^{n_1} \ldots o_k^{n_k}\}$ is semilinear and effectively constructible. In fact, L is either empty, or of the form $o_1^{m_1} \ldots o_k^{m_k}$, where $m_i = *$ or 0, $1 \leq i \leq k$.*

Proof. According to Lemma 4.3 (3), we can decide in polynomial time whether Π halts with input wx for every $x \in \{o_1\}^* \ldots \{o_k\}^*$. As a consequence, if Π does not halt on w, then $L = \emptyset$; otherwise, $L = \{o_1\}^{m_1} \ldots \{o_k\}^{m_k}$, where $m_i = *$ or 0, $1 \leq i \leq k$, and we can compute the m_i, $1 \leq i \leq k$, by proceeding inductively as follows, starting with $w_0 = w$: if $|w_{i-1}|_{o_i} \geq 1$, then we set $w_i = w_{i-1}$ and, based on Lemma 4.3 (2), take $m_i = *$; if $|w|_{o_i} = 0$, then we check whether Π halts with input wo_i, which according to Lemma 4.3 (3) can be decided in polynomial time. If Π halts, then we set $w_i = w_{i-1}o_i$ and, based on Lemma 4.3 (2), take $m_i = *$, otherwise we set $w_i = w_{i-1}$ and $m_i = 0$. Obviously, according to Lemma 4.4, the multiset w_k obtained in the last step of this procedure cannot be extended by any o_i for which m_i was set to 0, hence, the semilinear set L constructed in that way equals the set of inputs of the form $wo_1^{n_1} \ldots o_k^{n_k}$ on which Π halts. The whole procedure only needs polynomial time, hence, we have given an effective (and even quite efficient) construction for L.

We immediately infer the following result, which strengthens Theorem 4.8:

Corollary 4.7 *Deterministic catalytic P systems are not computationally complete.*

4.4.3 Prioritized Deterministic Catalytic P Systems

Already in the original version of catalytic P systems, a priority relation on the rules of the system was considered using a strict partial order (i.e. an irreflexive, asymmetric, and transitive relation) on the rules of the system: only rules that are maximal among the rules applicable in the corresponding region can be taken into a multiset of rules to be applied. In [18], the following variants of *prioritized* catalytic P systems were considered: whereas *weakly prioritized* catalytic P systems correspond with the original version of catalytic P systems with a priority relation on the rules, in *strongly prioritized* catalytic P systems a multiset of rules is applicable if no rule of a lower priority than those contained in this multiset can be applied even if the rules do not compete for objects. For establishing the computational completeness and the universality of non-deterministic P systems, such a priority relation on the rules was not necessary, yet for the universality of deterministic P systems such a feature is essential, i.e. *strongly prioritized* as well as *weakly prioritized* deterministic (purely) catalytic P systems are universal. In *statically prioritized* catalytic P systems, rules are divided into different priority groups, and if a rule in a higher priority group is applicable, then no rules from a lower priority group can be used, i.e. all rules in an applicable multiset are from the same priority group. This static condition turns out to be a very weak extension, i.e. statically prioritized deterministic catalytic P systems only accept semilinear sets.

4.4.4 k-Deterministic Catalytic P Systems

A careful look into the proof of Theorem 4.1 reveals the interesting fact that the non-determinism only arises from having to guess whether the register to be decremented is empty or not, but in case of a wrong choice, the trap symbol # will appear within a few steps. Hence, looking ahead a few transition steps allows us to make the correct choice. To capture this idea of looking ahead k steps, we introduce the notion of k-determinism as considered in [23].

To represent a computation in a catalytic P system, we will use the notion of a computation tree (see [25]). The *computation tree* of a catalytic P system is a labeled maximal tree with the root node of the tree corresponding to the initial configuration of the system. The children of a node are configurations that follow in a one-step transition. Nodes are labeled by configurations and edges are labeled by multisets of applicable rules. We say that a computation halts if it represents a finite branch in the computation tree.

To be more efficient in an implementation, we only consider catalytic P systems having the specific feature that we do not have to expand the complete computation tree during the simulation, but rather "look ahead" in the computation tree at most k steps for some fixed k to be able to exclude the paths which would lead to an infinite loop (because of containing a special symbol #, the trap symbol, for which rules always have to be present to guarantee that a configuration containing at least one such symbol can never be part of a halting computation) and choose the (only) path which may lead to a successful continuation (i.e. possibly being part of a halting computation).

For $k \geq 0$, the notion of *k-determinism* for catalytic P systems is defined as follows:

$k = 0$: A catalytic P system is called *deterministic* or *0-deterministic* if, at any step of a computation, there exists at most one configuration derivable from the current one. If at some stage of the computation the trap symbol # appears in the current configuration, we stop the computation without getting a result.

$k = 1$: A catalytic P system is called *1-deterministic* (is said to have a level of *look-ahead* 1) if, at any step of a computation, either:

- there is no configuration derivable from the current one, i.e. the computation halts (and yields a result) **or**
- all configurations derivable from the current one contain the trap symbol # (hence, we stop without continuing the computation, no result is obtained) **or**
- there exists exactly one configuration which is derivable from the current one and does not contain the trap symbol #. As configurations containing the trap symbol can never lead to a halting computation (and therefore can never yield a result), the only reasonable continuation is the configuration not containing the trap symbol.

$k > 1$: For $k > 1$, a catalytic P system is called *k-deterministic* (is said to have a *level of look-ahead k*) if the following condition holds: At any moment of a computation, either:

- the computation halts (and yields a result) **or**
- for any configuration derivable from the current one we make $k - 1$ further steps and for all these branches of depth $k - 1$ we end up with a configuration containing the trap symbol # (hence, we stop without continuing the computation, no result is obtained) **or**
- for exactly one configuration c derivable from the current one there is at least one branch of the computation tree which cannot be continued (and therefore will yield a result at the end of this branch) or is of depth $k - 1$, with this configuration being the root of this branch, and the trap symbol # does not appear in any of the configurations along this branch. This uniquely determined configuration c is chosen for continuing the computation.

In each step of a computation in a *k*-deterministic P system there exists exactly one uniquely determined multiset of rules to be applied for possibly continuing the computation: either this multiset is empty, i.e. we stop (getting a result only in the case that we halt), or we continue with applying a non-empty multiset yielding the uniquely determined configuration possibly being part of an eventually successful halting computation.

In the following, we shall denote the feature of k-determinism by inserting D_k in front of O in the notation for the families defined in Section 4.1. Inspecting the proofs elaborated in Section 4.2 we see that any partially recursive function $f : \mathbf{N}^\alpha \to \mathbf{N}^\beta$ can be computed by a 4-deterministic catalytic P system with only one membrane and with only $\alpha + 2$ ($\alpha + 3$ catalysts) which yields a refined version of Theorem 4.2:

Corollary 4.8 *For any* $\alpha, \beta \geq 1$,

$$F_{\alpha,\beta} D_4 O_{-c} P_1(cat_{\alpha+2}) = F_{\alpha,\beta} D_4 O_{-c} P_1(pcat_{\alpha+3}) = \Gamma_{\alpha,\beta} RE.$$

For accepting (purely) catalytic P systems, the result corresponding with Theorem 4.3 reads as follows:

Corollary 4.9 $Ps_a D_4 OP_1(cat_*) = Ps_a D_4 OP_1(pcat_*) = PsRE.$

4.5 Mobile Catalysts and Bistable Catalysts

In this section we consider variants of catalysts that are not totally resistant to changes. A *bistable catalyst* remains in the same region, but is allowed to switch between two states, i.e. it does not remain unchanged. A *mobile catalyst* remains the same symbol, but is allowed to change the region as other non-catalyst symbols through the application of an evolution rule. In both cases, the role of such a variant of a catalyst does not fulfill the condition of only having a catalytic function anymore, because it is changed anyway, although in a very restricted form. Hence, it is not surprising that mobile catalysts and bistable catalysts turn out to be even more powerful than non-changing catalysts.

In a *P system with bistable catalysts*, the catalysts may switch between two states, i.e. the catalytic rules are of the form $ca \to c'v$ and $c'a \to cv$ (in a more relaxed variant, also rules of the form $ca \to cv$ and $c'a \to c'v$ are allowed). From the results shown in [1], we immediately infer the following optimal result (*bicat* instead of *cat* indicates the use of bistable catalysts instead of normal ones):

Theorem 4.10 $NO_{-c} P_1(bicat_1) = NRE.$

The definition of a *P system with mobile catalysts* remains the same as that for catalytic P systems with the only exception: that in the catalytic rules we also allow any catalyst to have a target, i.e. the catalytic rules now are of the form $ca \to (c, tar)v$ with $c \in C$ and *tar* being the target of the catalyst c. In [21] the following results were shown (*mcat* instead of *cat* indicates the use of mobile catalysts instead of normal ones):

Theorem 4.11 $NOP_2(mcat_2) = NOP_3(mcat_1, tar) = NRE$.

The parameter *tar* indicates that instead of the target *in* the targets in_j are used in order to specify the exact region j where the object should enter through an inner membrane. The result $NOP_2(mcat_2) = NRE$ is weaker than that from Theorem 4.1, yet $NOP_3(mcat_1, tar) = NRE$ shows that when using more than one membrane and the mobility of the single catalyst, we need only this minimal number of one mobile catalyst to obtain computational completeness.

4.6 Variants of Transition Modes and Halting

In this section, we investigate the computational power of (purely) catalytic P systems working in other modes than the maximally parallel one. Besides the trivial modes, where any arbitrary number of rules may be applied in parallel (*asynchronous mode*) or exactly one rule is applied (*sequential mode*) in each transition step, the *minimally parallel mode* has become of special interest recently – starting from a specific partitioning of the set of rules, we require that, if possible, at least one rule from each partition has to be applied in each transition step.

The usual way of considering a computation in a P system to be successful is to require that it halts, i.e. that no rule can be applied any more in the whole system (we shall also use the notion *total halting* in the following). Taking a partitioning of the rule set (as for the minimally parallel mode), we may require that there exists an applicable multiset of rules containing one rule from each partition. A biological motivation for this variant of halting (*partial halting*) comes from the idea that a system may only survive as long as there are enough resources to give all subsystems the chance to continue their evolution. Computations also may be considered to be successful if at some moment a specific symbol appears in some membrane (*signal halting*) or if some specific pattern appears (*halting with states*). If the computation enters an infinite loop with a specific sequence of configurations appearing periodically, we speak of *adult halting*. We may even take every possible computation and extract a result from the output membrane in each reachable configuration (*unconditional halting*); this variant reveals the close connection of P systems with non-cooperative rules and Lindenmayer systems.

4.6.1 Sequential and Asynchronous Mode

Let $\Pi = (O, C, \mu, w_1, \ldots, w_m, R_1, \ldots, R_m, i_0)$ be a (catalytic) P system and let $Appl(\Pi, C_1)$ denote the set of multisets of rules applicable to the configuration C_1. Obviously, the set of multisets of rules applicable to the configuration C_1 in the *asynchronous mode*, $Appl(\Pi, C_1, asyn)$, equals $Appl(\Pi, C_1)$, and the set of multisets of rules applicable to C_1 in the *sequential mode* in a formal way is defined by

$$Appl(\Pi, C_1, sequ) = \{R' \mid R' \in Appl(\Pi, C_1) \text{ and } |R'| = 1\}.$$

By definition, we have $Appl(\Pi, C_1, sequ) \subseteq Appl(\Pi, C_1, asyn)$. As long as we consider the usual halting condition, every configuration reachable in the asynchronous mode is reachable in the sequential mode, too. Hence, the set of results obtained by a P system in the sequential mode equals the set of results obtained by a P system in the asynchronous mode. These observations not only hold true for (purely) catalytic P systems, but hold in general for any arbitrary variant of P systems.

Now let the families of sets of (vectors of) natural numbers generated by [purely] catalytic P systems with at most m membranes and at most k catalysts working in the asynchronous (*asyn*) and the sequential mode (*sequ*), be denoted by $XO_{-C}P_m([p]cat_*, Y)$, $X \in \{N, Ps\}$, $Y \in \{asyn, sequ\}$.

For (purely) catalytic P systems equipped with the asynchronous and the sequential mode, we obtain a characterization of semilinear sets:

Theorem 4.12 *For all* $X \in \{N, Ps\}$, $Y \in \{asyn, sequ\}$,

$$XO_{-C}P_*([p]cat_*, Y) = XREG.$$

Proof. As already argued above, we have $XO_{-C}P_*([p]cat_*, sequ) = XO_{-C}P_*([p]cat_*, asyn)$. Using the same construction as in the proof of Theorem 4.5, we get $XO_{-C}P_*(pcat_*, Y) \supseteq XREG$. Hence, it only remains to show that $XO_{-C}P_*([p]cat_*, Y) \subseteq XREG$. For proving $XO_{-C}P_1([p]cat_*, Y) \subseteq XREG$, we may again refer to the construction given in the proof of Theorem 4.5. Given a (purely) catalytic P system Π with an arbitrary membrane structure, we may apply the same encoding as used in Lemma 4.2 to obtain a (purely) catalytic P system Π' with only one membrane such that $h(X(\Pi')) = X(\Pi)$ where h is the projection erasing all symbols contained in a region different from the output membrane region. As $XREG$ is closed under projections, the assertion follows.

We would like to emphasize that a result similar to that established above holds true for many other variants of P systems, too.

4.6.2 Minimal Parallelism

For the *minimally parallel mode*, we need an additional feature for the set of rules R, i.e. we consider a partitioning of R into disjoint subsets $R_1, \ldots, R_h$. Usually, these partitions of R coincide with a specific assignment of the rules to the membranes, yet in this chapter we shall deal with an arbitrary partitioning of R, denoted by $\Theta(R)$. Hence, here we are dealing with P systems of the form $\Pi = (O, C, \mu, w_1, \ldots, w_m, R, \Theta, i_0)$ where R contains all rules to be applied in the regions of Π. For any set of rules $R' \subseteq R$, let $|R'|$ denote the number of rules in R'.

There are several possible interpretations of the minimally parallel mode which in an informal way can be described as applying multisets such that from every set R_j, $1 \leq j \leq h$, at least one rule – if possible – has to be used (e.g. see [6]). For the basic variant as defined in [14], which here we denote by *min′*, in each transition step we choose a multiset of rules R' from $Appl(\Pi, C_1, asyn)$ that cannot be extended to $R'' \in Appl(\Pi, C_1, asyn)$ with $R'' \supsetneq R'$ as well as $(R'' - R') \cap R_j \neq \emptyset$ and $R' \cap R_j = \emptyset$ for some j, $1 \leq j \leq h$, i.e. extended by a rule from a set of rules R_j from which no rule has been taken into R'. For this variant of minimal parallelism *min′*, we formally define the set of applicable multisets as follows:

$$Appl(\Pi, C_1, min') = \{R' \mid R' \in Appl(\Pi, C_1, asyn) \text{ and}$$
$$\text{there is no } R'' \in Appl(\Pi, C_1, asyn)$$
$$\text{with } R'' \supsetneq R', (R'' - R') \cap R_j \neq \emptyset$$
$$\text{and } R' \cap R_j = \emptyset \text{ for some } j, \ 1 \leq j \leq h\}.$$

In [14], further restricting conditions on the four basic modes *asyn*, *sequ*, *min′*, and *max*, especially interesting for *min′*, were considered. We here only define the variant where from each set of $\Theta(R)$ which contains an applicable rule a rule must be taken into any applicable multiset in the mode *amin*:

$$Appl(\Pi, C_1, amin) = \{R' \mid R' \in Appl(\Pi, C_1, asyn) \text{ and for all } j, 1 \leq j \leq h,$$
$$R_j \cap \bigcup_{X \in Appl(\Pi, C_1)} X \neq \emptyset \text{ implies } R_j \cap R' \neq \emptyset\}.$$

Based on the concept of minimal parallelism as introduced in [6], we now consider the set of all multisets of rules called to be *minimally applicable* in Chapter 1; as from each set R_j at most one rule is taken, we denote this mode by min_1:

$$Appl(\Pi, C_1, min_1) = \{R' \mid R' \in Appl(\Pi, C_1, min') \text{ and}$$
$$|R' \cap R_j| \leq 1 \text{ for all } j, \ 1 \leq j \leq h\}.$$

In the *minimimally parallel mode min*, any applicable multiset containing a multiset from $Appl(\Pi, C_1, min_1)$ can be applied, i.e. we define:

$$Appl(\Pi, C_1, min) = \{R' \mid R' \in Appl(\Pi, C_1, asyn) \text{ and}$$
$$R'' \subseteq R' \text{ for some } R'' \in Appl(\Pi, C_1, min_1)\}.$$

Whereas the computational power of (purely) catalytic P systems working in the minimally parallel mode with partitioning the rule set according to the underlying membrane structure μ (in this case, the partitioning is denoted by Θ_μ) has not been investigated so far, the following computational completeness result can be established for P systems with bistable catalysts:

Theorem 4.13 $NOP_2(bicat_*, (min, \Theta_\mu)) = NRE$.

Proof. Let Q be a set of natural numbers, take a universal register machine M_U (as U_{32} in [20]), and construct the corresponding register machine M_Q accepting Q with the encoding $code(M_Q)$ of M_Q as input for the second register and the input number put into the first register. Let m and n be the number of registers and subtract-instructions in $M_U = (m, H, l_0, l_h, I)$, respectively. Then we construct the P system Π_Q with bistable catalysts (for a catalyst c, the second state of c is denoted by $\bar{c}$) working in the minimally parallel mode with the standard membrane partitioning Θ_μ as follows:

$\Pi_Q = (O, C, [_1[_2]_2]_1, w_1, \lambda, R_1, R_2, 2)$,

$O = C \cup \overline{C} \cup \{A_j \mid 1 \leq j \leq m\} \cup \{a, E, F, \#\} \cup \{l, \bar{l} \mid l \in H\}$,

$C = \{c_i, c'_i \mid 0 \leq i \leq n\}$,

$w_1 = F c_0 c_1 \ldots c_n c'_0 c'_1 \ldots c'_n$,

$R_1 = R_{1,1} \cup R_{1,2} \cup R_{1,3} \cup R_{1,4}$,

$R_{1,1} = \{c_0 F \to \overline{c_0} F, \overline{c_0} F \to c_0 F(a, in) A_1, \overline{c_0} F \to c_0 E l_0 A_2^{code(M_Q)}\}$,

$R_{1,2} = \{c_0 l \to \overline{c_0} A_r \bar{l}, \overline{c_0} \bar{l} \to c_0 l', \overline{c_0} \bar{l} \to c_0 l'' \mid l : (ADD(r), l', l'') \in I\}$,

$R_{1,3} = \{c_i l \to \overline{c_i}, \overline{c_i} A_r \to c_i l', \overline{c_i} E \to c_i \#, c'_i l \to \overline{c'_i}(\bar{l}, in),$

$\overline{c'_i} A_r \to c'_0 \#, \overline{c'_i} \bar{l} \to c'_i l'' \mid l : (SUB(r), l', l'') \in I\}$,

$R_{1,4} = \{c'_0 l_h \to \overline{c'_0}, \overline{c'_0} E \to c'_0, c_0 \# \to \overline{c_0} \#, \overline{c_0} \# \to c_0 \#\}$,

$R_2 = \{\bar{l} \to (\bar{l}, out) \mid l \in H\}$.

By the rules in $R_{1,1}$, we either add one symbol a in region 2 and one symbol A_1 in region 1 or else start the simulation of M_Q with $code(M_Q)$ in the second register. The instructions of M_U then are simulated in Π_Q as follows: the rules in $R_{1,2}$ simulate the ADD-instructions in two steps using the catalyst c_0 with the second state $\overline{c_0}$. The subtract-instructions are simulated by the rules in $R_{1,3}$ each of them using its own pair of catalysts c_l and c'_l with their second states $\overline{c_l}$ and $\overline{c'_l}$; in that way, after having non-deterministically chosen to take c_l (assuming the corresponding register r to be non-empty) or c'_l (assuming the corresponding register r to be empty), in the second step we know which choice has been taken in the previous step—this short-term memory represents the power of bistable catalysts. Having activated $\overline{c_l}$, one symbol A_r must be present, otherwise we are forced to use the rule $\overline{c_l} E \to c_l \#$ introducing the trap symbol #. Whereas this simulation again takes only two steps, the correct simulation of the case assuming the register to be empty takes three steps: in the first step, the control symbol $\bar{l}$ is sent into region 2 (the only case where we need this region), thus forcing $\overline{c'_l}$ to use $\overline{c'_l} A_r \to c'_0 \#$ in case of a wrong choice. With having $\overline{c'_l}$ still being available in the third step after $\bar{l}$ having returned from region 2, we may correctly finish with applying $\overline{c'_l} \bar{l} \to c'_l l''$ in region 1. When M_U reaches l_h, we can eliminate the remaining non-catalytic symbols E and l_h by using the rules $c'_0 l_h \to \overline{c'_0}$ and $\overline{c'_0} E \to c'_0$ from $R_{1,4}$ and thus eventually stop the computation. Yet in case that at some moment we had taken the wrong choice and therefore had to introduce the trap symbol #, we end up in an infinite loop using $c_0 \# \to \overline{c_0} \#$ and $\overline{c_0} \# \to c \#$. In sum, we conclude $N(\Pi_Q) = Q$.

The construction elaborated in the proof above shows that P systems with bistable catalysts working in the minimally parallel mode with the standard partitioning Θ_μ are universal (with the number of catalysts bounded by $2(n+1)$ where n is the number of subtract-instructions in a universal register machine, e.g. Korec's U_{32}, see [20]).

If we do not take the standard partitioning Θ_μ as introduced in [6], then we may obtain even stronger results: let Θ_C denote the partitioning in purely catalytic P systems where a partition consists of all catalytic rules assigned to a catalyst in one region. Then all the results obtained in Section 4.2 for computational completeness with the maximally parallel mode now also hold for the minimally parallel mode with partitioning Θ_C, e.g. we obtain

$$NO_{-C} P_1(pcat_3, (min, \Theta_C)) = NRE.$$

More generally, we get the following results:

Corollary 4.9 *For any $m \geq 1$ and $k \geq 3$ as well as $X \in \{N, Ps\}$,*

$$XRE = XO_{-C} P_m(pcat_k, (min, \Theta_C)).$$

4.6.3 k-Restricted Minimal Parallelism

We now consider a restricted variant of the minimally parallel mode allowing only a bounded number of at most k rules to be taken from each set R_j, $1 \leq j \leq h$, of the partitioning $\Theta(R)$. For this k-restricted minimally parallel mode (min_k), we formally define

$$Appl(\Pi, C_1, min_k) = \{R' \mid R' \in Appl(\Pi, C_1, min) \text{ and}$$
$$|R' \cap R_j| \leq k \text{ for all } j, \ 1 \leq j \leq h\}.$$

If we now take the partitioning Θ_C in a purely catalytic P system, then we obviously obtain the same results with min_1 as with min, because a catalyst can only be used with one rule.

Corollary 4.10 *For any $m \geq 1$ and $k \geq 3$ as well as $X \in \{N, Ps\}$,*

$$XRE = XO_{-C}P_m(pcat_k, (min_1, \Theta_C)).$$

If we replace every catalytic rule $ca \to cv$ in a purely catalytic P system Π working in the mode min_1 with Θ_C by the corresponding non-cooperative rule $a \to v$, we obtain a P system Π' working in the mode min_1 with a partitioning according to the original Θ_C such that both generate the same result. Denoting an arbitrary partitioning with Θ and a special partitioning allowing for at most p partitions with Θ_p and the use of only non-cooperative rules by the abbreviation $ncoo$, we infer the following results:

Corollary 4.11 *For any $m \geq 1$, $k \geq 3$, and $p \geq 3$ as well as $X \in \{N, Ps\}$,*

$$XRE = XO_{-C}P_m(pcat_k, (min, \Theta_C)) = XO_{-C}P_m(pcat_k, (min_1, \Theta_C))$$
$$= XOP_m(ncoo, (min_1, \Theta_p)) \quad = XOP_m(cat_0, (min_1, \Theta_p)).$$

Obviously, cat_0 has the same meaning as $ncoo$, i.e. we only have non-cooperative rules of the form $a \to v$.

4.6.4 Partial Halting

Let us now consider the *partial halting*. As for the minimally parallel mode, we take a partitioning Θ of the rule set into sets $R_j \neq \emptyset$, $1 \leq j \leq h$, and for continuing a computation we require that there exists an applicable multiset of rules containing one rule from each set R_j, $1 \leq j \leq h$.

For P systems working in mode $X \in \{amin, asyn, sequ\}$ and with partial halting (in the following denoted by h), we only get Parikh sets of matrix languages (regular sets of non-negative integers) for a large class of rules as proved in [2] and [12]; we here only give a sketch of the proof for catalytic P systems, yet even for an arbitrary partitioning of the set of rules:

Theorem 4.14 *For every $Y \in \{amin, asyn, sequ\}$,*

$$PsO_E P_*([p]cat, Y, h) = PsMAT^\lambda \text{ and } NO_E P_*([p]cat, Y, h) = NREG.$$

For the mode amin, the underlying partitioning of the rules has to be the same as the partitioning for partial halting.

Proof. We only prove $PsO_E P_*([p]cat, Y, h, \Theta) \subseteq PsMAT^\lambda$. Hence, let us start with an extended catalytic P system

$$\Pi = (O, T, C, \mu, w_1, \ldots, w_m, R'_1, \ldots, R'_m, \Theta, i_0)$$

using (purely) catalytic rules, working with the mode X, and with Θ being a partitioning of the set of rules $R = \bigcup_{j=1}^{m} R'_j$ into sets of rules R_1 to R_d. The stopping condition h—partial halting—then guarantees that in order to continue a computation there must exist a sequence of rules $\langle r_1, \ldots, r_d \rangle$ with $r_i \in R_i$, $1 \leq i \leq d$, such that all these rules are applicable in parallel. We now consider all functions δ with $\delta(i, r) \in \{0, 1\}$ and $\delta(i, r) = 1$ if and only if the rule $r \in R_i$, $1 \leq i \leq d$, is assumed to be applicable to the current configuration, and we define a matrix grammar $G_M = (V_M, \overline{T}, S, M)$ generating representations of all possible configurations computable in the given P system Π with the representation of an object a in membrane i as (a, i). We start with the matrix $(S \to Kh(w_1, \ldots, w_m))$ where $h(w_1, \ldots, w_m)$ is a representation of the initial configuration $(w_1, \ldots, w_m)$. A transition step in Π then is simulated in G_M as follows:

(i) We non-deterministically choose some δ as described above, use the matrix $(K \to K(\delta))$ and then apply the matrix

$$(K(\delta) \to K'(\delta), s_1, t_1, \ldots, s_d, t_d, s'_1, t'_1, \ldots, s'_d, t'_d,).$$

Each pair of rules s_i, t_i, $1 \leq i \leq d$, checks for the applicability of a rule $r \in R_i$ with $\delta(i, r) = 1$ and marks the symbols consumed by this rule, and finally s'_i, t'_i re-mark the corresponding symbols again. For simulating $r : ca \to cv \in R_i \cap R'_j$, $1 \leq j \leq m$, we take $s_i = (c, j) \to \overline{(c, j)}$ and $t_i = (a, j) \to \overline{(a, j)}$. For simulating $r : a \to v \in R_i \cap R'_j$ we take $s_i = (a, j) \to \overline{(a, j)}$ and t_i as above. After having simulated the application of one rule from each R_i, $1 \leq i \leq d$, we remark all the symbols having been introduced before by using the sequence of rules $s'_1, t'_1, \ldots, s'_d, t'_d$: we take $s'_i = \overline{(c, j)} \to (c, j)$ and $t'_i = \overline{(a, j)} \to (a, j)$ for $s_i = (c, j) \to \overline{(c, j)}$ and $t_i = (a, j) \to \overline{(a, j)}$ as well as $s'_i = \overline{(a, j)} \to (a, j)$ and $t'_i = \overline{(a, j)} \to (a, j)$ for $s_i = (a, j) \to \overline{(a, j)}$ and $t_i = (a, j) \to \overline{(a, j)}$.

(ii) Finally, we take different matrices depending on the transition mode:

   1. In the sequential mode, we only have to take all possibilities of simulating the application of one rule from $R_i \cap R'_j$ with $\delta(i, r) = 1$: For $r : [c]a \to [c]v \in R_i \cap R'_j$, we take $(K'(\delta) \to K, (a, j) \to h_j(v))$ where h_j is the morphism mapping $(b, here)$ onto (b, j), (b, out) onto $(b, \hat{j})$ where $\hat{j}$ denotes the outer region of membrane j, and (b, in_l) onto (b, l) (we leave the details for simulating the target *in* to the reader).

2. In the asynchronous mode, we have to allow an arbitrary number of rules to be applied in parallel; we simulate the application of rules sequentially, marking the results such that they cannot be used immediately. Finally, if for the current transition step, the application of no further rule is intended, we can re-mark the result symbols to be available for the simulation of the next transition step. In sum, we use the matrices $(K'(\delta) \to K''(\delta))$, $(K''(\delta) \to K''(\delta), [(c, j) \to \overline{(c, j)},](a, j) \to \overline{h_j(v)})$ for simulating rules $r : [c]a \to [c]v \in R'_j$, as well as $(K''(\delta) \to \overline{K}(\delta))$, $(\overline{K}(\delta) \to \overline{K}(\delta), \overline{(b, j)} \to (b, j))$, $1 \le j \le m$, $b \in O$, and finally $(\overline{K}(\delta) \to K)$. We remark that here the sequential mode and the asynchronous mode do not coincide, because after having applied some rules sequentially the condition to continue the computation with respect to the partial halting condition may not be fulfilled any more.

3. For the mode *amin*, instead of $(K'(\delta) \to K''(\delta))$ as for the asynchronous mode, we simulate the application of a sequence of rules $\langle r_1, \ldots, r_d \rangle$ with $\delta(i, r_i) = 1$ and $r_i : [c_{r_i}]a_{r_i} \to [c_{r_i}]v_{r_i} \in R_i \cap R'_{j_i}$, $1 \le i \le d$, such that all these rules are applicable in parallel (this has been checked before), which is accomplished by the matrix

$$(K'(\delta) \to K''(\delta), \ldots, [(c_{r_i}, j_i) \to \overline{(c_{r_i}, j_i)},](a_{r_i}, j_i) \to \overline{h_{j_i}(v_{r_i})}, \ldots).$$

We remark that this simulation requires that the partitioning for the partial halting coincides with the partitioning for the mode *amin*, and, moreover, requires that we are able to apply a rule from every set R_i, $1 \le i \le d$, which is only guaranteed in the mode *amin*, but not in the mode *min* itself.

As a technical detail we have to mention that it does not matter whether all the marked symbols are re-marked again, this would just make them unavailable during the next steps. Any sentential form containing marked symbols is considered to be non-terminal, hence, it cannot contribute to $L(G_M)$.

Finally, we may stop the simulation of computation steps of Π and use the matrices $(K \to F)$, $(F \to F, (a, j) \to (\overline{a}, j))$ for every object a and every membrane $j, 1 \le j \le m$, and the final matrix $(F \to \lambda)$ for generating a terminal string of G_M.

Now, we have to extract the representations of final configurations from $L(G_M)$. For every possibility of choosing a sequence of rules $\langle r_1, \ldots, r_d \rangle$ with $r_i \in R_i$, $1 \le i \le d$, such that all these rules are applicable in parallel, we construct a regular set checking for the applicability of this sequence in any possible representation of configurations of Π; then we take the union of all these regular sets and take its complement thus obtaining a regular set R. In $L(G_M) \cap R$ we then find at least one representation for every final configuration of computations in Π, but no representation of a non-final configuration.

Finally, let g be a projection with $g((\bar{a}, j)) = \lambda$ for $j \neq i_0$, $g((\bar{a}, i_0)) = \lambda$ for $a \in O - T$, and $g((\bar{a}, i_0)) = a$ for $a \in T$. Due to the closure properties of MAT^λ, we obtain $\Psi(g(L(G_M) \cap R)) = Ps(\Pi) \in PsMAT^\lambda$.

4.6.5 Other Variants of Halting

Looking carefully into the proofs given in the previous sections showing computational completeness for several variants of (purely) catalytic P systems by simulating register machines, we realize that we may consider a computation to be successful if at some moment a specific symbol representing the final state of the simulated register machine appears, yet only if no trap symbol occurs (which would indicate that at some moment during the computation a wrong choice with respect to the contents of a register has been taken). Obviously, this situation can be described by a regular set, and such variants of *halting with states* especially are used in the area of P automata (e.g. see [7]). In the following chapter we shall investigate communication P systems that will allow us to stop with a signal like the appearance of a specific symbol in some membrane only (*signal halting*).

Another variant is to consider a computation to be successful when it enters an infinite loop with the same sequence of configurations appearing from some moment on; as no significant changes happen any more, this variant is called *adult halting* (in the literature, the length of the period often is required to be one, but this definition is too restrictive for P systems). All the proofs elaborated before showing computational completeness for some specific variant of (purely) catalytic P systems can easily be modified in such a way that successful computations end with adult halting.

As an example to elucidate the main idea, we give the necessary changes in the proof of Theorem 4.13: the set of rules

$$R_{1,4} = \{c'_0 l_h \to \overline{c'_0},\ \overline{c'_0} E \to c'_0,\ c_0\# \to \overline{c_0}\#,\ \overline{c_0}\# \to c_0\#\}$$

has to be replaced by the set of rules

$$R'_{1,4} = \{c'_0 l_h \to \overline{c'_0} \overline{l_h},\ \overline{c'_0} \overline{l_h} \to c'_0 l_h,\ c_0\# \to \overline{c_0}\#\#,\ \overline{c_0}\# \to c_0\#\#\}.$$

The rules $c_0\# \to \overline{c_0}\#\#$, $\overline{c_0}\# \to c_0\#\#$ in $R'_{1,4}$ now guarantee that the number of symbols in an infinite computation with the trap symbol # is not bounded anymore, whereas a computation reaching a configuration without # yet containing El_h with l_h being the final label now will cycle with a period of two (the configurations containing $E\overline{l_h}$ and El_h, respectively, in the skin membrane) and thus fulfills the condition for adult halting.

Finally, we may even take a result from the output membrane in each reachable configuration (*unconditional halting*, denoted by u):

Example 4.2 *Consider the (deterministic) P system from Example 4.1. In contrast to the case of normal halting where the generated language would be empty, now the reachable configurations $(a_1{}^{2^n})$ allow us to extract the corresponding results 2^n, i.e. $N_u(\Pi) = \{2^n \mid n \geq 0\}$.*

The computational power of extended P systems using noncooperative rules working in the maximally parallel mode with unconditional halting (where a result is only taken if the whole contents of the output membrane consists of terminal symbols) is equivalent to that of E0L systems:

Theorem 4.15 *For all $m \in \mathbf{N} \cup \{*\}$, $PsO_E P_m(ncoo, u) = PsE0L$.*

Proof. We first show that $PsE0L \subseteq PsO_E P_1(ncoo, u)$.

Let $G = (V, T, w, P)$ be an E0L system. Then the corresponding extended one-membrane P system (we omit the—empty—set of catalysts) is $\Pi = (V, T, [_1\ ]_1, w, P, 1)$. Due to the maximally parallel mode applied in the extended P system Π, the computations in Π directly correspond with the derivations in G. Hence, $Ps_u(\Pi) = \Psi(L(G))$.

By definition, we have $PsO_E P_1(ncoo, u) \subseteq PsO_E P_m(ncoo, u)$, for all $m \geq 1$; hence, it only remains to show that $PsO_E P_*(ncoo, u) \subseteq PsE0L$:

Let $\Pi = (V, T, w_1, \ldots, w_m, R_1, \ldots, R_m, i_0)$ be an extended P system generating $Ps_u(\Pi)$, i.e., with unconditional halting. Then we first construct the E0L system $G = (V \times [1..m], T_0, w, P)$ with $w = \sqcup_{i \in [1..m]} h_i(w_i)$ (where $\sqcup$ represents the union of multisets and $[1..m]$ the interval of natural numbers between 1 and m) and

$$T_0 = h_{i_0}(T) \cup \bigcup_{j \in [1..m], j \neq i_0} h_j(V),$$

where the $h_j : V^* \to \{(a, j) \mid a \in V\}^*$ are morphisms with $h_j(a) = (a, j)$ for $a \in V$ and $1 \leq j \leq m$, as well as $P = R \cup P'$ where P' contains the rule $(a, j) \to (a, j)$ for $a \in V$ and $1 \leq j \leq m$ if and only if R_j contains no rule for a (which guarantees that in P there exists at least one rule for every $b \in V \times [1..m]$) and R is the set of rules obtained from those from the R_i by replacing each object a supposed to be in or to go to region j by (a, j).

We now take the projection $h : T_0^* \to T^*$ with $h((a, i_0)) = a$ for all $a \in T$ and $h((a, j)) = \lambda$ for all $a \in V$ and $j \in [1..n]$, $j \neq i_0$. Due to the direct correspondence of computations in Π and derivations in G, respectively, we immediately obtain $\Psi(h(L(G))) = Ps_u(\Pi)$. As E0L is closed under morphisms and therefore $Ps_u(\Pi) = \Psi(L(G'))$ for some E0L system G', we finally obtain $Ps_u(\Pi) \in PsE0L$.

Without extensions, the membranes used in P systems yield additional computational power: as is well known, the finite language $\{a, aa\}$ cannot be generated by a 0L system. On the other hand, every finite set of natural numbers can be generated by a P system with non-cooperative rules:

Example 4.3 *Consider the deterministic P system*

$$\Pi = (\{a\}, \{a\}, [_1 \ldots [_n]_n \ldots]_1, a^{x_1}, \ldots, a^{x_n}, R_1, \ldots, R_n, n)$$

with $R_i = \{a \to (a, in)\}$ for $1 \leq i < n$ and $R_n = \emptyset$. In the first $n-1$ computation steps, the contents of the outer regions is shifted one region deeper in the linear membrane structure, and after n transition steps the contents of all regions $< n$ is empty and the computation stops. During the computation, at step i, $0 \leq i < n$, $a^{y_{i+1}}$ where $y_{i+1} = \sum_{j=0}^{i} x_{n-j}$, is found in output membrane n; hence, we obtain $N_u(\Pi) = \{y_i \mid 1 \leq i \leq n\}$.

As an immediate consequence of the constructions elaborated in Theorem 6.4 and the preceding example, we obtain the following results:

Theorem 4.16 $Ps[D]OP_1(ncoo, u) = Ps[D]0L \subsetneq Ps[D]OP_*(ncoo, u)$.

We conjecture that the families $Ps[D]OP_n(ncoo, u)$ form an infinite hierarchy with respect to the number of membranes; a formal proof of this conjecture is left as a challenge for the interested reader.

4.7 GENERATION OF LANGUAGES

In a very easy way, P systems in general and especially (purely) catalytic P systems can also be used as language generators: during a successful computation, all the symbols sent out through the skin membrane are taken as the symbols forming a string in just that sequence the symbols are sent out (we take all possible sequences of symbols that are sent out in one transition step as possible substrings to be concatenated with the string already generated by the preceding transition steps— thus, not just one string may result from a successful computation). The language generated by a (purely) catalytic P system Π in that way is denoted by $L(\Pi)$. The family of languages generated by [purely] catalytic P systems with at most m membranes and at most k catalysts is denoted by $LOP_m([p]cat_k)$.

For showing the computational completeness of (purely) catalytic P systems as language generating devices, we need the following result for register machines (see [22] and [9]):

Proposition 4.2 *For any partially recursive function $f : \mathbb{N} \to \mathbb{N}$ there exists a register machine M with two registers computing f in such a way that, when starting with 2^n in register 1 and 0 in register 2, M computes $f(n)$ by halting with $2^{f(n)}$ in register 1 and 0 in register 2. Moreover, in no configuration both registers are empty.*

The proof of the following result is based on the proposition above and the constructions elaborated in the proof of Theorem 4.1:

Theorem 4.17 $LOP_1(cat_2) = LOP_1(pcat_3) = RE$.

Proof. For a given language $L \in RE$, $L \subseteq T^*$ for some alphabet T with $card(T) = k$, we construct a purely catalytic P system

$$\Pi_L = (O_L, \{c_1, c_2, c_3\}, [\,_1\,]_1, c_1 c_2 c_3 p_0 \tilde{p}_0, R_L, 1)$$

with $L(\Pi_L) = L$ as follows: let $T = \{a_m \mid 1 \leq m \leq k\}$; then every symbol a_m in T can be interpreted as the digit m at base $k + 1$; hence, every string in T^* can be encoded as a natural number using the function $g_T : T^* \to \mathbb{N}$ inductively defined by $g_T(\lambda) = 0$, $g_T(a_m) = m$ for $1 \leq m \leq k$, and $g_T(wa) = g_T(w) * (k + 1) + g_T(a)$ for $a \in T$ and $w \in T^*$. We now describe the way how to iteratively generate w by sending the corresponding symbols of w to the environment, at the same time generating $o_1^{2^{g_T(w)}}$ inside the skin membrane.

Sending out a new symbol a_m and generating the corresponding number of symbols o_1 inside the skin membrane is accomplished by simulating a register machine procedure p_m generating $A^{2^{x(k+1)+m}}$ from A^{2^x}. According to Proposition 4.2, this task can be done with only two registers; thus, following the construction given in Theorem 4.1 we need not specify the details of simulating p_m in Π_L, yet we take $p_{(m,i)} \tilde{p}_{(m,i)}$ as the pair of symbols to represent the initial label (m, i) and $p_{(m,f)} \tilde{p}_{(m,f)}$ as the pair of symbols to represent the final label (m, f) of p_m for the simulation of this register machine procedure in Π_L following the construction given in the proof of Theorem 4.1.

In Π_L, we now start with $c_1 c_2 c_3 p_0 \tilde{p}_0$ in the skin membrane and generate an output w and $o_1^{2^{g_T(w)}}$ inside the skin membrane as follows: we may either add a new terminal symbol a_m, $1 \leq m \leq k$, by choosing the corresponding rules $c_1 \tilde{p}_0 \to c_1$ and $c_2 p_0 \to c_2 p_{(m,i)} \tilde{p}_{(m,i)} (a_m, out)$ or else finish the initial phase by applying $c_1 \tilde{p}_0 \to c_1$ and $c_2 p_0 \to c_2 p_{l_0} \tilde{p}_{l_0}$ where the label l_0 is the start label of the register machine M_L with two registers which halts when started with $2^{g_T(w)}$ in its first register if and only if w is in the given language L (e.g. see [22]). Following the constructions elaborated in the proof of Theorem 4.1, Π_L now halts, i.e. accepts $o_1^{2^{g_T(w)}}$, if and only if M_L halts, i.e. if and only if $w \in L$. Hence, we conclude $L(\Pi_L) = L$.

For bistable catalysts, the following optimal result was proved in [1]:

Theorem 4.18 $LOP_1(bicat_1) = RE$.

4.8 Conclusion

In this chapter, we have investigated the computational power of (purely) catalytic P systems working in the maximally parallel mode and have shown that they are able to generate any recursively enumerable set of vectors of natural numbers in only one membrane thereby needing at most two (three) catalysts. Catalytic P systems with no catalysts and purely catalytic P systems with one catalyst can only generate semilinear sets. The computational power of (purely) catalytic P systems with one (two) catalysts remains as a challenging open problem. Only when using the variant of one mobile catalyst, computational completeness could be shown.

Non-deterministic (purely) catalytic P systems are universal as accepting mechanisms, whereas the deterministic variants are not computationally complete. Computational completeness for accepting deterministic (purely) catalytic P systems can be regained by adding priority relations on the rules.

Using the idea of sending out symbols to the environment and concatenating these symbols to a string, (purely) catalytic P systems can also be used as language generating devices. Again, (purely) catalytic P systems are able to generate any recursively enumerable language with only one membrane, thereby needing at most two (three) catalysts, whereas the computational power of (purely) catalytic P systems as language generating devices with one (two) catalysts remains an open problem.

Equipped with the sequential or the asynchronous mode, (purely) catalytic P systems cannot generate more than semilinear sets. For the minimally parallel mode, only purely catalytic P systems using bistable catalysts could be shown to be computationally complete (needing two membranes) with the standard partitioning of rules assigning the partitions of rules to the membranes, whereas for (purely) catalytic P systems with normal catalysts working in the minimally parallel mode, this question is only solved for partitioning the set of rules with respect to the catalysts. On the other hand, using the 1-restricted minimally parallel mode together with a suitable partitioning of the rules in the skin membrane, we even do not need catalysts at all for obtaining computational completeness.

4.9 Bibliographic Notes

In the paper introducing P systems, [24], priorities were used to prove computational completeness for catalytic P systems. P. Sosík was the first author to realize that priorities on the rules are not necessary, see [26], [28], and [27]. The number of catalysts needed for establishing computational completeness for catalytic P systems was decreased to six in [13], and in [16], based on this result, it was observed that at most seven catalysts were needed for purely catalytic P systems. The best results known so far with respect to the number of catalysts (two for catalytic P systems and three for purely catalytic P systems, see Theorem 4.1) were established in [11] and some variants for purely catalytic P systems were discussed in [19]. In [17], similarities between purely catalytic P systems and vector addition systems were elucidated, and among other results, $NO_{-C} P_1(pcat_1) = NREG$ was shown.

In Section 4.4, the issue of "deterministic versus non-deterministic" in accepting catalytic P systems was discussed; the main results and proof techniques for the results established in Section 4.4 were taken from [18]. In [5], the divergence problem for catalytic P systems in the framework of well-structured transition systems was studied, yielding as a by-product an alternative proof for the non-universality of accepting deterministic catalytic P systems. Based on the theoretical results concerning the 4-determinism of (accepting) catalytic P systems, an effective implementation was described in [4].

Mobile catalysts were introduced in [21], where also the best result known at that time for bistable catalysts was to be found with one bistable and two normal catalysts in five membranes needed to obtain computational completeness. The optimal result with one bistable catalyst in one membrane (see Theorem 4.10) is derived from the result corresponding with Theorem 4.18 as established in [1].

The concept of minimal parallelism was already mentioned in [25], p. 84, and called *minimal synchronization:* "in each step, in each region where at least one rule *can* be used, at least one rule *must* be used." In fact, this can be interpreted as the mode *amin*. Theorem 4.13 therefore also holds true for the mode *amin* (a result already proved—in a different way—in [25], Theorem 3.4.7), yet as well for the more general variant of minimal paralellism *min'* as considered in [14].

The idea of partial halting was introduced in [12] and then further developed in [2]. A formal framework for different variants of halting as well as for various transition modes was developed in [14]. The variant of unconditional halting was investigated in [3]. The characterization of purely catalytic P systems by P systems with a suitable partitioning of non-cooperative rules in one membrane was established just recently in [15].

REFERENCES

[1] A. ALHAZOV: Number of protons/bi-stable catalysts and membranes in P systems. Time-freeness. *Membrane Computing. 6th Intern. Workshop WMC 2005* (R. Freund et al., eds.), Vienna, Austria, LNCS 3850, Springer, 2006, 79–95.

[2] A. ALHAZOV, R. FREUND, M. OSWALD, S. VERLAN: Partial versus total halting in P systems. *Proc. Fifth Brainstorming Week on Membrane Computing* (M. A. Gutiérrez-Naranjo et al., eds.), Sevilla, 2007, 1–20.

[3] M. BEYREDER, R. FREUND: (Tissue) P systems using noncooperative rules without halting conditions. *Pre-Proc. Ninth Workshop on Membrane Computing (WMC9)*, (P. Frisco et al. eds.), Edinburgh, 2008, 85–94.

[4] A. BINDER, R. FREUND, G. LOJKA, M. OSWALD: Implementation of catalytic P systems. *Pre-proc. CIAA 2004* (M. Domaratzky et al., eds.), Queen's University, Kingston, Ontario, Canada, 2004, 24–33.

[5] N. BUSI: Using well-structured transition systems to decide divergence for catalytic P systems. *Theoretical Computer Sci.*, 372 (2007), 125–135.

[6] G. CIOBANU, L. PAN, GH. PĂUN, M.J. PÉ REZ-JIMÉNEZ: P systems with minimal parallelism, *Theoretical Computer Sci.*, 378 (2007), 117–130.

[7] E. CSUHAJ-VARJÚ GY. VASZIL: P automata. *Membrane Computing* (Gh. Păun et al. eds.), LNCS 2597, Springer, 2003, 219–233.

[8] L.E. DICKSON: Finiteness of the odd perfect and primitive abundant numbers with n distinct prime factors. *Amer. Journal Math.*, 35 (1913), 413–422.

[9] H. FERNAU, R. FREUND, M. OSWALD, K. REINHARDT: Refining the nonterminal complexity of graph-controlled, programmed, and matrix grammars. *J. Automata, Languages and Combinatorics*, 12 (2007), 117–138.

[10] R. FREUND: Particular results for variants of P systems with one catalyst in one membrane. *Proc. the Fourth Brainstorming Week on Membrane Computing* (C. Graciani-Díaz et al., eds.), Fénix Editora, Sevilla, 2006, vol. II, 41–50.

[11] R. FREUND, L. KARI, M. OSWALD, P. SOSÍK: Computationally universal P systems without priorities: two catalysts are sufficient. *Theoretical Computer Sci.*, 330 (2005), 251–266.

[12] R. FREUND, M. OSWALD: Partial halting in P systems. *Intern. J. Foundations of Computer Sci.*, 18 (2007), 1215–1225.

[13] R. FREUND, M. OSWALD, P. SOSÍK: Reducing the number of catalysts needed in computationally universal systems without priorities. *Fifth Intern. Workshop Descriptional Complexity of Formal Systems* (E. Csuhaj-Varjú et al., eds.), Budapest, Hungary, July 2003, 102–113.

[14] R. FREUND, S. VERLAN: A formal framework for P systems. *Pre-proc. Membrane Computing, Intern. Workshop WMC8* (G. Eleftherakis et al., eds.), Thessaloniki, Greece, 2007, 317–330.

[15] R. FREUND, S. VERLAN: (Tissue) P systems working in the k-restricted minimally parallel derivation mode. *Intern. Workshop on Computing with Biomolecules* (E. Csuhaj-Varjú et al., eds.), Wien, Austria, 2008, 43–52.

[16] O.H. IBARRA, Z. DANG, O. EGECIOGLU, G. SAXENA: Characterizations of catalytic membrane computing systems. *28th Intern. Symp. Mathematical Foundations of Computer Sci., 2003* (B. Rovan, P. Vojtás, eds.), LNCS 2747, Springer, 2003, 480–489.

[17] O. IBARRA, Z. DANG, O. EGECIOGLU: Catalytic P systems, semilinear sets, and vector addition systems. *Theoretical Computer Sci.*, 312 (2004), 379–399.

[18] O. IBARRA H. YEN: Deterministic catalytic systems are not universal. *Theoretical Computer Sci.*, 363 (2006), 149–161.

[19] O. IBARRA, H. YEN, Z. DANG: The power of maximal parallelism in P systems. *Eighth Intern. Conf. Developments in Language Theory (DLT'04)*, LNCS 3340, Springer, 2004, 212–224.

[20] I. KOREC: Small universal register machines. *Theoretical Computer Sci.*, 168 (1996), 267–301.

[21] S.N. KRISHNA, A. PĂUN: Results on catalytic and evolution-communication P systems. *New Generation Comput.*, 22 (2004), 377–394.
[22] M.L. MINSKY: *Computation: Finite and Infinite Machines*. Prentice Hall, Englewood Cliffs, New Jersey, 1967.
[23] M. OSWALD: *P Automata*. PhD thesis, Faculty of Computer Science, Vienna University of Technology, Vienna, Austria, 2003.
[24] GH. PĂUN: Computing with membranes. *J. Computer and System Sci.*, 61 (2000), 108–143.
[25] GH. PĂUN: *Membrane Computing. An Introduction*. Springer, Berlin, 2002.
[26] P. SOSÍK: P systems versus register machines: two universality proofs. *Pre-Proc. Workshop on Membrane Computing (WMC-CdeA2002)*, Curtea de Argeş, Romania, 2002, 371–382.
[27] P. SOSÍK: The power of catalysts and priorities in membrane systems, *Grammars*, 6 (2003), 13–24.
[28] P. SOSÍK, R. FREUND: P Systems without priorities are computationally universal. *Membrane Computing. Intern. Workshop, WMC-CdeA 2002* (Gh. Păun et al, eds.), Curtea de Argeş, Romania, August 2002, LNCS 2597, Springer, 2003, 400–409.

CHAPTER 5

COMMUNICATION P SYSTEMS

RUDOLF FREUND

ARTIOM ALHAZOV

YURII ROGOZHIN

SERGEY VERLAN

5.1 INTRODUCTION

COMMUNICATION P systems are inspired by the idea of communicating substances through membrane channels of a cell. Molecules may go the same direction together—*symport*—or some of them may leave while at the same time other molecules enter the cell—*antiport*. Communicating objects between membrane regions is a powerful tool yielding computational completeness with one membrane using antiport rules or symport rules of size three, i.e. involving three objects, in the maximally parallel mode. Yet even with the minimally parallel mode (throughout this chapter, we only consider the special variant of minimal parallelism *amin* as defined in Chapter 4), we get computational completeness with two membranes. As register machines can be simulated in a deterministic manner, P systems with antiport rules or symport rules can accept any recursively enumerable set of (vectors of) natural numbers in a deterministic way.

In tissue P systems, the objects are communicated through channels between cells. In each transition step we apply only one rule for each channel, whereas at the level of the whole system we work in the maximally parallel way. Computational completeness can be obtained with a rather small number of objects and membranes or cells, in the case of tissue P systems even with copies of only one object. The computational power of P systems with antiport rules or symport rules involving copies of only one object remains one of the most challenging open questions.

The concept of P systems with antiport and/or symport rules can be generalized to systems using membrane rules evolving multisets of objects on both sides of the membrane even depending on permitting contexts (also called *promoters*) and/or forbidden contexts (also called *inhibitors*). Although these rules are very powerful, P systems applying rules without inhibitors in the minimally parallel mode *amin* and working under the condition of partial halting can only generate Parikh sets of matrix languages.

P systems with communication rules can also be used as language generators— we take the sequences of terminal objects sent out to the environment as the strings generated by the system. Another possibility for obtaining strings is to consider the sequences of configurations in a halting computation and to observe the occurrence of a specific object (*traveler*) in the membranes of these configurations.

5.2 P Systems with Symport/Antiport Rules

In this section we elucidate the computational power of the basic model of communication P systems, i.e. P systems using antiport and/or symport rules. Computational completeness can be obtained with rules of size (at most) 3 in only one membrane when working in the maximally parallel mode, whereas in the minimally parallel mode two membranes are needed. Using the sequential or the asynchronous mode, we only obtain Parikh sets of matrix languages. With partial halting, only Parikh sets of matrix languages are obtained even with the minimally parallel mode *amin*. For definitions of the transition and halting modes, we refer the reader to Chapter 4.

Definition 5.1 *A P system (of degree $d \geq 1$) with antiport and/or symport rules (in this section called P system for short) is a construct*

$$\Pi = (O, T, E, \mu, w_1, \ldots, w_d, R_1, \ldots, R_d, i_0) \text{ where}$$

1. *O is the alphabet of objects;*
2. *$T \subseteq O$ is the alphabet of terminal objects;*

3. $E \subseteq O$ is the set of objects occurring in an unbounded number in the environment;
4. μ is a membrane structure consisting of d membranes (usually labeled with i and represented by corresponding brackets $[_i$ and $]_i$, $1 \leq i \leq d$);
5. w_i, $1 \leq i \leq d$, are strings over O associated with the regions $1, 2, \ldots, d$ of μ; they represent multisets of objects initially present in the regions of μ;
6. R_i, $1 \leq i \leq d$, are finite sets of rules of the form $(u, out; v, in)$, with $u \neq \lambda$ and $v \neq \lambda$ (antiport rule), and (x, out) or (x, in), with $x \neq \lambda$ (symport rule);
7. i_0, $1 \leq i_0 \leq d$, specifies the output membrane of Π.

The antiport rule $(u, out; v, in)$ in R_i exchanges the multiset u inside membrane i with the multiset v outside membrane i; the symport rule (x, out) sends the multiset x out of membrane i and (x, in) takes x in from the region surrounding membrane i (if i is the skin membrane, then x has to contain at least one symbol not in E). The membrane structure μ and the multisets represented by w_i, $1 \leq i \leq d$, in Π constitute the *initial configuration* of the system.

In the *maximally parallel mode*, a transition from one configuration to another one is obtained by the application of a maximal multiset of rules. The system continues maximally parallel transition steps until there remain no applicable rules in any region of Π; then the system halts (*total halting*). We consider the vector of multiplicities of objects from T contained in the output membrane i_0 at the moment when the system halts as the *result* of the underlying computation of Π; observe that here we do not count the non-terminal objects present in the output membrane. The set of results of all halting computations possible in Π is denoted by $Ps(\Pi)$, respectively. The family of all sets of vectors of natural numbers computable by P systems with d membranes and using rules of type a is denoted by $Ps\,O_E\,P_d(a, max, H)$. When the parameter d is not bounded, it is replaced by $*$.

We consider variants of P systems using only rules of very restricted types a: $anti_k$ indicates that only antiport rules of weight at most k are used, where the *weight of an antiport rule* $(u, out; v, in)$ is defined as $\max\{|u|, |v|\}$; sym_k indicates that only symport rules of weight at most k are used, where the *weight of a symport rule* (x, out) or (x, in) is defined as $|x|$. For an antiport rule $(u, out; v, in)$, we may refine the weight by also considering the total number of symbols specified in the rule—the *size of an antiport rule* is defined as $|u| + |v|$; then $anti_k^s$ indicates that only antiport rules of weight at most k and size at most s are used. For example, $anti_2^3$ allows antiport rules of the forms $(a, out; b, in)$, $(ab, out; c, in)$, and $(c, out; ab, in)$, whereas $anti_2$ (coinciding with $anti_2^4$) also allows rules of the form $(ab, out; cd, in)$, with a, b, c, d being single objects. The *size of a symport rule* equals its weight, hence, we do not need additional notions.

When using the special *minimally parallel mode* amin, in each transition step we choose a multiset of rules from R in such a way that this chosen multiset includes at least one rule from every set of rules (in this chapter we only consider the partitioning of the rules according to the given membrane structure) containing applicable rules. In the *asynchronous* (*asyn*) and the *sequential mode* (*sequ*), in each transition step we apply an arbitrary number of rules/exactly one rule, respectively. The corresponding families of sets of vectors of natural numbers generated by P systems with d membranes and using rules of type a in the transition mode X are denoted by $Ps\,O_E\,P_d(a, X, H)$, $X \in \{amin, asyn, sequ\}$.

If instead of the total halting we take *partial halting*, i.e. computations halting as soon as no multiset of rules containing at least one rule from each set of rules from the partitioning of the rules is applicable anymore, the corresponding families are denoted by $Ps\,O_E\,P_d(a, X, h)$, $X \in \{max, amin, asyn, sequ\}$.

If at the end of a computation only a bounded number of at most k non-terminal objects remains in the output membrane, we replace the subscript E by $-k$; if we do not distinguish between terminal and non-terminal symbols, then we simply omit the subscript E.

All these variants of P systems with antiport and/or symport rules can also be considered as accepting devices, the input being given as the numbers of objects in the distinguished membrane i_0. In that case we write Ps_a instead of Ps; in the accepting case, the subscripts E and $-k$ to O are of no meaning and therefore omitted. The families specified by deterministic systems are denoted by adding D in front of O.

When we are only interested in the number of symbols in the output/input membrane, we replace Ps by N. In the case that only m objects are needed to generate/accept a set of natural numbers, for all $X \in \{max, amin, asyn, sequ\}$ and $Y \in \{H, h\}$ we obtain the families $NO_m P_d(a, X, Y)$ and $N_a O_m P_d(a, X, Y)$, respectively.

As the maximally parallel mode and total halting are the most common variants, in the following we shall sometimes omit the parameters *max* and H when denoting the corresponding families.

The following simple examples show that at least when considering only one single object in one membrane, the generating power of P systems with antiport and/or symport rules differs from the accepting power:

Example 5.1 *With only one object, we can only generate finite sets, i.e.* $NO_1 P_1(anti_*, sym_*) = NFIN$: on the one hand, consider a P system $\Pi = (\{b\}, \{b\}, E, [_1\,]_1, w_1, R_1, 1)$. Then $N(\Pi)$ is finite if E is empty, because no additional symbols can be brought in from the environment; if $E = \{b\}$, then by definition the rules in R_1 may only be of the form (b^k, out) or $(b^k, out; b^m, in)$. With symport rules (b^k, out) the number of objects in the skin membrane decreases, the same happens with antiport rules $(b^k, out; b^m, in)$ where $k > m$; yet these rules are the only ones which do not enforce infinite computations, as antiport rules $(b^k, out; b^m, in)$ where $k \leq m$ remain applicable as soon as the number of objects in the skin membrane is at least k. Hence, we conclude that in all cases $N(\Pi) \in NFIN$.

On the other hand, any non-empty set $M \in \mathbf{NFIN}$ is generated by the P system $\Pi = (\{b\}, \{b\}, \{b\}, [_1 \ ]_1, b^m, R_1, 1)$ where $m = \max(M) + 1$ and $R_1 = \{(b^m, out; b^j, in) \mid j \in M - \{0\}\} \cup \{(b^m, out) \mid 0 \in M\}$. The empty set is generated by the P system $\Pi = (\{b\}, \{b\}, \{b\}, [_1 \ ]_1, b, \{(b, out; b, in)\}, 1)$.

As in one membrane, partial halting has the same effect as total halting and all the arguments given above hold true for every transition mode $X \in \{max, amin, asyn, sequ\}$, we conclude $NO_1 P_1(anti_*, sym_*, X, Y) = \mathbf{NFIN}$, $Y \in \{H, h\}$.

Example 5.2 *The infinite set* $\mathbf{N}$ *is accepted by the P system*

$$\Pi = (\{b\}, \{b\}, \{b\}, [_1 \ ]_1, \lambda, \{(b, out)\}, 1).$$

Every computation for an input b^m *with* $m > 0$ *takes exactly one step in the maximally parallel mode and at most m steps in the other modes. Hence,* $\mathbf{N} \in N_a O_1 P_1(sym_1, X, Y)$ *for all* $X \in \{max, amin, asyn, sequ\}$ *and* $Y \in \{H, h\}$.

With more than one membrane, we may accept non-semilinear sets even when using copies of only one object:

Example 5.3 *The non-semilinear set* $\{2^n \mid n \in \mathbf{N}\}$ *is accepted by the P system with antiport rules*

$$\Pi = (\{b\}, \{b\}, \{b\}, [_1[_2[_3 \ ]_3]_2]_1, \lambda, \lambda, \lambda, R_1, R_2, R_3, 1),$$

$$R_1 = \{(bb, out; b, in)\},$$

$$R_2 = \{(b, in)\},$$

$$R_3 = \{(bb, out), (bb, in)\}.$$

Using the rule $(bb, out; b, in)$ *from* R_1 *the number m of objects b put into the skin membrane is divided by 2 in every maximally parallel transition step. If m has been of the form 2^n, then at the end of the computation the last symbol b enters membrane 2 by using* (b, in). *If this rule (b, in) was chosen or had to be used before that because of the contents of the skin membrane being an uneven number, then at the end of the computation at least two objects are in membrane two which starts an infinite computation with the rules from* R_3. *Hence, we conclude* $\{2^n \mid n \in \mathbf{N}\} \in N_a O_1 P_3(anti_2^3, sym_2)$.

5.2.1 Computational Completeness for the Maximally Parallel Mode

The following result shows that we only need one membrane to obtain computational completeness; in the case of accepting P systems, even deterministic systems are sufficient in contrast to the situation for catalytic P systems:

Theorem 5.1 $NO_{-1} P_1(anti_2^3) = N_a DO P_1(anti_2^3) = \mathbf{NRE}.$

Proof. Let $M = (3, H, l_0, l_h, I)$ be a deterministic register machine. We construct the P system generating the set $N_a(M)$ of numbers accepted by M as follows:

$$\Pi = (O, T, O, [_1\ ]_1, l_I, R_1, 1),$$
$$O = \{p, p', p'', \bar{p}, \breve{p} \mid p \in H\} \cup \{A_i \mid 1 \leq i \leq 3\} \cup \{l_I, l_{b_1}, b_1\},$$
$$T = \{b_1\},$$
$$R_1 = R_{1,I} \cup R_{1,A} \cup R_{1,S},$$
$$R_{1,I} = \{(l_I, out; A_1 l_{b_1}, in), (l_{b_1}, out; b_1 l_I, in)\} \cup \{(l_I, out; l_0, in)\},$$
$$R_{1,A} = \{(p, out; A_r q, in) \mid p : (\text{ADD}(r), q, q) \in I\},$$
$$R_{1,S} = \{(p, out; p'p'', in), (p''A_r, out; \bar{p}, in), (p', out; \breve{p}, in),$$
$$(\bar{p}\breve{p}, out; q, in), (\breve{p}p'', out; s, in) \mid p : (\text{SUB}(r), q, s) \in I\}.$$

The contents of register r is represented by the corresponding number of symbols A_r, $1 \leq r \leq 3$. First, we generate an arbitrary number n of symbols b_1 and A_1 by n times applying the rules $(l_I, out; A_1 l_{b_1}, in)$ and then $(l_{b_1}, out; b_1 l_I, in)$. With applying the rule $(l_I, out; l_0, in)$ we then start the simulation of M which accepts n if and only if $n \in N_a(M)$. An add-instruction $p : (\text{ADD}(r), q, q) \in I$ is simulated by using the rule $(p, out; A_r q, in)$. A subtract-instruction $p : (\text{SUB}(r), q, s) \in I$ is simulated by using the rules from $R_{1,S}$ starting with applying $(p, out; p'p'', in)$: in the next step, $(p', out; \breve{p}, in)$ is applied, and only if register r is not empty, $(p''A_r, out; \bar{p}, in)$ is applied in parallel; in the succeeding step, we either continue with $(\bar{p}\breve{p}, out; q, in)$ for this nonempty case and with $(\breve{p}p'', out; s, in)$ for the case that register r has been empty. When the final label l_h appears, the computation stops with the desired output of n symbols b_1 being found in the skin membrane together with the only additional symbol l_h.

If we consider acceptance, then the input n is given by A_1^n in the skin membrane. As the only non-determinism in the P system Π occurred in $R_{1,I}$, the P system $\Pi_a = (O_a, T_a, O_a, [_1\ ]_1, l_0, R_a, 1)$ with $O_a = O - \{l_I, l_{b_1}, b_1\}$, $T_a = \{A_1\}$, and $R_a = R_{1,A} \cup R_{1,S}$ is deterministic and accepts $N_a(M)$.

Adding only the very simple symport rule of size one (l_h, out) we can avoid the additional symbol l_h at the end of a computation of the P system Π constructed in the proof above, i.e. we obtain:

Corollary 5.1 $NOP_1(anti_2^3, sym_1) = NRE.$

The results established above for recursively enumerable sets of natural numbers can easily be extended to Parikh sets—we take $T = \{b_i \mid 1 \leq i \leq k\}$ and add l_{b_i}, b_i, $2 \leq i \leq k$, to O, and in $R_{1,I}$ we have to take all rules with b_i, $1 \leq i \leq k$, instead of b_1 only, i.e.

$$R_{1,I} = \{(l_I, out; A_i l_{b_i}, in), (l_{b_i}, out; b_i l_I, in) \mid 1 \leq i \leq k\} \cup \{(l_I, out; l_0, in)\}.$$

Moreover, we then have to simulate a deterministic register machine $M = (k + 2, H, l_0, l_h, I)$. In sum, we obtain:

Corollary 5.2 $Ps\,O_{-1}P_1(anti_2^3) = Ps_aDOP_1(anti_2^3) = PsRE.$

In the results stated above we can restrict ourselves even to antiport rules of size 3 exactly. This result is already optimal with respect to the size of the rules, because with antiport rules of size 2, i.e. being of the form $(b, out; c, in)$ with b and c being single objects, the number of objects in the system cannot be changed.

We now turn our attention to P systems with only symport rules; again rules of size 3 in one membrane are sufficient to obtain computational completeness:

Theorem 5.2 $Ps\,O_E P_1(sym_3) = Ps_aDOP_1(sym_3) = PsRE.$

Proof. Let $M = (k+2, H, l_0, l_h, I)$ be a deterministic register machine and construct the P system

$$\Pi = (O, T, E, [_1\ ]_1, w_1, R_1, 1),$$

$$O = \{p, p', \bar{p}, \bar{p}', \bar{p}'', \tilde{p}, \tilde{p}', \tilde{p}'', Z_p \mid p \in H\} \cup \{A_i \mid 1 \le i \le k+2\}$$
$$\cup \{X, l_I, l'_I\} \cup T,$$

$$T = \{b_i \mid 1 \le i \le k\},$$

$$E = O - (\{X, l_I, l'_I\} \cup \{p', \bar{p}, \bar{p}', Z_p \mid p \in H\}),$$

$$w_1 = \{X, l_I, l'_I\} \cup \{p', \bar{p}, \bar{p}', Z_p \mid p \in H\},$$

$$R_1 = R_{1,I} \cup R_{1,A} \cup R_{1,S},$$

$$R_{1,I} = \{(l_I l'_I X, out), (l_0 l'_I X, in), (l_I X, in)\} \cup \{(l'_I A_i b_i, in) \mid 1 \le i \le k\},$$

$$R_{1,A} = \{(pp', out), (A_r p'q, in) \mid p : (ADD(r), q, q) \in I\},$$

$$R_{1,S} = \{(pp', out), (p'\bar{p}\tilde{p}, in), (\bar{p}\tilde{p}' A_r, out), (\bar{p}'\bar{p}'', in), (\tilde{p}\tilde{p}', out),$$
$$(\tilde{p}'\tilde{p}'', in), (\bar{p}''\bar{p}'' Z_q, out), (\tilde{p}\tilde{p}'' Z_s, out) \mid p : (SUB(r), q, s) \in I\}$$
$$\cup \{(Z_p p, in) \mid p \in H\}.$$

Using the rules $(l_I l'_I X, out)$ as well as $(l_I X, in)$ and $(l'_I A_i b_i, in)$ from $R_{1,I}$, we are able to generate any number of symbols A_i and the same number of symbols b_i in the skin membrane. The application of the rule $(l_0 l'_I X, in)$ after $(l_I l'_I X, out)$ then starts the simulation of M. A deterministic add-instruction $p : (ADD(r), q, q) \in I$ is simulated by using the single copy of p' with the rules (pp', out) and $(A_r p'q, in)$. A subtract-instruction $p : (SUB(r), q, s) \in I$ is simulated by using the rules from $R_{1,S}$. The symbol p takes p' outside, which then returns together with $\bar{p}, \tilde{p}$; $\bar{p}$ goes to the environment with $\bar{p}'$ and returns transformed to $\bar{p}''$. At the same time, $\tilde{p}$ tries to decrement register r; if this is possible, it goes out together with $\tilde{p}'$ and a copy of A_r and returns back as $\tilde{p}''$ together with $\tilde{p}'$. If $\bar{p}''$ meets $\tilde{p}''$, then the new state q has to be chosen by sending out Z_q; on the other hand, if the register has been empty, i.e. if no symbol A_r had been present, then $\tilde{p}$ has remained in the skin membrane and the state s is chosen by sending out Z_s. When the final label l_h appears, the computation stops, with the garbage of symbols $(w_1 - \{l_I\}) \cup \{l_h\}$ remaining in the skin membrane.

Obviously, omitting the rules from $R_{1,I}$ in the generating P system Π, we obtain the corresponding P system Π_a with $T_a = \{A_i \mid 1 \le i \le k\}$, which is deterministic and accepts $N_a(M)$.

The number of garbage symbols remaining in the skin membrane at the end of a computation in the P system Π constructed in the proof above depends on the deterministic register machine to be simulated, yet using more sophisticated proof techniques, this number can be bounded by a constant (as shown in [43], this constant is at most seven).

Corollary 5.3 $NO_E P_1(sym_3) = N_a DOP_1(sym_3) = NRE$.

5.2.2 Variants of Transition and Halting Modes

In a similar way as exhibited in Chapter 4 for catalytic P systems, also for P systems with antiport and/or symport rules we can restrict ourselves to systems with one membrane by renaming the symbols in the different regions, yet now also taking into account the environment as region 0. In this way, we obtain a characterization of Parikh sets of matrix languages when using the sequential or the asynchronous mode:

Theorem 5.3 For every $X \in \{asyn, sequ\}$,

$$PsOP_*(anti_*, sym_*, X, H) = PsO_{-1}P_1(anti_2^3, X, H)$$
$$= PsO_E P_1(sym_3, X, H) = PsMAT^\lambda,$$
$$NOP_*(anti_*, sym_*, X, H) = NO_{-1}P_1(anti_2^3, X, H)$$
$$= NO_E P_1(sym_3, X, H) = NREG.$$

If we impose the condition of partial halting, then even with the minimally parallel mode *amin* we only get Parikh sets of matrix languages (where the partitioning of the rules for both the partial halting as well as the minimally parallel mode coincides with the assignment of the rules to the membranes):

Theorem 5.4 For every $X \in \{amin, asyn, sequ\}$,

$$PsOP_*(anti_*, sym_*, X, h) = PsO_{-1}P_1(anti_2^3, X, h)$$
$$= PsO_E P_1(sym_3, X, h) = PsMAT^\lambda,$$
$$NOP_*(anti_*, sym_*, X, h) = NO_{-1}P_1(anti_2^3, X, h)$$
$$= NO_E P_1(sym_3, X, h) = NREG.$$

As partial halting and total halting obviously have the same effect on P systems with only one membrane, if the partitioning of the rules for partial halting just has one partition, we immediately infer the following results from Corollary 5.2 and Theorem 5.2:

Corollary 5.4 $PsRE = PsO_{-1}P_1(anti_2^3, max, h) = PsO_E P_1(sym_3, max, h) = Ps_a DOP_1(anti_2^3, max, h) = Ps_a DOP_1(sym_3, max, h)$.

$ZOP_1(anti_*, sym_*, amin, Y) = ZOP_1(anti_*, sym_*, asyn, Y)$ obviously holds for every $Z \in \{N, Ps\}$ and $Y \in \{H, h\}$, therefore at least two membranes are needed to obtain computational completeness with the mode *amin*; hence, the following results are optimal with respect to the number of membranes (compare with Corollary 5.2 and Theorem 5.2 as well as Theorem 5.4):

Theorem 5.5 For every $Z \in \{N, Ps\}$, $ZO_E P_2(anti_2^3, amin, H) = ZO_E P_2(sym_3, amin, H) = ZRE$.

Proof. We only prove the result for the symport rules, because the basic ideas for the antiport rules are quite similar. Thus, let $M = (k+2, H, l_0, l_h, I)$ be a non-deterministic register machine and construct the P system

$\Pi = (O, T, E, [_1 [_2]_2]_1, w_1, w_2, R_1, R_2, 2)$,

$O = \{p, p', p'', p''', \bar{p}, \bar{p}', \bar{p}'', \bar{p}''', \hat{p}, \hat{p}', \hat{p}'', \hat{p}''' \mid p \in H\}$
$\quad \cup \{A_i \mid 1 \le i \le k+2\}$,

$T = \{A_i \mid 1 \le i \le k\}$,

$E = O - \{p', p'', \bar{p}'', \hat{p}, \hat{p}', \hat{p}''', \hat{p}' \mid p \in H\}$,

$w_1 = \{l_0\} \cup \{p', p'', \bar{p}'', \hat{p}, \hat{p}''' \mid p \in H\}$,

$w_2 = \{\bar{p}' \mid p \in H - \{l_h\}\}$,

$R_1 = R_{1,A} \cup R_{1,S}$,

$R_{1,A} = \{(pp', out), (p'A_r q, in), (p'A_r s, in) \mid p : (ADD(r), q, s) \in I\}$,

$R_{1,S} = \{(pp', out), (\bar{p}\bar{p}'p', in), (\bar{p}'\bar{p}''A_r, out), (\bar{p}''\bar{p}'''p''', in), (p'''\hat{p}, out),$
$\quad (\hat{p}p''\bar{p}''', out), (\hat{p}p''q, in), (\hat{p}\hat{p}'\bar{p}', out), (\hat{p}'\bar{p}'', in), (\hat{p}\hat{p}''\bar{p}''', out),$
$\quad (\hat{p}\bar{p}'''s, in) \mid p : (SUB(r), q, s) \in I\}$,

$R_2 = R_{2,A} \cup R_{2,S}$,

$R_{2,A} = \{(A_r, in) \mid 1 \le r \le k\}$,

$R_{2,S} = \{(\bar{p}\hat{p}, in), (\bar{p}\hat{p}', out), (\bar{p}'\bar{p}''', in), (\bar{p}'\bar{p}'', in),$
$\quad (\bar{p}\bar{p}''', out), (\bar{p}\bar{p}'', out) \mid p : (SUB(r), q, s) \in I\}$,

An add-instruction $p : (\text{ADD}(r), q, s)$ is simulated by sending out the label of the instruction p together with p' which returns with A_r as well as the label of the next instruction q or s to be simulated. In the case of a terminal register r, $1 \leq r \leq k$, A_r has to enter membrane 2 in the succeeding step by using (A_r, in) from $R_{2,A}$.

When simulating a subtract-instruction $p : (\text{SUB}(r), q, s)$, p' returns with $\bar{p}$ and $\bar{p}'$. Whereas $\bar{p}$ enters membrane 2 together with $\bar{p}$, $\bar{p}'$ gets the chance to take one copy of A_r out of region 1 using the rule $(\bar{p}'\bar{p}''A_r, out)$. Depending on whether this rule had to be applied or not, the simulation proceeds until finally the label of the corresponding instruction to be simulated next is brought in together with p'' or $\bar{p}'''$, respectively. We should like to mention that all the symbols from $O - E$ used during the simulation finally have returned to their original locations.

When the computation of the register machine stops, the label l_h appears and the computation in Π stops, too. The result of the computation is taken from the second membrane, which still contains the symbols from the initial multiset w_2. In order to avoid this garbage and to have the results in an elementary membrane, we would have to add a third membrane inside the second membrane and let all terminal symbols pass through this membrane, thereby using a set of rules for the third membrane being the same as $R_{2,A}$. Yet as long as we do not care about garbage symbols in the output membrane, two membranes as in the P system constructed above are sufficient.

5.3 Tissue P Systems with Antiport and Symport Rules

In tissue P systems, objects are communicated through membrane channels between cells as well as between the cells and the environment. In contrast to tree-like P systems, in tissue P systems every cell may have connection with the environment and every cell may have connection with any other cell.

Definition 5.2 *A tissue P system (of degree $m \geq 1$) with antiport and symport rules is a construct*

$$\Pi = (m, O, E, w_1, \ldots, w_m, ch, (R_{(i,j)})_{(i,j) \in ch}, i_0),$$

where m is the number of cells, O is the alphabet of objects, E is the set of objects occurring in an unbounded number in the environment, $w_1, \ldots, w_m$ are strings over O representing the initial multiset of objects present in the cells of the system (it is assumed that the m cells are labeled with $1, 2, \ldots, m$), $ch \subseteq \{(i, j) \mid i, j \in \{0, 1, 2, \ldots, m\}, (i, j) \neq (0, 0)\}$ is the set of links (or channels or synapses) between cells (0 indicates the environment), $R_{(i,j)}$ is a finite set of antiport and/or symport rules associated with the channel $(i, j) \in ch$, and i_0 is the output cell.

A rule in $R_{(i,j)}$ for the ordered pair (i, j) of cells is written as x/y $(xy \neq \lambda)$ and its application means moving the objects specified by x from cell i (from the environment, if $i = 0$; in that sense, we sometimes shall consider the environment as a cell, too) to cell j, at the same time moving the objects specified by y in the opposite direction. The rules with one of x, y being empty are *symport rules*, the others are *antiport rules*, but we do not always explicitly consider this distinction here. For short, we shall also speak of a *tissue P system* only when dealing with a *tissue P system with antiport and symport rules* as defined above.

The computation starts with the multisets specified by $w_1, \ldots, w_m$ in the m cells; in each time unit, a rule is used on each channel for which a rule can be used (if no rule is applicable for a channel, then no object passes over it). Therefore, the use of rules is sequential at the level of each channel, but it is parallel at the level of the system: all channels which can use a rule must do it (the system is synchronously evolving). The computation is successful if and only if it halts.

The result of a halting computation is the number described by the multiplicity of objects present in cell i_0 in the halting configuration. The set of all (vectors of) natural numbers computed in this way by the system Π is denoted by $N(\Pi)$ $(Ps(\Pi))$. The family of sets $N(\Pi)$ $(Ps(\Pi))$ of (vectors of) natural numbers computed as above by systems with at most n symbols and m cells is denoted by $NO_n t' P_m$ (resp. $Ps\, O_n t' P_m$). When any of the parameters m, n is not bounded, it is replaced by $*$.

In [27], only channels (i, j) with $i \neq j$ were allowed, and, moreover, for any i, j only one channel out of $\{(i, j), (j, i)\}$ was allowed, i.e. between two cells (or one cell and the environment) only one channel could be taken (as we shall see in the following, this technical detail may influence considerably the computational power). The family of sets $N(\Pi)$ $(Ps(\Pi))$ of (vectors of) natural numbers computed as above by such tissue P systems with at most n symbols and at most m cells is denoted by $NO_n t P_m$ (resp. $Ps\, O_n t P_m$).

Example 5.4 As tissue P systems of the form $(1, \{b\}, \{b\}, b^m, \{(1, 0)\}, R_{(1,0)}, 1)$ behave like P systems working in the sequential mode, from Example 5.1 we immediately infer $NO_1 t P_1 = NFIN$.

Now let $\Pi = (1, \{b\}, \{b\}, b^m, \{(0, 1), (1, 0)\}, R_{(0,1)}, R_{(1,0)}, 1)$. If $R_{(0,1)} \cup R_{(1,0)}$ is empty, then obviously $N(\Pi) = \{m\}$. Otherwise, let

$$n = \min\{j \mid b^j/b^i \in R_{(1,0)} \text{ or } b^i/b^j \in R_{(0,1)} \text{ for some } i\}.$$

Then a computation in Π can never halt as long as there are at least n objects in the cell. Therefore, $N(\Pi) \subseteq \{j \mid j < n\}$; hence, $N(\Pi) \in NFIN$.

As an immediate consequence of Example 5.4 we get the following results:

Theorem 5.6 $NO_1 t P_1 = NO_1 t' P_1 = NFIN$.

If we use two cells, we can already generate infinite sets (compare with Example 5.2):

Example 5.5 Consider the tissue P system

$$\Pi = (2, \{b\}, \{b\}, \lambda, b, \{(2,0), (2,1)\}, R_{(2,0)}, R_{(2,1)}, 1)$$

with $R_{(2,0)} = \{b^1/b^2\}$ and $R_{(2,1)} = \{b^1/\lambda, b^2/\lambda\}$. The second cell can provide the first one with one more object b in every step as long as its contents is not taken as a whole to cell 1 by one of the rules from $R_{(2,1)}$. Obviously, $N(\Pi) = \mathbf{N} - \{0\}$.

If we allow at least two objects in one cell, then all regular sets can be generated:

Example 5.6 Let M be a regular set of natural numbers, i.e. there exist finite sets of natural numbers M_0, M_1 and a natural number k such that $M = M_0 \cup \{i + jk \mid i \in M_1, j \in \mathbf{N}\}$. Then the tissue P system $\Pi = (1, \{b, p\}, \{b\}, pp, \{(1,0)\}, R_{(1,0)}, 1)$ with $R_{(1,0)} = \{pp/b^i \mid i \in M_0\} \cup \{pp/pb, pb/pb^{k+1}\} \cup \{pb/b^i \mid i \in M_1\}$ generates M as the number of symbols b in cell 1 in halting computations: initially, there are no objects b in the cell, so the system "chooses" between generating an element of M_0 in one step or exchanging pp by pb. In the latter case, there remains only one copy of p in the system. After an arbitrary number j of applications of the rule pb/pb^{k+1} a rule exchanging pb by b^i for some $i \in M_1$ is eventually applied, generating $jk + i$ symbols b. Hence, $N(\Pi) = M_0 \cup \{i + jk \mid i \in M_1, j \in \mathbf{N}\} = M$.

As shown in [5], with one channel between a single cell and the environment, we exactly get the regular sets:

Theorem 5.7 $NREG = NO_n t P_1$ for all $n \geq 2$.

With two channels between a single cell and the environment, we can already obtain computational completeness with five objects:

Theorem 5.8 $NRE = NO_n t' P_1$ for all $n \geq 5$.

Proof. Let us consider a register machine $M = (3, H, l_0, l_h, I)$; the main ideas for the construction of the tissue P system

$$\Pi = (1, \{a_1, a_2, a_3, p, q\}, \{a_1, a_2, a_3, p, q\}, w_1, \{(0, 1), (1, 0)\}, R_{(0,1)}, R_{(1,0)}, 1)$$

generating $N(M)$ are the following:

The objects a_1, a_2, and a_3 represent the three registers; the objects p and q are needed for encoding the instructions of M; q also has the function of a trap object, i.e. in case of the wrong choice for a rule to be applied we take in so many objects q that we can never get rid of them again and therefore get "trapped" in an infinite loop.

We use a combination of small numbers of objects p and q to be able to selectively check for the appearance of a_2 and a_3, i.e. for testing register 2 and register 3 for zero we take pq^2 and p^2q, respectively. We assume the labels from H to be $3i + 1$ for $0 \leq i \leq t - 1$, as well as $l_0 = 1$ and $l_h = 3(t - 1) + 1$, for some $t > 1$; these labels are encoded as suitable numbers of the object p by the function $c : \mathbf{N} \longrightarrow \mathbf{N}$ with $c(x) = 5x + 15t$ for $x \geq 0$. With $l_0 = 1$ we therefore obtain $c(l_0) = 15t + 5$ and $w_1 = p^{c(l_0)} = p^{15t+5}$.

The rules in the two channels between the cell and the environment are now defined as follows:

$$R_{(1,0)} = \{p^{c(l_h)}/\lambda\} \cup \{p^{c(l_1)}/p^{c(l_2)}a_i, \ p^{c(l_1)}/p^{c(l_3)}a_i \ | $$
$$l_1 : (\text{ADD}(i), l_2, l_3) \in I, 1 \leq i \leq 3\}$$
$$\cup \ \{p^{c(l_1)}a_k/p^{c(l_2)}, \ p^{c(l_1)}/p^{c(l_1+1)}p^{k-1}q^{4-k}, \ p^{c(l_1+1)}/p^{c(l_1+2)},$$
$$p^{c(l_1+2)}p^{k-1}q^{4-k}/p^{c(l_3)} \ | \ l_1 : (\text{SUB}(k), l_2, l_3) \in I, 2 \leq k \leq 3\},$$
$$R_{(0,1)} = \{q^6/q^3, q^6/p^3, q^6/pq^2a_2, q^6/p^2qa_3\}.$$

Throughout the whole computation in Π, the application of rules is directed by the code $p^{c(l)}$ for some $l \leq l_h$, and the corresponding rules in $R_{(1,0)}$ should guarantee the correct sequence of encoded rules. The rules in $R_{(0,1)}$ are only applied if something goes wrong, i.e. either superfluous objects p in case of a wrong choice of the rule from $R_{(1,0)}$ cause the application of the rule q^6/p^3 from $R_{(0,1)}$ or the failure of the zero test causes the application of the rule q^6/pq^2a_2 or q^6/p^2qa_3 in case a_2 or a_3 is present. The rule q^6/q^3 in $R_{(0,1)}$ acts as a trap rule, i.e. we inevitably enter an infinite loop with this rule, because at most two objects q can leave through channel $(1, 0)$ using a rule $p^{c(l_1+2)}p^{k-1}q^{4-k}/p^{c(l_3)}$.

As for the halting label l_h we take the rule $p^{c(l_h)}/\lambda$ from $R_{(1,0)}$, the work of Π stops exactly when the work of M stops (provided none of the trap rules from $R_{(0,1)}$ has been applied due to a wrong non-deterministic choice during the computation), and moreover, the final configuration of Π represents the final contents of the registers in M.

5.4 Tuning the Parameters

In this section we investigate the trade-off between several parameters in (tissue) P systems with antiport and symport rules, for example, between the number of membranes (cells) and the number of objects needed to obtain computational completeness.

5.4.1 Minimal Antiport and Minimal Symport

When using antiport or symport rules, we needed rules of size 3 to obtain computational completeness in only one membrane (see Theorem 5.1 and Corollary 5.3). These results are already optimal for systems with only one membrane; the situation changes completely if rules of size 2, called *minimal antiport* or *minimal symport* rules, are considered—in one membrane or cell, we only get finite sets:

Theorem 5.9 $NO[t]P_1(anti_1, sym_1) \cup N[t]OP_1(sym_2) \subseteq NFIN$.

Yet with two membranes or cells, in order to get computational completeness, we may already restrict ourselves to minimal symport and/or minimal antiport rules:

Theorem 5.10 $NRE = NO[t]P_2(anti_1, sym_1) = NO[t]P_2(sym_2)$.

The proof significantly differs if tissue or tree-like P systems are considered. In the tissue case, the proof is based on the possibility to reach a membrane from another one by two roads—directly or via the environment. In this way, a temporal de-synchronization of pairs of objects is obtained and it can be used to simulate the instructions of a register machine. For the tree-like case, the result in the previous theorem only holds true if we do not require the output membrane to be elementary, otherwise it were inevitable that some object remains in the output membrane at the end of a successful computation.

Moreover, in the tissue case, we have a deterministic construction for the acceptance of recursively enumerable sets. In the tree-like case it is not possible to use a similar technique, because only the root is connected to the environment, which considerably restricts the accepting power of deterministic P systems:

Theorem 5.11 *For any deterministic P system with rules of type sym_2 and $anti_1$, the number of objects present in the initial configuration of the system cannot be increased during halting computations.*

However, if non-deterministic systems are considered, then it is possible to reach computational completeness for the accepting case with two membranes: an initial pumping phase is performed to introduce a sufficient number of working objects needed to carry out the computation (a non-deterministic guess for the number of working objects is done). After that, the system simulates a register machine thereby consuming the number of working objects. In sum, we obtain the following results:

Theorem 5.12 *For $Z \in \{N, Ps\}$,*

$$ZRE = ZO[t]P_2(anti_1, sym_1) = ZO[t]P_2(sym_2) = Z_a OP_2(anti_1, sym_1)$$
$$= Z_a OP_2(sym_2) = Z_a DOtP_2(anti_1, sym_1) = Z_a DOtP_2(sym_2).$$

5.4.2 Generalized Minimal Communication

We can generalize the idea of minimal antiport and symport and introduce the concept of *minimal interaction tissue P systems* (or generalized communicating P systems). These are tissue P systems where at most two objects may interact, i.e. one object is moved with respect to another one. Such interactions can be described by rules of the form $(a, i)(b, j) \to (a, k)(b, l)$, which indicate that if symbol a is present in membrane i and symbol b is present in membrane j, then a will move to membrane k and b will move to membrane l. We may impose several restrictions on these interaction rules, namely by superposing several cells. Some of these restrictions directly correspond to antiport or symport rules of size 2.

Below we define all possible restrictions (modulo symmetry): let O be an alphabet and let $(a, i)(b, j) \to (a, k)(b, l)$ be an interaction rule with $a, b \in O$, $i, j, k, l \geq 0$. Then we distinguish the following cases:

1. $i = j = k \neq l$: the *conditional-uniport-out rule* sends b to membrane l provided that a and b are in membrane i.
2. $i = k = l \neq j$: the *conditional-uniport-in rule* brings b to membrane i provided that a is in that membrane.
3. $i = j \neq k = l$: the *symport2 rule* corresponds to the minimal symport rule, i.e. a and b move together from membrane i to k.
4. $i = l \neq j = k$: the *antiport1 rule* corresponds to the minimal antiport rule, i.e. a and b are exchanged in membranes i and k.
5. $i = k \neq j \neq l$: the *presence-move rule* moves the symbol b from membrane j to l, provided that there is a symbol a in membrane i.
6. $i = j \neq k \neq l$: the *separation rule* sends a and b from membrane i to membranes k and l, respectively.
7. $k = l \neq i \neq j$: the *joining rule* brings a and b together to membrane i.
8. $i = l \neq j \neq k$ or $i \neq j = k \neq l$: the *chain rule* moves a from membrane i to membrane k while b is moved from membrane j to membrane i, i.e. where a has previously been.
9. $i \neq j \neq k \neq l$: the *parallel-shift rule* moves a and b in independent membranes.

A minimal interaction tissue P system may have rules of several types as defined above. With respect to the computational power of such systems we immediately see that when only antiport1 rules or only symport2 rules are used, the number of objects in the system cannot be increased, hence, such systems can generate only finite sets of (vectors of) natural numbers. However, if we allow uniport rules (i.e. rules of the form $(a, i) \to (a, k)$ specifying that, whenever an object a is present in cell i, this may be moved to cell k), then minimal interaction tissue P systems with symport2 and uniport rules or with antiport1 and uniport rules become tissue P systems with minimal symport or minimal symport and antiport, respectively.

By combining conditional-uniport-in rules and conditional-uniport-out rules, computational completeness can be achieved by simulating a register machine. A register machine may be also simulated by using only the parallel-shift rule. In all other cases, when only one of the types of rules defined above is considered, it is not even clear whether infinite sets of natural numbers can be generated.

Another interesting problem is to investigate how an interaction rule may be simulated by some restricted variants. Such a study may lead to a formulation of sufficient conditions on how combinations of variants of rules $(a, i)(b, j) \to (a, k)(b, l)$ may guarantee that the system can be realized by using only specific restricted variants of rules in an equivalent minimal interaction tissue P system. After that, a system satisfying sufficient conditions of several restrictions may be automatically rewritten in terms of any corresponding restricted variants. A list of such results can be found in [47].

5.4.3 Number of Symbols

Not taking care of the complexity of the rules, yet instead regarding the number of objects, the main results for P systems with antiport (and symport) rules can be summarized in the following table:

In Table 5.1, the class of P systems indicated by A generates exactly *NFIN*, the class indicated by B generates at least *NREG*, in the case of C at least *NREG* can be generated and at least *NFIN* can be accepted, while a class indicated by a number d can simulate any d-register machine. A box around a number indicates a known computational completeness bound, (U) indicates a known unpredictability bound, and a number in boldface shows the diagonal where $m(s - 2)$ equals $(m - 1)(s - 1)$. The most interesting questions still remaining open are to characterize the families generated or accepted by P systems with only one symbol.

Table 5.1 Families $NO_m P_n$.

objects	membranes					
	1	2	3	4	5	m
1	A	B	B	B	B	B
2	C	1	2 (U)	[3]	4	$m - 1$
3	1	2 (U)	[4]	6	8	$2m - 2$
4	2 (U)	[4]	6	9	12	$3m - 3$
5	[3]	6	9	12	16	$4m - 4$
6	4	8	12	16	20	$5m - 5$
s	$s - 2$	$2s - 4$	$3s - 6$	$4s - 8$	$5s - 10$	$\max\{m(s - 2), (m - 1)(s - 1)\}$

Table 5.2 Families $NO_m t P_n$.

objects								
4	NREG	NRE	NRE	NRE	NRE	NRE	NRE	
3	NREG	A	NRE	NRE	NRE	NRE	NRE	
2	NREG	A	NRE	NRE	NRE	NRE	NRE	
1	NFIN	B	A	A	A	A	NRE	
	1	2	3	4	5	6	7	cells

The results for tissue P systems with only one channel between two cells and between a cell and the environment are listed in Table 5.2, the results for tissue P systems with two channels between cells as well as a cell and the environment are depicted in Table 5.3. In both tables, the entries NFIN, NREG, and NRE indicate the equality with the corresponding family $NO_m t P_n$ and $NO_m t' P_n$, respectively; A indicates that the corresponding family includes at least NREG, and B indicates that the corresponding family can generate more than NFIN.

The main open question concerns a characterization of the sets of natural numbers in $NO_2 t P_2$ and $NO_3 t P_2$. Further, it would be interesting to find the minimal number l such that $NO_1 t P_l$ contains all recursively enumerable sets of natural numbers, whereas the families $NO_1 t P_j$ with $j < l$ do not fulfill this condition. Finally, it remains to find characterizations of the sets of natural numbers in those families $NO_1 t P_j$ that do not contain all recursively enumerable sets of natural numbers.

The most interesting open problems for the families $NO_n t' P_m$ are to find the minimal number k as well as the minimal number l such that $NO_k t' P_1$ and $NO_1 t' P_l$, respectively, contain all recursively enumerable sets of natural numbers, whereas the families $NO_i t' P_1$ and $NO_1 t' P_j$ with $i < k$ and $j < l$, respectively, do not fulfill this condition. Moreover, it remains to find characterizations of the sets of natural numbers in those families $NO_n t' P_m$ that do not contain all recursively enumerable sets of natural numbers.

Table 5.3 Families $NO_m t' P_n$.

objects							
5	NRE	NRE	NRE	NRE	NRE	NRE	
4	A	NRE	NRE	NRE	NRE	NRE	
3	A	NRE	NRE	NRE	NRE	NRE	
2	A	A	NRE	NRE	NRE	NRE	
1	NFIN	A	A	A	A	NRE	
	1	2	3	4	5	6	cells

5.4.4 Number of Rules

Another complexity parameter investigated in the literature is the number of rules in a universal P system with antiport and symport rules. Such a bound can be obtained if we simulate a universal device for which a bound on the number of rules is already known. Since P systems with antiport and symport rules can easily simulate register machines, it is natural to consider simulations of register machines having a small number of instructions. An example of such a machine is the register machine U_{32} described in [36], which has 22 instructions (9 increment and 13 decrement instructions). The table below summarizes the best results known on this topic, showing the trade-off between the number of antiport rules and their size:

Table 5.4 P systems with small numbers of antiport rules.

number of rules	73	56	47	43	30	23
size of rules	3	5	6	7	11	19

5.5 A GENERAL MODEL OF P SYSTEMS WITH PERMITTING AND FORBIDDEN CONTEXTS

We now introduce a general model of P systems with permitting and forbidden contexts covering the most important models of communication P systems as well as evolution-communication P systems.

Definition 5.3 *A P system (of degree $d \geq 1$) with permitting and forbidden contexts is a construct*

$$\Pi = (O, T, E, \mu, w_1, \ldots, w_d, R, i_0)$$

where $O, T, E, \mu, w_1, \ldots, w_d, i_0$ are as in Definition 5.1, and R is a finite set of membrane rules with permitting and forbidden contexts over O associated with the membranes $1, 2, \ldots, d$ of μ; an evolution rule associated with membrane i is of the form

$$u^{w,\neg s}[_i^{z,\neg t} x \to v[_i y$$

where $w, z \in O^*/ s, t \in O^*$ are the permitting contexts (promoters)/ forbidden contexts (inhibitors) *in the region outside membrane i and inside membrane i, respectively, u outside membrane i is replaced by v and x inside membrane i is replaced by y.*

The rule $u^{w, \neg s} [_i^{z, \neg t} x \to v [_i y$ from R is applicable if and only if the multiset uw occurs in the region outside membrane i and s does not appear there as well as the multiset xz occurs in the region inside membrane i and t does not appear there. The application of this rule results in subtracting the multiset identified by u from the multiset in the region outside membrane i and adding v instead as well as subtracting x and adding y in the region inside membrane i. The definitions for the initial configuration of the system, the variants of transition modes and halting modes as well as for the families of sets of (vectors of) natural numbers generated or accepted by P systems with permitting/forbidden contexts are similar as for P systems with communication rules, yet now allowing more general types of rules.

5.5.1 General Results

Looking carefully into the definitions of the transition modes as well as the halting modes considered so far in this handbook, we immediately infer the following general results, as established in [24]:

Theorem 5.13 *Any variant of P systems yielding a family of sets of (vectors of) natural numbers F when working in the transition mode $X \in \{max, amin, asyn, sequ\}$, with only one set of rules assigned to a single membrane and stopping with total halting yields the same family F when working in the transition mode X with only one set of rules assigned to a single membrane when stopping with partial halting, too.*

Theorem 5.14 *Any variant of P systems yielding a family of sets of (vectors of) natural numbers F when working in the transition mode $X \in \{asyn, sequ\}$, with only one set of rules assigned to a single membrane and stopping with total or partial halting, respectively, yields the same family F when working in the minimally parallel mode amin and stopping with the corresponding halting mode, too.*

For any P system using rules of type α without forbidden contexts, with any transition mode $X \in \{amin, asyn, sequ\}$, and partial halting, we only get Parikh sets of matrix languages (regular sets of natural numbers):

Theorem 5.15 *For every $X \in \{amin, asyn, sequ\}$ and any type α of rules without forbidden contexts, $PsOP_*(\alpha, X, h) \subseteq PsMAT^\lambda$ and $NOP_*(\alpha, X, h) \subseteq NREG$.*

5.5.2 Results for P Systems with Specific Types of Rules

Based on the general model defined above, we now consider several variants of P systems covered by this general model, as there are P systems with antiport and/or symport rules, which have already been thoroughly investigated in this chapter, P systems with conditional uniport rules, P systems with boundary rules, which closely resemble P systems without forbidden contexts in our general model, and evolution-communication P systems.

5.5.2.1 P Systems with Antiport and Symport Rules

An *antiport rule* $(u, out; v, in) \in R_i$ in a P system is written as $v [_i u \to u [_i v$ in the general model defined above; the *symport rules* $(x, out) \in R_i$ and $(x, in) \in R_i$ are written as $[_i x \to x[_i$ and $x[_i \to [_i x$, respectively.

5.5.2.2 P Systems with Conditional Uniport Rules

A *conditional uniport rule* is a rule of one of the forms $ab[_i \to b[_i a, [_i ab \to a[_i b, a[_i b \to [_i ab, b[_i a \to ab[_i,$ with $a, b \in O$; in every case, the object a is moved across membrane i, whereas the object b stays where it is. Using only rules of that kind induces the type $uni_{1,1}$ which corresponds to the first two types from Subsection 5.4.2. Conditional uniport rules were introduced in [46] for the case of tissue P systems, showing computational completeness with maximal parallelism and total halting.

Using only conditional uniport rules of type $uni_{1,1}$, we again obtain computational completeness, even with the minimally parallel mode, together with total halting, whereas, as a direct consequence of Theorem 5.15, with partial halting we only get Parikh sets of matrix languages (regular sets of natural numbers) with the minimally parallel mode *amin*.

Theorem 5.16 *For every* $X \in \{amin, max\}$, $PsOP_*(uni_{1,1}, X, H) = PsRE$.

5.5.2.3 P Systems with Boundary Rules

P systems with boundary rules were first defined in [15], where evolution rules as well as communication rules with permitting contexts were considered. Non-cooperative evolution rules are of the form $b \to v$ with $b \in O$ and $v \in O^*$; a rule $b \to v \in R_i$ corresponds to $[_i b \to [_i v$ in our general notation. The communication rules considered in [15] were symport or antiport rules with permitting contexts, i.e. of the form $u^w[_i^z x \to x[_i u$. In [16], boundary rules of the form $u[_i x \to v[_i y$ were considered, i.e. rewriting on both sides of the membrane was allowed.

5.5.2.4 Evolution-Communication P Systems

In evolution-communication P systems as introduced in [17], we allow non-cooperative evolution rules (type *ncoo*) as well as antiport and symport rules. For evolution-communication P systems, the following results hold:

Theorem 5.17 *For* $X \in \{amin, max\}$, $PsRE = PsOP_2(ncoo, anti_1, sym_1, X, H) = PsOP_2(ncoo, sym_2, max, H) = Ps_aDOP_3(ncoo, anti_1, sym_1, X, H) = Ps_aDOP_3(ncoo, sym_2, max, H)$.

5.6 Generation of Languages

As already described in Chapter 4, Section 4.7, P systems can also be used as language generators: during a successful computation, all the terminal objects sent out through the skin membrane are taken as the objects forming a string in just that sequence the objects are sent out. If in one step more than one object is sent out—either by only one copy of a rule or by several rules from the applied multiset of rules—then we take all possible sequences of objects that are sent out in one transition step as possible substrings to be concatenated with the string already generated by the preceding transition steps, as we take all possible substrings from each transition step. Thus, not just one string may be the result of a successful computation. The language generated by a P system Π in that way is denoted by $L(\Pi)$. The family of languages generated by P systems with at most m membranes using communication (antiport, symport) rules of type a, using the transition mode $X \in \{max, amin, asyn, sequ\}$, and applying the halting condition $Y \in \{H, h\}$, are denoted by $LOP_m(a, X, Y)$.

For showing the computational completeness of P systems with communication rules as language generating devices, we may use the same techniques as elaborated in the proof of Theorem 4.17 in Chapter 4; based on the results established in Section 5.2 for how P systems with antiport or symport rules are able to simulate the actions of a register machine, we immediately infer the following results (which are optimal with respect to the number of membranes and the corresponding size of the communication rules):

Theorem 5.18 *For every* $Y \in \{H, h\}$, $RE = LOP_1(anti_2^3, max, Y) = LOP_1(sym_3, max, Y)$.

Moreover, for the minimally parallel mode we also get a characterization of RE with the minimal number of membranes being two (see Theorem 5.5):

Theorem 5.19 $RE = LOP_2(anti_2^3, amin, H) = LOP_2(sym_3, amin, H)$.

For evolution-communication P systems, in [2] the following results were shown; for both results of Theorem 5.20, it suffices to have cooperation with one special object (called *proton*) that only participates in communication rules; using four protons even yields time-free systems.

Theorem 5.20 $RE = LOP_2(ncoo, sym_2) = LOP_2(ncoo, anti_1, sym_1)$.

5.6.1 Traces

Another idea to generate strings is to follow the traces of a specific object during a successful computation, i.e. for a sequence of configurations representing a halting computation we consider the sequence of regions where this specific object ("traveler") is found in these configurations. If we assign letters or λ to the membrane labels (i.e. we specify a weak coding), the sequences of symbols obtained in this way observed during halting computations yield a string language. For languages over a one-letter alphabet $\{a_1\}$, we get similar results as in Theorem 5.18 by assigning a_1 to the label of the skin membrane. In general, for languages over the alphabet $V = \{a_i \mid 1 \leq i \leq n\}$ we need P systems with the membrane structure $[_1 [_2]_2 \ldots [_{n+1}]_{n+1}]_1$ and the weak coding $c_n : \{i \mid 0 \leq i \leq n+1\} \to \{a_i \mid 1 \leq i \leq n\}$ with $c_n(0) = c_n(1) = \lambda$ and $c_n(i+1) = a_i$ for $1 \leq i \leq n$. It is easy to see that taking such P systems with $n+1$ membranes we can generate every recursively enumerable language over the n-letter alphabet V with respect to the weak coding c_n by using antiport or symport rules of size 3 in the maximally or even in the minimally parallel mode.

5.7 CONCLUSION

The results established in this chapter have elucidated the computational power of pure communication. P systems with antiport and/or symport rules are computationally complete with rules of size 3 applied in the maximally parallel mode in only one membrane or applied in the minimally parallel mode in two membranes. With rules applied in the sequential and the asynchronous mode and total halting as well as with rules applied in the sequential, the asynchronous or the minimally parallel mode *amin* and partial halting we only get Parikh sets of matrix languages.

Tissue P systems with communication rules use one rule (if some rule is applicable) in every channel, but all channels have to work in parallel in each transition step. Computational completeness can already be obtained with one cell and two channels between the single cell and the environment as well as with two cells and only one channel between the cells and between the cells and the environment. Moreover, only a small number of membranes (cells) *and* objects are needed to obtain computational completeness with (tissue) P systems with antiport and symport rules. Some open questions are described in Section 5.4 concerning these features of descriptional complexity.

5.8 Bibliographic Notes

P systems with antiport and symport rules were introduced in [39], where the first results concerning computational completeness were established: $NRE = NOP_2(anti_2, sym_2) = NOP_5(anti_1, sym_2)$.

The computational completeness with one membrane was independently shown in [30] and [25] as well as in [21], where this result was obtained based on a more general model with channels through membranes. A deterministic simulation of register machines by P systems with antiport rules of weight 2 was first established in [26]. Tissue P systems were introduced in [37], and tissue-like P systems with channel states were investigated in [27]. There is another definition of tissue P systems that uses parallel channels, i.e. the maximal parallelism acts on the level of the whole system; such systems are a direct generalization of P systems and were considered in several places, we only mention [40], [12], [29], and [46].

P systems with minimal symport and antiport rules were first investigated in [14], where nine membranes were used to achieve computational completeness. This number was progressively decreased and finally established at two membranes in [11]. A deterministic proof using three cells for the tissue case was first presented in [45] and improved to two cells in [9].

The descriptional complexity of P systems with respect to the number of objects was first considered in [41], where three objects were shown to be sufficient for obtaining computational completeness in four membranes, then in [4] five objects in one membrane were shown to be enough, and for tissue P systems even with only one object computational completeness was established in [22]. The main results listed in Subsection 5.4.3 were established in [6] for P systems and in [5] for tissue P systems. Based on the universal register machine U_{32} in [36], universal P systems with a small number of antiport rules were described in [19], [23], and [13].

Example 5.3 first was published in [34]. $NOtP_1(anti_1, sym_1) \subseteq NFIN$ and $NtOP_1(sym_2) \subseteq NFIN$ from Theorem 5.9 was shown in [12] and [31], respectively. Evolution-communication P systems were introduced in [17] and investigated especially in [1].

In [10], the generation of languages by P systems with minimal antiport and symport rules in two membranes was investigated. Traces were introduced in [35], the optimal results described in Subsection 5.6.1 were established in [31].

In [44], another specific variant of communication P systems on a tree structure was considered: in terms of Section 5.4.2 it uses uniport, conditional-uniport-out and symport2 rules enriched with a possibility to bring a symbol from the environment. These systems are shown to be computationally complete, with respect to acceptance even in their deterministic variant.

The minimally parallel mode was introduced in [18], the special variant of minimal parallelism *amin* was considered in [29], where a formal framework for

P systems was developed. The universality proofs given in Subsection 5.2.2 and in Section 5.5 hold for the minimally parallel mode *min* as defined in Chapter 1 and for the more general variant of minimal parallelism as defined in [29].

Partial halting was investigated in [7] and [24]; the results shown in Subsection 5.5.1 were established in [8].

In [43], the history as well as open problems for the computational power of P systems with antiport and symport rules were described. The PhD thesis [3] investigates many variants of communication P systems.

Finally we have to mention that in this chapter we have restricted ourselves to (tissue) P systems with antiport and symport rules as generating and accepting devices. The automata variant of such P systems was introduced in [20] and, for example, also considered in [38]; a thorough discussion of P automata is carried out in Chapter 6.

REFERENCES

[1] A. ALHAZOV: On determinism of evolution-communication P systems. *J. Univ. Computer Sc.*, 10 (2004), 502–508.

[2] A. ALHAZOV: Number of protons/bi-stable catalysts and membranes in P systems. Time-freeness. In [28], 79–95.

[3] A. ALHAZOV: *Communication in Membrane Systems with Symbol Objects*. PhD Thesis, Tarragona, Spain, 2006.

[4] A. ALHAZOV, R. FREUND: P systems with one membrane and symport/antiport rules of five symbols are computationally complete. *Proc. of BWMC05* (M.A. Gutiérrez-Naranjo et al., eds.), Sevilla, 2005, 19–28.

[5] A. ALHAZOV, R. FREUND, M. OSWALD: Tissue P systems with antiport rules and a small number of symbols and cells. *Proc. of DLT 2005* (C. de Felice et al., eds.) Palermo, LNCS 3572, Springer, 2005, 100–111.

[6] A. ALHAZOV, R. FREUND, M. OSWALD: Symbol/membrane complexity of P systems with symport/antiport rules. In [28], 97–114.

[7] A. ALHAZOV, R. FREUND, M. OSWALD, S. VERLAN: Partial versus total halting in P systems. In [32], 1–20.

[8] A. ALHAZOV, R. FREUND, M. OSWALD, S. VERLAN: Partial halting in P systems using membrane rules with permitting contexts. *Proc. of MCU 2007* (J.O. Durand-Lose, M. Margenstern, eds.), Orléans, France, LNCS 4664, Springer, 2007, 110–121.

[9] A. ALHAZOV, R. FREUND, YU. ROGOZHIN: Some optimal results on symport/antiport P systems with minimal cooperation. In [32], 23–36.

[10] A. ALHAZOV, YU. ROGOZHIN: Generating languages by P systems with minimal symport/antiport. *The Computer Science Journal of Moldova*, 14 (2006), 299–323.

[11] A. ALHAZOV, YU. ROGOZHIN: Towards a characterization of P systems with minimal symport/antiport and two membranes. In [33], 135–153.

[12] A. ALHAZOV, YU. ROGOZHIN, S. VERLAN: Symport/antiport tissue P systems with minimal cooperation. In [32], 37–52.

[13] A. ALHAZOV, S. VERLAN: Minimization strategies for maximally parallel multiset rewriting systems. *TUCS Techical Report* 862, Turku, 2008.

[14] F. BERNARDINI, M. GHEORGHE: On the power of minimal symport/antiport. *Preproc. of the 3rd Workshop on Membrane Computing* (A. Alhazov, C. Martín-Vide, Gh. Păun, eds), Tarragona, 2003, 72–83.

[15] F. BERNARDINI, V. MANCA: P systems with boundary rules. In [42], 107–118.

[16] F. BERNARDINI, F.J. ROMERO-CAMPERO, M. GHEORGHE, M.J. PÉREZ-JIMÉNEZ, M. MARGENSTERN, S. VERLAN, N. KRASNOGOR: On P systems with bounded parallelism. *Proc. of the 7th Int. Symp. SYNASC*, Timişoara, Romania, 2005, IEEE Computer Society, Washington, USA, 399–406.

[17] M. CAVALIERE: Evolution-communication P systems. In [42], 134–145.

[18] G. CIOBANU, L. PAN, GH. PĂUN, M.J. PÉREZ-JIMÉNEZ: P systems with minimal parallelism. *Theoretical Computer Sci.*, 378 (2007), 117–130.

[19] E. CSUHAJ-VARJÚ, M. MARGENSTERN, GY. VASZIL, S. VERLAN: Small computationally complete symport/antiport P systems. *Theoretical Computer Sci.*, 372 (2007), 152–164.

[20] E. CSUHAJ-VARJÚ, GY. VASZIL: P automata or purely communicating accepting P systems. *Proc. of WMC-CdeA 2002* (Gh. Păun et al., eds.), LNCS 2597, Springer, 2003, 219–233.

[21] R. FREUND, M. OSWALD: P Systems with activated/prohibited membrane channels. In [42], 261–268.

[22] R. FREUND, M. OSWALD: Tissue P systems with symport/antiport rules of one symbol are computationally universal. In [32], 187–200.

[23] R. FREUND, M. OSWALD: Small universal antiport P systems and universal multiset grammars. *Proc. of BWMC06* (C. Graciani-Díaz et al., eds.), Vol. II, Fénix Editora, Sevilla, 2006, 51–64.

[24] R. FREUND, M. OSWALD: Partial halting in P systems. *Intern. J. Foundations of Computer Sci.*, 18 (2007), 1215–1225.

[25] R. FREUND, A. PĂUN: Membrane systems with symport/antiport: universality results. In [42], 270–287.

[26] R. FREUND, GH. PĂUN: On deterministic P systems. Manuscript, 2003, see [48].

[27] R. FREUND, GH. PĂUN, M.J PÉREZ-JIMÉNEZ: Tissue-like P systems with channel states. *Theoretical Computer Sci.*, 330 (2005), 101–116.

[28] R. FREUND, GH. PĂUN, G. ROZENBERG, A. SALOMAA, eds.: *Membrane Computing. 6th Int. Workshop WMC 2005*. Vienna, Austria, LNCS 3850, Springer, 2006.

[29] R. FREUND, S. VERLAN: A formal framework for static (tissue) P systems. *Proc. of WMC 2008* (G. Eleftherakis et al., eds.), Thessaloniki, Greece, Springer, 2007, LNCS 4860, 271–284.

[30] P. FRISCO, H.J. HOOGEBOOM: Simulating counter automata by P systems with symport/antiport. In [42], 288–301.

[31] P. FRISCO, H.J. HOOGEBOOM: P Systems with symport/antiport simulating counter automata. *Acta Informatica*, 41 (2004), 145–170.

[32] M.A. GUTIÉRREZ-NARANJO, GH. PĂUN, M.J. PÉREZ-JIMÉNEZ, eds.: *Cellular Computing (Complexity Aspects)*, ESF PESC Exploratory Workshop, Fénix Editorial, Sevilla, 2005.

[33] H. J. HOOGEBOOM, GH. PĂUN, G. ROZENBERG, A. SALOMAA: *Membrane Computing, 7th Int. Workshop, WMC 2006, Leiden, The Netherlands, 2006, Revised, Selected, and Invited Papers*, LNCS 4361, Springer, 2006.

[34] O. IBARRA, S. WOODWORTH: On symport/ antiport P systems with one or two symbols. *Proc. of the 7th Int. Symp. SYNASC*, Timișoara, Romania, 2005, IEEE Computer Society, Washington, USA, 431–439.

[35] M. IONESCU, C. MARTÍN-VIDE, A. PĂUN, GH. PĂUN: Unexpected universality results for three classes of P systems with symport/antiport. *Lecture Notes in Computer Sci.*, 2568 (2002), 281–290.

[36] I. KOREC: Small universal register machines. *Theoretical Computer Sci.*, 168 (1996), 267–301.

[37] C. MARTÍN-VIDE, J. PAZOS, GH. PĂUN, A. RODRÍGUEZ-PATÓN: Tissue P systems. *Theoretical Computer Sci.*, 296 (2003), 295 326.

[38] M. OSWALD: *P Automata*. PhD thesis, Faculty of Computer Science, Vienna University of Technology, Vienna, Austria, 2003.

[39] A. PĂUN, GH. PĂUN: The power of communication: P systems with symport/antiport. *New Generation Computing*, 20 (2002), 295–305.

[40] GH. PĂUN: *Membrane Computing. An Introduction*. Springer, 2002.

[41] GH. PĂUN, J. PAZOS, M.J. PÉREZ-JIMÉNEZ, A. RODRÍGUEZ-PATÓN: Symport/antiport P systems with three objects are universal. *Fundamenta Informaticae*, 64 (2005), 1–4.

[42] GH. PĂUN, G. ROZENBERG, A. SALOMAA, C. ZANDRON, eds.: *Membrane Computing. Intern. Workshop WMC 2002, Curtea de Argeș, Romania, Revised Papers*, LNCS 2597, Springer, Berlin, 2003.

[43] Y. ROGOZHIN, A. ALHAZOV, R. FREUND: Computational power of symport/antiport: history, advances, and open problems. In [28], 1–31.

[44] P. SOSÍK: P systems versus register machines: two universality proofs. *Pre-Proc. of WMC-CdeA2002*, Curtea de Argeș, 2002, 371–382.

[45] S. VERLAN: Tissue P systems with minimal symport/antiport. *Proc. of Developments in Language Theory, DLT 2004* (C. S. Calude et al., eds.), LNCS 3340, Springer, 418–430.

[46] S. VERLAN, F. BERNARDINI, M. GHEORGHE, M. MARGENSTERN: On communication in tissue P systems: conditional uniport. In [33], 521–535.

[47] S. VERLAN, F. BERNARDINI, M. GHEORGHE, M. MARGENSTERN: Generalized communicating P systems. *Theoretical Computer Sci.*, 404 (1-2) (2008), 170–184.

[48] *The Membrane Computing Web Page*: http://ppage.psystems.eu.

CHAPTER 6

P AUTOMATA

ERZSÉBET CSUHAJ-VARJÚ

MARION OSWALD

GYÖRGY VASZIL

6.1 Introduction

Since membrane computing was initiated with the aim of modeling the architecture and the functioning of the living cell [33, 34], studying variants of P systems which communicate with their environment is an especially important research area in P systems theory. One approach in this direction is represented by models where the environment is given as a supply of objects from which multisets of objects are imported into the system by the skin membrane.

Investigations of these constructs may aim at examining the effect of the consumed input multiset sequence on the behavior of the P system and describing the relation of the different classes of input multiset sequence sets and the corresponding classes of underlying P systems. *P automata*, constructs combining characteristics of classical automata and natural systems with distributed architecture, are models of these types. They resemble classical automata in the sense that in each computational step they receive an input from their environment, and this input has influence on their operation, that is, the input changes their configurations (states) and thus their functioning. The consumed input sequences can be distinguished as accepted or rejected multiset sequences.

Particularly important variants of P automata are the ones which are based on purely communicating P systems, as the most widely studied model, where the underlying membrane system is an antiport P system. One of the characteristics which makes these constructs different from classical automata is the property that the workspace that they can use for computation is provided by the objects of the already consumed input multisets. The objects which enter the system become, in some sense, part of the description of the machine, that is, the input, the object of the computation and the machine which executes the computation cannot be separated in the same way as in classical automata. This is a feature that can usually also be observed when we look at natural processes as "computations".

The first variant called *one-way P automaton* was introduced in [14, 15], and realized an accepting P system based on top-down symport rules with promoters (and implicitly inhibitors). Almost at the same time, a closely related notion, the *analyzing P system* was defined in [20] providing another concept of an automaton-like P system based on antiport P systems. Since that time, several variants of P automata have been introduced and investigated, which differ from each other in the main ingredients of these systems: the way of defining the acceptance, the way of communication with the environment, the types of the communication rules used by the regions, the types of the rules associated with the regions (whether or not evolution rules are allowed to be used), and whether or not the membrane structure changes in the course of the computation. Summaries on these constructs and their properties can be found in [30, 11].

Similarly to generating P systems, several variants of P automata are computationally complete, as powerful as Turing machines, even with limited size parameters. Although these models are of particular importance, since they offer alternative models for computing, P automata with less computational power are of interest as well, since for practical purposes, models of natural systems with low complexity are especially interesting. An adequate example of the latter system is the standard, generic variant of P automata, based on antiport rules with promoters or inhibitors, functioning with sequential rule application, and accepting with final states. By appropriately chosen mappings for defining the language of the P automaton, these constructs determine a language class with sub-logarithmic space complexity. By applying the rules in the maximally parallel manner, the class of languages computable in linear space, i.e. the family of context-sensitive languages is identified by these P automata variants. This representation provides a natural description of this well-known class of languages.

In the following sections we describe the most important variants of P automata and their properties, with the main emphasis on their computational power. We present proof techniques widely used in the literature, and demonstrate that these constructs have certain features which classical types of automata do not have. We also propose new topics and problems for future research.

6.2 BASIC DEFINITIONS

In the following we define the basic notions concerning P automata with unified terminology, and thus in a way which is slightly different from what can usually be found in the literature. To distinguish between strings of objects and sequences (strings) of multisets, we designate the set of finite multisets over a set V by V^*, and the set of their n-tuples by $(V^*)^n$. We also denote a multiset u by the corresponding string $a_1^{u(a_1)} a_2^{u(a_2)} \ldots a_t^{u(a_t)} \in V^*$, $V = \{a_1, a_2, \ldots, a_t\}$, or in the form $\{(a_1, u(a_1)), (a_2, u(a_2)), \ldots, (a_t, u(a_t))\}$.

As it was mentioned in the Section 6.1, the underlying P system of the generic variant of P automata is a P system with symport/antiport, possibly having promoters and/or inhibitors.

Remember that a symport rule is of the form (x, in) or (x, out), $x \in V^*$, and an antiport rule is of the form $(x, out; y, in)$, $x, y \in V^*$. These rules might be associated with a promoter or an inhibitor multiset, denoted as $(x, in)|_Z$, $(x, out)|_Z$, or $(x, out; y, in)|_Z$, $x, y \in V^*$, $Z \in \{z, \neg z \mid z \in V^*\}$. If $Z = z$, then the rule can only be applied if the region contains all objects of multiset z, and if $Z = \neg z$, then z must not be a sub-multiset of the multiset of objects present in the region. To simplify the notations, we denote symport and antiport rules with or without promoters/inhibitors as $(x, out; y, in)|_Z$, $x, y \in V^*$, $Z \in \{z, \neg z \mid z \in V^*\}$ where we also allow x, y, z to be the empty multiset. If $y = \lambda$ or $x = \lambda$, then the notation above denotes the symport rule $(x, in)|_Z$ or $(y, out)|_Z$, respectively, and if $Z = \lambda$, then the rules above are without promoters or inhibitors.

Definition 6.1 *A P automaton (with n membranes) is an $(n + 4)$-tuple, $n \geq 1$, $\Pi = (V, \mu, P_1, \ldots, P_n, c_0, \mathcal{F})$, where:*

- *V is a finite alphabet of objects,*
- *μ is a membrane structure of n membranes with membrane 1 being the skin membrane,*
- *P_i is a finite set of antiport rules with promoters or inhibitors associated to membrane i, $1 \leq i \leq n$,*
- *$c_0 = (w_1, \ldots, w_n)$ is called the initial configuration (or the initial state) of Π where each $w_i \in V^*$ is called the initial contents of region i, $1 \leq i \leq n$,*
- *$\mathcal{F}$ is a computable set of n-tuples $(v_1, \ldots, v_n)$ where $v_i \subseteq V^*$, $1 \leq i \leq n$; it is called the set of accepting configurations of Π.*

An n-tuple $(u_1, \ldots, u_n)$ of finite multisets of objects over V present in the n regions of the P automaton Π describes a *configuration* of Π; u_i, $1 \leq i \leq n$, is called the contents of region i in this configuration.

A P automaton changes its configurations as a standard symport/antiport P system (with promoters or inhibitors), the application of its rules takes place using a certain type of working mode known from the literature (e.g. the sequential rule application, the maximally parallel rule application, etc.). Sequential rule application for P automata was introduced in [14, 15]. In this case, the rule set to be applied is chosen in such a way that exactly one rule is applied in each region where the application of at least one rule is possible. This mode was also called 1-restricted minimally parallel in [25].

In the following, the set of the different types of working modes is denoted as $MODE$, and $seq, maxpar \in MODE$ stand for the *sequential* and the *maximally parallel* rule application, respectively, as the two most commonly studied variants in P automata theory.

Definition 6.2 *Let* $\Pi = (V, \mu, P_1, \ldots, P_n, c_0, \mathcal{F})$, $n \geq 1$, *be a P automaton working in the X-mode of rule application, where* $X \in MODE$. *The transition mapping of* Π *is defined as a partial mapping* $\delta_X : V^* \times (V^*)^n \to 2^{(V^*)^n}$ *as follows:*

For two configurations $c, c' \in (V^*)^n$, we say that $c' \in \delta_X(u, c)$ if Π enters configuration c' from configuration c by applying its rules in the X-mode while reading the input $u \in V^*$, i.e. if u is the multiset of objects that enter the skin membrane from the environment while the underlying P system changes configuration c to c' by applying its rules in mode X.

We define the sequence of multisets of objects accepted by a P automaton as the input sequence which is consumed by the skin membrane until the system reaches an accepting configuration. We shall denote these sequences in the form of strings.

Definition 6.3 *Let* $\Pi = (V, \mu, P_1, \ldots, P_n, c_0, \mathcal{F})$, $n \geq 1$, *be a P automaton. The set of input sequences accepted by* Π *with X-mode of rule application,* $X \in MODE$, *is defined as*

$$A_X(\Pi) = \{v_1 \ldots v_s \mid V_i \in V^*, \text{ there are } c_0, c_1, \ldots, c_s \in (V^*)^n \text{ such that}$$
$$c_i \in \delta_X(v_i, c_{i-1}), 1 \leq i \leq s, \text{ and } c_s \in \mathcal{F}\}.$$

We say that a P automaton Π, as above, accepts by final states if $\mathcal{F} = E_1 \times \cdots \times E_n$ for some $E_i \subseteq V^*, 1 \leq i \leq n$, where E_i is either a finite set of finite multisets or $E_i = V^*$. This means that a configuration $c = (u_1, \ldots, u_n)$ is final, if for all regions of Π, $u_i \in E_i, 1 \leq i \leq n$.

If Π accepts by halting, then $\mathcal{F}$ contains all halting configurations of Π, that is, configurations c with no $c' \in (V^*)^n$ such that $c' \in \delta_X(v, c)$ for some $v \in V^*$, $X \in MODE$.

Obviously, the accepted multiset sequences of a P automaton can be studied from several points of view. One possibility is to encode them to strings and investigate the properties of languages they define. In the case of sequential rule application, the set of multisets that may enter the system is finite and can easily be determined by observing the rules associated with the skin membrane. These multisets can obviously be encoded by a finite alphabet, thus any accepted input sequence can be considered as a string over a finite alphabet. In the case of parallel rule application, this fact does not necessarily hold, the number of objects which may enter the system in one step is not necessarily bounded by a constant. Therefore, in this case the accepted input sequences correspond to strings over infinite alphabets.

We first consider languages over finite alphabets, therefore we apply a mapping to produce a finite set of symbols from a possibly infinite set of multisets.

Definition 6.4 *Let $\Pi = (V, \mu, P_1, \ldots, P_n, c_0, \mathcal{F})$, $n \geq 1$, be a P automaton, Σ be a finite alphabet, and let $f : V^* \to \Sigma^*$ be a mapping. The language accepted by Π with respect to f using the X-mode rule application, where $X \in MODE$, is defined as*

$$L_X(\Pi, f) = \{f(v_1)\ldots f(v_s) \in \Sigma^* \mid v_1 \ldots v_s \in A_X(\Pi)\}.$$

The class of languages accepted by P automata with respect to a class of computable mappings $\mathcal{C}$ with X-mode rule application, $X \in MODE$, is denoted by $\mathcal{L}_{X,\mathcal{C}}(PA)$.

The reader can easily see both similarities and differences between P automata and classical variants of automata. For example, unlike conventional automata, P automata have no separate sets of states, their state sets are represented by the (possibly infinite) sets of configurations. But, similarly to conventional automata, several (restricted or generalized) variants of P automata can be defined. For example, P automata with only top-down symport rules (with or without promoters or inhibitors) can be distinguished as one-way P automata. Furthermore, if the change in the configuration of the P automaton has no influence on the input, we can speak of a read-only machine. Read-write versions of P automata can also be defined, if in the environment we specify an initial, finite supply of objects which may enter or leave the system.

We close the section with an example showing how a linear language can be described by a very simple P automaton with only one membrane.

Example 6.1 Consider the P automaton $\Pi = (\{a, b, p, q, r\}, [\ ]_1, P_1, (p), \{q\})$, where

$$P_1 = \{(p, out; pa, in), (p, out; qa, in), (qa, out; r, in), (r, out; sb, in),$$
$$(sb, out; q, in)\}.$$

Let us define $f(xy) = x$ where xy denotes a multiset over V with $x \in \{a, b, \lambda\}$, and $y \in \{p, r, s, q\}$. Then, starting from the initial configuration (p), an arbitrary number n of symbols a can be brought into the system by using the rule $(p, out; pa, in)$. In any moment, the rule $(p, out; qa, in)$ can be applied instead, importing at least one symbol a. From then on, the sequence of the rules $(qa, out; r, in)$, $(r, out; sb, in)$, and $(sb, out; q, in)$ takes care that for every symbol a that is sent out, a symbol b is taken in and sent out again immediately. Having accepted the multiset sequence $a^n b^n$, the computation halts successfully in the final state (q); hence, for $X \in \{seq, maxpar\}$, it holds that $L_X(\Pi, f) = \{a^n b^n \mid n \geq 1\}$.

6.3 ON THE POWER OF P AUTOMATA

One of the characteristics of P automata which makes them different from classical automata is the property that the workspace that they can use for the computations is provided by the objects of the already consumed input multisets. In the following, based on [12, 13], we show that if the rules of the P automata are applied sequentially, then we can obtain a class of languages accepted by a special, restricted variant of one-way Turing machines with a logarithmically bounded workspace. If the rules are applied in the maximally parallel manner, then appropriate variants of P automata determine the family of context-sensitive languages.

First we show how the amount of computational resources used by P automata corresponds to the standard notion of workspace used by Turing machine computations when the mapping f is non-erasing, that is, when $f : V^* \to \Sigma^*$ for some V, Σ with $f(u) = \lambda$ if and only if $u = \emptyset$.

Let us denote by **NSPACE**(S) the class of languages accepted by a non-deterministic Turing machine using a workspace which is bounded by a function $S : \mathbb{N} \to \mathbb{N}$ of the length of the input.

We say that $L \in \mathbf{r1NSPACE}(S)$ if there is a Turing machine which accepts L by reading the input from a read only input tape once from left to right, and for every accepted word of length n, there is an accepting computation during which the number of non-empty cells on the work-tape(s) is bounded in each step by $c \cdot S(d)$ where c is an integer constant, and $d \leq n$ is the number of input tape cells that have already been read, that is, the actual *distance* of the reading head from the left end of the one-way input tape.

Let $c = (u_1, \ldots, u_n)$ be a configuration of a P automaton. We denote by $|c|$ the number of objects present inside the membrane structure, that is, $|c| = \sum_{i=1}^{m} |u_i|$ where $|u_i|$ denotes the number of objects of $u_i \in V^*$.

Theorem 6.1 *Let Π be a P automaton, let $c_0, c_1, \ldots, c_m$ be a sequence of configurations during an accepting computation, and let $S : \mathbb{N} \to \mathbb{N}$, such that $|c_i| \leq S(d)$, $0 \leq d \leq i \leq m$, where $S(d)$ bounds the number of objects inside the system in the i-th step of functioning, $d \leq i$ being the number of transitions in which a non-empty multiset was imported into the system from the environment.*

If f is non-erasing and $f \in \mathbf{NSPACE}(S_f)$, then for any $X \in MODE$, $L_X(\Pi, f) \in \mathbf{r1NSPACE}(\log(S) + S_f)$.

Proof. Consider $\Pi = (V, \mu, P_1, \ldots, P_n, c_0, \mathcal{F})$, $n \geq 1$, a P automaton with $L_X(\Pi, f) \subseteq \Sigma^*$ where $f : V^* \to \Sigma^*$. In the following we present a possible approach towards the construction of a Turing machine M with a one-way read-only input tape which simulates the work of Π and satisfies the workspace requirements of the statement above.

Let M be given as $M = (t, Q, A \cup \Sigma, \Sigma, B, q_0, \{q_F\}, \delta_M)$ where $t \in \mathbf{N}$ is the number of work-tapes, Q is the set of internal states, $q_0, q_F \in Q$ are the starting and the final states, $A = \{0, \ldots, 9\} \cup \{B\}$ is the work-tape alphabet, Σ is the input alphabet, and $\delta_M : Q \times \Sigma \cup \{\lambda\} \times A^t \to Q \times A^t \times \{L, R\}^t$ is the transition function of M.

Let M have a work-tape assigned to each rule (occurrence) r_j of Π, $1 \le j \le \sum_{i=1}^n |P_i|$, a work-tape for each symbol and region pair $(a_i, j) \in V \times \{1, \ldots, n\}$, and an additional tape for each triple $(a_i, j, k) \in V \times \{0, \ldots, n\}^2$. Using the digits of the tape alphabet, $\{0, \ldots, 9\}$, M keeps track of the configurations of Π by having an integer written on each of the the work-tapes assigned to the symbol-region pair (a_i, j), denoting the number of objects a_i present in region j, that is, the situation where $a_{i,j}$ is the number recorded on the tape for (a_i, j), corresponds to the configuration $c = (u_1, \ldots, u_n)$ of Π with $u_j(a_i) = a_{i,j}$ for all $a_i \in V$, $1 \le j \le n$.

M simulates a configuration change of Π in the following way. First it guesses a rule collection to be used by writing an integer i on each tape corresponding to rule r_j, indicating that r_j will be used in i copies, and also makes sure that the promoters/inhibitors attached to the rules and the current configuration allow the application of the guessed rule collection. Then for each triple, (a_i, j, k), $a_i \in V$, $0 \le j, k \le n$, M examines the chosen rule set and writes the number of objects a_i leaving from region j to region k on the corresponding work-tape (where region 0 denotes the environment) decreasing the number on the tape for (a_i, j) accordingly. Now M makes sure that the configuration change conforms to the type of rule application of Π, and then creates the description of the new configuration by adding the values written on the tape for each (a_i, j, k) to the number stored on the tape for (a_i, k). Based on the contents of the tapes for $(a_i, 0, 1)$ where region 1 is the skin region, M also computes the input multiset v, the value $f(v) = x \in \Sigma^*$, and then reads x from the input tape.

To calculate the workspace used by the Turing machine M, notice that in any step of the computation, the work-tapes discussed above contain integers bounded by the number of objects inside Π during the corresponding computation step. This means that if the number of objects inside the P automaton in a configuration c during a computation is bounded by $S(d)$ where d is the number of non-empty multisets read so far from the environment, then the workspace used by M to record the configurations and calculate the configuration changes of Π (i.e. the number of digits necessary to record this information in the configuration corresponding to c), is also bounded by $t \cdot \log(S(d))$ (in base 10 since $A = \{0, \ldots, 9\}$). Constant t denotes the number of used work-tapes (that is, $t = \sum_{i=1}^n |P_i| + |V| \cdot n \cdot (n+1)^2$), and since f is non-erasing, d coincides with the number of input symbols read from the one-way input tape.

In addition, the transition function of M should also enable an initialization phase, and it should be able to make sure that the input is accepted by M if and only if it is accepted by Π. If the P automaton accepts by halting, then no additional space is required for the Turing machine to compute whether a simulated configuration is accepting or not, it only needs to check that no rule is applicable. If the P automaton accepts by final states, then the equality of the contents of the corresponding regions and elements of a finite set of multisets has to be checked, which can also be done within the given space bound.

We have to take into account the space necessary to compute the mapping f which maps the input multisets of the P automaton to the elements of the input alphabet of the Turing machine. The input symbol of M is computed from the input multiset of Π, so the multiset which is mapped by f does not depend on the length of the whole input sequence, but only on the actual input multiset. Therefore, we can conclude that if f is in $\mathbf{NSPACE}(S_f)$, then $L_X(\Pi, f) \in \mathbf{r1NSPACE}(\log(S) + S_f)$, $X \in MODE$.

Now we employ the theorem above to investigate the power of P automata. In the following, for the representation of Turing machines, we will use the notion of a *three-counter machine* which is a Turing machine with a one-way read only input tape and three work-tapes which can be used as three counters capable of storing any non-negative integer as the distance of the reading head from the only non-blank tape cell marked with the special symbol Z. Formally it can be given as $M = (Q, \{Z, B\} \cup \Sigma, \Sigma, B, q_0, \{q_F\}, \delta_M)$ where Q is a set of internal states containing the initial and the accepting states $q_0, q_F \in Q$ respectively, $\{Z, B\}$ is the work-tape alphabet, Σ is the input alphabet, and δ_M is the transition function which maps the 5-tuple of state, input symbol, and the fact that a counter is zero or non-zero (that is, whether Z or B is read on the work-tapes) to the quadruple of a new state and three instructions to increment, leave unchanged, or decrement the counters (that is, to move the reading heads away from, or towards the marked tape cell). As the work-tape of a general Turing machine can be simulated with two pushdown stacks, and two pushdown stacks can be simulated by three counters (see [18]), three-counter machines characterize the class of recursively enumerable languages. Moreover, if a Turing machine uses $S(n)$ tape cells for some $S : \mathbf{N} \to \mathbf{N}$ during a computation, where n is the length of the input, then there exists a simulating three-counter machine such that the integers stored in the counters during a simulating computation are bounded by $c^{S(n)}$ for a suitable constant c.

Theorem 6.2 (1) $\mathcal{L}_{seq,\mathcal{C}}(PA) = \mathbf{r1NSPACE}(\log(n))$ *for any class $\mathcal{C}$ of non-erasing mappings with a finite domain, and* (2) $\mathcal{L}_{maxpar,\mathcal{C}}(PA) = \mathbf{CS}$ *for any class $\mathcal{C}$ of non-erasing linear space computable mappings.*

Proof. Consider a P automaton Π, a non-erasing mapping f, and the sequence of configurations $c_0, c_1, \ldots, c_m$ of an accepting computation of Π. If the rules are applied sequentially, then $|c_i| \leq k \cdot d$ where k is a constant depending on the form of the rules, and $d \leq i$ is the number of non-empty multisets read by the P automaton up to reaching configuration c_i.

Since the number of possible input multisets is finite, the mapping $f \in \mathcal{C}$ maps a finite domain to a finite set of values, which means that its computation requires constant space. Therefore, by Theorem 6.1, we obtain that $L_{seq}(\Pi, f) \in$ **r1NSPACE**$(\log(k \cdot n) + k') = $ **r1NSPACE**$(\log(n))$.

If Π uses maximally parallel rule application, then the number of symbols inside the system is at most k^d where k is a suitable constant and d is as above. Therefore, if $f \in$ **NSPACE**(n), then by Theorem 6.1, we obtain that $L_{maxpar}(\Pi, f) \in$ **r1NSPACE**$(\log(k^n) + n) = $ **r1NSPACE**(n).

Now we prove that languages in **r1NSPACE**$(\log(n))$ and **r1NSPACE**(n) can be accepted by P automata using sequential and maximally parallel rule application, respectively. To see this, consider the three-counter machine $M = (Q, \{Z, B\} \cup \Sigma, \Sigma, B, q_0, \{q_F\}, \delta_M)$. If x is an element of the domain of δ_M, then the transitions are given by pairs of the form (x, y) where $y \in \delta_M(x)$. Let these pairs be labeled by elements of the finite set of labels $lab(\delta_M)$.

Let $\Pi = (V, \mu, P_1, P_2, P_3, P_4, P_5, c_0, \mathcal{F})$ be a P automaton with the membrane structure $\mu = [\,[\,]_2\,[\,]_3\,[\,]_4\,[\,]_5\,]_1$. It has a skin membrane, three regions, region 3, 4, and 5 for storing and manipulating symbols corresponding to the states of M, and maintaining the values of the counters, and one more membrane for an additional check of the correct functioning.

Let $V = \Sigma \cup \{A_1, A_2, F\} \cup \{\langle q \rangle, \langle t \rangle, \langle t \rangle_a \mid q \in Q,\ t \in lab(\delta_M),\ a \in \Sigma\}$. The symbols of $V - \Sigma$ govern the work of Π. The presence of $\langle q \rangle \in V$ in the skin membrane indicates that Π simulates a configuration of the counter machine when it is in state $q \in Q$. While simulating a transition from state q to state q' labeled by $t \in lab(\delta_M)$, Π needs extra steps for reading the input and manipulating the symbols which keep track of the counter values. During these steps the skin membrane contains copies of one of the symbols $\langle t \rangle$, $\langle t \rangle_a$, for some $a \in \Sigma$, and when the simulation of the transition is complete, $\langle q' \rangle \in V$ appears in the skin membrane.

The simulation starts in the initial state $c_0 = (w_1, w_2, w_3, w_4, w_5)$ with $w_1 = \langle q_0 \rangle \langle q_0 \rangle \langle q_0 \rangle a$, for some $a \in \Sigma$, $w_2 = A_1$, and for $3 \leq i \leq 5$ $w_i = \{(\langle q \rangle, 1), (\langle t \rangle, 1), (\langle t \rangle_a, 1) \mid q \in Q,\ q \neq q_0,\ t \in lab(\delta_M),\ a \in \Sigma\}$. The rule sets belonging to the regions are as follows. Let

$$P_1 = \{(x, out; y^k, in)|_{\langle t \rangle \langle t \rangle \langle t \rangle} \mid x, y \in \Sigma,\ t \in lab(\delta_M),\ y \text{ is read}$$

during the transition $t\} \cup \{((\langle q_F \rangle \langle q_F \rangle \langle q_F \rangle, out; A_2, in)\}$

$$\cup \{(x, out; FF, in)|_{a_1 a_2 a_3} \mid x \in \Sigma,\ a_i \in \{\langle q \rangle, \langle t \rangle, \langle t \rangle_a \mid q \in Q,$$

$t \in lab(\delta_M),\ a \in \Sigma\}$ such that $a_i \neq a_j$, for some $1 \leq i, j \leq 3\}$,

and $k \geq 1$ is a suitable constant. In the case of sequential rule application, if transition t is simulated, as indicated by the presence of $\langle t \rangle$, then k copies of the corresponding input symbol are read into the skin membrane, and one other symbol is sent out in the first simulating step. If the rules are applied in the maximally parallel manner, then after the first simulation step, the number of symbols in the skin region is k times as much as the number that was already present.

Before the simulation starts, that is, when a state symbol $\langle q \rangle$ is present, or in the later simulating steps, when the symbols $\langle t \rangle_a$, for some $a \in \Sigma$, are present, then nothing is read from the environment.

If the three symbols from $\{\langle q \rangle, \langle t \rangle, \langle t \rangle_a \mid q \in Q, t \in lab(\delta_M), a \in \Sigma\}$ which appear in the skin region do not belong to the same state or to the same transition with the same index, then the system imports a multiset of several objects F and, as we will see, never reaches a final state anymore.

In what follows, let us assume that $\Sigma = \{a_1, \ldots, a_m\}$. Now let

$$P_2 = \{(A_1, out; A_2, in), (A_1, out; F, in), (F, out; F, in)\},$$

and for $3 \leq i \leq 5$, let

$$P_i = \{(\langle t \rangle a_k, out; \langle q \rangle, in), (\langle t \rangle_{a_1}, out; \langle t \rangle, in), (\langle t \rangle_{a_{j+1}}, out; \langle t \rangle_{a_j}, in),$$
$$(\langle q' \rangle, out; \langle t \rangle_{a_m}, in) \mid 1 \leq k \leq m,\ 1 \leq j \leq m-1,\ t \in lab(\delta_M)$$
$$\text{is a transition from } q \text{ to } q' \text{ decreasing the } (i-2)\text{th counter } \}$$
$$\cup \{(\langle t \rangle, out; \langle q \rangle, in), (\langle t \rangle_{a_1}, out; \langle t \rangle, in)|_{Z_1}, (\langle q' \rangle, out; \langle t \rangle_{a_m} a, in),$$
$$(\langle t \rangle_{a_{j+1}}, out; \langle t \rangle_{a_j}, in)|_{Z_{j+1}} \mid 1 \leq j \leq m-1,$$

$t \in lab(\delta_M)$ is a transition from q to q', and

$a \in \Sigma$ if the $(i-2)$th counter is increased during t, or

$a = \lambda$ if the $(i-2)$th counter is not changed during t, or

$Z_k = \lambda$ for all $a_k \in \Sigma$ if the $(i-2)$th counter is non-empty before t, or

$Z_k = \neg a_k$ for all $a_k \in \Sigma$ if the $(i-2)$th counter must be empty before $t\}$.

Regions 3, 4, and 5 are responsible for keeping track of the simulated states and transitions while also maintaining the values stored in the counters of M. Together with moving the transition symbols $\langle t \rangle_a$, for some $a \in \Sigma$, they import and export the symbols originating from the input as necessary to maintain the correct counter contents. The emptiness of the counters are checked by the use of inhibitors.

If the work of these regions is not consistent, i.e. if they do not send the same symbols to the skin region, then a multiset containing FF is imported into the skin region, and the rules $(A_1, out; F, in)$ and $(F, out; F, in)$ are applied in region 2 making it impossible to halt, or to reach a final configuration.

To enable Π to accept with final states, we have $\mathcal{F} = (V^*, E_2, V^*, V^*, V^*)$ where $E_2 = \{A_2\}$. This means that the simulation of M can be finished if A_2 appears in the second region. This can happen if first three copies of $\langle q_F \rangle$, the object corresponding to the final state, are exported to the first region, then the rule $(\langle q_F \rangle \langle q_F \rangle \langle q_F \rangle, out; A_2, in)$ is applied, and the rule $(A_1, out; A_2, in)$ transfers A_2 to the second region. Reaching this configuration, the system halts, therefore the construction above also realizes a P automaton which accepts by halting.

Consider now the possible input multisets of the system entering during a successful, accepting computation. These are multisets containing some element of Σ in multiple number of copies, so if we use the mapping $f_1(x^i) = x$, $x \in \Sigma$, $i \geq 1$, $f_1(\emptyset) = \lambda$, to map V^* to Σ^*, the P automaton Π simulates the transitions of the counter machine M with both types of rule application, sequential and maximally parallel.

Let us compare now the possible values of the counters of M and the number of objects available inside the system Π for their representation. If Π uses its rules sequentially, the number of objects present in the system after reading the d-th non-empty input multiset is at most $c_1 \cdot d$ where c_1 is a suitable integer constant. Thus, the sum of the counter values of M which Π can simulate is $c_1 \cdot d$ where d is the number of input symbols read by M. This is sufficient for M to simulate a one-way non-deterministic Turing machine using $c_2 \cdot \log(d)$ non-empty cells on its worktape where c_2 is a constant and d is the number of symbols read by the reading head on the one-way input tape. Therefore, for all languages $L \in \mathbf{r1NSPACE}(\log(n))$, there is a P automaton with $L = L_{seq}(\Pi, f_1)$ for f_1 as above.

If Π uses its rules in the maximally parallel manner, then the number of objects present in the system after reading the d-th input multiset is at most c_3^d for some constant c_3. This is the bound on the sum of the counter values of M, where d is the number of input symbols read. With this bound, M can simulate a one-way non-deterministic Turing machine using a workspace bound of $c_4 \cdot d$ where c_4 is a constant and d is the number of symbols read by the reading head on the one-way input tape, that is, for all languages $L \in \mathbf{r1NSPACE}(n)$, there is a P automaton with f_1 as above, such that $L = L_{maxpar}(\Pi, f_1)$.

Note that systems with more complicated mappings can also be constructed, but as long as f is linear space computable, the power of those P automata would not increase. Of course, if f is more complex than linear space computable, then more complex language classes could be accepted, but in that case, the power would come from the function f and not from the P automaton itself.

To complete the proof, we show that $\mathbf{r1NSPACE}(n) = \mathbf{NSPACE}(n) = CS$. Clearly, $\mathbf{r1NSPACE}(n) \subseteq \mathbf{NSPACE}(n)$. To see the opposite inclusion, consider a language $L \in \mathbf{NSPACE}(n)$ accepted by some Turing machine M. We show that $L \in \mathbf{r1NSPACE}(n)$. Consider M', a modified version of M with an additional work-tape. M' starts its operation by copying the actual input string from the input tape to the additional work-tape. After this initial phase, it continues working exactly as M, but using the additional work-tape instead of the input tape. If M uses a linearly bounded workspace by the length of the input, then the workspace of M' is linearly bounded by d, the distance of the reading head from the left-end of the one-way input tape, thus, $L \in \mathbf{r1NSPACE}(n)$ which means that $\mathbf{NSPACE}(n) \subseteq \mathbf{r1NSPACE}(n)$, so our statement holds.

Restricted variants of P automata can be used for describing other well-known language classes, e.g. the regular language class [17] and the context-free language class [35].

Remark 6.1 By the above proof, it can be seen that if we allow arbitrary functions for mapping the input multisets of the P automaton to the alphabet of the accepted language, then we can obtain a characterization of the class of recursively enumerable languages even if we do not use mappings that are more complex than linear space computable, all we need to allow is the property that not only the empty multiset, but other non-empty multisets can also be mapped to the empty string.

We could generalize the construction used in the previous theorem to simulate an arbitrary three-counter machine. In the following, instead of providing the reader with details on the construction, we present another approach to describe language classes in terms of automata-like P systems.

6.4 Extended P Automata

The language of the P automaton is defined via a mapping which orders to any input multiset a string over an alphabet. A variant of this approach, which conforms with formal language theoretic constructs, is to distinguish input objects and auxiliary objects, i.e. terminal and non-terminal objects of the P automaton. The accepted language then is defined as the sequence of terminal strings of the input multisets during an accepting computation (where the set of terminal strings of a multiset consists of all permutations of its terminal symbols).

In the sequel, we call such a construct an *extended P automaton* and denote it by $\Pi = (V, \Sigma, \mu, P_1, \ldots, P_n, c_0, \mathcal{F})$, where $\Sigma \subseteq V$ is the set of terminal objects (symbols) and thus elements in $V - \Sigma$ are the non-terminal objects (symbols).

Any other notions concerning P automata, as the rule application mode, the configuration change, the accepting configurations, and the concept of an accepted multiset sequence, are extended to the new construct in a natural manner.

For an extended P automaton $\Pi = (V, \Sigma, \mu, P_1, \ldots, P_n, c_0, \mathcal{F})$, the language accepted by Π in the X mode of rule application, $X \in \text{MODE}$, is

$$L_X(\Pi) = \bigcup_{v_1 v_2 \ldots v_s \in A_X(\Pi)} f(v_1) f(v_2) \ldots f(v_s)$$

with $f : V^* \to 2^{\Sigma^*}$ is defined as $f(u) = strings(u')$ where $u' \in \Sigma^*$ is the multiset of terminal objects from u, that is, $u' \subseteq u$ such that if $u'' \subseteq u$ for some $u'' \in \Sigma^*$ then $u'' \subseteq u'$, and $strings(v) \subseteq \Sigma^*$ for a multiset $v \in \Sigma^*$ denotes the set of strings corresponding to v.

An example for an extended P automaton is the *analyzing P system*, introduced in [20], which has antiport rules (without promoters and without inhibitors), uses the maximally parallel rule application, and accepts by halting.

According to [20], these systems, even with very limited size parameters, are able to recognize any recursively enumerable language. In order to demonstrate a widely used proof technique in the theory of automata-like P systems, based on appropriate simulations of deterministic register machines, we present the proof of the following statement.

Theorem 6.3 *For any recursively enumerable language $L \subseteq \Sigma^*$, there exists an analyzing P system Π with one membrane, such that $L_{maxpar}(\Pi) = L$.*

Proof. Let $L \subseteq \Sigma^*$ be a recursively enumerable language. Let us introduce the encoding $g_z(w) \in \mathbf{N}$ for strings over the alphabet $\Sigma = \{a_1, \ldots, a_{z-1}\}$ being the integer value of $digits_z(w)$, a string of z-ary digits $(1), \ldots, (z-1)$, with $z = |\Sigma| + 1$ and $digits_z(a_i) = (i)$.

For any such L, there exists a deterministic register machine which halts after processing the input u_0 placed initially in its input register if and only if $u_0 = g_z(w)$ for some $w \in L$. Therefore, it is sufficient to show how to read the input string w, generate the encoding $g_z(w)$, and simulate the instructions of a register machine with an analyzing P system.

To simulate a register machine, the object alphabet of the P system contains an object l_j for each instruction label l_j, and an object a_i for representing the values of each register i. The system has one membrane, the presence of an instruction label indicates that the corresponding instruction is to be simulated, and the contents of the registers are represented by the multiplicities of the corresponding objects.

To simulate an add-instruction $l_i : (\text{ADD}(r), l_j, l_j)$ we need the antiport rule $(l_i, out; l_j a_r, in)$.

A subtract-instruction $l_i : (\text{SUB}(r), l_j, l_k)$ can be simulated by

- $(l_i, out; l'_i l''_i, in)$,
- $(l'_i a_r, out; \bar{l}_i, in)$,
- $(l''_i, out; \tilde{l}_i, in)$,
- $(\bar{l}_i \tilde{l}_i, out; l_j, in)$,
- $(l'_i \tilde{l}_i, out; l_k, in)$.

The halting instruction is simulated by having no rule with the label l_h of the halting instruction $l_h : \text{HALT}$.

To generate the encoding of an input string, we can read each terminal letter $a \in \Sigma$ by rules of the form $(l_{ini}, out; l_{ini,a}a, in)$ with l_{ini} and $l_{ini,a}$, $a \in \Sigma$, being "instruction" objects corresponding to the initial reading phase of the computation. If we have represented by $g_z(v)$ the encoding of the input sequence v, then the encoding $g_z(va)$ for some $a \in \Sigma$ is given by $z \cdot g_z(v) + g_z(a)$. This encoding step can be accomplished by an appropriate register machine program starting from the label $l_{ini,a}$, and ending in l_{ini} again, which can continue with $l_{ini,b}$ signaling the reading of another input symbol b, or with l_0, starting the computation of the register machine to evaluate the encoded input.

If the sequence of terminal symbols taken in is deterministically given in the environment (so that the choice which rule has to be applied depends on the actual terminal symbol), then the acceptance, according to the construction given above, is carried out in a deterministic way.

Furthermore, we note that the construction used in the proof of the previous theorem leads to a method for computing sets of integers or sets of vectors: we consider a set of non-negative integers or vectors of non-negative integers to be accepted by a P automaton if the multiset $w \in V^*$ to be analyzed is initially put into a specified (input) membrane.

As we can see, the "workspace", i.e. one type of the computational resources needed to obtain the computational completeness of the model, is due to non-terminal objects that can be available in a number not restricted by the length of the input string. One interesting question is whether or not the computational power changes if the system works with different types of bounded resources.

Such a model is the *exponential-space symport/antiport acceptor* introduced in [27], which is a particular variant of extended P automata. In this case, the set of terminal objects Σ contains a distinguished symbol \$, and the rules are of the following four types in the set P_1 corresponding to the skin region:

1. $(u, out; v, in), u, v \in (V - \Sigma)^*, |u| \geq |v|$,
2. $(u, out; va, in), u, v \in (V - \Sigma)^*, |u| \geq |v|$, and $a \in \Sigma$,
3. $(u, out; v, in)|_a, u, v \in (V - \Sigma)^*, a \in \Sigma$,
4. for every $a \in \Sigma$ there is at least one rule of the form $(a, out; u, in), u \in (V - \Sigma)^*$.

The other regions contain rules of the form $(u, out; v, in)$, $u, v \in (V - \Sigma)^*$.

Exponential-space symport/antiport systems function with maximally parallel application of rules and the final states are defined by halting.

The language accepted by an exponential-space symport/antiport acceptor Π is defined in a slightly different way from that used in the case of an extended P automaton. A string $a_1 \ldots a_n$, $a_i \in \Sigma - \{\$\}$, $1 \leq i \leq n$ is accepted, if the symbols $a_1, \ldots, a_n, \$$ are brought into the system from the environment in the required order (by rules of type 2) and after reading the end marker $\$$, the computation eventually halts.

The term "exponential-space symport/antiport acceptor" comes from the fact that due to the restricted form of the rules, the workspace which can be used by such a construct is not arbitrarily large, the membrane system contains no more than an exponential number of objects (up to some constant) at any time during the computation. Similarly to the generic variant of P automata using the maximally parallel rule application, these systems determine the class of context-sensitive languages [27].

Theorem 6.4 *A language L is accepted by an exponential-space symport/antiport acceptor with one membrane, if and only if L is context-sensitive.*

The proof, as for generic P automata, is done by simulating the corresponding restricted variant of Turing machines. The reader, however, may notice that the restricted computational power is due to the restrictions of the form of the antiport rules.

Further restrictions on descriptional complexity parameters, as the number of membranes and the number of non-terminals lead to infinite hierarchies in the computational power. A system is called a *2-restricted exponential-space symport/antiport acceptor* if it has a terminal alphabet $T = \{a, b, \$\}$, uses rules of type 3, and the difference between the number of objects which are sent out and which are imported is bounded in such a way that $2|u| \geq |v|$ holds. In [27] it is shown that the computational power is not decreased by these modifications in the sense that 2-restricted systems characterize the class of binary (two-letter) context-sensitive languages, but based on the number of membranes or the number of non-terminals infinite hierarchies are obtained. First, if we fix the number of membranes at some n, there is an infinite hierarchy in computational power with respect to the number of non-terminal objects, and second, fixing the number of non-terminal objects at some m, there is an infinite hierarchy in computational power with respect to the number of membranes.

We close the section by mentioning that extended P automata also provide very simple descriptions of the regular language class. In [27] it is shown that a language L is regular if and only if it can be accepted by an exponential-space symport/antiport acceptor using only rules of type 1 and type 2, and a similarly simple variant of analyzing P systems with one membrane, called *finite P automaton with antiport rules* is presented in [24].

6.5 P Automata and Infinity

One of the important characteristics of P automata is that the basic model is suitable for describing languages over infinite alphabets, without any extension or additional component added to the construct, and that finite and infinite runs (computations) can also be defined by the common frame in a natural manner.

6.5.1 P Automata Over Infinite Alphabets

As we mentioned above, P automata are devices providing the possibility for obtaining *descriptions of languages over infinite alphabets*. The idea comes very naturally if we recall that the language accepted by such a system corresponds to the sequence of multisets entering during a successful computation, and notice that the number of possible multisets which form this sequence, that is, the number of possible symbols which constitute the accepted string, is not fixed in advance but can be arbitrarily high.

An example of this approach is the notion of a *P finite automaton*, introduced in [17], a P automaton $\Pi = (V \cup \{a\}, \mu, P_1, \ldots, P_n, c_0, \mathcal{F})$ applying the rules in the maximally parallel manner, accepting by final states, where the object alphabet $V \cup \{a\}$ contains a distinguished symbol a; the set P_1 corresponding to the skin region contains rules of the form $(x, out; y, in)|_Z$ with $x \in (V \cup \{a\})^*$, $y \in \{a\}^*$, $Z \in \{z, \neg z\}$, $z \in V^*$; and if $i \neq 1$, the set P_i contains rules of the form $(x, out; y, in)|_Z$ with $Z \in \{z, \neg z\}$, $x, y, z \in V^*$. We also allow the use of rules of the form $(x, in)|_Z$ in the skin membrane in such a way that the application of any number of copies of the rule is considered "maximally" parallel.

Notice that the domain of the mapping f is infinite, so its range could also easily be defined to be infinite, as $f : \{a\}^* \to \Sigma \cup \{\lambda\}$ for an infinite alphabet $\Sigma = \{a_1, a_2, \ldots\}$ with $f(a^k) = a_k$ for any $k \geq 1$, and $f(\emptyset) = \lambda$.

The language accepted by a P finite automaton Π is $L(\Pi) = L_{maxpar}(\Pi, f)$ for f as above.

Theorem 6.5 *For any $L \subseteq \Sigma^*$ over a finite alphabet Σ, $L \in REG$ if and only if $L = L(\Pi)$ for some P finite automaton Π.*

Proof. Let $M = (Q, \Sigma, q_0, F, \delta)$ be a finite automaton over the input alphabet $\Sigma = \{a_1, \ldots, a_n\}$, with set of states Q, transition relation $\delta : Q \times \Sigma \to Q$, initial state $q_0 \in Q$, and set of final states $F \subseteq Q$. Let us assume that $q_0 \notin F$.

Let $TR = \{[q_1, a_i, q_2] \mid \delta(q_1, a_i) = q_2\}$ and let $TR' = \{[q_1, a_i, q_2]' \mid \delta(q_1, a_i) = q_2\}$, a primed and a non-primed set of triples corresponding to the transitions of M. Let us also denote for any $t' \in TR'$, by $next(t') \subseteq TR$ the set of those non-primed transition symbols which correspond to transitions that can follow the transition denoted by the primed symbol t', that is, $next(t') = \{[q_2, a_j, q_3] \in TR \mid t' = [q_1, a_i, q_2]'\}$.

Now we construct a P finite automaton Π simulating M.

Let $\Pi = (V \cup \{a\}, [\ [\ ]_2\ ]_1, P_1, P_2, c_0, \mathcal{F})$ where $V = TR \cup TR' \cup \{\#\}$ and $c_0 = (w_1, w_2)$ with $w_1 = a\#$, $w_2 = \{(t, 1), (t', 1) \mid t \in TR\}$.

$$P_1 = \{(a, out; a^i, in)|_t, (a^{i-1}, out)|_{t'} \mid t = [q_j, a_i, q_k]\}$$
$$\cup \{(a, out; a, in)|_t \mid t = [q_j, a_1, q_k]\},$$
$$P_2 = \{(t_0, out; \#, in) \mid \text{ for all } t_0 = [q_0, a_i, q], a_i \in \Sigma, q \in Q\}$$
$$\cup \{(t', out; t, in), (s, out; t', in) \mid t \in TR, s \in next(t')\},$$

and $\mathcal{F} = (V^*, E_2)$ with $E_2 = \{\{(\#, 1)\} \cup \{(t, 1), (t', 1) \mid t \in TR\} - \{(s', 1)\} \mid$ for all $s' \in TR'$ such that $s' = [q, a_i, q_f]'$, $q_f \in F\}$.

It is not difficult to see how Π simulates M. The rules of the second region are responsible for sending symbols representing the transitions of M into the first region in an order which is a legal transition sequence of M, and the rules of the first region import from the environment the necessary number of symbols a corresponding to the input symbol belonging to the simulated transition.

Therefore, $L(\Pi, f) = L(M)$, and thus, $L \in REG$ implies that L can be accepted by a P finite automaton. The proof of the reverse of this statement, involving a more complicated construction, can be found in [17].

So far we have considered finite alphabets, but it is clear that P finite automata can also be built in such a way that the number of input symbols entering the system in one step is unbounded. Since in the case when they work over finite input alphabets, P finite automata characterize the class of regular languages, the infinite alphabet language class they determine in their general form, might be considered as an extension of REG to infinite alphabets. This is a promising research direction since this class behaves differently from other infinite alphabet extensions of regular languages defined using ideas, such as, for example, the machine model called finite memory automata from [28], or the infinite alphabet regular expressions introduced in [32]. Given an infinite alphabet $\Sigma = \{a_1, a_2, \ldots\}$, P finite automata are able to describe, for example, the language $\{a_{2i} \mid i \geq 1\}$ which can be described by infinite alphabet regular expressions but cannot be accepted by finite memory automata, and also the language $\{a_i a_i \mid i \geq 1\}$ which is accepted by finite memory automata but cannot be captured by infinite alphabet regular expressions.

6.5.2 ω-P Automata

The generic model of P automata also provides possibilities of describing (possibly) infinite runs (sequences of configurations). This is of particular interest, since assuming a P automaton as a system being in interaction with its environment, we also should consider communication processes (functioning) not limited in time. Variants of P automata, motivated by these considerations, are the so-called ω-P automata [24].

When dealing with infinite strings (ω-words), the main problem that arises is the fact that now failing computations have to stop, whereas successful computations have to be infinite. Therefore, eventual acceptance can only be checked via the final states.

Let Σ^ω denote the space of infinite strings (ω-words) on a finite alphabet of cardinality ≥ 2, then subsets of Σ^ω are called ω-languages. For an ω-word $z \in \Sigma^\omega$, the i-th letter of z is denoted by $z(i)$.

We now consider Turing machines with a working tape and a separate input tape on which the read-only-head can only move to the right. Such an ω-Turing machine is a construct $M = (\Sigma, \Gamma, Q, q_0, \delta)$, where Σ is the input alphabet, Γ the work-tape alphabet, Q the finite set of internal states, q_0 the initial state, and $\delta \subseteq Q \times \Sigma \times \Gamma \times Q \times \{0, +1\} \times \Gamma \times \{-1, 0, +1\}$ defines the next configuration.

Let $x \in \Sigma^\omega$ be the input of the Turing machine M. A sequence $z \in Q^\omega$ of states that M runs through in (some of) its computation(s) with input x is called a *run of M on x*. An input sequence $x \in \Sigma^\omega$ is said to be accepted by M according to condition (mode) C if there is a run z of M on x such that z satisfies C.

In the literature, different *variants of accepting* infinite strings (ω-words) by Turing machines can be found (see e.g. [37], [38], [36], [9], [10]). Using the notation of [36], the following conditions were considered in [24] (also see [37]).

Let $a : Q^\omega \to 2^Q$ be a mapping which assigns to every ω-word $\zeta \in Q^\omega$ a subset $Q' \subseteq Q$, and let $R \subseteq 2^Q \times 2^Q$ be a relation between subsets of Q. A pair (M, Y), where $Y \subseteq 2^Q$, is said to accept an ω-word $x \in \Sigma^\omega$ if and only if $\exists Q' \exists z (Q' \in Y \land z$ is a run of M on $x \land (a(z), Q') \in R)$. (If Y consists of only one subset of Q, i.e., $Y = \{F\}$ for some $F \subseteq Q$, then one usually writes (M, F) instead of $(M, \{F\})$.)

For an ω-word $z \in Q^\omega$ let $ran(z) = \{v \mid v \in Q \land \exists i (i \in (\mathbf{N} - \{0\}) \land z(i) = v)\}$ be the *range* of z (considered as a mapping $z : (\mathbf{N} - \{0\}) \to Q$), i.e. the set of all letters occurring in z, and let $inf(z) = \{v \mid v \in Q \land z^{-1}(v)$ is infinite$\}$ be the *infinity set* of z, i.e. the set of all letters occurring infinitely often in z. As relations R we shall use $=, \subseteq$ and $\sqcap$ ($Z' \sqcap Z''$ means $Z' \cap Z'' \neq \emptyset$).

Hence, six types of acceptance (a, R) (for $a \in \{ran, inf\}$ and some $R \in \{\subseteq, \sqcap, =\}$) are obtained.

An ω-P automaton is a construct as defined above, but with the difference that now the sequence of terminal symbols taken in is infinite. These ω-words are accepted according to specific conditions defined via the final states, while in the case of non-acceptance, the system should halt.

In [24], it was shown that for any well-known variant of acceptance mode for ω-Turing machines, an ω-P automaton simulating the computations of the corresponding ω-Turing machine can effectively be constructed:

Theorem 6.6 *Let $L \subseteq \Sigma^\omega$ be an ω-language accepted by an ω-Turing machine in the acceptance mode (a, R), where $a \in \{ran, inf\}$ and $R \in \{\subseteq, \sqcap, =\}$. Then an ω-P automaton that simulates the actions of the Turing machine and accepts L in the same acceptance mode (a, R) can effectively be constructed.*

In the proof given in [24], it is shown in detail how the steps of an ω-Turing machine (represented by a 2-register machine) can be simulated by an ω-P automaton Π, that uses antiport rules and only two membranes (the second one is used for checking the corresponding acceptance condition defined via the final states). Hence, with respect to the number of membranes, this result is already optimal. Moreover, the construction would also work for deterministic variants of ω-Turing machines/ω-P automata as well as for ω-Turing machines describing functions on ω-languages.

6.6 FURTHER VARIANTS OF P AUTOMATA

During the years, several types of automata-like P systems were introduced with the aim of studying their boundaries as computational devices and exploring their relations to classical automata.

One of the motivations to develop the concept of the P automaton was to study properties of purely communicating accepting P systems. For this reason, the question whether or not any change in the underlying communicating P system implies changes in the power and the size complexity of the respective new class of P automata is of particular interest. This problem was approached from different points of view.

Additional constraints given by a partial binary relation were posed to the application of the communication rules of the generic model in the case of *P automata with priorities* in [5], where the rules with the highest priority must be applied in the configuration change. Based on a simulation of a two-counter machine, the computational completeness of this construct was proved. Two other variants, with conditional symport/antiport rules, are *P automata with membrane channels* [30, 21, 22], motivated by certain natural processes taking place in cells, and *P automata with conditional communication rules associated with the membranes* [30, 23]. Both models are computationally complete devices, and optimal results on their size parameters have also been obtained. The proofs of the corresponding statements were based on different techniques: simulation of deterministic register machines and graph-controlled grammars. The models and results above demonstrate that P automata with simple conditions for communication and small bounds on size parameters are as powerful as Turing machines.

The models that have been discussed so far have a static membrane structure, that is, the membrane structure is not altered during the functioning of the system. Considering P automata as models of complex biological systems, this condition is rather restrictive, since the architecture of natural systems may change in the course of their functioning. Moreover, P automata with varying membrane structure may utilize the distributed nature of their architecture more efficiently during the computational process.

A P automaton-like system working with a dynamically changing membrane structure is the *P automaton with marked membranes* ([16]), or P_{pp} automaton for short. The concept is motivated by several research areas, such as the theory of P systems, brane calculi [4], and traditional automata theory. The underlying P system models the situation when proteins are allowed to move through the membranes and to attach onto or to detach from the membranes, in such a way that their moves may also imply changes in the membrane structure. P automata with marked membranes are able to consume inputs from their environment, i.e. multisets of proteins, which might influence the behavior of the system, and correspond to the result of a computation if the P_{pp} automaton starts in the initial configuration and halts in a final configuration. As in the previous cases, the model is computationally complete. Its importance lies not only in the dynamic membrane structure, but also in building bridges between important research areas.

Another feature in which P automata differ from classical automata is the property that they have no separate internal state sets, the states are represented by the (possibly infinite) set of configurations. *P automata with states* attempted to make the generic concept resemble more to conventional automata [29]. In this model, both states and objects are considered, the states, together with the objects, govern the communication. As expected, the device is computationally complete, moreover, any recursively enumerable language can be described by these systems with very restricted form [19].

Although most of the variants of P automata realize a concept for modeling purely communicating, accepting P systems, they can also be used for *describing complex evolving systems*. Evolution-communication P automata, where both communication and evolution rules are allowed to be used are examples for such models [1]. The construct can be considered as a variant of extended P automata, and as expected, describes the class of recursively enumerable languages.

6.7 FURTHER ASPECTS AND RELATED MODELS

Another important research area is to investigate how concepts of classical automata theory can be related to concepts in the theory of P automata. An approach in this direction is presented in [6], where the so-called Mealy multiset automata and elementary Mealy membrane automata are studied, inspired by the concept of a Mealy automaton. An augmented version of the elementary Mealy membrane automaton, with extended communication capabilities, called simple P machine was examined in [7].

P systems with symport/antiport form the basis of the concept of the P transducer, which is basically an extended one-membrane P automaton working with input and output objects [8]. Four types of these machines were distinguished and studied, two of them are computationally complete, and two are incomparable to finite state sequential transducers. Iterating these latter P transducer classes, new characterizations of the recursively enumerable language class can be obtained.

Continuing the above line, another aspect to develop the theory of P automata is to study their usability in parsing languages. A variant of accepting P systems with dynamically changing membrane structure, called *active P automaton*, was proposed and used for parsing sentences of natural languages in [2, 3]. An active P automaton starts the computation with one membrane containing the string to be analyzed together with some additional information assisting the computation. It computes with the structure of the membrane system, using operations as membrane creation, division, and dissolution. There are also rules for extracting a symbol from the left-hand end of the input string and for processing assistant objects. The computation ends with acceptance when all symbols from the string are consumed and all membranes are dissolved. It was shown that the model is suitable for recognizing any recursively enumerable language, and with restrictions in the possible types of rules, also for identifying other well-known language classes, such as the regular language class and the class of context-sensitive languages. This special variant of accepting P systems resembles P automata since any symbol in the string can be considered as a multiset of objects with one element consumed from the environment.

A common feature of the models that we have discussed so far is the assumption that the environment contains an infinite supply of objects. This is in some sense artificial since usually only a finite collection of input objects are available for a natural system. A model called *restricted communicating P system* corresponding to this idea was discussed in [26]. In this case, the environment does not initially contain any object. The objects can be sent out to the environment and only those that were expelled earlier can enter the system again. The rules only allow the movement of the objects between the neighboring regions and importing objects from the environment into the skin membrane. It was shown that these constructs are equivalent to two-way multihead finite automata over bounded languages, and that the number of membranes induces an infinite hierarchy of the recognized language classes.

6.8 Topics for Future Research

The theory of P automata provides many promising directions for future research. In addition to studying the limits of the computational power and the size complexity of the different variants, a *systematic comparison of the different model classes to classes of conventional automata* would be of particular interest.

An interesting problem of this type is to characterize the exact power of P automata with non-erasing mappings and sequential rule application. The only result we have so far is that if we denote the language class accepted by one-way logarithmic space bounded Turing machines as $1\text{NSPACE}(\log(n))$, then $r1\text{NSPACE}(\log(n)) \subset 1\text{NSPACE}(\log(n))$.

Since P automata represent cells, i.e. complex systems being in interaction with their environments, system theoretic aspects of these constructs can also be an important and promising part of their future study. Concepts as *robustness, stability, equilibrium, periodical or aperiodical state transition sequence* should be developed and investigated in terms of P automata.

References

[1] A. ALHAZOV: Minimizing evolution-communication P systems and EC P automata. *Brainstorming Week on Membrane Computing* (M. Cavaliere et al., eds.), Technical Report GRLM 26/03, Rovira i Virgili University, Tarragona, Spain, 2003, 23–31.

[2] G. BEL-ENGUIX, R. GRAMATOVICI: Parsing with active P automata. *Lecture Notes in Computer Sci.*, 2933 (2004), 31–42.

[3] G. BEL-ENGUIX, R. GRAMATOVICI: Parsing with P automata. *Applications of Membrane Computing* (G. Ciobanu et al., eds.), Springer, 2006, 389–410.

[4] L. CARDELLI: Brane calculi. Interactions of biological membranes. *Lecture Notes in Computer Sci.*, 3082 (2005), 257–280.

[5] L. CIENCIALA, L. CIENCIALOVA: Membrane automata with priorities. *J. Computer Sci. and Technology*, 19 (2004), 89–97.

[6] G. CIOBANU, V.M. GONTINEAC: Mealy multiset automata. *Intern. J. Foundations of Computer Sci.*, 17 (2006), 111–126.

[7] G. CIOBANU and V.M. GONTINEAC: P machines: An automata approach to membrane computing. *Lecture Notes in Computer Sci.*, 4361 (2006), 314–329.

[8] G. CIOBANU, GH. PĂUN, GH. STEFĂNESCU: P transducers. *New Generation Computing*, 24 (2006), 1–28.

[9] R.S. COHEN, A.Y. GOLD: ω-computations on Turing machines. *Theoretical Computer Sci.* 6 (1978), 1-23.

[10] R.S. COHEN and A.Y. GOLD: On the complexity of ω-type Turing acceptors. *Theoretical Computer Sci.*, 10 (1980), 249–272.

[11] E. CSUHAJ-VARJÚ: P automata. *Lecture Notes in Computer Sci.*, 3365 (2005), 19–35.

[12] E. CSUHAJ-VARJÚ, O.H. IBARRA, GY. VASZIL: On the computational complexity of P automata. *Lecture Notes in Computer Sci.*, 3384 (2005), 77–90.

[13] E. CSUHAJ-VARJÚ, O.H. IBARRA, GY. VASZIL: On the computational complexity of P automata. *Natural Computing*, 5 (2006), 109–126.

[14] E. CSUHAJ-VARJÚ, GY. VASZIL: P automata. *Pre-Proc. Workshop on Membrane Computing WMC-CdeA 2002, Curtea de Argeş, Romania, August 19-23, 2002* (Gh. Păun, C. Zandron, eds.), Pub. No. 1 of MolCoNet-IST-2001-32008, 2002, 177–192.

[15] E. CSUHAJ-VARJÚ, GY. VASZIL: P automata or purely communicating accepting P systems. *Lecture Notes in Computer Sci.*, 2597 (2003), 219–233.

[16] E. CSUHAJ-VARJÚ, GY. VASZIL: (Mem)brane automata. *Theoretical Computer Sci.*, 404 (2008), 52–60.

[17] J. DASSOW, GY. VASZIL: P finite automata and regular languages over countably infinite alphabets. *Lecture Notes in Computer Sci.*, 4361 (2006), 367–381.

[18] P.C. FISCHER: Turing machines with restricted memory access. *Information and Control*, 9 (1966), 364–379.

[19] R. FREUND, C. MARTÍN-VIDE, A. OBTUŁOWICZ, GH. PĂUN: On three classes of automata-like P systems. *Lecture Notes in Computer Sci.*, 2710 (2003), 292–303.

[20] R. FREUND, M. OSWALD: A short note on analysing P systems. *Bulletin of the EATCS*, 78 (October 2002), 231–236.

[21] R. FREUND, M. OSWALD: P automata with activated/prohibited membrane channels. *Lecture Notes in Computer Sci.*, 2597 (2003), 261–269.

[22] R. FREUND, M. OSWALD: P automata with membrane channels. *Artificial Life and Robotics*, 8 (2004), 186–189.

[23] R. FREUND, M. OSWALD: P systems with conditional communication rules assigned to membranes. *J. Automata, Languages and Combinatorics*, 9 (2004), 387–397.

[24] R. FREUND, M. OSWALD, L. STAIGER: ω-P automata with communication rules. *Lecture Notes in Computer Sci.*, 2933 (2004), 203–217.

[25] R. FREUND, S. VERLAN: (Tissue) P systems working in the k-restricted minimally parallel derivation mode. *Intern. Workshop on Computing with Biomolecules, August 2008, Wien, Austria* (E. Csuhaj-Varjú et al., eds.), Österreichische Computer Gesellschaft, 2008, 43–52.

[26] O.H. IBARRA: On membrane hierarchy in P systems. *Theoretical Computer Sci.*, 334 (2005), 115–129.
[27] O.H. IBARRA, GH. PĂUN: Characterization of context-sensitive languages and other language classes in terms of symport/antiport P systems. *Theoretical Computer Sci.*, 358 (2006), 88–103.
[28] M. KAMINSKI, N. FRANCEZ: Finite-memory automata. *Theoretical Computer Sci.*, 134 (1994), 329–363.
[29] M. MADHU, K. KRITHIVASAN: On a class of P automata. *Intern. J. Computer Math.*, 80 (2003), 1111–1120.
[30] M. OSWALD: *P Automata*, PhD dissertation, Vienna University of Technology, 2003.
[31] M. OSWALD, R. FREUND: P Automata with membrane channels. *Proc. Eight Int. Symp. on Artificial Life and Robotics* (M. Sugisaka, H. Tanaka, eds.), Beppu, Japan, 2003, 275–278.
[32] F. OTTO: Classes of regular and context-free languages over countably infinite alphabets. *Discrete Applied Math.*, 12 (1985), 41–56.
[33] GH. PĂUN: Computing with membranes. *J. Computer and System Sci.*, 61 (2000), 108–143.
[34] GH. PĂUN: *Membrane Computing. An Introduction.* Springer, 2002.
[35] GY. VASZIL: A class of P automata characterizing context-free languages. *Proc. Fourth Brainstorming Week on Membrane Computing* (M.A. Gutiérrez-Naranjo et al., eds.), Fénix Editora, Sevilla, 2006, 267–276.
[36] J. ENGELFRIET, H.J. HOOGEBOOM: X-automata on ω-words. *Theoretical Computer Sci.*, 110 (1993), 1–51.
[37] L. STAIGER: ω-languages. *Handbook of Formal Languages* (G. Rozenberg, A. Salomaa, eds.), Springer, 1997, vol. 3, 339–387.
[38] K. WAGNER, L. STAIGER: Recursive ω-languages. *Lecture Notes in Computer Sci.*, 56 (1977), 532–537.

CHAPTER 7

P SYSTEMS WITH STRING OBJECTS

CLAUDIO FERRETTI

GIANCARLO MAURI

CLAUDIO ZANDRON

7.1 Introduction

IN the P systems basic model, membranes contain multisets of objects (represented by symbols of a given alphabet) which are meant to be atomic and hence only subject to be manipulated as a whole, i.e. they are consumed (disappear) to produce new, different, objects (that appear).

An interesting variant, proposed in [18], is obtained by taking into account structured objects and rules that can modify their structure. In particular, objects can be represented by strings over an alphabet. The evolution rules acting on this kind of objects will then be string processing rules of many types: rewriting, splicing, context adjoining rules, and so on. On the one hand, this corresponds to the biological reality, where a series of basic constituents (DNA, RNA, proteins) have a "natural" string structure. On the other hand, rewriting P systems can be immediately considered as language generating devices, and as such compared with the analogous devices from automata and formal languages theory. As a matter of fact, the system will transform the strings given in the initial configuration by using the rewriting rules contained in its different regions, and we can define the generated language as the set of strings collected in a particular output membrane, or sent outside the skin membrane during the computation.

The question that immediately arises concerns the characterization of the classes of languages that we can generate using P systems with different features and their comparison with the families in the Chomsky hierarchy. In particular, we are interested in studying the generating power of rewriting P systems with restrictions on the rewriting rules that can be used: only context-free rules will be allowed [18]. In this view, it is easy to show that context-free rewriting alone, even in the framework of a membrane structure, does not have enough power to generate the class of recursively enumerable (RE) languages, and hence we have to add other features that enhance this power.

Three main features have been introduced that are able to modify the generating power of rewriting P systems, as discussed in the next section. The first feature concerns the use of priority relations on the set of rules from each membrane. Another possibility to increase the generating power consists of modifying the membrane permeability [18, 26]: a membrane can be thin (in this case, strings will be allowed to pass through the membrane) or thick (no transmembrane communication allowed), and specific operators associated to the rules will modify its state, from thin to thick or vice versa; a thin membrane can also disappear. Finally, systems (called extended P systems) where the generated language is "filtered" through a terminal alphabet have been introduced.

In Section 7.3 the main results concerning the comparison among classes of languages generated by rewriting P systems with Chomsky classes, and with other families such as *MAT* and *ORD*, are reported, with particular attention to universality results, i.e. to the capability of generating the family of RE languages. Furthermore, many papers have dealt with the study of hierarchies of languages classes that can be obtained by using a different number of membranes, and some bounds on the minimum number of membranes needed to get the required power have been proved.

Other kinds of restrictions can be considered: for example, in [10] the case of imposing that each string is rewritten in the *leftmost* possible position has been considered, getting a characterization of recursively enumerable languages by systems with seven membranes.

In the same paper the use of *forbidding* conditions associated with rules, in the form of symbols which should not be present in the string to be rewritten, is considered. Somewhat surprisingly, the use of *permitting* conditions (certain symbols should be present in the rewritten string) turns out to be less powerful. In particular, forbidding/permitting conditions can be used to control communication, as described in Section 7.4.

Another aspect to be taken into account is the kind of parallelism we use in processing strings. We can just inherit the parallel processing mechanism used in P systems with atomic objects: all strings are processed in parallel, but each single string is rewritten by only one rule (the parallelism is maximal at the level of strings and rules, but the rewriting is sequential at the level of the symbols from each string). However, the string structure allows for stronger levels of parallelism, such as, for example, in Lindenmayer systems, where all the occurrences of a given symbol in the string are simultaneously rewritten. An intermediate situation is that of rewriting P systems with partial parallelism [12, 13], where each symbol of a string must be rewritten wherever possible, but if several occurrences of a symbol exist, then only one is rewritten. These different approaches to the use of parallelism in rewriting P systems are discussed in Section 7.5, and the resulting language families and hierarchies are compared.

7.2 Basic Definitions

Formally, *a rewriting P system* (or RP system) of degree $n \geq 1$, is a construct

$$\Pi = (V, \mu, M_1, \ldots, M_n, (R_1, \rho_1), \ldots, (R_n, \rho_n), i_0),$$

where:

- V is an alphabet,
- μ is a membrane structure consisting of n membranes (labeled with $1, 2, \ldots, n$),
- M_i, $1 \leq i \leq n$ are finite languages over V,
- R_i, $1 \leq i \leq n$, are finite sets of context free evolution rules. They are of the form $X \to v(tar)$ where X is a symbol of V and $v = v'$ or $v = v'\delta$ or $v = v'\tau$. v' is a string over V and δ, τ are special symbols not in V. $tar \in \{here, out\} \cup \{in_m \mid 1 \leq m \leq n\}$, represents the *target* membrane, i.e. the membrane where the string produced with this rule will go. Often, the indication "here" is omitted.
- ρ_i, $1 \leq i \leq n$, are partial order relations over R_i, representing priorities among rules,
- i_0 is the output membrane.

The membrane structure μ and the finite languages $M_1, \ldots, M_n$ constitute the initial configuration of the system. Membranes can have two different thickness levels. In the initial configuration all membranes are considered of thickness 1. The operators δ and τ produce thickness changes, as will be explained below. We can pass from one configuration to another by applying in parallel the evolution rules to all strings which can be rewritten, obeying the priority relations. The strings generated are collected in a designated membrane, the output one.

Note that each string is processed by one rule only; the parallelism refers to processing simultaneously all available strings by all applicable rules. If several rules can be applied to a string, then one of them is non-deterministically chosen, and we consider the obtained string as the next state of the object described by the string. It is important to highlight the fact that the strings do not evolve independently, but their evolutions are interrelated in two ways:

- If we have priorities, a rule r_1 applicable to a string x can forbid the use of another rule, r_2, of lower priority for rewriting another string y which is present at the same time in the same membrane.
- Even without priority, if a string can be rewritten forever, in a membrane or on an itinerary through several membranes, and this cannot be avoided, then all strings are lost, because the computation never stops.

Moreover, a rule specifies, as previously said, the target membrane. If a rule contains *here* as target membrane (or the target is not specified), it means that the string obtained after the rule is applied will remain in the same region where the rule is applied. If the specified target is *out* and the thickness of the membrane is 1, the string will be sent to the region placed immediately outside. Finally, if the specified target is in_m, where m is a label of a membrane adjacent to the region of the rule, and the membrane m has thickness 1, then the string is sent to that membrane. Notice that, as for the case of symbol object, systems with the generic indication *in* can also be considered. In this case, the string is sent to a membrane non-deterministically chosen among those immediately inside the region where the rule is applied.

If a rule contains the special symbol δ and the membrane where this rule is applied has thickness 1, then that membrane is dissolved and it is no longer recreated; the objects in the membrane become objects of the membrane placed immediately outside, while the rules of the dissolved membrane are removed. If the membrane has thickness 2, this symbol reduces the thickness to 1. If a rule contains the special symbol τ the thickness of the membrane where this rule is applied is increased; the thickness of a membrane of thickness 2 is not further increased. Notice that both these symbols could be introduced in the same region at the same time, when more rules are applied to different strings (one rule for each string) that contain different thickness operators. In this last case, the thickness of the membrane is not changed.

The transfer of objects has priority over the actions of δ and τ: if at the same step an object has to pass through a membrane and a rule changes the thickness of that membrane, then we first transmit the object, and after that we change the thickness.

We denote by $L(\Pi)$ the language generated by Π and by $RP_m(Pri, \delta, \tau)$, $m \geq 1$, the family of languages generated by rewriting P systems with at most m membranes, using priority over evolution rules and both actions indicated by δ and τ.

When one of the features $\alpha \in \{Pri, \delta, \tau\}$ is not used, we write $n\alpha$ instead of α.

We can consider systems in which a terminal alphabet (or output alphabet) $T \subseteq V$ is added. We obtain, in this case, *extended rewriting P systems* or *ERP* systems. The family of languages generated by ERP systems with at most m membranes is denoted by $ERP_m(\alpha, \beta, \gamma)$, $m \geq 1$, where, as in the previous case, $\alpha \in \{Pri, nPri\}$, $\beta \in \{\delta, n\delta\}$, and $\gamma \in \{\tau, n\tau\}$.

7.3 THE POWER OF REWRITING P SYSTEMS

Some preliminary results follow directly from the definitions:

Lemma 7.1 $RP_m(\alpha, \beta, \gamma) \subseteq ERP_m(\alpha, \beta, \gamma)$, $RP_m(\alpha, \beta, \gamma) \subseteq RP_{m+1}(\alpha, \beta, \gamma)$, $ERP_m(\alpha, \beta, \gamma) \subseteq ERP_{m+1}(\alpha, \beta, \gamma)$, for $m \geq 1$ and for all possible $\alpha \in \{Pri, nPri\}$, $\beta \in \{\delta, n\delta\}$, $\gamma \in \{\tau, n\tau\}$.

The computational power of rewriting membrane systems has been investigated in various research papers. We first consider RP systems without priority relations and without variable thickness of membranes. In [18] it is proved that the family of languages generated by rewriting P systems with a single membrane without making use of further features is exactly the family of CF languages. Using two membranes, one can generate non-CF languages:

Theorem 7.1 $CF = RP_1(nPri, n\delta, n\tau) \subset RP_2(nPri, n\delta, n\tau)$.

Thus, adding a membrane to systems with a single membrane increases the power of rewriting P systems. Nonetheless, allowing an unlimited number of membranes does not lead to universality, as shown in [10]:

Theorem 7.2 $MAT = ERP_4(nPri, n\delta, n\tau) = ERP_*(nPri, n\delta, n\tau)$.

Proof. To prove that $MAT \subseteq ERP_*(nPri, n\delta, n\tau)$, we consider a matrix grammar without appearance checking $G = (N, T, S, M)$, in the binary normal form. Each matrix $(X \to \lambda, A \to x)$ is replaced by $(X \to f, A \to x)$, where f is a new symbol.

Such a grammar can be simulated by a system

$$\Pi = (V, T, [_1[_2[_3[_4\ ]_4]_3]_2]_1, \{BZ\}, \emptyset, \emptyset, \emptyset, R_1, R_2, R_3, R_4),$$

with

$$V = N_1 \cup N_2 \cup T \cup \{f, f'\} \cup \{X' \mid X \in N_1\} \cup \{A' \mid A \in N_2\}$$
$$\cup \{X_{i,j} \mid 1 \leq i, j \leq k\} \cup \{A_{i,j} \mid 1 \leq i, j \leq k\},$$
$$R_1 = \{C \to C' \mid C \in N_2\}$$
$$\cup \{A \to (A_{i,1}, in) \mid m_i : (X \to a, A \to x), 1 \leq i \leq k\}$$
$$\cup \{f \to (\lambda, out)\},$$
$$R_2 = \{X \to (X_{i,1}, in) \mid m_i : (X \to a, A \to x), 1 \leq i \leq k\}$$
$$\cup \{X_{i,j} \to (X_{i,j+1}, in) \mid X \in N_1, 1 \leq j < i \leq k\}$$
$$\cup \{a' \to (a, out) \mid a \in N_1 \cup \{f\}\},$$
$$R_3 = \{A_{i,j} \to (A_{i,j+1}, out) \mid A \in N_2, 1 \leq j < i \leq k\}$$
$$\cup \{A_{i,i} \to (x, in) \mid m_i : (X \to a, A \to x), 1 \leq i \leq k\}$$
$$\cup \{a' \to (a', out) \mid a \in N_1 \cup \{f\}\},$$
$$R_4 = \{X_{i,i} \to (a', out) \mid m_i : (X \to a, A \to x), 1 \leq i \leq k\}$$
$$\cup \{C' \to C \mid C \in N_2\}.$$

Assume that we have a string of the form wX in membrane 1 (initially, we have the string BZ, for $(S \to ZB)$ being the start matrix of G). We can prime the non-terminals of w and in any moment we can use a rule $A \to (A_{i,1}, in)$, for some $1 \leq i \leq k$, and the string is sent to membrane 2. The only applicable rule in membrane 2 is $X \to (X_{j,1}, in)$, for some $1 \leq j \leq k$, and the string is sent to membrane 3. Here the second component of the subscript of A is increased by one and the string is sent to membrane 2. The string circulates between membranes 2 and 3 until one of the symbols gets a subscript of the form i, i or j, j. If $i < j$ or $j < i$, then the string gets stuck either in membrane 4 or 2, respectively. If $i = j$, then the string is sent to membrane 4 by the rule $A_{i,i} \to (x, in)$. Here, all primed non-terminals are replaced with the corresponding non-primed symbols, and then the string is sent back to membrane 1. The process can be iterated as long as we do not have $a = f$. The string can be sent out by using the rule $f \to (\lambda, out)$. If the string is terminal, it belongs to $L(\Pi)$. Consequently, $L(\Pi) = L(G)$.

To prove the inclusion $ERP_*(nPri, n\delta, n\tau) \subseteq MAT$, we consider an ERP system $\Pi = (V, T, \mu, M_1, \ldots, M_n, R_1, \ldots, R_n)$ generating the language $L(\Pi)$. We consider the skin membrane labeled with 1.

Clearly, each string present at any time in the system is rewritten independently with respect to the other strings. The only possible mutual influence is the fact that if a string can be rewritten forever, then no output of the computation is accepted. Thus, if $L(\Pi) \neq \emptyset$ (otherwise, the language is trivially in MAT), then there is an axiom w which leads to a string in T^* which exits the system, while other axioms either lead to strings which leave the system or to strings to which no rule can be applied. Thus, the axioms and their descendant strings can be assumed to evolve independently of each other, without taking care whether or not other strings can be rewritten forever.

We build a matrix grammar (without appearance checking) $G = (N, T, S, M)$, generating the same language as Π in the following way (h is the morphism defined by $h(a) = a', a \in V$):

$N = \{a' \mid a \in V\} \cup \{[i] \mid 1 \leq i \leq n\} \cup \{E, S\}$,

$M = \{(S \to [i]h(w)) \mid w \in M_i, 1 \leq i \leq n\}$

$\cup \{([i] \to [i], A' \to h(x)) \mid A \to (x, here) \in R_i, 1 \leq i \leq n\}$

$\cup \{([i] \to [j], A' \to h(x)) \mid A \to (x, in) \in R_i, 1 \leq i \leq n,$ and j is the label of a membrane placed directly inside membrane $i\}$

$\cup \{([i] \to [j], A' \to h(x)) \mid A \to (x, out) \in R_i, 2 \leq i \leq n,$ and j is the label of the membrane surrounding membrane $i\}$

$\cup \{([1] \to E, A' \to h(x)) \mid A \to (x, out) \in R_1\}$

$\cup \{(E \to E, a' \to a) \mid a \in T\}$

$\cup \{(E \to \lambda)\}$.

It is easy to see that the symbols $[i]$, $1 \leq i \leq n$, control the work of the grammar G in such a way that the computations in Π are correctly simulated: the symbols $[i]$ indicate the rules to be used as well as the membrane where the corresponding string is placed in the next configuration. In this way, all computations in Π can be simulated by derivations in G, working with primed symbols. When a string is to be sent out of the system, the symbol E is introduced, no further rule from Π can be simulated, and the symbols of T lose their primes. If no symbol from $V - T$ is present and if the unpriming is complete, then we get a string in $L(G)$. Consequently, $L(\Pi) = L(G)$, which completes the proof.

In order to obtain Turing powerful rewriting P systems we need to make use of other features, such as priority relations. In this case, universality can be obtained with a very simple membrane structure.

We denote by *ORD* the family of languages generated by context-free ordered grammars. Obviously, we have $RP_1(Pri, n\delta, n\tau) = ORD$. In [18] one shows that, using systems with priorities, three membranes suffice to generate all RE languages. The result has been improved in [11], where the following result is proved

Theorem 7.3 $RE = RP_2(Pri, n\delta, n\tau)$.

Proof. The inclusion $RP_2(Pri, n\delta, n_T) \subseteq RE$ follows directly from the Church–Turing thesis, or it can be proved (with a long and quite technical proof) by building the Turing machine simulating the system. We prove here the opposite inclusion.

Again, consider a matrix grammar with appearance checking $G = (N, T, S, P, F)$ in the binary normal form. Then consider a grammar G' in which we substitute each matrix of type 4 $(X \to \lambda, A \to x)$, where $x \in T^*$, with a matrix $(X \to X', A \to x)$ (which is considered of type 4'), and with the additional matrices $(X' \to \lambda)$, where X' is a symbol not in G associated with X. Clearly, $L(G) = L(G')$. We assume the matrices of the types 2, 3, and 4' labeled in a one-to-one manner, with $m_1, m_2, \ldots, m_k$. The following rewriting P systems with 2 membranes generates the same language as G':

$\Pi = (V, \mu, M_1, M_2, (R_1, \rho_1), (R_2, \rho_2), 2)$,

$V = N_1 \cup N_2 \cup \{E, Z, \dagger\} \cup T \cup \{X_i, X'_i \mid X \in N_1, 1 \le i \le k\}$,

$\mu = [_1 [_2]_2]_1$,

$M_1 = \{XAE \mid S \to XA \text{ is the rule of the matrix of type 1}\}$,

$M_2 = \emptyset$,

$R_1 = \{r_a : a \to a \mid a \in V - T, a \ne E\}$

$\cup \{r_0 : E \to \lambda(in_2), \dagger \to \dagger\}$

$\cup \{X \to Y_i(in_2) \mid m_i : (X \to Y, A \to z_i) \text{ type 2 or type 3 matrix},$

$z_i \in (N_2 \cup T)^* or\ z_i = \dagger, 1 \le i \le k\}$

$\cup \{X \to X'_i(in_2), X'_i \to \lambda \mid m_i : (X \to X', A \to x) \text{ type 4' matrix}\}$

$\cup \{Y_i \to Y, Y'_i \to Y \mid Y \in N_1, 1 \le i \le k\}$,

$\rho_1 = \{r_a > r_0 \mid a \in V - T, a \ne E\}$,

$R_2 = \{r_i : Y_i \to Y_i, r'_i : A \to x(out)$

$\mid m_i : (X \to Y, A \to x) \text{ type 2 matrix}\}$

$\cup \{p_i : Y_i \to Y'_i, p'_i : Y'_i \to Y_i, p''_i : A \to \dagger(out) \mid$

$m_i : (X \to Y, A \to \dagger) \text{ type 3 matrix}\}$

$\cup \{q_i : X'_i \to X'_i, q'_i : A \to x \mid m_i : (X \to X', A \to x) \text{ type4' matrix}\}$

$\cup \{s_0 : E \to E(out)\}$,

$\rho_2 = \{r_i > r'_j, \{p_i > p''_j, p'_i > p''_j, q_i > q'_j \mid i \ne j, \text{ for all } i, j\}$

$\cup \{r_i > p''_j, r_i > q'_j, p_i > r'_j, p_i > q'_j, p'_i > r'_j, p'_i > q'_j,$

$q_i > r'_j, q_i > p''_j \text{ for all } i, j\}$

$\cup \{r_i > s_0, p''_i > p_i > s_0, q_i > s_0 \mid \text{ for all } i\}$.

Consider the string XwE in membrane 1 with $w \in (N_2 \cup T)^*$ (initially we have XAE). We can apply a production $\{r_a : a \to a | a \in V - T, a \neq E\}$, and we get the same string XwE, or we can apply one of the rules that simulates the first rule of a matrix of type 2, 3, or 4. Let us examine the last three cases separately.

If we apply a production $X \to Y_i(in_2)$ for a type 2 matrix, we get Y_iwE, where i corresponds to a label of a matrix of type 2. This string is sent to membrane 2 where the we can apply a rule $r'_j : A \to m(out)$, and in particular the rule for $j = i$ because of the priority relations. If we apply this rule, we have correctly simulated a type 2 matrix and the obtained string is sent back in membrane 1. Notice that if the symbol A is not in the string, the computation will never halt, because we can forever apply the rule $Y_i \to Y_i$.

The simulation of a type 4 matrix proceeds in a similar way.

If we apply a production $X \to Y_i(in_2)$ for a type 3 matrix we get again a string Y_iwE where i corresponds to a label of a matrix of type 3, and the string is sent to membrane 2. Due to the priority relations, if the symbol A is in the string then we have to apply the rule $p''_k : A \to \dagger(out)$. The computation will never halt. If the symbol A is not in the string, we can apply the rule $Y_j \to Y'_j$, and we get the string Y'_jwE. On this string, we can apply a rule $Y'_j \to Y_j$, and we get again the string Y_jwE, or a rule $E \to E(out)$, to send the string in membrane 1.

In membrane 1 we can replace the symbols Y_i and Y'_i with Y and the symbol X'_i with λ. In this way, we can iterate the process of simulating the matrices of type 2, 3, and 4'. The rule $E \to \lambda(in_2)$ is used to send a terminal string in the output membrane. If any non-terminal symbol is present in the string, this rule cannot be applied, because we have to apply a rule $r_a : a \to a$ due to the priority relations.

In the output membrane, we collect the terminal strings generated by G', thus $L(G') = L(\Pi)$.

Universality can also be obtained without priority relations by making use of the variable thickness feature ([19]). The result was first proved in [25] and then improved in [9, 23]:

Theorem 7.4 $RE = RP_4(nPri, \delta, \tau)$.

Proof. The inclusion $RP(nPri, \delta, \tau) \subseteq RE$ follows directly from the Church–Turing thesis. We prove here the opposite inclusion.

We construct a rewriting P system without priority but with variable thickness that generates the language generated by a matrix grammar with appearance checking $G = (N, T, S, P, F)$ in the binary normal form, where each matrix of type 4 $(X \to \lambda, A \to x)$, is replaced with a matrix $(X \to X', A \to x)$ (which is considered of type 4'),

$$\Pi = (V, [_1[_2\ ]_2[_3\ ]_3[_4\ ]_4]_1, M_1, M_2, M_3, M_4, R_1, R_2, R_3, R_4).$$

The matrices of G without rules used in the appearance checking mode are simulated by rules in the skin membrane and by membrane 2, while the appearance checking control is performed in membranes 3 and 4.

The main idea behind the construction (which we only sketch here) is to have exactly two strings in the system: a string XwE (initially XAE) used to generate the string also generated by the matrix grammar, and a string F, used to control the application of the rules on the first string by means of the variable thickness feature of the membranes.

When a matrix is simulated, the presence of both strings at the same time in the same matrix is checked. This is done by applying a rule using a τ symbol to the first string XwE, and a rule with a δ symbol to the second string F.

If both strings are present at the same time in the same membrane, then both operations τ and δ are performed at the same time on the membrane, thus its thickness is not changed and the simulation can proceed correctly. Otherwise, the membrane where the first string XwE is present increases its thickness, due to the τ symbols, thus blocking the string inside it forever. At the same time, the membrane where the second string F is present is dissolved by the use of the symbol δ, inducing an infinite computation.

The changes of the thickness of membranes also allow the simulation of the appearance checking. Both strings are sent to the membrane used to simulate type 3 matrices, and its thickness is increased to 2, so that the strings cannot exit for the moment. While the appearance checking is performed on the string XwE, the second string F decreases the thickness again to 1. Now both strings can exit the membrane (without changing its thickness) and the simulation of a type 3 matrix is correctly done.

Partial results are known for membrane structures with less than four membranes:

Theorem 7.5 $RP_3(nPri, \delta, \tau) - MAT \neq \emptyset$; $MAT \subseteq RP_2(nPri, \delta, \tau)$; $RP_1(nPri, \delta, \tau) - CF \neq \emptyset$.

Different restrictions can be considered, for instance in the way the rewriting rules are applied to string objects. In [10] some restrictions which are well-known in formal language theory have been considered, for instance, leftmost derivations: any string is rewritten in the leftmost position which can be rewritten by a rule from its region. That is, the symbols of the string are examined one by one, from left to right; the first symbol which can be rewritten by a rule from the region where the string is present, is rewritten. If there are several rules with the same left hand symbol, then any of them is chosen.

We denote by $L_{left}(\Pi)$ the language generated by a system Π in this way and by $(E)RP_m(left)$, $m \geq 1$, we denote the family of all such languages, generated by (extended) systems with at most m membranes.

The following result was first produced in [10] and then improved in [15]:

Theorem 7.6 $RE = RP_4(left)$.

Again, only partial answers are known for rewriting P systems with less than four membranes, in particular (from [10, 15]) we have:

Theorem 7.7 $ET0L \subset ERP_3(left)$, $0L \subset RP_2(left)$, $E0L \subseteq ERP_2(left)$; $MAT \subset RP_2(left)$.

Two further restrictions, also considered in [10], are *forbidding conditions* and *permitting conditions*.

In the first case the rules are of the form $\langle X \to x; F \rangle$, where $F \subseteq V$, and the rule $X \to x$ can be applied only to the strings which do not contain any symbol from F (if $F = \emptyset$, then the rule is applied without any restriction). We denote by $L_{forb}(\Pi)$ the language generated by a P system Π using such rules, and by $(E)RP_m(forb)$, $m \geq 1$, we denote the family of all such languages, generated by (extended) systems with at most m membranes.

When we consider the application of the rules with *permitting conditions*, the rules are still of the form $\langle X \to x; F \rangle$, where $F \subseteq V$, but the rule $X \to x$ can be applied only to the strings which contain all symbols in F (again, if $F = \emptyset$, then no restriction applies). We denote by $L_{perm}(\Pi)$ the language generated by a P system Π using such rules and by $(E)RP_m(perm)$, $m \geq 1$, we denote the family of all such languages, generated by (extended) systems with at most m membranes.

As usual, if the degree of the systems is not bounded, then we replace the subscript m by $*$.

When we use forbidding conditions associated with rules, two membranes suffice to generate all RE languages, and this remains true also for non-extended systems:

Theorem 7.8 $RE = RP_2(forb) = RP_*(forb) = ERP_2(forb) = ERP_*(forb)$.

On the contrary, and quite surprisingly, the use of permitting conditions with extended rewriting P systems does not increase the computational power: the class of languages generated by such systems coincides with the class of languages generated by matrix grammars without appearance checking.

Theorem 7.9 $MAT = ERP_*(perm) = ERP_2(perm)$.

Proof. The inclusion $ERP_2(perm) \subseteq ERP_*(perm)$ is straightforward.

The inclusion $MAT \subseteq ERP_2(perm)$ can be proved by using a construction where rules of the form $\langle A \to (x, in); \{X_i\} \rangle$ are present to ensure that the second production of the matrix i is simulated only if the string contains the symbol X_i.

To prove the inclusion $ERP_*(perm) \subseteq MAT$, let us consider an ERP system with permitting conditions, $\Pi = (V, T, \mu, M_1, \ldots, M_n, R_1, \ldots, R_n)$ generating the language $L_{perm}(\Pi)$. We consider the skin membrane labeled with 1.

As already stated in the proof of Theorem 7.2, each string present at any time in the system evolves independently with respect to the other strings present in the system.

Thus, to simulate the system Π we build a matrix grammar (without appearance checking) $G = (N, T, S, M)$ similar to the one used for Theorem 7.2, generating the same language as Π. We only need to add some further rules to check the permitting conditions (h is the morphism defined by $h(a) = a', a \in V$):

$N = \{a' \mid a \in V\} \cup \{[i] \mid 1 \leq i \leq n\} \cup \{E, S\}$,

$M = \{(S \to [i]h(w)) \mid w \in M_i, 1 \leq i \leq n\}$

$\cup \{([i] \to [i], a_1' \to a_1', \ldots, a_k' \to a_k', A' \to h(x)) \mid$

$(A \to (x, here); \{a_1, \ldots, a_k\}) \in R_i, 1 \leq i \leq n\}$

$\cup \{([i] \to [j], a_1' \to a_1', \ldots, a_k' \to a_k', A' \to h(x)) \mid$

$(A \to (x, in); \{a_1, \ldots, a_k\}) \in R_i, 1 \leq i \leq n$, and j is the label

of a membrane placed directly inside membrane $i\}$

$\cup \{([i] \to [j], a_1' \to a_1', \ldots, a_k' \to a_k', A' \to h(x)) \mid$

$(A \to (x, out); \{a_1, \ldots, a_k\}) \in R_i, 2 \leq i \leq n$, and j is the label

of the membrane surrounding membrane $i\}$

$\cup \{([1] \to E, a_1' \to a_1', \ldots, a_k' \to a_k', A' \to h(x)) \mid$

$(A \to (x, out); \{a_1, \ldots, a_k\}) \in R_1\}$

$\cup \{(E \to E, a' \to a) \mid a \in T\}$

$\cup \{(E \to \lambda)\}$.

As previously said, the symbols $[i], 1 \leq i \leq n$, control the work of the grammar G in such a way that the computations in Π are correctly simulated. At the same time, the mode of working specific to matrix grammars makes possible the checking of the permitting conditions by using rules of the form $a' \to a'$, for a being an element of a permitting conditional set of a rule. In this way, all computations in Π can be simulated by derivations in G, working with primed symbols. When a string is to be sent out of the system, the symbol E is introduced, no further rule from Π can be simulated, and the symbols of T lose their primes. If no symbol from $V - T$ is present and if the unpriming is complete, then we get a string in $L(G)$. Consequently, $L_{perm}(\Pi) = L(G)$, which completes the proof.

Combining the results in Theorem 7.2 with this last theorem, we have the following result:

Corollary 7.1 $MAT = ERP_4(nPri, n\delta, n\tau) = ERP_*(nPri, n\delta, n\tau) = ERP_2(perm) = ERP_*(perm)$.

7.3.1 Normal Forms

Constraints can be imposed on the structure of P systems, for instance, in terms of how many membranes are allowed, or how many rules may be in each of them. Results are known about which kind of constraints still allow classes of P systems which are universal.

If the constraints state the exact number of allowed membranes or rules, and if the corresponding classes of P systems are universal, then that kind of structure may be considered a *normal form*.

Here, only constraints on the depth of the membrane structure and on the number of rules in each membrane are considered. Given such two parameters, a generalized definition of normal form for P systems can be given [16].

Definition 7.1 *A rewriting P system is in m-n-normal-form if its depth is exactly m and if in each membrane we have exactly n rewriting rules. If we put no restriction either on the depth or on the number of rewriting rules, then we replace the corresponding term (i.e. either m or n respectively) with *.*

Results in literature have considered normal forms of type m-$*$ [20, 16], m-n [26, 24, 16], or $*$-n [24], usually for rewriting systems without symbol parallelism, but some authors considered also occurrence parallelism [16]. Here we list the results.

Theorem 7.10 *Every recursively enumerable language can be generated by a P system of type $(Pri, n\delta, n\tau)$ in 2-2-normal-form.*

The next result relates the use of priorities in changing from one normal form to another.

Theorem 7.11 *Every language generated by a rewriting P system of degree m in k-*-normal-form, with or without priorities, can be generated by a rewriting P system of degree m in 2-*-normal-form with priorities.*

7.4 REWRITING P SYSTEMS WITH CONDITIONAL COMMUNICATION

In many reactions in the biological cell, many molecules can pass through membranes only if they are of a specific shape or if they contain some specific subsequences.

Starting from this idea, in [8] the possibility of controlthe communication of strings depending on the contents of the strings themselves was considered, based on the symbols which appear in a string or on the shape of the string.

Such a system is defined as in the basic variant of RP systems, where for each set of rules R_i, a set of *permitting conditions* P_i and a set of *forbidding conditions* F_i are also defined.

The conditions can be of various forms:

1. *Empty*: no restriction is imposed on strings, they either exit the current membrane or enter any of the directly inner membrane freely (but they cannot remain in the current membrane); an empty permitting condition is denoted by $(true, X)$, $X \in \{in, out\}$, and an empty forbidding condition by $(false, notX)$, $X \in \{in, out\}$.
2. *Symbols checking*: each P_i is a set of pairs (a, X), $X \in \{in, out\}$, for $a \in V$, and each F_i is a set of pairs $(b, notX)$, $X \in \{in, out\}$, for $b \in V$; a string w can go to a lower membrane only if there is a pair $(a, in) \in P_i$ with $a \in alph(w)$, and for each $(b, notin) \in F_i$, we have $b \notin alph(w)$; similarly, for sending the string w out of membrane i it is necessary to have $a \in alph(w)$ for at least one pair $(a, out) \in P_i$ and $b \notin alph(w)$ for all $(b, notout) \in F_i$.
3. *Substrings checking*: each P_i is a set of pairs (u, X), $X \in \{in, out\}$, for $u \in V^+$, and each F_i is a set of pairs $(v, notX)$, $X \in \{in, out\}$, for $v \in V^+$; a string w can go to a lower membrane only if there is a pair $(u, in) \in P_i$ with $u \in Sub(w)$, and for each $(v, notin) \in F_i$ we have $v \notin Sub(w)$; similarly, for sending the string w out of membrane i it is necessary to have $u \in Sub(w)$.
4. *Prefix/suffix checking*: exactly as in the case of substrings checking, with the checked string being a prefix or a suffix of the string to be communicated.
5. *Shape checking*: each P_i is a set of pairs (e, X), $X \in \{in, out\}$, where e is a regular expression over V, and each F_i is a set of pairs $(f, notX)$, $X \in \{in, out\}$, where f is a regular expression over V; a string w can go to a lower membrane only if there is a pair $(e, in) \in P_i$ with $w \in L(e)$, and for each pair $(f, notin) \in F_i$ we have $w \notin L(f)$; similarly, for sending the string w out of membrane i it is necessary to have $w \in L(e)$ for at least one pair $(e, out) \in P_i$ and $w \notin L(f)$ for all $(f, notout) \in F_i$.

These conditions are denoted by *empty*, *symb*, sub_k, $pref_k$, $suff_k$, *patt*, respectively, where k is the length of the longest string in all P_i, F_i; when no upper bound on this length is imposed we replace the subscript by $*$.

The family of languages computed by (extended) systems of degree at most $m \geq 1$ and with permitting conditions of type α and forbidding conditions of type β, is denoted by $(E)RP_m(\alpha, \beta)$, $\alpha, \beta \in \{empty, symb, sub_*, pref_*, suff_*, patt\} \cup \{sub_k, pref_k, suff_k \mid k \geq 1\}$.

When using both prefix and suffix checking, this is indicated by $prefsuff_k$. As usual, if the degree of the systems is not bounded, then the subscript m is replaced by $*$.

7.4.1 Computational Power

As expected, the possibility of controling the string movement according to the string properties is a powerful feature. The following results, that were first proved in [8] and then improved in [15, 21], show that in many cases we obtain universality by using a quite simple membrane structure.

Theorem 7.12 $RP_2(patt, empty) = RP_2(empty, sub_2) = RP_2(sub_2, symb) = RP_3 (sub_2, empty) = ERP_6(symb, empty) = RP_3(symb, symb) = RP_8(prefsuff_2, empty) = RE$.

Proof. We prove here the first result, to show the general technique used in the proofs. We refer the reader to the papers cited at the beginning of this section for the remaining proofs.

Let $G = (N, T, S, P)$ be a type-0 Chomsky grammar in the Kuroda normal form, and assume that all non-context-free rules in P are labeled in a one-to-one manner.

To prove $RP_2(patt, empty) = RE$, we construct the rewriting P system

$$\Pi = (V, V, [_1[_2\ ]_2]_1, S, \emptyset, R_1, P_1, F_1, R_2, P_2, F_2),$$

where:

$$V = T \cup N \cup \{(A, r), [B, r] \mid r : AB \to CD \in P\},$$
$$R_1 = \{A \to x \mid A \to x \in P\}$$
$$\cup \{A \to (A, r), B \to [B, r] \mid r : AB \to CD \in P\},$$
$$P_1 = \{(T^*, out)\} \cup \{((N \cup T)^*(A, r)[B, r](N \cup T)^*, in)$$
$$\mid r : AB \to CD \in P\},$$
$$F_1 = \{(false, notin), (false, notout)\},$$
$$R_2 = \{(A, r) \to C, [B, r] \to D \mid r : AB \to CD \in P\},$$
$$P_2 = \{((N \cup T)^*, out)\},$$
$$F_2 = \{(false, notout)\}.$$

Only one string is present in the system at every moment (initially, the string is the axiom of G). If a string contains at least a non-terminal symbol of G then it is either rewritten in the skin membrane, or it remains forever there, hence we get no output. A string present in the skin membrane can be sent to the membrane inside it only if it is of the form $x(A, r)[B, r]y$, for some rule $r : AB \to CD$ and $x, y \in (N \cup T)^*$, which ensures the correct simulation in membrane 2 of the rule. After replacing (A, r) with C and $[B, r]w$ with D, the string exits membrane. The context-free rules of P are simulated in the skin membrane. If a string contains only terminal symbols, then it can be sent out of the system. Consequently, $L(G) = L(\Pi)$.

For other families with very simple membrane structures only partial results are known.

Theorem 7.13 $RP_1(empty, symb) = CF$; $ERP_1(empty, empty) = RP_1(empty, empty) = FIN$; $RP_1(symb, empty) - LIN \neq \emptyset$; $ERP_1(symb, empty) \subseteq CF$; $RP_2(empty, empty) - CF \neq \emptyset$; $RP_2(empty, symb) - MAT \neq \emptyset$.

Theorem 7.14

1. All one-letter languages in $ERP_1(sub, empty)$ are finite.
2. The families $ERP_1(symb, empty)$ and $RP_1(symb, empty)$ are incomparable with REG, LIN, and strictly included in CF.
3. For each regular language $L \subseteq T^*$ and $c \notin T$, the language $L\{c\}$ is in $RP_1(symb, empty)$.

7.5 PARALLEL APPLICATION OF RULES

When considering rewriting P systems, it is quite natural to allow the application of several rules to a single string, as in many cases a cellular component is processed by many chemical reactions working at the same time on different sites. This was first considered in [13].

Considering the parallel application of the rewriting rules means that, at each step of a computation, more symbols of a string are replaced at the same time, in a single, atomic operation. The set of the rules that can be applied differs according to the chosen semantics of parallelism.

When we consider rewriting P systems with a parallel application of rules a problem arises: different rules applied on the same string can have different target indications. As a result, we have consistency problems for the communication of the resulting string, because there could be contradictory indications about the region where the string has to be sent. For instance, a rule could replace a symbol and have a target indication *here*, while another rule applied at the same time on a different symbol could have a target indication *out*.

Up to now, in the literature three types of P systems with parallel rewriting of strings have been proposed, each one facing the communication problem in a different way: P systems with partially parallel rewriting ([11, 12, 13]), P systems with deadlock ([2, 4, 1]), and P systems without target conflicts ([6, 3]).

We briefly recall now the definitions of main parallelism semantics which are standard in the formal language literature and the properties of the previous three types of parallel rewriting P systems, and then we summarize known results concerning various types of parallel rewriting P systems.

7.5.1 Parallel Rewriting Semantics

In this section we present the main parallel rewriting methods for context free rules. We assume that two or more rules are not allowed to rewrite a symbol at the same time, as in interactionless Lindenmayer systems.

Definition 7.2 *In a maximal parallelism rewriting step (M), all occurrences of all symbols (which can be the subject of a rewriting rule) are simultaneously rewritten by rules which are non-deterministically chosen in the set of all applicable rewriting rules. That is, if the string $w = x_1 a_1 x_2 a_2 x_3 a_3 \ldots x_n a_n x_{n+1}$, with $w \in V^+$, $a_1, \ldots, a_n \in V$ and $x_i \in (V - \{a_1, \ldots, a_n\})^*$, $i = 1, \ldots, n+1$, is such that there are no rules defined over symbols in the strings $x_1, \ldots, x_{n+1}$, and there are some rules $r_1 : a_1 \to \alpha_1, \ldots, r_m : a_m \to \alpha_m$, not necessarily distinct, then we obtain in one maximally parallel rewriting step the string $w' = x_1 \alpha_1 x_2 \alpha_2 x_3 \alpha_3 \ldots x_m \alpha_m x_{m+1}$.*

Definition 7.3 *A rewriting operation with unique parallelism (U) consists in the substitution of all occurrences of exactly one symbol according to exactly one rule, which is non-deterministically chosen between all rules that can be applied to that symbol. That is, given a string $w = x_1 a x_2 a x_3 \ldots x_n a x_{n+1}$ with $x_i \in (V - \{a\})^*$, $i = 1, \ldots, n+1$, and one context-free rule $r : a \to \alpha$, in one parallel rewriting step we obtain the string $w' = x_1 \alpha x_2 \alpha x_3 \ldots x_n \alpha x_{n+1}$.*

Definition 7.4 *When considering a symbol parallelism rewriting step (S), for each symbol that can be the subject of a rewriting rule, all of its occurrences are substituted according to the same rule. That is, given some distinct symbols $a_1, \ldots, a_n$ in V and a string $w = x_1 a'_1 x_2 a'_2 x_3 a'_3 \ldots x_m a'_m x_{m+1}$, with $a'_i \in \{a_1, \ldots, a_n\}$, $i = 1, \ldots, m$, $m \geq n$, and $x_j \in (V - \{a_1, \ldots, a_n\})^*$, $j = 1, \ldots, m+1$, and given one context-free rule for each symbol $r_1 : a_1 \to \alpha_1, \ldots, r_n : a_n \to \alpha_n$ (non-deterministically chosen between all rules which can be applied to each symbol), in one step we obtain the string $w' = x_1 \alpha'_1 x_2 \alpha'_2 x_3 \alpha'_3 \ldots x_m \alpha'_m x_{m+1}$, where $\alpha'_i \in \{\alpha_1, \ldots, \alpha_n\}$, $i = 1, \ldots, m$, and $\alpha'_i = \alpha_k$ in w' if and only if $a'_i = a_k$ in w for some $k = 1, \ldots, n$.*

Definition 7.5 *With an occurrence parallelism rewriting step (O), only one occurrence for each symbol in the string that can be rewritten is non-deterministically chosen to be rewritten. For example, given two symbols a_1, a_2 in V and a string $w = x_1 a_1 x_2 a_2 x_3 a_1 x_4 a_1 x_5 a_2 x_6$, with $x_j \in (V - \{a_1, a_2\})^*$, $j = 1, \ldots, 6$, and given* the context-free rules $r_1 : a_1 \to \alpha_1$, $r_2 : a_2 \to \alpha_2$, in one step we could obtain the string $w = x_1\alpha_1 x_2 x_3 a_1 x_4 a_1 x_5 \alpha_2 x_6$ (we non-deterministically choose the third occurrence of symbol a_1 and the first occurrence of symbol a_2).

Definition 7.6 We can also consider parallelism made of tables of rules (T), as in $(E)T0L$ systems: the set of rewriting rules is divided into subsets of rules. In this case, if we have a string w and some tables $t_1 : [r_1^1 : a_1 \to \alpha_1, \ldots, r_k^1 : a_k \to \alpha_{k_1}], \ldots, t_l : [r_1^l : a_1 \to \alpha_1, \ldots, r_k^l : a_k \to \alpha_{k_l}]$, then in one step only the rules from a table (which is non-deterministically chosen) can be applied, and these rules must be applied in parallel to all occurrences of all symbols in w, but not necessarily following the order the rules appear in the table. Moreover, if some rules in the chosen table are defined over symbols not in w, or if the number of rules in the table exceeds the length of the string, then we skip those (not defined or exceeding) rules without forbidding the application of the entire table.

Definition 7.7 In a partially parallel rewriting step, we apply exactly k rules $(P)_{=k}$ (for a fixed $k \geq 1$) in parallel when we have a string $w = x_1 a_1 x_2 a_2 x_3 a_3 \ldots x_k a_k x_{k+1}$ with $x_i \in V^*, i = 1, \ldots, k+1$, $a_i \in V, i = 1, \ldots, k$, and rules $r_1 : a_1 \to \alpha_1, \ldots, r_k : a_k \to \alpha_k$, not necessarily distinct, and in one step we obtain $w' = x_1 \alpha_1 x_2 \alpha_2 x_3 \alpha_3 \ldots x_k \alpha_k x_{k+1}$. So, in one step we substitute exactly k symbols in the string w, but if the string contains less than k symbols which can be the subject of a rewriting rule, than the parallel rewriting step is blocked.

Definition 7.8 In a partially parallel rewriting step, we apply at most k rules $(P)_{\leq k}$ (for a fixed $k \geq 1$) in parallel if and only if there exists a number k' such that $k' \leq k$ and we can apply exactly k' rules in parallel. In this case, up to k symbols are substituted in the string.

Definition 7.9 In a partially parallel rewriting step, we apply at least k rules $(P)_{\geq k}$ (for a fixed $k \geq 1$) in parallel if and only if there exists a number k' such that $k' \geq k$ and we can apply exactly k' rules in parallel. In this case, at least k symbols are substituted in the string.

7.5.2 Parallel Rewriting P Systems

Parallel rewriting P systems were first considered in [13, 12].

In particular, in [13] the maximally parallel semantics were considered: in one step all regions are processed simultaneously by using the rules in a non-deterministic and parallel manner. This means that in each region the objects to evolve and the rules to be applied to them are non-deterministically chosen, but all objects which can evolve should evolve.

The string is then communicated to the region corresponding to the target (if any) which appears the maximal number of times in the used rules. When several targets have occurred the same number of times and this is the maximum number, then one of them is non-deterministically chosen. In such a way, the problem of target consistency described before is solved.

The families of languages generated by extended parallel rewriting P systems is denoted by $EPRP_n(\pi)$, where n is the degree of the system and $\pi \in \{(M), (U), (S), (O), (T), (P)\}$ is the parallelism semantics adopted. If the number of membrane is not bounded, then the subscript n is replaced with $*$.

In [13] one shows that extended parallel rewriting P systems of this type are Turing-equivalent:

Theorem 7.15 $EPRP_*(M) = RE$.

Proof. Consider a matrix grammar with appearance checking $G = (N, T, S, P, F)$ in the binary normal form. We assume the matrices of the types 2, 3, and 4 labeled in a one-to-one manner, with $m_1, m_2, \ldots, m_n$.

We build a rewriting P system with $n + 1$ membranes which generates the same language as G. The system and the simulation is similar to the simulation presented in the proof of Theorem 7.3.

For each type 3 matrix $m_i : (X \to Y, A \to \dagger)$ the rule $\{X \to (Y_i, in_i)\}$ is added to the skin, which begins the simulation of such a matrix, sending the string to the corresponding membrane which operates the simulation.

In membrane i the rules $\{A \to (\dagger, out)\}$ and $\{Y_i \to (Y, out)\}$ are present. The last rule is used to conclude the simulation. At the same time, the first rule correctly simulates the appearance checking: if the string contains a symbol A, then the rule is applied in parallel to the second one, introducing in this way the trap symbol. Otherwise, the rule is not applied and the string is sent to the skin to continue the simulation of another matrix.

In [12] *occurrence parallelism* was also considered, and the following results were proved:

Theorem 7.16 $EPRP_*(O) = RE$; $EPRP_1(O) - CF \neq \emptyset$; $EPRP_4(O) - MAT \neq \emptyset$.

7.5.3 Parallel Rewriting P Systems with Deadlock

In a parallel P system with deadlock [2], when the set of rules applied have mixed target indications, then the strategy used to face and solve the communication problem is the following: the string is marked to be in a *deadlock state* inside the system. The string is not sent to outer or inner regions but it remains inside the current membrane, though it will not be processed anymore by any other rule; this choice does not mean that the indication *here* determines the target region, it means that further string processing and communications are stopped when a situation of deadlock arises for that string. Notice that other strings can enter the membrane, be processed by local rules, and even exit the region (if they are not in a deadlock state after the application of local rules).

We say that a generic configuration $C_t = (\mu, M_0^t, \ldots, M_n^t)$ is *free* if there are no deadlock states inside the system at that time. Otherwise, we say that the system is in a *deadlock configuration*, and we denote by $\langle M_j^t \rangle$ all multisets which contain at least a deadlocked string. A transition starting from a deadlock configuration will always reach another deadlock configuration, as it is not possible to remove deadlock states. So if $C_t = (\mu, M_0^t, \ldots, \langle M_j^t \rangle, \ldots, M_n^t)$ is a deadlock configuration, then for all $t' \geq t$ it holds that $C_{t'} = (\mu, M_0^{t'}, \ldots, \langle M_j^{t'} \rangle, \ldots, M_n^{t'})$, where other multisets besides $\langle M_j^{t'} \rangle$ could have reached a deadlock state. We remark that the multiset $\langle M_j^{t'} \rangle$ still represents a deadlock state even though it is not necessarily equal to $\langle M_j^t \rangle$, because other strings may have entered membrane j. A configuration where all multisets are in a deadlock state (that is, at least one string in each multiset is in a deadlock state) is said to be a *global deadlock* configuration; otherwise, we talk about *local deadlock* configurations.

In P systems with deadlock, any sequence of transitions of free and deadlock configurations forms a *computation*. If all the strings are in a deadlock configuration, then the computation halts, but produces no output.

We will denote by $(E)DPRP_n^k(\pi, \Delta)$ the family of languages generated by (extended) rewriting P systems of degree at most n and depth at most k, where $\pi \in \{(M), (U), (S), (O), (T), (P)\}$ denotes the used parallelism method and $\Delta \in \{D, nD\}$ denotes systems with or without the possibility of having deadlock states, respectively. We use the notation $EDPRP(\pi, \Delta)$ for systems whose depth or degree are not specified. If the depth is specified but the number of membranes is not limited, then the subscript n is replaced by $*$.

A first set of results from [2] compared the computational power of such systems with respect to standard classes of formal languages, such as languages generated by L systems or matrix grammars.

When we consider systems with maximal parallelism semantics and we have:

Theorem 7.17 $ET0L \subseteq EDPRP_3^2((M), D)$; $ET0L \subset EDPRP_4^4((M), D)$.

Proof. According to Theorem 1.3 in [22], for each language $L \in ET0L$ there exists an ET0L system G which generates L and contains only two tables, that is $G = (V, T, w, P_1, P_2)$. At the first step of a derivation, we use table P_1. After using table P_1, we either use again table P_1 or we use table P_2, and after each use of table P_2 we either use table P_1 or we stop the derivation.

Making use of this observation, we construct the P system $\Pi = (V', T, [_0[_1\ ]_1[_2\ ]_2]_0, M_0, \emptyset, \emptyset, R_0, R_1, R_2)$, such that $L(\Pi) \in EDPRP_3^2((M), D)$, where the alphabet is $V' = V \cup T \cup \{X, X_1, X_2, X_3, X_4, \dagger\}$, with $V \cap T \cap \{X, X_1, X_2, X_3, X_4, \dagger\} = \emptyset$, and the initial multiset is $M_0 = Xw$, with $w \in V^+$. The system contains the following sets of rules:

$$R_0 = \{X \to X_1(in), X_2 \to X_1(in), X_2 \to X_3(in),$$
$$X_4 \to X_1(in), X_4 \to \lambda(out)\},$$
$$R_1 = \{A \to xX_2(out) \mid A \to x \in P_1\} \cup \{X_1 \to \lambda(out), X_3 \to \dagger(out)\},$$
$$R_2 = \{A \to xX_4(out) \mid A \to x \in P_2\} \cup \{X_3 \to \lambda(out), X_1 \to \dagger(out)\}.$$

Each table P_i, $i = 1, 2$, is simulated in membrane m_i, $i = 1, 2$ respectively.

The computation starts in the skin membrane. Using the rule $X \to X_1(in)$, the string Xw can be sent to any of the inner membranes. If it enters membrane 2, then the trap symbol $\dagger$ is introduced and no output will be produced. In this way, we correctly simulate the fact that a derivation in G must always begin using the productions from table P_1. Otherwise, if the string enters membrane 1, we simulate the application of table P_1, the symbol X_1 is erased and some occurrences of the new symbol X_2 are introduced. The string then returns to membrane 0, where both rules $X_2 \to X_1(in)$ and $X_2 \to X_3(in)$ can be applied.

If both rules are simultaneously applied, in any case the trap symbol will be introduced in membrane 1 or 2, by means of the rule $X_3 \to \dagger(out)$ or $X_1 \to \dagger(out)$ respectively. Again, no output will be produced. Instead, if only one rule is applied in the skin membrane to all occurrences of X_2, then the string will enter membrane 1 or 2 and the computation can proceed. In membrane 1 we simulate the table P_1 once more, and the procedure can be iterated in the same way. In membrane 2 we simulate table P_2, in this case we erase the symbol X_3 and introduce some occurrences of the new symbol X_4, then the string returns to membrane 0.

At this moment, we can either choose to stop the simulation of G by using the rule $X_4 \to \lambda(out)$ (which deletes all occurrences of X_4), or to start another simulation of table P_1 by using the rule $X_4 \to X_1(in)$ (which rewrites all occurrences of X_4 into X_1). In any case, we correctly simulate the fact that after using table P_2 we cannot use it again. If $X_4 \to \lambda(out)$ is used, the string exits the system: if it is a terminal string, then it will be accepted, otherwise it will not contribute to the generated language. Observe that if both rules over X_4 are simultaneously applied in the skin membrane, then we have a deadlock state and no output will be produced.

It follows that $L(\Pi) = L(G)$.

For the second statement of the theorem, we consider the language $L = \{(ab^n)^m \mid m \geq n \geq 1\}$, which does not belong to $ET0L$, but can be generated by the system

$$\Pi = (V, T, [_0[_1[_2[_3\ ]_3]_2]_1]_0, \emptyset, \emptyset, \emptyset, M_3, R_0, \ldots, R_3),$$
$$V = \{A, B, B', C, a, b\}, T = \{a, b\}, M_3 = \{AB\},$$

$R_0 = \{B' \to \lambda(out), C \to \lambda(out)\}$;

$R_1 = \{A \to \lambda(in), A \to \lambda(out)\}$;

$R_2 = \{A \to \lambda(out), B' \to bB'(out), C \to A(out), C \to C(out)\}$;

$R_3 = \{A \to AC(here), A \to AC(out), B \to aB'B(here), B \to aB'(out)\}$.

The equality $L = L(\Pi)$ can be easily checked.

Using a similar proof, it is also possible to show that the family $ET0L$ is also included in the family of languages generated by systems which make use of table parallelism, with or without the use of deadlock:

Theorem 7.18 $ET0L \subseteq EDPRP_1^1((T), nD) \subseteq EDPRP_1^1((T), D)$.

Some results, from [1], concern systems using partial parallel rewriting systems with deadlock:

Theorem 7.19 1. $EDPRP((P)_{\leq k}, D) \subseteq EDPRP((P)_{=k}, D) \subseteq EDPRP((P)_{\geq k}, D)$, for all $k \geq 1$;
2. $EDPRP((P)_{=k}, D) = MAT$, for all $k \geq 2$.

Another natural research topic concerns the comparison between systems where the deadlock is allowed or not, in order to understand whether or not the possibility of generating deadlock states improves the power of parallel rewriting P systems. Of course, a system where deadlock states are not allowed can easily be simulated by a system with the same features and where, moreover, the deadlock is allowed. But is it possible to simulate parallel rewriting P systems with deadlock using systems where the deadlock is not allowed? The answer is positive, at least for some types of parallelism, even if it seems that we need a more complex membrane structure, as shown in [2, 4].

When we consider maximal parallelism semantics we have the following results:

Theorem 7.20 $EDPRP_n^k((M), nD) \subseteq EDPRP_n^k((M), D)$, and $EDPRP_n^k((M), D) \subseteq EDPRP_{7n}^{k+2}((M), nD)$, for all $n \geq 1, k \geq 1$.

Proof. The first inclusion directly follows from the definitions.

To show that $EDPRP_n^k((M), D) \subseteq EDPRP_{7n}^{k+2}((M), nD)$, for all $n \geq 1, k \geq 1$, consider a system $\Pi = (V, T, \mu, M_0, \ldots, M_{n-1}, R_0, \ldots, R_{n-1})$ such that $L(\Pi) \in EDPRP_n^k((M), D)$.

We build the P system $\Pi' = (V', T, \mu', M'_0, \ldots, M'_m, R'_0, \ldots, R'_m)$, such that $L(\Pi') \in EDPRP_m^{k+2}((M), nD)$ which generates the same language as Π.

Consider a generic membrane m_i of Π, for any $i = 0, \ldots, n-1$. Such a membrane contains a set of strings M_i, a set of rules R_i and, possibly, a set of other membranes; we will denote with $m_{i,1}, \ldots, m_{i,h}$ the membranes placed immediately inside m_i.

We build a corresponding membrane m'_i in Π' by replacing every string w in M_i with a string Xw, where X is a symbol not in V.

The set of rules of the membrane m'_i is

$$R'_i = \{X \to X_1(in),\ X_{here} \to X,\ X_{in} \to X(in),\ X_{out} \to X(out)\}.$$

(In the skin membrane ($i = 0$), the last rule is replaced with

$$X_{out} \to \lambda(out)).$$

Then, immediately inside m'_i, we add three new membranes denoted by $m_{i(here)}$, $m_{i(in)}$, $m_{i(out)}$. Each new membrane $m_{i(tar)}$, with $tar \in \{here, in, out\}$, will contain the following rules:

$$R_{i(tar)} = \{A \to \bar{y}(in) \mid A \to y(tar) \in R_i\}$$
$$\cup\ \{\bar{B} \to B(out) \mid \forall \bar{B} \in \bar{V}\} \cup \{X_2 \to X_{tar}(out),\ X \to \dagger(out)\}.$$

Moreover, each membrane $m_{i(here)}$, $m_{i(in)}$ and $m_{i(out)}$ will contain a single inner membrane, which will be denoted by $m_{i(here),check}$, $m_{i(in),check}$ and $m_{i(out),check}$ respectively. The rules belonging to $m_{i(tar),check}$, for each $tar \in \{here, in, out\}$, are:

$$R_{i(tar),check} = \{A \to \dagger(out) \mid A \to y(tar') \in R_i,\ tar' \in \{here, in, out\}\ \text{and}$$
$$tar' \neq tar\} \cup \{X_1 \to X_2(out)\}.$$

Finally, we add the rule $X_1 \to \dagger(out)$ in each membrane $m'_{i,1}, \ldots, m'_{i,h}$, which are all placed inside m'_i and correspond to the membranes $m_{i,1}, \ldots, m_{i,h}$ originally placed in membrane m_i.

From the construction of μ', it follows that we need seven membranes in Π' to simulate each membrane in Π, hence $m = 7n$, and that the depth is increased from the value k to the new value $k + 2$.

The idea behind the construction is that in membrane m'_i we always have to apply the rule $X \to X_1(in)$. Then, the obtained string $X_1 w$ is non-deterministically sent to one of the inner membranes. If the string reaches a membrane among $m'_{i,1}, \ldots, m'_{i,h}$ then an infinite computation is obtained by means of the rule $X_1 \to \dagger(out)$.

If the string reaches one of the added membranes $m_{i(here)}$, $m_{i(in)}$, $m_{i(out)}$, we first apply to w all rules with the corresponding target indication (e.g. in $m_{i(here)}$ we simulate all rules having target indication *here*). Then the string is sent to the inner membrane ($m_{i(here),check}$ in our example) to check if other rules with conflicting targets can also be applied. In this last case, an infinite computation is induced. Otherwise the string is sent back to the skin membrane to start the simulation of another computation step of Π.

Using a similar proof, it can be shown that the previous results are still valid even when we consider the symbol parallelism method:

Theorem 7.21 $EDPRP_n^k((S), nD) \subseteq EDPRP_n^k((S), D)$, and $EDPRP_n^k((S), D) \subseteq EDPRP_{7n}^{k+2}((S), nD)$, for all $n \geq 1$, $k \geq 1$.

Finally, some results from [4] concern the relations among different parallelism types. It is possible to show that, if we do not consider limits on the complexity of the membrane structure, systems which use the maximal parallelism or table parallelism have the same generative power. In fact:

Theorem 7.22 $EDPRP_n^k((M), \Delta) \subseteq EDPRP_n^k((T), \Delta)$, for $\Delta \in \{D, nD\}$; $EDPRP_n^k((T), D) \subseteq EDPRP_*^{k+1}((M), D)$; $EDPRP_n^k((T), nD) \subseteq EDPRP_*^{k+3}((M), nD)$.

Proof. The first inclusion can be easily proved by considering a generic P system Π such that $L(\Pi) \in EDPRP_n^k((M), \Delta)$. An equivalent P system Π' of type $EDPRP_n^k((T), \Delta)$ can be obtained by considering the same alphabets and membrane structure as Π and by putting all the rules of any membrane of Π inside a single table in the corresponding membrane of Π'.

The second and third inclusion can be proved by considering a system $\Pi = (V, T, \mu, M_0, \ldots, M_{n-1}, R_0, \ldots, R_{n-1})$ making use of table parallelism (with and without deadlock, respectively), and by building a corresponding system Π' which generates the same language by using maximal parallelism.

Consider a generic membrane m_i of Π, for any $i = 0, \ldots, n-1$, which can contain a set of strings M_i, a set of tables of rules R_i and, possibly, a set $\{m_{i,1}, \ldots, m_{i,h}\}$ of other membranes.

The membrane m_i' in Π', corresponding to m_i in Π, is obtained by replacing every string w in M_i with a string Xw, where X is a symbol not in V. Inside m_i' we add some new membranes, one for each table $t_r \in R_i$, for $r = 1, \ldots, s$, so that the various tables can be simulated by dividing the rules in Π' in different regions.

Adding the rules to control the movement of the string in Π' will lead to a system such that $L(\Pi') = L(\Pi)$.

As a consequence, the following corollary holds:

Corollary 7.2 $EDPRP((M), D) = EDPRP((M), nD) = EDPRP((T), D) = EDPRP((T), nD)$.

7.5.4 Decidable Properties

The result concerning the equivalence between partial parallel rewriting P systems with deadlock and matrix grammars without appearance of Theorem 7.19 has been used in [1] to prove the following important result, concerning the possibility of establishing if a given system will ever reach a deadlock configuration:

Theorem 7.23 *It is decidable whether or not a partial parallel rewriting P system (where exactly k rules are to be applied at each computing step) will ever reach a deadlock configuration.*

Proof. Consider a system $\Pi = (V, T, \mu, M_0, \ldots, M_n, R_0, \ldots, R_n)$ such that $L(\Pi) \in EDPRP((P)_{=k}, D)$. First, we construct a matrix grammar $G = (N, T, S, P)$, such that $L(G) = L(\Pi)$.

Then, we add to G the following *deadlock matrices*: $[\dagger \to \dagger, \bar{A} \to \lambda]$, for all $A \in V \setminus T$, and $[\dagger \to \dagger, \bar{a} \to a]$, for all $a \in T$.

We obtain a new grammar G' which generates the language $L(G)$ plus some new strings $w \in \{\dagger\}T^*$. In fact, after a simulation matrix (corresponding to a deadlock state in Π) has been used in G, by means of the deadlock matrices of G' we generate some new strings where all non-terminal overlined symbols are deleted, while the terminal overlined symbols are substituted with the corresponding non-overlined terminal symbols. Observe that the deadlock matrices do not modify any other original derivation in G, because they can be applied only over the strings which contain the symbol $\dagger$.

Consider now the language $\mathcal{L} = L(G') \cap L_\dagger$, where $L_\dagger = \{w = \dagger x \mid x \in T^*\}$. As $L_\dagger \in REG$ and the family MAT is closed under intersection with regular languages, it follows that also $\mathcal{L}$ belongs to MAT. The language $\mathcal{L}$ is non-empty if and only if the symbol $\dagger$ appears in at least one string of $L(G')$, that is if and only if at least one deadlock state occurs in the system. As the emptiness problem is decidable for the family of languages MAT, it holds that it is also decidable to state whether or not a partial parallel rewriting P system, where exactly k rules are to be applied at each computing step, ever reaches a deadlock configuration.

In [1] it is also shown that we are able to decide not only whether or not a deadlock configuration will occur, but also in which region of the membrane structure it will take place:

Corollary 7.3 *Given a partial parallel P system (where exactly k rules are to be applied at each computing step), it is decidable to determine in which membrane a deadlock state will ever occur.*

7.5.5 Parallel Rewriting P Systems Without Target Conflicts

Another possibility for facing the problem of applying parallel rewriting rules with different target indications was proposed in [6]; in this case, the idea was to avoid conflicts a priori, by allowing only subsets of rules with the same target indication to apply in parallel.

For every region $i = 0, \ldots, n$ of the membrane structure, the set R_i of evolution rules is divided into mutually disjoint subsets of rules which have the same target indications, that is $R_i = \mathcal{H}_i \cup \mathcal{O}_i \cup \mathcal{I}_i$, where $\mathcal{H}_i = \{r \in R_i \mid tar(r) = here\}$, $\mathcal{O}_i = \{r \in R_i \mid tar(r) = out\}$ and $\mathcal{I}_i = \{r \in R_i \mid tar(r) = in\}$.

Consider now some rules $r_1, \ldots, r_m$, for some $m \geq 2$, all of which can be applied to a common string w at the same time. To consider parallel rewriting P systems without target conflicts means that such rules will be applied to w only if no conflicts arise for the target indications, that is, only if it holds that (1) $r_1, \ldots, r_m \in \mathcal{H}_i$, or (2) $r_1, \ldots, r_m \in \mathcal{I}_i$, or (3) $r_1, \ldots, r_m \in \mathcal{O}_i$ (we apply in parallel only the rules which match on the target membrane). According to the chosen set of rules, the resulting string w' (obtained after the parallel application of $r_1, \ldots, r_m$) (1) remains inside the current region i, (2) is communicated to a (non-deterministically chosen) inner region, (3) is communicated to the outer region. If the string exits the system, it can never come back and it may contribute to the output of the computation.

The family of languages generated by extended rewriting P systems without target conflicts, of degree at most n, is denoted by $EWPRP_n(\pi, a)$, where $\pi \in \{(M), (U), (S), (O), (T), (P)\}$ denotes the used parallelism method, and $a \in \{Pri, nPri\}$ defines if the rules are applied following or not priority relations, as explained previously. When the number of membranes is not limited, then the subscript n is replaced by $*$.

The following results, from [4] and [3], show that this kind of system using table or maximal parallelism semantics is equivalent, and have the same generative power of ET0L systems. The proof is similar to the proof of Theorem 7.22.

Theorem 7.24 $EWPRP_2((M), nPri) = EWPRP_*((M), nPri) = EWPRP_1((T), nPri) = EWPRP_*((T), nPri) = ET0L$.

The fact that both families $EWPRP_*((M), nPri)$ and $EWPRP_*((T), nPri)$ collapse to ET0L means that these types of parallel rewriting P systems without target conflicts are strictly less powerful than Turing machines, as $ET0L \subset RE$.

Other results from [6] and [3] concern the use of unique parallelism semantics and symbol parallelism semantics. Also, these results can be proved using similar techniques to those used in the proof of systems which make use of table parallelism. In what concerns unique parallelism we have:

Theorem 7.25 $EDT0L \subseteq EWPRP_*((U), Pri)$; $ED0L \subseteq EWPRP_3((U), Pri)$; $EWPRP_1((U), nPri) - ED0L \neq \emptyset$; $EWPRP_n((U), Pri) \subseteq EWPRP_*((T), Pri)$, for every $n \geq 1$.

If we consider symbol parallelism, we have:

Theorem 7.26 $EDT0L = EWPRP_*((S), nPri)$; $EDoL \subset EWPRP_1((S), nPri)$.

From these results, it turns out that parallel rewriting P systems without target conflicts which use unique or symbol parallelism are equivalent, and they are strictly less powerful than systems with table or maximal parallelism:

Corollary 7.4 $EWPRP_*((U), Pri) = EWPRP_*((S), nPri) = EWPRP_3((S), nPri) = EDT0L \subset EWPRP_1((T), nPri) = EWPRP_2((M), nPri)$.

7.5.6 Relations Among the Three Types of Parallel P Systems

In the previous sections we have reported definitions and all known results about the three types of parallel rewriting P systems, stating that (i) Krishna–Rama P systems are able to generate every enumerable language (RE family), (ii) P systems with deadlock (which use maximal parallelism) strictly contain the $EToL$ family, (iii) P systems without target conflicts (with respect to any analyzed parallel rewriting method) have at most the same computational power as ETOL systems.

Hence, P systems without target conflicts are at the bottom of the *hierarchy of parallel rewriting P systems*, while Krishna–Rama systems are at its top. P systems with deadlock and (M)-parallelism, whose exact computational power is still unknown, are placed in the middle of the hierarchy.

Consequently, it can be easily established that, when maximal parallelism is used in each type of parallel rewriting P system, then the following hierarchy holds:

Theorem 7.27 $EWPRP_*(M, nPri) \subset EDPRP_4^4((M), D) \subseteq EPRP_*(M)$.

7.6 P Systems with Replicated Rewriting

The model for rewriting P systems proposed in [14] introduces a single variant to the base model: when applied, a single rule can be defined to produce multiple strings, starting from the same original. Formally, rules can have the form:

$$X \to v_1(tar_1) \| \ldots \| v_k(tar_k),$$

where X is a symbol of V, v is a string over V, and where $tar_i \in \{here, out, in_m\}$, $1 \leq m \leq n$, represents the target among the n available membranes in the usual way.

Rules with $k > 1$ operate a *replicated rewriting*, where a string $x_1 X x_2$ matched by the rule is transformed in a set of k strings $x_1 v_i x_2$. Each of them is sent to the respective tar_i membrane.

The family of languages generated by P systems with replicated rewriting, and with at most m membranes, is denoted by RRP_m. Again, we can define as usual the families of languages $ERRP_m$, or RRP_* and $ERRP_*$, when defining extended systems having a terminal alphabet $T \subseteq V$, or having an unbounded number of membranes, respectively.

Small systems with replicated rewriting can characterize RE [17]:

Theorem 7.28 $ERRP_6 = RRP_6 = RE$.

This theorem is proved by defining a P system which generates the same language of a given matrix grammar with appearance checking (in the strong binary normal form), and it is known that the family of such languages is equal to RE.

The P system has six membranes, with the structure $[_1[_2[_3[_4\ ]_4]_3]_2[_5\ ]_5[_6\ ]_6]_1$, and it has the further properties that no terminal alphabet has to be defined, and that its replication rules produce at most two strings.

It is also possible to prove the same result by using, in place of targets in_m, only targets in, which non-deterministically choose any directly lower membrane. If a string is sent to the wrong membrane, it finds rules of the form $A \to Z, Z \to Z$, where the "trap-symbol" Z is produced and therefore that non-deterministic branch of computation will never stop.

A more specific property of systems with replicated rewriting is to be able to solve well known NP-complete problems in linear time [14].

Theorem 7.29 *P systems with replicated rewriting can efficiently solve the following problems:*

- *SAT, in linear time in the number of variables and the number of clauses;*
- *HPP, in linear time in the number of nodes of the given graph.*

Proof. We consider here only SAT problem. For a given formula $C_1 \wedge \ldots \wedge C_m$ in conjunctive normal form over variables $x_1, \ldots, x_n$, we build the system:

$$\Pi = (V, V, \mu, M_1, \ldots, M_m, M_{m+1}, R_1, \ldots, R_m, R_{m+1}),$$

$$V = \{a_i, t_i, f_i \mid 1 \leq i \leq n\},$$

$$\mu = [_1 \ldots [_m[_{m+1}\ ]_{m+1}]_m \ldots]_1,$$

$$M_{m+1} = \{a_1\},$$

$$M_j = \{\lambda\}, 1 \leq j \leq m,$$

$$R_{m+1} = \{a_i \to t_i a_{i+1}(here) \| f_i a_{i+1}(here) \mid 1 \leq i \leq n-1\}$$

$$\cup \{a_n \to t_n(out) \| f_n(out)\},$$

$$R_j = \{t_i \to t_i(out) \mid x_i \text{ is present in } C_j, 1 \leq i \leq n\}$$

$$\cup \{f_i \to f_i(out) \mid \neg x_i \text{ is present in } C_j, 1 \leq i \leq n\}, 1 \leq j \leq m$$

The system works by producing inside membrane m in n steps, thanks to replication rules, all the 2^n truth assignments. In the following m steps, assignments are moved outward through membranes $j = m, \ldots, 1$, where they are checked against each clause C_j, and consequently selected. Therefore, the formula is satisfiable if and only if any string will exit the system, in $n + m$ steps.

Further results concerning small replicated rewriting P systems are as follows [17]

Theorem 7.30 $MAT \subset RRP_4$; $E0L \subset ERRP_4$; $ET0L \subset ERRP5$.

7.7 Conclusions

One of the motivations to study string rewriting P systems is that they have a familiar output, consisting of a usual language of strings, so that all the properties being studied can be compared to those in formal language theory. Second, strings give a structure to the objects moving across the membranes, a structure described by the exact sequence of symbols in them, which cannot be expressed when representing the content of membranes by multisets of atomic objects.

References

[1] D. Besozzi: *Computational and Modelling Power of P Systems*. PhD Thesis, Univ. degli Studi di Milano, 2004.

[2] D. Besozzi, C. Ferretti, G. Mauri, C. Zandron: Parallel rewriting P systems with deadlock. *Lecture Notes in Computer Sci.*, 2568 (2003), 302–314.

[3] D. Besozzi, G. Mauri, G. Vaszil, C. Zandron: Collapsing hierarchies of parallel rewriting P systems without target conflicts. *Lecture Notes in Computer Sci.*, 2933 (2004), 55–69.

[4] D. Besozzi, G. Mauri, C. Zandron: Hierarchies of parallel rewriting P systems. *Proc. Brainstorming Week on Membrane Computing* (M. Cavaliere et al., eds.), Tarragona, Rovira i Virgili Univ., Tech. Rep. No. 26, 2003, 61–74.

[5] D. Besozzi, G. Mauri, C. Zandron: Deadlock decidability in partial parallel P systems. *Lecture Notes in Computer Sci.*, 2943 (2004), 55–60.

[6] D. Besozzi, G. Mauri, C. Zandron: Parallel rewriting P systems without target conflicts. *Lecture Notes in Computer Sci.*, 2597 (2003), 119–133.

[7] D. Besozzi, G. Mauri, C. Zandron: Hierarchies of parallel rewriting P systems – A survey. *New Generation Computing*, 22 (2004), 331–347.

[8] P. Bottoni, A. Labella, C. Martin-Vide, Gh. Păun: Rewriting P systems with conditional communication. *Lecture Notes in Computer Sci.*, 2300 (2002), 325–353.

[9] R. Freund, C. Martín-Vide, Gh. Păun: From regulated rewriting to computing with membranes: Collapsing hierarchies. *Theoretical Computer Sci.*, 312 (2004), 143–188.

[10] C. Ferretti, G. Mauri, Gh. Păun, C. Zandron: On three variants of P systems with string-objects. *Theoretical Computer Sci.*, 301 (2003), 201–215.

[11] S.N. Krishna: *Languages of P Systems. Computability and Complexity*. PhD Thesis, IIT Madras, India, 2001.

[12] S. N. KRISHNA, R. RAMA: A note on parallel rewriting in P systems. *Bulletin of the EATCS*, 73 (February 2001), 147–151.

[13] S.N. KRISHNA, R. RAMA: On the power of P systems based on sequential/parallel rewriting. *Intern. J. Computer Math.*, 76 (2001), 317–330.

[14] S.N. KRISHNA, R. RAMA: P systems with replicated rewriting. *J. Automata, Languages, Combinatorics*, 6 (2001), 345–350.

[15] M. MADHU: Rewriting P systems: improved hierarchies. *Theoretical Computer Sci.*, 334 (2005), 161–175.

[16] M. MADHU, K. KRITHIVASAN: Generalized normal form for rewriting P systems. *Acta Informatica*, 38 (2002), 721–734.

[17] V. MANCA, C. MARTÍN-VIDE, GH. PĂUN: On the power of P systems with replicated rewriting. *J. Automata, Languages, Combinatorics*, 6 (2001), 359–374.

[18] GH. PĂUN: Computing with membranes. *J. Computer and System Sci.*, 61 (2000), 108–143.

[19] GH. PĂUN: Computing with membranes – A variant: P systems with polarized membranes. *Intern. J. Found. Computer Sci.*, 11 (2000), 167–182.

[20] I. PETRE: A normal form for P systems. *Bulletin of the EATCS*, 67 (1999), 165–172.

[21] H. RAMESH, R. RAMA: Rewriting P systems with conditional communication: improved hierarchies. *Pre-proc. Membrane Computing, Intern. Workshop, WMC8*, Thessaloniki, Greece, 2007, 527–538.

[22] G. ROZENBERG, A. SALOMAA: *The Mathematical Theory of L Systems*. Academic Press, 1980.

[23] C. ZANDRON: *A Model for Molecular Computing: Membrane Systems*. PhD Thesis, Universita' degli Studi di Milano, Dipartimento di Informatica, 2001.

[24] C. ZANDRON, C. FERRETTI, G. MAURI: Two normal forms for rewriting P systems. *Lecture Notes in Computer Sci.*, 2055 (2001), 153–164.

[25] C. ZANDRON, C. FERRETTI, G. MAURI: Using membrane features in P systems. *Romanian J. Information Science and Technology*, 4 (2001), 241–257.

[26] C. ZANDRON, G. MAURI, C. FERRETTI: Universality and normal forms on membrane systems. *Proc. Intern. Workshop Grammar Systems* (R. Freund, A. Kelemenova, eds.), Bad Ischl, Austria, July 2000, 61–74.

CHAPTER 8

SPLICING P SYSTEMS

SERGEY VERLAN
PIERLUIGI FRISCO

8.1 INTRODUCTION

A restriction enzyme is able to cut, with a high specificity, a double stranded DNA molecule. If DNA molecules and restriction enzymes are present in a test tube, then these enzymes can cut the DNA. If also ligase, another enzyme, is present in the tube, then 'new' DNA molecules (meaning molecules not initially present in the test tube) can be formed by recombination. This biological process is called *splicing*. Of course, also DNA molecules initially present in the tube can be formed (restored).

T. Head [11] was the first to formalize in mathematical terms biological splicing. Other formalizations of the same process were later presented in [22] and [30]. Splicing P systems were introduced in [25] and since then they have been widely investigated for their computational power, for the definition of (direct) universal P systems, etc. What differentiates splicing P systems from the vast majority of models of membrane computing is their use of sets of strings instead of multisets of symbols (objects).

8.2 BASIC DEFINITIONS

Definition 8.1 A splicing rule *over an alphabet V is a 4-tuple* (u_1, u_2, u_3, u_4) *where* $u_1, u_2, u_3, u_4 \in V^*$. *Splicing rules are frequently written as* $u_1\#u_2\$u_3\#u_4$, $\{\$, \#\} \notin V$. *The strings* u_1u_2 *and* u_3u_4 *are called splicing sites.*

Let V be an alphabet, $x, y \in V^*$, and let r be a splicing rule over V. We say that x *matches* rule r if x contains an occurrence of one of the two sites of r. We also say that x and y are *complementary* with respect to r if x contains one site of r and y contains the other one. In this case we also say that x or y may *enter* rule r. If $x = x_1u_1u_2x_2$, $y = y_1u_3u_4y_2$ with $x_1, x_2, y_1, y_2, u_1, u_2, u_3, u_4 \in V^*$, then the splicing rule $r = u_1\#u_2\$u_3\#u_4$ can be *applied* to x and y. The result of this application is w and z where $w = x_1u_1u_4y_2$ and $z = y_1u_3u_2x_2$. This is denoted by $(x, y) \vdash_r (w, z)$ and we say that x and y are spliced and w and z are the result of this splicing.

What was just defined is called *2-splicing* as two strings, w and z, are the result. If instead only the string w is the result, then *1-splicing* is performed. This is denoted by $(x, y) \vdash_r w$. In order to highlight the splicing sites we write: $(x_1u_1|u_2x_2, y_1u_3|u_4y_2) \vdash_r (x_1u_1u_4y_2, y_1u_3u_2x_2)$ where the vertical bar indicates where the splicing occurs.

The *diameter* of a splicing rule $r = u_1\#u_2\$u_3\#u_4$, denoted by $dia(r)$, is the vector $(|u_1|, |u_2|, |u_3|, |u_4|)$. It is possible to generalize this notation and say that a set of splicing rules R has the diameter (n_1, n_2, n_3, n_4), denoted by $dia(R)$, if $n_i = \max\{d_i \mid dia(r) = (d_1, d_2, d_3, d_4), r \in R\}, 1 \le i \le 4$.

The pair $\sigma = (V, R)$ where V is an alphabet and R is a set of splicing rules is called a *splicing scheme* or an *H-scheme*.

For a splicing scheme $\sigma = (V, R)$ and for a language $L \subseteq V^*$ we define:

$$\sigma_1(L) = \{w \in V^* \mid x, y \in L, (x, y) \vdash_r w, \text{ for some } r \in R\},$$
$$\sigma_2(L) = \{w, z \in V^* \mid x, y \in L, (x, y) \vdash_r (w, z), \text{ for some } r \in R\}.$$

We can now introduce *iterated splicing*. For $j \in \{1, 2\}$,

$$\sigma_j^0(L) = L,$$
$$\sigma_j^{i+1}(L) = \sigma_j^i(L) \cup \sigma_j(\sigma_j^i(L)), \ i \ge 0,$$
$$\sigma_j^*(L) = \cup_{i \ge 0} \sigma_j^i(L).$$

It is known that iterated splicing preserves the regularity of a language:

Theorem 8.1 [28] *Let* $L \subseteq V^*$ *be a regular language and let* $\sigma = (V, R)$ *be a splicing scheme. Then* $\sigma_1^*(L)$ *is a regular language.*

Definition 8.2 *[12] A Head splicing system (also called H system) is a pair* $H = (\sigma, A) = ((V, R), A)$ *also denoted by* $H = (V, A, R)$, *where V is an alphabet,* $A \subseteq V^*$ *is a set of strings, called* axioms, *and R is a set of splicing rules over V.*

We say that an H system $H = (V, A, R)$ is *finite* if A and R are finite.

The *language generated* by the system H based on 2-splicing (1-splicing) is $L(H) = \sigma_2^*(A)$ $(L(H) = \sigma_1^*(A))$.

We denote by $H_2(FIN, FIN)$ ($H_1(FIN, FIN)$) the family of languages generated by finite H systems. The following results are known:

Proposition 8.2 *[36, 33, 28]* $H_2(FIN, FIN) \subsetneq H_1(FIN, FIN)$.

Theorem 8.3 *[30, 12, 28]* $H_1(FIN, FIN) \subsetneq REG$.

We denote by $H_2(REG, FIN)$ ($H_1(REG, FIN)$) the family of languages generated by H systems having a regular set of axioms, finite set of splicing rules, and based on 2-splicing (1-splicing).

In the following sections of this chapter the use of 2-splicing is implicitly assumed while the use of 1-splicing is explicitly mentioned.

Definition 8.3 *An* extended H system *is a 4-tuple* $\gamma = (V, T, A, R)$, *where V is an alphabet,* $T \subseteq V$ *is called* terminal alphabet, $A \subseteq V^*$ *is a finite set of strings, called* axioms, *and R is a set of splicing rules over V.*

The language generated by an extended H system γ is $L(\gamma) = \sigma^*(A) \cap T^*$, where $\sigma = (V, R)$, that is, the set of all strings over the terminal alphabet T generated by the H system $H = (V, A, R)$.

We denote by $EH(FIN, FIN)$ the family of languages generated by finite extended H systems.

Extended H systems are more powerful than H systems:

Theorem 8.4 *[27, 28]* $EH(FIN, FIN) = REG$.

Various computationally complete models of H systems having a finite set of axioms and splicing rules have been proposed. These models were equipped with permitting/forbidding contexts for splicing rules, double splicing, multisets, etc.

We refer to [28] for an overview of different models based on splicing. Here we only define *time-varying distributed H systems*. This model uses different sets of rules for each splicing step.

Definition 8.4 *A* time-varying distributed H system, (TVDH system) *of degree n is an n + 3-tuple*

$$D = (V, T, A, R_1, R_2, \ldots, R_n),$$

where V is an alphabet, $T \subseteq V$ *is the terminal alphabet,* $A \subseteq V^*$ *is a finite set of axioms and* R_i, $1 \leq i \leq n$, *called* components, *are finite sets of splicing rules.*

At each moment $k = n \cdot j + i$, where $j \geq 0$, $1 \leq i \leq n$, only rules of component R_i are used to splice current strings. More precisely, we define

$L_1 = A$,
$L_{k+1} = \sigma_i(L_k)$, for $i \equiv k - 1 \pmod{n} + 1$, $k \geq 1$, $1 \leq i \leq n$, $\sigma_i = (V, R_i)$.

In each computational step k the strings in L_k are spliced using rules of the component R_i, $i \equiv k - 1 \pmod{n} + 1$ and only the result of this splicing forms the next set of strings L_{k+1}. No other string belongs to L_{k+1}.

The language generated by a TVDH system D consists of all strings over the terminal alphabet produced at some computational step.

$$L(D) = (\cup_{k \geq 1} L_k) \cap T^*.$$

We denote by VDH_n the family of languages generated by TVDH systems of degree at most n.

TVDH systems are of a great interest for splicing P systems because of the direct correspondences between these two systems. The following two theorems are basic to this research area:

Theorem 8.5 [15] $VDH_2 = RE$.

Theorem 8.6 [16] *For any type-0 grammar G there is a TVDH system D_G of degree 1 which simulates G and $L(G) = L(D_G)$. Moreover,*

- $L_k \cap L_{k+1} = \emptyset$, *i.e. strings produced at step k and $k + 1$ are different;*
- *a string from $L(D_G)$ is obtained by using once a rule from a fixed set $R_R \subseteq R$.*

8.3 SPLICING P SYSTEMS

Definition 8.5 *A (generating) splicing P system of degree $m \geq 1$ is a construct*

$$\Pi = (V, T, \mu, A_1, \ldots, A_m, R_1, \ldots, R_m), \text{ where}:$$

V is an alphabet with #, \$ $\notin V$;
$T \subseteq V$ is the terminal alphabet;
μ is a membrane structure with m membranes;
$A_i \subseteq V^*$, $i \in \{1, \ldots, m\}$, are finite languages associated with the compartments $1, \ldots, m$ of μ;
R_i for $i \in \{1, \ldots, m\}$ are sets of evolution rules of the form $(r; tar_1, tar_2)$ associated with the compartments $1, \ldots, m$ of μ, where r is a splicing rule over V and $tar_1, tar_2 \in \{here, in, out\}$ are target indicators.

A *configuration* of Π is an m-tuple $(M_1, \ldots, M_m)$ of languages over V. The m-tuple $(A_1, \ldots, A_m)$ is called *initial configuration*.

For two configurations $(M_1, \ldots, M_m)$, $(M'_1, \ldots, M'_m)$ of Π we denote by $(M_1, \ldots, M_m) \Rightarrow (M'_1, \ldots, M'_m)$ a *transition* from $(M_1, \ldots, M_m)$ to $(M'_1, \ldots, M'_m)$, that is, the application of splicing rules of Π in parallel to all possible strings that are in the corresponding compartments. When this happens the result of each splicing is distributed to other compartments according to the target indicators. Formally, for $i \in \{1, \ldots, m\}$, if $x = x_1 u_1 u_2 x_2$, $y = y_1 u_3 u_4 y_2 \in M_i$ and $(r = u_1 \# u_2 \$ u_3 \# u_4; tar_1, tar_2) \in R_i$, $x_1, x_2, y_1, y_2, u_1, u_2, u_3, u_4 \in V^*$, the splicing $(x, y) \vdash_r (z, w)$, $z, w \in V^*$ can take place in i. The strings z and w pass to the compartments indicated by tar_1 and tar_2, respectively, as explained in the following. For $j = 1, 2$:

if $tar_j = here$, then the string remains in compartment i;
if $tar_j = out$, then the string is sent to a compartment situated immediately above compartment i (if i is the skin membrane, then the string exits the system);
if $tar_j = in$, then the string is sent to a non-deterministically chosen compartment situated immediately below compartment i.

Since the strings are present in an arbitrary number of copies, after the application of rule r in compartment i, strings x and y are still present in the same compartment. Exceptions to this are presented in the following.

A *computation* in a splicing P system Π is a sequence of transitions between configurations of Π starting from the initial configuration $(A_1, \ldots, A_m)$. If the considered system has only one compartment, then we omit parentheses and we write $M_1 \Rightarrow M_2 \Rightarrow \ldots \Rightarrow M_k$ ($M_1 = A_1$) for a sequence of transitions.

The result of the computation consists of all strings over the terminal alphabet T sent outside the system at some moment of the computation.

We denote by $L(\Pi)$ the language generated by a splicing P system Π.

A splicing P system $\Pi = (V, T, \mu, A_1, \ldots, A_m, R_1, \ldots, R_m)$ is called *non-extended* if $V = T$ and it is said to have *global* rules if $R_1 = \ldots = R_m$.

The use of sets of strings instead of multisets for splicing P systems may lead to different semantics. For instance: what happens if there is a string w in compartment i and the same string w is produced by a certain rule r in compartment i indicating that w has to be sent away from i. As sets of strings are used, it is difficult to argue if the two strings w shall be distinguished or not. Hence, there are two solutions: the first one is to keep the old w in the compartment and to send a new copy of w away. The second solution is to send w away and to eliminate it from the compartment.

A similar problem arises in the case when there are two rules which produce the same string w one having the target indicator *here* and the other one having the target indicator *in* (or *out*). Also, both situations described above may happen at the same time.

The situations indicated in the above can be dealt with by several semantics.

Let w be a string which is produced by a rule r in compartment i and suppose that w is sent away from i by a target indicator *in* or *out*. Then:

1. The string w is sent away from the compartment and no copy of it remains in the same compartment.
2. The string w is sent away from the compartment, but a copy of it remains in the same compartment providing that one of the following conditions is satisfied;
   a. The string w is already present in the same compartment;
   b. There is another rule r' which produces w with the target indicator *here*;
   c. Both conditions (2.a) and (2.b) are satisfied;
   d. One of conditions (2.a) or (2.b) is satisfied.

We say that a splicing P system is of *type x* according to the above semantics x.

So, in type (1) all occurrences of string w are sent away of the compartment. In type (2.a) and (2.d), if w is produced in a compartment, then it remains there forever. Type (2.b) and (2.c) are situated somewhere between the above two extremes.

The distinction between these types is important only in the case of splicing P systems with one compartment as it leads to classes having different computational power. If two or more compartments are used, then it is always possible to define a system where the above situations never occur. Traditionally, the type (2.d) is used for splicing P systems and we assume that it will be used for systems having more than one compartment.

We denote by $ELSP_m(spl, in, x)$ the family of languages generated by splicing P systems of type x having a degree at most m. We omit the last parameter for systems of type (2.d) and denote the corresponding family by $ELSP_m(spl, in)$, or $LSP_m(spl, in)$ for the non-extended case. We also indicate the diameter of splicing rules and write $ELSP_m(spl, in, (n_1, n_2, n_3, n_4))$ for classes of splicing systems whose rules have the diameter at most (n_1, n_2, n_3, n_4).

Remark 8.1 If $ELSP_m(spl, in, (n_1, n_2, n_3, n_4)) = \mathcal{L}$ and if $\mathcal{L}$ is closed with respect to the mirror image, then $ELSP_m(spl, in, (n_{i_1}, n_{i_2}, n_{i_3}, n_{i_4})) = \mathcal{L}$ for all possible permutations of indices i_t, $1 \leq t \leq 4$, $1 \leq i_t \leq 4$, $i_k \neq i_m$ if $k \neq m$.

This follows from the fact that a language $L \in \mathcal{L}$ generated by a P system Π with the diameter (n_1, n_2, n_3, n_4) may be obtained by a P system with the diameter (n_3, n_4, n_1, n_2) by simply swapping the first and the second part of rules from Π. In order to generate L by a P system with the diameter (n_2, n_1, n_4, n_3) it is sufficient to take a P system generating the mirror image of L and to apply the same function to all its rules and axioms.

8.3.1 Splicing Tissue P Systems

Like for other models of P systems it is possible to consider a compartment structure described by a graph rather than a tree.

Definition 8.6 *A splicing tissue P system of degree $m \geq 1$ is a $(2m+3)$-tuple*

$$\Pi = (V, T, G, A_1, \ldots, A_m, R_1, \ldots, R_m),$$

where G is the underlying directed labeled graph of Π and the remaining components are as in Definition 8.5.

Configuration, transition, computation, and *result* of a splicing tissue P system are defined as before. For $k = 1, 2$, the application of evolution rules is such that:

if $tar_k = here$, then the corresponding string remains in node i;
if $tar_k = go_j$, then the string is sent to node j (it is clear that there must be an edge (i, j) in G);
if $tar_k = out$, the string is sent outside of the system.

It is also possible to write go instead of go_k and this means that the corresponding string may go to any of the connected compartments, non-deterministically.

We denote by $ELStP_m(spl, go)$ the family of languages generated by tissue splicing P systems having a degree at most m. The family of languages generated by non-extended systems with same parameters is denoted by $LStP_m(spl, go)$.

A *simple splicing tissue P system* is a splicing tissue P system such that for every evolution rule $(r; tar_1, tar_2)$ either $tar_1 = tar_2 = go_j$, or $tar_1 = tar_2 = out$. In this case, the rules are associated with edges, not with compartments. Specifically, a rule in R_i with targets go_j is associated with the edge (i, j). Then, a simple splicing tissue P system is denoted by $(V, T, G, A_1, \ldots, A_m, R)$, where V, T, G, and A_i, $1 \leq i \leq m$, have the same meaning as before and R is a set of splicing rules associated to edges. The family of languages generated by simple tissue splicing P systems having a degree at most m is denoted by $ELStP_m(spl, sgo)$.

8.4 Proof Techniques for Splicing

This section presents two common techniques used in proofs regarding formal systems based on splicing.

8.4.1 Rotate-and-simulate Technique

The *rotate-and-simulate* technique is the basic tool for universality proofs in the area of splicing.

Let S be a formal system based on splicing simulating a formal grammar G. Then, for any string $w \in (N \cup T)^*$ of G there are strings $Xw''Bw'Y$ ($X, Y, B \notin N \cup T$, $w = w'w''$) in S. These strings are called *rotational versions* of w. More precisely, the symbols X and Y flank w and the symbol B marks the beginning (or ending) of the string. It should be clear that w can be obtained from any of its rotational versions.

The system S simulates the formal grammar G as follows. For each production $u \to v$ of G the axiom ZvY and the splicing rule $\lambda\#uY\$Z\#vY$ are in S. The production of G is simulated as follows. The system S rotates the string Xw_1uw_2BY into Xw_2Bw_1uY and then it applies the corresponding splicing rule. Next, the resulting string Xw_2Bw_1vY is rotated into Xw_1vw_2BY. The rotation is made symbol by symbol, i.e. the string Xwa_iY, $a_i \in N \cup T \cup \{B\}$, is transformed into Xa_iwY after some computational steps.

The symbol by symbol rotation is often implemented as follows. The system starts with the string Xwa_iY. Later, this string becomes XwY_i, i.e. the presence of a_i at the right-hand end side is encoded by Y_i (it is also possible to use a unary encoding 1^iY). Next, the strings $X_ja_jwY_i$, $1 \leq j \leq n$, are generated, where n is the total number of symbols that may occur in the string. Afterwards, the system works in a cycle where indices i and j are decreased. If $j \neq i$, then the strings with these indices are removed. If $i = j$, the string $X_1a_iwY_1$ is obtained. Next, the string Xa_iwY is produced. Therefore the initial string Xwa_iY is rotated by one symbol.

8.4.2 The Method of Directing Molecules

Since the iterated splicing leads to regular languages only, various mechanisms that regulate the application of splicing rules were proposed. Most of them are inspired by the regulated rewriting and grammar systems theory. Usually the applicability of a splicing rule is controlled by external factors such as its spatial position, e.g. in a compartment, or some permitting/forbidding conditions.

The method of directing molecules uses the fact that splicing, contrary to rewriting, is a binary operation. Since a splicing operation involves two strings (we also say molecules), its application might be controlled by the presence or absence of one of the strings involved in the splicing. Hence, the evolution of the system is directed not only by the program (rules), but by the data (strings) which is currently processed. This property permits to refine the system by making a finer partition of rules. For example, by making a good choice of rules and axioms a TVDH system of degree 1 having two subsets of rules each working at odd or even steps might be obtained.

This method is well suited for computations where there is a clear distinction between strings representing the information and strings changing the information by splicing with above strings. More precisely, if it might be assumed that during the computation at each step i the configuration C_i of the system might be split into two parts, D_i representing the data and a fixed configuration M containing the strings that may be spliced with the data, and if there are no possible splicings between elements of D_i, for all i, then the method of directing molecules works by varying in time available elements of M. This can be done by marking by a number, the *state*, the corresponding strings and increasing this number, using new splicing rules, modulo k. Therefore, the presence of a string at some step of the computation will depend on its period k and on its initial state. This means that new subpartitions of the rules set are created, as some rules which belong to the same set of rules may not be always applied at the same time, but depending on the presence of directing strings.

The method of directing molecules works well in combination with rotate-and-simulate technique, because the last one fits in the scheme described above. More details can be found in [34] and [33].

8.5 Basic Universality Case

In this section we consider splicing P systems with two compartments and we show that these systems may generate any recursively enumerable language.

Theorem 8.7 $ELSP_2(spl, in) = RE$.

Proof. Consider a type-0 Chomsky grammar $G = (N, T, S, P)$. Consider a new symbol B and assume that $N \cup T \cup \{B\} = \{a_1, a_2, \ldots, a_n\}$ $(a_n = B)$.

In the construction below we assume that $1 \leq i \leq n, 0 \leq j \leq n-1, 1 \leq k \leq n$, $a \in N \cup T \cup \{B\}$.

We construct the following splicing P system:

$\Pi = (V, T, \mu, M_1, M_2, R_1, R_2)$,
$V = N \cup T \cup \{B\} \cup \{X, Y, X_i, Y_i, X_0, Y_0, X'_j, Y'_j, X', Y', Z, Z_X, Z_Y\}$,
$\mu = [_1 [_2 \ ]_2]_1$,
$M_1 = \{XSBY\} \cup \{ZvY_j \mid u \to va_j \in P\} \cup \{ZY_i, ZY_0, ZY'_j, X'Z, XZ, Z_X\}$,
$M_2 = \{X_i a_i Z, X'_j Z, X_j Z, ZY', ZY, Z_Y\}$,

R_1 contains the following rules:

   1.1 : $(\lambda\#uY\$Z\#vY_j; in, in)$, if $u \to va_j \in P$;

   1.1' : $(\lambda\#a_i uY\$Z\#Y_i; in, in)$, if $u \to \lambda \in P$;

   1.2 : $(\lambda\#a_i Y\$Z\#Y_i; in, in)$;   1.3 : $(a\#Y_k\$Z\#Y'_{k-1}; in, in)$;

   1.4 : $(a\#Y'_j\$Z\#Y_j; in, in)$;   1.5 : $(X_0\#a\$X'\#Z; in, in)$;

   1.6 : $(X'\#a\$X\#Z; in, in)$;   1.7 : $(X'\#a\$\lambda\#Z_X; out, out)$,

R_2 contains the following rules:

   2.1 : $(X\#a\$X_i a_i\#Z; out, out)$;   2.2 : $(X_k\#a\$X'_{k-1}\#Z; out, out)$;

   2.3 : $(X'_j\#a\$X_j\#Z; out, out)$;   2.4 : $(a\#Y_0\$Z\#Y'; out, out)$;

   2.5 : $(a\#Y'\$Z\#Y; out, out)$;   2.6 : $(a\#BY_0\$Z_Y\#\lambda; out, out)$.

We prove that $L(\Pi) = L(G)$ in the following way. First we show how the derivations of the grammar G can be simulated. In this way we prove that $L(G) \subseteq L(\Pi)$. At the same time we consider all other possible evolutions and we show that they do not lead to a terminal string. Thus, our assertion will be proved.

The simulation is based on the "rotate-and-simulate" method. The computation in Π follows the flow-chart shown in Fig. 8.1. The vertices of the flow-chart show a configuration of strings during the computation. We number every configuration and indicate the corresponding number in the upper right corner of each string. The numbering we use has the following property: configurations

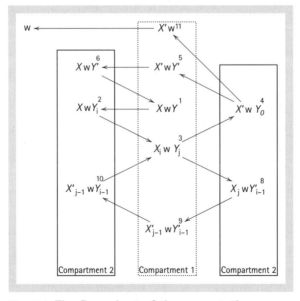

Fig. 8.1 The flow-chart of the computation.

having an odd (resp. even) number occur in the first (resp. second) compartment. In these configurations, the symbol w represents any string in the language $(N \cup T)^*\{B\}(N \cup T)^*$. We show that the computation follows the flow-chart from Fig. 8.1, i.e. all strings produced in one configuration will not be able to evolve anymore except strings from the next configuration.

We also note that the resulting string is immediately sent outside the compartment in which it was produced.

Rotation

We start with Xwa_kY in the first compartment (compartment 1).

$(Xw|a_kY, Z|Y_k) \vdash_{1.2} (XwY_k, Za_kY)$,

The string Za_kY cannot evolve anymore. The first string corresponds to configuration 2.

$(X|wY_k, X_ja_j|Z) \vdash_{2.1} (XZ, X_ja_jwY_k)$.

The string XZ is an axiom.

The second strings corresponds to configuration 3. There are 4 possible cases with respect to values of indices i and j:

a) There are strings of form $X_0a_kwY_i$, $i > 0$, in the first compartment. Then:

$(X_0|a_kwY_i, X'Z) \vdash_{1.5} (X_0Z, X'a_kwY_i)$.

None of these strings can evolve anymore as no rule in compartment two is applicable to them.

b) There are strings of form $X_ia_kwY_0$, $i > 0$, in the first compartment. No rule of the first compartment is applicable to these strings.

c) There are strings of form $X_ja_kwY_i$, $i, j > 0$, in the first compartment. Then we have the following computation:

$(X_ja_kw|Y_i, ZY'_{i-1}) \vdash_{1.3} (X_ja_kwY'_{i-1}, ZY_i)$.

The first string corresponds to configuration 8. The second string cannot evolve anymore.

$(X_j|a_kwY'_{i-1}, X'_{j-1}|Z) \vdash_{2.2} (X_jZ, X'_{j-1}a_kwY'_{i-1})$.

The second string corresponds to configuration 9. The first string cannot evolve anymore.

$(X'_{j-1}a_kw|Y'_{i-1}, Z|Y_{i-1}) \vdash_{1.4} (X'_{j-1}a_kwY_{i-1}, ZY'_{i-1})$.

The first string corresponds to configuration 10. The second string cannot evolve anymore.

$(X'_{j-1}|a_kwY_{i-1}, X_{j-1}|Z) \vdash_{2.3} (X'_{j-1}Z, X_{j-1}a_kwY_{i-1})$.

The second string corresponds to configuration 3. The first string cannot evolve anymore. Therefore, indices of X and Y were decreased simultaneously by 1. Another possible evolutions for configuration 10:

$(X'_{j-1}a_kw|Y_0, Z|Y') \vdash_{2.4} (X'_{j-1}a_kwY', ZY_0)$ and
$(X'_{j-1}a_kw|BY_0, Z_Y|) \vdash_{2.6} (X'_{j-1}a_kw, Z_YBY_0)$.

The string ZY_0 is an axiom. Other obtained strings cannot evolve anymore in compartment 1.

d) There are strings of form $X_0 a_k w Y_0$ in the first compartment. This gives us:
$(X_0|a_k w Y_0, X'|Z) \vdash_{1.5} (X_0 Z, X' a_k w Y_0)$.
The second string corresponds to configuration 4. The first string is an axiom.
(*) $(X' a_k w | Y_0, Z | Y') \vdash_{2.4} (X' a_k w Y', Z Y_0)$.
The first string corresponds to configuration 5. The second string is an axiom.
(**) $(X'|a_k w Y', X|Z) \vdash_{1.6} (X'Z, X a_k w Y')$.
The second string corresponds to configuration 6. The first string cannot evolve anymore.
$(X a_k w | Y', Z | Y) \vdash_{2.5} (X a_k w Y, Z Y')$
The first string corresponds to configuration 1. The second string cannot evolve anymore. So, letter a_k of the string $X w a_k Y$ was rotated. It is possible to apply rules 2.6 and 1.7 in cases labeled by (*) and (**), but these applications will be discussed later. There is another possible evolution in configuration 6:
$(X|a_k w Y', X_i a_i|Z) \vdash_{2.1} (XZ, X_i a_i a_k w Y'), i > 0$.
The string XZ is an axiom and the second string cannot evolve anymore in compartment 1.

Simulation of Productions of the Grammar

If string $X w u Y$ is present in the first compartment and if there is a production $u \to v$ in P, then it is possible to apply rule 1.1 or 1.1' in order to simulate the corresponding production of the grammar.

The Result

The word $X' a_k w Y_0$ is obtained in the second compartment (case (*)). If $w = w'B$, the following rules can be used:
$(X' a_k w'|B Y_0, Z_Y|) \vdash_{2.6} (X' a_k w', Z_Y B Y_0), (w = w'B)$.
The second string cannot evolve anymore. The first string may be involved in rule 1.7:
$(X'|a_k w', |Z_X) \vdash_{1.7} (X' Z_X, a_k w')$.
The string $X' Z_X$ is not terminal. If $a_k w' \in T^*$, then it will belong to the result. Instead of the previous rule it was also possible to apply rule 1.6:
$(X'|a_k w', X|Z) \vdash_{1.6} (X'Z, X a_k w')$.
The first string cannot evolve anymore. The second string may enter rule 2.1:
$(X|a_k w', X_i a_i|Z) \vdash_{2.1} (XZ, X_i a_i a_k w'), i > 0$.
The first string is an axiom and the second string cannot evolve anymore.

There are a few other possible evolutions, which lead to deadlock (to non-terminal strings which cannot evolve anymore), but we do not examine them here.

It is easy to see that by following the computation above all strings of $L(G)$ are generated and, conversely, the system does not produce other strings.

Remark 8.2 In rules above it is possible to use a target indicator *here* in the first position, for rules 1.1–1.4 and 2.4–2.5, and in the second position, for rules 1.5–1.6 and 2.1–2.3. The evolution of the system would remain the same.

Remark 8.3 It is worth noting that at any time only one rule is applicable. In the lower part of the flow-chart the first compartment is responsible for the transformations at the right hand side of a string $Y \to Y_i \to Y'_{i-1} \to Y_{i-1}$, while the second compartment contains rules permitting the left hand side to evolve $X \to X_i \to X'_{i-1} \to X_{i-1}$. In the upper part of the flow-chart the roles are changed and the first compartment performs the transformations on the left hand side: $X_0 \to X' \to X$ while the second compartment transforms the right hand side $Y_0 \to Y' \to Y$.

Let us observe that rules from R_1 cannot be used in compartment 2 because of the target indicator *in*, while using rules of R_2 in compartment 1 produces two outputs which exit the system and cannot evolve anymore. Thus, we get:

Corollary 8.1 *Every language $L \in RE$ can be generated by an extended splicing P system of degree 2 with global rules.*

8.5.1 Reducing the Diameter

The system presented in Theorem 8.7 has a diameter that depends on the size of the grammar productions. If we consider that the initial grammar is in Kuroda or Penttonen normal form, then the constructed system has the diameter $(1, 3, 2, 2)$ mainly because of rules 1.1, 1.1′, and 2.1. We show now how to replace these rules by a set of rules with a smaller diameter by moving some of the information to axioms.

Assume that that the grammar $G = (N, T, S, P)$ is in Penttonen normal form and construct a system Π as in the proof of Theorem 8.7. After that rules 1.1 and 1.1′ and the axioms ZvY_j are eliminated, and the following rules are added to R_1:

1.1 : $(\lambda \# AY \$ \bar{Z} \# Y; here, in)$, if $A \to \lambda \in P$,
1.1′ : $(A \# CY \$ Z \# Y_k; in, in)$, if $AC \to AD \in P$ and $D = a_k$,
1.1″ : $(\lambda \# AY \$ Z_l \# \lambda; in, in)$,   if $l : A \to x \in P$ and $1 \le |x| \le 2$.

In addition, axioms $\bar{Z}Y$ and $\{Z_l x Y_t \mid l : A \to xy \in P$ and $y = a_t, x \in N \cup T \cup \{\lambda\}\}$ are added to M_1.

The alphabet V is also updated to include symbols $\bar{Z}$ and $Z_l, l = 1, 2, \ldots, |P|$.

It is obvious that the rules above simulate productions of G. Indeed, the first rule simulates all productions $A \to \lambda$ just by erasing the letter A. The second rule simulates productions $AC \to AD$ by erasing the letter C and starting the rotation of D. The productions $A \to xy$, $1 \leq |xy| \leq 2$ are directly simulated by deleting AY and by inserting xY_t, $y = a_t$, at the end of the string.

The above transformations decrease the diameter of the system down to $(1, 2, 2, 1)$. Now we show how rule 2.1 can be eliminated. In order to do this, we replace it with the following rule:

2.1 : $(X\#a\$\lambda\#Z_i; out, out)$.

We also replace all strings $X_i a_i Z$ from M_2 by strings $X_i a_i Z_i$ and add Z_i, $1 \leq i \leq |V|$, to the alphabet.

It is clear that the obtained system evolves in the same way as the previous one, hence we prove the following theorem.

Theorem 8.8 $ELSP_2(spl, in, (1, 2, 1, 1)) = ELSP_2(spl, in, (1, 1, 1, 2)) = ELSP_2(spl, in, (2, 1, 1, 1)) = ELSP_2(spl, in, (1, 1, 2, 1)) = RE$.

These statements follow from Remark 8.1 and from the fact that RE is closed with respect to the mirror operation.

Remark 8.4 Rule 1.1 introduced above uses the target indicator *here*. This can be avoided by replacing rule 1.1 by the rule $(\lambda\#AY\$Z\#Y_{n+1}; in, in)$, where $n = |V|$ and by adding the axiom $X_{n+1} Z_{n+1}$ to the second compartment. In this way a new symbol having the index $n + 1$ is introduced and it can be used for the rotation which leads to the removal of A.

8.5.2 Translating to a TVDH System

In this section we present a procedure that permits to transform a splicing P system with two compartments satisfying certain conditions to a TVDH system of degree 2. Such a transformation is motivated by a simpler structure of the latter model and by the existence of a computer simulator for it [33].

Consider a splicing P system $\Pi = (V, T, M_1, M_2, R_1, R_2)$ of degree 2. Let $C_i = (M_1^{(i)}, M_2^{(i)})$ be a configuration of Π at step $i > 0$. Suppose that Π satisfies the following conditions:

1. Both targets for all rules are either *in*, or *out*.
2. There are two fixed sets of strings B_1 and B_2 over V^* such that for any two configurations of the system C_i and C_j, $i, j > 0$, if $w_1 \in M_1^{(i)} - B_1$ and $w_2 \in M_1^{(j)} - B_1$ (resp. $w_1 \in M_2^{(i)} - B_2$ and $w_2 \in M_2^{(j)} - B_2$), then w_1 and w_2 can never be complementary with respect to a rule from R_1 (resp. R_2).
3. All strings in C_i are non-terminal, for all i.
4. Strings sent outside of the system cannot evolve in the second compartment.

Then it is possible to construct a TVDH system D, of degree 2, that will generate the same language as Π. This system can be defined as follows: $D = (V', T, M'_1 \cup M'_2 \cup S_1 \cup S'_2, R_1 \cup R_A, R_2 \cup R_A)$, where M'_1, M'_2 and R_A permit to implement the method of directing molecules and make appear strings from B_1 at odd steps and strings from B_2 at even steps. The set S'_2 contains strings $w\bar{Z}, \bar{Z} / \forall$, for all $w \in S_2$ and there is also a rule $\lambda \# \bar{Z} \$ \bar{Z}' \# \lambda$ in R_A and a string $\bar{Z}'$ in M'_1. This will allow strings of S_2 to be generated in the next step.

We remark that the above method works for any splicing P system having a structure of a ring. Hence, any splicing P system of degree n with a ring structure satisfying conditions above may be transformed to a TVDH system of degree n by slightly modifying the construction presented before.

It is also possible to transform a TVDH system $D = (V, T, A, R_1, \ldots, R_n)$ into a splicing P system with a ring structure. Specifically, suppose that D satisfies the following conditions:

1. There is a fixed set of strings B over V^* such that for any two configurations of the system L_i and L_j, $i, j > 0$, $j = i \pmod{n}$, if $w_1 \in L_i - B$ and $w_2 \in L_j - B$, then w_1 and w_2 can never be complementary with respect to a rule from R_i.

2. A terminal string may be obtained only by a rule from $R_R \subseteq R_1$.

Then a splicing P system having a ring structure generating the same language as D may be constructed. This system is defined as follows: $\Pi = (V, T, G, M_1, \ldots, M_n, R_1, \ldots, R_n)$, rules from R_i have a target go, rules from R_R have the target out, and M_i contains strings from B that match a rule from R_i, $1 \leq i \leq n$.

8.6 Systems with One Compartment

In this section we concentrate on splicing P systems with one compartment. We will see that different variants of the definition lead to classes of systems having a different computational power. We start with systems of type (2.d).

8.6.1 Decidability Results

Let $\Pi = (V, T, [_1]_1, M_1, R_1)$ be a splicing P system of type (2.d) based on 1-splicing and such that $R_1 = R_1^h \cup R_1^o$, where R_1^h contains all rules having the target indicator *here* and R_1^o contains all rules having the target indicator *out*. We denote by $\sigma_{h_o} = (V \cup T, R_1^o)$ and $\sigma_{h_h} = (V \cup T, R_1^h)$ the corresponding H schemes and by $H_o = (\sigma_{h_o}, M_1)$ and $H_h = (\sigma_{h_h}, M_1)$ the corresponding H systems.

As in this section we deal with splicing P systems of degree 1, we indicate a configuration as M instead of (M). So, if $M = M^{(1)}$ is the initial configuration of a splicing P system of degree 1, then $M^{(1)} \Rightarrow M^{(2)} \Rightarrow \ldots \Rightarrow M^{(k)}$ for $k \geq 1$. This is denoted by $M^{(1)} \Rightarrow^k M^{(k)}$.

Lemma 8.1 $M^{(k+1)} = \sigma_{h_h}^k(M^{(1)})$.

Lemma 8.2 $L(\Pi) = \sigma_{h_o}(L(H_h)) \cap T^*$.

Proof. Let $w \in L(\Pi)$. Then $w \in T^*$, $w = x_1 u_1 u_4 y_2$, and there is a number $k \in \mathbb{N}$ and strings $x, y \in M^{(k+1)}$ such that $x = x_1 u_1 u_2 x_2$ and $y = y_1 u_3 u_4 y_2$. There is also a rule $r = (u_1 \# u_2 \$ u_3 \# u_4; out) \in R_1^o$ such that $(x, y) \vdash_r w$. From Lemma 8.1 we obtain $x, y \in \sigma_{h_h}^k(M^{(1)}) \subseteq L(H_h)$. Consequently, $w \in \sigma_{h_o}(L(H_h)) \cap T^*$.

Conversely, if $w \in \sigma_{h_o}(L(H_h)) \cap T^*$, then there is a number $k \in \mathbb{N}$, strings $x, y \in \sigma_{h_h}^k(M^{(1)})$ and a rule $r \in R_1^o$ such that $(x, y) \vdash_r w$. In this case $x, y \in M^{(k+1)}$, see Lemma 8.1, and w is sent outside of the system by r. Moreover, $w \in T^*$. This implies $w \in L(\Pi)$.

Corollary 8.2 $ELSP_1(spl, in, (2.d)) \subseteq REG$.

The converse inclusion is also true.

Lemma 8.3 $REG \subseteq ELSP_1(spl, in, (2.d))$.

Proof. An extended H system S can be simulated by a splicing P system Π of type $(2.d)$ of degree 1. This simulation can be done in the following way: for each splicing rule r of S there are two evolution rules $1 : (r; here, here)$ and $2 : (r; out, out)$ in Π. The evolution rule 1 allows to simulate the splicing rule r, the evolution rule 2 allows to send the result of the computation outside the compartment.

We then have:

Theorem 8.9 $ELSP_1(spl, in, (2.d)) = REG$.

8.6.2 Undecidability Results

In this subsection we show that systems corresponding to the family $ELSP_m(spl, in, (1))$ are computationally complete.

For a splicing rule $r = u_1 \# u_2 \$ v_1 \# v_2$ we define:

$$Out(r) = \{(u_1 \# u_2 \$ u_1 \# u_2; out, out), (v_1 \# v_2 \$ v_1 \# v_2; out, out)\}.$$

Now we show how it is possible to simulate a TVDH system of degree 1 by a splicing P system of type (1) having one compartment. The main idea is to simulate the application of a rule r of a TVDH system by rules $(r; here, here)$ and $Out(r)$.

The first rule allows to simulate the application of r, while the second set of rules removes initial strings.

In order to prove the theorem, first we prove the following lemma.

Lemma 8.4 *Let $D = (V, T, A, R \cup R_R)$ be a TVDH system of degree 1 such that:*

(i) $L_k \cap L_{k+1} = \emptyset$, i.e, strings produced at step k and $k + 1$ are different.
(ii) A string from $L(D)$ is obtained by using once a rule from a fixed set R_R.

Then, there is a splicing P system $\Pi \in ELSP_1(spl, in, (1))$, simulating D, such that $L(\Pi) = L(D)$.

We consider the following splicing P system $\Pi = (V, T, [_1]_1, A, R')$, where $R' = R_1 \cup R_2 \cup R_3$ with
$R_1 = \{(r; here, here), Out(r) \mid r \in R\}$;
$R_2 = \{(r; out, out), Out(r) \mid r \in R_R\}$;
$R_3 = \{(w\#\lambda\$w\#\lambda; out, out) \mid w \in A\}$.

It is important to notice that, by construction, if $M^{(k)}$ contains a string which cannot match any rule of Π, then this string remains without contributing to the generated language. For this reason in this proof we only consider strings matching at least one splicing rule in R'.

Based on this construction it is possible to prove

(I): $L_r - T^* = M^{(r)}$ for $r \geq 1$;
(II): $L_{r+1} \cap T^* = \sigma_{h_o}(M^{(r)}) \cap T^*$, where $\sigma_{h_o} = (V, R_o)$ and R_o contains all rules of R' having a target indicator *out*.

More details can be found in [32].

Theorem 8.10 $ELSP_1(spl, in, (1)) = RE$.

Proof. The TVDH system constructed in Theorem 1 from [16] (see also Theorem 6.3.1 from [33]) satisfies Lemma 8.4. We do not present here this system as it can be obtained by a transformation of the system from Theorem 8.7 to a TVDH system of degree 2, as described in Section 8.5.2, and after that by transforming that system to a TVDH system of degree 1, as described in Section 8.4.2.

8.6.3 Systems of Type (2.b) and (2.c)

In this section we show that splicing P systems of types (2.b) and (2.c) are computationally complete.

Theorem 8.11 *Let D be a TVDH system of degree 1. Then, there is a splicing P system Π of type (2.b) and of degree 1 such that $L(\Pi) = L(D)$.*

Proof. Let $D = (V, T, A, R)$. Consider $\Pi = (V, T, [_1]_1, A, R')$, with

$R' = \{(r; here, here); (r; out, out); Out(r) \mid r \in R\} \cup \{(w\#\lambda\$w\#\lambda; out, out) \mid w \in A\}$, where $Out(r)$ is defined as in Section 8.6.2.

It can be easily checked that the system Π simulates D.

As $VDH_1 = RE$ (see Theorem 8.6) we have:

Corollary 8.3 $ELSP_1(spl, in, (2.b)) = RE$.

For systems of type (2.c) we have:

Theorem 8.12 $ELSP_1(spl, in, (2.c)) = RE$.

This theorem can be regarded as a corollary of Theorem 8.10.

8.7 Restricted Models of Splicing P Systems

In this section we present some variants of the definition that impose restrictions on the structure of the system, types of rules, and alphabet.

8.7.1 One-way Communication

As suggested by its name, this variant implies that the communication graph is unidirectional, i.e. if an edge $i \to j$ exists in G, then there should not be an edge $j \to i$ in G. Because of this, in order to obtain non-trivial computations, the underlying graph G must have cycles.

We denote by $ELStP_m(spl, 1go, (d_1, d_2, d_3, d_4))$ the family of languages generated by splicing P systems with one-way communication of degree at most m and having rules of diameter at most (d_1, d_2, d_3, d_4).

It is possible to use results from Section 8.5 by adding an additional compartment between compartments 1 and 2 which sends all input strings to the next compartment. However, it is possible to use this third compartment in order to decrease the diameter of the system. More precisely, the following result holds.

Theorem 8.13 $ELStP_3(spl, 1go, (0, 2, 1, 0)) = RE$.

Proof. Consider a type-0 Chomsky grammar $G = (N, T, S, R)$ in Kuroda normal form. Consider a new symbol B and assume that $N \cup T \cup \{B\} = \{a_1, a_2, \ldots, a_n\}$ ($a_n = B$).

We construct the following splicing P system (with $1 \leq i, k \leq n$ and $0 \leq j \leq n-1$)

$\Pi = (V, T, G, M_1, M_2, M_3, R_1, R_2, R_3)$, where
G is the ring having 3 nodes and edges $1 \to 2, 2 \to 3, 3 \to 1$,
$V_1 = \{X, X', X_i, X_0\}$,
$V_2 = \{Y, Y', Y_i, Y_0\}$,
$V = T \cup V_1 \cup V_2 \cup \{Z_a \mid a \in V_1 \cup V_2\} \cup \{\bar{Z}_{r_i}, \bar{Z}_{X_i}, \tilde{Y}_i\}$
$M_1 = \{\bar{Z}_{Y_i} Y_i, Z_{X'_j} X'_j, Z_{Y_j} Y_j, Z_{Y'} Y'\} \cup \{\bar{Z}_{Y_i} a Y_k \mid a_i \to a a_k \in P, a \in T \cup \{\lambda\}\} \cup \{Z_{Y_i} Y_{n+1} \mid a_i \to \lambda \in P\} \cup \{\bar{Z}'_{Y_i} \tilde{Y}_i \mid A a_i \to CD \in P\} \cup \{Z_{A a_i} C Y_k \mid A a_i \to C a_k \in P\}$,
$M_2 = \{X_i a_i \bar{Z}_{X_i}, Z_{Y''_j} Y''_j, X_j Z_{X_j}, X Z_X\}$,
$M_3 = \{Z_{Y'_j} Y'_j, X''_j Z_{X''_j}, X' Z_{X'}, Z_Y Y\}$,

R_1 contains the following rules:

1.1 :$(\lambda \# a_i Y \$ \bar{Z}'_{Y_i} \# \lambda;\ here, in)$,   1.1' :$(\lambda \# a_t \tilde{Y}_i \$ Z_{a_t a_i} \# \lambda;\ go, go)$,

1.2 :$(\lambda \# a_i Y \$ \bar{Z}_{Y_i} \# \lambda;\ go, go)$,   1.3 :$(\lambda \# Z_{X'_{i-1}} \$ X_i \# \lambda;\ go, go)$,

1.4 :$(\lambda \# Y''_j \$ Z_{Y_j} \# \lambda;\ go, go)$,   1.5 :$(\lambda \# Y_0 \$ Z_{Y'} \# \lambda;\ go, go)$,

R_2 contains the following rules:

2.1 :$(\lambda \# \bar{Z}_{X_i} \$ X \# \lambda;\ go, go)$,   2.2 :$(\lambda \# Y'_j \$ Z_{Y''_j} \# \lambda;\ go, go)$,

2.3 :$(\lambda \# Z_{X_j} \$ X''_j \# \lambda;\ go, go)$,   2.4 :$(\lambda \# Z_X \$ X' \# \lambda;\ go, go)$,

R_3 contains the following rules:

3.1 :$(\lambda \# Y_i \$ Z_{Y'_{i-1}} \# \lambda;\ go, go)$,   3.2 :$(\lambda \# Z_{X''_j} \$ X'_j \# \lambda;\ go, go)$,

3.3 :$(\lambda \# Z_{X_0} \$ X_0 \# \lambda;\ go, go)$,   3.4 :$(\lambda \# Y' \$ Z_Y \# \lambda;\ go, go)$.

The systems works similarly to the system from the proof of Theorem 8.7, however, during the rotation, end symbols are primed twice. The simulation of a production $A a_i \to CD$ of G is done in two steps. First, a symbol a_i is replaced by symbol $\tilde{Y}_i$ and then the combination $A \tilde{Y}_i$ is replaced by $C Y_k$ where $D = a_k$.

Remark 8.5 All the rules of the system, except rule 1.1 have both target indicators *go*. It is possible to replace rule 1.1 by a similar rule but with both targets *go* if a new rule 3.1' : $(\lambda \# a_t \tilde{Y}_i \$ Z_{a_t a_i} \# \lambda;\ go, go)$ and new axioms $Z_{A a_i} C Y_{k-1}$ are added to compartment 3.

8.7.2 Immediate Communication

Another interesting variant of splicing (tissue) P systems are P systems with *immediate communication* which differ from the systems introduced in Section 8.3 by the fact that the result of each splicing shall move from the compartment in which it was produced. More exactly, we can see a splicing P system with immediate communication as a splicing P system whose rules do not contain the target indicator *here* and at the same time each splicing rule appears with the four possible combinations of target indicators *in* and *out* (or *go* in the tissue case).

We denote by $ELSP_m(spl, move)$ the family of languages generated by splicing P systems with immediate communication of degree at most m.

If we have only one compartment, then at each step we send outside of the system the same molecules which are the result of splicing the axioms, hence we have the following result:

Theorem 8.14 $ELSP_1(spl, move) = FIN$.

The inclusion $ELSP_1(spl, move) \supseteq FIN$ follows from the fact that any finite language $L = (a_1, \ldots, a_n)$ may be generated by a splicing P system with immediate communication having axioms a_i and rules $a_i\#\lambda\$a_i\#\lambda$, $1 \leq i \leq n$.

Two compartments are already enough in order to have maximal computational power. Theorem 8.7 is constructed in such a way that its result holds for splicing P systems with immediate communication. Indeed, strings from compartment 2 are sent to compartment 1, while all strings obtained in compartment 1 that normally are sent to compartment 2 contain symbols from the non-terminal alphabet, hence they will not be part of the result. Therefore, taking into account Theorem 8.8 and Remark 8.4 we obtain the following result.

Theorem 8.15 $ELSP_2(spl, move, (1, 2, 1, 1)) = ELSP_2(spl, move, (1, 1, 1, 2)) = ELSP_2(spl, move, (2, 1, 1, 1)) = ELSP_2(spl, move, (1, 1, 2, 1)) = RE$.

In the non-extended case the construction of Theorem 8.7 is not working anymore. However, such systems may generate non-regular languages:

Example 8.1 Consider the system

$$\Pi = (\{a, b, c, e\}, \{a, b, c, e\}, \{cabbc, cae\}, \{ebc\}, \{ca\#e\$c\#a\}, \{b\#c\$e\#bc\}).$$

Any string of form $ca^m b^n c$ present in the first compartment is spliced with cae producing $ca^{m+1}b^n c$ and ce, which are sent outside as well as in the inner compartment. In the second compartment the string $ca^{m+1}b^n c$ is spliced with ebc producing $ca^{m+1}b^{n+1}c$ and ec. Hence, only strings $ca^{m+k+1}b^{n+k}c$, $k \geq 0$, and ce will form the language $L(\Pi)$. Since we start with $cabbc$, we obtain $L(\Pi) = \{ca^n b^n \mid n \geq 2\} \cup \{ce\}$ which is not regular.

8.7.3 Non-terminal Alphabet

In this section we consider non-extended splicing P systems. The first theorem is a corollary of Theorem 8.9 and Theorem 8.3.

Theorem 8.16 $LSP_1(spl, in) \subsetneq REG$.

If two compartments are used, then the computational power increases.

Theorem 8.17 $LSP_2(spl, in) = RE$.

The proof of this result can be obtained by an appropriate modification of the system from the proof of Theorem 8.7. The main idea is the following. A rotation and the simulation of rules of the grammar may be performed as in Theorem 8.7. At the end of the simulation, when symbol B is at the end of the string, we substitute the markers X and Y by V and W. After that we perform the rotation step but we rotate only terminal letters of the grammar. When we have again the symbol B at the end of the string, we know that the considered string is terminal and that it is in a right form. At this moment we eliminate the symbols V and W and send the resulting string outside of the system. More technical details might be found in [32] and [33].

8.8 SMALL UNIVERSAL SPLICING TISSUE P SYSTEMS

In this section we show how to construct a universal simple splicing P system having only 8 evolution rules. Moreover, the obtained system has a very small descriptional complexity.

We need to introduce here some additional notions.

A *tag system* of degree $m > 0$ [2, 18], is a triple $TS = (m, V, P)$, where $V = \{a_1, \ldots, a_{n+1}\}$ is an alphabet and P is a set of productions of the form $a_i \to P_i$, $1 \le i \le n$, $P_i \in V^*$. The symbol a_{n+1} is called the *halting* symbol. A configuration of the system TS is a string w. A tag system passes from a configuration $w = a_{i_1} \ldots a_{i_m} w'$ to the next configuration z by erasing the first m symbols of w and by adding P_{i_1} to the end of the string: $w \Rightarrow z$, if $z = w' P_{i_1}$.

The computation of TS over the string $x \in V^*$ is a sequence of configurations $x \Rightarrow \ldots \Rightarrow y$, where either $y = a_{n+1} a_{i_1} \ldots a_{i_{m-1}} y'$, or $y' = y$ and $|y'| < m$, where $|w|$ is the length of string w. In this case we say that TS *halts* on x and that y' is the result of the computation of TS over x. We say that TS *recognizes* the language L if for all $x \in L$, TS halts on x, and TS halts only on strings from L.

Tag systems of degree 2 are able to recognize the family of recursively enumerable languages [2, 18]. Moreover, systems constructed in [2] have non-empty productions and halt only by reaching the symbol a_{n+1} in the first position.

We also define the notion of *input* for splicing P systems. An input string for a system Π is a string w over the non-terminal alphabet of Π. The computation of Π on input w is obtained by adding w to the axioms of A_1 and after that by evolving Π as usual.

Let $V = \{a_1, \ldots, a_n\}$ be an alphabet. Consider coding functions c and $\bar{c}$ defined as follows: $c(a_i) = \alpha^i \beta$ and $\bar{c}(a_i) = \beta \alpha^i$. We extend these functions to strings and put $c(w) = c(b_1) \ldots c(b_m)$ if $w = b_1 \ldots b_m$.

Theorem 8.18 *Let $TS = (2, V, P)$ be a tag system and let $w \in V^*$. Then, there is a splicing tissue P system $\Pi = (V', T, G, A_1, \ldots, A_m, R)$, having 8 rules, which given an encoding of TS and the string $Xc(w)Y$ as input, simulates TS on input w, i.e. such that:*

1. *for any string w on which TS halts producing the result w', the system Π produces a unique result $Xc(w')Y$;*
2. *for any string w on which TS does not halt, the system Π computes infinitely without producing any result.*

Proof. We construct the system Π as follows.

Let $|V| = n$, $V' = \{\alpha, \beta, X, Y, Z\}$, and $T = \{X, Y, \alpha, \beta\}$. The set of rules R consists of (the second column indicates the edge to which the rule is associated, i.e. for an edge $i \to j$ the rule is located in compartment i and its both target indicators are go_j):

1 : $\beta\#Y\$Z\#\alpha$ $1 \to 2$

2 : $\lambda\#\alpha Y\$Z\#Y$ $2 \to 3$

3 : $X\alpha\#\lambda\$X\#Z$ $3 \to 2$

4 : $\lambda\#\beta Y\$Z\#Y$ $2 \to 4$

5 : $X\beta\#\lambda\$X\#Z$ $4 \to 5$

6 : $X\alpha\#\alpha\$X\#Z$ $5 \to 5$

7 : $X\beta\#\alpha\$X\#Z$ $5 \to 1$

8 : $X\#c(a_{n+1})\$X\#Z$ $1 \to out$

The initial languages A_j are $A_1 = \{XZ\}$, $A_2 = \{ZY\}$, $A_3 = A_4 = A_5 = \{XZ\}$.

The tag system TS is encoded by the strings $\{Zc(P_i)\bar{c}(a_i)Y \mid a_i \to P_i \in P\}$ added as an input (to the first compartment).

The main idea of the construction is the following. The word $X\alpha^j \beta c(bw)Y$ in node 1 encodes the current configuration of TS. Using rule 1, a production P_i

and symbol a_i, $1 \leq i \leq n$, are concatenated at the end of the string (in this way a guess is made about the first symbol of the string). As a result, strings of the kind $Xa^j \beta c(bw P_i) \beta a^i Y$ are generated. After that, indices of the first and the last symbol are decreased simultaneously by taking off one a (since all symbols are coded in unary alphabet this decreases the index of the symbol). This is done by rules 2 and 3. When the same number of a is present at both ends of a string, i.e. the indices coincide, the system removes one more symbol and the string returns to node 1 where it may be processed again. The check for equality is made by rules 4 and 5. Since rule 4 (resp. 5) checks the presence of β at the end (resp. beginning) of the string, the test is successful if and only if both ends of the string contain β ($X\beta c(bw P_j)\beta Y$). Now the second symbol (b) is eliminated by rules 6 and 7. Rule 6 eliminates a's from $c(b)$ and when β is reached, i.e. b is erased, the resulting string $Xc(wP_j)Y$ is passed to node 1. It should be clear that Π simulates the productions of TS.

When a symbol a_{n+1} begins the string, rule 8 is used and the resulting string is sent outside of the system. Hence, Π simulates TS.

Conversely, it is clear that a successful computation in TS may be reconstructed from a successful computation in Π. For this it is enough to look at strings of the form $Xc(w)Y$ in node 1.

The alphabet V' can be reduced to two elements by encoding X, Y, Z, a, β with a binary alphabet.

Hence the system constructed above has 8 rules, 2 symbols and $n+5$ initial axioms. If we fix $c(a_{n+1}) = \beta\beta$, then the diameter of the whole system is (2, 2, 1, 1).

8.9 DIRECT UNIVERSAL SPLICING P SYSTEM

To prove that splicing P systems can simulate a computationally complete formal system S means to prove that also splicing P systems are computationally complete. If, moreover, S is universal (as Turing machines, grammars, etc.), then it also means that splicing P systems are universal. As a universal system S is able to simulate any splicing P system, then a splicing P system Π_u able to simulate S can also simulate any splicing P system Π, so Π_u is universal, too. The system Π_u would then get as input the encoding of S. For this reason we say that a splicing P system can simulate other splicing P systems in an *indirect* way (that is, through the encoding of another universal system). The system presented in the previous section is an example of such a simulation, the target system S being a tag system. If instead a splicing P system, Π_u could get as input the encoding of any other splicing P system Π and simulate it, then Π_u would perform a *direct* simulation.

Such a universal splicing P system Π_u would have a fixed alphabet, a fixed set of evolution rules, and a fixed compartment structure. One part of the axioms present in Π_u would be fixed, another part would encode the alphabet, axioms, evolution rules and compartment structure of any other splicing P system. The language generated by Π_u would then be an encoding of the language generated by Π.

In this section we sketch such a universal splicing P system

$$\Pi_u = (V_u, T_u, [_1[_2\ ]_2[_3\ ]_3]_1, L_{1_u}, L_{2_u}, L_{3_u}, R_{1_u}, R_{2_u}, R_{3_u})$$

simulating splicing (tissue) P systems. We do not completely describe Π_u, as such a description would be long and tedious. Readers interested in the complete description can refer to [10].

Let $\Pi = (V, T, \mu, L_1, \ldots, L_m, R_1, \ldots, R_m)$ be a splicing P system and let

$$C_L : V^* \to V_u^*;$$

$$C_R : V^*\#V^*\$V^*\#V^* \times TAR \times TAR \to V_u^*;$$

$$D : T_u^* \to T^*;$$

(coding) injective functions with $TAR = \{0, 1, \ldots, m\}$. The function C_L encodes strings in $L_1, \ldots, L_m$ into strings in L_{1_u}; C_R encodes evolution rules in $R_1, \ldots, R_m$ into strings in L_{1_u} and D encodes strings over T_u into strings over T.

To indicate that the system Π_u gets as input an encoding of Π we write $\Pi_u(\Pi) = (V_u, T_u, \mu_u, L_{1_u} \cup \bigcup_{x \in L_1 \cup \cdots \cup L_m} C_L(x) \cup \bigcup_{r \in R_1 \cup \cdots \cup R_m} C_R(r), L_{2_u}, L_{3_u}, R_{1_u}, R_{2_u}, R_{3_u})$. The language generated by $\Pi_u(\Pi)$ is denoted by $L(\Pi_u(\Pi))$. When the strings in this language are given as an input to the D function, then $L(\Pi)$, the language generated by Π, is obtained. So, we can write $\{D(y) \mid y \in L(\Pi_u(\Pi))\} = L(\Pi)$.

Let $N = V - T$, $N = \{v_1, \ldots, v_p\}$, and $T = \{v_{p+1}, \ldots, v_q\}$. Moreover, let $f : V^* \to \{0, 1, 2\}^*$ and $f' : V \to \{0, 1, 2\}^*$ be functions such that $f(v) = f'(v_{j_1}) \ldots f'(v_{j_{|v|}})$ and

$$f'(v_j) = \begin{cases} 20^j & \text{for } j \leq p, \\ 10^j & \text{for } p + 1 \leq j \leq q, \end{cases}$$

with $v = v_{j_1} \ldots v_{j_{|v|}}$, $v_{j_1}, \ldots, v_{j_{|v|}} \in V$. So, the function f encodes strings over V into strings over $\{0, 1, 2\}$.

The axioms in Π_u are of the kind $h_k v t_k$ with $v \in V_u^*$, $h_k, t_k \in V_u$, $k \in \mathbb{N}$. If $x \in L_i$, $1 \leq i \leq m$, then $C_L(x) = h_1 z^i s_2 z^i B f(x) s_1 t_1$ and these kind of strings are called *input-axioms*. Notice that $C_L(x)$ contains the encoding of both x and i.

Let $\mu = (Q, E)$ be the cell-tree underlying Π. In order to define C_R we have to introduce:

$bar : \{0, 1, 2, \#\}^* \to \{\bar{0}, \bar{1}, \bar{2}, \bar{\#}\}^*;$
$mi : \{0, 1, 2, \#\}^* \to \{0, 1, 2, \#\}^*;$
$t : V^*\#V^*\$V^*\#V^* \times \{here, out, in\} \times \{here, out, in\} \times \{1, 2\} \times \{1, \ldots, m\} \to \mathbb{N}_+,$

where $\mathbf{N}_+ = \{1, 2, \dots\}$. If $x_1, \dots, x_k \in \{0, 1, 2, \#\}$, then $bar(x_1 \dots x_k) = \bar{x}_1 \dots \bar{x}_k$ while $mi(x_1 \dots x_k) = x_k \dots x_1$. Let $r = (u_1 \# u_2 \$ u_3 \# u_4; tar_1, tar_2) \in R_j$, $1 \le j \le m$. Then,

$$t(r, l, j) = \begin{cases} j & \text{if } tar_l = here, \\ k & \text{if } tar_l = out \text{ and } (k, j) \in E, \\ k & \text{if } tar_l = in \text{ and } (j, k) \in E, \end{cases}$$

and

$$C_R(r) = h_3 s_1 bar(mi(f(u_1) \# f(u_2))) z^{t(r,1,j)} s_2 z^{t(r,1,j)} \bar{\$} z^{t(r,2,j)} s_2 z^{t(r,2,j)} s_1$$
$$bar(mi(f(u_3) \# f(u_4))) z^j t_3.$$

Strings of this kind are called *input-rules*.

Notice that $C_R(r)$ contains the encoding of the evolution rule (splicing rule and target indicators) and j.

The decoding function D is such that

$$D(\gamma) = \begin{cases} f^{-1}(\gamma) & \text{if } \gamma \in 10^+, \\ \lambda & \text{otherwise.} \end{cases}$$

In a splicing P system the strings present in a compartment can splice according to the evolution rules present in the same compartment. The generated strings are either sent to another compartment or they can remain in the same one. This is simulated by the universal splicing P system in the steps indicated in Fig. 8.2. We refer to [10] for details.

The splicing P system Π_u has 594 evolution rules.

1. Modify input-axioms;
2. Create pairs of input-axioms;
3. Check pairs of input-axioms;
4. Rotate pairs of input-axioms;
5. Join an input-rule with a pair of input-axioms; (a *working string* is obtained)
6. Check working strings;
7. Rotate working strings;
8. Match first splicing site;
9. Wrong match input-rule pair and input-axioms;
10. Mark first splicing site;
11. End of matching. Ordering;
12. Match second splicing site;
13. Wrong match input-rule pair of input-axioms;
14. Mark second splicing site;
15. End of matching. Ordering;
16. Simulate splicing;
17. End simulation;
18. Two cases.

Fig. 8.2 Steps followed by the universal splicing P system.

8.10 Bibliographical Notes

The splicing operation was introduced by T. Head in [11]. The definition of splicing as given in this chapter was given by Gh. Păun in [22]. A more general definition may be found in [30]. Theorem 8.3 first appeared in [3]. The proof uses complex arguments formulated in terms of semigroups of dominoes. A direct proof is presented in [30] and adapted to Păun's definition of splicing in [28]. Recently, a simpler proof of this theorem was given in [14] which uses a decomposition of splicing in three operations: cut, paste, and erase.

The rotate-and-simulate method was introduced in [23] and since then it is commonly used to prove results concerning systems based on splicing. The method of directing molecules was introduced in [34] and several results using it are present in [33].

TVDH systems were introduced in [24]. Theorems 8.5 and 8.6 are proved in [15] and [16], respectively. An overview of results on TVDH systems may be found in [33].

The operation of splicing in the context of P systems was considered for the first time in [25], but with a different definition (the splicing rules have an associated string that enters the splicing operation, hence the splicing is used as a unary operation). Splicing P systems as in Definition 8.5 were introduced in [29]. A precise version of the definition in the case of one compartment was given in [35] where the results from Section 8.6 may be found.

The universality of splicing P systems was obtained in [25] for four compartments; the result was improved to three compartments [29], and then to two compartments [21]. The last result was improved from the point of view of the diameter of splicing rules in [7] and included in [26]. The proof of Theorem 8.7 is a slight modification of the main theorem from [32]. The construction given in the theorem was checked by a transformation to TVDH and after that running a TVDH systems simulator described in [33]. It is worth noting that the obtained TVDH system is identical to the one from [15]. Global rules for splicing P systems were firstly considered in [19].

Splicing P systems with immediate communication were considered in [17]. In this article a universality proof is given having no bound on the number of compartments. The universality with two compartments was shown in [32]. Non-extended splicing P systems have been considered in [26] where the universality of such systems with 4 compartments was sketched. The article [32] improves this result to two compartments.

Splicing tissue P systems were introduced in [29] and the restriction to asymmetric graphs was studied in [7] where a result having same descriptional complexity parameters as in Theorem 8.13 is given. Theorem 8.13 is new and it was checked against errors by a transformation to TVDH and after that running a TVDH systems simulator.

The number of rules in a universal P system was firstly considered in [31] and Theorem 8.18 is taken from there. The complete proof of the direct universal splicing P system can be found in [10].

We would like to mention some other models of P systems based on splicing not included in this chapter. In [1] (also in [26]) splicing P systems with rule creation have been considered. These systems permit to activate or deactivate certain rules based on the applied splicing rule. A universality with one compartment is proved. Another model presented in [4] (also considered in a different form in [5]) uses the possibility of splitting the splicing operation into two independent operations: cut and recombination. More information on cut and recombination operations can be found in [6].

The PhD theses [37, 13, 20, 8, 33] contain chapters dedicated to splicing P systems. More information and open problems about splicing P systems can be found in [9].

References

[1] F. Arroyo, A.V. Baranda, J. Castellanos, Gh. Păun: Membrane computing: The power of (rule) creation. *J. Univ. Computer. Sci.*, 8 (2002), 369–381.

[2] J. Cocke, M. Minsky: Universality of tag systems with p=2. *Journal of the ACM*, 11 (1964), 15–20.

[3] K. Culik II, T. Harju: Splicing semigroups of dominoes and DNA. *Discrete Applied Math.*, 31 (1991), 261–277.

[4] F. Freund, R. Freund, M. Oswald, M. Margenstern, Y. Rogozhin, S. Verlan: P systems with cutting/recombination rules assigned to membranes. *Lecture Notes in Computer Sci.*, 2933 (2003), 191–202.

[5] R. Freund, F. Freund: Molecular computing with generalized homogeneous P-systems. *Lecture Notes in Computer Sci.*, 2054 (2000), 130–144.

[6] R. Freund, F. Wachtler: Universal systems with operations related to splicing. *Computers and Artificial Intelligence*, 15 (1996), 273–294.

[7] P. Frisco: On two variants of splicing super-cell systems. *Romanian J. Information Science and Technology*, 4 (2001), 89–100.

[8] P. Frisco: *Theory of Molecular Computing. Splicing and Membrane Systems*. PhD thesis, Leiden University, The Netherlands, 2004.

[9] P. Frisco: *Computing with Cells. Advances in Membrane Computing*. Oxford University Press, 2009.

[10] P. Frisco, H.J. Hoogeboom, P. Sant: A direct construction of a universal P system. *Fundamenta Informaticae*, 49 (2002), 103–122.

[11] T. Head: Formal language theory and DNA: an analysis of the generative capacity of specific recombinant behaviors. *Bulletin of Mathematical Biology*, 49 (1987), 737–759.

[12] T. HEAD, GH. PĂUN, D. PIXTON: Language theory and molecular genetics. Generative mechanisms suggested by DNA recombination. *Handbook of Formal Languages* (G. Rozenberg, A. Salomaa, eds.), Springer, 1997, vol 2, 295–360.

[13] S.-N. KRISHNA: *Languages of P Systems. Computability and Complexity*. PhD thesis, Indian Institute of Technology, Madras, India, 2002.

[14] V. MANCA: A proof of regularity for finite splicing. *Lecture Notes in Computer Sci.*, 2950 (2004), 309–317.

[15] M. MARGENSTERN, Y. ROGOZHIN, S. VERLAN: Time-varying distributed H systems of degree 2 can carry out parallel computations. *Lecture Notes in Computer Sci.*, 2568 (2002), 326–336.

[16] M. MARGENSTERN, Y. ROGOZHIN, S. VERLAN: Time-varying distributed H systems with parallel computations: the problem is solved. *Lecture Notes in Computer Sci.*, 2943 (2004), 48–53.

[17] C. MARTÍN-VIDE, GH. PĂUN, A. RODRÍGUEZ-PATÓN: P systems with immediate communication. *Romanian J. Information Science and Technology*, 4 (2001), 171–182.

[18] M. MINSKY: *Computations: Finite and Infinite Machines*. Prentice Hall, Englewood Cliffts, NJ, 1967.

[19] A. PĂUN: On P systems with global rules. *Lecture Notes in Computer Sci.*, 2340 (2001), 329–339.

[20] A. PĂUN: *Unconventional Models of Computation: DNA and Membrane Computing*. PhD thesis, Univ. of Western Ontario, Canada, 2003.

[21] A. PĂUN, M. PĂUN: On membrane computing based on splicing. *Where Mathematics, Computer Science, Linguistics and Biology Meet* (C. Martin-Vide, V. Mitrana, eds.), Kluwer, 2001, 409–422.

[22] GH. PĂUN: On the splicing operation. *Discrete Applied Math.*, 70 (1996), 57–79.

[23] GH. PĂUN: Regular extended H systems are computationally universal. *J. Automata, Languages, Combinatorics*, 1 (1996), 27–36.

[24] GH. PĂUN: DNA computing: distributed splicing systems. *Lecture Notes in Computer Sci.*, 1261 (1997), 351–370.

[25] GH. PĂUN: Computing with membranes. *J. Computer and System Sci.*, 1 (2000), 108–143.

[26] GH. PĂUN: *Membrane Computing. An Introduction*. Springer, 2002.

[27] GH. PĂUN, G. ROZENBERG, A. SALOMAA: Computing by splicing. *Theoretical Computer Sci.*, 168 (1996), 321–336.

[28] GH. PĂUN, G. ROZENBERG, A. SALOMAA: *DNA Computing: New Computing Paradigms*. Springer, 1998.

[29] GH. PĂUN, T. YOKOMORI: Membrane computing based on splicing. *DNA Based Computers V*, volume 54 of *DIMACS Series in Discrete Mathematics and Theoretical Computer Science* (E. Winfree, D.K. Gifford, eds.), American Mathematical Society, 1999, 217–232.

[30] D. PIXTON: Regularity of splicing languages. *Discrete Applied Math.*, 69 (1996), 101–124.

[31] Y. ROGOZHIN, S. VERLAN: On the rule complexity of universal tissue P systems. *Lecture Notes in Computer Sci.*, 3850 (2005), 356–362.

[32] S. VERLAN: About splicing P systems with immediate communication and non-extended splicing P systems. *Lecture Notes in Computer Sci.*, 2933 (2003), 369–382.

[33] S. VERLAN: *Head Systems and Applications to Bio-informatics*. PhD thesis, University of Metz, 2004.
[34] S. VERLAN: A boundary result on enhanced time-varying distributed H systems with parallel computations. *Theoretical Computer Sci.*, 344 (2005), 226–242.
[35] S. VERLAN, M. MARGENSTERN: About splicing P systems with one membrane. *Fundamenta Informaticae*, 65 (2005), 279–290.
[36] S. VERLAN, R. ZIZZA: 1-splicing vs. 2-splicing: separating results. *Proceedings of WORDS'03*, Turku, Finland, 2003, 320–331.
[37] C. ZANDRON: *A Model for Molecular Computing: Membrane Systems*. PhD thesis, Universita' degli Studi di Milano, Milano, Italy, 2002.

CHAPTER 9

TISSUE AND POPULATION P SYSTEMS

FRANCESCO BERNARDINI

MARIAN GHEORGHE

9.1 INTRODUCTION

MEMBRANE computing theory contains a number of computational models inspired by the structure and behavior of various biological entities. The cell-like P system model, the most developed and investigated, is inspired by the living cell with its compartments arranged in a hierarchical structure, each compartment containing other compartments and biochemical elements inside [30]. The tissue and population P systems, as well as P colony models, are inspired by more complex biological systems consisting of multiple individuals (cells, bacteria, social insects, birds, etc.) living and cooperating in a certain environment which is used to exchange information or chemical elements. Some of these systems have a well-defined fixed structure, like most of the tissue P systems, whereas others—tissue P systems with active membranes or population P systems—have a dynamic structure. For all these tissue P systems the underlying abstract model is a network of components. In this chapter we present tissue P systems and population P systems, together with variants or generalizations of them. P colonies are described in Chapter 23, Section 23.1.

Tissue P systems were first introduced as a generalization of cell-like P systems, by considering a more general structure than a tree [33], [32] and population P systems as an extension to the former. Both have now their specific theoretical developments and interesting applications.

In this chapter we present tissue P systems working with symbol objects in a pure communicative way, by using symport/antiport rules, or adopting an evolution-communication strategy with various combinations of rules, tissue P systems working on string objects, several variants of population P systems, and generalized communicating P systems. We end this chapter with a discussion about tissue P systems approaches inspired by other computational paradigms and briefly mention some applications of these models. Only tissue P systems working in a communicative manner will be presented with their definitions, an example, and a standard result with its associated proof. For the other types of systems only definitions and results will be stated.

9.2 Tissue P Systems with Communication Channels

Communication is one of the essential features of every P system that allows objects to move from one compartment (cell or environment) to another. One way of achieving communication in tissue P systems is by using channels between neighboring compartments which enable the movement of objects inside the system. The movement of different objects through communication channels is controlled by a set of communication rules which are associated with them; they determine the type of interactions between components. In biological terms, communication rules somehow define the "permeability" of the communication channel with respect to certain types of molecules.

Symport/antiport rules are used to build an abstract model of the active transport of molecules across the membrane. They have been widely used in various investigations regarding tissue P systems. The computational power of communication in the context of both cell-like and tissue P systems (see [30], [4], and also Chapter 5 of this handbook) has been intensively investigated. In this case a channel between cells i and j is modeled as an ordered pair (i, j) and has assigned a finite set of antiport rules. The simplest form of such a rule is (x/y), which describes that an occurrence of the multiset x belonging to cell i can be exchanged with an occurrence of the multiset y from cell j. A symport rule is then seen as a special antiport rule where $y = \lambda$, which means that through the channel defined by the ordered pair (i, j) an occurrence of the multiset x passes through from cell i to j. Moreover, a cell i can communicate with the environment, denoted usually by 0, by means of communication channels $(0, i)$ and/or $(i, 0)$ which have associated their own sets of symport/antiport rules. Specifically, one considers a purely communicative model of tissue P systems where rules cannot increase the number of objects inside the cells, but cells have to rely on an infinite environment to increase the number of objects in the system. This model of tissue P systems with symport/antiport rules was first investigated in [29] and then extended in [19] by adding a notion of state to communication channels—this state is modified every time a communication rule is applied and it determines which rules can be subsequently applied. Also, instead of using the maximal parallelism, the paper [19] has introduced the restriction that, in each step, all the communication channels evolve in parallel at the same time, but, for each channel, at most one rule is applied (maximal parallelism with sequential behavior on channels).

Formally, we give the following definition of tissue P systems with channel-states and symport/antiport rules working on symbol objects.

Definition 9.1 A *tissue P system with channel-states and symport/antiport rules* [19] is a construct

$$\Pi = (O, K, w_1, \ldots, w_m, E, ch, (s_{(i,j)})_{(i,j) \in ch}, (R_{(i,j)})_{(i,j) \in ch}, i_0),$$

where O *is the finite alphabet of* objects; K *is the alphabet of* states; $w_1, \ldots, w_m$ *are finite multisets of objects in* O *which are initially present inside the cells (labeled* $1, \ldots, m$*);* $E \subseteq O$ *is the set of symbol objects present in an arbitrary number of copies in the environment (labeled 0);* $ch \subseteq \{(i, j) \mid i, j \in \{0, 1, \ldots, m\}, i \neq j\}$ *is the set of* channels *between cells, and cells and environment (we call all these* components*);* $s_{(i,j)}$ *is the* initial state *of the channel* (i, j)*;* $R_{(i,j)}$ *is a finite set of* symport/antiport rules *of the form* $(s, x/y, s')$ *with* $s, s' \in K$, $x, y \in O^*$, *associated with channel* (i, j)*;* $i_0 \in \{1, \ldots, m\}$ *is the* output cell.

Thus, a tissue P system with channel-states and symport/antiport rules is a collection of m cells which initially contain multisets w_i, $1 \leq i \leq m$, and are pairwise connected through communication channels belonging to ch. Every channel (i, j) has associated an initial state $s_{(i,j)}$ and a finite set of rules $R_{i,j}$. The cells are immersed into the environment which contains arbitrarily many copies of E and these are never exhausted. A tissue P system has associated a *graph* structure which consists of *nodes* corresponding to cells and environment and *edges* for channels linking various components.

For a component i we will call *neighboring components* (or *neighboring cells*, if they do not involve the environment), all the components (or cells) j that are linked with i through channels (i, j).

A special class of tissue P systems with channel-states and symport/antiport rules is the class of tissue P systems where each channel has only one state which is never modified by the set of rules associated to it. These systems are simply called *tissue*

P systems with symport/antiport rules. In this case we drop from their definition all the notations regarding states (the set K and the initial state values) and for each channel (i, j), the rules in $R_{(i,j)}$ are given in the form (x/y), with x, y two finite multisets of symbol objects. Moreover, we say that a tissue P system with channel-states and symport/antiport rules with m cells uses the *single-channel restriction* if at most one channel is allowed between every two components. This means that, for every $0 \leq i, j \leq m$, if $R_{(i,j)}$ is not empty, then $R_{(j,i)}$ is empty.

A *computation* in a tissue P system with channel-states and symport/antiport rules, is a sequence of steps which start with the cells $1, \ldots, m$ containing the multisets $w_1, \ldots, w_m$ and where, in each step, one or more rules associated with certain channels are applied to the current multisets of symbol objects. We identify four strategies to apply these rules: *maximally parallel with sequential behavior on channels, maximally parallel, asynchronous,* and *sequential* modes. The *maximally parallel with sequential behavior on channels* mode (denoted by *maxsc*) [19] means that, in each step, for each channel, at most one rule is applied. Therefore, the use of the rules is sequential at the level of each channel, but it involves a maximal number of components. The *maximally parallel* mode (denoted by *max*) [30] means that, in each step, a maximal number of components are used and in each channel a maximal number of rules are applied, providing that all these rules change the current state of the channel to the same new one. However, since the environment contains an infinite number of symbol objects, maximal parallelism is defined only for systems satisfying the property that for each channel $(0, j)$ and each rule $(s, x/\lambda, s') \in R_{0,j}$, we have $|x|_a > 0$ for some $a \in O - E$ (otherwise an infinite number of symbol objects are brought into cell j). In the *asynchronous* case (denoted by *asyn*) [14], in each step, an arbitrary number of channels are non-deterministically chosen and for each of them, an arbitrary number of rules are non-deterministically chosen to be applied, providing they use the same states. If one of the components involved is the environment, then the restriction mentioned for maximal parallelism applies in this case as well. The *sequential* strategy (denoted by *seq*) [14] imposes that, in each step, a channel and a rule associated with are non-deterministically chosen.

A computation in a tissue P system with channel-states and symport/antiport rules is *successful* if and only if it halts. The result of such a computation is the Parikh vector which describes the multiplicities of every symbol object from O in the multiset occurring inside cell i_0. A successful computation may also generate non-negative numbers representing the size of the multiset of objects from O that occur in i_0 at the end of this computation. The set of Parikh vectors obtained by all successful computations in Π operating according to the strategy $r \in \{maxsc, max, asyn, seq\}$ is denoted by $Ps(\Pi, r)$; the set of non-negative numbers generated in these circumstances is denoted by $N(\Pi, r)$.

This definition of a successful computation, one that starts from the initial multisets of symbol objects, then evolves toward a halting stage, by following a certain strategy of applying the rules in each step and producing a result in an output cell, is valid for all the other variants of P systems used in this chapter.

The following example, adapted from [19], illustrates how tissue P systems with channel-states and symport/antiport rules work. In fact only symport rules are used.

Example 9.1 *Let us consider the following tissue P system:*

$$\Pi = (O, K, \lambda, \lambda, \lambda, E, ch, (s_{(i,j)})_{(i,j) \in ch}, (R_{(i,j)})_{(i,j) \in ch}, i_0),$$

$$O = \{a, b\},$$

$$K = \{s, s', s''\},$$

$$E = O,$$

$$ch = \{(0, 1), (1, 2), (1, 3)\},$$

$$R_{(0,1)} = \{(s, a/\lambda, s), (s, a/\lambda, s'), (s', b/\lambda, s'), (s', b/\lambda, s'')\},$$

$$R_{(1,2)} = \{(s, a/\lambda, s), (s, b/\lambda, s)\},$$

$$R_{(1,3)} = \{(s, b/\lambda, s'), (s', a/\lambda, s)\},$$

$$i_0 = 3.$$

In the case of maximal parallelism with sequential behavior on channels, the system, in state s, brings inside cell 1 from the environment $n \geq 0$ copies of a, by using n times the rule $(s, a/\lambda, s)$; then the channel $(0, 1)$ changes the state to s' when one further a is brought in by using $(s, a/\lambda, s')$. In state s' we can bring $m \geq 0$ copies of b and finally end up in state s'' and one more b is introduced in cell 1 using the rules $(s', b/\lambda, s')$ and $(s', b/\lambda, s'')$, respectively. Copies of a and b can either go to cell 2 or to cell 3. Those going to cell 2 are lost for the computation, whereas those selected by channel $(1, 3)$ will count toward the result of the computation. Therefore, the computation halts when all a's and b's introduced in cell 1 are consumed. In this case, in cell 3, either the number of a's and b's are equal or the number of b's is equal to the number of a's plus 1; this means that $Ps(\Pi, maxsc) = \{(n, n) \mid n \geq 1\} \cup \{(n, n + 1) \mid n \geq 1\}$ and $N(\Pi, maxsc) = \{n \mid n \geq 2\}$.

Moreover, the system Π operating in a sequential manner produces exactly the same multisets. Hence $Ps(\Pi, maxsc) = Ps(\Pi, seq)$ and $N(\Pi, maxsc) = N(\Pi, seq)$. The behavior of Π is defined neither for the maximal parallelism nor for the asynchronous strategy because the rules $(s, a/\lambda, s)$ and $(s', b/\lambda, s')$ instantaneously introduce inside cell 1 an infinite number of a's and b's, respectively.

We denote by $NO_n t' P_m(state_k, sym_{t_1}, anti_{t_2}, r)$, with $n, m, k, t_1, t_2 \geq 1$ and $r \in \{maxsc, max, asyn, seq\}$, the family of sets of non-negative numbers generated by tissue P systems with at most n symbols, m cells, and k states, and using symport rules of size at most t_1, antiport rules of size at most t_2 (the size of a rule

$(s, x/y, s')$ is $max\{|x|, |y|\}$), and mode r. When any of the values n, m, k, t_1, t_2 is not bounded, then it is replaced by $*$. Moreover, we use $NO_n t' P_m(sym_{t_1}, anti_{t_2}, r)$, with $n, m, t_1, t_2 \geq 1$ and $r \in \{maxsc, max, asyn, seq\}$ as above, to denote the families of sets of non-negative numbers generated by tissue P systems which do not use channel-states. We replace N by Ps whenever families of sets of Parikh vectors, instead of non-negative numbers, are considered and use t in place of t' when tissue P systems with single-channel restriction are investigated. When only symport or antiport rules are used, then we drop from the above notation $anti_{t_2}$ or sym_{t_1}, respectively.

The variant of tissue P systems with (channel-states and) symport/antiport rules, using maximal parallelism with sequential behaviour on channels is the most investigated class of tissue P systems using symport/antiport rules. The computational power and the complexity of the mechanisms used have been mainly studied ([4], [35], [5], [15], [1], [2], [19]). Subsequently we will refer to [4] (an excellent survey of the main developments regarding this variant of tissue P systems) to present the most significant outcomes.

The following results establish the computational power of these systems with respect to minimal cooperation, i.e. the minimum number of cells and the minimum size of the rules involved. It is shown that with 2 cells and using symport/antiport rules of size 1 or only symport rules of size 2, it is enough to achieve computational completeness.

Theorem 9.1 (i) $PsRE = PsO_*tP_m(state_k, sym_{t_1}, anti_{t_2}, maxsc) = PsO_*tP_m(sym_{t_1}, anti_{t_2}, maxsc)$, for all $m \geq 2$, $k, t_1, t_2 \geq 1$. (ii) $PsRE = PsO_*tP_m (state_k, sym_{t_1}, maxsc) = PsO_*tP_m(sym_{t_1}, maxsc)$, for all $m \geq 2$, $k \geq 1$, $t_1 \geq 2$.

The above results are in fact optimal with respect to the parameters considered, number of cells, and the size of the rules used. Indeed, this is proved by the following results that show that with 1 cell, we either obtain a characterization of the Parikh sets of matrix languages, when the size of the rules is at least 2, or only finite sets of numbers, when the size is at most 1.

Theorem 9.2 (i) $PsMAT = PsO_*tP_1(state_k, sym_{t_1}, anti_{t_2}, maxsc) = PsO_*tP_1 (sym_{t_1}, anti_{t_2}, maxsc)$, for $k \geq 1$, $t_1, t_2 \geq 2$. (ii) $NO_*tP_1(state_k, sym_{t_1}, anti_{t_2}, maxsc) = NO_*tP_1(sym_{t_1}, anti_{t_2}, maxsc) \subseteq NFIN$, for $k \geq 1$, $t_1, t_2 \geq 1$. (iii) $NO_*tP_1(state_k, sym_{t_1}, maxsc) = NO_*tP_1(sym_{t_1}, maxsc) \subseteq NFIN$, for $k \geq 1$, $t_2 \geq 2$.

In the above results although the number of cells and the size of the rules are kept at minimal values, the number of symbol objects is arbitrarily large. We will next present some results where the values corresponding to the number of cells and symbols used by these tissue P systems are small. Point (ii) of the following theorem also shows that by dropping the single-channel restriction one might sometimes save one cell.

Theorem 9.3 (i) $NRE = NO_n tP_m(state_k, sym_*, anti_*, maxsc) = NO_n tP_m (sym_*, anti_*, maxsc)$, for $(n, m) \in \{(4, 2), (2, 3), (1, 7)\}, k \geq 1$. (ii) $NRE = NO_n t'P_m (state_k, sym_*, anti_*, maxsc) = NO_n t'P_m(sym_*, anti_*, maxsc)$, for $(n, m) \in \{(5, 1), (3, 2), (2, 3), (1, 6)\}, k \geq 1$.

Tissue P systems with channel-states and symport/antiport rules are studied in the case of maximal parallelism using the same complexity principles, i.e. number of symbols, of cells, and the size of the rules involved ([30], [29], [39]). If maximal parallelism is allowed, then one cell can be saved: tissue P systems with one cell that use symport/antiport rules of size 2 are computationally complete [21]. The proof in [21] makes use of a special version of register machines called counter automata. Here we present a proof that is based on the simulation of the same type of register machine which is considered in the proof of Theorem 9.1 (see [19]).

Theorem 9.4 $PsRE = PsO_* tP_m(state_k, sym_{t_1}, anti_{t_2}, max) = PsO_* t'P_m(state_k, sym_{t_1}, anti_{t_2}, max)$, for $m, k \geq 1, t_1, t_2 \geq 2$.

Proof. We prove the result only for tissue P systems using single-channel restriction; obviously this will remain true for arbitrary tissue P systems with channel-states and symport/antiport rules. In fact, we only prove the inclusion $PsRE \subseteq PsO_* t P_m(state_k, sym_{t_1}, anti_{t_2}, max)$, for all $m, k \geq 1, t_1, t_2 \geq 2$.

We consider a register machine $M = (n, H, l_0, l_h, I)$ generating a set of vectors $N(M) \subseteq \mathbf{N}^p$, for some $p \geq 1$, and construct the following tissue P system with symport/antiport rules (there are no states in this system):

$$\Pi = (O, l_0, O, \{(1, 0)\}, R_{(1,0)}, 1), \text{ where:}$$

$$O = \{a_i \mid 1 \leq i \leq n\} \cup \{l, l', l'', l''' \mid l \in H\} \cup \{e\},$$

and the set $R_{(1,0)}$ is built as described below.

1. For each addition-instruction, $l_1 : (\text{ADD}(r), l_2, l_3) \in I$, $(l_1/l_2 a_r)$ and $(l_1/l_3 a_r)$ are added to $R_{(1,0)}$. These rules simulate an addition by bringing from the environment into cell 1 an object a_r together with either an object l_2 or an object l_3. This means that the value of the register r has been increased by 1 and the next instruction to be simulated is either labeled by l_2 or by l_3.
2. For each subtraction-instruction, $l_1 : (\text{SUB}(r), l_2, l_3) \in I$, the rules indicated in the table below are introduced in $R_{(1,0)}$. These rules simulate in three steps a subtraction-instruction.

Step	$R_{(1,0)}$
1	$(l_1/l'_1\,l''_1)$
2	(l''_1/l'''_1) and $(l'_1\,a_r/e)$
3	$(l'''_1\,e/l_2)$ or
	$(l'''_1\,l'_1/l_3)$

The label l_1 in cell 1 is replaced by the multiset $l'_1\,l''_1$. In the second step, if (at least) a copy of a_r is present in cell 1, then the object l'_1 is sent to the environment together with a copy of a_r and the auxiliary object e is brought into cell 1; if no copy of a_r exists inside cell 1, then l'_1 remains there. Simultaneously, l'''_1 is brought into cell 1 and l''_1 is sent to the environment. In the third step, l'''_1 checks what happened in cell 1 in the previous step: if e is inside cell 1 (i.e. a_r was present), then the multisets $l'''_1\,e$ is removed from cell 1 and the new label l_2 is introduced in cell 1, thus completing the simulation of the subtraction, when this is possible. Otherwise, if l'_1 is still in cell 1, then the multiset $l'''_1\,l'_1$ is exchanged with l_3 from the environment, thus completing the simulation of the subtraction, when the register r is empty and the operation fails.

3. $R_{(1,0)}$ also contains the rule (l_h/λ). Thus, the work of Π will stop exactly when the halting label of M is introduced in cell 1.

Therefore, we conclude that $N(M) = Ps(\Pi, max)$. This proof works under the assumption that all the registers, but the first k defining the output, are emptied before the end of the computation.

The proof above shows how to trade cells and channels when maximal parallelism is used: two rules from the same channel connecting the first cell to the environment are applied in parallel, hence we can avoid using the second cell to control the simulation of the register machine [19]. It follows that if we allow two channels between a cell and the environment, i.e. (0, 1) and (1, 0), then a similar result can be obtained to the previous one by using maximal parallelism with sequential behavior on channels. Indeed, all the rules, but $(l'_1\,a_r/e)$ rules, will be kept associated to (1, 0) channel. These rules will be transformed into $(e/l'_1\,a_r)$ and associated to (0, 1) channel. This way the rules (l''_1/l'''_1) and $(e/l'_1\,a_r)$ will run in parallel, but on different channels.

Theorem 9.5 $PsRE = PsO_*t'P_m(state_k, sym_{t_1}, anti_{t_2}, maxsc)$, for m, $k \geq 1$ and $t_1, t_2 \geq 2$.

A result similar to that stated in Theorem 9.2 holds for tissue P systems with channel-states and symport/antiport rules working either in sequential or asynchronous mode [14].

Theorem 9.6 $PsMAT = PsO_*tP_1(state_k, sym_{t_1}, anti_{t_2}, r) = PsO_*tP_1(sym_{t_1}, anti_{t_2}, r)$, for $k \geq 1$, $t_1, t_2 \geq 2$ and $r \in \{seq, asyn\}$.

9.3 Evolution-Communication Models

In this section we will refer to tissue P systems that use evolution (rewriting) as well as communication rules and where the environment no longer plays the role of a source of objects as these are now generated by evolution rules. For this reason the environment will be either dropped from the next definitions or utilized to collect the elements sent out of the system. Three types of such systems will be presented.

9.3.1 Neural-like Networks of Membranes

A first class of evolution-communication (in short, EC) tissue P systems we consider are inspired by the way neurons work together in order to process impulses in a complex network where they are linked through synapses; this class of tissue P systems is called *neural-like networks of membranes* ([30], [27]).

Definition 9.2 A *neural-like network of membranes* is a construct

$$\Pi = (O, \sigma_1, \ldots, \sigma_m, ch, i_0),$$

where O, ch, i_0 have the meaning provided in Definition 9.1; for each cell i, $1 \leq i \leq m$, $\sigma_i = (Q_i, s_{i,0}, w_i, R_i)$ consists of a set of states Q_i, an initial state $s_{i,0}$, an initial multiset w_i, and a set of EC rules R_i.

Comparing this with Definition 9.1, we notice that the states and rules are local to cells and not to channels. For each cell i the set of evolution-communication rules, R_i, contains rules of the form $s_1 w \to s_2 x y_{go} z_{out}$, where s_1, s_2 are states and w, x, y, z are multisets of symbol objects. When such a rule is applied to a multiset w, this is replaced by x which stays in the current cell, y which goes to neighboring cells, and z which is sent out of the system. In fact, the multiset z is non-empty only for the output cell.

For these neural-like networks of membranes three modes of processing their objects and three ways of communicating them between cells are defined.

The three *processing modes* are: the *minimal* mode (denoted by *min*), which requires that in every cell only one occurrence of the multiset from the left-hand side of a rule is processed; the *parallel* mode (denoted by *par*), which asks for as many occurrences of the multiset defining the left-hand side of a chosen rule to be processed as possible; the *maximal* (denoted by *max*) mode, which requires that a maximal number of multisets are processed with respect to all rules. The rules involved in these transformations use the same states.

The communication modes refer to what happens with the symbol objects occurring in the multiset y_{go} of a rule $s_1 w \to s_2 x y_{go} z_{out}$ that is applied to a multiset w of a given cell. There are three *communication strategies*: the *replicative* (denoted by *repl*) manner (each symbol object occurring in y_{go} is sent to all the neighboring cells); the *unique destination* (denoted by *one*) manner (all the symbol objects occurring in y_{go} are sent to the same neighboring cell, non-deterministically chosen); and the *non-deterministic distribution* (denoted by *spread*) (the symbols in y_{go} are non-deterministically distributed among the cells that are linked with the current cell).

Two *types* of such systems are defined: *cooperative* (denoted by *coo*) systems, if at least one rule $s_1 w \to s_2 x y_{go} z_{out}$ has $|w| > 1$, and *non-cooperative* (denoted by *ncoo*) systems, otherwise.

In the rest of this section we will use the following notations: $Proc = \{min, par, max\}$, $Comm = \{repl, one, spread\}$ and $Type = \{coo, ncoo\}$.

We denote by $NOnP_{m,r}(x, \alpha, \beta)$ the family of sets of non-negative numbers computed by neural-like networks of membranes of type $x \in Type$, using at most $m \geq 1$, cells and $r \geq 1$, states, applying the processing mode $\alpha \in Proc$, and following the communication strategy $\beta \in Comm$.

The results below show that irrespective of the type of the neural-like networks of membranes used and the communication strategy adopted, in most of the processing modes, with no more than 2 cells and 2 to 5 states, computational completeness is achieved.

Theorem 9.7 (i) $NRE = NOnP_{2,5}(ncoo, min, \beta) = NOnP_{4,4}(ncoo, min, \beta)$, for $\beta \in Comm$. (ii) $NRE = NOnP_{2,2}(coo, \alpha, \beta)$, for $\alpha \in Proc$, $\beta \in Comm$.

The above results have been further improved [25] decreasing by 2 the number of states of each of the family involved in point (i).

9.3.2 EC P Systems with Receptors

In the model introduced in Subsection 9.3.1 the evolution and communication aspects came out of the same rules, which are in fact evolution rules with an indication regarding which multiset will be communicated. In this subsection we will refer to an EC model of tissue P systems where these two aspects are separated and distinct rules are associated to them [8].

We define these systems [8] by adapting Definition 9.2 as follows:

- the states are dropped from the model;
- for each cell i, the set of rules is $R_i = C_i \cup E_i$, with C_i a finite set of *communication rules* of the forms $(x; y, in)$ or $(x, exit)$, for $x, y \in O^+$, two finite non-empty multisets of symbol objects, and E_i a finite set of *evolution rules* of the form $x \to y$, for $x, y \in V^+$, two finite non-empty multisets of symbol objects.

In an EC tissue P system defined above, the channels do not have a direction of the communication process associated with it and are denoted by unordered pairs $\{i, j\}$, with $1 \leq i \neq j \leq m$. When the direction between cells will be considered then the corresponding systems will be called *unidirectional*.

A communication rule $(x; y, in) \in C_i$, $1 \leq i \leq m$, may be applied in cell i when a multiset x is present in this cell and it leads to receiving a multiset y from one of its neighboring cells, chosen in a non-deterministic way. In this case x acts as a receptor that is used to recognize the signal y. Moreover, a rule of the form $(x, exit) \in C_i$ allows the cell i to release a multiset x outside of the system. In fact, the objects are lost because we do not provide any specific notion of environment.

An evolution rule of the form $x \to y \in E_i$, $1 \leq i \leq m$, allows consumption of a multiset x in order to produce a new multiset y inside cell i.

Next, we introduce a variant of EC tissue P systems that use a special type of objects called *receptors* for communication purposes [8]. These systems, called EC tissue P systems with receptors, have the following characteristics with respect to those introduced in the previous subsection: for each object $a \in O$ a receptor $\bar{a}$ is defined; initially each cell i contains a finite multiset of objects w_i and a finite multiset of receptors ρ_i; each cell i contains a finite set of rules, R_i, consisting of evolution rules $x \to yz$, with $x, y \in O^+$ and z denoting a finite multiset of receptors, and communication rules $(u, exit)$, with $u \in O^+$. The receptors replace the communication rules $(x; y, in)$ used by general EC tissue P systems.

In each step of a computation, in each cell i, a multiset of objects x can be transformed by a rule $x \to yz \in R_i$ or sent to a neighboring cell j when a receptor $\bar{x}$ is present within this cell or sent out of the system if R_i contains a rule $(x, exit)$; each receptor $\bar{x}$ which is used in a communication to receive an x is removed from the cell and new receptors z produced by rules $x \to yz$ are added into.

Evolution rules can be either cooperative or non-cooperative. Like in the previous case the rules, either evolution or communication rules, are applied in a non-deterministic and maximally parallel manner.

$NO[U]TP_m(s, r, f)$, for $m \geq 1$, $s, r \geq 0$, $f \in \{ncoo, coo\}$, denotes the family of sets of non-negative numbers that are generated by [unidirectional] EC tissue P systems with at most m cells, using communication rules of type (s, r) and evolution rules of type f; a communication rule $(x; y, in)$ is of type (s, r) when $|x| \leq s$, $|y| \leq r$; a communication rule $(x, exit)$ is of type (s, r) when $|x| \leq s$, for any $r \geq 0$. $NO[U]RTP_{m,k}(r, f)$, for $m \geq 1$, $k, r \geq 0$, $f \in \{ncoo, coo\}$, denotes the family of sets of non-negative numbers that are generated by (unidirectional) tissue P systems that use at most m receptors of size at most r, rules of the form $(x, exit)$ with $|x| \leq r$, and evolution rules of type f; each cell has initially at most k receptors.

The next result proves the universality of [unidirectional] EC tissue P systems (with receptors) when cooperative evolution rules are used.

Theorem 9.8 $NRE = NO[U]TP_1(1, 0, coo) = NO[U]T\mathcal{R}P_{1,0}(1, coo)$.

This is a direct consequence of the universality result obtained for the basic model of P systems with cooperative evolution rules [30]. Indeed, in our case we use only one cell and rely on evolution and $(a, exit)$ rules, with $a \in O$.

In the case of non-cooperative rules we obtain again computational completeness in slightly modified conditions.

Theorem 9.9 $NRE = NOTP_2(1, 1, ncoo) = NOUTP_3(1, 1, ncoo)$.

9.3.3 EC Tissue P Systems with Terminal Communication

In this subsection we present a class of EC tissue P systems that contain only evolution rules, but in a certain state some objects are communicated among the neighboring cells. This state, we will see, is associated with a terminal process and for this reason these devices are called *evolution tissue P systems with terminal communication* [6]. We refer to Definition 9.2 and the constraints imposed in Subsection 9.3.2 to define these systems; the only difference consists in the way the sets of rules are defined. In this case the sets R_i contain only evolution rules $a \to v$, $a \in O, v \in O^*$.

For each cell i, we will denote by $T_i \subseteq O$, the set of all the objects that can no longer be transformed by applying rules from R_i. The communication is defined in three different ways: *terminal at the level of objects* (denoted by $tObj$), which means that every time an object $a \in T_i$, is produced inside cell i, it will then be sent to one of the neighboring cells; *terminal at the level of cells* (denoted by $tCell$), which means that the communication takes place from cell i if all the objects in i are in T_i (in this case all the objects are sent to one of the neighboring cells); *terminal at the level of the system* (denoted by $tSys$) is the most general communication mode which happens only when the entire system reaches a terminal state, i.e. no rule is applicable across the system (in this case each cell will communicate its objects to one of the neighbors, non-deterministically chosen).

The communication processes described above do not affect the output cell, i_0, which is not engaged in any communication that involves sending objects out of i_0.

We will denote by $NtOP_m(x)$, the family of non-negative numbers computed by evolution tissue P systems with terminal communication, where at most $m \geq 1$ cells are used in the terminal mode $x \in \{tObj, tCell, tSys\}$.

The following results have been established for these systems in [6].

Theorem 9.10 (i) $NCF = NtOP_m(tObj)$, for $m \geq 4$. (ii) $NET0L = NtOP_m(tSys)$, for $m \geq 4$. (iii) $NET0L \subseteq NtOP_m(tCell)$, for $m \geq 4$.

9.4 Tissue P Systems with Strings

The use of string objects in membrane computing is motivated by both its theoretical roots in formal language theory and the inspiration coming from cellular biology. Indeed, the key data structures in formal languages are sets of strings, whereas in living cells, apart from simple biochemical elements, like ions, water molecules, described by symbol objects, there are more complex macromolecules, like DNA, RNA, proteins, which are depicted by string objects. Some of the operations (rules) that are applied to symbol objects may be used or reformulated for strings, but new ones can be also defined. In this section, tissue P systems using EC rules will be presented. (Chapter 8 has also discussed the case of the splicing operation.)

Indeed, EC tissue P systems defined for symbol objects can also be considered for string objects [26]. Definition 9.2 given in Subsection 9.3.1, for neural-like networks of membranes, provides the key elements for an EC tissue P system working with string objects [26].

Definition 9.3 An *EC tissue P system working with string objects* is a construct

$$\Pi = (O, T, \sigma_1, \ldots, \sigma_m, ch, i_0),$$

where O, ch, i_0 have the meaning provided in Definition 9.1; $T \subseteq O$ is the set of terminal symbols. For each $1 \leq i \leq m$, $\sigma_i = (Q_i, s_{i,0}, L_{i,0}, R_i)$, where Q_i is the set of states, $s_{i,0}$ is the initial state, $L_{i,0}$ is the initial set of strings over O^*, and R_i is the set of EC rules belonging to cell i.

Each set R_i contains EC rules of the form $s_1 X \to s_2 x(tar)$, where $X \in (O - T)$, $x \in O^*$, and $tar \in \{here, go, out\}$. These rules are also called rewriting rules. The string objects in a cell are processed in parallel, but each string object is rewritten by only one rule. If a rule $s_1 X \to s_2 x(tar)$ is applied to a string $x_1 X x_2$ in cell i, then it produces $y = x_1 x x_2$ and if $tar = here$ then y remains in i, if $tar = go$ then y goes to one of the neighboring cells, non-deterministically chosen, and if $tar = out$ then y is removed from the system. Like in the case of neural-like networks of membranes, the use of *out* target is only allowed for the output cell. The processing modes and communication strategies applied for neural-like networks of membranes are used in this case as well. The language computed by such a system consists of all the string objects over T obtained in the output cell at the end of a halting computation.

The family of languages generated by EC tissue P systems with string objects using at most $m \geq 1$ cells, each of which containing at most $r \geq 1$ states, is denoted by $RStP_{m,r}(\alpha, \beta)$, where $\alpha \in Proc, \beta \in Comm$ (see Subsection 9.3.1 for these notations).

These mechanisms are computationally complete as the following theorem shows [26].

Theorem 9.11 $RE = RStP_{1,5}(max, \beta) = RStP_{2,4}(\alpha, \beta)$, $\alpha \in Proc$, $\beta \in Comm$.

An EC tissue P system with hybrid rules, both context-free and contextual, is considered in [24]. Again these devices achieve computational completeness [24].

A special class of EC tissue P systems with string objects is that of *(extended) gemmating tissue P systems* [18]. In order to introduce such systems the Definition 9.1 is slightly changed as follows:

- the states are dropped;
- there is no output cell, but the environment will be used instead to collect the results;
- the sets R_i contain EC rules having the form $a \to @_j x$ or $a \to x@_j$, where j indicates a target cell, $a \in O$ and $x \in O^*$.

The rule $a \to @_j x$ is applicable to string objects aw and the second, $a \to x@_j$, to string objects wa, where a is replaced by x, leading to xw and wx, respectively. The strings are then sent to the target cell, j. So, in this case the target cell is known and is no longer arbitrarily chosen from the list of neighboring cells. The rules are applied at the very ends of a string object. In a step of a computation the maximal parallelism is applied at the system level and each string object is rewritten by one single rule. The parameters considered in computing the generative power of these mechanisms are given by the number of cells and the size of the right-hand side of the rules (i.e. for $a \to @_j x$ or $a \to x@_j$, the size is $|x|$). The language generated is obtained as the set of strings collected in the environment at the end of a halting computation.

The family of languages generated by (extended) gemmating tissue P systems with string objects using at most $m \geq 1$ cells and using rules of size at most $r \geq 1$ is denoted by $[E]StGP_{m,r}$.

It is proved that such systems achieve universality for size 2 (minimum), either using 1 cell when extended gemmating tissue P systems are considered or 2 cells in the more general case of gemmating systems (when there is no terminal alphabet).

Theorem 9.12 (i) $RE = EStGP_{m,r}$, for $m \geq 1, r \geq 2$. (ii) $RE = StGP_{m,r}$, for $m \geq 2, r \geq 2$.

9.5 Population P Systems

The tissue P systems presented so far exhibit a fixed structure with respect to the number of cells and their links. However, in the literature of P systems, there have been introduced P systems with active membranes where cells can divide or disappear from the system [30]. In this section we consider a more general type of

P system with a dynamic structure where not only cell division and death aspects are envisaged, but the entire structure of the system can be changed during the evolution of the system. This model abstracts from the behavior of different biological systems, like tissues evolving in time, bacterium colonies, or social insects, where the structure of such communities evolves in time and space. As this model is not only accounting for various tissues, but for a more general class of biological systems, we call the basic units of such systems *components* instead of cells. The system itself will be called *population P system*.

We remember from Section 9.2 that a tissue P system is associated with a certain graph. In the case of population P systems the underlying graph structure will be part of the model as we need to refer to its structure at any given moment in time. This structure is continuously changing through specific rules, called *bond making rules* [7].

Definition 9.4 A *population P system* is a construct

$$\Sigma = (O, \gamma, \alpha, w_e, \sigma_1, \ldots, \sigma_m, i_0),$$

where O is the alphabet of symbol objects; $\gamma = (\{1, \ldots, m\}, A)$, *with* $A \subseteq \{\{i, j\} | 1 \leq i \neq j \leq m\}$, *is the undirected graph defining the initial links between the m components;* α *is the finite set of bond making rules of the form* $(i, x; y, j)$, *with* $1 \leq i \neq j \leq m, x, y \in O^*$; $w_e \in O^*$ *is the finite multiset of symbol objects initially assigned to the environment;* $\sigma_i = (w_i, R_i), 1 \leq i \leq m$, *with* w_i, *the finite multiset of symbol objects initially assigned to component i and* R_i *the finite set of rules associated to i (the set* R_i *consists of a set* E_i *of evolution rules of the form* $a \to v$, *for* $a \in O, v \in O^+$, *and a set* C_i *of communication rules of the forms* $(a; b, enter)$, $(a; b, in)$, $(b, exit)$, *for* $a \in O \cup \{\lambda\}$, $b \in O$); $i_0 \in \{1, \ldots, m\}$ *is the output component.*

A population P system Σ consists of m components in an environment. These components are initially linked, via some undirected communication channels, as specified by γ. Inside a component i we can either apply evolution rules $a \to v$ (a is replaced by v) or communication rules. Applying a rule $(a; b, in)$ means that if the component i contains an a then a b is brought in from a neighboring component j (i.e. $\{i, j\} \in A$); similarly $(a; b, enter)$ is used to bring in b from the environment; in both cases, when $a = \lambda$, an element b is freely introduced into component i; $(a, exit)$ means releasing a into the environment. Please note that in this case the environment is not an infinite supplier of objects as in the case of tissue P systems with symport/antiport rules, but an additional component with no rules associated, which allows it to bring in objects and send them. The bond making rules allow the creation of new links (bonds) between components. A rule $(i, x; y, j)$ allows the creation of an undirected communication channel between components i and j when x is in component i and y in j.

In every step evolution and communication rules are applied in a maximally parallel manner. Then all the links are destroyed, i.e. $A = \emptyset$, and the bond making rules from α are used to create new links (bonds); so, a new set A is obtained.

A number of parameters are considered when a computation takes place. R denotes the use of restricted communication rules, i.e. $(\lambda; b, in)$, $(\lambda; b, enter)$, $(b, exit)$; nR means arbitrary (non-restricted) communication rules; α_t, $t \geq 0$, means all the bond making rules have the property $(i, x; y, j)$, $|x| \leq t$, $|y| \leq t$; na means no bond making rules (i.e. the system will not change its structure).

We denote by $NOPP_{m,k}(x, y)$ the family of sets of non-negative numbers generated by population P systems using at most $m \geq 1$ components and in every step of the computation the number of components in every connected component of the graph defining the structure of the system is at most k, $n \geq k \geq 1$, with $x \in \{R, nR\}$, $y \in \{na\} \cup \{\alpha_t \mid t \geq 0\}$.

First, we consider the case when there are no bond making rules. This case combined with restricted communication rules leads to mechanisms that generate no more than the length set of context-free languages. When unrestricted communication rules are used then either with 2 components linked between them or 4 separate components, computational completeness is achieved.

Theorem 9.13 (i) $NCF = NOPP_{*,*}(R, na) = NOPP_{1,1}(R, na)$.
(ii) $NRE = NOPP_{2,2}(nR, na)$. (iii) $NRE = NOPP_{4,1}(nR, na)$.

When bond making rules are allowed, even for systems using restricted communication rules, universality is obtained.

Theorem 9.14 $NRE = NOPP_{6,2}(R, \alpha_2)$.

The population P system approach has been successfully used as a basic framework in defining more specialized modeling paradigms. The current stochastic modeling developments based on the use of Gillespie approach in specifying various interactions at molecular level in different biological systems reported in Chapter 18, and the agent based software engineering platforms that deal with dynamic systems exhibiting a dynamic structure, created around X-machines, reported in Chapter 23, Section 23.4, are all based on this concept.

We refer now to another type of population P systems, called *quorum sensing P systems*. They are inspired by some specific phenomena occurring in bacterium colonies. The bacteria that colonize an environment release some signaling molecules in this space. When the environment is densely populated, the concentration of these molecules increases and they penetrate back inside bacteria producing a shift in the bacteria behavior by specific gene expression changes leading to a high rate production of the signaling molecule. This process is called quorum sensing [38].

The model consists of a number of cells and multiple environments, called environment elements. The cells do not communicate directly, but only through their local environment elements; there are no communication channels between environment elements. Each cell contains a specific symbol characterizing this state and a non-negative number denoting the level of the signal molecule. Each environment element can contain only a certain amount of signal molecules and has a limited capacity. The system evolves either by producing signal molecules or diffusing them to and from environment elements. The model does not provide explicit links between cells and environment elements, but the rules identify precisely these relationships [9].

Definition 9.5 A *quorum sensing P system* is a construct

$$\Sigma = (Q, C_1, \ldots, C_m, E_1, \ldots, E_k, R, I),$$

where Q is the set of *states*; $C_i = (q_i, s_i)$, $1 \leq i \leq m$, with q_i, *the* initial state *of cell i (at every moment in time only one state value occurs within cell i) and s_i, a non-negative integer value, the* initial amount of signal molecule *available in cell i*; $E_j = (c_j, t_j)$, $1 \leq j \leq k$, with c_j, *the* capacity of the environment element j and t_j, *the* current amount of *the* initial amount of signal molecule *in j, $t_j \leq c_j$; the set R of* rules *contains:* evolution rules $(q, s)_i \rightarrow (q', s')_i$, *where $1 \leq i \leq m$, $q, q' \in Q$ are states, $s, s' \leq c_j$ are non-negative integer values (if the cell i is in state q with at least s signal molecules, it replaces q by q' and extracts s from the current signal molecule value and then adds s' to it)*, releasing rules $(q, s)_i [\ ]_j \rightarrow (q', s')_i [s'']_j$, *where $1 \leq i \leq m$, $1 \leq j \leq k$, $q, q' \in Q$ are states, s, s', s'' are non-negative integer values (if the cell i is in state q with at least s signal molecules, it replaces q by q', extracts s from the current signal molecule value and then adds s', releases s'' in the environment element j which is added to the current value, keeping the new value under the capacity c_j)*, diffusion rules $(s)_i [t]_j \rightarrow (s')_i [t']_j$, *where $1 \leq i \leq m$, $1 \leq j \leq k$, s, t, s', t' are non-negative integer values, $t, t' \leq c_j$ and $s + t = s' + t'$ (if in cell i there are at least s signal molecules and in the environment element j there is at least t signal molecules, then s is extracted from the current amount in i and then s' is added to it, whereas t is extracted from the amount available in the environment element j and then t' is added to)*; $I = (i_1, \ldots, i_h)$ *denotes the h output cells.*

The system starts from the set of initial states associated with cells and non-negative values associated with states and environment elements and in any step, it evolves as many cells as possible and inside every cell only a rule is applied, if at all possible. The result of a halting computation is obtained in the h output cells as a vector of h non-negative integer values.

$N^h QSP_{m,k}(\delta, \sigma)$ denotes the family of vectors of $h \geq 1$ non-negative integer values generated by quorum sensing P systems with at most $m \geq h$ cells, exactly h output cells, at most $k \geq 1$ environment elements, the capacity of every environment at most δ, and using at most α states.

The following result shows the power of these mechanisms and the linear dependency of m, k, δ on the length of the vector of numbers, with an unbounded number of states.

Theorem 9.15 $N^h RE = N^h QSP_{2h+6, h+3}(3h + 8, *)$, for $h \geq 1$.

This model is not only interesting and powerful as a novel computational paradigm, but is shown to be very effective in modeling real quorum sensing behavior of *Vibrio fischeri* bacterium [9].

Finally, we briefly mention another special case of population P systems, tissue P systems using unit rules and energy assigned to membranes [3]. In this case the cells, like in the quorum sensing P systems, communicate only with the environment by using some specific EC rules. Each such rule has the form $(\alpha : a, e, b)$, where $\alpha = in_i$ or $\alpha = out$, a, b are symbols from the current alphabet, and e is an integer number. Every membrane has associated an integer value, its energy. When a rule $(out : a, e, b)$ is applied in a cell i, a is replaced by b, which is sent out into the environment and the energy of the membrane is changed by adding e to the current energy value, which will increase, if $e \geq 0$, or will decrease, when $e < 0$. When the rule is $(in_i : a, e, b)$, then an object a from the environment is brought into cell i and transformed into b, and the energy value of its membrane will be changed by adding e to it. The rules are applied either in parallel—maximal parallelism for cells and at most one rule applied in each cell, or sequentially—one rule in the entire system is used. The sequential mode may be supplemented by some priority feature which means that if more than a rule can be applied then the rule is selected with the maximum value of $|e|$. It is proved that these mechanisms are computationally complete when maximally parallel mode or sequential mode enforced with priorities are used [3].

9.6 Generalized Communicating P Systems

In biological systems the parallelism occurs at various levels and involves different components and chemical elements. In all the variants of P systems this characteristic occurs either when certain (all) components are running in the same time or multiple transformations take place in each compartment simultaneously. Moreover, through different intercellular communication processes, elements in two adjacent compartments are involved at the same time in a certain transformation. In this respect there have been introduced and studied (tissue) P systems using communication rules of symport/antiport type [28], conditional uniport

[36] or different evolution-communication paradigms (see Section 9.3). In all these examples the rules used involve objects from one or two adjacent compartments. In this section we present a model involving four adjacent compartments, two acting as inputs and two as outputs. In this case we end up with a network of compartments linked by various channels, like in a tissue system, in such a way that these more general communication rules may be applied. This model, called *generalized communicating P system* [37], relying entirely on communication rules, will again use the environment as a supplier of objects for the system.

Definition 9.6 A *generalized communicating P system* is a construct [37]

$$\Pi = (O, w_1, \ldots, w_m, E, R, i_0),$$

where O is the finite alphabet of symbol objects; $w_1, \ldots, w_m$ are finite multisets over O, which are initially present inside the cells (labeled $1, \ldots, m$) of the system; $E \subseteq O$ is the set of symbol objects present in arbitrary number of copies in the environment (labeled 0); R is a finite set of interaction rules of the form $(a, i \to k; b, j \to l)$, with $a, b \in O, 0 \leq i, j, k, l \leq m$; $i_0 \in \{1, \ldots, m\}$ is the output cell.

When a rule $(a, i \to k; b, j \to l)$ is applied, a is moved from cell i to cell j and b is moved from cell k to cell l at the same time. The rules are applied according to the maximally parallel strategy. The family of non-negative numbers computed by a generalized communicating P system using at most $m \geq 1$ cells is denoted by $NGCPS_m$.

An interesting characteristic of these systems consists in their ability to generate in a flexible way different macros based on basic interaction rules or previously defined macros. We will restrict below to only two macros that use two inputs and two outputs each, and where the relationship between inputs and outputs is well-defined, but more general definitions can be provided instead.

The first macro we define is meant to produce the following behavior: when an object a appears in cell 1, then it is moved to cell 2 and a new copy of b is brought into cell 3 from the environment. We denote this macro by $(a, 1 \to 2; \delta b, 0 \to 3)$; δb means that b is brought into cell 3 with a delay, i.e. after a arrives to 2.

This macro is relatively obvious to be implemented; it requires two additional cells, an additional object, I_a, present in one of these two cells. It is simulated by two basic interaction rules: $(a, 1 \to 2; I_a, 5 \to 6), (b, 0 \to 3; I_a, 6 \to 5)$.

The second macro has the following behavior: in the presence of an object a in cell 1, all objects b that are present in cell 3 are moved, one by one, to cell 4; at the end a is moved to cell 2. This macro is denoted by $(a, 1 \to 2; b^+, 3 \to 4)$; b^+ means that all b's from cell 3 are brought into cell 4. It is implemented by using six interaction rules and requires seven more cells and three objects, two X's and one I_a present in three of them. The six interaction rules are: $(a, 1 \to 6; I_a, 5 \to 9)$, $(a, 6 \to 7; b, 3 \to 4)$, $(X, 11 \to 11; I_a, 9 \to 8)$, $(a, 7 \to 6; X, 10 \to 10)$, $(b, 3 \to 4; I_a, 8 \to 9)$, $(a, 6 \to 2; I_a, 8 \to 5)$.

As an illustration, we now define a generalized communicating P system that computes n^2. In defining this system we will use an alphabet of four elements, s, a, b, c, one interaction rule, five macros, and 10 cells, denoted 1 to 10 (we do not count those hidden by the macros used). In order to compute n^2 we will start with n copies of a in cell 1, a copy of b in cell 5 and a copy of s in cell 2; at the end of the computation we obtain n^2 copies of c in the output cell, 8. Indeed, in order to obtain this computation we use the interaction rule $(s, 2 \to 3; a, 1 \to 0)$ and the following five macros: $(s, 3 \to 4; b^+, 5 \to 6)$, $(s, 4 \to 9; b^+, 7 \to 5)$, $(b, 6 \to 7; \delta c, 0 \to 8)$, $(s, 9 \to 10; \delta b, 0 \to 5)$, $(s, 10 \to 2; \delta b, 0 \to 5)$.

We leave the reader to observe that with this set of transformations, one rule and five macros, n^2 c's from o (the environment) are correctly brought into cell 8; all the details are provided in [37].

This approach is very effective in reducing the size of any specification based on generalized communicating P systems and provides further ways to define more general and powerful macros.

These mechanisms are also very powerful, achieving computational completeness with 19 cells. Perhaps this result is not optimal with respect to the number of cells used, but its proof (see [37]) uses again the macros defined above, plus another that helps simulating the decrementing rule of a register machine.

Theorem 9.16 $NRE = NGCPS_m$, for $m \geq 19$.

9.7 Tissue P Systems and Other Formalisms

The numerous variants of P systems have inevitably many features in common with other computational models or paradigms. More than this, sometimes, concepts from other models are considered in the context of P systems. So far, various concepts from grammar systems and DNA computing theories or Petri nets, brane calculus, and X-machines formalisms, have been translated into the P systems framework, with specific adaptations and studied from different perspectives. In this section we present some developments in the area of tissue P systems that have been influenced by such computational paradigms.

Membrane computing and brane calculus have both been based on phenomena that occur on, or around membranes, as well as inside the compartments delimited by them. Some initial investigations have studied classes of P systems inspired by operations defined in the context of brane calculus [10], [11]. A similar approach has been also applied for tissue P systems and specific classes have been appropriately defined—we refer to [17] for details.

Another field of research that has provided interesting concepts to the study of different variants of tissue P systems, is that of grammar systems. In [31] a systematic approach to investigating the cross-fertilization between membrane computing and grammar systems is suggested and preliminary results are provided. Some parallel communicating grammar systems using multisets are considered; tissue-like P systems able to solve in linear time NP-complete problems are suggested. In [16] a special class of tissue P systems with channel states and symport/antiport rules are studied in order to model classes of distributed-cooperating grammar systems.

9.8 MISCELLANEOUS

In the previous sections we have analyzed the power of different variants of tissue, population, or generalized communicating P systems. Many of these are also very effective in solving various problems. Some of them have been used in addressing ways to find solutions to NP-complete problems. In [31] a class of tissue-like P systems is introduced as an extension to the class of parallel communicating grammar systems working on multisets of strings; they provide linear solutions to the satisfiability problem (SAT). SAT and the Hamiltonian Path Problem are solved in linear time by using rewriting tissue P systems [26].

From a slightly different perspective, SAT and 3-coloring problems are studied in [23], [13], as decidability problems. Again linear time solutions are provided for these problems.

Tissue P systems are developed to tackle more specific problems. In [12] a certain class of tissue-like P systems with active membranes is defined to model picture languages. In [34] the three processing modes and three transmitting modes investigated in [30], [27], and extended in [26] to string rewriting tissue P systems, are used to model Boolean circuits. Other applications of the models presented in this chapter are developed in this handbook in Chapters 18 and 21.

9.9 FINAL REMARKS

The concepts and results presented in this chapter represent a well-defined topic of membrane computing with specific theoretical problems and powerful applications. Some of the variants of tissue or population P systems mentioned here define classes of systems that need further investigation. Generalized communicating P systems introduced in [37] require a systematic investigation of the way well-known variants of (tissue) P systems can be obtained from or simulated with them in the light of the general framework defined in [20]. The interactions between various classes of (tissue) P systems and other computational formalisms are only in the first stage of investigation when a transfer from one area to another has been studied, but this type of investigation opens up various other more fruitful research topics, like the study of more general formalisms that cover the characteristics of those initially considered, the development of more case studies that show the benefits and limitations of these approaches, and the identifications of more specific classes of systems that suit better various application problems.

References

[1] A. Alhazov, R. Freund, M. Oswald: Cell/symbol complexity of tissue P systems with symport/antiport. *Intern. J. Found. Computer Sci.*, 17 (2006), 3–26.

[2] A. Alhazov, R. Freund, M. Oswald: Tissue P systems with antiport rules and small numbers of symbols and cells. In [22], 7–22.

[3] A. Alhazov, R. Freund, A. Leporati, M. Oswald, C. Zandron: (Tissue) P systems with unit rules and energy assigned to membranes. *Fundamenta Informaticae*, 74 (2006), 391–408.

[4] A. Alhazov, R. Freund, Y. Rogozhin: Computational power of symport/antiport: History, advances, and open problems. *Lecture Notes in Computer Sci.* 3850 (2006), 1–30.

[5] A. Alhazov, Y. Rogozhin, S. Verlan: Symport/antiport tissue P systems with minimal cooperation. In [22], 37–52.

[6] F. Bernardini, R. Freund: Tissue P systems with communication modes. *Lecture Notes in Computer Sci.*, 4361 (2006), 170–182.

[7] F. Bernardini, M. Gheorghe: Population P systems. *J. Univ. Computer Sci.*, 10 (2004), 509–539.

[8] F. Bernardini, M. Gheorghe: Cell communication in tissue P systems: Universality results. *Soft Computing*, 9 (2005), 640–649.

[9] F. Bernardini, M. Gheorghe, N. Krasnogor: Quorum sensing P systems. *Theoretical Computer Sci.*, 371 (2007), 20–33.

[10] D. Besozzi, N. Busi, G. Franco, R, Freund, Gh. Păun: Two universality results for (mem)brane systems. *Proc. Fourth Brainstorming Week on Membrane Computing*, Sevilla, 2006, RGNC Report 02/2006, 49–62

[11] L. Cardelli, Gh. Păun: A universality result for a (mem)brane calculus based on mate/drip operations. *Intern. J. Found. Computer Sci.*, 17 (2006), 49–68.

[12] R. Ceterchi, R. Gramatovici, N. Jonoska: Tiling rectangular pictures with P systems. *Workshop on Membrane Computing, WMC4, Tarragona, 2003* (C. Martin-Vide et al., eds.), LNCS 2933, Springer, 2004, 88–103.

[13] D. Díaz-Pernil, M.A. Gutiérrez-Naranjo, M.J. Pérez-Jiménez, A. Riscos-Núñez: A uniform family of tissue P systems with cell division solving 3-COL in a linear time. *Theoretical Computer Sci.*, 404 (2008), 76–87.

[14] R. Freund: Asynchronous P systems. *Lecture Notes in Computer Sci.*, 3365 (2005), 36–62.

[15] R. Freund, M. Oswald: Tissue P systems with symport/antiport rules of one symbol are computationally universal. In [22], 187–200.

[16] R. Freund, M. Oswald: Modelling grammar systems by tissue P systems working in the sequential mode. *Fundamamenta Informaticae*, 76 (2007), 305–323.

[17] R. Freund, M. Oswald: Tissue P systems and (mem)brane systems with mate and drip operations working on strings. *Electronic Notes in Theoretical Computer Sci.*, 171 (2007), 105–115.

[18] R. Freund, M. Oswald, A. Păun: Optimal results for the computational completeness of gemmating (tissue) P systems. *Intern. J. Found. Computer Sci.*, 16 (2005), 929–942.

[19] R. Freund, Gh. Păun, M.J. Pérez-Jiménez: Tissue-like P systems with channel-states. *Theoretical Computer Sci.*, 330 (2005), 101–116.

[20] R. Freund, S. Verlan: A formal framework for static (tissue) P systems. *Lecture Notes in Computer Sci.*, 4860 (2007), 271–284.

[21] P. Frisco, H.J. Hoogeboom: Simulating counter automata by P systems with symport/antiport. *Lecture Notes in Computer Sci.*, 2597 (2003), 288–301.

[22] M.A. Gutiérrez-Naranjo, Gh. Păun, M.J. Pérez-Jiménez, eds.: *Proc. ESF Exploratory Workshop on Cellular Computing (Complexity Aspects)*, Fénix Editora, Seville, 2005.

[23] M.A. Gutiérrez-Naranjo, M.J. Pérez-Jiménez, A. Riscos-Núñez, F.J. Romero-Campero: Cell-like solution versus tissue-like solution for the SAT problem. *Intern. J. Intelligent Systems*, 24 (2009), 747–765.

[24] S.N. Krishna, K. Lahshmanan, R. Rama: Tissue P systems with contextual and rewriting rules. *Lecture Notes in Computer Sci.*, 2597 (2002), 339–351.

[25] S.N. Krishna, R. Rama: On the power of tissue P systems working in the minimal mode. *Lecture Notes in Computer Sci.*, 2509 (2002), 176–190.

[26] M. Madhu, V.J. Prakash, K. Krithivasan: Rewriting tissue P systems. *J. Univ. Computer Sci.*, 10 (2004), 1250–1271.

[27] C. Martin-Vide, Gh. Păun, J. Pazos, A. Rodriguez-Paton: Tissue P systems. *Theoretical Computer Sci.*, 296 (2003), 295–326.

[28] A. Păun, Gh. Păun: The power of communication: P systems with symport/antiport rules. *New Generation Computing*, 20 (2002), 295–305.

[29] A. Păun, Gh. Păun, G. Rozenberg: Computing by communication in networks of membranes. *Intern. J. Found. Computer Sci.*, 13 (2002), 779–798.

[30] Gh. Păun: *Membrane Computing. An Introduction*, Springer, 2002.

[31] Gh. Păun: Grammar systems vs. membrane computing: A preliminary approach. *Proc. Grammar Systems Week Workshop, Budapest, 2004* (E. Csuhaj-Varjú, G. Vaszil, eds.), 255–275.

[32] Gh. Păun, Y. Sakakibara, T. Yokomori: P systems on graphs of restricted forms. *Publicationes Mathematicae*, 60 (2002), 635–660.

[33] Gh. Păun, T. Yokomori: Computing based on splicing. *Pre-Proc. Fifth Intern. Meeting on DNA Based Computers*, MIT, 1999, 213–227.

[34] V.J. Prakash, K. Krithivasan: Simulating Boolean circuits with tissue P systems. *Workshop on Membrane Computing, WMC5, Milan, 2004* (G. Mauri et al., eds.), 343–359.

[35] S. Verlan: Tissue P systems with minimal symport/antiport. *Lecture Notes in Computer Sci.*, 3340 (2004), 418–430.

[36] S. Verlan, F. Bernardini, M. Gheorghe, M. Margenstern: Computational completeness of tissue P systems with conditional uniport. *Lecture Notes in Computer Sci.*, 4361 (2006), 521–535.

[37] S. Verlan, F. Bernardini, M. Gheorghe, M. Margenstern: Generalized communicating P systems. *Theoretical Computer Sci.*, 404 (2008), 170–184.

[38] K. Winzer, K.R. Hardie, P. Williams: Bacterial cell-to-cell communication: Sorry, can't talk now – gone to lunch. *Current Opinion in Microbiology*, 5 (2002), 216–222.

[39] X. Xu: Tissue P systems with minimal symport/antiport. *Proc. Conf. on Bio-Inspired Computing - Theory and Applications, BIC-TA, Wuhan, 2006, Membrane Computing Section* (L. Pan, Gh. Păun, eds.), 168–177.

CHAPTER 10

CONFORMON P SYSTEMS

PIERLUIGI FRISCO

10.1 INTRODUCTION

A theoretical model of the living cell based on the concept of *conformon* has been the inspiration for *conformon P systems*. Conformons are sequence-specific mechanical strains embedded in the (macro)molecules present in cells. DNA supercoils and proteins are examples of conformational deformations that provide both the information and the energy needed for biopolymers to drive molecular processes. The concept of *conformon* was introduced independently in [12] and [21]. Following the definition given in [12] conformons and conformon-like entities have been classified into 10 families according to their biological functions [14].

Conformons are at the base of the *Bhopalator* [13, 15], a theoretical model of the living cell in which the processes present in a cell are described using conformons arranged in space and time with appropriate force vectors. In this model conformons interact exchanging all or part of their free-energy.

The mathematical abstraction of a biological conformon is an ordered pair of a symbol in an alphabet and a number. The symbol represents the information while the number represents the energy of a conformon. The interaction between two such mathematical conformons becomes then the passage of part of the number from one pair to another.

Compartments define the locality of these interactions. The passage of conformons is possible between different compartments and this passage is solely based on the number present in a conformon. A conformon can pass from one compartment to another only if its number is in between certain boundaries.

Conformon P systems can be regarded as a type of tissue P systems with the difference that several edges can be present between two compartments and that the passage of conformons from one compartment to another is not dictated by a rule but depends on some features of the conformons. Moreover, differently than the vast majority of membrane systems, conformon P systems do not operate under *maximal parallelism*. Instead, they are *asynchronous*: the transition from one configuration to another is such that at least one rule is applied.

After introducing in Sections 10.2, 10.3, and 10.4 some necessary formalism and notation, in Sections 10.6, 10.7, and 10.8 we describe in details the known results on the computing power of conformon P systems, while in Section 10.9 we give an overview on how these devices have been successfully used to model the dynamics of HIV infection.

10.2 Conformon P Systems

The model of membrane system considered in the present chapter was introduced and initially studied in its computational aspects in [5]. Further studies in the same directions were reported in [6, 7]. A *conformon* is an element of $V \times \mathbb{N}$, denoted by $[\Phi, a]$. We refer to Φ as the *name* (that is, the information) of the conformon $[\Phi, a]$ and to a as its *value* (that is, the energy). If, for instance, $V = \{A, B, C, \ldots, Z\}$, then $[A, 5], [C, 0], [Z, 14]$ are conformons, while $[AB, 21]$ and $[D, 0.5]$ are not. Provided an unambiguous context, the sole name identifies the conformon itself.

Two conformons can interact according to an *interaction rule*. An interaction rule is of the form $\Phi \xrightarrow{e} Y$, $\Phi, Y \in V$, and $e \geq 1$, and it says that a conformon with name Φ can give e from its value to the value of a conformon having name Y. If, for instance, there are conformons $[G, 5]$ and $[R, 9]$ and the rule $G \xrightarrow{3} R$, one application of this rule leads to $[G, 2]$ and $[R, 12]$. As for the moment we consider only conformons whose value cannot be a negative number, then the rule $G \xrightarrow{3} R$ cannot be applied to $[G, 2]$.

The compartments present in a conformon P system have a label, every label being different. Membrane compartments present in a conformon P system can be unidirectionally connected to each other and for each connection there is a *predicate*. A predicate is an element of the set $pred(\mathbb{N}) = \{\geq n, \leq n \mid n \in \mathbb{N}\}$. If, for instance, there are two compartments (with labels) m_1 and m_2 and there is a connection from m_1 to m_2 having predicate ≥ 4, then conformons having value greater than or equal to 4 can pass from m_1 to m_2. These are *unidirectional connections*: m_1 connected to m_2 does not imply that m_2 is connected to m_1. Moreover, each connection has its own predicate. If, for instance, m_1 is connected to m_2 and m_2 is connected to m_1, the two connections can have different predicates.

The interaction with another conformon and the passage to another compartment are the only *operations* that can be performed by a conformon.

We are going to study a few models of conformon P systems. Here we introduce one of these models.

Definition 10.1 *A conformon P system is a construct*

$$\Pi = (V, \mu, \omega_z, ack, L_1, \ldots, L_m, R_1, \ldots, R_m), \text{ where:}$$

V is an alphabet;

$\mu = (Q, E)$ *is a directed labeled graph underlying Π with*

$Q \subset \mathbf{N}$ *contains vertices. For simplicity we define $Q = \{1, \ldots, m\}$. Each vertex in Q defines a compartment of Π;*

$E \subseteq Q \times Q \times pred(\mathbf{N})$ *defines directed labeled edges between vertices, indicated by $(i, j, pred(n))$, $i, j \in Q, i \neq j$, where for each $n \in \mathbf{N}$ we consider $pred(n) \in \{\geq n, \leq n\}$;*

ω_z *with $\omega \in \{input, output\}$ and $z \in Q$ indicates if Π is an accepting ($\omega = input$) or generating ($\omega = output$) device. The compartment z contains the input or output, respectively;*

$ack \in Q$ *indicates the* acknowledge *compartment;*

$L_i : (V \times \mathbf{N}) \to \mathbf{N} \cup \{+\infty\}$, $i \in Q$, *are multisets of conformons initially associated with the vertices in Q; $L_{ack} = \emptyset$;*

R_i *are finite sets of interaction rules associated with the vertices $i \in Q$.*

Let M_i be the multiset of conformons present in compartment $i \in Q$. Two conformons in M_i can interact according to a rule in R_i such that the multiset of conformons M_i changes into M'_i. If, for instance, $[\Phi, a], [Y, b] \in M_i$, $\Phi \xrightarrow{e} Y \in R_i$, and $a \geq e$, then $M'_i = (M_i - \{[\Phi, a], [Y, b]\}) \cup \{[\Phi, a-e], [Y, b+e]\}$.

A conformon $[\Phi, a]$ present in i can *pass* to compartment j if $(i, j, pred(n)) \in E$ and $pred(n)$ holds on a. That is, if $pred(n)$ is $\leq n$, then $a \leq n$; if $pred(n)$ is $\geq n$, then $a \geq n$. This passage changes the multisets of conformons M_i and M_j into M'_i and M'_j, respectively. In this case $M'_i = M_i - \{[\Phi, a]\}$ and $M'_j = M_j \cup \{[\Phi, a]\}$.

At the moment we do not assume any requirement on the application of operations. If a conformon can pass to another compartment or interact with another conformon according to a rule, then one of the two operations or none of them is non-deterministically chosen. That is, conformon P systems are *asynchronous* and *non-deterministic*. Non-determinism can also arise from the configurations of a conformon P system if in a compartment a conformon can interact with more than one conformon and also from the graph underlying Π if a compartment has edges with the same predicate going to different compartments.

A *configuration* of Π is an m-tuple $(M_1, \ldots, M_m)$ of multisets over $V \times \mathbf{N}$. The m-tuple $(L_1, \ldots, L_m)$ is called *initial configuration* (remember that in the initial configuration the acknowledge compartment does not contain any conformon) while any configuration having $M_{ack} \neq \emptyset$ is called *final configuration*. In a final configuration no operation is performed even if it could.

For two configurations $(M_1, \ldots, M_m)$, $(M'_1, \ldots, M'_m)$ of Π we write $(M_1, \ldots, M_m) \Rightarrow (M'_1, \ldots, M'_m)$ indicating a *transition* from $(M_1, \ldots, M_m)$ to $(M'_1, \ldots, M'_m)$, that is, the application of one operation to at least one conformon. In other words, in any configuration in which $M_{ack} \neq \emptyset$ any conformon present in a compartment can either interact with another conformon present in the same compartment or pass to another compartment or remain in the same compartment unchanged. If no operation is applied to a multiset M_i, then no transition takes place. The reflexive and transitive closure of $\Rightarrow$ is indicated by $\Rightarrow^*$.

A *computation* is a finite sequence of transitions between configurations of a system Π starting from $(L_1, \ldots, L_m)$.

In case Π is an accepting device ($\omega = input$), then the input is given by the number of conformons (counted with their multiplicity) present in L_z. The input is accepted by Π if it reaches a configuration in which any conformon is present in ack, *halting* in this was the computation:

$$N(\Pi) = \{|L_z| \mid (L_1, \cdots, L_m) \Rightarrow^* (M'_1, \cdots, M'_m) \Rightarrow (M_1, \cdots, M_m),$$
$$M'_{ack} = \emptyset, M_{ack} \neq \emptyset\}.$$

In case Π is a generating device ($\omega = output$), then $L_z = \emptyset$. The result of a computation is given by M_z when any conformon is present in ack. When this happens the computation is *halted* and the number of conformons (counted with their multiplicity) present in M_z defines the *number generated* by Π, indicated by $N(\Pi)$:

$$N(\Pi) = \{|M_z| \mid (L_1, \cdots, L_m) \Rightarrow^* (M'_1, \cdots, M'_m) \Rightarrow (M_1, \cdots, M_m),$$
$$M'_{ack} = \emptyset, M_{ack} \neq \emptyset\}.$$

If instead Π reaches a configuration in which no conformon is present in ack and no operation can be applied, then we say Π *stops*.

We also consider *conformon P systems where interaction rules have priority on passage rules*. In these systems, if in a configuration a conformon can be subject to an interaction rule and to a passage rule, then the interaction rule is applied. In other words, in each compartment a passage rule is applied to a conformon only if no interaction rule can be applied to that conformon.

We do not provide formal definitions for most of the conformon P systems considered in this chapter. Figures depicting them are provided instead.

10.3 ABOUT THE FIGURES

Figures representing conformon P systems have compartments represented by rectangles with their label written in **bold** on their right-upper corner. Conformons and interaction rules related to a compartment are written inside a rectangle.

Conformons present in the initial configuration of a system are written in **bold** inside a rectangle while the ones written in normal font are present in that compartment in one of the possible configurations of the system. A slash (/) between values in a conformon indicates that a conformon can have any of the indicated values. For instance, $[A, 3/5/10]$ indicates that there are configurations in which $[A, 3]$, $[A, 5]$ or $[A, 10]$ can be present in a compartment. If m conformons $[\Phi, a]$ can be present in a compartment, then $([\Phi, a], m)$ is indicated. If an unbounded number of conformons $[\Phi, a]$ can be present in a compartment, then $([\Phi, a], +\infty)$ is indicated.

If in a compartment the interaction rules $\Phi \xrightarrow{e} Y$ and $Y \xrightarrow{e} \Phi$ are present, then $\Phi \xleftrightarrow{e} Y$ is indicated.

Directed edges between compartments are represented as arrows with their predicate indicated close to them. Several edges connecting two compartments are depicted as just one edge with different predicates separated by a slash (/). For instance, an edge with predicate $\leq 2/ \geq 5$ indicates two edges, one with predicate ≤ 2 and the other with predicate ≥ 5.

Further explanations related to the figures in this chapter are present in the next section.

Example 10.1 *A conformon P system accepting any positive even number.*

The conformon P system related to this example is depicted in Fig. 10.1. In order to have a simple system we slightly change the condition of acceptance into: the input is accepted if the system reaches a configuration in which any conformon is present in the acknowledge compartment and the input compartment is empty.

We only give an informal definition of this system, because a formal definition would be tedious. This system is called Π_1, its input compartment is 1 and its acknowledge compartment is 11.

In the initial configuration $p \in \mathbb{N}$ occurrences of $[A, 0]$ (the input) are present in compartment 1, $[B, 3]$ and $[C, 11]$ are present in compartment 2. The system works as follows: pairs of occurrences of $[A, 0]$ are taken out from compartment 1. This is performed with the help of conformons C and B. If when any conformon is present in the acknowledgement compartment the input compartment is empty, then there was a positive even number of $[A, 0]$ in this compartment. The system is non-deterministic.

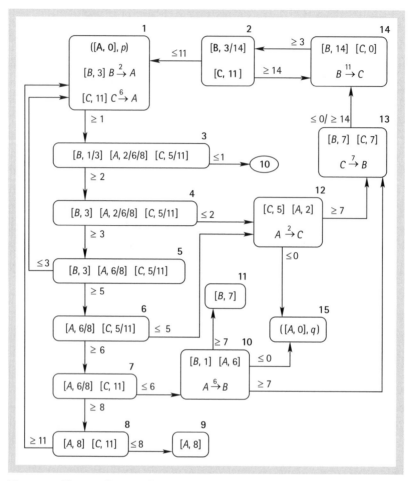

Fig. 10.1 The conformon P system in Example 10.1.

The conformons $[B, 3]$ and $[C, 11]$, initially present in compartment 2, can pass to compartment 1. Here:

i) the B or C conformon interact with two different occurrences of $[A, 0]$;
ii) the B or C conformon interact with the same occurrence of $[A, 0]$;
iii) the B or C conformon pass to compartment 3.

If item i in the previous list takes place, then $[B, 1]$, $[A, 2]$, $[C, 5]$, and $[A, 6]$ are generated and then they can pass to compartment 3. The compartments from 3 to 8 can be regarded as filtering conformons depending on their value (see Definition 10.2). From these compartments conformons with value 1 and 6 (B and A, respectively) pass to compartment 10, while the ones with value 5 and 2 (C and A, respectively) pass to compartment 12. This sequence of compartments lets conformons with value 3 and 11 to pass back to compartment 1 (this is related to item iii in the previous list).

If in compartment 10 $[B, 1]$ and $[A, 6]$ are present, then for sure two occurrences of $[A, 0]$ have been removed from compartment 1. One of these occurrences became $[A, 6]$ while the other interacted with the B conformon so that $[B, 1]$ is generated.

In compartment 10 the B and A conformons can interact so that $[A, 0]$ and $[B, 7]$ are generated. The conformon $[A, 0]$ can pass to compartment 15, while the conformon $[B, 7]$ can either pass to compartment 11 (the acknowledge one) or to compartment 13. What happens in compartment 12 between $[C, 5]$ and $[A, 2]$ is similar to that just described for compartment 10.

If $[B, 7]$ passes to the acknowledge compartment, the computation halts and the input is accepted only if compartment 1 is now empty. If instead $[B, 7]$ passes to compartment 13, then it stays here until $[C, 7]$ (generated in compartment 12) arrives. When $[B, 7]$, $[C, 7]$ are both in compartment 13 they can interact so that $[B, 14]$ and $[C, 0]$ are generated. These two last conformons can pass to compartment 14 and interact so that $[B, 3]$ and $[C, 11]$ are generated and they can pass to compartment 2.

A configuration similar to the initial one is recreated: two occurrences of $[A, 0]$ have been removed from compartment 1 and passed to compartment 10, 12, or 15.

Item ii in the previous list sees $[A, 8]$ to be generated. This conformon can only pass to compartment 9. In this case a halting configuration is never reached.

Item iii in the previous list has been discussed in the above.

If in the initial configuration there is an odd number of $[A, 0]$ conformons in compartment 1, then at a certain point either $[B, 1]$ or $[A, 6]$ are present in compartment 10. In this case the system stops.

In case in the initial configuration there are no occurrences of $[A, 0]$ in compartment 1, then $[B, 3]$ and $[C, 11]$ keep looping through compartments 1, 3, ..., 8.

We can say that Π_1 accepts $\{2n \mid n \geq 1\}$.

10.4 MODULES OF CONFORMON P SYSTEMS

In the following we use the concept of a *module*: a group of compartments with conformons and interaction rules in a conformon P system able to perform a specific task.

As a module is supposed to be part of a bigger system it has edges coming from or going to compartments not defined in the module itself. These compartments are depicted with a dashed line in the figures representing detailed modules.

258   CONFORMON P SYSTEMS

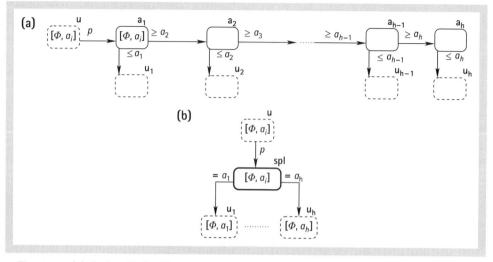

Fig. 10.2 (a) A detailed splitter and (b) its module representation, $1 \leq i \leq h$.

Some modules are depicted with a rectangle with a thicker line. Such modules have a *label* written in **bold** on its right upper corner indicating the kind of module. A subscript is added to differentiate labels referring to the same kind of module present in one system.

Definition 10.2 *A splitter is a module that can select conformons depending on their value. Specifically, consider the conformons* $[\Phi, a_i]$, $1 \leq i \leq h$, *and* $a_j < a_{j+1}$ *for* $1 \leq j \leq h-1$. *If* $[\Phi, a_{i_0}]$, $1 \leq i_0 \leq h$, *is present in a specific compartment of the splitter, then this conformon can pass to another specific compartments, precisely identified by* a_i.

Figure 10.2.a represents a detailed splitter.

No conformon and no rule is present in the initial configuration of this module; h compartments with labels a_i, $1 \leq i \leq h$, are present. From compartment u, external to the module, a conformon $[\Phi, a_i]$, $1 \leq i \leq h$, can reach compartment a_1 (there can be any predicate for this edge; in Fig. 10.2 we indicated p as a generic predicate). Each of the a_i compartments has edges $(a_i, a_{i+1}, \geq a_{i+1})$ for $1 \leq i \leq h-1$, and $(a_i, u_i, \leq a_i)$ for $1 \leq i \leq h$, where u_i are compartments external to the module. If a conformon $[\Phi, a_i]$ is present in compartment a_j, $1 \leq j \leq h$, then (considering what just described, that $1 \leq i \leq h$ and $a_j < a_{j+1}$ for $1 \leq j \leq h-1$) it can pass to compartment u_k only if $a_i = a_k$. Otherwise $a_i > a_k$ and $[\Phi, a_i]$ passes to compartment a_{k+1}.

Considering what was just stated, the operation performed by a splitter can be indicated in a more convenient way by more specific predicates on the edges outgoing the module. Figure 10.2.b illustrates the module representation of a splitter having **spl** as label and edges with predicates $= a_i$, $1 \leq i \leq h$, indicating that

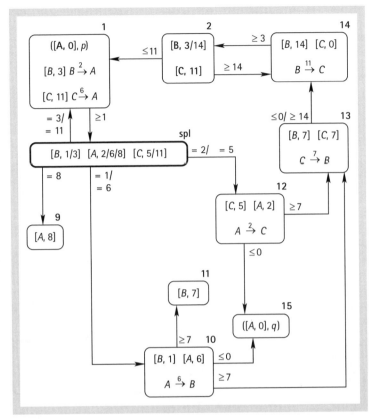

Fig. 10.3 The conformon P system in Example 10.1, with a splitter.

a conformon $[\Phi, a_i]$ can pass from compartment u to compartment u_k only if $a_i = a_k$. It is important to notice that the module representations of splitters can miss the edges having labels $= a_1$ and $= a_h$ and having one outgoing edge with label $\leq a_1$ and one outgoing edge with label $\geq a_h$.

The compartments from 3 to 8 in Fig. 10.1 define a splitter. This means that the conformon P system Π_1 can be depicted as done in Fig. 10.3.

Definition 10.3 *A separator is a module that can select conformons depending on their name. This module is such that when any conformon of type $[\Phi_i, a]$, $1 \leq i \leq h$, is present in a specific compartment of it, then it can pass to a compartment precisely identified by Φ.*

Figure 10.4.a represents a detailed separator.

The number of compartments in this module and the conformons present in them depend on h and a. From compartment u, external to the module, a

260 CONFORMON P SYSTEMS

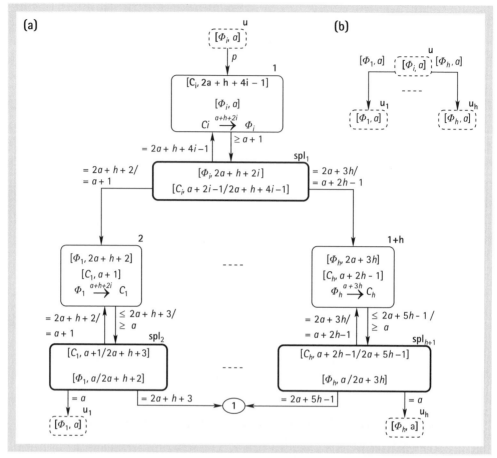

Fig. 10.4 (a) A detailed separator and (b) its shorthand representation.

conformon $[\Phi_i, a]$ can reach compartment 1. There can be any predicate for this edge; in Fig. 10.4.a we indicated p as a generic predicate. In the initial configuration compartment 1 contains the conformons $[C_i, 2a + h + 4i - 1]$, $C_i \neq \Phi_j$, $1 \leq i, j \leq h$. In order to differentiate the conformons $[\Phi_i, a]$, $a + h + 2i$ is added to the value of them and subtracted from the value of $[C_i, 2a + h + 4i - 1]$. In this way, $[\Phi_i, 2a + h + 2i]$ and $[C_i, a + 2i - 1]$ are created and they can pass to spl_1. The pairs of conformons that interacted in compartment 1 can now be selected according to their value. Conformons with values $2a + h + 2i$ and $a + 2i - 1$ pass to specific compartments where the rule $\Phi_i \xrightarrow{a+h+2i} C_i$ is present. In these compartments $[\Phi_i, a]$ and $[C_i, 2a + h + 4i - 1]$ are created and they can pass to compartments u_i and 1, respectively.

The operation performed by a separator can be indicated in a more convenient way by a label on an edge. In Fig. 10.4.b a conformon P system having edges with

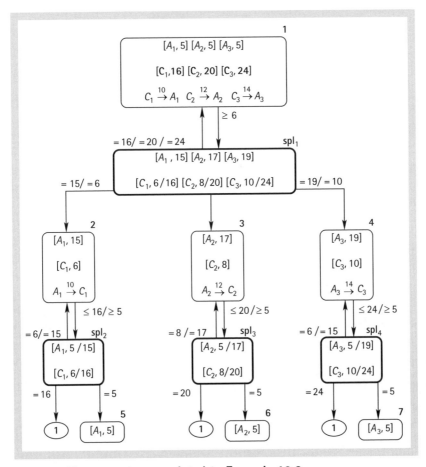

Fig. 10.5 The separator associated to Example 10.2.

labels $[\Phi_i, a]$, $1 \leq i \leq h$, is depicted. This indicates that a conformon $[\Phi_j, a]$ can pass from compartment u to compartment u_i only if $j = i$.

Example 10.2 *A separator for $[A_1, 5]$, $[A_2, 5]$, and $[A_3, 5]$.*

If we consider the definition of a separator, then we have $a = 5$ and $h = 3$. The C_i conformons initially present in compartment 1 are: $[C_1, 16]$, $[C_2, 20]$, and $[C_3, 24]$. The interactions rules present in compartment 1 are: $C_1 \xrightarrow{10} A_1$, $C_2 \xrightarrow{12} A_2$, and $C_3 \xrightarrow{14} A_3$. These rules allow the conformons $[A_1, 15]$, $[C_1, 6]$, $[A_2, 17]$, $[C_2, 8]$, $[A_3, 19]$, and $[C_3, 10]$ to be generated. These conformons can pass (through spl_1) to compartments 2, 3, and 4. In compartment 2, for instance, only the conformons $[A_1, 15]$ and $[C_1, 6]$ can pass. Here they interact such that $[A_1, 5]$ and $[C_1, 16]$ are generated. These two conformons can then pass to compartments 5 and 1, respectively.

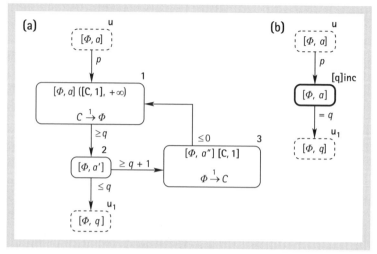

Fig. 10.6 (a) A detailed increaser and (b) its module representation.

In the pictorial representations of conformon P systems several conformons separated by a slash (/) close to an edge indicate several shorthand representations of separators.

Definition 10.4 *An increaser is a module that can increase the value of conformons until a specific amount. This module is such that when a conformon $[\Phi, a]$ is present in a specific compartment of it, then its value can increase until q, $q \geq 1$, so that $[\Phi, q]$ can pass to another specific compartment.*

In Fig. 10.6.a a detailed increaser is depicted. From compartment u, external to the module, a conformon $[\Phi, a]$ can reach compartment 1. There can be any predicate for this edge; in Fig. 10.6.a we indicated p as a generic predicate. In the initial configuration an unbounded number of conformons $[C, 1]$ is present in compartment 1 and one occurrence of $[C, 1]$ is present in compartment 3. In compartment 1 the value of Φ can be increased to any value but only when the value of this conformon is $\geq q$, then it can pass to compartment 2.

In this compartment the Φ conformon can pass to compartment u_1, external to the module, only if its value is $\leq q$, that is, if it is equal to q. Otherwise, this conformon can pass to compartment 3 where its value can be decreased (and added to the one of the C conformon present in this compartment). When the value of Φ is equal to zero, then it can pass back to compartment 1.

In Fig. 10.6.b the module representation of an increaser is depicted. The label of this module is $[q]$inc.

Definition 10.5 *A decreaser is a module that can decrease the value of conformons until a specific amount. This module is such that when a conformon $[\Phi, a]$ is present*

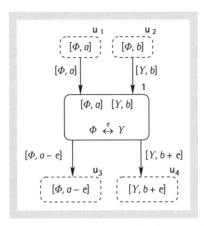

Fig. 10.7 A detailed strict interaction.

in a specific compartment of it, then its value can decrease until q, $q \geq 1$, so that $[\Phi, q]$ can pass to another specific compartment.

This module can be easily obtained starting from an increaser. If we consider Fig. 10.6, then it is enough to remove the edge from compartment u to compartment 1 and to add an edge from compartment u to compartment 3. The module representation of such a module is similar to the one of an increaser with the difference that the label is $[q]$dec.

The combination of separators, decreasers, and increasers allows us to define *strict interaction* rule: $\Phi^{(a)} \stackrel{e}{\rightarrow} Y_{(b)}$ where $a, b, e \in \mathbb{N}$, meaning that a conformon with name Φ can interact with Y passing just e only if the value of Φ and Y before the interaction is a and b, respectively. The detailed module for strict interaction is depicted in Fig. 10.7.

Interactions of the kind $\Phi \stackrel{e}{\rightarrow} Y_{(b)}$ (before the interaction Φ can have any value while Y must have b as value) and $\Phi^{(a)} \stackrel{e}{\rightarrow} Y$ (before the interaction Φ must have a as value while Y can have any value) can be defined, too.

10.5 A Technical Remark

The proofs in the following sections are based on register machines, partially blind register machines, and restricted register machines.

A *restricted register machine* is a register machine which can increase the value of a register, say β, only if it decrease the value of another register, say γ at the same time.

So, restricted register machines have only one kind of instruction: $s : (SUB(\gamma), ADD(\beta), v, w)$ with s, v, w states and γ, β different register of the restricted register machine. If when in state s the content of register γ can be decreased of 1, then the one of register β is increased by 1 and the machine goes into state v, otherwise no operation is performed on the registers and the machine goes into state w. Two registers γ and β are *connected* if there is an instruction in which both registers are present.

It is known that restricted register machines with $n+1$ registers are more powerful than the ones with n registers. A consequence of this theorem is that the number of registers induces an infinite hierarchy among families of computed sets of numbers.

In order to have similar constructions both in the accepting and the generating cases, we consider register machines with ADD instructions of the form $l_i : (ADD(r), l_j)$, with the non-determinism moved to the labeling, which is allowed not to be injective: the same l_i can label two or more different instructions. Clearly, for deterministic register machines such a case does not appear. For non-deterministic register machines it is possible to pass from instructions of the form $l_i : (ADD(r), l_j, l_k)$ to instructions of the form $l_i : (ADD(r), l_s)$ with $s \in \{j, k\}$. Conversely, the instructions $l_i : (OP(r_1), labels_1)$, $l_i : (OP'(r_2), labels_2)$, where OP and OP' can be ADD or SUB while $labels_1$ ($labels_2$) is l_j if OP = ADD (OP' = ADD) and $labels_1$ ($labels_2$) is l_j, l_k if OP = SUB (OP' = SUB), can be simulated by the instructions:

$$l_i : (ADD(2), l'_i, l''_i),$$

$$l'_i : (SUB(2), l'''_i, l''''_i), \; l'''_i : (OP(r_1), labels_1),$$

$$l''_i : (SUB(2), l''''_i, l'''''_i), \; l'''''_i : (OP(r_2), labels_2).$$

where $l'_i, \ldots, l'''''_i$ are not used for any other instruction. Note that register 1 is not used and that no other register except r_1, r_2 is modified.

In the following the labels of the instructions are called *states*, and when a register machine M passes to execute an instruction $s : (OP(r), labels)$ we say that M is in state s.

10.6 ONE AND TWO UNBOUNDED ELEMENTS

Let us call *total value* the sum of all the values of the conformons present in a conformon P system. In the following theorem we consider conformon P systems such that all the conformons present in an unbounded number of copies have 0 as value. In these systems the total value is finite.

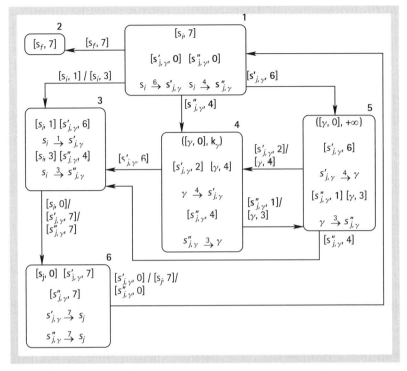

Fig. 10.8 The conformon P system related to Theorem 10.1.

Theorem 10.1 *The family of sets of numbers generated by conformon P systems coincides with the one generated by partially blind register machines.*

Proof. (I) Consider a partially blind register machine $M = (m, H, s_1, s_f, I)$ with $H = \{s_1, \ldots, s_f\}$; we construct the conformon P system Π depicted in Fig. 10.8.

In the initial configuration of Π for each register γ of M there is an unbounded number of occurrences of $[\gamma, 0]$ in compartment 5. Compartment 4 is the input compartment and initially it contains k_γ copies of $[\gamma, 0]$, where k_γ is the value of γ in the initial configuration of the simulated machine. The addition of one to γ is simulated moving one occurrence of $[\gamma, 0]$ from compartment 5 to compartment 4; the subtraction of one is simulated with the passage of one occurrence of $[\gamma, 0]$ from compartment 4 to compartment 5. The acknowledge compartment is 2. The system stops if there is no $[\gamma, 0]$ in compartment 4 while the system tried to simulate a subtraction from γ.

For each $s \in H$, the initial configuration of Π has $[s_j, 0]$ in compartment 6. For each instruction $s : (\text{ADD}(\gamma), v) \in I$, the initial configuration of Π has $[s'_{j,\gamma}, 0]$ in compartment 1; for each instruction $s : (\text{SUB}(\gamma), v) \in I$ the initial configuration of Π has $[s''_{j,\gamma}, 0]$ in compartment 1.

At most one conformon of the kind $[s_i, 7]$, $1 \leq i \leq f$, can be present in compartment 1 (then, $[s_i, 0]$ is not present in compartment 6). Initially it is the one related to the initial state of M.

If $s_i : (\text{ADD}(\gamma), s_j) \in I$, then $s_i \xrightarrow{6} s'_{j,\gamma} \in R_1$. The conformons $[s'_{j,\gamma}, 6]$ and $[s_i, 1]$ are created when this rule is applied, then they can pass to compartments 5 and 3, respectively. In compartment 5 $[s'_{j,\gamma}, 6]$ is needed to let an occurrence of $[\gamma, 0]$ pass from this compartment to the final one. This is performed first increasing to 4 the value of an occurrence $[\gamma, 0]$. When this happens, $[s'_{j,\gamma}, 2]$ and $[\gamma, 4]$ can pass to compartment 4. Here these two conformons can interact by $\gamma \xrightarrow{4} s'_{j,\gamma}$ so that $[\gamma, 0]$ and $[s'_{j,\gamma}, 6]$ are created. Conformon $[\gamma, 0]$ remains in this compartment, while $[s'_{j,\gamma}, 6]$ can pass to compartment 3. Here, $[s'_{j,\gamma}, 6]$ and $[s_i, 1]$ can interact so to create $[s'_{j,\gamma}, 7]$ and $[s_i, 0]$. These last two conformons can pass to compartment 6 where $[s'_{j,\gamma}, 7]$ interacts with $[s_j, 0]$. The result of this interaction creates $[s'_{j,\gamma}, 0]$ and $[s_j, 7]$ and both these conformons can pass to compartment 1.

If $s_i : (\text{SUB}(\gamma), s_j) \in I$, then $s_i \xrightarrow{4} s''_{j,\gamma} \in R_1$. The passage of one occurrence of $[\gamma, 0]$ from compartment 4 to compartment 5 is performed in a way similar to what just described. In case no conformon $[\gamma, 0]$ is present in compartment 4, then $[s'_{j,\gamma}, 4]$ can only remain in compartment 4.

When $[s_f, 7]$ (s_f is the final state of the register machine) is present in compartment 1, then it can only pass to the acknowledge compartment, compartment 2 in the figure, halting in this way the computation.

(II) Let $\Pi = (V, \mu = (N, E), fin, ack, L_1, \cdots, L_m, R_1, \ldots, R_m)$ be a conformon P system. We know that the number of membranes, the different conformons, and the total amount t of values are finite. We define $M = (m, H, s_0, f, I)$ a partially blind register machine simulating Π. We say that M correctly simulated Π if M reaches the final state with all its registers empty. Each register of M will be labeled O_{X_x} where O indicates a membrane, X the name content, and x the value content of a conformons associated with Π. There will be no counters representing infinite conformons (with o as value) associated with a membrane. The number recorded in a counter O_{X_x} indicates how many copies of the conformon $[X, x]$ are associated with membrane O. The simulation is composed by three phases:

1) creation of the initial configuration of the simulated system;
2) simulation of the possible operations of the simulated system;
3) check of the accepted number.

We now describe these three steps in detail.

Phase 1. Starting from state s_0 the initial configuration of Π is recorded into the registers. For each conformon $[X, x]$, present in k occurrences in membrane O in the initial configuration of Π, there will be a sequence of k instructions $s_1 : (\text{ADD}(O_{X_x}), s_2)$, $s_2 : (\text{ADD}(O_{X_x}), s_3), \ldots, s_k : (\text{ADD}(O_{X_x}), r)$ increasing the value of the counter O_{X_x} to k. If more that one conformon is associated with the initial configuration, then the state reached at the end of a sequence of instructions related to one conformon (r in the previous sequence) is the first state of another sequence related to another conformon. The state reached by M at the end of Phase 1 is $\bar{s}$; when this state is reached Phase 2 starts. If, for instance, the conformons $([A, 5], 3)$ and $[B, 2]$ associated with membrane O represent the initial configuration of Π, then the instructions $s_0 : (\text{ADD}(O_{A_5}), s_{[O_{A_5+},2]})$, $s_{[O_{A_5+},2]} : (\text{ADD}(O_{A_5}), s_{[O_{A_5+},3]})$, $s_{[O_{A_5+},3]} : (\text{ADD}(O_{A_5}), s_{[O_{B_2+},1]})$, $s_{[O_{B_2+},1]} : (\text{ADD}(O_{B_2}), \bar{s})$ are present in I.

Phase 2. The passage of a conformon $[X, x]$ from membrane U to membrane Q is represented by removing one unit from the counter U_{X_x} and adding one unit to the counter Q_{X_x}. The interaction between two conformons is represented in a similar way. The several operations possible in the simulated system are simulated in a non-deterministic way. In case M tries to simulate an operation that is not allowed in Π, then M stops. The only way for M to reach a final state is to add one unit to one of the counters O_{X_x} where O is the label of the acknowledge membrane of Π. Now we describe the simulation in more detail; for a better understanding instructions are numbered.

The simulation of the passage of a conformon $[X, x]$ from membrane U to membrane Q considering that $(U, Q, pred(n)) \in E$ and $pred(n)$ holds on x, is performed in the following way: for each register U_{X_x} such that $pred(n)$ holds on x there are instructions:

1) $\bar{s} : (\text{SUB}(U_{X_x}), s_{[U_{X_x-}, Q_{X_x+}]})$;
2) $s_{[U_{X_x-}, Q_{X_x+}]} : \text{ADD}(\hat{s}), Q_{X_x+})$.

Instruction 1 decreases the value of register U_{X_x} and changes into $s_{[U_{X_x-}, Q_{X_x+}]}$ the state. In case the content of U_{X_x} is 0, then M stops. If not, then the content of register Q_{X_x} is increased by one by the application of instruction 2. If $Q = ack$, then $\hat{s} = \bar{\bar{s}}$ (so that phase 3 starts), otherwise $\hat{s} = \bar{s}$. In case the conformon $[X, x]$, with $x = 0$, is associated to membrane Q with infinite multiplicity, then instruction 2 is not present and $s_{U_{X_x-}, Q_{X_x+}} = \bar{s}$.

The simulation of the rule $A \xrightarrow{e} B$ between $[A, a]$ and $[B, b]$, both associated with membrane O, is performed in the following way: for each $e \leq a \leq t$ there are instructions:

3) $\bar{s} : (\text{SUB}(O_{A_a}), s'_{[A\xrightarrow{e} B, a, O]})$;
4) $s'_{[A\xrightarrow{e} B, a, O]} : (\text{ADD}(O_{A_{a-e}+}), s''_{[A\xrightarrow{e} B, a, O]})$;
5) $s''_{[A\xrightarrow{e} B, a, O]} : (\text{SUB}(O_{B_b-}), s'''_{[A\xrightarrow{e} B, b, O]})$;
6) $s'''_{[A\xrightarrow{e} B, b, O]} : (\text{ADD}(O_{B_{b+e}+}), \bar{s})$.

Instruction 3: a conformon associated with membrane O having value $a \geq e$ is randomly chosen and its content is decreased by one. If its content is 0, this means that in that moment of the simulation membrane O does not contain any conformon $[A, a]$ so that the rule cannot be simulated and M stops. If its content is at least 1, then one unit is subtracted from it and the machine changes state into $s'_{[A\xrightarrow{e}B,a,O]}$.

Instruction 4: if the rule $A \xrightarrow{e} B$ is applied in membrane O, then the subtraction of e to the value of $[A, a]$ implies the creation in membrane O of one occurrence of $[A, a - e]$. This instruction simulates this and it is present only if $[A, a - e]$ is not one of the conformons with 0 energy and infinite multiplicity associated with membrane O. If this is the case this instruction is not present and $s'_{[A\xrightarrow{e}B,a,O]} = s''_{[A\xrightarrow{e}B,a,O]}$.

Instruction 5: in order to simulate the rule $A \xrightarrow{e} B$ in membrane O a conformon $[B, b]$ has to be associated to this membrane. This instruction subtracts one from the value of the register associated to a conformon with name B randomly chosen between the ones in membrane O. In case membrane O contains infinitely many occurrences of a conformon with name B and $b = 0$, then this instruction is not present, and $s''_{[A\xrightarrow{e}B,a,O]} = s'''_{[A\xrightarrow{e}B,b,O]}$.

Instruction 6: the content of the register related to the conformon $[B, b + e]$ is increased by one.

Phase 3. When M is in state $\bar{s}$ it means that the simulation of the passage of one conformon to the acknowledge membrane has been simulated. From this state all registers related to conformons associated with membranes different than the final one are randomly decreased. The ones related to conformons associated with the final membrane are decreased together with the content of register l_1: for one unit removed from one of these registers one unit is subtracted from l_1. If in this process the machine tries to subtract one unit from a register whose content is 0, then the machine stops in a non-final state. At any stage of this process the machine may go into the final state f. If then all counters are empty we say that M correctly simulated the system Π accepting the number y.

If priorities are added to the conformon P systems considered in the previous theorem, then computationally complete systems are obtained.

Theorem 10.2 *The family of sets of numbers generated by conformon P systems where interactions rules have priority on passage rules coincides with that generated by register machines.*

Proof. The conformon P system depicted in Fig. 10.9 is very similar to the one related to Theorem 10.1 and depicted in Fig. 10.8. We only describe the differences with respect to this proof.

Let $M = (m, H, s_1, s_f, I)$ be a register machine. For $s_i : (\text{SUB}(\gamma), s_j, s_k) \in I$ the rule $s_i \xrightarrow{4} s''_{j,\gamma}$ is in R_1. If the conformon $[s_i, 7]$ is present in this compartment, then the conformons $[s_i, 3]$ and $[s''_{j,\gamma}, 4]$ are created and they can pass to compartments 3 and 4, respectively.

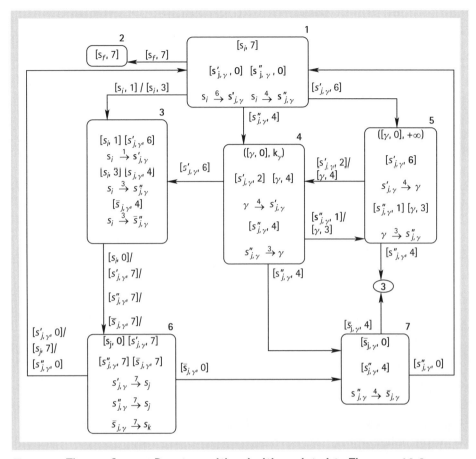

Fig. 10.9 The conformon P system with priorities related to Theorem 10.2.

If in compartment 4 there is an instance of $[\gamma, 0]$, then the priorities force the application of $s''_{j,\gamma} \xrightarrow{3} \gamma$, otherwise $[s''_{j,\gamma}, 4]$ can pass to compartment 7. What was just described is the only use of priorities in the conformon P system.

In case $[s''_{j,\gamma}, 4]$ passes to compartment 7 it can interact with $[\bar{s}_{j,\gamma}, 0]$ initially present in this compartment. The conformons $[s''_{j,\gamma}, 0]$ and $[\bar{s}_{j,\gamma}, 4]$ are created and they can pass to compartments 1 and 3, respectively. In compartment 3 $[\bar{s}_{j,\gamma}, 4]$ can interact with $[s_j, 3]$. This creates $[\bar{s}_{j,\gamma}, 7]$ and $[s_j, 0]$ and they can both pass to compartment 6. Here $[\bar{s}_{j,\gamma}, 7]$ and $[s_k, 0]$ can interact so to create $[\bar{s}_{j,\gamma}, 0]$ and $[s_k, 7]$. These two conformons can pass to compartment 1.

In [8, 9] it is proved that the presence of priorities can be regarded as an unbounded quantity.

We now consider conformon P systems with an unbounded number of conformons and these conformons can have a positive integer as value. In this way the total value becomes unbounded too.

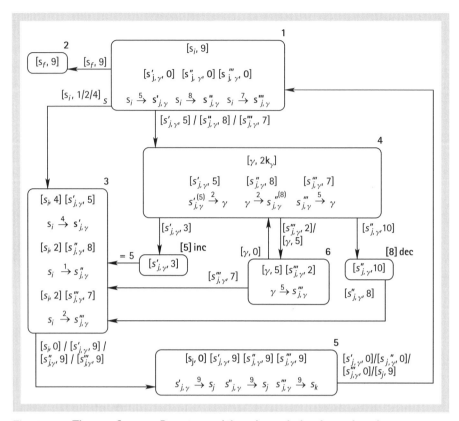

Fig. 10.10 The conformon P system with unbounded value related to Theorem 10.3.

Theorem 10.3 *The family of sets of numbers generated by conformon P systems with unbounded value coincides with the one generated by register machines.*

Proof. Figure 10.10 represents a conformon P system Π with unbounded value simulating a register machine $M = (m, H, s_1, s_f, I)$, $H = \{s_1, \ldots, s_f\}$.

For each state $s_j \in H$, $1 \leq j \leq f$, the initial configuration of Π has $[s_j, 0]$ in compartment 5. For each instruction $s_i : (\text{ADD}(\gamma), s_j) \in I$ the initial configuration of Π has $[s'_{j,\gamma}, 0]$ in compartment 1; for each instruction $s_i : (\text{SUB}(\gamma), s_j, s_k) \in I$ the initial configuration of Π has $[s''_{j,\gamma}, 0]$ and $[s'''_{j,\gamma}, 0]$ in compartment 1. Moreover, if k_γ is the initial value of the counter γ in M, then the conformon $[\gamma, 2k_\gamma]$ is present in compartment 4 in the initial configuration. The simulation of the addition of 1 to a counter corresponds to the addition of 2 to the value of the related conformon; the simulation of subtraction of 1 from a counter corresponds to the subtraction of 2 from the value of the related conformon. Considering that initially empty counters are represented by conformons having 0 as value we have that the value of any γ conformon is an even number.

At most one conformon of the kind $[s_i, 9]$ can be present in compartment 1. Initially it is the one related to s_1 the initial state of M.

If $s_i : (\text{ADD}(\gamma), s_j) \in I$, then the interaction rule $s_i \xrightarrow{5} s'_{j,\gamma}$ is in R_1. The conformons $[s_i, 4]$ and $[s'_{j,\gamma}, 5]$ are created when this rule is applied and then they can pass to compartments 3 and 4, respectively.

In compartment 4 the conformons $[s'_{j,\gamma}, 5]$ and $[\gamma, 2k_\gamma]$ can strictly interact: only when the value of the former conformon is 5 it can pass 2 units of its value to the latter conformon. When this happens $[s'_{j,\gamma}, 3]$ can pass to an increaser that will change the value of this conformon into 5. The resulting conformon can then pass to compartment 3 and interact with $[s_i, 4]$ so that $[s'_{j,\gamma}, 9]$ and $[s_i, 0]$ are created. These last two conformons can pass to compartment 5. Here $[s'_{j,\gamma}, 9]$ can interact with $[s_j, 0]$. The outcome of this interaction, conformons $[s'_{j,\gamma}, 0]$ and $[s_j, 9]$, can pass to compartment 1.

The simulation of the instruction $s_i : (\text{SUB}(\gamma), s_j, s_k) \in I$ is performed by "gambling" (see [8]). For each of these instructions two interaction rules $s_i \xrightarrow{8} s''_{j,\gamma}$ and $s_i \xrightarrow{7} s'''_{j,\gamma}$ are in R_1. In case the former interaction rule is applied, then the conformon P system "gambles" that the counter is not empty. This is performed by the strict interaction between conformons γ and $s''_{j,\gamma}$: the former can pass 2 units of its value to the latter only if the value of the latter is 8. When this happens $[s''_{j,\gamma}, 10]$ can pass to a decreaser that will bring its value to 8. From here $[s''_{j,\gamma}, 8]$ can pass to compartment 3.

In case the system Π "gambles" that the counter is empty, then $[s'''_{j,\gamma}, 7]$ is created in compartment 1 and it can pass to compartment 4. Here it can interact with $[\gamma, 2k_\gamma]$ so that $[s'''_{j,\gamma}, 2]$ and $[\gamma, 2k_\gamma + 5]$ are created. The former conformon can pass to compartment 6 while the latter one can pass to this compartment only if $k_\gamma = 0$. If this does not happen, then Π never halts. It is relevant to note that the value of the γ conformon can be 5 (an odd number) only because of this sequence of configurations. If k_γ was 0, then $[\gamma, 5]$ can pass to compartment 6 and here interact with $[s'''_{j,\gamma}, 2]$ such that $[\gamma, 0]$ and $[s'''_{j,\gamma}, 7]$ are created. The former conformon can pass back to compartment 4, while the latter passes to compartment 3.

When $[s_f, 7]$ (s_f is the final state of M) is present in compartment 1, then it can only pass to the acknowledge compartment, with label 2 in the figure, ending in this way the computation.

10.7 Negative Values

What if the value of a conformon can become negative? In this case we have *conformon P systems with negative values*. If, for instance, we consider the conformons

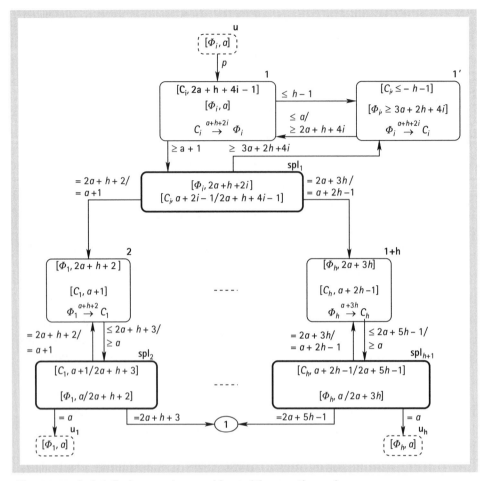

Fig. 10.11 A detailed separator working with negative values.

$[G, 5]$ and $[R, 4]$ and the rule $G \xrightarrow{3} R$, then the rule can be applied forever leading to $[G, 5 - 3n]$ and $[R, 4 + 3n]$, $n \geq 1$. Despite the fact that the previous proofs do not consider this possibility, they can be easily be adapted to this new feature.

As splitters do not contain interaction rules, they are not directly effected by possible negative values. Separators, on the other hand, are strongly based on the fact that some interactions occur only once. A separator also working with negative values is depicted in Fig. 10.11.

It works essentially in the same way as the separator depicted in Fig. 10.4 with the difference that conformons which interacted in compartment 1 more than once pass (through $\mathbf{spl_1}$) to compartment $1'$ where the reverse interactions can take place. Considering what we said in Section 10.4 in relation to splitters, the edge from $\mathbf{spl_1}$ to compartment $1'$ having label $\geq 3a + 2h + 4i$ in Fig. 10.11 is allowed.

The definition of separators working with negative values simplifies the description of conformon P systems with negative values. Increasers and decreasers also work for negative values.

Theorem 10.4 *Conformon P systems with negative values can simulate partially blind register machines.*

Proof. Such a P system can be obtained from the one in Part I of the proof of Theorem 10.1 where for each interaction rule its reverse is present.

Theorem 10.5 *The family of sets of numbers generated by conformon P systems with negative values and unbounded value coincides with the one generated by register machines.*

Proof. Such a P system can be obtained from the one indicated in the proof of Theorem 10.3 where for each interaction rule its reverse is present.

10.8 Infinite Hierarchies

The problem of finding classes of P systems which induce infinite hierarchies of the computed sets of numbers (for example, with respect to the number of membranes, of symbols, etc.) was investigated for various types of P systems. In this section we consider this problem also for conformon P systems. To this aim, we introduce some restricted classes of such systems.

A-restricted conformon P systems can have more than one input compartment and they have only one conformon with a distinguished name, let us say D, encoding the input. The formal definition of such conformon P systems changes then into:

$$\Pi = (V, \mu, input_{z_1}, \ldots, input_{z_n}, L_1, \ldots, L_m, R_1, \ldots, R_m),$$

where $z_1, \ldots, z_n \in Q$ indicate the input compartments. In the initial configuration the input compartments contain only D conformons and no other compartment contains D conformons. The definitions of configuration, transition, computation, halt, stop, and set of numbers accepted follow from the ones given in Section 10.2.

Lemma 10.1 *A-restricted conformon P systems can simulate restricted register machines with two counters.*

Proof. The conformon P system related to this proof is depicted in Fig. 10.12. It is very similar to the one depicted in Fig. 10.10. The only difference between the two figures is that the increaser and decreaser modules present in Fig. 10.10 have been substituted by compartment 7 in Fig. 10.12. This proofs is then similar to the one of Theorem 10.2.

274 CONFORMON P SYSTEMS

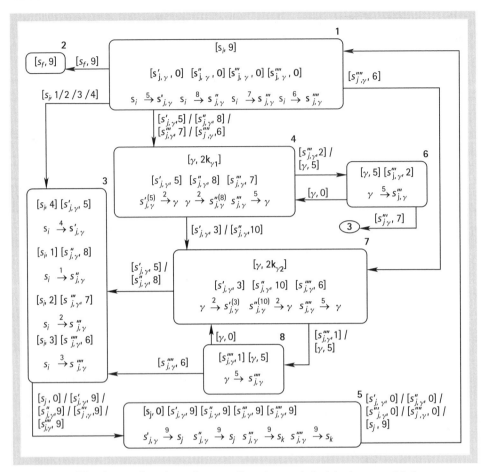

Fig. 10.12 The A-restricted conformon P system related to Lemma 10.1.

Let $M = (m, H, s_1, s_f, I)$ be a restricted register machine with two registers, γ_1 and γ_2. The only instructions present in this register machine are of the kind $s_i : (\text{SUB}(\alpha), \text{ADD}(\beta), s_j, s_k)$ with $\alpha, \beta \in \{\gamma_1, \gamma_2\}$, $\alpha \neq \beta$. If the initial values of these counters is k_{γ_1} and k_{γ_2}, respectively, then the value of the γ conformons initially present in compartments 4 and 7, the input compartments of the conformon P system, is $2k_{\gamma_1}$ and $2k_{\gamma_2}$, respectively. We refer with $\gamma_{[4]}$ to the conformon $[\gamma, 2k_{\gamma_1}]$ initially present in membrane 4 and with $\gamma_{[7]}$ to the conformon $[\gamma, 2k_{\gamma_2}]$ initially present in membrane 7. These conformons are the only ones present in compartments 4 and 7 in the initial configuration, so γ is the distinguished conformon name encoding the input. This is similar to what happens to $\gamma_{[4]}$ and $\gamma_{[7]}$.

The conformons with name $s'_{j,\gamma}$ are used to increase the value of $\gamma_{[4]}$ and then to decrease, if possible, the one of $\gamma_{[7]}$. The conformons with name $s''_{j,\gamma}$ are use to decrease, if possible, the value of $\gamma_{[4]}$ and then to increase the one of $\gamma_{[7]}$. The conformons with name $s'''_{j,\gamma}$ and $s''''_{j,\gamma}$ test if the value of respectively $\gamma_{[4]}$ and $\gamma_{[7]}$ is 0. If any of these operations cannot be performed, then the system never halts.

The system halts when the conformons $[s_f, 9]$ passes to compartment 2, the acknowledge compartment.

The just given constructive proof can be used to create A-restricted conformon P systems simulating restricted register machines with any number of counters.

Let us assume that a given restricted register machine has n counters $\{\gamma_1, \ldots, \gamma_n\}$ each with an initial value. Then it is possible to build an A-restricted conformon P system Π' having conformons with name γ initially present in n different input compartments. Considering the proof of Lemma 10.1 this seems to be a must, as collecting more than one conformon with name γ in the same compartment would not allow the system Π' to perform a simulation on the restricted register machine. The system Π' would be such that every time the value of a γ conformon is increased (decreased), then the one of its connected counter (for the particular simulated instruction) is decreased (increased, respectively) by the same amount. The subscripts or the value of the s', s'', s''', and s'''' conformons can be used to pass these conformons to specific membranes in order to interact with the appropriate γ conformons.

So we can state:

Corollary 10.1 *A-restricted conformon P systems can simulate restricted register machines.*

Here is the reverse of this inclusion:

Lemma 10.2 *Restricted register machines with n counters can simulate A-restricted conformon P systems having n input compartments.*

Proof. In the initial configuration the value stored in the n counters is the value of the D conformons present in the initial configuration of the conformon P system. The rest of the description of the conformon P system is encoded into the finite control of the restricted register machine.

The increase/decrease of the value of the D conformons is split into a sequence of increases/decreases of 1. Every time 1 is subtracted by the value of a D conformon, then the value of the associated counter is decreased by 1 and that of the connected counter is increased by 1. Chains of coupled increase/decrease of connected counters simulate the passage of value between different conformons. Similarly when the value of a D conformon is increased.

When the simulation of a conformon passing to the acknowledge membrane is performed, then the register machine reaches the final state.

Knowing from Section 10.5 that restricted register machines induce an infinite hierarchy on the number of counters, we can state:

Theorem 10.6 *A-restricted conformon P systems induce an infinite hierarchy on the number of compartments.*

Another class of conformon P systems we consider here is that of *B-restricted conformon P systems*, which have only one input compartment and the input can be encoded by conformons with names from a given finite set of names (they are called *input conformons*).

An initial configuration is such that some input conformons are present in the input compartment, no input conformon is present in the remaining compartments, and the acknowledge compartment is empty. The definition of configuration, transition, computation, and set of numbers accepted follow from those given in Section 10.2.

Lemma 10.3 *B-restricted conformon P systems with two input conformons can simulate restricted register machine with two counters.*

Then, an extension to register machines with any number of registers can be obtained in a way similar to the way Corollary 10.1 follows from Lemma 10.1. Therefore, we have:

Corollary 10.2 *B-restricted conformon P systems can simulate restricted register machines.*

Here is the reverse of this inclusion whose proof is similar to that of Lemma 10.2:

Lemma 10.4 *Restricted register machines with n counters can simulate B-restricted conformon P systems having n input conformons.*

Because restricted register machines induce an infinite hierarchy on the number of counters, we have:

Theorem 10.7 *B-restricted conformon P systems induce an infinite hierarchy on the number of input conformons.*

Several open problems related to conformon P systems can be found in [8].

10.9 Modeling with Conformon P Systems

Conformon P systems have been successfully used as a modeling platform for biological systems [11, 1, 10]. When the activity of these systems is simulated by a computer, then probabilities are associated with interaction and passage rules. The

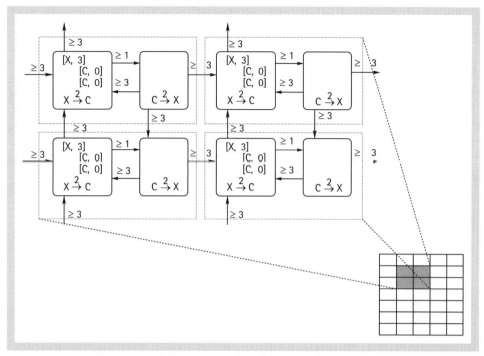

Fig. 10.13 A grid of conformon P systems. © Reprinted from [1], Page 5, Copyright (2008), with permission from Elsevier.

presence of probabilities, together with the other features of conformon P systems, let these devices be classified as discrete probabilistic formal models characterized by locality of interaction (that is, spatial heterogeneity) and parallel evolution of parts such that mechanistic processes can be explicitly or implicitly (phenomenologically) described.

Let us classify as *simple* all the kinds of conformon P systems considered in the previous sections of the present chapter. A *grid of conformon P systems* is a conformon P system composed by *cells*, each cell being a simple conformon P system. In Fig. 10.13 a grid of conformon P systems is depicted.

This way to create P systems resembles cellular automata [2, 20]. In [1] the differences between grids of conformon P systems and cellular automata have been discussed.

A *simulator* for (*grids* of) conformon P systems is available from [19]. Such a simulator accepts as input an XML file describing the systems to be simulated together with other files describing parameters of the simulation (as, for instance, the size of the grid, if the grid is a torus or not, etc.). This simulator has been used to model glycolysis [11] and the dynamics of HIV infection [1, 10]. In the following we give an overview of the latter model.

Several mathematical models for the dynamics of HIV infection (see, for instance, [17, 22, 16]) have been proposed but all of them fail to describe the dynamics of the infection in a complete way. Typically, these models do not maintain biological plausibility together with a qualitative match of the known dynamics of the concentrations of healthy, infected, and dead cells. Such dynamics occur in two distinct time scales: one over the first weeks and another over several years.

Our model is based on the one using cellular automata described in [3] and uses a grid of conformon P systems.

The conformons and rules associated with each cell in the considered grid can be logically divided in two parts. In the following we describe only the conformons and rules belonging to *part* 1. Later on we extend this description to *part* 2.

In *part* 1 the *state* conformons with names H, A, AA, PD, and D, each present with occurrence 1 in each cell, are used to identify a cell being *healthy*, *A-infected*, *AA-infected*, *pre-dead*, and *dead*, respectively. Cells in the grid are such that in any configuration only one (any one) of the state conformons present in them can have a value bigger than 0. So, for instance, if in a configuration a cell contains $[H, 1]$, $[A, 0]$, $[AA, 0]$, $[PD, 0]$, and $[D, 0]$, then the state of that cell is *healthy*. Moreover, each cell has an unbounded number of occurrences of $[R, 1]$, $[V, 10]$, $[E, 0]$, and $[W, 0]$. The conformons with name R and W do not have any direct relationship with any aspect of HIV infection. In broad terms, the conformons with name R can be regarded as 'food' molecules needed by a cell in a certain state to perform an action. The conformons with name W can be regarded as "waste" molecules to which some conformons can pass part of their value. Conformons $[V, 11]$ represent viruses produced by *A-infected* cells (notice that a $[V, 10]$ conformon does not represent a virus). Conformons of the kind $[E, 1]$ can be generated by *AA-infected* cells.

In the following we list the actions implemented by the rules present in each cell. The numbers present in between parentheses refer to the rules numbers indicated in the Appendix.

(1, 2) an *A-infected* cell can generate a virus;
(3, 4, 5) a *healthy* cell can become *A-infected* if it contains a virus;
(6, 7) an *AA-infected* cell can generate $[E, 1]$;
(8, 9) $[E, 1]$ conformons can generate $[E, 4]$;
(10, 11, 12) a *healthy* cell can become *A-infected* if it contains $[E, 4]$;
(13) an *A-infected* cell can become *AA-infected*;
(14) an *AA-infected* cell can become *pre-dead*;
(15, 16, 17, 18, 20, 21, 22) a *pre-dead* cell removes viruses and E conformons present in it;
(19) a *pre-dead* cell can become a *dead* cell;
(23) a *dead* cell can become *2-healthy*.

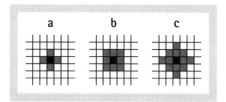

Fig. 10.14 The considered neighborhoods: (a) von Neumann, (b) Moore, (c) diamond. © Reprinted from [1], page 12, Copyright (2008), with permission from Elsevier.

The state conformons with name $H2$, $A2$, $AA2$, $PD2$, and $D2$ are associated with *phase 2* and they indicate a cell being in state 2*healthy*, 2*A-infected*, 2*AA-infected*, 2*pre-dead*, and 2*dead*, respectively, in this phase (where the "2" indicates that the state is related to *part* 2). The rules used in *phase* 2, indicated in the Appendix, are similar to the ones in *phase* 1 but some have a lower probability.

In both phases the conformons $[V, 11]$ and $[E, 1]$ can pass from one cell to another according to the considered neighborhood. We run our tests using the three neighborhoods depicted in Fig. 10.14, where a black cell can pass a conformon to any of the grey cells.

The simulation performed was based on a 50 × 50 toroidal grid, used all the three neighborhoods, run 10 times with different random number sequences, and started with either 5% or 0.04% *infected* cells.

The vast majority of the tests we ran showed dynamics very similar to that observed *in vivo*. A typical outcome is depicted in Fig. 10.15.

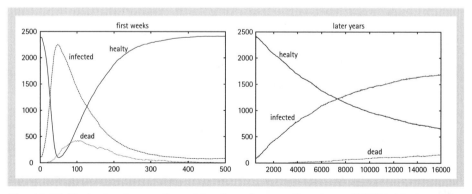

Fig. 10.15 Typical outcome for grids of conformon P systems. © With kind permission of Springer Science and Business Media [10].

Appendix: Rules, links, and probabilities

	part 1			part 2	
label	rule	prob.	label	rule	prob.
1	$R \xrightarrow{1} A_{(1)}$	0.7071	24	$R \xrightarrow{1} A2_{(1)}$	0.7071
2	$A^{(2)} \xrightarrow{1} V_{(10)}$	0.7071	25	$A2^{(2)} \xrightarrow{1} V_{(10)}$	0.7071
3	$V \xrightarrow{11} H_{(1)}$	0.79	26	$V \xrightarrow{11} H2_{(1)}$	0.79
4	$H^{(12)} \xrightarrow{12} A_{(0)}$	0.79	27	$H2^{(12)} \xrightarrow{12} A2_{(0)}$	0.0001
5	$A^{(12)} \xrightarrow{11} W_{(0)}$	0.79	28	$A2^{(12)} \xrightarrow{11} W_{(0)}$	0.79
6	$R \xrightarrow{1} AA_{(1)}$	0.7071	29	$R \xrightarrow{1} AA2_{(1)}$	0.7071
7	$AA^{(2)} \xrightarrow{1} E_{(0)}$	0.7071	30	$AA2^{(2)} \xrightarrow{1} E_{(0)}$	0.7071
8	$E^{(1)} \xrightarrow{1} E_{(1)}$	0.07071			
9	$E^{(2)} \xrightarrow{2} E_{(2)}$	0.07071			
10	$E \xrightarrow{4} H_{(1)}$	0.79	31	$E \xrightarrow{4} H2_{(1)}$	0.79
11	$H^{(5)} \xrightarrow{5} A_{(0)}$	0.79	32	$H2^{(5)} \xrightarrow{5} A2_{(0)}$	0.0001
12	$A^{(5)} \xrightarrow{4} W_{(0)}$	0.79	33	$A2^{(5)} \xrightarrow{4} W_{(0)}$	0.79
13	$A^{(1)} \xrightarrow{1} AA_{(0)}$	0.04	34	$A2^{(1)} \xrightarrow{1} AA2_{(0)}$	0.0001
14	$AA^{(11)} \xrightarrow{1} PD_{(0)}$	0.1	35	$AA2^{(11)} \xrightarrow{1} PD2_{(0)}$	0.00075
15	$V^{(11)} \xrightarrow{1} PD_{(1)}$	0.7071	36	$V^{(11)} \xrightarrow{1} PD2_{(1)}$	0.7071
16	$E \xrightarrow{1} PD_{(1)}$	0.7071	37	$E \xrightarrow{1} PD2_{(1)}$	0.7071
17	$E \xrightarrow{2} PD_{(1)}$	0.7071	38	$E \xrightarrow{2} PD2_{(1)}$	0.7071
18	$E \xrightarrow{4} PD_{(1)}$	0.7071	39	$E \xrightarrow{4} PD2_{(1)}$	0.7071
19	$PD^{(1)} \xrightarrow{1} D_{(0)}$	0.2	40	$PD2^{(1)} \xrightarrow{1} D2_{(0)}$	0.001
20	$PD^{(2)} \xrightarrow{1} W_{(0)}$	0.7071	41	$PD2^{(2)} \xrightarrow{1} W_{(0)}$	0.7071
21	$PD^{(3)} \xrightarrow{2} W_{(0)}$	0.7071	42	$PD2^{(3)} \xrightarrow{2} W_{(0)}$	0.7071
22	$PD^{(5)} \xrightarrow{4} W_{(0)}$	0.7071	43	$PD2^{(5)} \xrightarrow{4} W_{(0)}$	0.7071
23	$D^{(1)} \xrightarrow{1} H2_{(0)}$	0.1	44	$D2^{(1)} \xrightarrow{1} H2_{(0)}$	0.001

© With kind permission of Springer Science and Business Media [10].

Links:
$[V, 11]$ can pass with probability 1 from any cell to any of its neighbors; $[E, 1]$ can pass with probability 0.01 from any cell to any of its neighbors.

References

[1] D.W. CORNE, P. FRISCO: Dynamics of HIV infection studied with cellular automata and conformon-P systems. *BioSystems*, 91 (2008), 531–544.

[2] A. DEUTSCH, S. DORMANN: *Cellular Automaton Modeling of Biological Pattern Formation: Characterization, Applications, and Analysis.* Birkäusen, Boston, 2004.

[3] R.M. DOS SANTOS, S. COUTINHO: Dynamics of HIV infection: a cellular automata approach. *Physical Review Letters*, 87 (2001), 168102.

[4] G. ELEFTHERAKIS, P. KEFALAS, GH. PĂUN, G. ROZENBERG, A. SALOMAA, eds.: *Membrane Computing. 8th International Workshop, WMC 2007, Thessaloniki, Greece, June 2007*, LNCS 4860, Springer, 2007.

[5] P. FRISCO: The conformon-P system: A molecular and cell biology-inspired computability model. *Theoretical Computer Sci.*, 312 (2004), 295–319.

[6] P. FRISCO: Infinite hierarchies of conformon-P systems. *Lecture Notes in Computer Sci.*, 4361 (2006), 395–408.

[7] P. FRISCO: Conformon-P systems with negative values. In [4], 331–344.

[8] P. FRISCO: *Computing with Cells. Advances in Membrane Computing.* Oxford University Press, 2009.

[9] P. FRISCO: A hierarchy of computational processes. Technical report, Heriot-Watt University, 2008. HW-MACS-TR-0059.

[10] P. FRISCO, D.W. CORNE: Modeling the dynamics of HIV infection with conformon-P systems and cellular automata. In [4], 21–32.

[11] P. FRISCO, R.T. GIBSON: A simulator and an evolution program for conformon-P systems. *SYNASC 2005, 7th Intern. Symp. Simbolic and Numeric Algorithms for Scientific Computing*, Timisoara, Romania, 2005, IEEE Computer Society, 2005, 427–430.

[12] D.E. GREEN, S. JI: The electromechanical model of mitochondrial structure and function. *Molecular Basis of Electron Transport* (J. Schultz, B.F. Cameron, eds.), Academic Press, New York, 1972, 1–44.

[13] S. JI: The Bhopalator: a molecular model of the living cell based on the concepts of conformons and dissipative structures. *J. Theoretical Biology*, 116 (1985), 395–426.

[14] S. JI: Free energy and information contents of conformons in proteins and DNA. *BioSystems*, 54 (2000), 107–214.

[15] S. JI: The Bhopalator: an information/energy dual model of the living cell (II). *Fundamenta Informaticae*, 49 (2002), 147–165.

[16] C. KAMP, S. BORNHOLDT: From HIV infection to AIDS: a dynamically induced percolation transition? *Proc. Royal Society B: Biological Sciences*, 269 (2002), 2035–2040.

[17] A.S. PERELSON, P.W. NELSON: Mathematical analysis of HIV-1 dynamics in vivo. *SIAM Review*, 41 (1999), 3–44.

[18] GH. PĂUN: *Membrane Computing. An Introduction.* Springer, 2002.

[19] The P Systems Webpage. http://ppage.psystems.eu/.

[20] T. TOFFOLI, N. MARGOLUS: *Cellular Automata Machines: A New Environment for Modeling.* MIT Press, 1987.

[21] M.V. VOLKENSTEIN: The conformon. *J. Theoretical Biology*, 34 (1972), 193–195.

[22] D. WODARZ, M.A. NOWAK: Mathematical models of HIV pathogenesis and treatment. *Bioessays*, 24 (2002), 1178–1187.

CHAPTER 11

ACTIVE MEMBRANES

GHEORGHE PĂUN

11.1 INTRODUCTION

In the previous chapters, the evolution of a P system was defined mainly by rules making evolve the multisets of objects (or the sets or multisets of strings) present in the compartments of the membrane structure, while the membranes themselves remained the same during the evolution/computation, maybe only disappearing through dissolution. However, both from a mathematical and computational point of view and also from a biological point of view, this is a too restricted case. First, it is natural to also evolve the membrane structure, not only by removing membranes through dissolution, but also by increasing the number of membranes. Many possibilities exist: creating new membranes, budding them from existing membranes, dividing membranes, splitting their objects according to a given criterion and enclosing the new classes of objects in new membranes, and so on. Many of these possibilities are also supported by biological facts, while biology brings another strong motivation for making the rules depend directly on the membranes, and thus involving them in evolution: many reactions taking place in the compartments of a cell are controlled by enzymes bound on membranes ("visible" on only one side of a membrane or going across the membrane, thus controlling reactions on both sides of it).

All these observations were the starting point of introducing P systems with *active membranes*, having rules which directly involve the membranes where the objects evolve and also making the membranes themselves evolve.

We start here by introducing a basic version of P systems with active membranes, those with rules for (a) evolution of objects in compartments, (b) sending objects inside and (c) outside a membrane, for (d) membrane dissolution, and (e) membrane division (under the control of objects). A new feature of these systems is the fact that the membranes are *polarized*, they have one of three possible "electrical charges": +, −, 0. For such systems, we look for universality results, in most restricted setups, in what concerns the types of rules used and the number of polarizations. Later, we consider various variants of these rules, restrictions or extensions, as well as other types of related rules, also mentioning for them results concerning computing power. A comprehensive bibliography ends the chapter.

We stress the fact that we are here concerned only with the computing power of the many classes of P systems with active membranes, and not with the possibility of using them for devising efficient solutions to computationally hard problems, a subject which will be tackled in Chapter 12.

11.2 Basic Types of Rules

We work with cell-like membrane structures (hence with membranes hierarchically arranged), with multisets of symbol objects placed in the compartments. As usual, we consider an alphabet O of objects and the membranes are represented by labeled parentheses, with the right squared parenthesis also marked with a polarization, an element $e \in \{+, -, 0\}$. For instance, a membrane with label h and polarization e is represented as $[\ ]_h^e$.

The five basic types of rules with active membranes are the following ones (H is a set of labels for membranes and s is the label of the skin membrane):

(a) $[a \to v]_h^e$, for $h \in H, e \in \{+, -, 0\}, a \in O, v \in O^*$
(object evolution rules, associated with membranes and depending on the label and the charge of the membranes, but not directly involving the membranes, in the sense that the membranes are neither taking part to the application of these rules nor are they modified by them);

(b) $a[\ ]_h^{e_1} \to [b]_h^{e_2}$, for $h \in H, e_1, e_2 \in \{+, -, 0\}, a, b \in O$
(*in* communication rules; an object is introduced in the membrane, maybe modified during this process; also the polarization of the membrane can be modified, but not its label);

(c) $[a]_h^{e_1} \to [\ ]_h^{e_2} b$, for $h \in H, e_1, e_2 \in \{+, -, 0\}, a, b \in O$
(*out* communication rules; an object is sent out of the membrane, maybe modified during this process; also the polarization of the membrane can be modified, but not its label);

(d) $[\ a\ ]_h^e \to b$, for $h \in H - \{s\}$, $a \in \{+, -, 0\}$, $a, b \in O$
(dissolving rules; in reaction with an object, a membrane can be dissolved, while the object specified in the rule can be modified);

(e) $[\ a\ ]_h^{e_1} \to [\ b\ ]_h^{e_2} [\ c\ ]_h^{e_3}$, for $h \in H - \{s\}$, $e_1, e_2, e_3 \in \{+, -, 0\}$, $a, b, c \in O$
(division rules for elementary membranes; in reaction with an object, the membrane is divided into two membranes with the same label, maybe of different polarizations; the object specified in the rule is replaced in the two new membranes by possibly new objects; all objects different from a are duplicated in the two new membranes).

Let us note that in all rules only one object is specified in the left-hand side (that is, the objects do not directly interact—but they do it indirectly, by means of the membranes and their polarizations) and that, with the exception of rules of type (a), single objects are always transformed into single objects (the two objects produced by a division rule of type (e) are placed in two different regions). Also, it is important to note that, as formulated above, rules of type (e) refer to elementary membranes, but a direct extension is possible, allowing such rules to be applied both to elementary and to non-elementary membranes. If also in rules of type (a) we impose an object-to-object transformation (i.e. the rules are of the form $[\ a \to b]_h^e$, for $a, b \in O$), then we say that the system is in the *one-normal form* and we denote that type of rules with (a_1). Also, if in any type of rules (b)–(e) the objects from the right-hand side of the rule can be missing (e.g. a rule of type (b) is of the form $a[\ ]_h^{e_1} \to [\ ]_h^{e_2}$, for $a \in O$), hence removing the object from the left-hand of the rule in cases (b)–(d), then we write (a_λ) instead of (a), for each $a \in \{b, c, d, e\}$.

These rules are applied in the usual non-deterministic maximally parallel manner, with the following details: Any object can be subject of only one rule of any type and any membrane can be subject of only one rule of types (b)–(e); rules of type (a) are not counted as applied to membranes, but only to objects (when a rule of type (a) is applied, the membrane can also evolve by means of a rule of another type); as usual, if a membrane is dissolved, then all the objects (and membranes, in the case that the division rule is applied to a non-elementary membrane) in its region are left free in the surrounding region; if a rule of type (e) is applied to a membrane, and its inner objects and membranes evolve at the same step, then it is assumed that first the inner objects and membranes evolve and then the division takes place, so that the result of applying rules inside the original membrane is replicated in the two new membranes (in short, the rules are applied in a bottom-up manner). Of course, the rules associated with a membrane h are used for all copies of this membrane, irrespective of whether or not the membrane is an initial one or it is obtained by division. The skin membrane cannot be dissolved or divided, but it can be the subject of in/out operations; because the environment is empty at the beginning, only objects which were expelled from the system can be present there, hence only such objects can be brought back to the system.

A *P system with active membranes* is then obtained as usual, as a construct

$$\Pi = (O, H, \mu, w_1, \ldots, w_m, R, i_0),$$

where O is the alphabet of objects, H is the alphabet of labels for membranes, μ is the initial membrane structure, of degree m, with all membranes labeled with elements of H and having neutral polarization, $w_1, \ldots, w_m$ are strings over O specifying the multisets of objects present in the compartments of μ, R is a finite set of rules of the forms mentioned above, and $i_0 \in \{0, 1, \ldots, m\}$ indicates the region where the result of a computation is obtained (0 represents the environment).

The configurations of Π identify the current membrane structure and the multisets of objects present in its compartments; $(\mu, w_1, \ldots, w_m)$ is the initial configuration. The transitions, computations, and successful (i.e. halting) computations are defined in the usual way. Also the result of a (halting) computation is defined as customary in membrane computing; here we only consider the number of objects present in region i_0 (not their vector); thus, with a system Π we associate the set $N(\Pi)$ of numbers computed in this way.

The family of all such sets $N(\Pi)$ of natural numbers computed by systems Π with at most n_1 initial membranes, with at most n_2 membranes simultaneously present during a computation, using rules as in *list-of-rules* is denoted by $NOP_{n_1,n_2}(act_3, \textit{list-of-rules})$; act_3 indicates the fact that we use rules with active membranes with three polarizations; when only two polarizations are used, we write act_2; in case no polarization is used (which amounts to using only one electrical charge, never changed), we write act_1. If no bounds on one of the parameters n_1, n_2 are imposed, then we replace it with $*$.

11.3 Universality for Restricted Cases

The natural issue of investigating the computing power of P systems with active membranes has been addressed already in the initial paper where these systems were introduced, [34], and, as expected, universality results were obtained. Initially, all types of rules were used (even one further type, which will be introduced in Section 11.6, rules for dividing non-elementary membranes under the control of inner membranes of different polarizations) and the number of membranes was not bounded (the proof started from a matrix grammar with appearance checking in the binary normal form and the number of membranes depended on the number of matrices). This result was repeatedly improved, in three directions: types of rules, number of membranes, number of polarizations. For instance, in [35], one recalls the following result:

Theorem 11.1 $NOP_{3,3}(act_3, (a), (b), (c)) = NRE$.

We give here improvements on this result, in the number of membranes or in the number of polarizations, but first we present the basic construction, with a rather simple proof, for a slight variation of Theorem 11.1: we use one further membrane, but only two polarizations.

Theorem 11.2 $NOP_{4,4}(act_2, (a), (b), (c)) = NRE$.

Proof. Consider a register machine $M = (m, H, l_0, l_h, I)$ working in the generative mode. Because three registers ensure the universality, we assume that $m = 3$. We construct the P system

$$\Pi = (O, B, [\ [\ ]_1^0[\ ]_2^0[\ ]_3^0]_s^0, l_0, \lambda, \lambda, \lambda, R, 1), \text{ with}$$

$$O = \{a\} \cup \{l, l', l'' \mid l \in H\},$$

$$B = \{s, 1, 2, 3\},$$

and the set of rules obtained as follows (with each register $r = 1, 2, 3$ of M we associate a membrane with label r; the number of copies of object a present in membrane r represents the value of the corresponding register r):

1. For each ADD instruction $l_i : (\text{ADD}(r), l_j, l_k)$, $r = 1, 2, 3$, of M we introduce in R the rules

$$l_i[\ ]_r^0 \to [\ l_i']_r^0,$$
$$[\ l_i' \to l_j a]_i^0,$$
$$[\ l_i' \to l_k a]_i^0,$$
$$[\ l_g]_r^0 \to [\ ]_r^0 l_g, \ g \in \{j, k\}.$$

The object l_i enters the correct membrane r, produces there one further copy of a, and exits in the form of a label l_j or l_k, as requested by the ADD instruction.

2. For each SUB instruction $l_i : (\text{SUB}(r), l_j, l_k)$, $r = 1, 2, 3$, of M we introduce in R the rules

$$l_i[\ ]_r^0 \to [\ l_i']_r^+,$$
$$[\ a]_r^+ \to [\ ]_r^0 a,$$
$$[\ l_i' \to l_i'']_r^+,$$
$$[\ l_i'']_r^0 \to [\ ]_r^0 l_j,$$
$$[\ l_i'']_r^+ \to [\ ]_r^0 l_k.$$

The interplay of primed versions of l_i and the polarizations ensures the correct simulation of the SUB instruction. The label-object l_i enters the correct membrane r, changing its polarization to +. In a membrane with positive charge, both a copy of a can exit, thus returning the polarization to 0 (and, because of the maximal parallelism, this should happen if at least one copy of a is present), and, in parallel, l'_i becomes l''_i. This last object plays the role of a checker: if a was present (hence the polarization is again neutral), l''_i exits transformed in l_j, otherwise (if the polarization remained +), l''_i exits transformed in l_k (and, at the same time, the polarization is returned to neutral). If copies of a exist, only one can exit. The simulation of the SUB instruction is correct.

The observation that the label l_h remains idle in the skin region, hence when it appears the computation stops, concludes the proof (the contents of membrane 1 clearly corresponds to the value of register 1 of M, hence $N(M) = N(\Pi)$).

In the previous construction we have not used the fact that the skin membrane can also be involved in communication rules. Making use of this feature, we can save one membrane:

Theorem 11.3 $NOP_{3,3}(act_2, (a), (b), (c)) = NRE$.

Proof. We start again from a register machine $M = (m, H, l_0, l_h, I)$ with three registers and we construct the P system

$$\Pi = (O, B, [\ [\ ]_1^0 [\ ]_2^0]_3^0, \lambda, \lambda, l_0, R, 1), \text{ with}$$

$$O = \{a_i \mid i = 1, 2, 3\} \cup \{l, l', l'', l''', l^{iv}, l^v, l^{vi} \mid l \in H\},$$

$B = \{1, 2, 3\}$, with 3 being the skin membrane,

and the set of rules obtained as follows; this time, with each register $r = 1, 2, 3$ of M we associate a membrane with label r and also an object a_r:

1. For each ADD instruction $l_i : (ADD(r), l_j, l_k)$, $r = 1, 2$, of M we introduce in R rules as in the previous proof, and for $r = 3$ we consider the rules

$$[\, l_i \to l_g a_3]_3^0, \ g \in \{j, k\}.$$

The simulation of the ADD instruction is obvious.

2. For each SUB instruction $l_i : (SUB(r), l_j, l_k)$, $r = 1, 2$, of M we introduce in R rules as in the previous proof, with a replaced with the corresponding a_r,

$r = 1, 2$, while for $r = 3$ we consider the following rules:

1. $[\, l_i \to l_i' l_i'' \,]_3^0$,
2. $[\, l_i' \to l_i''' \,]_3^0$,
   $[\, l_i'' \,]_3^0 \to [\, \,]_3^+ l_i^{iv}$,
3. $[\, a_3 \,]_3^+ \to [\, \,]_3^0 a_3$,
   $[\, l_i''' \to l_i^v \,]_3^+$,
4. $[\, l_i^v \to l_j \,]_3^0$,
   $[\, l_i^v \,]_3^+ \to [\, \,]_3^+ l_i^{vi}$,
5. $l_i^{vi} [\, \,]_3^+ \to [\, l_k \,]_3^0$.

The simulation of SUB instructions for register 3 is slightly more difficult than for other registers: In the first step we introduce two objects, one for producing the "checker" l_i^v, and one for changing the polarization of the skin membrane. With the polarization changed to $+$, an object a_3 can be removed (due to the maximal parallelism, it must be removed if it exists), and this is recorded in changing back the polarization to neutral. Now, the "checker" l_i^v introduces the correct object l_j or l_k, depending on the membrane polarization (returning this polarization to neutral in the case the register was empty, and this requires two steps, for sending out and bringing back primed versions of l_i).

The end of a computation in M coincides with the halting of a computation in Π and $N(M) = N(\Pi)$, which concludes the proof.

For the price of using three polarizations, we can save one further membrane: as in the previous proof, we use different objects a_r for different registers r, and the same membrane for $r = 1$ and $r = 2$. When simulating a SUB instruction for $r = 1$, we use as above the polarization $+$, but when simulating a SUB instruction for $r = 2$ we can use the negative polarization. The details are left to the reader, we mention here only the result:

Theorem 11.4 $NOP_{2,2}(act_3, (a), (b), (c)) = NRE$.

The previous results can be further improved. For instance, in [2] one can find proofs (significantly more complex than the previous ones and based on simulating graph-controlled grammars) for the following equalities:

Theorem 11.5 $NOP_{1,1}(act_2, (a), (c_\lambda)) = NOP_{2,2}(act_2, (a), (c)) = NRE$.

Thus, rules of type (b) (sending-in) are avoided and at the same time one uses only two membranes and two polarizations.

11.4 Avoiding Polarizations

Polarizations are a strong feature of P systems with active membranes, as they constitute a sort of "memory" of the membrane, similar to the states of an automaton, passing information from one step of a computation to the next step. On the other hand, electrical charges, as involved in rules introduced above, are not very realistic from a biological point of view. That is why the problem of avoiding using polarizations was a concern of many researches in membrane computing.

Some ideas are rather obvious. For instance, we can trade polarizations for labels, by allowing rules which also change the label of a membrane, not only its polarization. This can be done only for rules of types (b), (c), (e), which can then be of the following forms (the prime indicates the possibility of changing the labels of membranes; because we use only one electrical charge, this plays no role in the evolution of the system and hence it is omitted when writing the membranes and the rules):

(b′) $a + [\ ]_{h_1} \to [\ b\ ]_{h_2}$, where $a, b \in O$ and $h_1, h_2 \in H$,

(c′) $[\ a\ ]_{h_1} \to [\ ]_{h_2} b$, where $a, b \in O$ and $h_1, h_2 \in H$,

(e′) $[\ a\ ]_{h_1} \to [\ b\ ]_{h_2} [\ c\ ]_{h_3}$, where $a, b, c \in O$ and $h_1, h_2, h_3 \in H$.

The power of using rules of these types was examined in [6], where the following equalities were obtained:

Theorem 11.6 $NOP_{2,2}(act_1, (a), (b'), (c)) = NOP_{2,2}(act_1, (a), (b), (c')) = NOP_{2,*}(act_1, (a), (c), (e')) = NRE.$

Note that one uses only one type of rule which is allowed to change the labels of membranes.

At the price of using rules of all five types (and starting from a larger number of membranes, five), we can avoid polarizations and still preserve the universality even using rules in the one-normal form (always having only one object in the right-hand of rules of type (a) and no erasing in any type of rules). Because this result is rather strong and its proof uses an idea different from those used in the previous proofs, we recall here some details from [23].

Theorem 11.7 $NOP_{5,*}(act_1, (a_1), (b_1), (c_1), (d_1), (e_1)) = NRE.$

Proof. Let us consider a register machine $M = (3, H, l_0, l_h, I)$ and construct the P system

$$\Pi = (O, B, \mu, w_s, w_h, w_1, w_2, w_3, R, 0), \text{ with}$$

$$O = \{d_i \mid 0 \leq i \leq 5\} \cup \{g, \#, \#', p, p', p'', c, c', c''\} \cup H \cup \{l' \mid l \in H\}$$

$\cup \{l_{iu} \mid l_i$ is the label of an ADD instruction in $I, 1 \leq u \leq 4\}$

$\cup \{l_{iu0} \mid l_i$ is the label of a SUB instruction in $I, 1 \leq u \leq 4\}$

$\cup \{l_{iu+} \mid l_i$ is the label of a SUB instruction in $I, 1 \leq u \leq 6\}$,

$B = \{s, h, 1, 2, 3\}$,

$\mu = [\,[\,]_1[\,]_2[\,]_3[\,]_h\,]_s$,

$w_s = l_0 d_0$, $w_a = \lambda$, for all $a \in B - \{s\}$,

and with the rules constructed as described below.

The value of a register $r = 1, 2, 3$ of M is represented by the number of copies of the membrane with label r plus one (if the value of the register r is k, then the system contains $k + 1$ membranes with label r). The membrane with label h is auxiliary, it is used for controlling the correct simulation of instructions of M by computations in Π. Each step of a computation in M, i.e. using an ADD or a SUB instruction, corresponds to six steps of a computation in Π. We start with all membranes being empty, except for the skin region, which contains the initial label l_0 of M and the auxiliary object d_0. If the computation in M halts, that is, the object l_h appears in the skin region of Π, then we pass to producing and sending to the environment one object c for each membrane with label 1 present in the system, except one; in this way, the number of copies of c sent to the environment represents the correct result of the computation in M.

We first indicate the rules used in each of the six steps of simulating instructions ADD and SUB, and then we present the rules for producing the result of the computation; in tables below, only the membranes and the objects relevant for the simulation of the respective instruction are specified. In particular, we ignore the "garbage" object g, because once introduced it remains idle for the whole computation.

The **simulation of an instruction** $l_i : (\text{ADD}(r), l_j, l_k)$ uses the rules from Table 11.1.

The label object l_i enters into the correct membrane r (even if the register r is empty, there is at least one membrane with label r) and in the next step divides it. The object l_{i2} exits the newly produced membrane, but g remains inside; l_{i2} will evolve three further steps, just to synchronize with the evolution of the auxiliary objects $d_u, u \geq 0$, and the auxiliary membrane h, so that in the sixth step we end with the label l_j or l_k of the next instruction to be simulated present in the skin

Table 11.1 The simulation of an ADD instruction.

Step	Main rules	Auxiliary rules	Configuration
–	–	–	$[\, l_i d_0 [\,]_r \ldots [\,]_h]_s$
1	$l_i [\,]_r \to [\, l_i']_r$	$[\, d_0 \to d_1]_s$	$[\, d_1 [\, l_i']_r \ldots [\,]_h]_s$
2	$[\, l_i']_r \to [\, l_{i1}]_r [\, g]_r$	$[\, d_1 \to d_2]_s$	$[\, d_2 [\, l_{i1}]_r [\, g]_r \ldots [\,]_h]_s$
3	$[\, l_{i1}]_r \to [\,]_r l_{i2}$	$d_2 [\,]_h \to [\, d_3]_h$	$[\, l_{i2} [\,]_r [\,]_r \ldots [\, d_3]_h]_s$
4	$[\, l_{i2} \to l_{i3}]_s$	$[\, d_3]_h \to [\,]_h d_4$	$[\, l_{i3} d_4 [\,]_r [\,]_r \ldots [\,]_h]_s$
5	$[\, l_{i3} \to l_{i4}]_s$	$[\, d_4 \to d_5]_s$	$[\, l_{i4} d_5 [\,]_r [\,]_r \ldots [\,]_h]_s$
6	$[\, l_{i4} \to l_t]_s, t \in \{j, k\}$	$[\, d_5 \to d_0]_s$	$[\, l_t d_0 [\,]_r [\,]_r \ldots [\,]_h]_s$

membrane, together with d_0; note that the number of copies of membrane r was increased by one.

The auxiliary objects d_u, $u \geq 0$, and the auxiliary membrane h are used in the simulation of SUB instructions, as we will see immediately; in step 4, there also are other rules to be used in membrane h, introducing the trap-object #, but using such rules will make the computation never halt, hence, they will not produce an unwanted result.

The **simulation of an instruction** $l_i : (\text{SUB}(r), l_j, l_k)$ is done with the help of the auxiliary membrane h. The object l_i enters a membrane r (there is at least one copy of it) and divides it, and on this occasion makes a non-deterministic choice between trying to continue as having register r non-empty or as having it empty. If the guess has been correct, then the correct action is done (decrementing the register in the first case, doing nothing in the second case) and the correct next label is introduced, i.e. l_j or l_k, respectively. If the guess has not been correct, then the trap object # is introduced. For all membranes x of the system we consider the rules $[\, \# \to \#']_x$, $[\, \#' \to \#]_x$, hence, the appearance of # will make the computation last forever.

In Table 11.2 we present the rules used in the case of guessing that the register r is not empty. As in the case of simulating an ADD instruction, the auxiliary objects and membranes do not play any role in this case.

In step 2 we divide the membrane r containing the object l_i' and the objects l_{i1+}, l_{i2+} are introduced in the two new membranes. One of them is immediately dissolved, thus the number of copies of membrane r remains unchanged; the objects l_{i3+}, l_{i4+} are introduced in this step. In the next step (the fourth one of the simulation), objects l_{i3+}, l_{i4+} look for membranes r in the skin region. If both of them find such membranes—and this is the correct/desired continuation corresponding to a non-empty register r—then both of them enter such membranes; l_{i3+} becomes g and l_{i4+} becomes l_{i5+}. If only one of them finds a membrane r, then the other one has to evolve to object # in the skin membrane and the computation never halts.

Table 11.2 The simulation of a SUB instruction, guessing that register r is not empty.

Step	Main rules	Auxiliary rules	Configuration
–	–	–	$[\, l_i d_0 \,[\,]_r \ldots [\,]_h\,]_s$
1	$l_i [\,]_r \to [\, l_i' \,]_r$	$[\, d_0 \to d_1 \,]_s$	$[\, d_1 \,[\, l_i' \,]_r \ldots [\,]_h\,]_s$
2	$[\, l_i' \,]_r \to [\, l_{i1+} \,]_r [\, l_{i2+} \,]_r$	$[\, d_1 \to d_2 \,]_s$	$[\, d_2 [\, l_{i1+} \,]_r [\, l_{i2+} \,]_r \ldots [\,]_h\,]_s$
3	$[\, l_{i1+} \,]_r \to l_{i3+}$	$d_2 [\,]_h \to [\, d_3 \,]_h$	
	$[\, l_{i2+} \,]_r \to [\,]_r l_{i4+}$		$[\, l_{i3+} l_{i4+} [\,]_r \ldots [\, d_3 \,]_h\,]_s$
4	$l_{i3+} [\,]_r \to [\, g \,]_r$	$[\, d_3 \,]_h \to [\,]_h d_4$	
	$l_{i4+} [\,]_r \to [\, l_{i5+} \,]_r$		$[\, d_4 [\, l_{i5+} \,]_r [\, g \,]_r \ldots [\,]_h\,]_s$
	or $[\, l_{i3+} \to \# \,]_s$,		
	$[\, l_{i4+} \to \# \,]_s$		$[\, d_4 \# [\,]_r \ldots [\,]_h\,]_s$
5	$[\, l_{i5+} \,]_r \to l_{i6+}$	$[\, d_4 \to d_5 \,]_s$	$[\, d_5 l_{i6+} [\,]_r \ldots [\,]_h\,]_s$
6	$[\, l_{i6+} \to l_j \,]_s$	$[\, d_5 \to d_0 \,]_s$	$[\, l_j d_0 [\,]_r \ldots [\,]_h\,]_s$

If we had enough membranes r and l_{i3+}, l_{i4+} went there, then in the next step l_{i5+} dissolves one membrane r and is changed to l_{i6+}. In the sixth step, l_{i6+} becomes l_j (simultaneously with introducing d_0), and this completes the simulation. Thus, the computation continues without having # present in the system if and only if the guess made in step 2 has been correct, i.e. if register r has been non-empty. Because in step 5 a membrane r was dissolved, the number of these membranes was decreased by one, and this corresponds to subtracting one from register r.

However, in step 2, instead of the rule $[\, l_i' \,]_r \to [\, l_{i1+} \,]_r [\, l_{i2+} \,]_r$ we can use the rule $[\, l_i' \,]_r \to [\, l_{i10} \,]_r [\, l_{i20} \,]_r$, with the intention to simulate the subtract instruction in the case when the register r is empty—that is, only one membrane with label r is present in the system. The rules used in the six steps of the simulation are given in Table 11.3.

In this case, the auxiliary objects d_u, $u \geq 0$, and the auxiliary membrane h play an essential role.

Until step 3 we proceed exactly as above, but now we introduce the objects l_{i30}, l_{i40}. The first one can enter the available membrane r, there evolving to g. If there is a second membrane r, i.e. if the guess has been incorrect, then the rule $l_{i40} [\,]_r \to [\, \# \,]_r$ should be used simultaneously (step 4), and the computation never ends. If there is no second membrane r, then in step 4 l_{i40}, can also enter membrane h, but then the trap object is produced here by the rule $[\, d_3 \to \# \,]_h$. The only way not to introduce the object # is (i) not to have a second membrane r, (ii) to use the rule $[\, d_3 \,]_h \to [\,]_h d_4$, thus preventing the use of the rule $l_{i40} [\,]_h \to [\, l_k' \,]_h$, and (iii) not sending l_{i40} to the unique membrane r. This means that during step 4 the object l_{i40} should wait unchanged in the skin region.

Table 11.3 The simulation of a SUB instruction, guessing that register r is empty.

Step	Main rules	Auxiliary rules	Configuration
–	–	–	$[\, l_i d_0 \,[\,]_r \ldots [\,]_h\,]_s$
1	$l_i[\,]_r \to [\, l'_i\,]_r$	$[\, d_0 \to d_1\,]_s$	$[\, d_1 \,[\, l'_i\,]_r \ldots [\,]_h\,]_s$
2	$[\, l'_i\,]_r \to [\, l_{i10}\,]_r [\, l_{i20}\,]_r$	$[\, d_1 \to d_2\,]_s$	$[\, d_2 \,[\, l_{i10}\,]_r [\, l_{i20}\,]_r \ldots [\,]_h\,]_s$
3	$[\, l_{i10}\,]_r \to l_{i30}$ $[\, l_{i20}\,]_r \to l_{i40}[\,]_r$	$d_2[\,]_h \to [\, d_3\,]_h$	$[\, l_{i30} l_{i40} \,[\,]_r \ldots [\, d_3\,]_h\,]_s$
4	$[\, l_{i30}\,]_r \to [\, g\,]_r$, l_{i40} waits or $l_{i40}[\,]_r \to [\, \#\,]_r$ or $l_{i40}[\,]_h \to [\, l'_k\,]_h$	$[\, d_3\,]_h \to d_4[\,]_h$ $[\, d_3 \to \#\,]_h$	$[\, l_{i40} d_4 [\, g\,]_r \ldots [\,]_h\,]_s$ $[\, d_4 [\, \#\,]_r \ldots [\,]_h\,]_s$ $[[\, g\,]_r \ldots [\, l'_k \#\,]_h\,]_s$
5	$l_{i40}[\,]_h \to [\, l'_k\,]_h$	$[\, d_4 \to d_5\,]_s$	$[\, d_5 [\,]_r \ldots [\, l'_k\,]_h\,]_s$
6	$[\, l'_k\,]_h \to l_k[\,]_h$	$[\, d_5 \to d_0\,]_s$	$[\, l_k d_0 [\,]_r \ldots [\,]_h\,]_s$

In the next step, l_{i40} can enter membrane h (or the unique membrane r, but it becomes # there). In the next step, l_k is released from membrane h, at the same time producing d_0, hence, the simulation is completed correctly and the system can pass to simulating another instruction.

For both guesses, the simulation of the SUB instruction works correctly, and the process can be iterated.

If the computation in M halts, i.e. if the object l_h is introduced in the skin region, we can start to use the following rules:

$$l_h[\,]_1 \to [\, l'_h\,]_1,$$

$$[\, l'_h\,]_1 \to p,$$

$$p[\,]_1 \to [\, p'\,]_1,$$

$$[\, p'\,]_1 \to [\, p''\,]_1 [\, c'\,]_1,$$

$$[\, p''\,]_1 \to p,$$

$$[\, c'\,]_1 \to c'',$$

$$[\, c''\,]_s \to [\,]_s c,$$

$$p[\,]_h \to [\, p'\,]_h,$$

$$[\, p'\,]_h \to p.$$

The object l_h dissolves one membrane with label 1 (thus, the remaining membranes with this label are now as many as the value of register 1 of M), and gets transformed into p. This object enters each membrane with label 1, divides it, reproduces itself (after passing through p' and p'') and also produces a copy of the object c''; this happens when dissolving the two membranes with label 1 obtained by division (rule $[\,p'\,]_1 \to [\,p''\,]_1[\,c'\,]_1$), hence, the number of copies of the membrane with label 1 has decreased by one. The object c'' immediately exits the system, changed into c.

At any time, the object p can also enter membrane h and then dissolves it. The computation can stop only after having dissolved all membranes with label 1 (i.e. a corresponding number of copies of c have been sent out) and after having dissolved the membrane h, hence, the evolution of objects d_u, $u \geq 0$, also stops.

Consequently, $N(M) = N(\Pi)$, and this concludes the proof.

11.5 USING MINIMAL PARALLELISM

Besides the much investigated maximal parallelism, other strategies of using the rules were considered in membrane computing: sequential (exactly one rule is used in each membrane), asynchronous (each membrane evolves or not, thus the time is ignored), with bounded parallelism (at most a given number of rules is used in each step in each membrane), etc. One of the most interesting strategies is that introduced in [18] under the name of *minimal parallelism*. The idea is that each membrane which can evolve in a given step should do so by using at least one rule; otherwise stated, if at least one rule *can* be used, then at least one *must* be used; one, two, or more rules, any variant is permitted, hence a considerable degree of non-determinism is introduced, which sometimes makes the proofs more difficult.

This way of using the rules can be defined for all types of P systems, with some additional care being necessary in the case when rules from different membranes interact, like in systems with symport/antiport rules and in the case of P systems with active membranes. Using a rule from a region can make impossible the use of a rule from a neighboring region and vice-versa. As usual, in such cases the difficulty is solved by invoking the non-determinism: if a region (a membrane) can use at least one rule, we say that it is *alive*; from each alive region (membrane) we choose one or zero rules; the combination of these rules is said to be *acceptable* if no rule from the alive regions (membranes) where zero rules were chosen can be added; then, this combination of rules can be applied or any other obtained by choosing further rules from alive regions (membranes) where one rule was previously chosen.

Somewhat surprisingly, this way of using the rules also leads to universality in most, if not all, cases where maximal parallelism leads to universality. This is the case, for instance, for P systems with polarized membranes using the basic five types of rules; the following counterpart of Theorems 11.1, 11.2, 11.3 was proved already in [18] (the fact that one uses the minimal parallelism is indicated by the superscript *min*):

Theorem 11.8 $NOP_{3,3}^{min}(act_3, (a), (b), (c)) = NRE.$

Similarly, in [23], a counterpart of Theorem 11.7 is obtained:

Theorem 11.9 $NOP_{7,*}^{min}(act_1, (a_1), (b_1), (c_1), (d_1), (e_1)) = NRE.$

Note that in both cases there is a price to pay when passing from the maximal parallelism to the minimal one: three polarizations in the first case and two more membranes in the latter one. It is highly possible that these results can be improved from these points of view.

11.6 Further Types of Rules

The five types of rules discussed in the previous sections are the basic ones that are most investigated—mainly when looking for polynomial (even linear) time solutions to computationally hard problems (typically, NP-complete problems). This direction of research will be presented in Chapter 12, where it will be shown that rules of types (d) and (e), i.e. for membrane dissolution and membrane division, are essential (this was not the case in the universality results from the previous sections). Besides these five basic types of rules, there are several others where the membranes are directly involved. We briefly present here some of these types of rules, but details and many other operations can be found in the papers mentioned at the end of this chapter and, of course, in the general bibliography of membrane computing from http://ppage.psystems.eu.

First, let us consider a type of rule for membrane division already introduced in [33], [34]: dividing membranes which contain membranes of opposite polarizations. They are known as rules of type (f).

(f) $[\,[\ ]_{h_1}^{e_1} \ldots [\ ]_{h_k}^{e_1} [\ ]_{h_{k+1}}^{e_2} \ldots [\ ]_{h_n}^{e_2}\,]_{h_0}^{e_0}$
$\to [\,[\ ]_{h_1}^{e_3} \ldots [\ ]_{h_k}^{e_3}\,]_{h_0}^{e_5} [\,[\ ]_{h_{k+1}}^{e_4} \ldots [\ ]_{h_n}^{e_4}\,]_{h_0}^{e_6},$

for $k \geq 1, n > k, h_i \in H, 0 \leq i \leq n$, and $e_0, \ldots, e_6 \in \{+, -, 0\}$ with $\{e_1, e_2\} = \{+, -\}$; if the membrane with the label h_0 contains other membranes than those with the labels $h_1, \ldots, h_n$ specified above, then they must have neutral charges in order to make this rule applicable (division of non-elementary membranes; this is possible only if a membrane contains two immediately lower membranes of opposite polarization, + and $-$; the membranes of opposite polarizations are separated in the two new membranes, but their polarization can change; always, all membranes of opposite polarizations are separated by applying this rule; all membranes of neutral polarization are duplicated and then are part of the contents of both copies of the membrane h_0).

Such rules are very powerful, as they are applied to non-elementary membranes, hence by division both objects and membranes are replicated.

A related type of rule was introduced in [42] under the name of *strong division rules for non-elementary membranes*. They are a combination of rules of types (e) and (f), but without using polarizations: the division is performed under the control of two inner membranes, somewhat similar to the case of rules of type (e), where the division is controlled by inner objects. These rules are of the form

$$[[\ ]_{h_1}[\ ]_{h_2}]_{h_3} \to [[\ ]_{h_1}]_{h_3}[[\ ]_{h_2}]_{h_3},$$

where h_1, h_2, h_3 are labels of membranes. Thus, if a membrane with label h_3 contains a membrane with label h_1 and one with label h_2, then these membranes are separated (like the membranes with opposite polarizations in rules of type (f)), placed in two copies of membrane h_3; all membranes and objects placed inside membranes h_1 and h_2 or in membrane h_3 outside membranes h_1, h_2 are replicated in the new copies of membrane h_3. As usual, the application of rules takes place in a bottom-up way, each membrane can be the subject of only one rule, and the skin membrane cannot be divided.

Another way of controlling the division of a membrane is by taking into account all the objects it contains, and *separating* these objects according to a specified criterion. The most natural form of such separation rules is the following one.

Consider a subset Q of the set of objects O. A separation rule with respect to Q is of the form

$$[\ O\ ]_h \to [\ Q\ ]_h[\ O - Q\ ]_h,$$

with the obvious meaning that the elementary membrane with label h is separated (hence divided) in two membranes, with the same label, the first one containing all objects of the initial membrane which belong to Q and the second membrane contains the objects not in Q. The rule can be applied only if the initial membrane h contains at least one object from Q and at least one object from $O - Q$. A variant can be to allow that the labels of the new membranes are not necessarily the same as the label of the initial membrane.

Related rules and their properties (especially the usefulness in solving computationally hard problems) were investigated in [3], [24], [25].

A systematic study of rules for handling membranes—implicitly, rules explicitly involving membranes, hence "with active membranes"—was started in [20]. Besides rules of types (a)–(e) as above, but without polarizations, one considers rules of the following types:

$$(mer) \quad [a]_h[b]_{h'} \rightarrow [c]_{h''} - \text{merge},$$

$$(endo) \quad [a]_h[b]_{h'} \rightarrow [[d]_{h'}c]_h - \text{endocytosis},$$

$$(exo) \quad [a[b]_{h'}]_h \rightarrow [c]_h[d]_{h'} - \text{exocytosis},$$

$$(cre) \quad a \rightarrow [b]_h - \text{create}.$$

In all cases, a, b, c, d are objects and h, h', h'' are labels of membranes. Actually, rules of these types have been considered in various papers, sometimes with slight variations (changing labels, using polarizations, involving only one object in the left hand side of (endo), (exo) rules, etc.). For instance, merging rules are investigated in [25] (see also the bibliography of this thesis), endo- and exocytosis rules were investigated in [26], [27], while rules for membrane creation were introduced in [8].

Then, using rules for membrane creation, we can start from systems with only one membrane and, in a first phase of the computation, we can construct any given membrane structure by handling in a suitable way the objects introduced in the new membranes; starting from a given P system Π, we can construct it from a "seed system", including the multisets of Π. Thus, universality results for P systems with membrane creation can be obtained from existing universality results (e.g. for catalytic systems), even for systems with only one initial membrane.

Universality can also be obtained for other combinations of rules (the notations in the next theorem are similar to those used in the previous sections; note that we return to the maximal parallelism—which is also indicated by the absence of the superscript *min* from now on):

Theorem 11.10 $NOP_4(act_1, (a), (endo), (exo)) = NRE$.

It is worth noting that the (endo), (exo) operations mentioned above are formally different from those considered in Chapter 14 (in papers [15], [36]), which can also be considered as performed by rules with active membranes.

Returning to [20], the approach started there is not concerned with the computational power or the efficiency of P systems with various combinations of rules, but with constructing "membrane calculi" based on various operations. The focus is not on computations, but on configurations and their processing. Several natural issues can be raised in this context: normal forms, reachability, edit distances between configurations (with respect to a given set of operations), reproducibility of a system, etc. For instance, it is shown that the edit distance is not a metric, but a weak-metric (it is not symmetric in general), which in many cases is not computable (just because for many combinations of rules we obtain classes of P systems which are computationally complete, hence undecidable). A sort of AFL (abstract family of languages) theory can be constructed in this area (dealing with "abstract families of configurations"), looking for closure properties, generators, etc. This direction of research looks rather promising.

Let us now also return to rules for creating membranes, of the form $a \to [\, b\,]_h$, where a, b are objects and h is a label. It is written in biology books that nature creates membranes not only for delimiting "protected reactors" and surfaces where reactants are bound, but also to ensure that these "reactors" are small enough so that the reactants swimming in water can meet each other frequently enough (stirred by Brownian motion). Let us read this last idea in the sense that "smaller is faster", and generalize/exaggerate it to "the reactions take place twice as fast in a membrane than in its surrounding membrane". This directly recalls the idea of accelerated Turing machines, which perform the first step of a computation in one time unit, the next one in half of a time unit (the machine "learns" how to execute steps of computation), and so on, with any step lasting half of the previous one. This means that the machine performs infinitely many steps in two (external) time units. This is one of the tricks for devising computing devices which compute beyond Turing (see [19] for details and for similar ideas of passing beyond the "Turing barrier"). This was exactly the strategy used in [14], where it is proved that P systems with membrane creation, with the children membranes working twice as fast as their parent membranes, can compute Turing non-computable sets of numbers. Some details can be also found in Subchapter 23.2.

We end this section by mentioning one further class of P systems with a dynamical membrane structure, using so-called *gemmation rules*. Such a rule produces membranes with a specified address, reminding the vesicles produced by the Golgi apparatus in a cell: objects of a membrane i can introduce special objects $@_j$ (by so-called *pre-dynamical* evolution rules), these objects determine the budding of *mobile membranes*, carrying the "address" j, which leave the membrane i and travel across membranes towards the target membrane j; when this membrane is reached, the mobile membrane fuses with it, leaving its contents inside membrane j. We do not enter into details, but we refer to papers [11], [12], [13], to Chapters 7 and 9, and to their bibliographies.

11.7 Concluding Remarks

This chapter presented only basic ideas, types of rules, and results dealing with active membranes. Many other issues were considered in the literature. We mention just one of them in order to prove the interest of this area of research, of dealing with P systems with a dynamical membrane structure, namely, focusing the attention on the membrane structure itself, not on the multisets of objects processed in its compartments. In the previous sections, the goal was to compute numbers, using the membrane structure as an "additional" stuff, the "hardware" of the system; it is natural to switch the perspective, and to take the membrane structure (the tree describing it) as the goal of the computation, and to use the multisets of objects as "additional". P systems with rules for handling membranes can then be interpreted as tree processing devices; we only mention [21] for details. Then, membrane division and other similar operations can also be considered for tissue-like P systems, which makes possible the generation (more general, processing) of more complex structures than trees—we refer to [17] for steps in this direction.

Thus, besides being well motivated biologically, the classes of P systems with a dynamical membrane structure are worth investigating also from a mathematical and computational point of view, and, although many results have been obtained already, many issues are still open to research.

References

[1] A. Alhazov, R. Freund, Gh. Păun: P systems with active membranes and two polarizations. *Proc. Second Brainstorming Week on Membrane Computing*, Sevilla, 2004, RGNC Report 01/2004, 20–36.

[2] A. Alhazov, R. Freund, Gh. Păun: Computational completeness of P systems with active membranes and two polarizations. *Lecture Notes in Computer Sci.*, 3354 (2005), 82–92.

[3] A. Alhazov, R. Freund, A. Riscos-Núñez: Membrane division, restricted membrane creation and object complexity in P systems. *Intern. J. Computer Math.*, 83 (2006), 529–548.

[4] A. Alhazov, T.-O. Ishdorj: Membrane operations in P systems with active membranes. *Proc. Second Brainstorming Week on Membrane Computing*, Sevilla, 2004, RGNC Report 01/2004, 37–44.

[5] A. Alhazov, L. Pan: Polarizationless P systems with active membranes. *Grammars*, 7 (2004), 141–159.

[6] A. Alhazov, L. Pan, Gh. Păun: Trading polarizations for labels in P systems with active membranes. *Acta Informatica*, 41 (2005), 111–144.

[7] B. Aman, G. Ciobanu: Reachability problem in mobile membranes. *Proc. WMC8*, Thessaloniki, June 2007, 111–122.

[8] F. Arroyo, A. Baranda, J. Castellanos, Gh. Păun: Membrane computing: The power of (rule) creation. *J. Universal Computer Sci.*, 8 (2002), 369–381.

[9] F. Bernardini, M. Gheorghe: Language generating by means of P systems with active membranes. *Brainstorming Week on Membrane Computing*, Tarragona, February 2003, TR 26/03, URV, 2003, 46–60.

[10] F. Bernardini, M. Gheorghe: P systems with 7 active membranes are computationally complete. *Recent Results in Natural Computing* (M.J. Pérez-Jiménez et al., eds.), Univ. Sevilla, 2004, 7–16.

[11] D. BESOZZI, E. CSUHAJ-VARJU, G. MAURI, C. ZANDRON: Size and power of extended gemmating P systems. *Soft Computing*, 9 (2005), 650–656.

[12] D. BESOZZI, G. MAURI, GH. PĂUN, C. ZANDRON: Gemmating P systems: collapsing hierarchies. *Theoretical Computer Sci.*, 296 (2003), 253–267.

[13] D. BESOZZI, C. ZANDRON, G. MAURI, N. SABADINI: P systems with gemmation of mobile membranes. *Lecture Notes in Computer Sci.*, 2202 (2001), 136–153.

[14] C. CALUDE, GH. PĂUN: Bio-steps beyond Turing. *BioSystems*, 77 (2004), 175–194.

[15] L. CARDELLI, GH. PĂUN: An universality result for a (mem)brane calculus based on mate/drip operations. *Intern. J. Found. Computer Sci.*, 17 (2006), 49–68.

[16] M. CAVALIERE, M. IONESCU, T.-O. ISHDORJ: Inhibiting/de-inhibiting P systems with active membranes. *Cellular Computing. Complexity Aspects* (M.A. Gutiérrez-Naranjo et al., eds.), Fenix Editora, Sevilla, 2005, 117–129.

[17] R. CETERCHI, R. GRAMATOVICI, N. JONOSKA, K.G. SUBRAMANIAN: Tissue-like P systems with active membranes for picture generation. *Fundamenta Informaticae*, 56 (2003), 311–328.

[18] G. CIOBANU, L. PAN, GH. PĂUN, M.J. PÉREZ-JIMÉNEZ: P systems with minimal parallelism. *Theoretical Computer Sci.*, 378 (2007), 117–130.

[19] B.J. COPELAND: Hypercomputation. *Minds and Machines*, 12 (2002), 461–502.

[20] E. CSUHAJ-VARJÚ, A. DI NOLA, GH. PĂUN, M.J. PÉREZ-JIMÉNEZ, G. VASZIL: Editing configurations of P systems. *Fundamenta Informaticae*, 82 (2008), 29–46.

[21] R. FREUND, M. OSWALD, A. PĂUN: P systems generating trees. *Proc. Second Brainstorming Week on Membrane Computing*, Sevilla, 2004, RGNC Report 01/2004, 221–232.

[22] R. FREUND, A. PĂUN: P systems with active membranes and without polarizations. *Soft Computing*, 9 (2005), 657–663.

[23] R. FREUND, GH. PĂUN, M.J. PÉREZ-JIMÉNEZ: Polarizationless P systems with active membranes working in the minimally parallel mode. *Proc. Sixth Brainstorming Week on Membrane Computing*, Sevilla, 2007, RGNC Report 01/2007, 131–155.

[24] M. IONESCU, T.-O. ISHDORJ: Replicative-distribution rules in P systems with active membranes. *Lecture Notes in Computer Sci.*, 3407 (2005), 68–83.

[25] T.-O. ISHDORJ: *Membrane Computing, Neural Inspirations, Gene Assembly in Ciliates*. PhD Thesis, Sevilla University, 2007.

[26] S.N. KRISHNA: On the efficiency of a variant of P systems with mobile membranes. *Ramanujan Math. Soc. Lecture Notes Series*, 3 (2007), 171–178.

[27] S.N. KRISHNA, GH. PĂUN: P systems with mobile membranes. *Natural Computing*, 4 (2005), 255–274.

[28] M. MARGENSTERN, C. MARTIN-VIDE, GH. PĂUN: Computing with membranes; Variants with an enhanced membrane handling. *Proc. 7th Intern. Meeting on DNA Based Computers* (N. Jonoska, N.C. Seeman, eds.), Tampa, Florida, USA, 2001, 53–62.

[29] C. MARTÍN-VIDE, GH. PĂUN, A. RODRIGUEZ-PATÓN: On P systems with membrane creation. *Computer Science J. of Moldova*, 9 (2001), 134–145.

[30] L. PAN, A. ALHAZOV, T.-O. ISHDORJ: Further remarks on P systems with active membranes, separation, merging, and release rules. *Soft Computing*, 9 (2005), 686–690.

[31] L. PAN, T.-O. ISHDORJ: P systems with active membranes and separation rules. *J. Universal Computer Sci.*, 10 (2004), 630–649.

[32] A. PĂUN: On P systems with active membranes. *Unconventional Models of Computation* (I. Antoniou et al., eds.), Springer, 2000, 187–201.

[33] GH. PĂUN: Computing with membranes (P Systems); Attacking NP-complete problems. *Unconventional Models of Computation* (I. Antoniou et al., eds.), Springer, 2000, 94–115.

[34] GH. PĂUN: P systems with active membranes: Attacking NP-complete problems. *J. Automata, Languages, Combinatorics*, 6 (2001), 75–90.

[35] GH. PĂUN: *Membrane Computing. An Introduction.* Springer, 2002.

[36] GH. PĂUN: One more universality result for P systems with objects on membranes. *Intern. J. of Computers, Communication and Control*, 1 (2006), 44–51.

[37] GH. PĂUN, M.J. PÉREZ-JIMÉNEZ, A. RISCOS-NÚÑEZ: Tissue P systems with cell division. *Proc. Second Brainstorming Week on Membrane Computing*, Sevilla, 2004, RGNC Report 01/2004, 380–386.

[38] GH. PĂUN, Y. SUZUKI, H. TANAKA, T. YOKOMORI: On the power of membrane division in P systems. *Theoretical Computer Sci.*, 324 (2004), 61–85.

[39] M.J. PÉREZ-JIMÉNEZ, F.J. ROMERO-CAMPERO: Trading polarizations for bi-stable catalysts in P systems with active membranes. *Proc. Second Brainstorming Week on Membrane Computing*, Sevilla, 2004, RGNC Report 01/2004, 327–342.

[40] A. RODRIGUEZ-PATÓN, P. SOSÍK: P systems with active membranes characterize PSPACE. *Lecture Notes in Computer Sci.*, 4287 (2007), 33–46.

[41] CL. ZANDRON, CL. FERRETTI, G. MAURI: Solving NP complete problems using P systems with active membranes. *Unconventional Models of Computation* (I. Antoniou et al., eds.), Springer, 2000, 289–301.

[42] C. ZANDRON, A. LEPORATI, C. FERRETTI, G. MAURI, M.J. PÉREZ-JIMÉNEZ: On the computational efficiency of polarizationless recognizer P systems with strong division and dissolution. *Fundamenta Informaticae*, 87 (2008), 79–91.

CHAPTER 12

COMPLEXITY— MEMBRANE DIVISION, MEMBRANE CREATION

MARIO J. PÉREZ-JIMÉNEZ
AGUSTÍN RISCOS-NÚÑEZ
ÁLVARO ROMERO-JIMÉNEZ
DAMIEN WOODS

12.1 INTRODUCTION

MEMBRANE systems are very flexible and versatile devices. There are many different approaches, depending on the way the computations are interpreted. Is a number or a string expected as output? Does the interest lie in observing the "behavior" of the system? This chapter addresses how real-life problems could be solved using P systems. To this aim, notions from classical computational complexity theory are adapted for the membrane computing framework. As mentioned in Chapter 3, the purpose of computational complexity theory is to provide bounds on the amount of resources necessary for any mechanical procedure (algorithm) that solves a problem.

If one accepts that the class **P** contains all the feasible problems, or at least excludes most of the unfeasible ones, then from the point of view of modeling biological phenomena it can be argued that P systems which solve **NP**-hard problems are beyond our modeling capabilities on realistic computer hardware. However, on the one hand, if we show that the computational power of some P system is bounded above by **P**, or some subclass of **P**, then we can consider the possibility of directly modeling this P system on a computer. Consequentially we can consider computer simulations of aspects of biological phenomena that are faithfully modeled with the P system. On the other hand, if we wish to build real membrane computers in the laboratory, then it stands to reason that it is more realistic to use membrane systems that are realized by a feasible amount of (e.g. biological) material.

Usually, computational complexity theory deals with decision problems which are problems that require a "*yes*" or "*no*" answer. A *decision problem*, X, is a pair (I_X, θ_X) such that I_X is a language over a finite alphabet (whose elements are called *instances*) and θ_X is a total Boolean function (that is, a predicate) over I_X.

Of course, many abstract problems are not decision problems. For example, in combinatorial optimization problems some value must be optimized (minimized or maximized). In order to deal with such problems, they can be transformed into roughly equivalent decision problems by supplying a target/threshold value for the quantity to be optimized, and then asking whether this value can be attained.

A natural correspondence between decision problems and languages can be established as follows. Given a decision problem $X = (I_X, \theta_X)$, its associated language is $L_X = \{w \in I_X \mid \theta_X(w) = 1\}$. Conversely, given a language L, over an alphabet Σ, its associated decision problem is $X_L = (I_{X_L}, \theta_{X_L})$, where $I_{X_L} = \Sigma^*$, and $\theta_{X_L} = \{(x, 1) \mid x \in L\} \cup \{(x, 0) \mid x \notin L\}$.

The solvability of decision problems is defined through the recognition of the languages associated with them. Let M be a Turing machine with working alphabet Γ and L a language over Γ. Assume that the result of any halting computation of M is *yes* or *no*. If M is a *deterministic* device, then we say that M *recognizes* or *decides* L whenever, for any string u over Γ, if $u \in L$, then the answer of M on input u is *yes* (that is, M accepts u), and the answer is *no* otherwise (that is, M rejects u). If M is a *non-deterministic* device, then we say that M *recognizes* or *decides* L if for any string u over Γ, $u \in L$ if and only if there exists a computation of M with input u such that the answer is *yes*.

Throughout this chapter, it is assumed that each abstract problem has an associated fixed *reasonable encoding scheme* that describes the instances of the problem by means of strings over a finite alphabet. We do not define *reasonable* in a formal way, however, following [11], instances should be encoded in a concise manner, without irrelevant information, and where relevant numbers are represented in binary (or any fixed base other than 1). It is possible to use multiple reasonable encoding schemes to represent instances, but it is proved that the input sizes differ at most by a polynomial. The *size* $|u|$ of an instance u is the length of the string associated with it, in some reasonable encoding scheme.

P systems take multisets as input, usually in a unary fashion. Hence, it is important to be careful when asserting that a problem is polynomial-time solvable by membrane systems. In this context, polynomial-time solutions to **NP**-complete problems in the framework of membrane computing can be considered as *pseudopolynomial* time solutions in the classical sense (see [11] and [42] for details).

12.2 RECOGNIZER P SYSTEMS

Membrane systems—usually called P systems—are distributed (maximally) parallel computing devices inspired by the structure and functioning of living cells. They have several *syntactic* ingredients: a *membrane structure* consisting of a hierarchical arrangement of membranes embedded in a *skin* membrane, and delimiting *regions* or compartments where multisets of *objects* are placed and evolve according to sets of *rules*.

Also, P systems have two main *semantic* ingredients: their inherent *parallelism* and *non-determinism*. The objects inside the membranes can evolve according to given rules in a synchronous (in the sense that a global clock is assumed), parallel, and non-deterministic manner.

Definition 12.1 *A P system of degree* $q \geq 1$ *is a tuple of the form* $\Pi = (\Gamma, H, \mu, w_1, \ldots, w_q, R, h_o)$, *where:*

1. Γ *is a working alphabet of objects, and* H *is a finite set of labels;*
2. μ *is a membrane structure (a rooted tree) consisting of* q *membranes injectively labeled with elements of* H;
3. $w_1, \ldots, w_q$ *are strings over* Γ *describing the initial multisets of objects placed in the* q *initial regions of* μ;
4. R *is a finite set of developmental rules;*
5. $h_o \in H$ *or* $h_o = env$ *indicates the output region: in the case* $h_o \in H$, *for a computation to be successful there must be exactly one membrane with label* h_o *present in the halting configuration; in the case* $h_o = env$, h_o *is usually omitted from the tuple.*

Many variants of P systems can be obtained depending on the kind of *developmental rules* and the semantics which are considered. The *length of a rule* is the number of symbols necessary to write it, both its left and right sides.

If h is the label of a membrane, then $f(h)$ denotes the label of the father of the membrane labeled with h. We adopt the convention that the father of the skin membrane is the environment (*env*).

Definition 12.2 *A P system with input membrane is a tuple* (Π, Σ, h_i), *where: (a)* Π *is a P system; (b)* Σ *is an (input) alphabet strictly contained in* Γ *such that the initial multisets are over the alphabet* $\Gamma \setminus \Sigma$*; and (c)* h_i *is the label of a distinguished (input) membrane.*

The difference between P systems with and without input membrane is not related to their computations, but only to their initial configurations.

Definition 12.3 *A P system* Π *without input has a single initial configuration* $(\mu, \mathcal{M}_1, \ldots, \mathcal{M}_q)$. *A P system* (Π, Σ, h_i) *with input has many initial configurations: for each multiset* $m \in \Sigma^*$, *the initial configuration associated with* m *is* $(\mu, \mathcal{M}_1, \ldots, \mathcal{M}_{h_i} \cup m, \ldots, \mathcal{M}_q)$.

In order to solve decision problems, we define *recognizer P system*.

Definition 12.4 *A recognizer P system is a P system such that: (a) the working alphabet contains two distinguished elements* yes *and* no*; (b) all computations halt; and (c) if* $\mathcal{C}$ *is a computation of the system, then either object* yes *or object* no *(but not both) must have been sent to the output region of the system, and only in the last step of the computation.*

For recognizer P systems, a computation $\mathcal{C}$ is said to be an *accepting computation* (respectively, *rejecting computation*) if the object *yes* (respectively, *no*) appears in the output region associated with the corresponding halting configuration of $\mathcal{C}$.

For technical reasons all computations are required to halt, but this condition can often be removed without affecting computational efficiency.

Throughout this chapter, $\mathcal{R}$ denotes an arbitrary class of recognizer P systems.

12.2.1 Uniform Families of P Systems

Many formal machine models (e.g. Turing machines or register machines) have an infinite number of memory locations. On the other hand, P systems, or logic circuits, are computing devices of finite size and they have a finite description with a fixed amount of initial resources (number of membranes, objects, gates, etc.). For this reason, in order to solve a decision problem a (possibly infinite) family of P systems is considered.

The concept of solvability in the framework of P systems also takes into account the pre-computational process of (efficiently) constructing the family that provides the solution. In this paper, the terminology *uniform family* is used to denote that this construction is performed by a *single* computational machine.

In the case of P systems with input membrane, the term uniform family is consistent with the usual meaning for Boolean circuits: a family $\Pi = \{\Pi(n) \mid n \in \mathbf{N}\}$ is uniform if there exists a deterministic Turing machine which constructs the system $\Pi(n)$ from $n \in \mathbf{N}$ (that is, which on input 1^n outputs $\Pi(n)$). In such a family, the P system $\Pi(n)$ will process all the instances of the problem with numerical parameters (reasonably) encoded by n—the common case is that $\Pi(n)$ processes all instances of size n. Note that this means that for these families of P systems with input membrane further pre-computational processes are needed in order to (efficiently) determine which P system (and from which input) deals with a given instance of the problem. The concept of *polynomial encoding* introduced below tries to capture this.

In the case of P systems without input membrane a new notion arises: a family $\Pi = \{\Pi(w) \mid w \in I_X\}$ associated with a decision problem $X = (I_X, \theta_X)$ is *uniform* (some authors [28, 51, 55] use the term *semi-uniform* here) if there exists a deterministic Turing machine which constructs the system $\Pi(w)$ from the instance $w \in I_X$. In such a family, each P system usually processes only one instance, and the numerical parameters and syntactic specifications of the latter are part of the definition of the former.

It is important to point out that, in both cases, the family should be constructed in a efficient way. This requisite was first included within the term uniform family (introduced in [37]), but nowadays it is preferred to use the term *polynomially uniform by Turing machines* to indicate a uniform (by a single Turing machine) and effective (in polynomial time) construction of the family.

Definition 12.5 *A family $\Pi = \{\Pi(w) \mid w \in I_X\}$ (respectively, $\Pi = \{\Pi(n) \mid n \in \mathbf{N}\}$) of recognizer membrane systems without input membrane (resp., with input membrane) is* polynomially uniform by Turing machines *if there exists a deterministic Turing machine working in polynomial time which constructs the system $\Pi(w)$ (resp., $\Pi(n)$) from the instance $w \in I_X$ (resp., from $n \in \mathbf{N}$).*

Some papers [54, 55] present an (apparently) different concept of uniform families of P systems: each member of the family is said to be constructed by a specific deterministic Turing machine. That is, different P systems of the family can be constructed by different Turing machines. The authors of this chapter believe that this is a rather non-standard notion of uniformity which can lead to unnatural complexity classes, so they will stick to Definition 12.5.

12.2.2 Confluent P Systems

In order for recognizer P systems to capture the true algorithmic concept, a condition of *confluence* is imposed, in the sense that all possible successful computations must give the same answer. This contrasts with the standard notion of accepting computations for non-deterministic (classic) models.

Definition 12.6 *Let $X = (I_X, \theta_X)$ be a decision problem, and $\Pi = \{\Pi(w) \mid w \in I_X\}$ be a family of recognizer P systems without input membrane.*

- *Π is said to be* sound *with respect to X if the following holds: for each instance of the problem, $w \in I_X$, if there exists an accepting computation of $\Pi(w)$, then $\theta_X(w) = 1$.*
- *Π is said to be* complete *with respect to X if the following holds: for each instance of the problem, $w \in I_X$, if $\theta_X(w) = 1$, then every computation of $\Pi(w)$ is an accepting computation.*

The concepts of soundness and completeness can be extended to families of recognizer P systems with input membrane in a natural way. However, an efficient process of selecting P systems from instances must be made precise.

Definition 12.7 *Let $X = (I_X, \theta_X)$ be a decision problem, and $\Pi = \{\Pi(n) \mid n \in \mathbb{N}\}$ a family of recognizer P systems with input membrane. A* polynomial encoding *of X in Π is a pair (cod, s) of polynomial time computable functions over I_X such that for each instance $w \in I_X$, $s(w)$ is a natural number (obtained by means of a reasonable encoding scheme) and $cod(w)$ is an input multiset of the system $\Pi(s(w))$.*

Polynomial encodings are stable under polynomial time reductions [45].

Proposition 12.1 *Let X_1, X_2 be decision problems, r a polynomial time reduction from X_1 to X_2, and (cod, s) a polynomial encoding from X_2 to Π. Then $(cod \circ r, s \circ r)$ is a polynomial encoding from X_1 to Π.*

Next, the concepts of soundness and completeness are defined for families of recognizer P systems with input membrane.

Definition 12.8 *Let $X = (I_X, \theta_X)$ be a decision problem, $\Pi = \{\Pi(n) \mid n \in \mathbb{N}\}$ a family of recognizer P systems with input membrane, and (cod, s) a polynomial encoding of X in Π.*

- *Π is said to be* sound *with respect to (X, cod, s) if the following holds: for each instance of the problem, $w \in I_X$, if there exists an accepting computation of $\Pi(s(w))$ with input $cod(w)$, then $\theta_X(w) = 1$.*
- *Π is said to be* complete *with respect to (X, cod, s) if the following holds: for each instance of the problem, $w \in I_X$, if $\theta_X(w) = 1$, then every computation of $\Pi(s(w))$ with input $cod(w)$ is an accepting computation.*

Note that if a family of recognizer P systems is sound and complete, then every P system of the family is confluent, in the sense previously mentioned.

12.2.3 Semi-Uniform Solutions versus Uniform Solutions

The first results showing that membrane systems could solve computationally hard problems in polynomial time were obtained using P systems without input membrane. In that context, a specific P system is associated with each instance of the problem. In other words, the syntax of the instance is part of the description of the associated P system. Thus this P system can be considered *special purpose*.

Definition 12.9 *A decision problem X is* solvable in polynomial time *by a family of recognizer P systems without input membrane $\Pi = \{\Pi(w) \mid w \in I_X\}$, denoted by $X \in \mathbf{PMC}^*_{\mathcal{R}}$, if the following holds:*

- *The family Π is polynomially uniform by Turing machines.*
- *The family Π is polynomially bounded; that is, there exists a natural number $k \in \mathbf{N}$ such that for each instance $w \in I_X$, every computation of $\Pi(w)$ performs at most $|w|^k$ steps.*
- *The family Π is sound and complete with respect to X.*

The family Π is said to provide a *semi-uniform solution* to the problem X.

Next, recognizer P systems with input membrane are defined to solve problems in a *uniform* way in the following sense: all instances of a decision problem of the same *size* (via a given reasonable encoding scheme) are processed by the same system, to which an appropriate input is supplied.

Definition 12.10 *A decision problem $X = (I_X, \theta_X)$ is* solvable in polynomial time *by a family of recognizer P systems with input membrane $\Pi = \{\Pi(n) \mid n \in \mathbf{N}\}$, denoted by $X \in \mathbf{PMC}_{\mathcal{R}}$, if the following holds:*

- *The family Π is polynomially uniform by Turing machines.*
- *There exists a polynomial encoding (cod, s) of X in Π such that:*
  - *The family Π is polynomially bounded with respect to (X, cod, s); that is, there exists a natural number $k \in \mathbf{N}$ such that for each instance $w \in I_X$, every computation of the system $\Pi(s(w))$ with input $cod(w)$ performs at most $|w|^k$ steps.*
  - *The family Π is sound and complete with respect to (X, cod, s).*

The family Π is said to provide a *uniform solution* to the problem X.

As a direct consequence of working with recognizer membrane systems, these complexity classes are closed under complement. Moreover, they are closed under polynomial time reductions [45].

Obviously, every uniform solution of a decision problem provides a semi-uniform solution using the same amount of computational resources. That is, $\mathbf{PMC}_{\mathcal{R}} \subseteq \mathbf{PMC}^*_{\mathcal{R}}$, for any class $\mathcal{R}$ of recognizer P systems.

Remark: It is interesting to distinguish the concept of *polynomially uniform by Turing machines* from the concepts of *semi-uniform* and *uniform* solutions. The first concept is related with the resources required to construct the family of P systems solving a decision problem. The last two refer to the way in which the family processes the instances. In semi-uniform solutions, every instance is processed by a special purpose P system, while in uniform solutions, each P system processes all instances of a given size.

12.3 Efficiency of Basic Transition P Systems

In this section, the computational efficiency of P systems whose membrane structure does not increase is studied.

A *basic transition* P system is a P system with only evolution, communication, and dissolution rules, which do not increase the size of the membrane structure. Let $\mathcal{T}$ denote the class of recognizer basic transition P systems.

In [17], an efficient simulation of deterministic Turing machines by recognizer basic transition P systems is given.

Proposition 12.2 *(Sevilla Theorem) Every deterministic Turing machine working in polynomial time can be simulated in polynomial time by a family of recognizer basic transition P systems with input membrane.*

In the same paper it is also proved that each confluent basic transition P system can be (efficiently) simulated by a deterministic Turing machine. As a consequence, these P systems efficiently solve most tractable problems.

Proposition 12.3 *If a decision problem is solvable in polynomial time by a family of recognizer basic transition P systems with input membrane, then there exists a deterministic Turing machine solving it in polynomial time.*

These results are also verified for recognizer basic transition P systems without input membrane. Therefore, the following holds.

Theorem 12.1 $\mathbf{P} = \mathbf{PMC}_{\mathcal{T}} = \mathbf{PMC}^*_{\mathcal{T}}$.

Thus, the ability of a P system in $\mathcal{T}$ to create exponential workspace (in terms of number of objects) in polynomial time (e.g. via evolution rules of the type $[a \to a^2]_h$) is not enough to efficiently solve **NP**–complete problems (unless $\mathbf{P} = \mathbf{NP}$). Theorem 12.1 provides a tool to attack the conjecture $\mathbf{P} = \mathbf{NP}$ in the framework of membrane computing.

Corollary 12.1 P $\neq$ NP *if and only if every, or at least one, NP-complete problem is not in* $PMC_T = PMC_T^*$.

12.4 MEMBRANE DIVISION

Two ways of producing new membranes in living cells are *mitosis* (membrane division) and *autopoiesis* (membrane creation). The abstraction of both processes has given rise to the models of P systems with membrane division and P systems with membrane creation. Both models have been proved to be universal, but up to now there has been no theoretical result proving that they simulate each other in polynomial time.

P systems with active membranes having associated electrical charges with membranes were introduced in [38].

Definition 12.11 *A P system with active membranes of degree* $q \geq 1$ *is a tuple* $\Pi = (\Gamma, H, \mu, w_1, \ldots, w_q, R, h_o)$, *where* $\Gamma, H, w_1, \ldots, w_q, h_o$ *are as in a usual P system,* μ *is a membrane structure consisting of q membranes injectively labeled with elements of H and with electrical charges* $(+, -, 0)$ *associated with them, and R is a finite set of rules, of the following forms:*

1. $[a \to u]_h^\alpha$, *for* $h \in H, \alpha \in \{+, -, 0\}$, $a \in \Gamma, u \in \Gamma^*$ (object evolution rules).
2. $a [\ ]_h^{\alpha_1} \to [b]_h^{\alpha_2}$, *for* $h \in H$, $\alpha_1, \alpha_2 \in \{+, -, 0\}$, $a, b \in \Gamma$ (send-in communication rules).
3. $[a]_h^{\alpha_1} \to [\ ]_h^{\alpha_2} b$, *for* $h \in H$, $\alpha_1, \alpha_2 \in \{+, -, 0\}$, $a, b \in \Gamma$ (send-out communication rules).
4. $[a]_h^\alpha \to b$, *for* $h \in H$, $\alpha \in \{+, -, 0\}$, $a, b \in \Gamma$ (dissolution rules).
5. $[a]_h^{\alpha_1} \to [b]_h^{\alpha_2} [c]_h^{\alpha_3}$, *for* $h \in H$, $\alpha_1, \alpha_2, \alpha_3 \in \{+, -, 0\}$, $a, b, c \in \Gamma$ (division rules for elementary membranes).
6. $[[\ ]_{h_1}^{\alpha_1} \ldots [\ ]_{h_k}^{\alpha_1} [\ ]_{h_{k+1}}^{\alpha_2} \ldots [\ ]_{h_n}^{\alpha_2}]_h^\alpha \to [[\ ]_{h_1}^{\alpha_3} \ldots [\ ]_{h_k}^{\alpha_3}]_h^\beta [[\ ]_{h_{k+1}}^{\alpha_4} \ldots [\ ]_{h_n}^{\alpha_4}]_h^\gamma$, *for* $k \geq 1, n > k, h, h_1, \ldots, h_n \in H$, $\alpha, \beta, \gamma, \alpha_1, \ldots, \alpha_4 \in \{+, -, 0\}$ *and* $\{\alpha_1, \alpha_2\} = \{+, -\}$ (division rules for non-elementary membranes).

These rules are applied as usual (see [37] for details).

Note that these P systems have some important features: (a) they use three electrical charges; (b) the polarization of a membrane, but not the label, can be modified by the application of a rule; and (c) they do not use cooperation nor priorities.

In the framework of P systems without input membrane, in [56] it is proved that confluent recognizer P systems, with active membranes making use of no membrane division rule, can be efficiently simulated by a deterministic Turing machine.

Proposition 12.4 *(Milano Theorem) A deterministic P system with active membranes but without membrane division can be simulated by a deterministic Turing machine with a polynomial slowdown.*

Let $\mathcal{NAM}$ be the class of recognizer P systems with active membranes which do not make use of division rules. As a consequence of the previous result, the following holds:

Corollary 12.2 $\text{PMC}^*_{\mathcal{NAM}} \subseteq \text{P}$.

In [50] one provides a simple proof of each tractable problem able to be solved (in a semi-uniform way) by a family of recognizer P systems with active membranes (without polarizations) operating in exactly one step and using only send-out communication rules. That proof can be easily adapted to uniform solutions.

Proposition 12.5 $\text{P} \subseteq \text{PMC}_{\mathcal{NAM}}$.

Thus, we have a version of Theorem 12.1 for the class $\mathcal{NAM}$.

Theorem 12.2 $\text{P} = \text{PMC}_{\mathcal{NAM}} = \text{PMC}^*_{\mathcal{NAM}}$.

The first efficient solutions to **NP**-complete problems by using P systems with active membranes were given in a *semi-uniform* way (where the P systems of the family depend on the syntactic structure of the instance) for Hamiltonian Path and Vertex Cover in [23] and [36], for SAT in [32], [38, 39], and [56].

Let $\mathcal{AM}(+n)$ (respectively, $\mathcal{AM}(-n)$) be the class of recognizer P systems with active membranes using division rules for elementary and non-elementary membranes (respectively, only for elementary membranes).

In the framework of $\mathcal{AM}(-n)$, efficient *uniform* solutions to weakly **NP**-complete problems (Knapsack [44], Subset Sum [43], Partition [14]), and strongly **NP**-complete problems (SAT [49], Clique [5], Bin Packing [47], Common Algorithmic Problem [46]) have been obtained.

In order to illustrate these results, a family of recognizer P systems with active membranes solving SAT in a uniform way and in a linear time is shown next.

A Uniform Solution to SAT using Membrane Division

The SAT problem is the following: *Given a Boolean formula in conjunctive normal form (CNF), determine whether or not it is satisfiable, that is, whether there exists an assignment to its variables on which it evaluates true.*

First we consider the polynomial-time computable function $\langle m, n \rangle = ((m + n)(m + n + 1)/2) + m$ (the *pair* function). It is a primitive recursive and bijective function from $\mathbf{N}^2$ onto $\mathbf{N}$.

For each $m, n \in \mathbf{N}$, the P system of degree 2 with input and with external output $\Pi(\langle m, n \rangle) = (\Gamma(m, n), \Sigma(m, n), H, \mu, w_1, w_2, R(m, n), h_i)$, is defined as follows:

- $\Sigma(m, n) = \{x_{i,j}, \bar{x}_{i,j} \mid 1 \leq i \leq m, 1 \leq j \leq n\}$.
- $\Gamma(m, n) = \Sigma(m, n) \cup \{c_k \mid 1 \leq k \leq m + 2\}$
  $\cup \{d_k \mid 1 \leq k \leq 2m + 3n + 3\}$
  $\cup \{r_{i,k} \mid 0 \leq i \leq m, 1 \leq k \leq 2n\}$
  $\cup \{e, t\} \cup \{Yes, No\}$
- $H = \{1, 2\}$.
- $\mu = [\,[\,]_2\,]_1$ (we call *internal membranes* to membranes with label 2).
- $w_1 = \lambda$ and $w_2 = d_1$.
- $h_i = 2$.
- The list of rules in $R(m, n)$ is given below, together with some short comments on their purpose and functioning.

  (a) $\{[d_k]_2^0 \to [d_k]_2^+ [d_k]_2^- \mid 1 \leq k \leq n\}$.
  By using a rule of (a), an internal membrane is divided into two membranes with opposite polarizations. These rules allow us to duplicate, in one step, the total number of internal membranes.

  (b) $\{[x_{i,1} \to r_{i,1}]_2^+, [\bar{x}_{i,1} \to r_{i,1}]_2^- \mid 1 \leq i \leq m\}$.
  $\{[x_{i,1} \to \lambda]_2^-, [\bar{x}_{i,1} \to \lambda]_2^+ \mid 1 \leq i \leq m\}$.
  The rules of (b) try to implement a process allowing the internal membranes to encode the *assignment* of a variable and, simultaneously, to check the value of all clauses by this assignment, in such a way that an object $r_{i,1}$ will appear in the membrane only if the i-th clause is true. Otherwise, the object encoding the variable will disappear.

  (c) $\{[x_{i,j} \to x_{i,j-1}]_2^+, [x_{i,j} \to x_{i,j-1}]_2^- \mid 1 \leq i \leq m, 2 \leq j \leq n\}$.
  $\{[\bar{x}_{i,j} \to \bar{x}_{i,j-1}]_2^+, [\bar{x}_{i,j} \to \bar{x}_{i,j-1}]_2^- \mid 1 \leq i \leq m, 2 \leq j \leq n\}$.
  This checking process is performed with respect to the *first* variable appearing in the internal membrane. Hence, the rules of (c) take care of making a cyclic path through the variables to get (initially, the first variable is x_1, then x_2, and so on).

  (d) $\{[d_k]_2^+ \to [\,]_2^0 d_k, [d_k]_2^- \to [\,]_2^0 d_k \mid 1 \leq k \leq n\}$.
  $\{d_k[\,]_2^0 \to [d_{k+1}]_2^0 \mid 1 \leq k \leq n - 1\}$.
  The rules of (d) are used as controllers of the generating process of the assignments and the encoding of the satisfied clauses: the objects d_k are sent out to the skin, at the same time the checking is made and they come back to the internal membranes to perform a further division of these membranes.

(e) $\{[r_{i,k} \to r_{i,k+1}]_2^0 \mid 1 \leq i \leq m, \ 1 \leq k \leq 2n-1\}$.

The use of objects $r_{i,2n}$ in the rules (i), (j), and (k) makes it necessary to perform a rotation of the indices of these objects.

(f) $\{[d_k \to d_{k+1}]_1^0 \mid n \leq k \leq 3n-3\}$; $[d_{3n-2} \to d_{3n-1}e]_1^0$.

Through the counter-objects d_k, the rules of (f) *control* the rotation of the indices of the elements $r_{i,k}$ in the internal membranes.

(g) $e[\ ]_2^0 \to [c_1]_2^+$; $[d_{3n-1} \to d_{3n}]_1^0$.

The application of the rules of (g) show that the system is ready to check which clauses are made true by the assignment encoded by each internal membrane.

(h) $\{[d_k \to d_{k+1}]_1^0 \mid 3n \leq k \leq 2m+3n+2\}$.

The rules of (h) supply counters in the skin through objects d_k, in such a way that, if objects d_{2m+3n} appear, then they indicate the end of the checking of the clauses. The objects d_k, with $2m+3n+1 \leq k \leq 2m+3n+3$, will control the final stage of the computation.

(i) $[r_{1,2n}]_2^+ \to [\ ]_2^- r_{1,2n}$.

The applicability of the rule (i) indicates the fact that an internal membrane makes true the clause *represented* by the object $r_{1,2n}$ through a change of its polarization.

(j) $\{[r_{i,2n} \to r_{i-1,2n}]_2^- \mid 1 \leq i \leq m\}$.

Because of the previous rule, we must re-label the values of r representing the different clauses.

(k) $r_{1,2n}[\ ]_2^- \to [r_{0,2n}]_2^+$.

By using the rule (k) the task of making explicit the assignments that make true the clause encoded in that moment of the execution by the object $r_{1,2n}$ is ended.

(l) $\{[c_k \to c_{k+1}]_2^- \mid 1 \leq k \leq m\}$.

The presence of objects c_k (with $2 \leq k \leq m+1$) in the internal membranes shows that the assignments making true all clauses are being determined.

(m) $[c_{m+1}]_2^+ \to [\ ]_2^+ c_{m+1}$.

The rule (m) sends to the skin the objects c_{m+1} appearing in the internal membranes.

(n) $[c_{m+1} \to c_{m+2}t]_1^0$.

By using the rule (n) the objects c_{m+1} in the skin evolve to objects c_{m+2} and t. The objects t in the skin are produced simultaneously with the appearance of the objects $d_{2m+3n+2}$ in the skin, and will show that there exists some assignment making the formula true.

(o) $[t\ ]_1^0 \to [\ ]_1^+ t$.

The rule (o) sends out of the system an object t changing the polarization of the skin to positive (then no object t remaining in the skin is able to evolve). This allows that an object c_{m+2} can exit the skin producing an object *Yes* in the environment through the rule (p), telling us that the formula is satisfiable, and the computation stops.

(p) $[c_{m+2}]_1^+ \to [\ ]_1^- Yes$.

The application of the rule (p) changes the polarization of the skin to negative in order that the objects c_{m+2} remaining in it are not able to continue evolving.

(q) $[d_{2m+3n+3}]_1^0 \to [\ ]_1^+ No$.

Rule (q) is only applicable when the object $d_{2m+3n+3}$ is present in the skin having a neutral charge (this is the case when the formula is not satisfiable). Then the system will evolve sending to the environment an object *No* and changing the polarization of the skin to positive, in order that the objects $d_{2m+3n+3}$ remaining in the skin do not evolve.

Finally, let (cod, s) be a polynomial encoding of SAT in $\Pi = \{\Pi(\langle m, n\rangle) : m, n \in \mathbb{N}\}$ defined as $s(\varphi) = \langle m, n\rangle$ and $cod(\varphi) = \{x_{i,j} : x_j \in C_i\} \cup \{\bar{x}_{i,j} : \neg x_{i,j} \in C_i\}$, for any Boolean formula in CNF, $\varphi = C_1 \wedge \cdots \wedge C_m$ such that $Var(\varphi) = \{x_1, \ldots, x_n\}$.

The family Π provides a uniform solution to the SAT problem according to Definition 12.10. Although the proof is omitted here (see [49] for more details), let us highlight that the necessary resources to build the family are of the order $\Theta(m \cdot n)$. Moreover, the designed P systems are deterministic and the total number of steps performed by $\Pi(s(\varphi))$ with input $cod(\varphi)$ is $2m + 5n + 3 \in \Theta(m + n)$.

From the above we have the following result.

Proposition 12.6 SAT $\in$ **PMC**$_{\mathcal{AM}(-n)}$.

Since **PMC**$_\mathcal{R}$ is closed under complement and polynomial time reductions, for any class $\mathcal{R}$ of recognizer P systems, the following result is obtained.

Proposition 12.7 **NP** $\cup$ **co-NP** $\subseteq$ **PMC**$_{\mathcal{AM}(-n)}$.

In the framework of $\mathcal{AM}(+n)$, in [55] one gives a efficient *semi-uniform* solution to QBF-SAT (satisfiability of quantified propositional formulas), a well known PSPACE-complete problem [11]. Hence the following is deduced.

Proposition 12.8 **PSPACE** $\subseteq$ **PMC**$^*_{\mathcal{AM}(+n)}$.

This result has been extended in [6], showing that QBF-SAT can be solved in a linear time and in a *uniform* way by a family of recognizer P systems with active membranes (without using dissolution rules) and using division rules for elementary and non-elementary membranes.

Proposition 12.9 **PSPACE** $\subseteq$ **PMC**$_{\mathcal{AM}(+n)}$.

A (deterministic and efficient) algorithm simulating a single computation of any confluent recognizer P system with active membranes and without input is described in [51]. Such P systems can be simulated by a deterministic Turing machine working with exponential space, and spending a time of the order $O(2^{p(n)})$, for some polynomial $p(n)$. Thus,

Proposition 12.10 $\text{PMC}^*_{\mathcal{AM}(+n)} \subseteq \textbf{EXP}$.

Therefore, $\textbf{PMC}_{\mathcal{AM}(+n)}$ and $\textbf{PMC}^*_{\mathcal{AM}(+n)}$ are two membrane computing complexity classes between **PSPACE** and **EXP**.

Corollary 12.3 $\textbf{PSPACE} \subseteq \textbf{PMC}_{\mathcal{AM}(+n)} \subseteq \textbf{PMC}^*_{\mathcal{AM}(+n)} \subseteq \textbf{EXP}$.

The reverse inclusion of Proposition 12.8 holds as well, [54]. Nevertheless, the concept of *uniform family* of P systems considered in that paper is different from that of Definition 12.5, although maybe the proof can be adapted to fit into the framework presented in this chapter. In this case the following would hold: **PSPACE** = $\textbf{PMC}^*_{\mathcal{AM}(+n)}$. Thus, the class of recognizer P systems with active membranes, with electrical charges, using division for elementary and non-elementary membranes, would be computationally equivalent to standard parallel machine models as PRAMs or alternating Turing machines.

The previous results show that the usual framework of P systems with active membranes for solving decision problems is too powerful from the computational complexity point of view. Therefore, it would be interesting to investigate weaker models of P systems with active membranes able to characterize classical complexity classes below **NP** and providing borderlines between efficiency and non-efficiency.

Efficient (semi-uniform and/or uniform) solutions to computationally hard problems have been obtained within different apparently weaker variants of P systems with active membranes:

- P systems with separation rules instead of division rules, in two different cases: first, using polarizations without changing membrane labels; and second, without polarizations but allowing change of membrane labels (SAT, uniform solution [34]).
- P systems using division for elementary membranes, without changing membrane labels, without polarizations, but using bi-stable catalysts (SAT, uniform solution [48]).
- P systems using division for elementary membranes, without label changing, but using only two electrical charges (SAT, uniform solution [3], Subset Sum, uniform solution [52]).
- P systems without polarizations, without label changing, without division, but using three types of membrane rules: separation, merging, and release (SAT, semi-uniform solution [33]).
- P systems without dissolution nor polarizations, but allowing the labels of membranes to change in division rules (SAT, uniform solution [4]).

- P systems without dissolution nor polarizations, but allowing the labels of membranes to change in send-out rules (SAT, uniform solution [4]).
- P systems without polarizations, but using division for elementary and non-elementary membranes (SAT, semi-uniform solution [4]).

12.4.1 Avoiding Polarizations

Next, several classes of recognizer P systems with active membranes without electrical charges and with different kinds of membrane division rules are studied from a computational complexity point of view.

Definition 12.12 *A polarizationless P system with active membranes of degree $q \geq 1$ is a P system with active membranes $\Pi = (\Gamma, H, \mu, w_1, \ldots, w_q, R, h_o)$ without using polarizations, hence with the rules in R of the following forms:*

(a) $[a \to u]_h$, *for $h \in H, a \in \Gamma, u \in \Gamma^*$ (object evolution rules).*
(b) $a[\ ]_h \to [b]_h$, *for $h \in H, a, b \in \Gamma$ (send-in communication rules).*
(c) $[a]_h \to [\ ]_h b$, *for $h \in H, a, b \in \Gamma$ (send-out communication rules).*
(d) $[a]_h \to b$, *for $h \in H, a, b \in \Gamma$ (dissolution rules).*
(e) $[a]_h \to [b]_h [c]_h$, *for $h \in H, a, b, c \in \Gamma$ (weak division rules for elementary or non-elementary membranes).*
(f) $[[\ ]_{h_1} \ldots [\ ]_{h_k} [\ ]_{h_{k+1}} \ldots [\ ]_{h_n}]_h \to [[\ ]_{h_1} \ldots [\ ]_{h_k}]_h [[\ ]_{h_{k+1}} \ldots [\ ]_{h_n}]_h$, *where $k \geq 1, n > k, h, h_1, \ldots, h_n \in H$ (strong division rules for non-elementary membranes).*

These rules are applied as usual in P systems.

Notice that in this polarizationless framework there is no cooperation, priority, nor changes of the labels of membranes. Also, throughout this chapter, rules of type (f) are used only for $k = 1, n = 2$, that is, rules of the form (f) $[[\ ]_{h_1}[\ ]_{h_2}]_h \to [[\ ]_{h_1}]_h [[\ ]_{h_2}]_h$. They can also be restricted to the case where they are controlled by the presence of a specific membrane, that is, rules of the form (g) $[[\ ]_{h_1}[\ ]_{h_2}[\ ]_p]_h \to [[\ ]_{h_1}[\ ]_p]_h [[\ ]_{h_2}[\ ]_p]_h$.

The class of recognizer polarizationless P systems with active membranes (resp., which do not make use of division rules) is denoted by $\mathcal{AM}^0$ (resp., $\mathcal{NAM}^0$), and $\mathcal{AM}^0(\alpha, \beta, \gamma, \delta)$, where $\alpha \in \{-d, +d\}$, $\beta \in D = \{-n, +nw, +ns, +nsw, +nsr\}$, $\gamma \in \{-e, +e\}$, and $\delta \in \{-c, +c\}$, denotes the class of all recognizer P systems with polarizationless active membranes such that:

(a) if $\alpha = +d$ (resp., $\alpha = -d$) then dissolution rules are permitted (resp., forbidden);
(b) if $\beta = +nw$ or $+ns$ (resp., $\beta = +nsw$) then division rules for elementary and non-elementary membranes, weak or strong (resp., weak and strong) are permitted; if $\beta = +nsr$ then division rules of the types (e), (f) and (g) are

permitted; if $\beta = -n$ then only division rules for elementary membranes are permitted.
(c) if $\gamma = +e$ (resp., $\gamma = -e$) then evolution rules are permitted (resp., forbidden);
(d) if $\delta = +c$ (resp., $\delta = -c$) then communication rules are permitted (resp., forbidden).

Proposition 12.5 can be adapted to polarizationless P systems with active membranes which do not make use of division nor evolution rules, providing a lower bound about their efficiency.

Proposition 12.11 $P \subseteq PMC_{\mathcal{NAM}^0(-d,-e,+c)}$.

12.4.2 A Conjecture of Păun

At the beginning of 2005, Gh. Păun (problem F from [40]) wrote:

My favorite question (related to complexity aspects in P systems with active membranes and with electrical charges) is that about the number of polarizations. Can the polarizations be completely avoided? The feeling is that this is not possible—and such a result would be rather sound: passing from no polarization to two polarizations amounts to passing from non-efficiency to efficiency.

This so-called Păun's conjecture can be formally formulated in terms of membrane computing complexity classes as follows:

$$P = PMC^{[*]}_{\mathcal{AM}^0(+d,-n,+e,+c)}$$

where the notation $PMC^{[*]}_{\mathcal{R}}$ indicates that the result holds for both $PMC_{\mathcal{R}}$ and $PMC^*_{\mathcal{R}}$.

Let Π be a recognizer polarizationless P system with active membranes which do not make use of dissolution rules. A directed graph can be associated with Π verifying the following property: every accepting computation of Π is characterized by the existence of a path in the graph between two specific nodes.

Each rule of Π can be considered as a *dependency relation* between the object triggering the rule and the object(s) produced by its application. We can consider a general pattern for rules of types $(a), (b), (c), (e)$ in the form $(a, h) \to (a_1, h')(a_2, h') \ldots (a_s, h')$, where the rules of type (a) correspond to the case $h = h'$, the rules of type (b) correspond to the case $h = f(h')$ and $s = 1$, the rules of type (c) correspond to the case $h' = f(h)$ and $s = 1$, and the rules of type (e) correspond to the case $h = h'$ and $s = 2$. A formal definition of the *dependency graph* associated with a P system can be found in [16].

Note that a P system can dynamically evolve according to its rules, but the dependency graph associated with it is static. Further, rules of the kind (f) and (g) do not provide any node nor arc to the dependency graph.

Let Δ_Π be the set of all pairs $(a, h) \in \Gamma \times H$ such that there exists a path (within the dependency graph) from (a, h) to (yes, env)—the environment is considered to be the output region, although the results obtained are also valid for any h_o. In [16] the following results are shown.

Proposition 12.12 *Let Π be a recognizer polarizationless P systems with active membranes not using dissolution rules, and where every kind of division rules is permitted. Then,*

- *There exists a Turing machine that constructs the dependency graph associated with Π in a time bounded by a polynomial function depending on the total number of rules and the maximum length of the rules.*
- *There exists a Turing machine that constructs the set Δ_Π in a time bounded by a polynomial function depending on the total number of rules and the maximum length of the rules.*

Given a family $\Pi = \{\Pi(n) \mid n \in \mathbf{N}\}$ of recognizer P systems solving a decision problem in a uniform way (with (cod, s) being the associated polynomial encoding), the acceptance of a given instance of the problem, w, can be characterized by using the set $\Delta_{\Pi(s(w))}$ associated with $\Pi(s(w))$.

Let $\overline{\mathcal{M}_j} = \{(a, j) \mid a \in \mathcal{M}_j\}$, for $1 \leq j \leq q$ and $\overline{m} = \{(a, h_i) \mid a \in m\}$, for each input multiset m over Σ (recall that h_i is the label of the input membrane). Then, the following holds [16]:

Proposition 12.13 *Let $X = (I_X, \theta_X)$ be a decision problem, and $\Pi = \{\Pi(n) \mid n \in \mathbf{N}\}$ a family of recognizer polarizationless P systems and not using dissolution rules solving X in a uniform way. Let (cod, s) be a polynomial encoding associated with that solution. Then, for each instance w of the problem X the following statements are equivalent:*

(a) $\theta_X(w) = 1$ *(that is, the answer to the problem is yes for w).*

(b) $\Delta_{\Pi(s(w))} \cap \left(\overline{cod(w)} \cup \bigcup_{j=1}^{q} \overline{\mathcal{M}_j} \right) \neq \emptyset$, *where $\mathcal{M}_1, \ldots, \mathcal{M}_q$ are the initial multisets of $\Pi(s(w))$.*

A similar result holds for semi-uniform solutions [16] and the following theorem can be deduced.

Theorem 12.3 $\mathbf{P} = \mathbf{PMC}^{[*]}_{\mathcal{AM}^0(-d,\beta,+e,+c)}$, where $\beta \in D$.

Thus, polarizationless P systems with active membranes which do not make use of dissolution rules are non–efficient in the sense that their cannot solve **NP**-complete problems in polynomial time (unless **P=NP**).

Let us now consider polarizationless P systems with active membranes making use of dissolution rules. Will it be possible to solve NP-complete problems in that framework?

In [28] a negative answer is given for the case that division rules are used only for elementary membranes and being *symmetric*, in the following sense $[a]_h \to [b]_h[b]_h$.

Theorem 12.4 $P = PMC_{\mathcal{AM}^0(+d,-n(sym),+e,+c)}^{[*]}$.

Several authors [4, 16] gave a positive answer when division for non-elementary membranes, in the strong sense, is permitted. The mentioned papers provide semi-uniform solutions in a linear time to SAT and Subset Sum, respectively. Thus, we have the following result:

Proposition 12.14 $NP \cup co\text{-}NP \subseteq PMC_{\mathcal{AM}^0(+d,+ns,+e,+c)}^{*}$.

As a consequence of Theorems 12.3 and 12.14, a *partial negative* answer to Păun's conjecture is given: assuming that $P \neq NP$ and making use of dissolution rules and division rules for elementary and non-elementary membranes, computationally hard problems can be efficiently solved avoiding polarizations. The answer is partial because efficient solvability of NP-complete problems by polarizationless P systems with active membranes making use of dissolution rules and division *only* for elementary membranes is unknown.

Theorem 12.14 was improved in [2] by giving a family of recognizer polarizationless P systems with active membranes using dissolution rules and division for elementary and (strong) non-elementary membranes solving QBF-SAT in a *uniform* way and in a linear time. Then,

Proposition 12.15 $PSPACE \subseteq PMC_{\mathcal{AM}^0(+d,+ns,+e,+c)}$.

Next, we present some results about the efficiency of polarizationless P systems with active membranes when evolution rules and/or communication rules are forbidden.

First, one can adapt a solution given in [4] to provide a semi-uniform solution to SAT in a linear time by a family of recognizer polarizationless P systems with active membranes by using evolution, dissolution and division rules for elementary and non-elementary membranes (both in the strong and weak versions), and avoiding communication rules. That is, we have the following:

Proposition 12.16 $NP \cup co\text{-}NP \subseteq PMC_{\mathcal{AM}^0(+d,\beta,+e,-c)}^{*}$, where $\beta \in \{+nw, +ns\}$.

Evolution and communication rules can be avoided without loss of efficiency. Indeed, in [57] a semi-uniform solution to 3-SAT in a linear time by a family of polarizationless recognizer P systems with active membranes by using only dissolution rules and division rules for elementary and non-elementary membranes of the types (e) and (f), is presented. Thus, the following holds:

Proposition 12.17 $NP \cup co\text{-}NP \subseteq PMC_{\mathcal{AM}^0(+d,+nsw,-e,-c)}^{*}$.

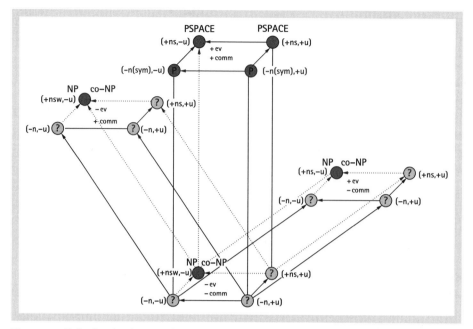

Fig. 12.1 Polarizationless active membranes *using dissolution rules*.

Moreover, Proposition 12.17 can be extended when non-elementary membrane division controlled by the presence of a membrane is allowed. In [24] it was presented a semi-uniform solution to QBF-3-SAT in a linear time by a family of polarizationless recognizer P systems with active membranes by using only dissolution rules and division rules of the types (e), (f), and (g). Thus, the following holds:

Proposition 12.18 $\text{PSPACE} \subseteq \text{PMC}^*_{\mathcal{AM}^0(+d,+nsr,-e,-c)}$.

Figure 12.1 graphically summarizes the results known related with complexity classes associated with polarizationless P systems with active membranes making use of dissolution rules. In the picture, $-u$ (resp. $+u$) means semi-uniform (resp. uniform) solutions, $-n$ (resp. $+ns$) means using division only for elementary membranes (resp. division for elementary and non-elementary membranes in the *strong* version). A standard class inside (respectively, over) a dark node means that the corresponding membrane computing class is equal (resp., is a lower bound) to the standard class.

12.4.3 Complexity and Uniformity Below P

In this section we look at recent results concerning polarizationless P systems with active membranes that use uniformity conditions that are significantly tighter than the usual polynomial time (P) uniformity conditions (Definition 12.5).

If we are solely interested in whether a P system solves intractable (e.g. NP- or PSPACE-hard) problems then P-uniformity (see Section 12.2.1) is a sufficient notion. However if we are interested in membrane systems whose computational efficiency is equal to, or less than P, then we should consider tighter (more restricted) uniformity. Otherwise, polynomial time membrane systems can (trivially) solve any P problem: in the definition of P uniformity (with input membrane), the function *cod* can (efficiently) solve any P problem (for P semi-uniform solutions the same is true for the function $w \to \Pi(w)$). This is quite different from circuit families, where we define the uniformity solely with respect to the input length n.

How tight should the uniformity condition be? This depends on the efficiency of the P system. For example, L-uniformity is sufficient if we can show (for example) that the system characterizes NL or P. As is the case elsewhere in complexity theory, it is desirable that results should hold under uniformity conditions that are significantly tighter than P.

This work was initiated in [28, 29], and more thoroughly developed in [30].

The readers are assumed to be familiar with the complexity classes L (logspace), NL (non-deterministic logspace), NC [35]. Let $AC = \bigcup_{h \in \mathbb{N}} AC^k$, where AC^k is the class of problems accepted by unbounded fan-in, uniform circuits of depth $O(\log n)^k$, and polynomial size (unbounded fan-in AND and OR gates have an arbitrary (polynomial) number of inputs). Classes within P to define uniformity conditions on membrane families are used: the "smallest" such class being AC^0.

Naturally our AC^0 circuit families should themselves be uniform [22, 35]. Here concurrent random access machines (CRAMs) that run in constant time are used, since they characterize uniform AC^0 [27]. The following inclusions are well known: $AC^0 \subsetneq NC^1 \subseteq L \subseteq NL \subseteq NC = AC \subseteq P$ [35, 7, 13, 22].

12.4.3.1 Uniformity for membrane systems

Here L and AC^0 uniform families of membrane systems are described. Rather than giving full definitions, we refer the reader to [30] and merely observe that in order to get L-uniformity, Definition 12.5 is modified so that "polynomial time" is replaced by "logspace". More precisely, in the case of systems with input membrane (L uniform solutions), each of the three functions $cod, s, n \to \Pi(n)$ are computable in deterministic space $O(\log n)$. Analogously, for membrane systems without input membrane (L-semi-uniform solutions), Definition 12.5 is modified so that the function $w \to \Pi(w)$ is computable in deterministic space $O(\log n)$. Similarly, to get AC^0-uniformity, we modify the definitions so that the relevant functions are constant time CRAM computable, instead of being from **P**. Uniformity conditions utilizing other complexity classes would be defined analogously.

Essentially, the idea is to use uniformity conditions that are as tight as is necessary to see the true efficiency of the membrane system itself, as opposed to observing the efficiency of the functions cod, s, $n \to \Pi(n)$, or $w \to \Pi(w)$. Note that in [30] the notion of acceptance differs from that in Definition 12.4(c), and the function s is the identity.

Notation (AC^0)–$PMC_{\mathcal{R}}^{[*]}$ means that in the definition of the corresponding complexity class AC^0 uniformity is considered instead of **P** uniformity.

12.4.3.2 Active membranes with and without dissolution

We begin by looking at some results on active membranes without polarizations and without dissolution. Let $\beta \in D$. Theorem 12.3 states that $\mathbf{P} = \mathbf{PMC}_{\mathcal{AM}^0(-d,\beta,+e,+c)}^{[*]}$. However in [30] it was shown that under tighter uniformities the actual efficiency of the membrane system is significantly weakened.

Proposition 12.19 (AC^0)–$PMC_{\mathcal{AM}^0(-d,\beta,+e,+c)}^{[*]} \subseteq \mathbf{NL}$.

For the case of semi-uniform solutions, the converse inclusion has also been shown [30], giving a characterization of **NL**:

Theorem 12.5 $\mathbf{NL} = (AC^0)$–$PMC_{\mathcal{AM}^0(-d,\beta,+e,+c)}^{*}$.

It is an open problem to find a characterization for the case of uniform solutions. If we could prove that such systems characterize a class conjectured to be a strict subset of **NL**, then it would be the first time we have shown a strong difference in the efficiency of P systems with and without input.

So some membrane systems that characterize **P**, when **P** uniformity is used, seemingly are less efficient under tighter uniformity. Others characterize **PSPACE**.

It turns out that the remaining systems, i.e. with dissolution and elementary division, have a **P** lower bound under tight uniformity [30].

Proposition 12.20 $\mathbf{P} \subseteq (AC^0)$–$PMC_{\mathcal{AM}^0(+d,\beta,+e,+c)}^{[*]}$.

Of the systems that fall under the conjecture of Păun, Theorem 12.4 (symmetric division) describes the most general that has been shown to characterize **P** under tight uniformity conditions [28]. Also, the known **PSPACE** characterizations (Section 12.4.2) hold under AC^0 uniformity [30].

12.5 MINIMALLY PARALLEL MODE

In [12] a more relaxed strategy of using the rules is considered, the so-called *minimal parallelism*: in each region where at least a rule can be applied, at least one rule must be applied (if there is no conflict with the objects), without any other restriction. This introduces an additional degree of non-determinism in the system evolution.

P systems with active membranes working in the minimally parallel mode means the following:

- All the rules of any type involving a membrane h form the set R_h, this means all the rules of the form $[a \to v]_h^\alpha$, $a[\ ]_h^\alpha \to [b]_h^\beta$, $[a\ ]_h^\alpha \to [\ ]_h^\beta b$, $[a]_h^\alpha \to z$ and $[a]_h^\alpha \to [b]_h^\beta[c]_h^\gamma$, with the same h, constitute the set R_h.
- If a membrane h appears several times in a given configuration of the system, then for each occurrence of the membrane we consider different copies of the set R_h.
- Then, in each step, from each set R_h associated with each membrane labeled by h, from which at least a rule *can* be used, at least one rule *must* be used, provided that there is no conflict with the objects. For example, if we have only an object a in membrane h and we have an evolution rule $[a \to b]_h$ and a send-in rule $a[\ ]_{h'} \to [c]_{h'}$, where h' is the label of a membrane immediately inside membrane h, then we can apply at least a rule from R_h and from $R_{h'}$, but we will apply only one between these two rules.

Of course, as usual for P systems with active membranes, each membrane and each object can be involved in only one rule, and the choice of rules to use and of objects and membranes to evolve is done in a non-deterministic way. In each step, the use of rules is done in the bottom-up manner (first the inner objects and membranes evolve, and the result is duplicated if any surrounding membrane is divided).

In [12] efficient semi-uniform solutions to SAT were obtained in the new framework by using P systems with active membranes and with polarizations.

Theorem 12.6 *The SAT problem can be solved in a semi-uniform way and in a linear time by P systems with active membranes, without dissolution rules and using (weak) division for non-elementary membranes, and working in the minimally parallel mode.*

Next, we define new classes of P systems related to $\mathcal{AM}^0$. Let $\alpha \in \{-d, +d\}$, $\beta \in D = \{-n, +nw, +ns, +nsw, +nsr\}$, $\gamma \in \{-e, +e\}$, $\delta \in \{-c, +c\}$, and $\epsilon \in \{m, M, md, Md\}$. Then we denote by $\mathcal{AM}^0(\alpha, \beta, \gamma, \delta, \epsilon)$ the class of recognizer polarizationless P systems with active membranes such that:

- the values of α, β, γ, δ with the same meaning as in Section 12.4.1.
- $\epsilon = m$: working in the minimally parallel mode.

- $\epsilon = md$: working in the deterministic minimally parallel mode.
- $\epsilon = M$: working in the maximally parallel mode.
- $\epsilon = Md$: working in the deterministic maximally parallel mode.

Next result is a consequence of working with confluent P systems.

Proposition 12.21 *The following two inclusions hold:*

(1) $\text{PMC}_{\mathcal{AM}^0(a,\beta,\gamma,\delta,m)} \subseteq \text{PMC}_{\mathcal{AM}^0(a,\beta,\gamma,\delta,M)}$.
(2) $\text{PMC}^*_{\mathcal{AM}^0(a,\beta,\gamma,\delta,m)} \subseteq \text{PMC}^*_{\mathcal{AM}^0(a,\beta,\gamma,\delta,M)}$.

A version of Păun's conjecture for P systems working in the minimally parallel mode can be proposed.

The class of all decision problems solvable in polynomial time by polarizationless P systems with active membranes using evolution, communication, dissolution, and division rules for elementary membranes (working in the minimally parallel mode*) is equal to the class* **P**.

Next, we study partial answers to the new version of Păun's conjecture.

The concept of *dependency graph* associated with a P system can be extended to P systems working in the minimally parallel mode.

In each transition step of such a kind of P system, it can be thought that objects are assigned to rules, non-deterministically choosing the rules and the objects assigned to each rule, according to the semantic of the minimally parallel mode. The objects which remain unassigned are left where they are, and they are passed unchanged to the next configuration (and belonging to the same membrane because dissolution rules are not permitted). So, in this kind of P system working in the minimally parallel mode (without using dissolution rules) we can pass from a node (a, h) to another node (a', h') in the dependency graph if and only if there exists a transition step producing (a', h') from (a, h).

Thus, Theorem 12.3 has the following version in the new context.

Theorem 12.7 $\mathbf{P} = \text{PMC}^{[*]}_{\mathcal{AM}^0(-d,\beta,\gamma,\delta,md)} = \text{PMC}^{[*]}_{\mathcal{AM}^0(-d,\beta,\gamma,\delta,m)}$.

In [26] a linear time semi-uniform solution to SAT through polarizationless P systems with active membranes making use of dissolution rules and working in the *minimally parallel mode* is given. Furthermore, the solution presented is deterministic.

Proposition 12.22 *The SAT problem can be solved in a semi-uniform way and in a linear time by polarizationless P systems with active membranes, using evolution, communication, dissolution, and (weak) division rules for non-elementary membranes, and working in the* deterministic minimally parallel mode.

Thus, we have the following:

Proposition 12.23 $\mathbf{NP} \cup \mathbf{co\text{-}NP} \subseteq \text{PMC}^*_{\mathcal{AM}^0(+d,+nw,+e,+c,md)}$.

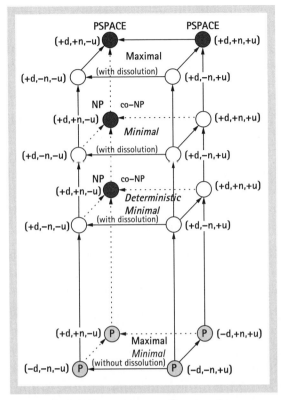

Fig. 12.2 Polarizationless active membranes using dissolution rules.

Figure 12.2 describes the results known related with Păun's conjecture in both modes, where $-u$ (resp. $+u$) means semi-uniform (resp. uniform) solutions, $-ne$ (resp. $+ne$) means using division only for elementary membranes (resp. division for elementary and non-elementary membranes, *strong* in the maximal parallelism and *weak* in the minimal parallelism). Through this graph, we try to specify whether or not it is possible to solve computationally hard problems by recognizer P systems of the class associated with each node. The direction of each arrow shows a relation of inclusion, and each blank node provides an open question.

12.6 MEMBRANE CREATION

In this section we investigate the behavior of P systems with active membranes (from the computational complexity point of view) when we substitute membrane division rules for other rules able to produce new membranes from objects (under the influence of existing chemical substances).

The *membrane creation* was first considered in [21] and [31], inspired by the fact that when a compartment becomes too large, it often happens that new membranes appear inside it, more or less spontaneously or during biological evolution [37].

The first definition of P systems with membrane creation is developed considering transition P systems with rules for membrane creation of the form $[a \rightarrow [u]_{h_1}]_{h_2}$, where a is an object, u is a string representing a multiset of objects, and h_1, h_2 are labels. When such a rule is applied in a membrane labeled by h_2, the object a is replaced by a new membrane, with the label h_1 and the contents as indicated by u. Thus, by applying a membrane creation rule there is no replication of objects into the new membrane, which becomes a daughter of the original membrane (the depth of the membrane structure can increase).

Definition 12.13 *A polarizationless P system with membrane creation of degree $q \geq 1$ is a tuple $\Pi = (\Gamma, H, \mu, w_1, \ldots, w_q, R, h_o)$, where all components are as in a general P system and R is a finite set of rules of the following forms:*

(a) $[a \rightarrow u]_h$, for $h \in H$, $a \in \Gamma$, $u \in \Gamma^*$ (object evolution rules).
(b) $a[\ ]_h \rightarrow [b]_h$, for $h \in H$, $a \in \Gamma$, $b \in \Gamma \cup \{\lambda\}$ (send-in communication rules).
(c) $[a]_h \rightarrow [\ ]_h b$, for $h \in H$, $a \in \Gamma$, $b \in \Gamma \cup \{\lambda\}$ (send-out communication rules).
(d) $[a]_h \rightarrow b$, for $h \in H$, $a \in \Gamma$, $b \in \Gamma \cup \{\lambda\}$ (dissolution rules).
(e) $[a \rightarrow [u]_{h_1}]_{h_2}$, for $h_1, h_2 \in H$, $a \in \Gamma$, and $u \in \Gamma^*$. *In reaction with an object, a new membrane is created. This new membrane is placed inside the same membrane where the object which triggers the rule was, and has associated an initial multiset and a label* (creation rules).

These rules are applied according to usual principles of polarizationless P systems (see [15] for details).

We denote by $\mathcal{MC}^0$ the class of recognizer polarizationless P systems with membrane creation, and we denote by $\mathcal{MC}^0(\alpha, \gamma, \delta)$, where $\alpha \in \{-d, +d\}$, $\gamma \in \{-e, +e\}$ and $\delta \in \{-c, +c\}$, the class of all recognizer polarizationless P systems with membrane creation such that:

(a) if $\alpha = +d$ (resp. $\alpha = -d$) then dissolution rules are permitted (resp. forbidden);
(b) if $\gamma = +e$ (resp. $\gamma = -e$) then evolution rules are permitted (resp. forbidden);
(c) if $\delta = +c$ (resp. $\delta = -c$) then communication rules are permitted (resp. forbidden).

Efficient *uniform* solutions to computationally hard problems have been given in the framework of polarizationless P systems with membrane creation: Subset Sum [15], SAT [20].

COMPLEXITY: MEMBRANE DIVISION/CREATION

In order to illustrate these results, a family of recognizer P systems with membrane creation solving SAT in a uniform way and in a linear time is shown next.

12.6.1 A Uniform Solution to SAT using Membrane Creation

The solution is addressed via a brute force algorithm which consists of the following phases:

- Generation and Evaluation Stage: Using membrane creation we generate membranes representing all possible assignments associated with the formula and evaluate it on each one.
- Checking Stage: In each membrane we check whether or not the formula evaluates true on the assignment associated with it.
- Output Stage: The system sends out to the environment the right answer according to the previous stage.

For each $m, n \in \mathbb{N}$, the P system of degree 2 with input and with external output $\Pi(\langle m, n \rangle) = (\Gamma(m, n), \Sigma(m, n), H, \mu, \mathcal{M}_a, \mathcal{M}_t, R(m, n), h_i)$, is defined as follows:

- $\Sigma(m, n) = \{x_{i,j}, \overline{x}_{i,j} \mid 1 \leq i \leq m, 1 \leq j \leq n\}$.
- $\Gamma(m, n) = \Sigma(m, n) \cup \{q, k_0, k_1, k_2, t_0, t_1, t_2, t_3\}$
  $\cup \{x_{i,j,t}, x_{i,j,f}, \overline{x}_{i,j,t}, \overline{x}_{i,j,f} \mid 1 \leq i \leq n, 1 \leq j \leq m\}$
  $\cup \{r_j, r_{j,t}, r_{j,f}, d_j \mid 1 \leq j \leq m\}$
  $\cup \{z_i, z_{i,t}, z_{i,f} \mid 0 \leq i \leq n\}$
  $\cup \{yes, no, yes_i, no_j \mid 0 \leq i \leq 9, 0 \leq j \leq 2n + 11\}$,
- $H = \{a, t, f, 1, \ldots, m\}$,
- $\mu = [\ [\ ]_a\ ]_t$,
- $\mathcal{M}_a = no_0$ and $\mathcal{M}_t = z_{0,t}\, z_{0,f}$,
- $h_i = t$.
- The list of rules in $R(m, n)$ is given below, together with some short comments on their purpose and functioning.

  (a) $\left.\begin{array}{l}[z_{j,t} \to [z_{j+1}\, k_0]_t]_l \\ [z_{j,f} \to [z_{j+1}\, k_0]_f]_l\end{array}\right\}$ for $\begin{array}{l}l = t, f \\ j = 0, \ldots, n - 2\end{array}$

  The goal of these rules is to create one membrane for each assignment to the variables of the formula. The new membrane with label t, where the object z_{j+1} is placed, represents the assignment $x_{j+1} = true$; on the other hand the new membrane with label f, where the object z_{j+1} is placed represents the assignment $x_{j+1} = false$.

  (b) $\left.\begin{array}{l}[x_{ij} \to x_{i,j,t} x_{i,j,f}]_l \\ [\overline{x}_{i,j} \to \overline{x}_{i,j,t} \overline{x}_{i,j,f}]_l \\ [r_i \to r_{i,t} r_{i,f}]_l \\ [z_k \to z_{k,t} z_{k,f}]_l\end{array}\right\}$ for $\begin{array}{l}l = t, f \\ k = 0, \ldots, n - 1 \\ i = 1, \ldots, m \\ j = 1, \ldots, n.\end{array}$

These rules duplicate the objects representing the variables, so they can be evaluated on the two possible assignments, $x_j = true$ ($x_{i,j,t}, \bar{x}_{i,j,t}$) and $x_j = false$ ($x_{i,j,f}, \bar{x}_{i,j,f}$). The objects r_i are also duplicated ($r_{i,t}, r_{i,f}$) in order to keep track of the clauses that evaluate true on the previous assignments to the variables. Finally the objects z_k produce the objects $z_{k,t}$ and $z_{k,f}$ which will create the new membranes representing the two possible assignments for the next variable.

(c) $\left.\begin{array}{l} x_{i,1,t}[\,]_t \to [r_i]_t, \quad \bar{x}_{i,1,t}[\,]_t \to [\lambda]_t \\ x_{i,1,f}[\,]_f \to [\lambda]_f, \quad \bar{x}_{i,1,f}[\,]_f \to [r_i]_f. \end{array}\right\}$ for $i = 1, \ldots, m$.

According to these rules the formula is evaluated in the two possible assignments for the variable that is being analyzed. The objects $x_{i,1,t}$ (resp. $\bar{x}_{i,1,f}$) get into the membranes labeled by t (resp. f) being transformed into the objects r_i representing that the clause number i evaluates true on the assignment $x_j = true$ (resp. $x_j = false$). On the other hand the objects $\bar{x}_{i,1,t}$ (resp. $x_{i,1,f}$) get into the membranes labeled by f (resp. t) producing no objects. This represents that these objects do not make the clause true in the assignment $x_{j+1} = true$ (resp. $x_{j+1} = false$).

(d) $\left.\begin{array}{l} x_{i,j,t}[\,]_t \to [x_{i,j-1}]_t, \quad \bar{x}_{i,j,t}[\,]_t \to [\bar{x}_{i,j-1}]_t \\ x_{i,j,f}[\,]_f \to [x_{i,j-1}]_f, \quad \bar{x}_{i,j,f}[\,]_f \to [\bar{x}_{i,j-1}]_f \\ r_{i,t}[\,]_t \to [r_i]_t, \quad\quad\quad\; r_{i,f}[\,]_f \to [r_i]_f \end{array}\right\}$ for $\begin{array}{l} i = 1, \ldots, m \\ j = 2, \ldots, n. \end{array}$

In order to analyze the next variable the second subscripts of the objects $x_{i,j,l}$ and $\bar{x}_{i,j,l}$ are decreased when they are sent into the corresponding membrane labeled by l. Moreover, following the last rule, the objects $r_{i,l}$ get into the new membranes to keep track of the clauses that evaluate true on the previous assignments.

(e) $\left.\begin{array}{l} [k_s \to k_{s+1}]_l \\ [k_2]_l \to \lambda \end{array}\right\}$ for $\begin{array}{l} l = t, f \\ s = 0, 1. \end{array}$

The objects k_i for $i = 0, 1, 2$ are counters that dissolve membranes when they are not useful anymore during the rest of the computation.

(f) $\left.\begin{array}{l} [z_{n-1,t} \to [z_n]_t]_l, \quad [z_{n-1,f} \to [z_n]_f]_l \\ [z_n \to d_1 \ldots d_m q]_l \end{array}\right\}$ for $l = t, f$.

At the end of the generation stage the objects $z_{n-1,l}$ create two new membranes where the formula will be evaluated on the two possible assignments for the last variable x_n. The object z_n is placed in both membranes and will produce the objects $d_1, \ldots, d_m$ and yes_0, which will take part in the checking stage.

(g) $\left.\begin{array}{l} [d_i \to [t_0]_i]_l \\ r_{i,t}[\,]_i \to [r_i]_i, \quad [r_i]_i \to \lambda \\ [t_s \to t_{s+1}]_i, \quad [t_2]_i \to t_3 \end{array}\right\}$ for $\begin{array}{l} i = 1, \ldots, m \\ s = 0, 1. \end{array}$

Following these rules each object d_i creates a new membrane with label i where the object t_0 is placed; this object will act as a counter. The object r_i gets into the membrane labeled by i and dissolves it preventing the counter, t_i, from reaching the object t_2. The fact that the object t_2 appears in a membrane with label i means that there is no object r_i, that is, the clause number i does not evaluate true on the assignment associated with the membrane; therefore the formula evaluates false on the associated assignment.

(h) $\left.\begin{array}{l} [q \to [yes_0]_a]_l \\ t_3[\,]_a \to [t_3]_a \quad\quad [t_3]_a \to \lambda \\ [yes_h \to yes_{h+1}]_a, \quad [yes_5]_a \to yes_6 \\ [yes_6]_l \to yes_7[\,]_l \end{array}\right\}$ for $\begin{array}{l} l = t, f \\ h = 0, \ldots, 4. \end{array}$

The object q creates a membrane with label a where the object yes_0 is placed. The object yes_h evolves to the object yes_{h+1}; at the same time the objects t_3 can get into the membrane labeled by a and dissolve it preventing the object yes_6 from being sent out from this membrane.

(i) $\left.\begin{array}{l} [no_p \to no_{p+1}]_a, \quad [no_{2n+10}]_a \to no_{2n+11} \\ [no_{2n+11}]_t \to no[\,]_t \\ yes_7[\,]_a \to [yes_8]_a, \quad [yes_8]_a \to yes_9 \\ [yes_9]_t \to yes[\,]_t \end{array}\right\}$ for $p = 0, \ldots, 2n + 9$.

From the beginning of the computation the object no_p evolves to the object no_{p+1} inside the membrane labeled by a. If any object yes_7 is produced during the computation, which means that the formula evaluates true on some assignment to its variables, it gets into this membrane and dissolves it producing the object yes_9 that will send out to the environment the object yes. On the other hand if no object yes_7 appears in the skin the object no_{2n+10} will dissolve the membrane labeled by a producing the object no_{2n+11} that will send out to the environment the object no.

Finally, let (cod, s) be a polynomial encoding of SAT in $\Pi = \{\Pi(\langle m, n\rangle) \mid m, n \in \mathbb{N}\}$ defined as $s(\varphi) = \langle m, n\rangle$ and $cod(\varphi) = \{x_{i,j} \mid x_j \in C_i\} \cup \{\overline{x}_{i,j} \mid \neg x_{i,j} \in C_i\}$, for any Boolean formula in CNF, $\varphi = C_1 \wedge \cdots \wedge C_m$ such that $Var(\varphi) = \{x_1, \ldots, x_n\}$.

The family Π provides a uniform solution to the SAT problem according to Definition 12.10. Although the proof is omitted here (see [20] for more details), let us highlight that the necessary resources to build the family are of the order $\Theta(m \cdot n)$. Moreover, the designed P systems are confluent and the total number of steps performed by any computation of $\Pi(s(\varphi))$ with input $cod(\varphi)$ is $2n + 11$ if the output is yes and $2n + 12$ if the output is no.

From the above we have the following result.

Proposition 12.24 SAT $\in$ PMC$_{\mathcal{MC}^0(+d, +e, +c)}$.

Since $\text{PMC}_\mathcal{R}$ is closed under complement and polynomial time reductions, for any class $\mathcal{R}$ of recognizer P systems, the following result is obtained.

Proposition 12.25 $\text{NP} \cup \text{co-NP} \subseteq \text{PMC}_{\mathcal{MC}^0(+d,+e,+c)}$.

Moreover, an improved result was obtained in [18] by providing a uniform solution to QBF-SAT.

Proposition 12.26 $\text{PSPACE} \subseteq \text{PMC}_{\mathcal{MC}^0(+d,+e,+c)}$.

12.7 Non-Deterministic Complexity Classes

P systems are non-deterministic computational devices but in order to capture a true algorithmic concept, a restrictive definition of acceptance has been considered: an input is accepted (in a confluent way) by a P system if and only if every computation answers yes.

In this section polynomial complexity classes associated with P systems with the same acceptance policy that non-deterministic Turing machines have are studied. That is, an input is accepted (in a non-confluent way) by a P system if there is *some* computation of the P system that answers yes.

In this new context, non-deterministic complexity classes in P systems without requiring them to be confluent are considered.

Definition 12.14 *A decision problem $X = (I_X, \theta_X)$ is non-deterministically solvable in polynomial time by a family of recognizer P systems without input membrane $\Pi = \{\Pi(w) \mid w \in I_X\}$, and we denote it by $X \in \text{NPMC}^*_\mathcal{R}$, if the following holds:*

- *The family Π is polynomially uniform by Turing machines.*
- *The family Π is polynomially bounded, that is, there is a natural number $k \in \mathbb{N}$ such that the minimum length of the accepting computations of $\Pi(w)$ is, at most, $|w|^k$.*
- *For each instance of the problem $w \in I_X$, $\theta_X(w) = 1$ if and only if there exists an accepting computation of $\Pi(w)$.*

The family Π is said to provide a *semi-uniform solution* to the problem X (in a non-deterministic or non-confluent manner).

In contrast to the corresponding definition for standard membrane computing complexity classes, in this definition we only demand that for each instance w with affirmative answer there exists at least *one* accepting computation of the system

$\Pi(w)$, instead of demanding *every* computation of the system to be an accepting one.

The class $\mathbf{NPMC}^*_{\mathcal{R}}$ is closed under polynomial time reduction, but notice that it is not necessarily closed under complement.

A family of recognizer basic transition P systems solving HPP (the directed version of the Hamiltonian Path Problem with two distinguished nodes) in a *linear time*, in a non-deterministic manner, is provided in [49]:

Proposition 12.27 HPP $\in \mathbf{NPMC}^*_{\mathcal{T}}$, and NP $\subseteq \mathbf{NPMC}^*_{\mathcal{T}}$.

In a similar way, non-deterministic complexity classes for recognizer membrane systems with input can be defined.

Definition 12.15 *A decision problem $X = (I_X, \theta_X)$ is non-deterministically solvable in polynomial time by a family of recognizer P systems with input membrane $\Pi = \{\Pi(n) \mid n \in \mathbf{N}\}$, and we denote it by $X \in \mathbf{NPMC}_{\mathcal{R}}$, if the following holds:*

- *The family Π is polynomially uniform by Turing machines.*
- *There exists a polynomial encoding (cod, s) of X in Π such that:*
  - *The family Π is polynomially bounded with respect to (X, cod, s), that is, there is a natural number $k \in \mathbf{N}$ such that the minimum length of the accepting computations of $\Pi(s(w))$ with input $cod(w)$ is, at most, $|w|^k$.*
  - *For each instance of the problem $w \in I_X$, $\theta_X(w) = 1$ if and only if there exists an accepting computation of $\Pi(s(w))$ with input $cod(w)$.*

The family Π is said to provide a *uniform solution* to the problem X (in a non-deterministic or non-confluent manner).

The class $\mathbf{NPMC}_{\mathcal{R}}$ is closed under polynomial time reduction, but it does not have to be closed under complement.

In [49] a family of recognizer basic transition P systems solving SAT in *constant time* (specifically, in *two steps*), in a non-deterministic manner, is presented. That is:

Proposition 12.28 SAT $\in \mathbf{NPMC}_{\mathcal{T}}$, and NP $\subseteq \mathbf{NPMC}_{\mathcal{T}}$.

12.8 CONCLUSIONS

In this chapter we have described some motivations, the basic concepts, and the main results that pertain to pioneering computational complexity in the membrane computing field.

We conclude by presenting new research directions within membrane computing complexity theory by listing some of the current open questions.

(A) Are there significant differences between uniform and semi-uniform solutions? Namely, is there some class $\mathcal{R}$ of recognizer P systems such that the inclusion $\mathbf{PMC}_\mathcal{R} \subseteq \mathbf{PMC}^*_\mathcal{R}$ is strict? We do not yet have a lower bound on the power of membrane systems with input membrane, without dissolution, and of course with uniformity that is tighter than **P**. These systems seem quite weak, and it might well be possible to characterize some (conjectured strict) subset of **NL**. If so, this would be the first time uniformity and semi-uniformity were shown to have different computational power.

(B) Class **P** is characterized by the complexity classes $\mathbf{PMC}^{[*]}_\mathcal{T}$ and $\mathbf{PMC}^{[*]}_{\mathcal{NAM}}$. It would be interesting to study non-deterministic versions of these results. That is, determine whether or not $\mathbf{NP} = \mathbf{NPMC}^{[*]}_\mathcal{T}$ and/or $\mathbf{NP} = \mathbf{NPMC}^{[*]}_{\mathcal{NAM}}$.

(C) Efficient uniform solutions to **NP**-complete problems have been given by models of $\mathcal{AM}(-n)$. Is it possible to efficiently solve **PSPACE**-complete problems by using families of P systems from $\mathcal{AM}(-n)$?

(D) What is the efficiency of P systems with active membranes and electrical charges where evolution and communication rules are forbidden? Are there some relations with the results obtained for polarizationless P systems?

(E) Dissolution rules provide a borderline between tractability and intractability in the framework of polarizationless P systems with active membranes making use of division rules for elementary and non-elementary membranes. What happens if division for only elementary membranes is allowed? We already have a **P** lower bound, is it possible to find a **P** upper bound? Is $\mathbf{P} = \mathbf{PMC}^{[*]}_{\mathcal{AM}^0(+d,-n,+e,+c)}$ true?

(F) It is well known that $\mathbf{PSPACE} \subseteq \mathbf{PMC}^*_{\mathcal{AM}^0(+d,+nsr,-e,-c)}$. Determine an upper bound for that membrane computing complexity class.

(G) In the framework $\mathcal{AM}^0(+d, +n)$ evolution and communication rules can be avoided without loss of efficiency. Can it be extended to P systems belonging to $\mathcal{MC}(+d)$? That is, does $\mathbf{NP} \subseteq \mathbf{PMC}^{[*]}_{\mathcal{MC}^0(+d,-e,-c)}$ hold?

(H) There are already a number of **P** and **PSPACE** characterizations. Using uniformity that is tighter than **P**, it turns out that some systems characterize **NL**. Can we characterize other interesting complexity classes besides **NL**, **P**, and **PSPACE**?

(I) Complexity classes can be extended in a natural way to other areas of membrane systems (tissue P systems, spiking neural P systems, etc.). Tissue P systems with symport/antiport rules which solve a decision problem can be efficiently simulated by a family of basic transition P systems solving the same problem [9]. **NP**-complete problems can be efficiently solved by tissue P systems with cell division [10] and spiking neural P systems [25].

Borderlines between efficiency and non-efficiency in these new areas should be provided.

(J) The idea of using as tight as possible uniformity conditions can be immediately applied to existing complexity results for different types of membrane systems (tissue P systems, spiking neural P systems, etc.). This could give rise to new and interesting characterizations (within P) of other complexity classes in membrane systems.

ACKNOWLEDGEMENTS

The authors acknowledge the support of the project TIN2006-13425 of the Ministerio de Educación y Ciencia of Spain, co-financed by FEDER funds, and the support of the project of excellence TIC-581 of the Junta de Andalucía. Damien Woods thanks his co-author Niall Murphy who worked on key ideas for the definitions and results discussed in Section 12.4.3.

REFERENCES

[1] A. ALHAZOV, R. FREUND: On efficiency of P systems with active membranes and two polarizations. *Lecture Notes in Computer Sci.*, 3365 (2005), 81–94.

[2] A. ALHAZOV, M.J. PÉREZ–JIMÉNEZ: Uniform solution of QSAT using polarizationless active membranes. *Lecture Notes in Computer Sci.*, 4664 (2007), 122–133.

[3] A. ALHAZOV, R. FREUND, GH. PĂUN: P systems with active membranes and two polarizations. *Second Brainstorming Week on Membrane Computing* (Gh. Păun et al. eds.), Report RGNC 01/04, 2004, 20–35.

[4] A. ALHAZOV, L. PAN, GH. PĂUN: Trading polarizations for labels in P systems with active membranes. *Acta Informaticae*, 41 (2004), 111–144.

[5] A. ALHAZOV, C. MARTÍN-VIDE, L. PAN: Solving graph problems by P systems with restricted elementary active membranes. *Lecture Notes in Computer Sci.*, 2950 (2004), 1–22.

[6] A. ALHAZOV, C. MARTÍN-VIDE, L. PAN: Solving a PSPACE–complete problem by recognizing P systems with restricted active membranes. *Fundamenta Informaticae*, 58 (2003), 67–77.

[7] J.L. BALCÁZAR, J. DÍAZ, J. GABARRÓ: *Structural Complexity, Vols I and II*. Springer, 1988.

[8] D. DÍAZ-PERNIL, M.A. GUTIÉRREZ-NARANJO, M.J. PÉREZ-JIMÉNEZ, A. RISCOS-NÚÑEZ: A logarithmic bound for solving Subset Sum with P systems. *Lecture Notes in Computer Sci.*, 4860 (2007), 257–270.

[9] D. DÍAZ-PERNIL, M.A. GUTIÉRREZ-NARANJO, M.J. PÉREZ-JIMÉNEZ, A. ROMERO-JIMÉNEZ: Efficient simulation of tissue-like P systems by transition cell-like P systems. *Natural Computing*, to appear.

[10] D. DÍAZ-PERNIL, M.A. GUTIÉRREZ-NARANJO, M.J. PÉREZ-JIMÉNEZ, A. RISCOS-NÚÑEZ: A uniform family of tissue P systems with cell division solving 3–COL in a linear time. *Theoretical Computer Sci.*, 404 (2008), 76–87.

[11] M.R. GAREY, D.S. JOHNSON: *Computers and Intractability. A Guide to the Theory of NP-completeness*. W.H. Freeman and Company, New York, 1979.

[12] G. CIOBANU, L. PAN, GH. PĂUN, M.J. PÉREZ-JIMÉNEZ: P systems with minimal parallelism. *Theoretical Computer Sci*, 378 (2007), 117–130.

[13] R. GREENLAW, H.J. HOOVER, W.L. RUZZO: *Limits to Parallel Computation: P-completeness Theory*. Oxford University Press, Oxford, 1995.

[14] M.A. GUTIÉRREZ-NARANJO, M.J. PÉREZ-JIMÉNEZ, A. RISCOS-NÚÑEZ: A fast P system for finding a balanced 2-partition. *Soft Computing*, 9 (2005), 673–678.

[15] M.A. GUTIÉRREZ-NARANJO, M.J. PÉREZ-JIMÉNEZ, F.J. ROMERO-CAMPERO: A linear solution of Subset Sum problem by using membrane creation. *Lecture Notes in Computer Sci.*, 3561 (2005), 258–267.

[16] M.A. GUTIÉRREZ-NARANJO, M.J. PÉREZ-JIMÉNEZ, A. RISCOS-NÚÑEZ, F.J. ROMERO-CAMPERO: On the power of dissolution in P systems with active membranes. *Lecture Notes in Computer Sci.*, 3850 (2006), 224–240.

[17] M.A. GUTIÉRREZ-NARANJO, M.J. PÉREZ-JIMÉNEZ, A. RISCOS-NÚÑEZ, F.J. ROMERO-CAMPERO, A. ROMERO-JIMÉNEZ: Characterizing tractability by cell–like membrane systems. *Formal Models, Languages and Applications* (K.G. Subramanian et al., eds.), World Scientific, Singapore, 2006, 137–154.

[18] M.A. GUTIÉRREZ-NARANJO, M.J. PÉREZ-JIMÉNEZ, F.J. ROMERO-CAMPERO: A linear time solution for QSAT with membrane creation. *Lecture Notes in Computer Sci.*, 3850 (2006), 241–252.

[19] M.A. GUTIÉRREZ-NARANJO, M.J. PÉREZ-JIMÉNEZ, A. RISCOS-NÚÑEZ, F.J. ROMERO-CAMPERO: Computational efficiency of dissolution rules in membrane systems. *Intern. J. Computer Math.*, 83 (2006), 593–611.

[20] M.A. GUTIÉRREZ-NARANJO, M.J. PÉREZ-JIMÉNEZ, F.J. ROMERO-CAMPERO: A uniform solution to SAT using membrane creation. *Theoretical Computer Sci.*, 371 (2007), 54–61.

[21] M. ITO, C. MARTÍN-VIDE, GH. PĂUN: A characterization of Parikh sets of EToL languages in terms of P systems. *Words, Semigroups and Transducers* (M. Ito et al., eds.), World Scientific, Singapore, 2001, 239–254.

[22] D.S. JOHNSON: A catalog of complexity classes. *Handbook of Theoretical Computer Science*, Elsevier, Amsterdam, 1990.

[23] S.N. KRISHNA, R. RAMA: A variant of P systems with active membranes: Solving NP-complete problems. *Romanian J. Information Science and Technology*, 2 (1999), 357–367.

[24] A. LEPORATI, C. FERRETTI, G. MAURI, M.J. PÉREZ-JIMÉNEZ, C. ZANDRON: Complexity aspects of polarizationless membrane systems. *Natural Computing*, to appear.

[25] A. LEPORATI, G. MAURI, C. ZANDRON, GH. PAUN, M.J. PÉREZ-JIMÉNEZ: Uniform solutions to SAT and Subset Sum by spiking neural P systems. *Natural Computing*, online version (http://dx.doi.org/10.1007/s11047-008-9091-y).

[26] G. MAURI, M.J. PÉREZ-JIMÉNEZ, C. ZANDRON: On a Păun's conjecture in membrane systems. *Lecture Notes in Computer Sci.*, 4527 (2007), 180–192.

[27] D.A. MIX BARRINGTON, N. IMMERMAN, H. STRAUBING: On uniformity within NC^1. *J. Computer and System Sci.*, 41 (1990), 274–306.

[28] N. MURPHY, D. WOODS: Active membrane systems without charges and using only symetric elementary division characterise **P**. *Lecture Notes in Computer Sci.*, 4860 (2007), 367–384.

[29] N. MURPHY, D. WOODS: A characterisation of **NL** using membrane systems without charges and dissolution. *Technical Report NUIM-CS-TR-2008-01, Dept. of Computer Science, NUI Maynooth*, Ireland, Jan. 2008.

[30] N. MURPHY, D. WOODS: A characterisation of **NL** using membrane systems without charges and dissolution. *Lecture Notes in Computer Sci.*, 5204 (2008), 164–176.

[31] M. MUTYAM, K. KRITHIVASAN: P systems with membrane creation: Universality and efficiency. *Lecture Notes in Computer Sci.*, 2055 (2001), 276–287.

[32] A. OBTULOWICZ: Deterministic P systems for solving SAT problem. *Romanian J. Information Science and Technology*, 4 (2001), 551–558.

[33] L. PAN, A. ALHAZOV, T.-O. ISHDORJ: Further remarks on P systems with active membranes, separation, merging, and release rules. *Second Brainstorming Week on Membrane Computing* (Gh. Păun et al. eds.), Report RGNC 01/04, 2004, 316–324.

[34] L. PAN, T.-O. ISHDORJ: P systems with active membranes and separation rules. *J. Univ. Computer Sci.*, 10 (2004), 630–649.

[35] C.H. PAPADIMITRIOU: *Computational Complexity*. Addison-Wesley, Massachussetts, 1995.

[36] A. PĂUN: On P systems with membrane division. *Unconventional Models of Computation* (I. Antoniou et al., eds.), Springer, 2000, 187–201.

[37] GH. PĂUN: *Membrane Computing. An Introduction*. Springer, 2002.

[38] GH. PĂUN: P systems with active membranes: Attacking **NP**-complete problems. *J. Automata, Languages and Combinatorics*, 6 (2001), 75–90.

[39] GH. PĂUN: Computing with membranes: Attacking **NP**-complete problems. *Unconventional Models of Computation* (I. Antoniou et al., eds.), Springer, 2000, 94–115.

[40] GH. PĂUN: Further twenty six open problems in membrane computing. *Third Brainstorming Week on Membrane Computing* (M.A. Gutiérrez-Naranjo et al. eds.), Fénix Editora, Sevilla, 2005, 249–262.

[41] M.J. PÉREZ-JIMÉNEZ, A. ROMERO-JIMÉNEZ, F. SANCHO-CAPARRINI: The P versus NP problem through cellular computing with membranes. *Lecture Notes in Computer Sci.*, 2950 (2004), 338–352.

[42] M.J. PÉREZ-JIMÉNEZ: An approach to computational complexity in membrane computing. *Lecture Notes in Computer Sci.*, 3365 (2005), 85–109.

[43] M.J. PÉREZ-JIMÉNEZ, A. RISCOS-NÚÑEZ: Solving the Subset-Sum problem by active membranes. *New Generation Computing*, 23 (2005), 367–384.

[44] M.J. PÉREZ-JIMÉNEZ, A. RISCOS-NÚÑEZ: A linear-time solution to the Knapsack problem using P systems with active membranes. *Lecture Notes in Computer Sci.*, 2933 (2004), 250–268.

[45] M.J. PÉREZ-JIMÉNEZ, A. ROMERO-JIMÉNEZ, F. SANCHO-CAPARRINI: A polynomial complexity class in P systems using membrane division. *J. Automata, Languages and Combinatorics*, 11 (2006), 423–434.

[46] M.J. PÉREZ-JIMÉNEZ, F.J. ROMERO-CAMPERO: Attacking the Common Algorithmic Problem by recognizer P systems. *Lecture Notes in Computer Sci.*, 3354 (2005), 304–315.

[47] M.J. PÉREZ-JIMÉNEZ, F.J. ROMERO-CAMPERO: An efficient family of P systems for packing items into bins. *J. Univ. Computer Sci*, 10 (2004), 650–670.

[48] M.J. PÉREZ-JIMÉNEZ, F.J. ROMERO-CAMPERO: Trading polarizations for bi-stable catalysts in P systems with active membranes. *Lecture Notes in Computer Sci.*, 3365 (2005), 373–388.

[49] M.J. PÉREZ-JIMÉNEZ, A. ROMERO-JIMÉNEZ, F. SANCHO-CAPARRINI: Complexity classes in cellular computing with membranes. *Natural Computing*, 2 (2003), 265–285.

[50] A.E. PORRECA: *Computational Complexity Classes for Membrane Systems*. Master Degree Thesis, Universita' di Milano-Bicocca, Italy, 2008.

[51] A.E. PORRECA, G. MAURI, C. ZANDRON: Complexity classes for membrane systems. *Informatique théorique et applications*, 40 (2006), 141–162.

[52] A. RISCOS-NÚÑEZ: *Cellular Programming: Efficient Resolution of NP-complete Numerical Problems*. PhD Thesis, University of Sevilla, Spain, 2004.

[53] A. ROMERO-JIMÉNEZ: *Complexity and Universality in Cellular Computing Models*. PhD Thesis, University of Sevilla, Spain, 2003.

[54] P. SOSÍK, A. RODRÍGUEZ-PATÓN: Membrane computing and complexity theory: A characterization of PSPACE. *J. Computer and System Sci.*, 73 (2007), 137–152.

[55] P. SOSÍK: The computational power of cell division. *Natural Computing*, 2 (2003), 287–298.

[56] C. ZANDRON, C. FERRETTI, G. MAURI: Solving NP-complete problems using P systems with active membranes. *Unconventional Models of Computation* (I. Antoniou et al., eds.), Springer, 2000, 289–301.

[57] C. ZANDRON, A. LEPORATI, C. FERRETTI, G. MAURI, M.J. PÉREZ-JIMÉNEZ: On the computational efficiency of polarizationless recognizer P systems with strong division and dissolution. *Fundamenta Informaticae*, 87 (2008), 79–91.

CHAPTER 13

SPIKING NEURAL P SYSTEMS

OSCAR H. IBARRA

ALBERTO LEPORATI

ANDREI PĂUN

SARA WOODWORTH

13.1 INTRODUCTION

NEURONS are arguably one of the most interesting cell types in the human body. A large number of neurons working in a cooperative manner are able to perform tasks that are not yet matched by the tools we can build with our current technology. Some of these tasks are thought, self-awareness, intuition, etc. Coming closer to computer science, even "simple" tasks that have been studied extensively, such as pattern matching, are performed much faster and more reliably by our brains using the "technology" of neurons than by our computers, which are several orders of magnitude faster in their information processing capabilities.

We believe the distributed manner in which the brain processes information is important in obtaining better performance, thus we are interested in the emerging area of spiking neural P systems defined as a computational model in a seminal paper [14], and investigated in more than three dozen papers, many of which are available at http://ppage.psystems.eu/. In particular, see [1, 2, 3, 4, 9, 11, 13, 24, 25].

SN P systems incorporate ideas from spiking neurons into membrane computing [23], see, e.g. [20, 21].

Spiking neural P systems are computing models inspired by the way the neurons communicate by means of electrical impulses of identical shape, called *spikes*. In short, neurons (which can be regarded as one membrane cells) are placed in the nodes of a graph whose arcs represent the synapses; inside neurons there are spikes, represented as copies of a specified symbol, which are processed by spiking rules and forgetting rules; the spikes produced in a neuron are sent—with a specified delay—to all neurons linked by a synapse to the emitting neuron; in each time unit (the system is synchronized by means of a global clock) each neuron which can use a rule should use a rule (hence the process is parallel at the level of the system and sequential at the level of each neuron). The applicability of a rule is usually determined by a regular expression associated with it, against which the number of spikes that occur in the neuron is checked. Certain neurons are designated as *input* neurons, and one of them is designated as the *output* neuron. *Spike trains* (sequences of moments in time when spikes are introduced in or emitted from the system) are processed (generated, accepted, computed, translated) by SN P systems starting in an initial configuration, in which an initial number of spikes is assigned to each neuron. At each time moment, a computation step is executed by applying a rule in all the neurons in which this is possible.

If the computation halts, then the result of the computation is obtained from the output neuron in one of many possible ways: for example, as the number of spikes contained in the neuron, as the number of spikes emitted to the environment during the computation, or as the interval of time (number of steps) elapsed between the emission of the first two spikes. In this way an SN P system computes a function $f : \mathbf{N} \to \mathbf{N}$.

Other possibilities exist, as we will see in the course of this chapter: SN P systems may be used as language generators, as language recognizers, and as computation devices that solve computationally difficult (NP-complete) decision problems.

13.2 SN P Systems: Definition and Example

Spiking neural P systems (in short SN P systems) were introduced in [14] with the aim of defining P systems based on ideas specific to spiking neurons, recently much investigated in neural computing.

In SN P systems the cells (also called *neurons*) are placed in the nodes of a directed graph, called the *synapse graph*. The contents of each neuron consist of a number of copies of a single object type, called the *spike*. Every cell may also contain a number of *firing* and *forgetting* rules. Firing rules allow a neuron to send information to other neurons in the form of electrical impulses (also called *spikes*) which are accumulated at the target cell. The applicability of each rule is determined by checking the contents of the neuron against a regular set associated with the rule. In each time unit, if a neuron can use one of its rules, then one of such rules must be used. If two or more rules could be applied, then only one of them is non-deterministically chosen. Thus, the rules are used in the sequential manner in each neuron, but neurons function in parallel with each other. Observe that, as usually happens in membrane computing, a global clock is assumed, marking the time for the whole system, and hence the functioning of the system is synchronized. When a cell sends out spikes it becomes "closed" (inactive) for a specified period of time, that reflects the refractory period of biological neurons. During this period, the neuron does not accept new inputs and cannot "fire" (that is, emit spikes). Another important feature of biological neurons is that the length of the axon may cause a time delay before a spike arrives at the target. In SN P systems this delay is modeled by associating a delay parameter to each rule which occurs in the system. If no firing rule can be applied in a neuron, there may be the possibility to apply a *forgetting rule*, that removes from the neuron a predefined number of spikes.

Formally, a *spiking neural P system* (SN P system, for short) of degree $m \geq 1$, as defined in [14], is a construct of the form

$$\Pi = (O, \sigma_1, \sigma_2, \ldots, \sigma_m, syn, in, out), \text{ where:}$$

1. $O = \{a\}$ is the singleton alphabet (a is called *spike*);
2. $\sigma_1, \sigma_2, \ldots, \sigma_m$ are *neurons*, of the form $\sigma_i = (n_i, R_i)$, $1 \leq i \leq m$, where:
   (a) $n_i \geq 0$ is the *initial number of spikes* contained in σ_i;
   (b) R_i is a finite set of *rules* of the following two forms:
      (1) *firing* (also *spiking*) rules $E/a^c \rightarrow a;d$, where E is a regular expression over a, and $c \geq 1$, $d \geq 0$ are integer numbers. If $E = a^c$, then it is usually written in the simplified form $a^c \rightarrow a;d$. If $d = 0$, then sometimes it is omitted when writing the rule;
      (2) *forgetting* rules $a^s \rightarrow \lambda$, for $s \geq 1$, with the restriction that for each rule $E/a^c \rightarrow a;d$ of type (1) from R_i, we have $a^s \notin L(E)$ (where $L(E)$ is the regular language defined by E);
3. $syn \subseteq \{1, 2, \ldots, m\} \times \{1, 2, \ldots, m\}$, with $(i, i) \notin syn$ for $1 \leq i \leq m$, is the directed graph of *synapses* between neurons;
4. $in, out \in \{1, 2, \ldots, m\}$ indicate the *input* and the *output* neurons of Π.

A firing rule $E/a^c \rightarrow a;d \in R_i$ can be applied in neuron σ_i if it contains $k \geq c$ spikes, and $a^k \in L(E)$. The execution of this rule removes c spikes from σ_i (thus leaving $k - c$ spikes), and prepares one spike to be delivered to all the neurons σ_j such that $(i, j) \in syn$. If $d = 0$, then the spike is immediately emitted, otherwise it is emitted after d computation steps of the system. As stated above, during these d computation steps the neuron is *closed*, and it cannot receive new spikes (if a neuron has a synapse to a closed neuron and tries to send a spike along it, then that particular spike is lost), and cannot fire (and even select) rules. A *forgetting* rule $a^s \to \lambda$ can be applied in neuron σ_i if it contains *exactly* s spikes, and no firing rules are applicable. The execution of this rule simply removes all the s spikes from σ_i.

The *initial configuration* of the system is described by the numbers $n_1, n_2, \ldots, n_m$ of spikes present in each neuron, with all neurons being open. During the computation, a configuration is described by both the contents of each neuron and its *state*, which can be expressed as the number of steps to wait until it becomes open (zero if the neuron is already open). Thus, $\langle r_1/t_1, \ldots, r_m/t_m \rangle$ is the configuration where neuron σ_i contains $r_i \geq 0$ spikes and it will be open after $t_i \geq 0$ steps, for $i = 1, 2, \ldots, m$; with this notation, the initial configuration of the system is $C_0 = \langle n_1/0, \ldots, n_m/0 \rangle$.

A *computation* starts in the initial configuration. In order to compute a function $f : \mathbb{N} \to \mathbb{N}$, a positive integer number is given as input to a specified *input neuron*. In the original model, as well as in some early variants, the number is encoded as the interval of time steps elapsed between the insertion of two spikes into the neuron. To pass from a configuration to another one, for each neuron a rule is chosen among the set of applicable rules, and it is executed. Generally, a computation may not halt. However, in any case the spikes emitted by the *output neuron* are also sent to the environment. The moments of time when a spike is emitted by the output neuron are marked with 1, the other moments are marked with 0. This binary sequence is called the *spike train* of the system—it might be infinite if the computation does not halt. In the spirit of spiking neurons, the result of a computation is encoded in the distance between consecutive spikes sent into the environment by the (output neuron of the) system. For example, we can consider only the distance between the first two spikes of a spike train, or the distances between the first k spikes, the distances between all consecutive spikes, taking into account all intervals or only intervals that alternate, all computations or only halting computations, etc.

The previous definitions cover many types of systems/behaviors. By neglecting the output neuron we can define *accepting* SN P systems, in which a natural number n is introduced in the input neuron, specified as the interval of time between two consecutive spikes, and is accepted if the computation halts. On the other hand, by ignoring the input neuron (and thus starting from a predefined input configuration) we can define *generative* SN P systems. We can also consider both input and output neurons and then an SN P system can work as a transducer.

Languages on arbitrary alphabets can be obtained by generalizing the form of rules: take rules of the form $E/a^c \to a^p; d$, with the meaning that, provided that the neuron is covered by E, c spikes are consumed and p spikes are produced, and sent to all connected neurons after d steps (such rules are called *extended*). Then,

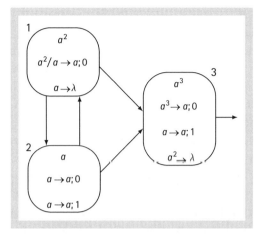

Fig. 13.1 An SN P system generating all natural numbers greater than 1.

with a step when the system sends out i spikes, we associate a symbol b_i, and thus we get a language over an alphabet with as many symbols as the number of spikes simultaneously produced. Another natural extension is to consider several output neurons, thus producing vectors of numbers, not only single numbers.

We now give a simple example of a generative SN P system—hence the input neuron in is omitted. We give it first in a formal manner (if a rule $E/a^c \to a; d$ has $L(E) = \{a^c\}$, then we write it in the simplified form $a^c \to a; d$):

$$\Pi_1 = (O, \sigma_1, \sigma_2, \sigma_s, syn, out), \text{ with}$$

$$O = \{a\},$$

$$\sigma_1 = (2, \{a^2/a \to a; 0, \ a \to \lambda\}),$$

$$\sigma_2 = (1, \{a \to a; 0, \ a \to a; 1\}),$$

$$\sigma_3 = (3, \{a^3 \to a; 0, \ a \to a; 1, \ a^2 \to \lambda\}),$$

$$syn = \{(1, 2), (2, 1), (1, 3), (2, 3)\},$$

$$out = 3.$$

This system is represented in a graphical form in Fig. 13.1 and it functions as follows. All neurons can fire in the first step, with neuron σ_2 choosing non-deterministically between its two rules. Note that neuron σ_1 can fire only if it contains two spikes; one spike is consumed, the other remains available for the next step.

Both neurons σ_1 and σ_2 send a spike to the output neuron, σ_3; these two spikes are forgotten in the next step. Neurons σ_1 and σ_2 also exchange their spikes; thus, as long as neuron σ_2 uses the rule $a \to a; 0$, the first neuron receives one spike, thus completing the needed two spikes for firing again.

However, at any moment, starting with the first step of the computation, neuron σ_2 can choose to use the rule $a \to a; 1$. On the one hand, this means that the spike of neuron σ_1 cannot enter neuron σ_2, it only goes to neuron σ_3; in this way, neuron σ_2 will never work again because it remains empty. On the other hand, in the next step neuron σ_1 has to use its forgetting rule $a \to \lambda$, while neuron σ_3 fires, using the rule $a \to a; 1$. Simultaneously, neuron σ_2 emits its spike, but it cannot enter neuron σ_3 (it is closed this moment); the spike enters neuron σ_1, but it is forgotten in the next step. In this way, no spike remains in the system. The computation ends with the expelling of the spike from neuron σ_3. Because of the waiting moment imposed by the rule $a \to a; 1$ from neuron σ_3, the two spikes of this neuron cannot be consecutive, but at least two steps must exist in between. Thus, we conclude that Π computes/generates all natural numbers greater than or equal to 2.

13.3 NORMAL FORMS FOR SN P SYSTEMS

In this section we will describe the major results from the paper [9] with respect to normal forms for SN P Systems.

With each computation (halting or not) in an SN P system we have associated a spike train, describing the sequence of steps when the output neuron spikes. Several numbers can be associated with such a spike train. The first possibility investigated was to consider the number of steps elapsed between the first two spikes emitted by the output neuron; the set of all such numbers associated with all computations of a system Π is denoted by $N_2(\Pi)$. If the spiking train of an SN P System Π produces exactly two spikes, then we call that the *strong* case, and the corresponding sets of numbers is denoted by $N_{\underline{2}}(\Pi)$. Two further cases are also of interest: (1) Taking only *halting* computations; this makes sense only for $N_2(\Pi)$, where the respective subsets are denoted by $N_2^h(\Pi)$. (2) Considering *strong halting* computations: halting computations as described above, with the extra condition that when the system halts, no spike is in the whole system. The respective set of computed numbers is denoted by $N_{\underline{2}}^h(\Pi)$.

We denote by $Spik_a^\beta P_m(rule_k, cons_p, forg_q)$ the family of sets $N_a^\beta(\Pi)$, for all systems Π with at most m neurons, each neuron having at most k rules, each of the spiking rules consuming at most p spikes, and each forgetting rule removing at most q spikes; then, $\alpha \in \{2, \underline{2}\}$, and β is either omitted or it belongs to the set $\{h, \underline{h}\}$. As usual, a parameter m, k, p, q is replaced with $*$ if it is not bounded.

In the above notation, we add to the list of features mentioned between parentheses the following two: $dley_r$, meaning that we use SN P systems whose rules $E/a^c \to a;d$ have $d \leq r$ (the delay is at most r), and $outd_s$, meaning that the outdegree of the synapse graph has the outdegree at most s. We also write $rule_k^*$ if the firing rules are of the form $E/a^c \to a;d$ with the regular expression of one of the forms $E = a^c$ or $E = a^*$ – in the first case, the rule is written simply as $a^c \to a;d$.

We are now ready to give the normal form results from [9]:

The first result shows that such systems do not lose their universality power even when the delays are not used, furthermore we see that we obtained even upper bounds for the number of rules in each neuron (at most 3), each spiking rule consuming at most 4 spikes, each forgetting rule removing either 4 or 5 spikes (depending on the halting condition) and with systems having the outdegree of the synapses at most 2:

Theorem 13.1 $Spik_2^\beta P_*(rule_3, cons_4, forg_a, dley_0, outd_2) = NRE$, where $\beta \in \{h, \underline{h}\}$ or β is omitted, and $a = 5$ for $\beta = \underline{h}$, otherwise $a = 4$.

If we consider the case of acceptors (rather than generators), then some of the constants in the previous theorem can be decreased:

In the accepting mode we can impose the restriction that in each neuron, in each time unit at most one rule can be applied, hence that the system behaves deterministically. A counterpart of Theorem 13.1 is then true (the notation of the respective families are obvious):

Theorem 13.2 $DSpik_{2acc}^\beta P_*(rule_2, cons_3, forg_2, dley_0, outd_2) = NRE$, where $\beta \in \{h, \underline{h}\}$ or β is omitted.

Another property observed was the fact that the forgetting rules are not necessary for the universality of these systems:

Theorem 13.3 $Spik_2 P_*(rule_k, cons_p, forg_0) = NRE$ for all $k \geq 2, p \geq 3$.

When considering the outdegree of the systems, we can have such systems without delays still universal even with outdegree 2:

Theorem 13.4 $Spik_2^\beta P_*(rule_3, cons_4, forg_4, dley_0, outd_2) = NRE$, where either $\beta = h$ or β is omitted.

One of the strongest results of the paper [9] is the one obtained about the simplification of the regular expressions. The following theorem required the design of a completely new proof construction: *dynamical circulation of spikes*:

Theorem 13.5 $Spik_2^\beta P_*(rule_2^*, cons_2, forg_1, dley_2, outd_2) = NRE$, where either $\beta = h$ or β is omitted.

For the rather involved proof construction the interested reader can consult [9].

If we consider the case of the strong halting acceptance, then the previous result can be almost matched even in the new case (the only exception is the maximum number of spikes forgotten).

Theorem 13.6 $Spik_2^h P_*(rule_3^*, cons_2, forg_{15}, dley_2, outd_2) = NRE$.

Finally, when considering the same question for acceptors we obtain:

Theorem 13.7 $DSpik_{2acc}^\beta P_*(rule_2, cons_2, forg_1, dley_2, outd_2) = NRE$ where $\beta = h$ or β is omitted.

The proofs of the universality results mentioned above are based on simulating register machines by means of SN P systems; we omit the respective constructions and refer to the bibliography for details.

13.4 EXTENDED SN P SYSTEMS

We have mentioned before that a generalization of SN P systems, introduced in [3, 22], is to use *extended rules*, of the form $E/a^j \to a^p; d$, with $p \geq 0$. Such a rule operates in the same manner as before except that firing sends p spikes along each outgoing synapse (and these p spikes are received simultaneously by each neighboring neuron). Clearly, when $p = 1$ the extended rules reduce to the standard (or non-extended) rules in the original definition. Note also that forgetting rules are just a special case of firing rules, i.e. when $p = 0$.

We will consider systems with three types of neurons:

1. A neuron is *bounded* if every rule in the neuron is of the form $a^i/a^j \to a^p; d$, where $1 \leq j \leq i$, $p \geq 0$, and $d \geq 0$. There can be several such rules in the neuron. These rules are called *bounded rules*.
2. A neuron is *unbounded* if every rule in the neuron is of the form $a^i(a^k)^*/a^j \to a^p; d$, where $i \geq 0$, $k \geq 1$, $j \geq 1$, $p \geq 0$, $d \geq 0$. Again, there can be several such rules in the neuron. These rules are called *unbounded rules*.
3. A neuron is *general* if it can have *general rules*, i.e. bounded as well as unbounded rules.

One can allow rules like $a_1 + \cdots + a_n \to a^p; d$ in the neuron, where all a_i's have bounded (resp., unbounded) regular expressions as defined above. But such a rule is equivalent to putting n rules $a_i \to a^p : d$ $(1 \leq i \leq n)$ in the neuron. It is known that any regular set over a 1-letter symbol a can be expressed as a finite union of regular sets of the form $\{a^i(a^j)^k \mid k \geq 0\}$ for some $i, j \geq 0$. Note that such a set is finite if $j = 0$. We can define three types of SN P systems:

1. *Bounded SN P system*: a system in which every neuron is bounded.
2. *Unbounded SN P system*: a system in which every neuron is either bounded or unbounded.
3. *General SN P system*: a system with general neurons (i.e., each neuron can contain both bounded and unbounded rules).

Let $k \geq 1$. A *k-output SN P system* has k output neurons, $O_1, \ldots, O_k$. We say that the system generates a k-tuple $(n_1, \ldots, n_k) \in \mathbf{N}^k$ if, starting from the initial configuration, there is a sequence of steps such that each output neuron O_i generates (sends out to the environment) exactly n_i spikes and then the system eventually halts.

We will consider systems with delays and systems without delays (i.e. $d = 0$ in all rules).

13.5 ASYNCHRONOUS GENERAL SN P SYSTEMS

The standard model of SN P systems is synchronized, meaning that all neurons fire at each step of the computation whenever they are fireable. This synchronization is quite powerful. It is known that a set $Q \subseteq \mathbf{N}$ is recursively enumerable if and only if it can be generated by a 1-output general SN P system (with or without delays) [14, 9]. This result holds for systems with standard rules or extended rules, and it generalizes to systems with multiple outputs. Thus, such systems are universal.

In [1] the computational power of SN P systems that operate in an asynchronous mode was introduced and studied. In an *asynchronous SN P system*, we do not require the neurons to fire at each step. During each step, any number of fireable neurons are fired (including the possibility of firing no neurons). When a neuron is fireable it may (or may not) choose to fire during the current step. If the neuron chooses not to fire, it may fire in any later step as long as the rule is still applicable. (The neuron may still receive spikes while it is waiting, which may cause the neuron to no longer be fireable.) Hence there is no restriction on the time interval for firing a neuron. Once a neuron chooses to fire, the appropriate number of spikes are sent out after a delay of exactly d time steps and are received by the neighboring neurons during the step when they are sent.

The following result was recently shown in [1]. It says that SN P systems which operate in an asynchronous mode of computation are still universal, provided that the neurons are allowed to use extended rules.

Theorem 13.8 *A set $Q \subseteq \mathbf{N}^k$ is recursively enumerable if and only if it can be generated by an asynchronous k-output general SN P system with extended rules. The result holds for systems with or without delays.*

It remains an open question whether the above result holds for the case when the system uses only standard (i.e. non-extended) rules.

13.6 ASYNCHRONOUS UNBOUNDED SN P SYSTEMS

In this section we will examine *unbounded* SN P systems, again assuming the use of extended rules. Recall that these systems can only use bounded and unbounded neurons (i.e. no general neurons are allowed). In contrast to Theorem 13.8, these systems can be characterized by partially blind register (we mainly say multi-counter) machines (PBCMs).

A *partially blind k-output multicounter machine* (k-output PBCM) [7] is a k-output CM, where the registers/counters cannot be tested for zero. The output counters are non-decreasing. The other counters can be incremented by 1 or decremented by 1, but if there is an attempt to decrement a zero counter then the computation aborts (i.e. the computation becomes invalid). By definition, a successful generation of a k-tuple requires that the machine enters an accepting state with all non-output counters set to zero.

It is known that k-output PBCMs can be simulated by k-dimensional vector addition systems, and vice-versa [7]. Hence, such counter machines are not universal. In particular, a k-output PBCM can generate the reachability set of a vector addition system.

In [1], asynchronous unbounded SN P systems without delays were investigated. The systems considered in [1] are restricted to halt in a pre-defined configuration. Specifically, a computation is valid if, at the time of halting, the numbers of spikes that remain in the neurons are equal to pre-defined values; if the system halts but the neurons do not have the pre-defined values, the computation is considered invalid and the output is ignored. These systems were shown to be equivalent to PBCMs in [1]. However, it was left as an open question whether the "pre-defined halting" requirement was necessary to prove this result. It was recently shown in [12] that this condition is, in fact, not necessary. Note that for these systems, firing zero or more neurons at each step is equivalent to firing one or more neurons at each step (otherwise, since there are no delays, the configuration stays the same when no neuron is fired).

Theorem 13.9 *A set $Q \subseteq \mathbf{N}^k$ is generated by a k-output PBCM if and only if it can be generated by an asynchronous k-output unbounded SN P system without delays. Hence, such SN P systems are not universal.*

Note that by Theorem 13.8, if we allow both bounded rules and unbounded rules to be present in the neurons, SN P systems become universal.

Again, it remains an open question whether the above theorem holds for the case when the system uses only standard rules.

It is known that PBCMs with only one output counter can only generate semilinear sets of numbers. Hence:

Corollary 13.1 *Asynchronous 1-output unbounded SN P systems without delays can only generate semilinear sets of numbers.*

The results in the following corollary can be obtained using Theorem 13.9 and the fact that they hold for k-output PBCMs.

Corollary 13.2

1. *The family of k-tuples generated by asynchronous k-output unbounded SN P systems without delays is closed under union and intersection, but not under complementation.*
2. *The membership, emptiness, infiniteness, disjointness, and reachability problems are decidable for asynchronous k-output unbounded SN P systems without delays; but containment and equivalence are undecidable.*

In Theorem 13.9, we showed that restricting an asynchronous SN P system without delays to contain only bounded and unbounded neurons gives us a model equivalent to a PBCM. However, it is possible that allowing delays would give additional power. For asynchronous unbounded SN P systems with delays, we can no longer assume that firing zero or more neurons at each step is equivalent to firing one or more neurons at each step.

Note that not every step in a computation has at least one neuron with a fireable rule. In a given configuration, if no neuron is fireable but at least one neuron is closed, we say that the system is in a *dormant* step. If there is at least one fireable neuron in a given configuration, we say the system is in a *non-dormant* step. (Of course, if a given configuration has no fireable neuron, and all neurons are open, we are in a halting configuration.) Thus, an SN P system with delays might be dormant at some point in the computation until a rule becomes fireable. However, the clock will keep on ticking.

Interestingly, the addition of delays does not increase the power of the system.

Theorem 13.10 *A set $Q \subseteq \mathbf{N}^k$ is generated by a k-output PBCM if and only if it can be generated by an asynchronous k-output unbounded SN P system with delays.*

This result contrasts the result in [9] which shows that synchronous unbounded SN P systems with delays and standard rules (but also standard output) are universal.

13.7 Asynchronous Bounded SN P Systems

In this section we consider *asynchronous* SN P systems, where the neurons can only use *bounded* rules. We show that these bounded SN P systems with extended rules generate precisely the semilinear sets.

A *k-output monotonic* CM is a non-deterministic machine with k counters, all of which are output counters. The counters are initially set to zero and can only be incremented by 1 or 0 (they cannot be decremented). When the machine halts in an accepting state, the k-tuple of values in the k-counter is said to be generated by the machine. Clearly, a k-output monotonic CM is a special case of a PBCM, where all the counters are output counters and all the instructions are addition instructions.

It is known that a set $Q \subseteq \mathbf{N}^k$ is semilinear if and only if it can be generated by a k-output monotonic CM [8]. We can show the following:

Theorem 13.11 *$Q \subseteq \mathbf{N}^k$ can be generated by a k-output monotonic CM if and only if it can be generated by a k-output asynchronous bounded SN P system with extended rules. The result holds for systems with or without delays.*

At present, we do not know whether Theorem 13.11 holds when the system is restricted to use only standard (non-extended) rules. However, the result holds for synchronous bounded SN P systems using only standard rules.

13.8 Sequential SN P Systems

Sequential SN P systems are another closely related model introduced in [13]. These are systems that operate in a sequential mode. This means that at every step of the computation, if there is at least one neuron with at least one rule that is fireable, we only allow one such neuron and one such rule (both non-deterministically chosen) to be fired. If there is no fireable rule, then the system is dormant until a rule becomes fireable. However, the clock will keep on ticking. The system is called *strongly sequential* if at every step, there is at least one neuron with a fireable rule.

Unlike for asynchronous systems (considered in the previous section), where the results relied on the fact that the systems use extended rules, the results here hold for systems that use standard rules as well as for systems that use extended rules.

Theorem 13.12 *The following results hold for systems with delays.*

1. *Sequential k-output unbounded SN P systems with standard rules and strongly sequential k-output general SN P systems with standard rules are universal.*
2. *Strongly sequential k-output unbounded SN P systems with standard rules and k-output PBCMs are equivalent.*

The above results also hold for systems with extended rules.

Item 2 in the above theorem improves the result found in [13] which required a special halting configuration similar to the halting configuration in [1]. In fact, this halting requirement is not necessary [12].

13.9 CLOSURE AND DECIDABLE PROPERTIES

We note that SN P systems that are characterized by PBCMs share the same closure and decidable properties of PBCMS. For example, the following results are known for PBCMs and therefore also hold for SN P systems equivalent to them.

Theorem 13.13

1. *(Union, intersection, complementation) The sets of k-tuples generated by k-output PBCMs are closed under union and intersection, but not under complementation.*
2. *(Membership) It is decidable to determine, given a k-output PBCM M and a k-tuple α (of integers), whether M generates α.*
3. *(Emptiness) It is decidable to determine, given a k-output PBCM, whether it generates an empty set of k-tuples.*
4. *(Infiniteness) It is decidable to determine, given a k-output PBCM, whether it generates an infinite set of k-tuples.*
5. *(Disjointness) It is decidable to determine, given two k-ouput PBCMs, whether they generate a common k-tuple.*
6. *(Containment, equivalence) It is undecidable to determine, given two k-output PBCMs, whether the set generated by one is contained in the set generated by the other (or whether they generate the same set).*
7. *(Reachability) It is decidable to determine, given a PBCM with k output counters and m auxiliary counters (thus a total of k + m counters) and configurations*

$\alpha = (i_1, \ldots, i_k, j_1, \ldots, j_m)$ and $\beta = (i'_1, \ldots, i'_k, j'_1, \ldots, j'_m)$ *(the first k components correspond to the output), whether α can reach β.*

13.10 SEQUENTIALITY INDUCED BY SPIKE NUMBER

In this section we consider sequential SN P systems where the sequentiality of the system is induced by a simple choice: the neuron with the maximum number of spikes out of the neurons that can spike at one step will fire. This corresponds to a global view of the whole network that makes the system sequential. We describe the properties of this restriction without giving the details of the proofs. The interested reader can consult the proofs in [10].

Several authors have recently noticed that the maximal parallelism way of rule application (which is widely used in membrane computing) is rather nonrealistic in some cases. This fact motivated the consideration of various other "strategies" for rule application in membrane systems (or neuron firing in SN P systems).

Here we consider the spiking restriction on neurons in the following way: if at any step there are more than one neuron that can spike (according to their rules) then only the neuron(s) containing the maximum number of spikes (among the currently "active" neurons) will fire. This is contrasting with the maximal parallel application of the rules case, in which case all the "active" neurons will fire at that step. To exemplify the firing mechanism of the new strategy, let us consider four neurons, with labels n_1, n_2, n_3, n_4 that are the only neurons that can fire at this step (according to their internal rules and the contents of spikes for each of them). In such a case we would find the maximum number of spikes stored in n_1 through n_4, say we have the values 5, 3, 7, 1, thus the neuron n_3 holds the maximum number of spikes, thus n_3 will fire at the next step. After n_3 fires, we update the number of the spikes in the whole system according to this neuron's spiking, and at the next step the neurons n_1, n_2, n_4 together with n_3 will be checked if they can fire (in the new conditions as they may be rendered inactive by an increment in their number of spikes stored). If there is a tie for the maximum number of spikes stored in the active neurons, then all the neurons containing the maximum will fire.

The main motivation behind this spiking strategy is the observation that in a population of cells of the same type (neurons in this case) which have similar types of rules (the spiking rules in our case) one can notice that the cells containing larger numbers of a specific molecule species are more active (spike faster/more frequently) than the cells containing lower numbers of the same molecule. Another observation is the fact that the neurons that receive a large number of spikes are more likely to spike than the neurons that do not receive many spikes. The same modeling path was taken also when the *integrate-and-fire* models were defined for neurons, which leads to the neurons that receive more spikes to fire faster than the neurons that receive lower numbers of spikes.

The restriction proposed above makes the spiking of the neurons in the system almost sequential (more than one neuron can spike only in the special case when there is a tie in the number of spikes contained). Because of this, we will call this application strategy *max pseudo-sequential*. One can also consider the sequential strategy, which resolves the ties by choosing for the next spiking neuron non-deterministically one of the neurons containing the maximum number of spikes at that moment (out of the active neurons). This second strategy will be called in the following *max sequential*.

We will consider also the difference between these devices from the point of view of generators versus acceptors, specifically, we notice a major difference between systems with deterministic neurons working as generators as opposed to acceptors. We see that the acceptors are universal whereas the generators are only able to generate one single value (thus are non-universal).

We start with the case of SN P systems based on strongly max sequentiality with delays. The next theorem shows the universality of such systems as opposed to the result in [13], where the strongly sequential systems were shown to be not universal.

Theorem 13.14 *Unbounded SN P systems with delays working in the max sequential mode are universal, even in the strongly sequential setting.*

If we consider the case of extended systems, then we can easily remove the delay (again we refer the interested reader to [10] for the details of the proof).

Theorem 13.15 *Extended SN P systems with max sequentiality, unbounded and without delays (thus strongly sequential) are universal.*

Concerning max pseudo-sequentiality, we start by noticing that there is no non-determinism at the level of the system: from each configuration to the next, we know for sure which neurons fire (this was not the case with the max sequentiality discussed in the previous section).

Theorem 13.16 *Unbounded SN P systems working in the max pseudo-sequential mode, without delays, are universal.*

An interesting observation is the fact that if one considers deterministic neurons (neurons in which the regular languages associated with each rule are disjoint), then such a system cannot produce non-determinism. Thus we have the following result:

Theorem 13.17 *A system of deterministic neurons working in a maximal pseudo-sequential manner (as a generator) is non-universal.*

This result contrasts the fact that if such devices are used in the accepting mode, then they are universal:

Theorem 13.18 *A system of deterministic neurons working in a maximal pseudo-sequential manner (as an acceptor) is universal.*

Finally, one can extend the notion of max sequentiality to the case of considering the minimum number (rather than the maximum) of spikes. This induces immediately the notion of *min-sequentiality*.

As opposed to Theorem 13.14, we obtain the universality without delays for the case in which we consider the minimum number of spikes in neurons rather than the maximum. This is made possible by considering that the output is realized through accumulation of spikes in the output neuron, rather than the time elapsed between the two spikes of the output neuron.

Theorem 13.19 *Unbounded SN P systems without delays, working in the min sequential mode (hence strongly sequential), are universal. In this case we consider the output to be given as the number of spikes in a given neuron.*

The topic described in the current section is an active research field and we expect more results to be reported in the future about the sequentiality induced by the spike number. Thus the reader should consider the current section as an introductory material to this area rather than a comprehensive treatise.

13.11 SN P Systems as Language Generators

In this section we use SN P systems as language generators, as described in two recent papers [2, 4]. Consider an SN P system Π with output neuron, σ_{out}, which is bounded. Interpret the output as follows. At times when 0 spikes, a is interpreted to be 1, and at times when it does not spike, interpret the output to be σ_{out}. We say a binary string $x = a_1 \ldots a_n$, where $n \geq 1$, is generated by Π if starting in its initial configuration, it outputs x and halts. We assume that the SN P systems use standard rules.

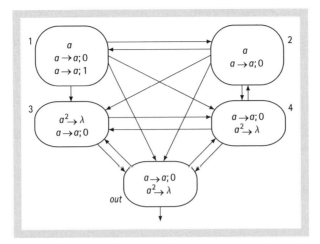

Fig. 13.2 A bounded SN P system generating the language $F1 = 0^+1$.

13.11.1 Regular Languages

It was recently shown in [2] that for any finite binary language F, the language $F1$ (i.e. with a supplementary suffix of 1) can be generated by a bounded SN P system. However, this result does not hold when F is an infinite regular language.

Observation 13.1 *Let $F = 0^*$. Then $F1$ cannot be generated by a bounded SN P system.*

The proof of this assertion is left to the reader.

It is interesting to note that by just modifying the previous language 0^* to always begin with at least one zero (so $F = 0^+$) we can generate $F1$.

Observation 13.2 *Let $F = 0^+$. Then $F1$ can be generated by a bounded SN P system.*

We give a bounded SN P system which generates $F1$ (shown in Fig. 13.2). Here the output neuron initially contains no spikes, guaranteeing that it will not spike during the first step. Both neurons $sigma_1$ and σ_2 will fire during the first step. If neuron σ_1 chooses to fire the rule $a \to a; 0$ then two spikes are received by neurons σ_3, σ_4, and σ_{out}. These spikes are forgotten during the next step causing the output to be σ_{out}. This is repeated until neuron σ_1 chooses to fire rule $a \to a; 1$. This will cause neurons σ_3, σ_4, and σ_{out} to receive one spike at this time step. This will cause all three neurons to fire during the next time step (when neuron σ_1 also fires). This leaves neurons σ_2, σ_3, σ_4, and σ_{out} with three spikes. No further neuron is fireable, causing the system to halt after producing 0^+1.

In contrast to Observation 13.1, we have:

Observation 13.3 *Let $F = 0^*$. Then $1F$ can be generated by a bounded SN P system.*

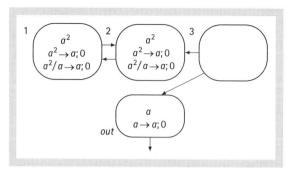

Fig. 13.3 A bounded SN P system generating the language $1F = 10^*$.

A bounded SN P system which generates $1F$ is shown in Fig. 13.3.
Observation 13.3 actually generalizes to the following rather surprising result:

Theorem 13.20 *Let $L \subseteq (0 + 1)^*$. Then the language $1L$ (i.e. with a supplementary prefix 1) can be generated by a bounded SN P system if and only if L is regular. (The result holds also for $0L$, i.e. the supplementary prefix is 0 instead of 1.)*

13.11.2 Another Way of Generating Languages

Now we define a new way of "generating" a string. Under this new definition, various classes of languages can be obtained. We say a binary string $x = a_1 \ldots a_n$, where $n \geq 0$, is generated by Π if it outputs $1x10^d$, for some d which may depend on x, and halts. Thus, in the generation, Π outputs a 1 before generating x, followed by 10^d for some d. (Note that the prefix 1 and the suffix 10^d are not considered as part of the string.) The set $L(\Pi)$ of binary strings generated by Π is called the *language* generated by Π. We can show the following:

1. When there is no restriction, Π can generate *any unary recursively enumerable* language. Generalizing, for $k \geq 1$, any recursively enumerable language $L \subseteq 0^*10^* \ldots 10^*$ (k occurrences of 0^*s) can be generated by an unrestricted SN P system.
   There are variants of the above result. For example, for $k \geq 1$, let $w_1, \ldots, w_k$ be fixed (not necessarily distinct) non-null binary strings. Let $L \subseteq w_1^* \ldots w_k^*$ be a bounded language. Then L can be generated by an unrestricted SN P system if and only it is recursively enumerable.
2. There are non-bounded binary languages, e.g. $L = \{xx^r \mid x \in \{0, 1\}^+\}$ (the set of even-length palindromes), that cannot be generated by unrestricted SN P systems. However, interestingly, as stated in item 5 below, the complement of L can be generated by a very simple SN P system.

3. Call the SN P system *linear spike-bounded* if the number of spikes in the unbounded neurons at any time during the computation is at most $O(n)$, where n is the length of the string generated. Define *polynomial spike-bounded* SN P systems in the same way. Then linear spike-bounded SN P systems and polynomial spike-bounded SN P systems are weaker than unrestricted SN P systems. In fact, $S(n)$ spike-bounded SN P systems are weaker than unrestricted SN P systems, for any recursive bound $S(n)$.

4. Call the SN P system *1-reversal* if every unbounded neuron satisfies the property that once it starts "spiking" it will no longer receive future spikes (but may continue computing). Generalizing, an SN P system is *r-reversal* for some fixed integer $r \geq 1$, if every unbounded neuron has the property that the number of times its spike size changes values from non-increasing to non-decreasing and vice-versa during any computation is at most r. There are languages generated by linear spike-bounded SN P systems that cannot be generated by r-reversal SN P systems. For example, $L = \{1^{k^2} \mid k \geq 1\}$ can be generated by the former but not by the latter.

5. Let $w_1, \ldots, w_k$ be fixed (not necessarily distinct) non-null binary strings. For every string x in $w_1^* \ldots w_k^*$, let $\psi(x) = (\#(x)_1, \ldots, \#(x)_k)$ in $\mathbf{N}^k$, where $\#(x)_i =$ number of occurrences of w_i in x. Thus, $\psi(x)$ is the Parikh map of x with respect to $(w_1, \ldots, w_k)$. For $L \subseteq w_1^* \ldots w_k^*$, let $\psi(L) = \{\psi(x) \mid x \in L\}$. A 1-reversal (resp. r-reversal) SN P system can generate L if and only if $\psi(L)$ is semilinear.

    Similarly, a 1-reversal (resp., r-reversal) SN P system can generate a language $L \subseteq 0^*10^* \ldots 10^*$ (k occurrences of 0^*s) if and only if the set $\{(n_1, \ldots, n_k) \mid 0^{n_1}1 \ldots 10^{n_k} \in L\}$ is semilinear.

    While the language $L = \{xx^r \mid x \in \{0, 1\}^+\}$ cannot be generated by an unrestricted SN P system (by item 2), its complement can be generated by a 1-reversal SN P system.

6. 1-reversal (resp., r-reversal) SN P system languages are closed under some language operations (e.g. union), but not under other operations (e.g. complementation). Many standard decision problems (e.g. membership, emptiness, disjointness) for 1-reversal (resp., r-reversal) SN P systems are decidable, but not other questions (e.g. containment and equivalence).

13.12 SOLVING NP-COMPLETE PROBLEMS

Besides generating and recognizing languages, SN P systems can also be used to solve *decision problems*, that is, problems whose answer is either YES or NO.

Formally, a decision problem Q is a pair (I_Q, Y_Q), where I_Q is a set of *instances* and $Y_Q \subseteq I_Q$ is the set of *positive* instances. The instances of I_Q are encoded by means of a *"reasonable" encoding scheme*, in the sense given in [6, p. 21]. NP-complete problems constitute an important class of computationally hard decision problems. For an introduction to the theory of NP-completeness, as well as an overview of the main results and techniques of the theory, we refer the reader to [6].

The use of SN P systems to solve NP-complete problems has been investigated in few papers [19, 18, 17, 15, 16], where both *semi-uniform* and *uniform* solutions to the well known problems SAT, 3-SAT, and SUBSET SUM have been proposed. When solving a problem Q in the semi-uniform setting, for each specified instance $\mathcal{I} \in I_Q$ we build in a polynomial time (with respect to the size of $\mathcal{I}$) an SN P system $\Pi_{Q,\mathcal{I}}$, whose structure and initial configuration depend upon $\mathcal{I}$, that outputs a specified number of spikes in a given interval of time if and only if $\mathcal{I} \in Y_Q$. On the other hand, a uniform solution of Q consists of a family $\{\Pi_Q(n)\}_{n \in \mathbb{N}}$ of SN P systems such that for all $n \in \mathbb{N}$ the system $\Pi_Q(n)$ is able to solve all the instances $\mathcal{I} \in I_Q$ of size n. The specific instance $\mathcal{I}$ to be solved is specified by sending a polynomial (in n) number of spikes—that is, an *encoding* of $\mathcal{I}$—to a designated (set of) input neuron(s) of $\Pi_Q(n)$; then, the system will emit a specified number of spikes in a given interval of time if and only if $\mathcal{I} \in Y_Q$. The preference for uniform solutions over semi-uniform ones is due to the fact that they are more strictly related to the structure of the problem. Indeed, in the semi-uniform setting we do not even need any input neuron, as the instance of the problem can be embedded into the structure (number of spikes, graph of neurons, rules) of the system from the very beginning.

A first result that we should mention is the following theorem, proved in [19]. Recall that the *description size* of an SN P system is defined as the number of bits which are needed to describe it, whereas an SN P system is *deterministic* if, for every neuron that occurs in the system and for any possible contents of the neuron, at most one of the rules that occur in the neuron may be applied.

Theorem 13.21 *Consider a (possibly universal) deterministic accepting SN P system Π, of degree $m \geq 1$, in which all the regular expressions are of the following restricted forms: a^i, with $i \leq 3$, or $a(aa)^+$. Then, any t steps of computation of Π can be simulated by a deterministic Turing machine in a time which is polynomial with respect to t and to the description size of Π.*

This result shows that polynomial size (with respect to the problem instance size) *deterministic* SN P systems cannot solve NP-complete problems in polynomial time unless **P = NP**, a very unlikely situation. Hence efficient solutions to NP-complete (or harder) problems can only be obtained by introducing features which enhance the efficiency (pre-computed resources, ways to exponentially grow the workspace during the computation, non-determinism, and so on). The theorem holds even if the system Π applies its rules in the maximally parallel way, provided that it is deterministic and that the regular expressions that occur in the firing rules are simple. In fact, we will see in a few moments that if arbitrary regular expressions are allowed then choosing whether a rule may be applied or not in a given neuron may be equivalent to solving an instance of the SUBSET SUM problem.

In the next three subsections we briefly review the main results obtained up to now on solving the NP-complete problems SAT, 3-SAT, and SUBSET SUM with SN P systems.

13.12.1 Solving SAT and 3-SAT

Let us start from the NP-complete decision problem SAT [6, p. 39]. The instances depend upon two parameters: the number n of variables, and the number m of clauses. We recall that a *clause* is a disjunction of literals, occurrences of x_i or $\neg x_i$, built on a given set $X = \{x_1, x_2, \ldots, x_n\}$ of Boolean variables. Without loss of generality, we can avoid the clauses in which the same literal is repeated or both the literals x_i and $\neg x_i$, for any $1 \leq i \leq n$, occur. In this way, a clause can be seen as a *set* of at most n literals. An *assignment* of the variables $x_1, x_2, \ldots, x_n$ is a mapping $a : X \to \{0, 1\}$ that associates to each variable a truth value. The number of all possible assignments to the variables of X is 2^n. We say that an assignment *satisfies* the clause C if, assigned the truth values to all the variables which occur in C, the evaluation of C (considered as a Boolean formula) gives 1 (*true*) as a result. The SAT problem can thus be formally stated as follows. An *instance* is a set $C = \{C_1, C_2, \ldots, C_m\}$ of clauses, built on a finite set $X = \{x_1, x_2, \ldots, x_n\}$ of Boolean variables. The *question* to be answered is: is there an assignment to the variables $x_1, x_2, \ldots, x_n$ that satisfies all the clauses in C? The 3-SAT problem is defined just like SAT, the only difference being that each clause contains exactly three literals. In what follows, by SAT(n, m) we denote the set of instances of SAT which have n variables and m clauses, whereas 3-SAT(n) indicates the set of instances of 3-SAT which can be built using n variables.

As shown in [6, p. 48], every instance γ of SAT can be transformed in polynomial time (with respect to n and m) into an instance γ' of 3-SAT, in such a way that γ is satisfiable if and only if γ' is satisfiable.

The first solution to these problems by means of SN P systems is described in [19], where it is shown that *non-deterministic* SN P systems are able to solve in the *semi-uniform* setting any instance of 3-SAT in *constant* time. The systems are acyclic (they operate like Boolean circuits), and are composed of four layers. In the first layer, a non-deterministic choice between two firing rules allows guessing for each variable the correct assignment that (possibly) satisfies the instance. The next two layers apply the generated assignment first to the literals and then to the 3-clauses that compose the instance. Finally, the last layer operates like an AND gate, emitting one spike during the fourth computation step if and only if all the clauses are satisfied.

A completely different approach is followed in [17], where a *uniform* family $\{\Pi_{SAT}(\langle n, m\rangle)\}_{n,m\in \mathbf{N}}$ of SN P systems is defined, such that for all $n, m \in \mathbf{N}$ the system $\Pi_{SAT}(\langle n, m\rangle)$ solves all the instances of SAT(n, m) in a number of steps which is linear in n and independent of m. The notation $\langle n, m\rangle$ indicates the positive integer number obtained by applying an appropriate bijection (for example, Cantor's pairing) from $\mathbf{N}^2$ to $\mathbf{N}$. All the systems $\Pi_{SAT}(\langle n, m\rangle)$ are non-deterministic and use only standard rules. However, both the size and the structure of the system depend upon the number m of clauses, and thus can be exponential with respect to n. In the same paper a *uniform* solution to 3-SAT is also presented, which uses non-deterministic systems $\Pi_{3SAT}(n)$ of polynomial size to solve all the instances of 3-SAT(n). $\Pi_{3SAT}(n)$ contains one subsystem for each possible 3-clause of n variables; the instance of 3-SAT to be solved is encoded as a sequence of $\Theta(n^3)$ bits, where a 1 means that the corresponding clause is part of the instance. The basic construction of $\Pi_{3SAT}(n)$ uses either extended rules or forgetting rules that compete with firing rules (a feature not present in the standard definition of SN P systems). However, simple modifications allow to obtain standard systems, that produce one spike during the sixth computation step if and only if the guessed assignment satisfies the instance.

13.12.2 Solving SUBSET SUM

SUBSET SUM is one of the most famous NP-complete decision problems [6, p. 223]. We can state it as follows: Given a (multi)set $V = \{v_1, v_2, \ldots, v_n\}$ of positive integer numbers, and a positive integer number S, is there a sub(multi)set $B \subseteq V$ such that $\sum_{b \in B} b = S$?

An important aspect of this problem is that it is a *numerical* and *pseudo–polynomial* NP-complete problem: the difficulty of solving it depends upon the magnitude of the integer numbers that occur in the instance. Denoting $K = max\{v_1, v_2, \ldots, v_n, S\}$, the instance size of SUBSET SUM is $\Theta(n \log K)$, since each number requires $\log K$ bits to be represented in binary form. A well known algorithm based on dynamic programming [5] solves any instance of SUBSET SUM in $\Theta(nK)$ steps, an exponential amount with respect to the instance size; hence this algorithm can be used only for small values of K (for example, when K is polynomially bounded in n). Note that if we should represent the instances of SUBSET SUM in *unary* form, then the instance size would be $\Theta(nK)$ and the above mentioned algorithm could be used to solve them in a polynomial time. Indeed, SUBSET SUM is not *strongly* NP-complete, meaning that it does not remain NP-complete when its instances are represented in unary form. The relevance of SUBSET SUM in the study of the computational efficiency of SN P systems becomes clear if we consider that *uniform* standard SN P systems read the instances from the environment *in unary form*, in a time which is proportional to K.

The first solution of SUBSET SUM by means of SN P systems is given in [18], where a *non-deterministic, semi-uniform* family of SN P systems solves any instance of SUBSET SUM in two steps. A drawback of this solution is that firing and forgetting rules compete for the same number of spikes, a feature not present in standard SN P systems. Moreover, the systems contain an exponential number of spikes in their initial configurations, and manipulate an exponential number of spikes in one step. A *uniform* family of SN P systems is also presented in [18]; here the systems are deterministic and use only standard rules, but their execution time is in general exponential with respect to the instance size (whereas the size of the systems is linear in n). The instances are encoded in *binary* form, and *maximal parallelism* is needed to convert them in the usual unary form before solving the problem. An interesting feature of these systems is that their computations halt if and only if the instance they are solving is positive.

Perhaps the simplest non-deterministic *semi-uniform* solution to the SUBSET SUM problem by means of SN P systems is given in [19, 17]. Given the instance ($V = \{v_1, v_2, \ldots, v_n\}$, S) of SUBSET SUM, let $E = (a^{v_1} \cup \lambda)(a^{v_2} \cup \lambda) \ldots (a^{v_n} \cup \lambda)$ be the corresponding regular expression built using the values $v_1, v_2, \ldots, v_n$ from V. The system is composed of a single neuron, that contains only the firing rule $E/a^S \to a; 0$, and is initialized with S spikes. According to the standard definition of SN P systems, the rule can fire (thus sending a single spike to the environment) if and only if $a^S \in L(E)$, that is, if and only if there exists a subset of V whose elements sum up to S. Hence, as stated above, in some situations determining whether a firing rule can be applied may be equivalent to solving an instance of SUBSET SUM. Note that this solution does not use extended rules nor forgetting rules in competition with spiking rules. However it uses complicated regular expressions, that involve strings whose length is proportional to the values v_i contained in V (and thus, possibly exponential with respect to the instance size of SUBSET SUM).

13.12.3 Using Pre-computed Resources

A completely different approach to solve SAT, 3-SAT, and SUBSET SUM in a polynomial time is exposed in [15, 16], where some (possibly exponentially large) pre-computed resources are given in advance. In particular, [15] describes a *uniform* family of (exponential size) *deterministic* SN P systems that solves all the instances of SAT(n, m) in a time which is quadratic in n and linear in m, as well as a *uniform* family of *deterministic* SN P systems that solve all the instances of 3-SAT(n) in a time which is cubic in n. In [16], a *semi-uniform* and a *uniform* solution to SUBSET SUM are illustrated, by means of (exponential size) *deterministic* SN P systems that work in polynomial time with respect to the instance size. In these papers it is not specified how the pre-computed resources could be built: it is only required that such systems have a structure which is as regular as possible (a requirement which is related to the concept of *uniformity*, used in the theory of Circuit Complexity to measure the difficulty of constructing families of Boolean circuits) and they do not contain neither "hidden information" that simplify the solution of specific instances, nor an encoding of all possible solutions (that is, an exponential amount of information that allows cheating while solving the instances of the problem). Pre-computed resources could be built by deterministic Turing machines whose computational power has been augmented by adding to their set of instructions some form of controlled duplication, that replicates (possibly substituting some pieces of the structure) part of the output it has built up to that moment. However, it is an open problem to precisely determine how this controlled duplication should work, and hence what kind of exponential size pre-computed resources may be considered acceptable.

13.13 Final Remarks

SN P systems have been intensely investigated in the past two years since their definition in [14]. This can be justified from two perspectives: they are both simple/transparent and model the processes at the level of a collection of neurons. One can notice an increased interest in the recent period for the sequentiality and the properties of systems working under a sequential (or even asynchronous) firing strategy. This can be motivated by the fact that the sensing technology for neurons is developing rapidly, allowing us to differentiate between spikes starting even fractions of seconds apart. Since the spikes are electrical charges they are sent with much faster speeds than the cellular reactions, thus one can consider the spiking at the level of a network of neurons as a sequential process.

We have reviewed several results in the area, starting with normal forms for general spiking systems. Later on in the chapter we have reviewed the conditions under which an asynchronous unbounded SN P system becomes universal. It is interesting to note that *all* of the following are needed to gain universality: (1) delays, (2) dormant steps, and (3) at least one neuron must fire at every step if the system is not in a dormant state. Clearly we must have (1) in order to have (2). If we just have (1) and (2), Theorem 13.10 shows the system is not universal. If we just have (1) and (3), it can be shown that the system is not universal.

An interesting open question is whether these results hold if we restrict the SN P system to use only standard (i.e. non-extended) rules.

For *sequential* SN P systems, we were able to obtain similar results with the use of only standard rules. If the system is *strongly sequential* with only bounded and unbounded neurons, the model is only as powerful as PBCMs. However, if we allow either general neurons or dormant steps, the system becomes universal.

Finally, we presented some results concerning sequentiality induced by the spikes number, the language generating capability of SN P systems, and their capability to solve computationally hard (**NP**-complete) decision problems. Many topics for future research remain to be examined—several of them pointed out in the previous sections.

References

[1] M. CAVALIERE, O. EGECIOGLU, O. H. IBARRA, M. IONESCU, GH. PĂUN, S. WOODWORTH: Asynchronous spiking neural P systems; decidability and undecidability. *Proc. DNA 2007*, 246–255.

[2] H. CHEN, R. FREUND, M. IONESCU, GH. PĂUN, M.J. PÉREZ-JIMÉNEZ: On string languages generated by spiking neural P systems. *Proc. 4th Brainstorming Week on Membrane Computing*, 2006, 169–194.

[3] H. CHEN, M. IONESCU, T.-O. ISHDORJ, A. PĂUN, GH. PĂUN, M.J. PÉREZ-JIMÉNEZ: Spiking neural P systems with extended rules: universality and languages. *Natural Computing* (special issue devoted to DNA 12 Conference), *Natural Computing*, 7 (2008), 147–166.

[4] H. CHEN, M. IONESCU, A. PĂUN, GH. PĂUN, B. POPA: On trace languages generated by (small) spiking neural P systems. *Pre-proc. 8th Workshop on Descriptional Complexity of Formal Systems*, June 2006.

[5] T.H. CORMEN, C.H. LEISERSON, R.L. RIVEST: *Introduction to Algorithms*. MIT Press, 1990.

[6] M.R. GAREY, D.S. JOHNSON: *Computers and Intractability. A Guide to the Theory on NP-Completeness*. W.H. Freeman and Company, 1979.

[7] S. GREIBACH: Remarks on blind and partially blind one-way multicounter machines. *Theoretical Computer Sci.*, 7 (1978), 311–324.

[8] T. HARJU, O. IBARRA, J. KARHUMAKI, A. SALOMAA: Some decision problems concerning semilinearity and commutation. *J. Computer and System Sci.*, 65 (2002), 278–294.

[9] O.H. IBARRA, A. PĂUN, GH. PĂUN, A. RODRÍGUEZ-PATÓN, P. SOSIK, S. WOODWORTH: Normal forms for spiking neural P systems. *Theoretical Computer Sci.*, 372 (2007), 196–217.

[10] O.H. IBARRA, A. PĂUN, A. RODRÍGUEZ-PATÓN: Sequentiality induced by spike number in SN P systems, *Pre-proc. Fourteenth meeting on DNA Computing* (DNA14), June 2008, Prague, Czech Republic, 36–47.

[11] O.H. IBARRA, S. WOODWORTH: Characterizations of some restricted spiking neural P systems. *Lecture Notes in Computer Sci.*, 4361 (2006), 424–442.

[12] O.H. IBARRA, S. WOODWORTH: Spiking neural P systems: some characterizations. *Lecture Notes in Computer Sci.*, 4639 (2007), 23–37.

[13] O.H. IBARRA, S. WOODWORTH, F. YU, A. PĂUN: On spiking neural P systems and partially blind counter machines. *Lecture Notes in Computer Sci.*, 4135 (2006), 113–129.

[14] M. IONESCU, GH. PĂUN, T. YOKOMORI: Spiking neural P systems. *Fundamenta Informaticae*, 71 (2006), 279–308.

[15] T.-O. ISHDORJ, A. LEPORATI: Uniform solutions to SAT and 3-SAT by spiking neural P systems with pre-computed resources. *Natural Computing*, to appear. *Natural Computing*, 7 (2008), 519–534.

[16] A. LEPORATI, M.A. GUTIÉRREZ-NARANJO: Solving SUBSET SUM by spiking neural P systems with pre-computed resources. *Fundamenta Informaticae*, 87 (2008), 61–77.

[17] A. LEPORATI, G. MAURI, C. ZANDRON, GH. PĂUN, M.J. PÉREZ-JIMÉNEZ: Uniform solutions to SAT and SUBSET SUM by spiking neural P systems. *Natural Computing*, to appear.

[18] A. LEPORATI, C. ZANDRON, C. FERRETTI, G. MAURI: Solving numerical NP-complete problems with spiking neural P systems. *Lecture Notes in Computer Sci.*, 4860 (2007), 336–352.

[19] A. LEPORATI, C. ZANDRON, C. FERRETTI, G. MAURI: On the computational power of spiking neural P systems. *Intern. J. Unconventional Computing*, 5 (2009), 459–473.

[20] W. MAASS: Computing with spikes. *Special Issue on Foundations of Information Processing of TELEMATIK*, 8 (2002), 32–36.

[21] W. MAASS, C. BISHOP, eds.: *Pulsed Neural Networks*. MIT Press, 1999.

[22] A. PĂUN, GH. PĂUN: Small universal spiking neural P systems. *BioSystems*, 90 (2007), 48–60.

[23] GH. PĂUN: *Membrane Computing – An Introduction*. Springer, 2002.

[24] GH. PĂUN, M.J. PÉREZ-JIMÉNEZ, G. ROZENBERG: Spike trains in spiking neural P systems. *Intern. J. Found. Computer Sci.*, 17 (2006), 975–1002.

[25] GH. PĂUN, M.J. PÉREZ-JIMÉNEZ, G. ROZENBERG: Infinite spike trains in spiking neural P systems. Submitted, 2006.

[26] The P Systems Web Page: http://ppage.psystems.eu/.

CHAPTER 14

P SYSTEMS WITH OBJECTS ON MEMBRANES

MATTEO CAVALIERE

SHANKARA NARAYANAN KRISHNA

ANDREI PĂUN

GHEORGHE PĂUN

14.1 INTRODUCTION

IN "standard" membrane computing, objects are placed in compartments defined by membranes and these objects evolve by means of rules corresponding to reactions taking place in compartments of a cell, or by other types of rules such as the symport/antiport ones; in turn, membranes evolve under the control of objects placed in compartments.

However, this covers only part of the biological reality, where most of the reactions taking place in a cell are controlled by proteins bound on membranes. They can be of two types: *peripheral* proteins, placed on one side of a membrane, internal or external, or *integral* proteins (also called transmembrane proteins), which have parts of the molecule on both sides of the membrane (it is estimated that in the animal cells, the proteins constitute about half of the mass of the membranes).

Thus, it is a natural challenge for membrane computing to take care of this situation and, indeed, several types of P systems were investigated where membranes carry objects.

One of the first papers where this aspect was addressed is [16], where a compartment was considered between the two phospholipid layers, with the objects supposed to be bound on the membrane being placed in this compartment. In what follows we recall other approaches, somewhat more faithful to the reality, both in what concerns the place of objects and the rules to process them.

Three approaches are discussed. The first is inspired from [5], where six basic operations were introduced, counterparts of biological operations with membranes and two calculi were defined on this basis: pino/exo/phago calculus and mate/drip/bud calculus. Part of these operations were rephrased in terms of membrane computing and used in defining classes of P systems where objects are placed only on membranes. We call *(mem)brane systems* the obtained P systems. Ideas and results from [6], [2], [20] will be recalled in Section 14.2.

In the above mentioned systems we do not have objects also in compartments, as in usual P systems. Two models taking care of this aspect are what we call *the Ruston model* and *the Trento model*. In both cases, objects are placed both on the membranes and in their compartments, with the difference that in the former case the objects do not change their places, those bound on membranes remain there, while in the latter model the objects can move from compartments to membranes and back. The first type of system was introduced in [17] and then investigated in a series of papers among which we mention [18], [19], [21], and [14]. The second type of system was introduced in [3] and then investigated in [7], [8], [9], etc.

Besides main definitions, we recall a few typical results, with only hints about the proofs. Details can be found in the papers mentioned above and in other related papers listed in the bibliography which ends the chapter.

14.2 (Mem)brane Systems

We start with some elements of brane calculi, then we introduce the P systems based on the respective operations, and after that we recall some results about the computing power of these systems.

14.2.1 The Pino/Exo/Phago and Mate/Drip/Bud Calculi

Instead of giving preliminary details, we pass directly to presenting the six basic operations, *pino, exo, phago* and *mate, drip, bud* (grouped in this way because distinct calculi were defined based on them, respectively. In all cases, **P, Q** are arbitrary subsystems and s, t, u, v, w are arbitrary patches (sequences of actions); in certain cases, the actions have also complementary actions; the notations are those of process algebra.

(pino) $[\,\mathbf{P}\,]u|(pino(s).t) \to [\,\mathbf{P}\,[\,]\,s\,]u|t,$

(exo) $[\,[\,\mathbf{P}\,]u|(exo.t)\mathbf{Q}\,]w|(co\text{-}exo.v) \to \mathbf{P}[\,\mathbf{Q}\,]u|w|t|v,$

(phago) $[\,\mathbf{P}\,]u|(phago.t)[\,\mathbf{Q}\,]w|(co\text{-}phago(s).v) \to [\,[\,[\,\mathbf{P}\,]u|t\,]s\mathbf{Q}\,]w|v,$

(drip) $[\,\mathbf{P}\,]u|(drip(s).t) \to [\,\mathbf{P}\,]u|t[\,]s,$

(mate) $[\,\mathbf{P}\,]u|(mate.t)[\,\mathbf{Q}\,]w|(co\text{-}mate.v) \to [\,\mathbf{PQ}\,]u|t|wv,$

(bud) $[\,[\,\mathbf{P}\,]u|(bud.t)\mathbf{Q}\,]w|(co\text{-}bud(s).v) \to [\,[\,\mathbf{P}\,]u|t\,]s[\,\mathbf{Q}\,]w|v.$

The action $pino(s)$ creates an empty membrane inside the membrane where the pino action is placed; the patch on the empty bubble so created, s, is a parameter to pino. The exo operation models the merging of two nested membranes; note that the subsystem **P** gets expelled to the outside, and all the residual patches of the two membranes, u, w, t, v, become contiguous. The phago operation models a membrane (the one with **Q**) "eating" another membrane (the one with **P**). In turn, *drip* produces an empty bubble (like *pino*), but outside a membrane, *mate* merges two membranes (like *exo*), but the membranes are not initially nested, while *bud* expels a membrane from inside another one (the opposite of *phago*).

14.2.2 Pino/Exo and Mate/Drip as Membrane Computing Operations

Only the four operations *pino, exo, mate, drip* were used in defining P systems, that is why we introduce here only their membrane computing counterparts, with two possible variants for *pino, exo*.

Let us consider an alphabet A of "proteins" (usually, we use the general term "objects"). A membrane which has associated a multiset $u \in A^*$ (as customary, we represent multisets by strings) is represented by $[\ ]_u$; we also use to say that u *marks* the membrane. When necessary, we can also use labels for identifying membranes, but in most cases the objects marking them are sufficient. The following operations with membranes can be defined:

$$pino_i : [\ ]_{uav} \to [[\ ]_{ux}]_v, \tag{1}$$

$$exo_i : [[\ ]_{ua}]_v \to [\ ]_{uxv}, \tag{2}$$

$$pino_e : [\ ]_{uav} \to [[\ ]_v]_{ux}, \tag{3}$$

$$exo_e : [\,[\ ]_u\,]_{av} \to [\ ]_{uxv}, \tag{4}$$

$$mate : [\ ]_{ua}[\ ]_v \to [\ ]_{uxv}, \tag{5}$$

$$drip : [\ ]_{uav} \to [\ ]_{ux}[\ ]_v, \tag{6}$$

where $a \in V$ and $u, v, x \in A^*$. The length of the string uxv (i.e. the total multiplicity of the multiset represented by this string) from each rule is called the *weight* of the rule.

In each case, multisets of objects are transferred from membranes appearing in the left-hand side of rules to membranes appearing in the right-hand side of rules, with protein a evolved into the multiset x. It is important to note that the multisets u, v and the object a marking the left-hand membranes of these rules correspond to the multisets u, v, x from the right-hand side of the rules; specifically, in (1), (3), (6), the multiset uxv resulting when applying the rule is precisely split into ux and v, with these two multisets assigned to the two new membranes.

The rules are applied to given membranes if they are marked with multisets of objects which include the multisets mentioned in the left-hand side of rules. All objects placed on membranes but not involved in the rules are not affected by the use of rules; in the case of *pino* and *drip*, these objects are randomly distributed to the two resulting membranes. In the case of the other operations, the result is uniquely determined. From a pair of membranes $[\,[\ ]_{z_1ua}\,]_{z_2v}$, by an exo_i rule we obtain the membrane $[\ ]_{z_1z_2uxv}$, and from $[\,[\ ]_{z_1u}\,]_{z_2av}$, by an exo_e rule we obtain the membrane $[\ ]_{z_1z_2uxv}$. In turn, from a pair of membranes $[\ ]_{z_1ua}[\ ]_{z_2v}$, by means of a *mate* rule we obtain the membrane $[\ ]_{z_1z_2uxv}$. In all cases, z_1, z_2 are arbitrary multisets over A. Thus, the rules of types (1), (3), and (6) introduce non-determinism in the evolution of the membranes, while the rules of types (2), (4), and (5) remove the non-determinism introduced by the former rules—providing that they are applied to the pair of membranes produced by the first rules.

The contents of membranes involved in these operations (i.e. the membranes possibly placed inside) are transferred from the input membranes to the output membranes in the same way as in brane calculi. Denoting these contents (empty or consisting of other membranes) by P, Q as in Subsection 14.2.1, we can indicate the effect of the six operations as follows:

$$pino_i : [\ P\ ]_{uav} \to [\,[\ ]_{ux}\ P\ ]_v,$$

$$exo_i : [\,[\ P\ ]_{ua}\ Q\ ]_v \to P\,[\ Q\ ]_{uxv},$$

$$pino_e : [\ P\ ]_{uav} \to [\,[\ ]_v\ P\ ]_{ux},$$

$$exo_e : [\,[\ P\ ]_u\ Q\ ]_{av} \to P\,[\ Q\ ]_{uxv},$$

$$mate : [\ P\ ]_{ua}[\ Q\ ]_v \to [\ PQ\ ]_{uxv},$$

$$drip : [\ P\ ]_{uav} \to [\ ]_{ux}[\ P\ ]_v.$$

Some of these operations are used in the membrane computing framework as indicated below.

14.2.3 P Systems Using the Mate, Drip Operations

Typically, rules as above can be used in a P system of the form

$$\Pi = (A, \mu, u_0, u_1, \ldots, u_m, R), \text{ where :}$$

1. A is an alphabet (its elements are called *objects* or *proteins*);
2. μ is a membrane structure with at least two membranes (hence $m \geq 1$), labeled with $0, 1, \ldots, m$, where 0 is the skin membrane;
3. $u_0, u_1, \ldots, u_m \in A^*$ are multisets of objects bound to the membranes of μ at the beginning of the computation, with $u_0 = \lambda$;
4. R is a finite set of *mate, drip* rules, of the forms specified in the previous subsection, using objects from the set A.

Note that the skin membrane has no protein associated, because it cannot enter any rule, it is only meant to delimit the system from its environment.

As usual in membrane computing, the evolution of the system proceeds through *transitions* among *configurations*, based on the *non-deterministic maximally parallel* use of rules. In each step, each membrane and each object can be involved in only one rule. A membrane remains unchanged if no rule is applied to it. The skin membrane never evolves.

A computation which starts from the initial configuration (the one described by μ and multisets $u_0, u_1, \ldots, u_m$) is *successful* only if (i) it halts, that is, it reaches a configuration where no rule can be applied, and (ii) in the halting configuration there are only two membranes, the skin (marked with the empty multiset) and an inner one. The *result* of a successful computation is the number of proteins which mark the inner membrane in the halting configuration.

The set of all numbers computed in this way by Π is denoted by $N(\Pi)$. The family of all sets $N(\Pi)$ computed by P systems Π using at any moment during a halting computation at most m membranes, and *mate, drip* rules of weight at most p, q, respectively, is denoted by $NOP_m(mate_p, drip_q)$.

14.2.4 Universality for the Mate/Drip Case

The following result is proved in [1], improving in all parameters m, p, q a previous result from [6].

Theorem 14.1 $NRE = NOP_m(mate_p, drip_q)$ for all $m \geq 5$ and $p, q \geq 4$.

Proof. Only the inclusion $NRE \subseteq NOP_5(mate_4, drip_4)$ needs a proof and to this aim one uses the characterization of NRE by means of register machines. Taking a register machine $M = (m, H, l_0, l_h, I)$ we construct the P system

$$\Pi = (A, [\,[\,]_1[\,]_2]_0, \lambda, hl_0, bc, R), \text{ with}$$
$$A = \{a_r \mid 1 \leq r \leq m\} \cup \{l, l', l'' \mid l \in H\}$$
$$\cup \{d_i \mid 1 \leq i \leq 8\} \cup \{b, c, h, h', s, \#\},$$

and the set R of rules constructed as suggested below.

The system Π starts with one inner membrane marked with the label l_0 corresponding to the initial instruction of M, together with the helping object h, and the second inner membrane marked with the auxiliary (control) objects b and c. The contents of a register r of M will be represented by the number of copies of object a_r present in the system.

Both ADD and SUB instructions of M are simulated in eight steps of a computation in Π, with the membranes produced from $[\,]_{bc}$ evolving in the same way in the two cases. The simulation of a SUB instruction is more complex. Specifically, in order to simulate an instruction $l_i : (\mathrm{SUB}(r), l_j, l_k)$ one uses the rules from Table 14.1 (in the configurations presented in the right-hand column we do not

Table 14.1 Rules simulating a SUB instruction $l_i : (\mathrm{SUB}(r), l_j, l_k)$; $l_i^? \in \{l_i', l_i''\}$, if $l_i^? = l_i''$, then $l_a = l_j$, and if $l_i^? = l_i'$, then $l_a = l_k$.

Step	Rules	Types	Configuration
initial			$[\,[\,]_{hl_i}[\,]_{bc}\,]_\lambda$
1	$[\,]_{bc} \to [\,]_{d_1 d_2 d_3}[\,]_c$	drip	$[\,[\,]_{hl_i}[\,]_{d_1 d_2 d_3}[\,]_c\,]_\lambda$
2	$[\,]_{d_1 d_2 d_3} \to [\,]_s[\,]_{d_2 d_3}$	drip	$[\,[\,]_{hl_i}[\,]_s[\,]_{d_2 d_3}[\,]_c\,]_\lambda$
3	$[\,]_{hl_i}[\,]_s \to [\,]_{hl_i's}$	mate	$[\,[\,]_{hl_i's}[\,]_{d_4}[\,]_{d_3}[\,]_c\,]_\lambda$
	$[\,]_{d_2 d_3} \to [\,]_{d_4}[\,]_{d_3}$	drip	
4	$[\,]_{l_i' a_r hs} \to [\,]_{l_i'}[\,]_{hs}$	drip	$[\,[\,]_{l_i'}[\,]_{hs}[\,]_{d_5 d_6 d_3}[\,]_c\,]_\lambda$
	$[\,]_{d_4}[\,]_{d_3} \to [\,]_{d_5 d_6 d_3}$	mate	or $[\,[\,]_{l_i' sh}[\,]_{d_5 d_6 d_3}[\,]_c\,]_\lambda$
5	$[\,]_{l_i'}[\,]_{sh} \to [\,]_{l_i'' sh}$	mate	$[\,[\,]_{hl_i^? s}[\,]_{d_7}[\,]_{d_5 d_6}[\,]_c\,]_\lambda$
	$[\,]_{d_3 d_5 d_6} \to [\,]_{d_7}[\,]_{d_5 d_6}$	drip	
6	$[\,]_{hl_i^? s}[\,]_{d_7} \to [\,]_{hl_a sd_7}$	mate	$[\,[\,]_{hl_a sd_7}[\,]_{d_8}[\,]_{d_5}[\,]_c\,]_\lambda$
	$[\,]_{d_6 d_5} \to [\,]_{d_8}[\,]_{d_5}$	drip	
7	$[\,]_{hsl_a d_7} \to [\,]_h[\,]_{l_a d_7}$	drip	$[\,[\,]_h[\,]_{l_a d_7}[\,]_{bc}[\,]_{d_5}\,]_\lambda$
	$[\,]_{d_8}[\,]_c \to [\,]_{bc}$	mate	
8	$[\,]_{l_a d_7}[\,]_h \to [\,]_{l_a h}$	mate	$[\,[\,]_{l_a h}[\,]_{bc}\,]_\lambda$
	$[\,]_{d_5}[\,]_{bc} \to [\,]_{bc}$	mate	

specify the other objects present on the respective membranes, but only those involved in the rules from the left-hand column).

If a copy of the symbol a_r exists (which means that register r is not empty), then it is removed in step 4 and the computation continues with the symbol l'_i (corresponding to the l_k instruction); if a_r does not exist, then the computation continues with the symbol l''_i (corresponding to the l_j instruction); in both cases the configuration $[\ [\ ]_{hl^?_i s}[\ ]_{d_7}[\ ]_{d_5 d_6}[\ ]_c\ ]_\lambda$ is reached after step 5.

Due to the subscripts of objects d_i, $1 \leq i \leq 8$, the simulation proceeds almost deterministically: in each step there is only one choice of rules to apply. The only exceptions are the possibilities of using (i) the rule $[\ ]_{d_2 d_3} \to [\ ]_{d_4}[\ ]_{d_3}$ already in step 2, and not in step 3 as indicated in Table 14.1, (ii) the rule $[\ ]_{d_6 d_5} \to [\ ]_{d_8}[\ ]_{d_5}$ already in step 5, and not in step 6 as indicated in Table 14.1, and (iii) the rule $[\ ]_{bc} \to [\ ]_{d_1 d_2 d_3}[\ ]_c$ in step 8 instead of the rule $[\ ]_{d_5}[\ ]_{bc} \to [\ ]_{bc}$.

In order to prevent the halting of the computation in these cases, we introduce in R the additional rules:

1. $[\ ]_h[\ ]_{d_5} \to [\ ]_{\#\# d_5}$,
2. $[\ ]_{\#\#} \to [\ ]_\#[\ ]_\#$,
3. $[\ ]_\#[\ ]_\# \to [\ ]_{\#\#}$.

By means of these rules, if we do not use exactly the rules indicated in Table 14.1 for the eight steps of the simulation, then the computation never halts. If we use the rules as indicated, then the simulation of the SUB instruction is correctly completed and we return to a configuration similar to the one we have started with, hence with two inner membranes, one marked with hl_α, for $\alpha \in \{l_j, l_k\}$, and one marked with bc.

The simulation of an ADD instruction $l_i : (\mathrm{ADD}(r), l_j, l_k)$ is done (also in eight steps) by the rules given in Table 14.2.

We do not describe in detail the use of these rules.

The computation in Π halts if and only if the computation in M halts. However, in order to have a successful computation in Π we need to keep only one membrane inside the system and to remove all other objects than a_1, the object whose multiplicity represents the contents of register 1 of M. To this aim, we use the rules from Table 14.3, which can be applied deterministically after having the object l_h present in the system.

The maximum number of membranes used is 5 (after step 2 of simulating ADD and SUB instructions) and we never use more than 4 objects in controlling the rules (in steps 4, 6, 7 from Table 14.1, in steps 6, 7 from Table 14.2, and in step 1 from Table 14.3). This concludes the proof.

Remark 14.1 If one considers as output the number of proteins marking the membrane just below the skin membrane, without restricting the number of membranes

Table 14.2 Rules simulating an ADD instruction $l_i : (ADD(r), l_j, l_k)$; $l_a \in \{l_j, l_k\}$.

Step	Rules	Types	Configuration
initial			$[\,[\,]_{hl_i}[\,]_{bc}\,]_\lambda$
1	$[\,]_{bc} \to [\,]_{d_1 d_2 d_3}[\,]_c$	drip	$[\,[\,]_{hl_i}[\,]_{d_1 d_2 d_3}[\,]_c\,]_\lambda$
2	$[\,]_{d_1 d_2 d_3} \to [\,]_s[\,]_{d_2 d_3}$	drip	$[\,[\,]_{hl_i}[\,]_s[\,]_{d_2 d_3}[\,]_c\,]_\lambda$
3	$[\,]_{hl_i}[\,]_s \to [\,]_{hl'_i a r s}$	mate	$[\,[\,]_{hl'_i a r s}[\,]_{d_4}[\,]_{d_3}[\,]_c\,]_\lambda$
	$[\,]_{d_2 d_3} \to [\,]_{d_4}[\,]_{d_3}$	drip	
4	$[\,]_{l'_i hs} \to [\,]_{l'_i}[\,]_{hs}$	drip	$[\,[\,]_{l'_i}[\,]_{hs}[\,]_{d_5 d_6 d_3}[\,]_c\,]_\lambda$
	$[\,]_{d_4}[\,]_{d_3} \to [\,]_{d_5 d_6 d_3}$	mate	
5	$[\,]_{l'_i}[\,]_{sh} \to [\,]_{l''_i sh}$	mate	$[\,[\,]_{hl''_i s}[\,]_{d_7}[\,]_{d_5 d_6}[\,]_c\,]_\lambda$
	$[\,]_{d_3 d_5 d_6} \to [\,]_{d_7}[\,]_{d_5 d_6}$	drip	
6	$[\,]_{hl''_i s}[\,]_{d_7} \to [\,]_{hl_a s d_7}$	mate	$[\,[\,]_{hl_a s d_7}[\,]_{d_8}[\,]_{d_5}[\,]_c\,]_\lambda$
	$[\,]_{d_6 d_5} \to [\,]_{d_8}[\,]_{d_5}$	drip	
7	$[\,]_{hsl_a d_7} \to [\,]_h[\,]_{l_a d_7}$	drip	$[\,[\,]_h[\,]_{l_a d_7}[\,]_{bc}[\,]_{d_5}\,]_\lambda$
	$[\,]_{d_8}[\,]_c \to [\,]_{bc}$	mate	
8	$[\,]_{l_a d_7}[\,]_h \to [\,]_{l_a h}$	mate	$[\,[\,]_{l_a h}[\,]_{bc}\,]_\lambda$
	$[\,]_{d_5}[\,]_{bc} \to [\,]_{bc}$	mate	

to two, then we have an improvement of the above result. Specifically, it has been proved in [13] that $PsRE = PsOP_m(mate_p, drip_q)$, for all $m \geq 4$ and $p, q \geq 3$. This proof is based on simulating matrix grammars with appearance checking in the strong binary normal form.

Whether or not the above results are optimal in what concerns the number of membranes and the weight of the used rules remains as an *open problem*.

Table 14.3 Rules ending the computation.

Step	Rules	Types	Configuration
initial			$[\,[\,]_{hl_h}[\,]_{bc}\,]_\lambda$
1	$[\,]_{bc}[\,]_{hl_h} \to [\,]_{hl_h b}$	mate	$[\,[\,]_{hl_h b}\,]_\lambda$
2	$[\,]_{l_h h b} \to [\,]_{l_h h'}[\,]_b$	drip	$[\,[\,]_{l_h h'}[\,]_b\,]_\lambda$
3	$[\,]_{h' l_h}[\,]_b \to [\,]_{h' b}$	mate	$[\,[\,]_{h' b}\,]_\lambda$
4	$[\,]_{h' b a_1} \to [\,]_{h'}[\,]_{a_1}$	drip	$[\,[\,]_{h'}[\,]_{a_1}\,]_\lambda$
5	$[\,]_{h'}[\,]_{a_1} \to [\,]_{a_1}$	mate	$[\,[\,]_{a_1}\,]_\lambda$

14.2.5 P Systems with Projective Operations

In the previous set-up, the proteins are of the integral type. The case of peripheral proteins leads to what in brane calculi was called *projective* operations, see [10].

The fact that a multiset u is placed on the internal side of a membrane and a multiset v on its external side is denoted as $[\ _u]_v$, hence with the right-hand bracket having both left and right subscripts.

A projective counterpart of Theorem 14.1 can be obtained, but at the price of using one further operation, *exo*. In the projective form, these three operations are defined as follows:

$$mate : [\ ]_{ua}[\ ]_v \to [\ ]_{uxv},$$
$$drip : [\ ]_{uav} \to [\ ]_{ux}[\ ]_v,$$
$$exo : [[\ ]_v{}_{au}] \to [\ _{uxv}],$$

where $a \in V, u, v, x \in A^*$. At the syntactic level, there is no difference between the projective and the standard *mate, drip* operations, they use only proteins placed on the external side of membranes. However, the proteins present on the membranes entering these operations and not used by the rules are distributed on one of the sides of the resulting membranes as suggested below (x_1, x_2, x_3, x_4 are generic multisets, and P, Q indicates the contents of the respective membranes, i.e. the possible membranes present inside it):

$$mate : [\ P\ _{x_1}]_{uax_2}[\ Q\ _{x_3}]_{vx_4} \to [\ PQ\ _{x_1x_3}]_{uxvx_2x_4},$$
$$drip : [\ Q\ _{x_1x_2}]_{uavx_3x_4} \to [\ _{x_1}]_{uxx_3}[\ Q\ _{x_2}]_{vx_4},$$
$$exo : [[\ Q\ _{x_1}]_{vx_2}{}_{aux_3}]_{x_4} \to [\ _{uxvx_2x_3}]_{x_1x_4}Q.$$

Note that in the case of *exo*, the proteins change the inside-outside position.

A P system using operations as above is defined in the usual way, with the only difference that for each membrane we have to specify two multisets, the one marking it from inside and the one marking it from outside. Formally, we write $\Pi = (A, \mu, (u_1, v_1), \ldots, (u_m, v_m), R)$, with the meaning that u_i is the internal marking of membrane i and v_i is the external marking of this membrane.

The projective approach makes possible the definition of the result of a computation in a more "realistic" way: instead of reading the result on an inner membrane, we can define now the result of a computation as the number of objects which mark the external side of the skin membrane of the system in the halting configuration—the skin membrane is allowed to enter *exo* operations.

The family of all sets $N(\Pi)$ computed by P systems Π using at any moment during a halting computation at most m membranes, and projective *mate, drip, exo* rules of weight at most p, q, r, respectively, is denoted by $NOP_m(pmate_p, pdrip_q,$

Table 14.4 Rules for the last two steps of the computation in the projective case.

Step	Rules	Types	Configuration
7			$[[[\ _{f_1}]_{bh'a_1 a_1^{n-1}}\ _{f_2}]_\lambda\ _{f_3}]_\lambda$
8	$[[\ ]_{h'a_1}\ _{f_2}] \to [\ _{h'a_1}]$	exo	$[[\ _{h'a_1^n}]_{f_1}\ _{f_3}]_\lambda$
9	$[[\ ]_{f_1}\ _{f_3}] \to [\ _{f_1}]$	exo	$[\ _{f_1}]_{h'a_1^n}$

$pexo_r$). When numbers 0 and 1 are ignored, we denote by $1NOP_m(pmate_p, pdrip_q, pexo_r)$ the corresponding families.

The proof of the following result is very much similar to the proof of Theorem 14.1. We start from a register machine M and construct a P system with the initial configuration

$$[\ [\ [\ _{f_1}]_{l_0 h}\ [\ _\lambda]_{bc}\ _{f_2}]_\lambda\ _{f_3}]_\lambda$$

(some membranes are marked on both sides, and one further membrane is added around the system); the new objects f_1, f_2, f_3 are added to the alphabet of objects.

The simulation of ADD and SUB instructions use the projective version of the rules from Tables 14.1 and 14.2, and they do not use the external membranes, those with internal markings f_2 and f_3.

In what concerns the final sequence of steps, we replace the last two rows from Table 14.3 with the two steps indicated in Table 14.4—we recall also the configuration obtained after step 3, including the multiset a_1^n marking the inner membrane, $n \geq 1$.

The two *exo* steps move the result of the computation on the external side of the unique membrane present in the system in the halting configuration. Besides the copies of a_1 corresponding to a value $n \in N(M)$, we also have here the object h', that is why numbers 0 and 1 should be ignored.

Therefore, we obtain:

Theorem 14.2 $1NRE = 1NOP_m(pmate_p, pdrip_q, pexo_r)$ for all $m \geq 6$, $p, q \geq 4$, and $r \geq 3$.

14.2.6 P Systems with Pino, Exo Operations

The study of P systems with other combinations of rules inspired in brane calculi is still a research topic. In what follows we use rules with the same syntax as *pino, exo* operations, but with a different interpretation of what concerns the contents of the handled membranes, that is why we call them *cre, dis* operations (for "membrane creation" and "membrane dissolution").

Specifically, we change the interpretation of the *pino, exo* rules as suggested below:

$$cre_i : [\ P\ ]_{uav} \to [[\ P\ ]_{ux}]_v,$$
$$dis_i : [[\ P\ ]_{ua}\ Q\ ]_v \to [\ P Q\ ]_{uxv},$$
$$cre_e : [\ P\ ]_{uav} \to [[\ P\ ]_v]_{ux},$$
$$dis_e : [[\ P\ ]_u\ Q\ ]_{av} \to [\ P Q\ ]_{uxv}.$$

That is, when a membrane is created inside an existing membrane, the new membrane contains all previously existing membranes, and while dissolving a membrane, its contents remain inside the membrane where it was placed before the operation. The interpretation of the latter operation is rather similar to the usual dissolution operation in membrane computing, while the membrane creation is understood as doubling the existing membrane, with a distribution of the multiset marking the initial membrane to the two new membranes.

Using rules of these types, we can define a P system in the standard way. For such a system Π we denote by $N(\Pi)$ the set of numbers computed as in the previous sections, and by $NOP_m(cre_p, dis_q)$ the family of sets $N(\Pi)$ computed by systems Π using at any moment during a computation at most m membranes, and cre_i, dis_i rules of weight at most p, q, respectively. (P systems based on cre_e, dis_e rules were not considered yet.)

A proof of the following result can be found in [20]. We only mention that this proof is based on simulating matrix grammars with appearance checking by means of P systems using *cre, dis* rules; finding a proof based on simulating register machines (maybe improving the parameters appearing in the theorem below) remains an *open problem*.

Theorem 14.3 $1NRE = 1NOP_m(cre_p, dis_q)$ *for all* $m \geq 7$ *and* $p, q \geq 4$.

14.3 THE RUSTON MODEL

We pass now to the case where we can have objects both in compartments, as in usual P systems, and on membranes, but with the latter objects never leaving their place and used mainly to control the evolution of other objects. Actually, the objects evolve by means of a sort of antiport rules, controlled by "protein" objects—with the notice that we consider minimal rules: one object inside the compartment and one outside evolve (they may not only interchange their place, but also change into other objects) with the help of one object placed on the membrane (which remains unchanged or may also evolve).

We present immediately this new type of rule, after specifying some notation. In order to distinguish the objects placed on membranes from those from the compartments, we call the former *proteins* and the latter simply *objects*. The fact that a multiset u of proteins is placed on a membrane (with label) i is written in the form $[_i u|$. Then, a membrane structure can contain both proteins and objects, which leads to expressions of the form

$$[_1 u_1 | u_2 [_2 u_3 | u_4]_2]_1.$$

We have here two membranes, with labels 1 and 2; the inner membrane 2 contains the multiset u_4 inside its region and the multiset of proteins u_3 on it, while in region 1 we have the objects of the multiset u_2 and the respective membrane is marked with the multiset of proteins u_1.

14.3.1 New Types of Rules

We can consider several types of rules for handling the objects and the proteins; we recall here those introduced in [17] (see also [21]). In all of them, a, b, c, d are objects, p, p' are proteins, and i is a label ("res" stands for "restricted", and "cp" stands for "change protein").

In the first group of rules, the protein only plays the role of a catalyst, it is not changed when applying the rule:

Type	Rule	Effect
1res	$[_i p\|a \to [_i p\|b$ $a[_i p\| \to b[_i p\|$	modify an object, but not move
2res	$[_i p\|a \to a[_i p\|$ $a[_i p\| \to [_i p\|a$	move an object, but not modify
3res	$[_i p\|a \to b[_i p\|$ $a[_i p\| \to [_i p\|b$	modify and move one object
4res	$a[_i p\|b \to b[_i p\|a$	interchange two objects
5res	$a[_i p\|b \to c[_i p\|d$	interchange and modify two objects

A generalization is to allow rules of the forms below, where the protein can also change:

Type	Rule	Effect (besides changing also the protein)
1cp	$[_i p\|a \to [_i p'\|b$ $a[_i p\| \to b[_i p'\|$	modify an object, but not move
2cp	$[_i p\|a \to a[_i p'\|$ $a[_i p\| \to [_i p'\|a$	move an object, but not modify
3cp	$[_i p\|a \to b[_i p'\|$ $a[_i p\| \to [_i p'\|b$	modify and move one object
4cp	$a[_i p\|b \to b[_i p'\|a$	interchange two objects
5cp	$a[_i p\|b \to c[_i p\|d$	interchange and modify two objects

Rules of type *cp* become of type *res* as soon as $p = p'$. An intermediate case can be that of changing proteins, but in a restricted manner, by allowing at most two states for each protein, p and $\bar{p}$, and the rules either as in the first table (without changing the protein), or changing from p to $\bar{p}$ and back (like the bistable catalysts). Rules with such flip-flop proteins are denoted by *nff*, $n = 1, 2, 3, 4, 5$ (note that in this case we allow both rules which do not change the protein and rules which switch from p to $\bar{p}$ and back).

Clearly, rules of type 3 are more general than those of type 2 and rules of type 5 are more general than those of type 4; in turn, rules *cp* are more general than rules *ff*, which are more general than rules *res*.

Then, it is easy to see that any rule $[_i p | a \to [_i p' | b$ of type 1*cp* can be simulated, in two steps, by rules of type 3*cp*: $[_i p | a \to \bar{b} [_i \bar{p} |,\ \bar{b} [_i \bar{p} | \to [_i p' | b$.

The rules of type 2*res* correspond to uniport rules, while rules of type 4*res* correspond to minimal antiport rules. Is is important however to note that in our case the number of proteins never change, hence at a given step the number of rules which can be used is bounded by the number of proteins (hence the parallelism is restricted in this way).

14.3.2 The Power of P Systems with Rules as Above

Rules as defined above can be used in a natural way for defining P systems where proteins placed on membranes are also present. Because we cannot create objects in the system (the rules are "conservative" from this point of view), we need a supply of objects in the environment, as in symport/antiport systems.

Formally, we can consider systems as follows:

$$\Pi = (O, P, \mu, w_1/z_1, \ldots, w_m/z_m, E, R_1, \ldots, R_m, i_0),$$

where O is the set of objects, P is the set of proteins (with $O \cap P = \emptyset$), μ is the membrane structure of degree m, $w_1, \ldots, w_m$ are the multisets of objects present in the m regions of μ, $z_1, \ldots, z_m$ are the multisets of proteins present on the m membranes of μ, $E \subseteq O$ is the set of objects present in the environment (in an arbitrarily large number of copies each), $R_1, \ldots, R_m$ are finite sets of rules associated with the m membranes of μ, and i_0 is the output membrane, an elementary one in μ.

The rules can be of the forms specified above. If we use only rules which change the protein (for instance, we do not mix flip-flop rules with non-flipping rules), then we say that the system is *pure*. Also the set of *cp* rules can be *pure*: all of them change the protein. These cases are indicated by writing *ffp*, *cpp* instead of *ff*, *cp*, respectively.

The rules are used in the maximally parallel way; as usual, each object and each protein can be involved in the application of only one rule, but the membranes are not considered as involved in the rule applications, hence the same membrane can appear in any number of rules at the same time (this resembles the case of evolution rules in P systems with active membranes). With a halting computation we associate a result, in the form of the multiplicity of objects present in region i_0 in the halting configuration. We denote by $N(\Pi)$ the set of numbers computed in this way by a given system Π and by $NOP_m(pro_r;list\text{-}of\text{-}types\text{-}of\text{-}rules)$ the family of sets of numbers $N(\Pi)$ generated by systems Π with at most m membranes, using rules as specified in the list-of-types-of-rules, and with at most r proteins present on a membrane.

The current best results concerning the computing power of P systems with proteins on membranes are as presented below. Proofs can be found in [17], [14]:

Theorem 14.4 $NOP_1(pro_2;2cpp) = NOP_1(pro_2;2res, 4cpp) =$
$NOP_1(pro_2;2res, 1cpp) = NOP_1(pro_6;3ffp) =$
$NOP_1(pro_6;2ffp, 4ffp) = NOP_1(pro_6;2ffp, 5ffp) =$
$NOP_1(pro_9;1res, 2ffp) = NOP_1(pro_6;1ffp, 2ffp) =$
$NOP_1(pro_8;1ffp, 2res) = NOP_1(pro_8;2ffp, 3res) =$
$NOP_1(pro_7;1ffp, 3res) = NOP_1(pro_8;3res, 4ffp) =$
$NOP_1(pro_7;2ffp, 5res) = NRE.$

We give here only a few hints about the proof of the universality of the first family, $NOP_1(pro_2; 2cpp)$.

As usual, we only have to prove the inclusion $NRE \subseteq NOP_1(pro_2; 2cpp)$ and to this aim we use the characterization of NRE by register machines.

Consider such a machine $M = (m, H, l_0, l_h, I)$ (without loss of the generality we may assume that in ADD instructions $l_i : (\text{ADD}(r), l_j, l_k)$ we have $l_i \neq l_j$ and $l_i \neq l_k$). We construct the system

$$\Pi = (O, P, [\,_1\,]_1, \lambda/l_0 p, E, R_1, 1)$$

with the following components

$$O = \{a_r \mid 1 \leq r \leq m\} \cup \{c_l \mid l \in H\} \cup \{c, d\},$$
$$P = \{l, l', l'' \mid l \in H\} \cup \{p, p', p''\} \cup \{p_l \mid l \in H\},$$
$$E = \{a_r \mid 1 \leq r \leq m\} \cup \{c_l \mid l \in H\} \cup \{c, d\},$$

and the set R_1 containing the following rules.

1. For an ADD instruction $l_i : (\text{ADD}(r), l_j, l_k) \in I$, we consider the rules

$$a_r[\,_1 l_i| \to [\,_1 l_g | a_r, \ g \in \{j, k\}.$$

The correct simulation of the ADD instruction is obvious.

2. For a SUB instruction $l_i : (\text{SUB}(r), l_j, l_k) \in R$ we consider the rules from the following table (we also specify the proteins present on the membrane):

Step	Proteins	Rules
1	l_i, p	$c_{l_i}[_1 l_i| \to [_1 l'_i|c_{l_1}$
2	l'_i, p	$[_1 p|c_{l_i} \to c_{l_i}[_1 p_{l_i}|$ and $d[_1 l'_i| \to [_1 p'|d$
3	p', p_{l_i}	$[_1 p_{l_i}|a_r \to a_r[_1 l''_j|$, if a_r exists, and $c[_1 p'| \to [_1 p''|c$
4	l''_j or p_{l_i} and p''	$[_1 l''_j|c \to c[_1 l_j|$ or $[_1 p_{l_i}|c \to c[_1 l_k|$, and $[_1 p''|d \to d[_1 p|$

In the presence of protein l_i, the rule $c_{l_1}[_1 l_i| \to [_1 l'_i|c_{l_i}$ is applied. In the second step we apply in parallel both rules indicated in the previous table. The first rule sends object c_r out and changes protein p into p_{l_i}; the second rule brings object d inside and changes protein l'_i into p'. At step 3 we bring inside the object c and we change protein p' into p''. If we have at least one copy of object a_r inside the region, we can also apply the rule $[_1 p_{l_i}|a_r \to a_r[_1 l''_j|$, which sends out object a_r and changes the protein p_{l_i} into l''_j. In the last step we send out object d and change the protein p'' into its original form p. If at step 3 we have sent out a copy of object a_r, then we can apply the rule $[_1 l''_j|c \to c[_1 l_j|$, otherwise we still have protein p_{l_i} on the membrane, and we apply the rule $[_1 p_{l_i}|c \to c[_1 l_k|$. Therefore, we change the protein l_i into l_j or l_k depending whether we can send out an object a_r or not, and this way we correctly simulate the SUB instruction.

When the halt label l_h is present on the membrane, no further instruction can be simulated, and the number of copies of a_1 in membrane 1 is equal to the value of register 1 of M.

It is worth noting that in all cases considered in Theorem 14.4 we use at least two proteins. Actually, P systems with any type of rules as above, but using only one protein (and only one membrane), generate only semilinear sets of numbers. More specifically, the following result was proved in [17] (obviously, $PsOP_m(\ldots)$ denotes the family of vectors of natural numbers generated by P systems with at most m membranes and with proteins and rules as indicated in the notation):

Theorem 14.5 $PsOP_1(pro_1; any\text{-}rules) \subseteq PsMAT$.

14.3.3 Other Types of Rules

Let us first consider a more restrictive type of rule, namely with the possibility of changing the protein, but not moving or changing objects. Thus, we somewhat reverse the role of objects and proteins: this time the objects assist the proteins in evolving. Specifically, rules of the following forms are considered:

$$[_i p|a \to [_i p'|a, \quad a[_i p| \to a[_i p'|,$$

where a is an object and p, p' are proteins. We say that these rules are of type $0cp$.

Using such rules we cannot do too much, that is why we have to combine them with rules as in Subsection 14.3.1. Moreover, we can also consider evolution rules of the form

$$[a \to u]_i,$$

where a is an object and u is a multiset (hence, non-cooperative evolution rules as in P systems with active membranes). Evolution rules can be combined with all types of rules as in Subsection 14.3.1. A systematic investigation of these possibilities remains as a research topic, but universality results are expected in many (restricted) cases. For instance, we have (see [19]):

Theorem 14.6 $NOP_2(pro_*; ncoo, 2res, 0cp) = NRE$.

Note that this time the number of proteins is not bounded.

14.3.4 Accepting and Computing Systems

In the previous subsections, the systems are used in the generative mode: we start from the initial configuration, we proceed non-deterministically, and in the end of halting computations we get a result which is included in the computed set of numbers (or of vectors). Naturally, we can also consider *accepting* or *computing* systems. In the first case, a number is introduced in a membrane, in the form of the multiplicity of a specified object, and this number is accepted if the computation halts. In the computing case, starting with such an input, we also collect a result, like in a generative system.

Counterparts of the equalities in Theorem 14.4 can be obtained also for the accepting mode of using P systems. Details can be found in [21]. What is not systematically examined is the question whether or not the universality can be reached also by deterministic systems (remember that deterministic accepting register machines are universal, which is a good starting point for proofs).

Using a P system in the computing mode suggests the question of building *universal* systems, fixed systems which are able to simulate any particular system (from a specified class) after introducing a code of the particular system in the universal one. Building such universal systems can start from universal register machines (as constructed, e.g. in [12]). For instance, if we directly apply the construction from the proof of the inclusion $NRE \subseteq NOP_1(pro_2; 2cpp)$ sketched above (after Theorem 14.4), then we obtain a universal system with two proteins, using rules of type $2cpp$, with 113 rules. This number can be reduced by some "code optimization" (saving rules when simulating couples of instructions of the register machine which are applied one after the other one), but a "reasonably small" universal P system is not known.

Furthermore, accepting P systems can be used for solving decidability problems, as discussed in detail in Chapter 12. To this aim (to trade space for time), we need tools for producing an exponential workspace in polynomial (if possible, linear) time, and the usual way to do it is by considering rules for membrane division. This is particularly natural in the case of P systems with objects on membranes, because we can consider rules for the division of a membrane with the division controlled by proteins placed on the membrane itself (not inside the delimited region, as in P systems with active membranes). Specifically, we can consider rules of the form

$$[_i p | \;]_i \rightarrow [_i p' | \;]_i [_i p'' | \;]_i,$$

where p, p', p'' are proteins. Under the influence of protein p, membrane i is divided into two copies with the same label, with protein p replaced by p' and p'', respectively.

Using such rules and rules as in Subsection 14.3.1, solutions to SAT can be obtained in a time polynomial with respect to the number of variables and the number of clauses. We refer to [18] for details.

14.4 THE TRENTO MODEL

In the models considered in Section 14.2 the proteins are bound to the membranes and cannot change their place. An attractive possibility is to also consider *the movement of the proteins*, in particular the ability for the proteins to attach/de-attach to/from the membranes, and to be transported across the regions of the system. This feature can be then coupled with membranes operations (this model is presented in Section 14.4.1) or with standard evolution rules (presented in Section 14.4.3).

14.4.1 Transport of Proteins and Membranes Operations

We review a model (introduced in [3]) that considers marked membranes, operations to evolve the membranes, and the possibility of proteins changing their place.

We use the same notations as in Section 14.2. A membrane is represented by a pair of square brackets, [ ]. With each membrane a multiset of proteins u (over a certain alphabet A) can be associated that marks the membrane and this is denoted by $[\ ]_u$. The objects of A are called proteins or simply objects. The contents of a membrane can consist of proteins and/or other membranes.

The considered operations for the evolution of the membranes (simply called *membrane operations*) are the $pino_i$, $pino_e$, and *drip* defined as in Section 14.2; moreover we also use restrictions on these operations, by considering a non-cooperative version of these operations. This is obtained by imposing $uv = \lambda$ and we add the prefix (*ncoo*) to denote it. For instance, $(ncoo)pino_i : [\ ]_a \to [\ [\ ]_x\ ]$ is a non-cooperative $pino_i$ rule.

We also use rules that take care of the movement of the proteins, and, in particular that can attach/de-attach proteins to/from the membranes, and rules to move the proteins through the membranes of the system.

The *protein movement rules* have the following forms:

$$attach : [a]_u \to [\ ]_{ua}, \quad [\ ]_u a \to [\ ]_{ua},$$
$$de\text{-}attach : [\ ]_{ua} \to [a]_u, \quad [\ ]_{ua} \to [\ ]_u a,$$
$$move_{in} : [\ ]_u a \to [a]_u,$$
$$move_{out} : [a]_u \to [\ ]_u a,$$

with $a \in A$, $u \in A^*$.

The effect of the *attach* rules is simply to attach the protein a to the corresponding membrane if the marking of the membrane includes u.

The rules $move_{out}$ ($move_{in}$) move the protein a outside (inside, resp.) if the marking of the corresponding membrane includes u. In what follows *prot* is used to denote the set of protein movement rules.

A *membrane system with marked membranes, membrane operations, and protein movement rules* (in short, a *PP system*) is a construct $\Pi = (A, \mu, u_1, \ldots, u_m, R, F)$, where A a finite set of proteins (also referred as A_Π), μ is the membrane structure (with $m \geq 1$ membranes), $u_1, \ldots, u_m \in A^*$ are the markings of the membranes of μ at the beginning of the computation (the *initial markings* of Π), R is a finite set of membrane operations and protein movement rules, $F \subseteq A$ is the set of *protein-flags*, simply called *flags*, marking the *output membranes*. A *configuration* of Π consists of a membrane structure, the markings of the membranes, and the multisets of proteins present inside the regions. We suppose that in the *initial configuration* the regions are empty, thus the initial configuration is defined by μ and $u_1, \ldots, u_m$. A *transition* of Π from a configuration to a new one is performed by applying, to each membrane of the system, either (i) the protein movement rules in the non-deterministic maximally parallel manner (each protein can only be used by one rule), or (ii) one of the membrane operations. The choice between using protein movement rules or using a membrane operation, for each membrane, is done in a non-deterministic way if both types of rules can be applied for a given membrane. In this way, a membrane remains unchanged (only) if no rules can be applied to it. As usual, a sequence of transitions forms a *computation* that is *successful* if it starts from the initial configuration and halts—it reaches a *halting configuration*, where no rule can be applied anywhere in the system. In the halting configuration we consider the *output membranes* that are membranes whose markings contain at least one flag from F. The *result* of a successful computation is the set of vectors describing the multiplicities of proteins in the markings of the output membranes (notice that this is different from the output in the previous considered models, where we had only one output vector).

Collecting all the results, for all possible successful computations, we get the set of vectors generated by Π, denoted by $Ps(\Pi)$. We denote then by $\mathcal{PP}_m(\alpha, prot)$, with $\alpha \in \{pino_i, pino_e, drip, (ncoo)pino_i, (ncoo)pino_e, (ncoo)drip\}$, $m \geq 1$, the class of PP systems using membrane operations of type α, protein movement rules, and at most m membranes (α or $prot$ are removed if the corresponding rules are not used). The family of sets of vectors generated by PP systems from $\mathcal{PP}_m(\alpha, prot)$ is denoted by $PsPP_m(\alpha, prot)$. As usual, m is substituted by $*$ if the number of employed membranes is arbitrary. A configuration of a PP system that can be reached by a (possibly empty) sequence of transitions, starting from the initial configuration, is called *reachable*. A multiset w of proteins is then a *reachable marking* for Π if there exists a reachable configuration of Π which contains a membrane marked by w.

14.4.2 Computational Power and Decision Problems

We review the results from [3], [4] concerning the computational power of PP systems and several decision problems.

PP systems using non-cooperative pino and drip rules are quite limited as computational power: the generated family of sets of vectors is strictly included in the family of Parikh images of context-free languages.

Theorem 14.7 $PsPP_*((ncoo)\alpha) \subset PsCF$, $\alpha \in \{pino_i, pino_e, drip\}$.

In particular, in [4] it is shown that there is no PP system from $\mathcal{PP}_*((ncoo)\alpha)$ with $\alpha \in \{pino_i, pino_e, drip\}$ that can generate the Parikh image of the regular language $\{a^{2n} \mid n \geq 1\}$.

The computational power of these PP systems increases when cooperative membrane operations are used.

Theorem 14.8 $PsMAT \subseteq PsPP_*(\alpha)$, $\alpha \in \{pino_i, pino_e, drip\}$.

Proof. We give here only a sketch of the proof. We use the equality $MAT=PR$ and consider a programmed grammar $G = (N, T, P, S)$ without appearance checking, with productions written in the form $(i : A \rightarrow x, E_i)$. Without loss of generality, we can suppose that in G there exists an unique initial production (with label l_0) and an unique final production $Z \to \lambda$ (with label l_h). $Lab(G)$ denotes the set of labels used by G.

We construct a PP system $\Pi = (A, [\, [\,]\,], \lambda, E\, Sl_0, T)$ from $\mathcal{PP}_*(pino_i)$ that generates exactly the Parikh image of $L(G)$ with:

$$A = N \cup T \cup \{E\} \cup Lab(G) \cup \{\#\},$$

and the pino rules in R are defined as follows.

1. (simulation of the programmed grammar productions)
   $[\,]_{E\,Ai} \to [\, [\,]_{Exj}\,]_i$, for $(i : A \to x, E_i) \in P$, $j \in E_i$, $i \neq l_h$,
2. (used when a production of G cannot be applied)
   $[\,]_A \to [\, [\,]_{E\#\#}]$, $A \in N$,
3. (used for making a computation non-halting)
   $[\,]_{E\#\#} \to [\, [\,]_{E\#\#}\,]_\#$,
4. (used to keep the symbols from a sentential form on the same membrane)
   $[\,]_{Xi} \to [\, [\,]_{E\#\#}\,]_i$, for $X \in (N \cup T)$, $i \in Lab(G)$,
5. (used to halt the computation)
   $[\,]_{E\,Zl_h} \to [\, [\,]\,]_{El_h}$.

The simulation of the application of a production in G is done by using one of the rules in group 1. The label j of the next production is also produced while the old one, i, is stored on the created external membrane (the structure of the system contains at any time a unique innermost membrane that is marked by the objects corresponding to the sentential form. Rules of group 4 guarantee that the sentential form is always kept on such innermost membrane, that is identified by the object E). If a production cannot be applied (the corresponding non-terminal is not present), then the computation does not halt and this is ensured by the rules in groups 2 and 3.

The computation must halt only when the sentential form is composed by terminal objects (i.e. from T) and to this aim we use the rules of group 5 that remove the symbol E and l_h from the innermost membrane and delete the non-terminal object Z. If some non-terminal is still present, a rule of group 2 is used and the computation does not halt. Therefore, any successful derivation of G producing w can be simulated by a successful computation in Π halting in a configuration containing a unique innermost membrane—the unique output membrane which is marked by the multiset $\Psi_V(w)$.

On the other hand, unsuccessful derivations of G can be of the type $S \Rightarrow^* w$, for $w \in Z(N \cup T)^*$, or of the type $S \Rightarrow^* w_1 \Rightarrow w_2$, with $w_1 \in Z(N \cup T)^* N(N \cup T)^*$ and $w_2 \in (N \cup T)^* N(N \cup T)^*$. The simulation in Π of these two types of derivations results in non-halting computations.

Therefore $Ps(\Pi)$ is exactly the Parikh image of the language generated by the grammar G.

Table 14.5 Computational power for PP systems using $pino_i$ and protein movement rules (*prot*). The same table holds also for $pino_e$ and *drip* operations.

	w/o prot	prot
w/o $pino_i$		PsFIN
(ncoo)$pino_i$	$\subset$ PsCF	$\supseteq$ PsCF
$pino_i$	$\supseteq$ PsPR	PsRE

A similar construction can be given by using only $pino_e$ or using only *drip* rules. Therefore the theorem follows.

If PP systems are equipped with cooperative membrane operations and protein movement rules, then they are computationally complete.

Theorem 14.9 $PsPP_*(a, prot) = PsRE$, $a \in \{pino_i, pino_e, drip\}$.

Informally, it seems that the ability to move the proteins (in a controlled way) through the regions of the system is important for reaching computational completeness. On the other hand, it is interesting to note that the generative power of PP systems using exclusively protein movement rules is very restricted, since they can generate only finite families of sets of vectors.

The result of Theorem 14.9 can be obtained by simulation of programmed grammars with appearance checking, that are known to be universal. Intuitively, this can be done by adding to the rules employed in the proof of Theorem 14.8 a group of rules used to simulate the appearance checking mechanism present in the programmed grammars.

The current knowledge about the computational power of PP systems is summarized in Table 14.5 where several interesting cases are still *open*: which of the inclusions $PsCF \subseteq PsPP_*((ncoo)a, prot)$, $PsPP_*((ncoo)a, prot) \subseteq PsRE$, $PsPR \subseteq PsPP_*(a)$, $PsPP_*(a) \subseteq PsRE$, $a \in \{pino_i, pino_e, drip\}$, are strict?

Another line of research consists in finding classes of PP systems in which relevant properties can be algorithmically checked (i.e. decided).

From the universality result of Theorem 14.9 one can easily obtain:

Theorem 14.10 *It is undecidable whether or not, for an arbitrary PP system Π and an arbitrary multiset w of proteins, w is a reachable marking of Π.*

If PP systems use only protein movement rules, only pino rules, or only drip rules, then the above considered problem becomes decidable.

Theorem 14.11 *It is decidable whether or not, for an arbitrary PP system Π from $PP_*(a)$, $a \in \{prot, pino_i, pino_e, drip\}$, and an arbitrary multiset w of proteins over A_Π, w is a reachable marking of Π.*

The case $\alpha = prot$ follows from the fact that, when only protein movement rules are used, the number of distinct reachable configurations is finite. The other cases can be proved by showing that the set of strings representing all the reachable markings for $\Pi \in \mathcal{PP}_*(\alpha)$, $\alpha \in \{pino_i, pino_e, drip\}$, can be generated by a programmed grammar G without appearance checking (hence, with the membership problem decidable).

However, it is not yet known if it is possible to decide whether or not an arbitrary multiset of proteins is a reachable marking for an arbitrary PP system from $\mathcal{PP}_*((ncoo)\alpha, prot)$, with $\alpha \in \{pino_i, pino_e, drip\}$.

Looking back to the original biological motivations of the model, other relevant properties concern the reachability of an arbitrary configuration and the boundedness of an arbitrary system.

Theorem 14.12 *It is decidable whether or not, for an arbitrary PP system Π and an arbitrary configuration C of Π, C is a reachable configuration of Π.*

This result simply derives from the fact that one can generate in a systematic fashion all reachable configurations of Π containing no more than r membranes (each application of a pino or drip rule increases the number of membranes and this generation process takes a bounded number of steps).

We say that a PP system Π is *bounded* if there exists an integer k, such that any reachable configuration of Π has less than k membranes.

Theorem 14.13 *It is decidable whether or not an arbitrary PP system Π from $\mathcal{PP}_*(\alpha)$, $\alpha \in \{pino_i, pino_e, drip\}$, is bounded.*

The result can be obtained by constructing a programmed grammar G without appearance checking such that $L(G)$ consists of strings corresponding to all the reachable markings of Π.

The computational power of a variant of PP systems using operations $endo_i$, $endo_e$, bud, $wrap$, exo_i, exo_e, $mate$, and $pino_i$, $pino_e$ was studied also in [15]. The model as considered in [15] differs from PP systems in the following ways: (i) the de-attach operation is more general, i.e. is done by rules as follows $[\ ]_{ua} \to [\ ]_u a$, $[\ ]_{ua} \to [a]_u$, $[\ ]_a u \to [\ ] ua$, $[u]_a \to [ua]$, (ii) the output is interpreted as the set of Parikh vectors of objects marking (externally only) the membrane(s) just below the skin.

Several universality results are proved in [15], but we omit here the details.

14.4.3 Transport of Proteins and Evolution Rules

The model presented in Section 14.4.1 is obtained by joining the transport of proteins with membrane operations. An alternative model, proposed in [7], consists in coupling the transport of proteins with standard evolution rules. In this way, one can capture some of the basic features present in cellular processes: (i) biochemical reactions in the compartments; (ii) transport of molecules across the regions of the system involving the proteins attached to the membranes (on one or possibly both sides) and (iii) attachment/de-attachment of objects to/from the sides of the membranes.

As earlier, a membrane is represented by a pair of square brackets, [ ] and, as in Section 14.2.5, we associated to each topological side of a membrane multisets u and v (over a particular alphabet A of proteins/objects) and this is denoted by $[\ _u]_v$. We say that the membrane is marked by u and v, where v is the *external marking* and u is the *internal marking*. An object is called *free* if it is not attached to the sides of a membrane, so is not part of a marking.

Moreover, each membrane has also associated a label that is written as superscript of the square brackets. The set of all labels is then denoted by *Lab*.

We consider rules for the attachment, de-attachment and movement of the objects by extending the definitions given in Section 14.4.1:

$$\text{attach}: [\ a\ _u]_v^i \to [\ _{ua}]_v^i, \quad a[\ _u]_v^i \to [\ _u]_{va}^i,$$

$$\text{de-attach}: [\ _{ua}]_v^i \to [a\ _u]_v^i, \quad [\ _u]_{va}^i \to [\ _u]_v^i a,$$

$$\text{move}_{in}: a[\ _u]_v^i \to [\ a\ _u]_v^i,$$

$$\text{move}_{out}: [\ a\ _u]_v^i \to a[\ _u]_v^i,$$

with $a \in A$, $u, v \in A^*$ and $i \in Lab$.

As done in Section 14.4.1, the rules of attach, de-attach, move$_{in}$, move$_{out}$ are called *protein movement rules* and here we consider several restrictions.

Protein movement rules for which $|uv| \geq 2$ are called *cooperative* protein movement rules (in short, coo$_{prot}$). Protein movement rules for which $|uv| = 1$ are called *non-cooperative* protein movement rules (in short, ncoo$_{prot}$). Protein movement rules for which $|uv| = 0$ are called *simple* protein movement rules (in short, sim$_{prot}$).

We also use *evolution rules* $evol: [u \to v]^i$, with $u \in A^+$, $v \in A^*$ and $i \in Lab$, as considered in Section 14.3.3. An evolution rule is called *cooperative* (in short, coo$_e$) if $|u| > 1$, otherwise the rule is called *non-cooperative* (ncoo$_e$).

The semantics of the rules should be already clear after Section 14.4.1 and Section 14.3.3.

We consider a *membrane system with peripheral proteins* (in short, a *PR system*) as the construct $\Pi = (A, \mu, (u_1, v_1), \ldots, (u_m, v_m), w_1, \ldots, w_m, R)$ where A is the alphabet of proteins (objects) (also referred as A_Π), $(u_1, v_1), \ldots, (u_m, v_m) \in A^* \times A^*$ are the markings associated at the beginning to the membranes $1, \ldots, m$, respectively (*initial markings* of Π; the first element of each pair specifies the internal marking, the second one the external marking), $w_1, \ldots, w_m$ denotes the multisets of objects present in the m regions of the membrane structure μ (*initial contents*) of the regions, and R is a finite set of evolution and protein movement rules. A *configuration* of system Π consists of a membrane structure, the markings of the membranes (internal and external), and the multisets of free objects present inside the regions. The *initial configuration* consists of the membrane structure μ, the initial markings, and the initial contents of the regions. The set of all possible configurations of Π is denoted by $\mathcal{C}(\Pi)$. A *transition* from a configuration to a new one is obtained by applying the rules of the system to the objects present in the current configuration. We consider two different ways of assigning the objects, hence two types of transitions: *free-parallel transitions* and *maximal-parallel transitions*. In the free-parallel case, in each region and for each marking, an *arbitrary number* of applicable rules is executed. A single object (free or not) may only be assigned to a single rule. In the maximal-parallel case, as usual, objects are assigned to rules in a maximal, parallel, and non-deterministic manner. Also in this case, a single object (free or not) may only be assigned to a single rule. We denote by $\mathcal{C}_R(\Pi, fp)$ [$\mathcal{C}_R(\Pi, mp)$] the set of all configurations of Π reached, starting from the initial configuration, and applying free-parallel transitions [maximal-parallel transitions, resp.]. We then denote by $\mathcal{M}_R(\Pi, fp)$ [$\mathcal{M}_R(\Pi, mp)$] the set of all markings (internal and external) associated to the membranes present in the configurations in $\mathcal{C}_R(\Pi, fp)$ [$\mathcal{C}_R(\Pi, mp)$, resp.]. We also denote by $\mathcal{PR}_m(\alpha, \beta)$ with $\alpha \in \{coo_e, ncoo_e\}$, $\beta \in \{coo_{prot}, ncoo_{prot}, sim_{prot}\}$, the class of PR systems with evolution rules of type α, protein movement rules of type β, and at most m membranes ($*$ is used if m is unbounded and rules not used are omitted).

14.4.4 Reachability Problems

The model introduced in Section 14.4.3 was originally meant to describe cellular processes (in [9]) and several biologically-motivated problems were then investigated in [7] and [8]. Specifically the reachability of configurations and of markings was analyzed. In what follows, we survey the obtained results, by giving hints of the proofs (the reader can find details in [7] and [8]).

Theorem 14.14 *It is decidable whether or not, for an arbitrary PR system Π from $\mathcal{PR}_*(coo_e, coo_{prot})$ and an arbitrary configuration C of Π, $C \in \mathcal{C}_R(\Pi, fp)$.*

The proof of this result is based on the fact that all reachable configurations of a system Π from $\mathcal{PR}_*(coo_e, coo_{prot})$, working in the free-parallel way, can be generated by a pure matrix grammar without appearance checking, for which the membership problem is decidable. In a similar manner, one can prove that there is an algorithm to decide the reachability of an arbitrary marking.

Theorem 14.15 *It is decidable whether or not, for an arbitrary PR system Π from $\mathcal{PR}_*(coo_e, coo_{prot})$ and an arbitrary pair of multisets (u, v) over A_Π, $(u, v) \in \mathcal{M}_R(\Pi, fp)$.*

The situation is different when the evolution of a PR system is obtained by using maximal-parallel transitions. In this case, the reachability of an arbitrary configuration is decidable when either the evolution rules are non-cooperative and the protein movement rules are simple or the system uses only protein movement rules (even cooperative).

Theorem 14.16 *It is decidable whether or not:*

- *For an arbitrary PR system Π from $\mathcal{PR}_*(coo_{prot}) \cup \mathcal{PR}_*(ncoo_e, sim_{prot})$ and an arbitrary configuration C of Π, $C \in \mathcal{C}_R(\Pi, mp)$.*
- *For an arbitrary PR system Π from $\mathcal{PR}_*(coo_{prot}) \cup \mathcal{PR}_*(ncoo_e, sim_{prot})$ and an arbitrary pair of multisets u, v over A_Π, $(u, v) \in \mathcal{M}_R(\Pi, mp)$.*

However, the reachability of an arbitrary configuration becomes undecidable in systems combining non-cooperative evolution rules and cooperative protein movement rules.

Theorem 14.17 *It is undecidable whether or not, for an arbitrary PR system Π from $\mathcal{PR}_*(ncoo_e, coo_{prot})$ and an arbitrary configuration C of Π, $C \in \mathcal{C}_R(\Pi, mp)$.*

The proof of the above result is based on the fact that an arbitrary programmed grammar with appearance checking can be simulated by a PR system from $\mathcal{PR}_*(ncoo_e, coo_{prot})$.

The decidability of reachability of an arbitrary configuration (and marking) for PR systems using non-cooperative evolution rules, non-cooperative protein movement rules, in the maximal-parallel case, remains an *open problem*.

References

[1] D. Besozzi, N. Busi, G. Franco, R. Freund, Gh. Păun: Two universality results for (mem)brane systems. *Proc. Fourth Brainstorming Week on Membrane Computing*, Sevilla, 2006, RGNC Report 02/2006, 49–62.

[2] D. Besozzi, G. Rozenberg: Extended P systems for the analysis of (trans)membrane proteins. *Pre-proc. WMC7, 2006*, 8–9.

[3] R. Brijder, M. Cavaliere, A. Riscos-Núñez, G. Rozenberg, D. Sburlan: Membrane systems with marked membranes. *Electronic Notes in Theoretical Computer Sci.*, 171 (2007), 25–36.

[4] R. Brijder, M. Cavaliere, A. Riscos-Núñez, G. Rozenberg, D. Sburlan: Membrane systems with proteins embedded in membranes. *Theoretical Computer Sci.*, 404 (2008), 26–39.

[5] L. Cardelli: Brane calculi – interactions of biological membranes. *Lecture Notes in Computer Sci.*, 3082 (2005), 257–280.

[6] L. Cardelli, Gh. Păun: An universality result for a (mem)brane calculus based on mate/drip operations. *Intern. J. Found. Computer Sci.*, 17 (2006), 49–68.

[7] M. CAVALIERE, S. SEDWARDS: Membrane systems with peripheral proteins: transport and evolution. *Electronic Notes in Theoretical Computer Sci.*, 171 (2007), 37–53.

[8] M. CAVALIERE, S. SEDWARDS: Decision problems in membrane systems with peripheral proteins, transport and evolution, *CoSBi Technical Report* 12/2006, www.cosbi.eu, and *Theoretical Computer Sci.*, 404 (2008), 40–51.

[9] M. CAVALIERE, S. SEDWARDS: Modelling cellular processes using membrane systems with peripheral and integral proteins, *Lecture Notes in Bio-Informatics*, 4210 (2006), 108–126.

[10] V. DANOS, S. PRADALIER: Projective brane calculus. *Lecture Notes in Computer Sci.*, 3082 (2005), 134–148.

[11] R. FREUND, M. OSWALD: Tissue P systems with mate and drip operations. *Proc. MeCBIC 2006* (N. Busi, C. Zandron, eds.), Venice, 2006.

[12] I. KOREC: Small universal register machines. *Theoretical Computer Sci.*, 168 (1996), 267–301.

[13] S.N. KRISHNA: Universality results for P systems based on brane calculi operations. *Theoretical Computer Sci.*, 371 (2007), 83–105.

[14] S.N. KRISHNA: On the computational power of flip-flop proteins on membranes. *Lecture Notes in Computer Sci.*, 4497 (2007), 695–704.

[15] S.N. KRISHNA: Membrane computing with transport and embedded proteins. *Theoretical Computer Sci.*, 410 (2009), 355–375.

[16] T.Y. NISHIDA: Simulations of photosynthesis by a K-subset transforming system with membranes. *Fundamenta Informaticae*, 49 (2002), 249–259.

[17] A. PĂUN, B. POPA: P systems with proteins on membranes. *Fundamenta Informaticae*, 72 (2006), 467–483.

[18] A. PĂUN, B. POPA: P systems with proteins on membranes and membrane division. *Lecture Notes in Computer Sci.*, 4036 (2006), 292–303.

[19] A. PĂUN, A. RODRIGUEZ-PATÓN: On flip-flop membrane systems with proteins on membranes. *Lecture Notes in Computer Sci.*, 4860 (2007), 424–427.

[20] GH. PĂUN: One more universality result for P systems with objects on membranes. *Intern. J. of Computers, Communication and Control*, 1 (2006), 44–51.

[21] B. POPA: *Membrane Systems with Limited Parallelism*. PhD Thesis, Louisiana Tech. Univ., Ruston, USA, 2006.

CHAPTER 15

PETRI NETS AND MEMBRANE COMPUTING

JETTY KLEIJN

MACIEJ KOUTNY

15.1 INTRODUCTION

MEMBRANE systems, or P systems are a computational model inspired by the compartmentalization of living cells and the effect this has on how they function. The biochemical reactions taking place in the compartments of a cell are abstracted to rules specifying which and how many new objects (molecules) can be produced from objects of a certain kind and quantity, possibly involving a transfer to a neighboring compartment. The dynamic aspects of a membrane system and its potential behavior (its computations) derive from these evolution (or reaction) rules. Thus membrane computing is essentially concerned with multiset rewriting.

Petri nets ([19, 43, 47]) on the other hand are an operational model for concurrent systems directly generalizing state machines by their distributed states and local actions. Thus, a Petri net is essentially a bipartite directed graph consisting of two kinds of nodes, usually called *places* and *transitions*. Places determine the distribution of the system's states over local states, whereas transitions are the local actions which when they occur affect only the information provided by adjacent local states. In Place/Transition nets, or PT-nets, a typical and prominent Petri net model, system states indicate the local availability of resources by marking each place with a number of *tokens*. When a transition occurs at a state, it consumes some tokens from each of its input places and produces new tokens in each of its output places. Hence multiset calculus is also basic for transforming the token distribution in PT-nets. This key connection between the two models provides the basis for the faithful translation proposed in [36, 37], from the basic membrane systems to PT-nets.

This translation forms the starting point of this chapter as a structural link between the two models bringing to light the concurrent nature of the behavior of P systems. Each different kind of object in a compartment of the P system is represented by a different place and each evolution rule corresponds directly to a transition together with input and output places. Thus the operational properties of the net reflect the dependencies and independencies (causality, conflict, and concurrency) between the executions of evolution rules. As a consequence, tools and techniques developed for Petri nets become available for the description, analysis, and verification of behavioral properties of membrane systems, and in particular for the investigation of the structure of ongoing behavior of P systems. This complements and broadens the standard approach to the analysis of P systems which concentrates primarily on the ultimate results of "successful" or halting computations of P systems and on the computational power, including aspects of complexity, of different variants of the model.

Given the occurrence rule for single transitions, potential concurrency or synchronicity in the behavior of a PT-net is captured in a natural way by step semantics in terms of arbitrary combinations (multisets or *steps*) of concurrently occurring transitions. Both (singleton and arbitrary step) semantics are standard definitions for the dynamic behavior of PT-nets. Membrane systems however are often assumed to evolve in a *fully synchronous* fashion. This means that, at each tick of a global clock common to all compartments, the current configuration of the system is transformed by a maximally concurrent application of the rules, i.e. no further rules could have been applied within the same time unit. Thus in this case one would consider for the corresponding PT-nets maximally concurrent steps (as investigated in [27]).

In addition, their role as a semantic model for P systems has led to a novel extension of Petri nets. In the PT-net constructed from a P system, each transition is given a *locality* to reflect the compartmentalization of the original system. These localities are first of all a modeling tool—co-located transitions correspond to evolution rules in the same compartment—allowing one to identify the active parts of the system during a computation. But note that the places of PT-nets obtained from P systems, support by construction the local aspects of the consumption and production of resources by the evolution rules represented by the transitions. As a consequence, the localities assigned to transitions are not needed for a proper interpretation of fully synchronous behavior of P systems. They play no role in the maximal concurrent step semantics nor in the singleton or arbitrary step semantics. Localities are however important in the case of *local synchronicity* or locally maximal steps, when at each tick of the global clock and for each locality active during that tick, as many transitions belonging to this locality as possible are executed. This locally maximally concurrent semantics is a new and interesting feature for Petri nets. In fact, PT-nets with localities (or PTL-nets) executing locally maximal steps are a model well-suited for the analysis of systems exhibiting a mix of synchronous and asynchronous executions, i.e. *globally asynchronous locally synchronous* (GALS) systems (see [50]). In particular, they provide a framework for the investigation of P systems evolving according to the natural assumption that synchronicity is restricted to the compartments of the system as delineated by the membranes.

For the investigation of concurrent runs and in particular of the relations between transition occurrences during such runs, the notion of *process* has been developed for Petri nets. Already in [42], Petri nets are unfolded into labeled acyclic nets to obtain an explicit representation of causality and concurrency. For various classes of Petri nets, ongoing (potentially infinite) system behavior can be investigated using such (infinite) processes (see, e.g. [7, 8, 21, 29, 30, 42, 48]), a possibility also of interest from a biological point of view. Through the translation of basic P systems to PT-nets, processes could provide a convenient, formal, tool for analyzing the causality and (a)synchrony in the behavior of P systems and shed light on the causal relationships between the reactions taking place during a computation.

Unfortunately, the strategy for unfolding PT-nets into labeled occurrence nets does not work for PTL-nets subject to the locally maximal step semantics. The resulting occurrence nets are not true representations of possible net behavior, due to a lack of information concerning potential executability of transitions which affects the local maximality of executed steps. A possible solution to this problem is to register explicitly the existence of potential alternative occurrences of transitions. This strategy has led in [37] to the addition of so-called barb-events (or barbs) to processes. Here we will exemplify this approach and briefly explain why the resulting barb-processes are sound.

The last part of this chapter is concerned with extensions of basic membrane systems. First we consider membrane systems in which molecules are not only consumed and produced, but may also act as triggers or inhibitors of certain reactions [9]. To model this effect we use PTL-nets extended with *range* arcs (or PTRL-nets) [32]. Range arcs generalize activator and inhibitor arcs, both already existing extensions of the Petri net model (see, e.g. [12, 43, 26, 51]). These arcs give transitions the possibility to test for the presence and absence, respectively, of objects (tokens) in specific places without consuming them. We demonstrate that the translation from basic P systems to PT-nets with localities is robust enough to be generalized to the level of these extended models while retaining the direct correspondence between occurrences of transitions and evolution rules. Thus also in this more general setting, the synchrony and causal dependence in computations of the P system have corresponding interpretations in the PT-net. Moreover, as shown in [33], again barb-events can be used to extend the existing process semantics for PT-nets with activator and inhibitor arcs to a sound process notion for the locally maximal step semantics.

Next, we consider P systems with rules which may have an additional effect on the membranes, dissolving or thickening them. If this happens the membrane structure changes, the mobility of objects is affected and some rules may be prevented from ever occurring again. Evolution rules in these dynamically changing membrane systems cannot be related in a one-to-one correspondence to Petri net transitions. Still, there is a way of modeling their characteristics with special control structures using activator and inhibitor arcs. These control structures correspond uniquely to evolution rules (in a dynamically changing environment) and consequently, no new Petri net features are needed. So, perhaps surprisingly, the behavior of these P systems can be described and analyzed using the barb-processes developed for PT-nets with localities and activator and inhibitor arcs.

We conclude this chapter with a summary of the presented approach, a discussion of related work, and possible directions for future research. This chapter is based on our work reported in [36, 37, 31, 33, 34].

15.1.1 Notation

A multiset over a (finite, in this chapter) set X is a function $m : X \to \mathbb{N} = \{0, 1, 2, \ldots\}$. The cardinality of m is $|m| \stackrel{df}{=} \sum_{x \in X} m(x)$.

For two multisets m and m' over X, the sum $m + m'$ is the multiset given by $(m + m')(x) \stackrel{df}{=} m(x) + m'(x)$ for all $x \in X$, and if $k \in \mathbb{N}$ then $k \cdot m$ is the multiset given by $(k \cdot m)(x) \stackrel{df}{=} k \cdot m(x)$ for all $x \in X$. We denote $m \leq m'$ whenever $m(x) \leq m'(x)$ for all $x \in X$, and if $m' \leq m$, then the difference $m - m'$ is $(m - m')(x) \stackrel{df}{=} m(x) - m'(x)$ for all $x \in X$.

As usual in membrane computing, we represent multisets over X by strings over X.

The set of all (closed) intervals of natural numbers is denoted by INT.

15.2 Basic Membrane Systems and PTL-nets

15.2.1 Basic Membrane Systems

We start from the most basic definition of P systems.

Definition 15.1 *[membrane structure] A membrane structure μ (of degree $m \geq 1$) is given by a rooted tree with m nodes identified with the integers $1, \ldots, m$. We will write $(i, j) \in \mu$ or $i = parent(j)$ to mean that there is an edge from i (parent) to j (child) in the tree of μ, and $i \in \mu$ to mean that i is a node of μ.*

The nodes of a membrane structure represent nested membranes which in turn determine compartments. Compartment j is enclosed by membrane j and lies in-between j and its children (if any). Figure 15.1 shows a membrane structure (with $m = 5$) together with the corresponding compartments. Note that 1 is the root node, $(1, 2) \in \mu$ and $3 = parent(5)$.

In the rest of this section, we will use notions specified in a previous, numbered definition (like μ above) directly in subsequent definitions.

Each distribution of objects over its compartments determines a configuration of a membrane system.

Definition 15.2 *[basic membrane system and its configurations] Let V be a finite alphabet of names of objects (molecules). A (basic) membrane system (P system) over μ is a tuple $\Pi \stackrel{df}{=} (V, \mu, w_1^0, \ldots, w_m^0, R_1, \ldots, R_m)$ such that, for every membrane i, w_i^0 is a multiset of objects, and R_i is a finite set of evolution rules r of the form $lhs^r \to rhs^r$, where $lhs^r \neq \lambda$, is a multiset over V, and rhs^r is a multiset over*

$$V \cup \{a_{out} \mid a \in V\} \cup \{a_{in_j} \mid a \in V \text{ and } (i, j) \in \mu\}$$

such that if i is the root of μ then no a_{out} occurs in rhs^r.
A configuration of Π is a tuple $C \stackrel{df}{=} (w_1, \ldots, w_m)$ of multisets of objects, and $C_0 \stackrel{df}{=} (w_1^0, \ldots, w_m^0)$ is the initial configuration.

We refer to lhs^r as the left-hand side of the rule r, and rhs^r as its right-hand side. The left-hand side of a rule specifies which and how many objects are needed as input for a single execution of this rule and its right-hand side specifies which and how many new objects are produced as a consequence. A symbol a_{in_j} represents an object a that is sent to a child node (compartment) j and a_{out} means that a is sent to the parent node. Fig. 15.2 shows a basic P system over the membrane structure from Fig. 15.1. This is our running example. In this particular case, $V = \{a, b, c\}$,

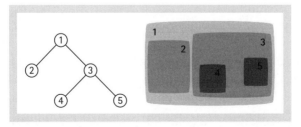

Fig. 15.1 Membrane structure and compartments for the running example.

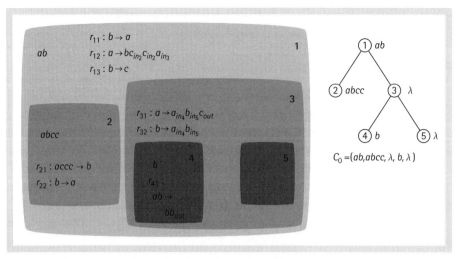

Fig. 15.2 A basic P system with initial configuration.

$lhs^{r_{21}} = accc$, $rhs^{r_{12}} = b\, c_{in_2}c_{in_2}a_{in_3}$, $w_1^0 = ab$ and $w_5^0 = \lambda$. The initial configuration is depicted also on the right, by placing the string w_i^0 next to the node i of the membrane structure.

A membrane system evolves from configuration to configuration as a consequence of the application (or execution) of evolution rules. Having said that, there is more than one strategy in which this can be done, ranging from applying a single rule at a time, to maximal parallelism. We distinguish four such *execution modes*, all based on the notion of a vector multi-rule.

Definition 15.3 *[vector multi-rule]* A vector multi-rule of Π is a tuple $\mathbf{r} \stackrel{df}{=} \langle \mathbf{r}_1, \ldots, \mathbf{r}_m \rangle$ where, for each membrane i of μ, $\mathbf{r}_i$ is a multiset of rules from R_i.

We lift the notion of left and right-hand sides of rules to vector multi-rules. For a vector multi-rule $\mathbf{r}$ as above, we denote by $lhs_i^{\mathbf{r}}$ the multiset $\sum_{r \in R_i} \mathbf{r}_i(r) \cdot lhs^r$ in which all objects in the left-hand sides of the rules in $\mathbf{r}_i$ are accumulated, and by $rhs_i^{\mathbf{r}}$ the multiset $\sum_{r \in R_i} \mathbf{r}_i(r) \cdot rhs^r$ of all (indexed) objects in the right-hand sides. The first multiset specifies per compartment how many objects are needed for the simultaneous execution of a combination of evolution rules.

Definition 15.4 *[enabled vector multi-rule]* A vector multi-rule $\mathbf{r}$ of Π is free-enabled at a configuration C if $lhs_i^{\mathbf{r}} \leq w_i$, for each i. A free-enabled $\mathbf{r}$ is: min-enabled if $|\mathbf{r}_1| + \cdots + |\mathbf{r}_m| = 1$; max-enabled *if no $\mathbf{r}_i$ can be extended to yield a vector multi-rule which is free-enabled at C; and* lmax-enabled *if no non-empty $\mathbf{r}_i$ can be extended to yield a vector multi-rule which is free-enabled at C.*

Referring to the running example and considering its initial configuration, $\langle \lambda, \lambda, r_{31}, \lambda, \lambda \rangle$ is not free-enabled, $\langle r_{11}, \lambda, \lambda, \lambda, \lambda \rangle$ is min-enabled but not lmax-enabled, $\langle r_{11}r_{12}, \lambda, \lambda, \lambda, \lambda \rangle$ is lmax-enabled but not max-enabled, and $\langle r_{11}r_{12}, r_{22}, \lambda, \lambda, \lambda \rangle$ is max-enabled.

If **r** is *free-enabled* (*free*) at a configuration C, then C has in each membrane i enough copies of objects for the application of the multiset of evolution rules r_i. Maximal concurrency (*max*) requires that adding any extra rule makes **r** demand more objects than C can provide. Locally maximal concurrency (*lmax*) is similar but in this case only those compartments which have rules in **r** cannot enable even more rules; in other words, each compartment either uses no rule, or uses a maximal multiset of rules. Minimal enabling (*min*) allows only a single copy of just one rule to be applied at any time.

We now describe the effect of the application of the rules which is independent of the mode of execution $\mathfrak{m} \in \{free, min, max, lmax\}$.

Definition 15.5 [*computations*] *A vector multi-rule* **r** *which is* $\mathfrak{m}$-*enabled at* C *can* $\mathfrak{m}$-*evolve to a configuration* $C' = (w'_1, \ldots w'_m)$ *such that, for each i and object a:*

$$w'_i(a) = w_i(a) - lhs^{\mathbf{r}}_i(a) + rhs^{\mathbf{r}}_i(a) + rhs^{\mathbf{r}}_{parent(i)}(a_{in_i}) + \sum_{i=parent(j)} rhs^{\mathbf{r}}_j(a_{out})$$

where $rhs^{\mathbf{r}}_{parent(i)} \stackrel{df}{=} \lambda$ *if i is the root of μ. We denote this by* $C \stackrel{\mathbf{r}}{\Longrightarrow}_{\mathfrak{m}} C'$. *An* $\mathfrak{m}$-*computation is a sequence of* $\mathfrak{m}$-*evolutions starting from* C_0.

Note that the evolution of C is non-deterministic in the sense that there may be different vector multi-rules applicable to C as described above. Figure 15.3 shows a two-stage lmax-evolution for the running example.

15.2.2 Place/Transition Nets with Localities

We now introduce the class of Petri nets to be used for a direct, behavior preserving translation from basic P systems. To model membrane systems, multisets of places are used to represent the availability of molecules within the compartments, while transitions correspond to evolution rules. As a consequence, each transition is

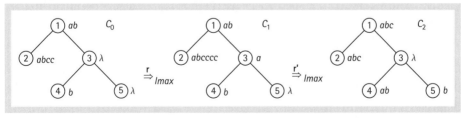

Fig. 15.3 An lmax-computation for the running example with the vector multi-rules r = $\langle r_{11}r_{12}, \lambda, \lambda, \lambda, \lambda \rangle$ and r' = $\langle \lambda, r_{21}r_{22}, r_{31}, \lambda, \lambda \rangle$.

associated with a compartment and this information is explicitly added to the model through the concept of the locality of a transition.

Definition 15.6 *[PTL-net] A PT-net with localities (or PTL-net) is a tuple* $NL \stackrel{df}{=} (P, T, W, \mathfrak{D}, M_0)$ *where P and T are finite disjoint sets of respectively the* places *and* transitions, $\mathfrak{D} : T \to \mathbf{N}$ *is a* locality mapping, $W : (T \times P) \cup (P \times T) \to \mathbf{N}$ *is the* weight function, *and* $M_0 : P \to \mathbf{N}$ *is the* initial marking *(in general, any multiset of places is a marking). We assume that, for every transition t there is a place p such that* $W(p, t) \neq 0$.

In diagrams, like that in Fig. 15.4, places are drawn as circles, and transitions as boxes. If $W(x, y) \geq 1$ for some $(x, y) \in (T \times P) \cup (P \times T)$, then (x, y) is an *arc* leading from x to y. An arc is annotated with its weight if the latter is greater than one. A marking M is represented by drawing in each place p exactly $M(p)$ tokens (small black dots). Boxes representing transitions belonging to the same localities either have the same shade or are displayed on a grey background of the same shade. The locality mapping $\mathfrak{D}$ partitions the transition set by associating with each transition a locality, given by an integer, for correspondence with compartments.

The role of vector multi-rules is now played by multisets of simultaneously executed transitions.

Definition 15.7 *[step] A* step U *of NL is a multiset of transitions. Its* pre-multiset *and* post-multiset *of places,* PRE(U) *and* POST(U), *are respectively given by* $\sum_{t \in U} U(t) \cdot W(p, t)$ *and* $\sum_{t \in U} U(t) \cdot W(t, p)$, *for each place p.*

As for the basic P systems, four modes of execution for PTL-nets (from sequential to fully synchronous) can be defined and analyzed.

Definition 15.8 *[enabled step] A step* U *is* free-enabled *at a marking M if* PRE(U) $\leq M$. *A free-enabled* U *is:* min-enabled *if* $|U| = 1$; max-enabled *if* U *cannot be extended to yield a step which is free-enabled at M; and* lmax-enabled *if* U *cannot be extended by a transition t satisfying* $\mathfrak{D}(t) \in \mathfrak{D}(U)$ *to yield a step which is free-enabled at M.*

Step U is enabled at a marking M if in each place there are enough tokens for the specified multiple occurrence of each of its transitions (note that each transition t needs to consume from each place p exactly $W(p, t)$ tokens which cannot be shared with any other transition). For the PTL-net in Fig. 15.4 and its initial marking, r_{31} is not free-enabled, r_{11} is min-enabled but not lmax-enabled, $r_{11}r_{12}$ is lmax-enabled but not max-enabled, and $r_{11}r_{12}r_{22}$ is max-enabled.

Definition 15.9 *[step sequences] A step* U *which is* m-enabled *at a marking M can be* m-executed *leading to the marking M′ given by* $M' \stackrel{df}{=} M - \text{PRE}(U) + \text{POST}(U)$. *We denote this by* $M[U\rangle_m M'$.
An m-step sequence *is a sequence of* m-executions *starting from* M_0.

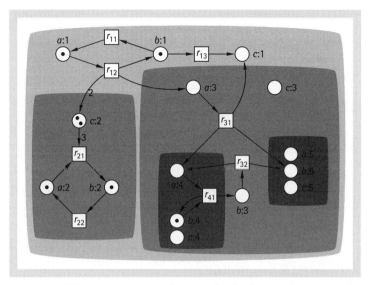

Fig. 15.4 PTL-net corresponding to the basic running example.

A possible two-stage lmax-computation for the PTL-net in Fig. 15.4 is:

$$M_0[r_{11}r_{12}\rangle_{lmax}M'[r_{21}r_{22}r_{31}\rangle_{lmax}M \quad (\dagger)$$

where M is as M_0 except that $M(c:2) = M(c:1) = M(a:4) = M(b:5) = 1$.

15.2.3 From Basic Membrane Systems to PTL-nets

To model a basic P system as a PTL-net, we introduce a separate place (a, j) for each molecule a and membrane j. For each rule r associated with a compartment i we introduce a separate transition t_i^r with locality i. If the transformation described by a rule r of compartment i consumes k copies of molecule a from compartment j, then we introduce a k-weighted arc from place (a, j) to transition t_i^r, and similarly for molecules being produced. Finally, assuming that, initially, compartment j contained n copies of molecule a, we introduce n tokens into place (a, j). This idea is formalised as follows.

Definition 15.10 *[translating P systems into PTL-nets] The PTL-net corresponding to Π is* $NL_\Pi \stackrel{df}{=} (P, T, W, \mathfrak{D}, M_0)$, *where:*

- $P \stackrel{df}{=} V \times \{1, \ldots, m\}$ and $M_0(p) \stackrel{df}{=} w_j(a)$ for every place $p = (a, j)$.
- $T \stackrel{df}{=} T_1 \cup \ldots \cup T_m$ where each T_i consists of distinct transitions $t = t_i^r$ for every evolution rule $r \in R_i$, with locality $\mathfrak{D}(t) \stackrel{df}{=} i$; and for each place $p = (a, j)$ the weight of the incoming arc is $W(p, t) \stackrel{df}{=} lhs^r(a)$ if $i = j$, and otherwise $W(p, t) \stackrel{df}{=} 0$ and of the outgoing one it is $W(t, p) \stackrel{df}{=} rhs^r(a)$ if $i = j$, $W(t, p) \stackrel{df}{=}$

$rhs^r(a_{out})$ if $j = parent(i)$, $W(t, p) \stackrel{df}{=} rhs^r(a_{in_j})$ if $i = parent(j)$, and otherwise $W(t, p) \stackrel{df}{=} 0$.

Figure 15.4 shows the translation for the running example, where each place (x, i) is denoted as $x{:}i$ and each transition t_i^r as r.

To capture the very tight correspondence between the P system Π and the PTL-net NL_Π, we introduce a straightforward bijection between configurations of Π and markings of NL_Π, based on the correspondence of object locations and places as well as that of vector multi-rules and steps.

Definition 15.11 [*mapping configurations & vector multi-rules*] *The marking $\nu(C)$ corresponding to configuration $C = (w_1, \ldots, w_m)$ is defined by $\nu(C)(a, i) \stackrel{df}{=} w_i(a)$, for every place (a, i).*

The step $\rho(\mathbf{r})$ corresponding to vector multi-rule $\mathbf{r} = (\mathbf{r}_1, \ldots, \mathbf{r}_m)$ is defined by $\rho(\mathbf{r})(t_i^r) \stackrel{df}{=} \mathbf{r}_i(r)$, for every transition t_i^r.

Fact 15.1 $\nu(C_0) = M_0$.

For the lmax-computations given in Fig. 15.3 and (†), we have $\rho(\mathbf{r}') = r_{21}r_{22}r_{31}$ and $\nu(C_2) = M$.

For any translation from P systems to Petri nets to be useful, it is essential to ensure that the latter can provide a faithful representation of the behavior of the former. Here, in fact, it is possible to establish the desired relationship between (the operation of) P systems and Petri nets at the system level. The fundamental link between the dynamics of a membrane system and that of its corresponding PTL-net is formulated next.

Fact 15.2 $C \stackrel{\mathbf{r}}{\Longrightarrow}_\mathfrak{m} C'$ *if and only if $\nu(C)$ $[\rho(\mathbf{r})\rangle_\mathfrak{m}$ $\nu(C')$ in NL_Π, for each mode of execution $\mathfrak{m}$.*

Together with Fact 15.1, this immediately implies that the (finite and infinite) $\mathfrak{m}$-computations of Π coincide with $\mathfrak{m}$-step sequences of NL_Π. For example, the lmax-evolution of the running example given in Fig. 15.3 coincides with that in (†) of the corresponding PTL-net.

15.2.4 PTL-nets and Membrane System Analysis

A central reason for defining Petri net models of P systems is that there is a variety of analytical and verification techniques developed over the years for Petri nets which could be applicable to P systems. Some techniques can be useful to assert properties of reachable configurations; for example, one can show (via the invariant analysis based on linear algebra [49] of the PTL-net in Fig. 15.4) that for the basic membrane system in Fig. 15.2 we have $0 \leq M(a:4) + M(b:3) - M(c:1) \leq 2$, for every marking reachable from the initial one.

Another way of deploying the analytical power of Petri nets is to use their causality semantics based on occurrence nets. This kind of analysis and verification would be oriented more towards dealing with entire computations rather than concentrating solely on the properties of reachable configurations. In a nutshell, an occurrence net associated with a PTL-net NL is an abstract record of a single step sequence of NL in which only information about causality and concurrency between executed transitions and visited local states is represented. Together with the natural requirement that causality is acyclic, this means that the underlying structure of an occurrence net is that of a partial order.

Consider, for example, the PTL-net NL in Fig. 15.4 and its lmax-step sequence which executes consecutively the three steps: $U_1 = r_{11}r_{12}$, $U_2 = r_{21}r_{22}r_{31}$ and $U_3 = r_{41}$. The construction of an occurrence net ON corresponding to such an lmax-step sequence can be explained as follows.

Referring to Fig. 15.5, we first create 7 places corresponding to the 7 tokens in the initial marking of NL and labeled accordingly; these places are the only places marked in the default initial marking of ON (with one token each). We then take $|U_1| = 2$ fresh transitions, corresponding (through their labels) to the transitions in U_1 and provide each with the correct inputs from the places of the initial marking of ON. The execution of U_1 creates new tokens; these are again represented as labeled places of ON but, crucially, these new places are completely fresh. In particular, even though r_{12} produces tokens in $c:2$, we do not draw an arc from r_{12} to one of the $c:2$-labeled places in the initial marking of ON; rather, we create two brand new $c:2$-labeled places. A similar construction is carried out for the remaining two steps U_2 and U_3, resulting in the net shown in Fig. 15.5 (with one exception explained below).

Essentially, ON traces the changes of markings due to steps being executed along some legal behavior of the original PTL-net, and in doing so records which resources are consumed and produced. As a result, the original net is unfolded into a labeled acyclic net, called a *process*, representing the structure of the given step sequence. As is standard in causality analysis of nets, such a representation should allow one to talk about important behavioral aspects of NL_Π, such as:

- *Causality.* The causality relationships among the executed transitions can be read-off by following directed paths in ON.
- *Concurrency.* Executed transitions without a directed path between their representations in ON are concurrent.
- *Executability.* Executions of ON define m-executions of NL.
- *Representation.* The m-execution of NL on basis of which ON is constructed can be executed by ON.

The third property expresses a soundness condition and implies an obligation to define an execution semantics for processes consistent with the chosen execution

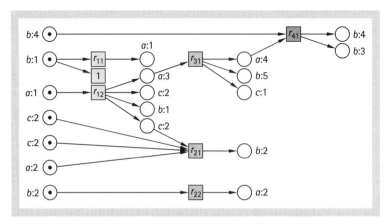

Fig. 15.5 Causal process for the PTL-net in Fig. 15.4.

step semantics of the original PTL-net (for a general semantical framework for Petri net processes, see [29]). In case of $m \in \{\textit{free}, \textit{min}, \textit{max}\}$ one can use the same step semantics for PTL-nets and their processes, and the Executability and Representation properties will hold [29]. However, as was argued in [36, 37], the relatively straightforward unfolding construction given above may fail to satisfy the Executability property in case of lmax-step sequences.

To make things work properly in the sense of the theory expounded in [29], one needs to suitably modify the standard process construction, as first outlined in [36]. Firstly, to have an lmax-step sequence semantics for the processes, all their transitions will have a locality assigned, corresponding to the original locality in the PTL-net. Secondly, the processes are augmented with information about the existence of transitions that could have been chosen for execution but were not, for example, due to being in conflict with some of the other executed transitions. In other words, barb-events (barbs) are used to signal *potential* executability of transitions in lmax-executions of PTL-nets. In the example, the transition labeled with 1 and connected to the b:1-labeled place in the initial marking is such a barb-event. It indicates that with the initial token in place $(b, 1)$, at this point of the step sequence, some non-specified transition at location 1 was enabled (but did not occur). The lmax-enabling rule is now adjusted to use barbs as devices enforcing lmax-execution within the process net, but they are not meant to be executed. In this way, barbs prevent that lmax-step sequences of the process net postpone the enabling of process transitions when that would imply that other transitions of the original net (with the same location) should have been executed.

The complete set of computations of a P system combined with all possible non-deterministic choices at reachable configurations can be represented by a single object, its *reachability graph*, an initialized directed graph with configurations as its nodes and edges labeled with vector multi-rules leading from configuration to configuration. Moreover, thanks to the close correspondence between basic

P systems and their associated PTL-nets outlined in the previous section, for the purpose of analysis of a basic P system we can also investigate the reachability graph of its PTL-net as this is isomorphic to the reachability graph of the P system. Since reachability graphs combine step sequences and reachable states (markings), they are useful for the analysis and verification of behavioral properties. Researching properties of reachability graphs has a long tradition in the field of Petri nets, and has produced over the years several fundamental insights. One of the crucial results is that the reachability problem for PT-nets (under the min-step or free-step sequence semantics) is decidable [41, 38]. In turn, this immediately implies that the problem of deciding whether a basic membrane system has a free- or min-computation leading to a given configuration (desired or non-desired) can be decided, and this result does not depend on the number of reachable configurations which may be infinite.

Another relevant property which one might want to verify for a basic P system is whether the concentration of specific molecule(s) in specific compartment(s) can grow unboundedly during one of its possible computations. This problem, referred to as *boundedness*, can also be tackled within the Petri net domain where it is shown to be decidable using the *coverability tree* construction introduced in [28] and then investigated, among others, in [23, 43]. What is more, coverability trees can also provide a tool for deciding many other relevant behaviorial problems, such as mutual exclusion, i.e. to show that two kind of molecules cannot be simultaneously present in the same compartment.

15.3 EXTENSIONS

Motivated by natural phenomena in biological systems, a whole range of different extensions of the basic membrane system have been introduced, often in the form of more sophisticated evolution rules. For some of these extensions, like catalysts and symport/antiport rules, the basic translation to PTL-nets can be used as before. For others, like i/o communication and rule creation/consumption, the correspondence between evolution rules and Petri net transitions is slightly more involved, but the resulting nets are again PTL-nets (see [31]). However, not all interesting variants of P systems can be modeled using no more than the PTL-net model. In this section, we discuss two such models and sketch how they can be properly captured in a suitably extended PTL-net model.

15.3.1 Rules with Promoters and Inhibitors

First we consider an extended version of P systems for modeling how reactions may be triggered or blocked in the presence of certain molecules. The role of such molecules differs from that of catalysts which actively take part in reactions and are returned afterwards. An example is object b which acts as a catalyst in evolution rule $r_{41} : ab \to b\, b_{out}$ of our running example (see Fig. 15.2). To model the subtle effect that the presence of molecules may have, P systems have evolution rules r of the form $lhs^r \to rhs^r|_{pro^r, inh^r}$ where pro^r and inh^r are multisets over V specifying respectively the *promoters* and *inhibitors* of r. The intuition behind pro^r and inh^r is that they test respectively for the presence and absence of certain objects inside a compartment, but without consuming them. As a consequence, any number of rules can test for the presence of a single object at the same time. In order for r to occur there must be *at least* $pro^r(a)$ copies of each symbol a in its associated compartment, and *less than* $inh^r(a)$ copies of each symbol a which occurs in inh^r. (Thus if $inh^r = \lambda$, the empty multiset, no object inhibits by its presence the occurrence of rule r.) It is assumed without loss of generality, that $pro^r(a) < inh^r(a)$ for all rules r and symbols $a \in inh^r$. We retain all definitions introduced for basic P systems in Subsection 15.2.1 with only one change regarding the notion of a free-enabled vector multi-rule $\mathbf{r}$. This is strengthened by additionally requiring that, for each i and $r \in \mathbf{r}_i$, we have $pro_i^r \le w_i$ and, moreover, if $a \in inh^r$ then $w_i(a) < inh^r(a)$.

PTL-nets are not expressive enough to model inhibitors and promoters because arcs between transitions and places indicate consumption and production of tokens (objects) rather than testing for their presence or absence. A possible way out is to use PTL-nets extended with *range* arcs [32]. Each such arc links a place to a transition and is specified by an interval (possibly infinite) of non-negative integers. This interval indicates the *range* (in the set **INT**) for the number of tokens that should be present in the place to enable the occurrence of the transition. Clearly, like pro^r and inh^r, range arcs can be used to model certain forbidden/required concentrations of molecules in a compartment.

A *PT-net with range arcs and localities* (or PTRL-net) is a tuple $NRL \stackrel{df}{=} (P, T, W, R, \mathfrak{D}, M_0)$ where the components other than R are as in Definition 15.6, and $R : P \times T \to \mathbf{INT}$ defines the *range arcs*. Let $R(p, t) = [k, l]$ be the interval associated as a range arc from place p to transition t. In diagrams, $R(p, t)$ is drawn as an arrow from p to t with a small grey circle as arrowhead and annotated with (k, l). There are two important special cases:

- $k = 0 = l$. Then the range arc is an *inhibitor* arc, p is an *inhibitor place of t*, and t can only be executed if p does not contain any tokens (we draw an arrow from p to t with a small open circle as arrowhead).
- $k = 1$ and $l = \infty$. Then the range arc is an *activator* arc, p is an *activator place of t*, and t can only be executed if p contains at least one token (we draw an arrow from p to t with a small black circle as arrowhead).

As far as the behavior is concerned, all we need to do is to stipulate that a step U is *free-enabled* at a marking M if, in addition to the requirements in Definition 15.8, for every place $p \in P$ and transition $t \in U$, we have $M(p) \in R(p, t)$.

The translation from membrane systems to PTRL-nets proceeds as in Definition 15.10 for the case of basic membrane system. The only additional feature is that for each transition $t = t_i^r$ and place $p = (a, j)$, we introduce a range arc such that $R(p, t) \stackrel{df}{=} [0, \infty]$ whenever $i \neq j$, and otherwise $R(p, t) \stackrel{df}{=} [pro^r(a), inh^r(a) - 1]$ if $a \in inh^r$ and $R(p, t) \stackrel{df}{=} [pro^r(a), \infty]$ if $a \notin inh^r$. Figure 15.6 illustrates the modified translation. It may be of interest to compare in Fig. 15.4 and 15.6 the different translations of the role of b in rule r_{41}.

In [33], it has been shown that key properties of the modified translation are very similar to those obtained in the basic case; in particular, Definition 15.11 and Fact 15.2 can simply be restated. Moreover, the treatment of causality developed for PTL-nets can be extended to PTRL-nets after allowing activator arcs in the occurrence nets (see [32, 33]).

When it comes to the properties which might be expressed or investigated using reachability or coverability graphs, the situation changes dramatically when we move from PTL-nets to PTRL-nets. The reason is that PTRL-nets allow one to test for the absence of resources (zero-testing). Nets with this kind of relationship between places and transitions (i.e. inhibitor arcs) have been considered in [1, 25, 43]. The extension with inhibitor arcs gives the resulting model of Petri nets the expressive power of Turing machines and decidability for certain important behavioral properties, such as reachability and boundedness, is lost [24, 1].

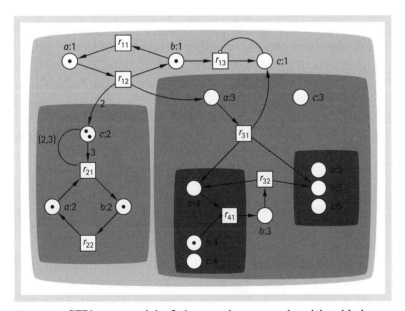

Fig. 15.6 PTRL-net model of the running example with added promoter and inhibitor constraints. The membrane system is as in Fig. 15.2 except that $r_{13} : b \to c|_{\lambda, c}$, $r_{21} : accc \to b|_{cc, cccc}$ and $r_{41} : a \to b_{out}|_{b, \lambda}$.

Although boundedness (like reachability) is undecidable for general PT-nets with inhibitor arcs, partial solutions have been proposed by restricting the class of nets under consideration as, for example, in [10].

15.3.2 Thickening and Dissolving Membranes

What may come as a surprise, is that PTRL-nets are also robust enough to model in a faithful way, P systems with a dynamic structure due to rules which may thicken or dissolve membranes. Evolution rules are now of the form $lhs^r \to rhs^r \mid^\tau$ or $lhs^r \to rhs^r \mid^\delta$.

The first rule is annotated with the special symbol τ which indicates that, when executed, r causes its immediately enclosing membrane to become "thick", and no object can pass through it anymore. The translation to PTRL-nets can proceed as before (Definition 15.10), but to capture the thickening of membranes in addition, for each membrane i a single special initially empty *control* place $\tau{:}i$ is introduced as an extra output place for all transitions which correspond to rules which make membrane i thick (thus there may be several transitions inserting tokens into $\tau{:}i$). Then each transition corresponding to a rule transferring objects through membrane i, is connected to $\tau{:}i$ by an inhibitor arc. As a result, as long as no transition which produces a token in $\tau{:}i$ is executed, the execution of transitions transferring objects through membrane i is not affected. However, once there is at least one token in $\tau{:}i$, these transitions can no longer be executed because no transition ever removes a token from $\tau{:}i$. Figure 15.7 illustrates the modified translation for the running example in which r_{21} is now a thickening rule. As in the case of membrane systems with inhibitors and promoters, the key properties of the resulting PTRL-nets are similar to those obtained in the basic case.

It is, however, more complicated to deal with rules which may lead to the dissolution of membranes. The special symbol δ in $lhs^r \to rhs^r \mid^\delta$ indicates that this rule, when executed in some compartment i enclosed by a membrane different from the root membrane, causes the membrane to "dissolve". Moreover, all objects present in that compartment are incorporated into compartment $parent(i)$, and all rules formerly associated to compartment i are prevented from ever occurring in the future. The basic translation to PTL-nets is now modified in the following way. To model evolution rules from compartment $parent(i)$ as transitions having access to the tokens stored in the places representing objects in the former compartment i, we use copies of transitions and activator and inhibitor arcs to signal which membranes have been dissolved. For each membrane i that can be dissolved, a membrane status control place $\delta{:}i$ is introduced. Each transition representing a dissolving rule associated with compartment i has $\delta{:}i$ as an additional output place. All transitions associated with membrane i are connected with an inhibitor arc to place $\delta{:}i$. Hence once the membrane has been dissolved, they can never occur again. Rules associated

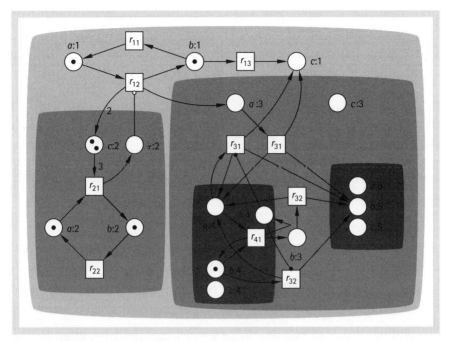

Fig. 15.7 PTRL-net model of the running example with added membrane thickening and dissolving features. The membrane system is as in Fig. 15.2 except that $r_{21} : accc \to b|^\tau$ and $r_{41} : ab \to bb_{out}|^\delta$. There are now two copies of the transitions for r_{31} and r_{32}. Note that only the essential additional places are shown.

with a membrane with a nested sequence of dissolving descendant membranes will be represented by transitions for each possible combination of access to resources. This can be modeled by a combination of activator and inhibitor arcs (see [34]). Consider, for example, a membrane system with three membranes such that $1 = parent(2)$ and $2 = parent(3)$. Assume further that there are two dissolution rules, r' within membrane 2 and r'' within 3. Then we add two membrane status places, $\delta{:}2$ and $\delta{:}3$, each place being initially empty and having an incoming directed arc from transition $t_2^{r'}$ and $t_3^{r''}$, respectively. Then, given a rule $r : a \to aa$ associated with membrane 2, we construct two transitions with locality 2 and a directed arc of weight 2 to place $a{:}2$, viz. $t^{r,2}$ with a directed arc from $a{:}2$, and $t^{r,3}$ with a directed arc from $a{:}3$. We also add an inhibitor arc between $\delta{:}2$ and each of these two transitions, as well as an activator arc between $t^{r,3}$ and $\delta{:}3$. As a result, $t^{r,2}$ can be executed only if r' has not occurred whereas $t^{r,3}$ only if r' has not occurred but r'' has occurred.

Figure 15.7 illustrates the construction for the running example in which r_{41} is now a dissolving rule.

It has been shown in [34] that key properties of the modified translation are similar to those obtained in the basic case; however, Definition 15.11 and Fact 15.2 cannot simply be restated. The reason is that more than one transition can correspond to a single rule, as in the case of $t^{r,2}$ and $t^{r,3}$ above. As a consequence, when comparing the computations of a membrane system with dissolving rules with the step sequences of a corresponding PTRL-net, we need to treat executions of transitions like $t^{r,2}$ and $t^{r,3}$ as if they were the same transition. This is a standard approach in Petri net theory when one deals with nets with labeled transitions with equal labeling identifying semantically equivalent transitions. As demonstrated in [34], having more transitions in the PTRL-net than rules in the original membrane system cannot be avoided if the step sequences should correspond to computations. Finally, the constructions outlined above can be adapted to simulate also simultaneous executions of thickening and dissolving of membranes.

15.4 Concluding Remarks

The faithful translation of *basic* P systems into PT-nets (with localities) has been the starting point of this chapter. The main contribution of the work presented is its potential for further development and exploitation of this structural and systematic link between membrane computing and Petri nets. For membrane systems this means that results, techniques, and tools from the Petri net world become available, like linear algebra techniques (invariants), and coverability and reachability analysis including decidability results and verification techniques (model checking). Moreover, with the process semantics of Petri nets, it becomes possible to describe the concurrent behavior of P systems in terms of causal dependencies between executions of evolution rules. For Petri nets the direct connection with P systems has led to new, interesting notions like locality (PTL-nets) and locally maximal step sequence semantics, necessitating research into a new process semantics suitable for PT-nets modeling GALS systems. Moreover, it has been pointed out how also P systems with more involved features like promoters and inhibitors could be related to PTL-nets with range arcs (to test the concentration of molecules inside compartments). The observation that even dynamic structures, resulting from membrane thickening and dissolving, can be hard-coded within the PTRL-net model seems to indicate that Petri nets in general, and PTRL-nets in particular, provide a robust and uniform framework which can be employed as a convenient and powerful tool for analyzing the computations of membrane systems.

15.4.1 Related Work

The interpretation of the evolution (reaction) rules of a P system as Petri net transitions is a quite natural idea and appears in various places in the literature, e.g.

in [16, 45] which both also regard the connection of (variants of) P systems to Petri nets as an opportunity to describe behavioral aspects of P systems and make results from the realm of Petri nets available for the analysis of membrane computations. In [20] on the other hand, Petri nets are used to describe features capturing the computational completeness of P systems. In [5] the relationship between P systems and Petri nets is investigated on the basis of two different P system interpretations of the producer/consumer example. A completely different approach is followed by [3] for P systems with symport/antiport rules. The hierarchy of a membrane structure is reflected in an hierarchy of agents, each modeled by a Petri net with nets as tokens. Molecules are represented by unstructured (ordinary) tokens and exchanging molecules is modeled by agents exchanging subagents. Actually, also the containment relation of higher-level agents can change dynamically when transitions are executed, which makes it possible to also investigate in the future P systems with active membranes. It is shown in [18] how PB systems (P systems with boundary rules) and membrane dissolution and creation can be "weakly" simulated by Petri nets as a means to investigate the decidability status of reachability and boundedness for such systems. This extends the work for PB systems with static membrane structure from [16]. In an attempt to relate P systems to the Petri net model for simulation and decidability purposes, [22, 44] propose an approach based on a new Petri net model with dynamic rewriting of descriptive expressions.

Other related work is concerned with the synthesis of Petri nets with localities and inhibitor arcs and activator arcs from behavioral descriptions in the form of step transition systems (i.e. state graphs with step-labeled edges). More precisely, given a step transition system one is asked to construct a net with an isomorphic reachability graph. This problem was first investigated for the class of elementary net systems in [39, 40] and the solution was extended to other classes of nets, including PTRL-nets, in [17].

Next to Petri nets, other models for concurrent and mobile systems have also been applied as a framework for the investigation of the behavior of membrane systems. In particular, calculi with membranes have been considered including ways to describe through such bioinspired calculi the causal dependencies among the reactions taking place in membrane systems (see, e.g. [11, 2, 14, 13, 15, 46]). In [6] the modeling methods of membrane computing, π-calculus, Petri nets, and two tools are combined and compared.

15.4.2 Further Outlook

The translation given in Definition 15.10 provides a fundamental connection between P systems and Petri nets. It essentially relates evolution rules to Petri net transitions at a most basic level with the local aspects of consumption and production of resources taken care of by the correspondence between places and located objects. This is immediately clear from Fact 15.2 for the execution mode min (single rules correspond to single transitions) and then also for the free and max modes. So, localities of transitions play no role in these semantics. Actually, since the localities of the transitions as defined by the translation can be deduced immediately from the location of their input places, even the lmax mode could have been defined without explicitly specifying localities. It can be argued that what appears to be most robust in the modeling approach described here is the idea of representing molecules in different compartments using dedicated places. In particular, there is no reason why the localities of a PTL-net should be ordered in a hierarchical (tree-like) structure. Consequently, as already pointed out in [31], the current scheme can also be used to model, e.g. tissue systems as Petri nets. It would be interesting to investigate what aspects of this Petri net modeling approach would still be relevant if other, increasingly sophisticated, classes of P systems were to be considered.

Having transitions associated with specific spatially disjoint locations is, however, important from a modeling point of view as they allow one to specify what happens where. The lmax-step sequences of PTL-nets provide a framework for the investigation of membrane systems evolving according to the natural assumption that synchronicity is restricted to the compartments of the system as delineated by the membranes. Note that the localities of transitions in PTL-nets do not necessarily depend on the specific input and output places per transition. Still, one could see localities as providing information on relationships between membranes or compartments within a single cell. If one abandons the idea that the "localities" of different cells are disjoint, then one needs to look for Petri net models more expressive than PTL-nets or PTRL-nets. For example, in the *networks of cells* introduced in [4, 5] evolution rules are not by definition tied to individual compartments or borders between two membranes. As a result, PTRL-nets may then not be able to properly reflect the lmax-execution rule defined for networks of cells. It appears that this is due to the fact that the locality mapping implicitly splits the space where the evolution rules operate onto disjoint fragments, and so a possible solution could be a more sophisticated locality mapping. Such a solution would require a re-definition of lmax-enabledness, perhaps using the general framework of *firing policies* introduced in [17] which subsumes under a single definition a variety of execution modes, including all those considered in this chapter.

Localities are also important for the formalization of dynamically changing membrane structures. They provide a means to model which evolution rules should be "switched on" or "switched off" depending on the presence of an enclosing membrane. The construction discussed here to model the effect of thickening and dissolving the membranes fits neatly in the PTRL-net set-up and should still be useful for more complex classes of P systems. For example, a simple modification using general k weighted inhibitor and activator arcs should be enough to capture the behavior of membranes with variable permeability.

Another direction for research aimed at providing support for dealing with behavioral properties of P systems using their established connection with Petri nets is to investigate specialized analysis techniques that can be used in dealing with PTRL-nets. As remarked before, properties like reachability and boundedness are undecidable for these nets. In [35] it is investigated to what extent the classical coverability tree construction might be adapted for Petri nets with inhibitor arcs and a step sequence semantics.

PT-nets with localities and range arcs provide a uniform common framework for the systematic investigation of many different variants of P systems. To understand concurrency and dependence relations in the behaviors of these nets (and hence of P systems) one should develop a causal semantics appropriate for the chosen operational semantics. The process semantics outlined here is no more than a first step on the way to understanding the role and effect of localities in the behavior of such systems. It is not yet known what the typical properties of such processes are and, closely related to this, what the abstract concurrency structures underlying such processes are (causal partial orders or even the extended causality structures of [26, 29] are insufficient to capture the intricacies of the interplay between local synchrony and causality). A related objective would be to investigate the notions of conflict and choice in relation to the barb-events leading to a suitable notion of unfolding for PTRL-nets that could be the basis for efficient verification procedures.

Finally, the relation between Petri nets and membrane computing as formalized here through the concept of PTRL-nets should be compared and combined with fundamental insights and ideas gained in other investigations in the general area of natural computing.

Acknowledgment

We are indebted to Grzegorz Rozenberg for introducing us to the area of membrane computing and many inspiring discussions.

References

[1] T. AGERWALA: *A Complete Model for Representing the Coordination of Asynchronous Processes*. Hopkins Computer Research Report 32, Johns Hopkins University, 1974.

[2] B. AMAN, G. CIOBANU: Translating mobile ambients into P systems. *Electronic Notes in Theoretical Computer Sci.*, 171 (2007), 11–23.

[3] L. BERNARDINELLO, N. BONZANNI, M. MASCHERONI, L. POMELLO: Modeling symport/antiport P systems with a class of hierarchical Petri nets. *Lecture Notes in Computer Sci.*, 4860 (2007), 124–137.

[4] F. BERNARDINI, M. GHEORGHE, M. MARGENSTERN, S. VERLAN: Networks of cells and Petri nets. *Proc. 5th Brainstorming Week on Membrane Computing*, Sevilla, 2007, 33–62.

[5] F. BERNARDINI, M. GHEORGHE, M. MARGENSTERN, S. VERLAN: Producer/consumer in membrane systems and Petri nets. *Lecture Notes in Computer Sci.*, 4497 (2007), 43–52.

[6] F. BERNARDINI, M. GHEORGHE, F.J. ROMERO-CAMPERO, N. WALKINSHAW: A hybrid approach to modeling biological systems. *Lecture Notes in Computer Sci.*, 4860 (2007), 138–159.

[7] E. BEST, R. DEVILLERS: Sequential and concurrent behavior in Petri net theory. *Theoretical Computer Sci.*, 55 (1988), 87–136.

[8] E. BEST, C. FERNÁNDEZ: *Nonsequential Processes. A Petri Net View*. Springer, 1988.

[9] P. BOTTONI, C. MARTÍN-VIDE, GH. PĂUN, G. ROZENBERG: Membrane systems with promoters/inhibitors. *Acta Informatica*, 38 (2002), 695–720.

[10] N. BUSI: Analysis issues in Petri nets with inhibitor arcs. *Theoretical Computer Sci.*, 275 (2002), 127–177.

[11] N. BUSI: Causality in membrane systems. *Lecture Notes in Computer Sci.*, 4860 (2007), 160–171.

[12] N. BUSI, G.M. PINNA: Process semantics for place/transition nets with inhibitor and read arcs. *Fundamenta Informaticae*, 40 (1999), 165–197.

[13] L. CARDELLI: Brane calculi. *Lecture Notes in Computer Sci.*, 3082 (2005), 257–278.

[14] L. CARDELLI, A. GORDON: Mobile ambients. *Lecture Notes in Computer Sci.*, 1378 (1998), 140–155.

[15] G. CIOBANU, D. LUCANU: Events, causality and concurrency in membrane systems. *Lecture Notes in Computer Sci.*, 4860 (2007), 209–227.

[16] S. DAL ZILIO, E. FORMENTI: On the dynamics of PB systems: a Petri net view. *Lecture Notes in Computer Sci.*, 2933 (2004), 153–167.

[17] PH. DARONDEAU, M. KOUTNY, M. PIETKIEWICZ-KOUTNY, A. YAKOVLEV: Synthesis of nets with step firing policies. *Lecture Notes in Computer Sci.*, 5062 (2008), 112–131.

[18] G. DELZANNO, L. VAN BEGIN: On the dynamics of PB systems with volatile membranes. *Lecture Notes in Computer Sci.*, 4860 (2007), 240–256.

[19] J. DESEL, W. REISIG, G. ROZENBERG, eds.: *Lectures on Concurrency and Petri Nets*. LNCS 3098, Springer, 2004.

[20] P. FRISCO: P systems, Petri nets, and program machines. *Lecture Notes in Computer Sci.*, 3850 (2005), 209–223.

[21] U. GOLTZ, W. REISIG: The non-sequential behavior of Petri nets. *Information and Control*, 57 (1983), 125–147.

[22] E. GUȚULEAC: Descriptive timed membrane Petri nets for modelling of parallel computing. *Int. J. Computers, Communications and Control*, 1 (2006), 33–39.

[23] M. HACK: *Decision Problems for Petri Nets and Vector Addition Systems*. Technical Memo 59, Project MAC, MIT, 1975.

[24] M. HACK: *Petri Net Languages*. Technical Report 159, MIT, 1976.

[25] M. HACK: *Decidability Questions for Petri Nets*. PhD Thesis, MIT, 1976.
[26] R. JANICKI, M. KOUTNY: Semantics of inhibitor nets. *Information and Computation*, 123 (1995), 1–16.
[27] R. JANICKI, P.E. LAUER, M. KOUTNY, R. DEVILLERS: Concurrent and maximally concurrent evolution of nonsequential systems. *Theoretical Computer Sci.*, 43 (1986), 213–238.
[28] R.M. KARP, R.E. MILLER: Parallel program schemata. *J. Comput. Syst. Sci.*, 3 (1969), 147–195.
[29] H.C.M. KLEIJN, M. KOUTNY: Process semantics of general inhibitor nets. *Information and Computation*, 190 (2004), 18–69.
[30] H.C.M. KLEIJN, M. KOUTNY: Infinite process semantics of inhibitor nets. *Lecture Notes in Computer Sci.*, 4024 (2006), 282–301.
[31] J. KLEIJN, M. KOUTNY: Synchrony and asynchrony in membrane systems. *Lecture Notes in Computer Sci.*, 4361 (2006), 66–85.
[32] J. KLEIJN, M. KOUTNY: Processes of Petri nets with range testing. *Fundamenta Informaticae*, 80 (2007), 199–219.
[33] J. KLEIJN, M. KOUTNY: Processes of membrane systems with promoters and inhibitors. *Theoretical Computer Sci.*, 404 (2008), 112–126.
[34] J. KLEIJN, M. KOUTNY: A Petri net model for membrane systems with dynamic structure. *Natural Computing*, to appear.
[35] J. KLEIJN, M. KOUTNY: Steps and coverability in inhibitor nets. *Perspectives in Concurrency Theory* (K. Lodaya, M. Mukund, R. Ramanujan, eds.), Universities Press, Hyderabad, India. 2008, 264–295.
[36] J. KLEIJN, M. KOUTNY, G. ROZENBERG: Towards a Petri net semantics for membrane systems. *Lecture Notes in Computer Sci.*, 3850 (2006), 292–309.
[37] J. KLEIJN, M. KOUTNY, G. ROZENBERG: Process semantics for membrane systems. *J. Automata, Languages, Combinatorics*, 11 (2006), 321–340.
[38] S.R. KOSARAJU: Decidability of reachability in vector addition systems. *STOC*, 1982, 267–281.
[39] M. KOUTNY, M. PIETKIEWICZ-KOUTNY: Transition systems of elementary net systems with localities. *Lecture Notes in Computer Sci.*, 4137 (2006), 173–187.
[40] M. KOUTNY, M. PIETKIEWICZ-KOUTNY: Synthesis of elementary net systems with context arcs and localities. *Lecture Notes in Computer Sci.*, 4546 (2007), 1–300.
[41] E.W. MAYR: An algorithm for the general Petri net reachability problem. *SIAM J. Comput.*, 13 (1984), 441–460.
[42] M. NIELSEN, G. PLOTKIN, G. WINSKEL: Petri nets, event structures and domains, Part I. *Theoretical Computer Sci.*, 13 (1980), 85–108.
[43] J.L. PETERSON: *Petri Net Theory and the Modeling of Systems*. Prentice Hall, 1981.
[44] A. PROFIR, E. GUŢULEAC, E. BOIAN: Encoding continuous-time P systems with descriptive times Petri nets. *TAPS'05*, IEEE Computer Press, 2005, 91–94.
[45] Z. QI, J. YOU, H. MAO: P systems and Petri nets. *Lecture Notes in Computer Sci.*, 2933 (2004), 286–303.
[46] A. REGEV, E.M. PANINA, W. SILVERMAN, L. CARDELLI, E.Y. SHAPIRO: BioAmbients: an abstraction for biological compartments. *Theoretical Computer Sci.*, 325 (2004), 141–167.
[47] W. REISIG, G. ROZENBERG, eds.: *Lectures on Petri Nets*. LNCS 1491 and 1492, Springer, 1998.

[48] G. ROZENBERG, J. ENGELFRIET: Elementary net systems. *Lecture Notes in Computer Sci.*, 1491 (1998), 12–121.

[49] M. SILVA, E. TERUEL, J.M. COLOM: Linear algebraic and linear programming techniques for the analysis of place/transition net systems. *Lecture Notes in Computer Sci.*, 1491 (1998), 309–373.

[50] C. STAHL, W. REISIG, M. KRSTIĆ: Hazard detection in a GALS wrapper: a case study. *ACSD'05*, IEEE Computer Society, 2005, 234–243.

[51] W. VOGLER: Partial order semantics and read arcs. *Theoretical Computer Sci.*, 286 (2002), 33–63.

[52] Membrane computing web page: http://ppage.psystems.eu/

CHAPTER 16

SEMANTICS OF P SYSTEMS

GABRIEL CIOBANU

16.1 INTRODUCTION

OPERATIONAL semantics provides a way of describing rigorously the evolution of a computing system. Configurations are states of a transition system, and a computation consists of a sequence of transitions between configurations terminating (if it terminates) in a final configuration. Structural operational semantics (SOS) provide a framework for defining a formal description of a system. It is intuitive and flexible, and it became more attractive over the years by the developments presented by G. Plotkin [21], G. Kahn [13], and R. Milner [16]. In basic P systems, a computation is regarded as a sequence of parallel applications of rules in various membranes, followed by a communication step and a dissolving step. An SOS of the P systems emphasizes the deductive nature of membrane computing by describing the transition steps by using a set of inference rules. Considering a set $\mathcal{R}$ of inference rules of the form $\frac{premises}{conclusion}$, the evolution of a P system can be presented as a deduction tree.

A sequence of transition steps represents a *computation*. A computation is successful if this sequence is finite, namely there is no rule applicable to the objects present in the last committed configuration. In a halting committed configuration, the result of a successful computation is the total number of objects present either in the membrane considered as the output membrane, or in the outer region.

16.2 CONFIGURATIONS AND TRANSITIONS

First we present an inductive definition of the membrane structure, the sets of configurations for a P system, and an intuitive definition for the transition system is given by considering each of the transition steps: maximally parallel rewriting, parallel communication, and parallel dissolving. The operational semantics of P systems is implemented by using the rewriting system called Maude. The relationship between the operational semantics of P systems and Maude rewriting is given by certain operational correspondence results.

Let O be a finite alphabet of objects over which we consider the *free commutative monoid* O_c^*, whose elements are *multisets*. The empty multiset is denoted by *empty*.

Objects can be enclosed in messages together with a target indication. We have *here* messages of typical form $(w, here)$, *out* messages (w, out), and *in* messages (w, in_L). For the sake of simplicity, hereinafter we consider that the messages with the same target indication merge into one message:

$$\prod_{i \in I}(v_i, here) = (w, here),$$

$$\prod_{i \in I}(v_i, in_L) = (w, in_L),$$

$$\prod_{i \in I}(v_i, out) = (w, out),$$

with $w = \prod_{i \in I} v_i$, I a non-empty set, and $(v_i)_{i \in I}$ a family of multisets over O.

We use the mappings rules and priority to associate with a membrane label the set of evolution rules and the priority relation : $\text{rules}(L_i) = R_i$, $\text{priority}(L_i) = \rho_i$, and the projections L and w which return from a membrane its label and its current multiset, respectively.

The set $\mathcal{M}(\Pi)$ *of membranes for a P system* Π, and *the membrane structures* are inductively defined as follows:

- if L is a label, and w is a multiset over $O \cup (O \times \{here\}) \cup (O \times \{out\}) \cup \{\delta\}$, then $\langle L \mid w \rangle \in \mathcal{M}(\Pi)$; $\langle L \mid w \rangle$ is called *simple (or elementary) membrane*, and it has the structure $\langle\rangle$;
- if L is a label, w is a multiset over $O \cup (O \times \{here\}) \cup (O \times \{in_{L(M_j)} \mid j \in [n]\}) \cup (O \times \{out\}) \cup \{\delta\}$, $M_1, \ldots, M_n \in \mathcal{M}(\Pi)$, $n \geq 1$, where each membrane M_i has the structure μ_i, then $\langle L \mid w ; M_1, \ldots, M_n \rangle \in \mathcal{M}(\Pi)$; $\langle L \mid w ; M_1, \ldots, M_n \rangle$ is called *a composite membrane* having the structure $\langle \mu_1, \ldots, \mu_n \rangle$.

We conventionally suppose the existence of a set of sibling membranes denoted by *NULL* such that $M, NULL = M = NULL, M$ and $\langle L \mid w ; NULL \rangle = \langle L \mid w \rangle$.

The use of *NULL* significantly simplifies several definitions and proofs. Let $\mathcal{M}^*(\Pi)$ be the free commutative monoid generated by $\mathcal{M}(\Pi)$ with the operation $(\_,\_)$ and the identity element *NULL*. We define $\mathcal{M}^+(\Pi)$ as the set of elements from $\mathcal{M}^*(\Pi)$ without the identity element. Let M_+, N_+ range over non-empty sets of sibling membranes, M_i over membranes, M_*, N_* range over possibly empty multisets of sibling membranes, and L over labels. The membranes preserve the initial labeling, evolution rules, and priority relation among them in all subsequent configurations. Therefore, in order to describe a membrane we consider its label and the current multiset of objects together with its structure.

A *configuration* for a P system Π is a skin membrane which has no messages and no dissolving symbol δ, i.e. the multisets of all regions are elements in O_c^*. We denote by $\mathcal{C}(\Pi)$ the set of configurations for Π.

An *intermediate configuration* is an arbitrary skin membrane in which we may find messages or the dissolving symbol δ. We denote by $\mathcal{C}^\#(\Pi)$ the set of intermediate configurations. We have $\mathcal{C}(\Pi) \subseteq \mathcal{C}^\#(\Pi)$.

Each P system has an initial configuration which is characterized by the initial multiset of objects for each membrane and the initial membrane structure of the system. For two configurations C_1 and C_2 of Π, we say that there is a *transition* from C_1 to C_2, and write $C_1 \Rightarrow C_2$, if the following *steps* are executed in the given order:

1. *maximally parallel rewriting step*: each membrane evolves in a maximally parallel manner;
2. *parallel communication of objects through membranes*, consisting in sending and receiving messages;
3. *parallel membrane dissolving*, consisting in dissolving the membranes containing δ.

The last two steps take place only if there are messages or δ symbols resulting from the first step, respectively. If the first step is not possible, then neither are the other two steps; we say that the system has reached a *halting configuration*.

16.3 OPERATIONAL SEMANTICS

We present an operational semantics of P systems, considering each of the three steps.

16.3.1 Maximally Parallel Rewriting Step

Here we formally define the maximally parallel rewriting $\overset{mpr}{\Longrightarrow}_L$ for a multiset of objects in one membrane, and we extend it to maximally parallel rewriting $\overset{mpr}{\Longrightarrow}$ over several membranes. Some preliminary notions are required.

Definition 16.1 *The irreducibility property w.r.t. the maximally parallel rewriting relation for multisets of objects, membranes, and for sets of sibling membranes is defined as follows:*

- *a multiset of messages and the dissolving symbol δ are L-irreducible;*
- *a multiset of objects w is L-irreducible iff there are no rules in rules(L) applicable to w with respect to the priority relation priority(L);*
- *a simple membrane $\langle L \mid w \rangle$ is **mpr-irreducible** iff w is L-irreducible;*
- *a non-empty set of sibling membranes $M_1, \ldots, M_n$ is **mpr-irreducible** iff M_i is mpr-irreducible for every $i \in [n]$; NULL is **mpr-irreducible**;*
- *a composite membrane $\langle L \mid w ; M_1, \ldots, M_n \rangle$ is **mpr-irreducible** iff w is L-irreducible, and the set of sibling membranes $M_1, \ldots, M_n$ is mpr-irreducible.*

The priority relation is a form of control on the application of rules. In the presence of a priority relation, no rule of a lower priority can be used during the same evolution step when a rule with a higher priority is used, even if the two rules do not compete for the same objects. We formalize the conditions imposed by the priority relation on rule applications in the definition below.

Definition 16.2 *Let M be a membrane labeled by L, and w a multiset of objects. A non-empty multiset $R = (u_1 \to v_1, \ldots, u_n \to v_n)$ of evolution rules is (L, w)-**consistent** if:*

$-R \subseteq \text{rules}(L)$,
$-w = u_1 \ldots u_n z$, so each rule $r \in R$ is applicable on w,
$-(\forall r \in R, \forall r' \in \text{rules}(L))$ r' applicable on w implies $(r', r) \notin \text{priority}(L)$ $((r_1, r_2) \in \text{priority}(L)$ iff $r_1 > r_2)$,
$-(\forall r', r'' \in R)$ $(r', r'') \notin \text{priority}(L)$,
$-$*the dissolving symbol δ has at most one occurrence in the multiset $v_1 \ldots v_n$.*

Maximally parallel rewriting relations $\overset{mpr}{\Longrightarrow}_L$ and $\overset{mpr}{\Longrightarrow}$ are defined by the following inference rules:

For each $w = u_1 \ldots u_n z \in O_c^+$ such that z is L-irreducible, and (L, w)-consistent rules $(u_1 \to v_1, \ldots, u_n \to v_n)$,

$$(R_1) \; \frac{}{u_1 \ldots u_n z \overset{mpr}{\Longrightarrow}_L v_1 \ldots v_n z}$$

For each $w \in O_c^+$, $w' \in (O \cup Msg(O) \cup \{\delta\})_c^+$, and mpr-irreducible $M_* \in \mathcal{M}^*(\Pi)$,

$$(R_2) \frac{w \overset{mpr}{\Longrightarrow}_L w'}{\langle L \mid w\,;\, M_* \rangle \overset{mpr}{\Longrightarrow} \langle L \mid w'\,;\, M_* \rangle}$$

For each L-irreducible $w \in O_c^*$, and $M_+, M'_+ \in \mathcal{M}^+(\Pi)$,

$$(R_3) \frac{M_+ \overset{mpr}{\Longrightarrow} M'_+}{\langle L \mid w\,;\, M_+ \rangle \overset{mpr}{\Longrightarrow} \langle L \mid w\,;\, M'_+ \rangle}$$

For each $w \in O_c^+$, $w' \in (O \cup Msg(O) \cup \{\delta\})_c^+$, $M_+, M'_+ \in \mathcal{M}^+(\Pi)$,

$$(R_4) \frac{w \overset{mpr}{\Longrightarrow}_L w',\, M_+ \overset{mpr}{\Longrightarrow} M'_+}{\langle L \mid w\,;\, M_+ \rangle \overset{mpr}{\Longrightarrow} \langle L \mid w'\,;\, M'_+ \rangle}$$

For each $M, M' \in \mathcal{M}(\Pi)$, and $M_+, M'_+ \in \mathcal{M}^+(\Pi)$,

$$(R_5) \frac{M \overset{mpr}{\Longrightarrow} M',\, M_+ \overset{mpr}{\Longrightarrow} M'_+}{M, M_+ \overset{mpr}{\Longrightarrow} M', M'_+}$$

For each $M, M' \in \mathcal{M}(\Pi)$, and mpr-irreducible $M_+ \in \mathcal{M}^+(\Pi)$,

$$(R_6) \frac{M \overset{mpr}{\Longrightarrow} M'}{M, M_+ \overset{mpr}{\Longrightarrow} M', M_+}$$

We note that $\overset{mpr}{\Longrightarrow}$ for simple membranes can be described by rule (R_2) with $M_* =$ NULL.

Remark 16.1 M is mpr-irreducible if there does not exist M' such that $M \overset{mpr}{\Longrightarrow} M'$.

Proposition 16.1 *Let Π be a P system. If $C \in \mathcal{C}(\Pi)$ and $C' \in \mathcal{C}^\#(\Pi)$ such that $C \overset{mpr}{\Longrightarrow} C'$, then C' is mpr-irreducible.*

The proof of Proposition 16.1 follows by structural induction on C.

The formal definition of $\overset{mpr}{\Longrightarrow}$ given above corresponds to the intuitive description of the maximal parallelism. The non-determinism is given by the associativity and commutativity of the concatenation operation over objects used in R_1. The parallelism of the evolution rules in a membrane is also given by R_1: $u_1 \ldots u_n z \overset{mpr}{\Longrightarrow}_L v_1 \ldots v_n z$ saying that the rules of the multiset $(u_1 \to v_1, \ldots, u_n \to v_n)$ are applied simultaneously. The fact that the membranes are evolving in parallel is described by rules $R_3 - R_6$.

16.3.2 Parallel Communication of Objects

We say that a multiset w is *here-free/out-free/in_L-free* if it does not contain any *here/out/in_L* messages, respectively. For w a multiset of objects and messages, we introduce the operations obj, here, out, and in_L as follows:

$\text{obj}(w)$ is obtained from w by removing all messages,

$$\text{here}(w) = \begin{cases} empty & \text{if } w \text{ is } here\text{-free,} \\ w'' & \text{if } w = w'(w'', here) \wedge w' \text{ is } here\text{-free;} \end{cases}$$

$$\text{out}(w) = \begin{cases} empty & \text{if } w \text{ is } out\text{-free,} \\ w'' & \text{if } w = w'(w'', out) \wedge w' \text{ is } out\text{-free;} \end{cases}$$

$$\text{in}_L(w) = \begin{cases} empty & \text{if } w \text{ is } in_L\text{-free,} \\ w'' & \text{if } w = w'(w'', in_L) \wedge w' \text{ is } in_L\text{-free.} \end{cases}$$

We consider the extension of the operator w (previously defined over membranes) to non-empty sets of sibling membranes by setting $\text{w}(NULL) = empty$ and $\text{w}(M_1, \ldots, M_n) = \text{w}(M_1) \ldots \text{w}(M_n)$.

We recall that the messages with the same target merge in one larger message.

Definition 16.3 *The **tar-irreducibility** property for membranes and for sets of sibling membranes is defined as follows:*

- *a simple membrane $\langle L \mid w \rangle$ is **tar-irreducible** iff w is here-free and $L \neq Skin \vee (L = Skin \wedge w$ out-free);*
- *a non-empty set of sibling membranes $M_1, \ldots, M_n$ is **tar-irreducible** iff M_i is tar-irreducible for every $i \in [n]$; NULL is **tar-irreducible**;*
- *a composite membrane $\langle L \mid w; M_1, \ldots, M_n \rangle$, $n \geq 1$, is **tar-irreducible** iff: w is here-free and $in_{L(M_i)}$-free for every $i \in [n]$, $L \neq Skin \vee (L = Skin \wedge w$ is out-free), $\text{w}(M_i)$ is out-free for all $i \in [n]$, and the set of sibling membranes $M_1, \ldots, M_n$ is tar-irreducible.*

Notation. We treat messages of the form $(w', here)$ as a particular communication inside a membrane, and we substitute $(w', here)$ by w'. We denote by $\overline{w}$ the multiset obtained by replacing $(\text{here}(w), here)$ with $\text{here}(w)$ in w. For instance, if $w = a\,(bc, here)\,(d, out)$ then $\overline{w} = abc\,(d, out)$, where $\text{here}(w) = bc$. We note that $\text{in}_L(\overline{w}) = \text{in}_L(w)$, and $\text{out}(\overline{w}) = \text{out}(w)$.

Parallel communication relation $\stackrel{tar}{\Longrightarrow}$ is defined by the following inference rules:

For each tar-irreducible $M_* \in \mathcal{M}^*(\Pi)$ and multiset w such that here(w) $\neq$ empty, or $L = Skin \wedge$ out(w) $\neq$ empty, or it exists $M_i \in M_*$ with $\mathrm{in}_{L(M_i)}(w)\mathrm{out}(w(M_i)) \neq$ empty,

$$(C_1)\frac{}{\langle L \mid w\,;\, M_* \rangle \stackrel{tar}{\Longrightarrow} \langle L \mid w'\,;\, M'_* \rangle}$$

where

$$w' = \begin{cases} \mathrm{obj}(\overline{w})\,\mathrm{out}(w(M_*)) & \text{if } L = Skin, \\ \mathrm{obj}(\overline{w})\,(\mathrm{out}(w), out)\,\mathrm{out}(w(M_*)) & \text{otherwise}, \end{cases}$$

and
$$w(M'_i) = \mathrm{obj}(w(M'_i))\,\mathrm{in}_{L(M_i)}(w), \text{ for all } M_i \in M_*$$

For each $M_1, \ldots, M_n, M'_1, \ldots, M'_n \in \mathcal{M}^+(\Pi)$, and multiset w,

$$(C_2)\frac{M_1, \ldots, M_n \stackrel{tar}{\Longrightarrow} M'_1, \ldots, M'_n}{\langle L \mid w\,;\, M_1, \ldots, M_n \rangle \stackrel{tar}{\Longrightarrow} \langle L \mid w''\,;\, M''_1, \ldots, M''_n \rangle}$$

where

$$w'' = \begin{cases} \mathrm{obj}(\overline{w})\,\mathrm{out}(w(M'_1, \ldots, M'_n)) & \text{if } L = Skin, \\ \mathrm{obj}(\overline{w})\,(\mathrm{out}(w), out)\,\mathrm{out}(w(M'_1, \ldots, M'_n)) & \text{otherwise}, \end{cases}$$

and each M''_i is obtained from M'_i by replacing its resources with
$$w(M''_i) = \mathrm{obj}(\overline{w(M'_i)})\,\mathrm{in}_{L(M'_i)}(w), \text{ for all } i \in [n]$$

For each $M, M' \in \mathcal{M}(\Pi)$, and tar-irreducible $M_+ \in \mathcal{M}^+(\Pi)$,

$$(C_3)\frac{M \stackrel{tar}{\Longrightarrow} M'}{M, M_+ \stackrel{tar}{\Longrightarrow} M', M_+}$$

For each $M \in \mathcal{M}(\Pi)$, $M_+ \in \mathcal{M}^+(\Pi)$,

$$(C_4)\frac{M \stackrel{tar}{\Longrightarrow} M',\, M_+ \stackrel{tar}{\Longrightarrow} M'_+}{M, M_+ \stackrel{tar}{\Longrightarrow} M', M'_+}$$

Remark 16.2 M is tar-irreducible iff there does not exist M' such that $M \stackrel{tar}{\Longrightarrow} M'$.

Proposition 16.2 *Let Π be a P system. If $C \in C^\#(\Pi)$ with messages and $C \stackrel{tar}{\Longrightarrow} C'$, then C' is tar-irreducible.*

The proof of Proposition 16.2 is done by structural induction on C.

16.3.3 Parallel Membrane Dissolving

If the special symbol δ occurs in the multiset of objects of a membrane labeled by L, that membrane is dissolved, its evolution rules and the associated priority relation are lost, and its contents (objects and membranes) are added to the contents of the surrounding membrane. We say that a multiset w is δ-*free* if it does not contain the special symbol δ.

Definition 16.4 *The δ-irreducibility property for membranes and for sets of sibling membranes is defined as follows:*

- *a simple membrane is δ-irreducible if it has no messages;*
- *a non-empty set of sibling membranes $M_1, \ldots, M_n$ is δ-irreducible if every membrane M_i is δ-irreducible, for $1 \le i \le n$; NULL is δ-irreducible;*
- *a composite membrane $\langle\, L \mid w\,;\, M_+ \,\rangle$ is δ-irreducible if w has no messages, M_+ is δ-irreducible, and $w(M_+)$ is δ-free.*

Parallel dissolving relation $\overset{\delta}{\Longrightarrow}$ is defined by the following inference rules:
For each $M_* \in \mathcal{M}^*(\Pi)$, δ-irreducible $\langle\, L_2 \mid w_2\delta\,;\, M_* \,\rangle$, and label L_1,

$$(D_1)\frac{}{\langle\, L_1 \mid w_1\,;\, \langle\, L_2 \mid w_2\delta\,;\, M_* \,\rangle \,\rangle \overset{\delta}{\Longrightarrow} \langle\, L_1 \mid w_1 w_2\,;\, M_* \,\rangle}$$

For each $M_+ \in \mathcal{M}^+(\Pi)$, $M'_* \in \mathcal{M}^*(\Pi)$, δ-free multiset w_2, multisets w_1, w'_2, and labels L_1, L_2

$$(D_2)\frac{\langle\, L_2 \mid w_2\,;\, M_+ \,\rangle \overset{\delta}{\Longrightarrow} \langle\, L_2 \mid w'_2\,;\, M'_* \,\rangle}{\langle\, L_1 \mid w_1\,;\, \langle\, L_2 \mid w_2\,;\, M_+ \,\rangle \,\rangle \overset{\delta}{\Longrightarrow} \langle\, L_1 \mid w_1\,;\, \langle\, L_2 \mid w'_2\,;\, M'_* \,\rangle \,\rangle}$$

For each $M_+ \in \mathcal{M}^+(\Pi)$, $M'_* \in \mathcal{M}^*(\Pi)$, multisets w_1, w_2, w'_2, and labels L_1, L_2

$$(D_3)\frac{\langle\, L_2 \mid w_2\delta\,;\, M_+ \,\rangle \overset{\delta}{\Longrightarrow} \langle\, L_2 \mid w'_2\delta\,;\, M'_* \,\rangle}{\langle\, L_1 \mid w_1\,;\, \langle\, L_2 \mid w_2\delta\,;\, M_+ \,\rangle \,\rangle \overset{\delta}{\Longrightarrow} \langle\, L_1 \mid w_1 w'_2\,;\, M'_* \,\rangle}$$

For each $M_+ \in \mathcal{M}^+(\Pi)$, $M'_*, N'_* \in \mathcal{M}^*(\Pi)$, δ-irreducible $\langle\, L \mid w\,;\, N_+ \,\rangle$, and multisets w', w'',

$$(D_4)\frac{\langle\, L \mid w\,;\, M_+ \,\rangle \overset{\delta}{\Longrightarrow} \langle\, L \mid w'\,;\, M'_* \,\rangle}{\langle\, L \mid w\,;\, M_+, N_+ \,\rangle \overset{\delta}{\Longrightarrow} \langle\, L \mid w'\,;\, M'_*, N_+ \,\rangle}$$

$$(D_5)\frac{\langle\, L \mid w\,;\, M_+ \,\rangle \overset{\delta}{\Longrightarrow} \langle\, L \mid ww'\,;\, M'_* \,\rangle \quad \langle\, L \mid w\,;\, N_+ \,\rangle \overset{\delta}{\Longrightarrow} \langle\, L \mid ww''\,;\, N'_* \,\rangle}{\langle\, L \mid w\,;\, M_+, N_+ \,\rangle \overset{\delta}{\Longrightarrow} \langle\, L \mid ww'w''\,;\, M'_*, N'_* \,\rangle}$$

Remark 16.3 M is δ-irreducible iff there does not exist M' such that $M \overset{\delta}{\Longrightarrow} M'$.

Proposition 16.3 *Let Π be a P system. If $C \in \mathcal{C}^\#(\Pi)$ is tar-irreducible and $C \overset{\delta}{\Longrightarrow} C'$, then C' is δ-irreducible.*

The proof of Proposition 16.3 follows by a structural induction on C.

It is worth noting that $C \in \mathcal{C}(\Pi)$ iff C is tar-irreducible and δ-irreducible. According to the standard description in membrane computing, a *transition step* between two configurations $C, C' \in \mathcal{C}(\Pi)$ is given by: $C \Rightarrow C'$ iff C and C' are related by one of the following relations:

either $C \stackrel{mpr}{\Longrightarrow}; \stackrel{tar}{\Longrightarrow} C'$, or $C \stackrel{mpr}{\Longrightarrow}; \stackrel{\delta}{\Longrightarrow} C'$, or $C \stackrel{mpr}{\Longrightarrow}; \stackrel{tar}{\Longrightarrow}; \stackrel{\delta}{\Longrightarrow} C'$.

The three alternatives in defining $C \Rightarrow C'$ are given by the existence of messages and dissolving symbols along the system evolution. Starting from a configuration without messages and dissolving symbols, we apply the "mpr" rules and get an intermediate configuration which is mpr-irreducible; if we have messages, then we apply the "tar" rules and get an intermediate configuration which is tar-irreducible; if we have dissolving symbols, then we apply the dissolving rules and get a configuration which is δ-irreducible. If the last configuration has no messages or dissolving symbols, then we say that the transition relation $\Rightarrow$ is well-defined as an evolution step between the first and last configurations.

Proposition 16.4 *The relation $\Rightarrow$ is well-defined over the entire set $\mathcal{C}(\Pi)$ of configurations.*

Examples of inference trees, as well as the proofs of the results, are presented in [3]. We have briefly presented the operational semantics just to give sense to the implementations of P systems into rewriting logic.

16.4 IMPLEMENTING P SYSTEMS USING MAUDE

Generally, by using a rewriting engine called Maude [10], a formal specification of a system can be automatically transformed into an interpreter. Moreover, Maude provides a useful search command, a semi-decision procedure for finding failures of safety properties, and also a model checker. Since the P systems combine the power of parallel rewriting in various locations (compartments) with the power of local and contextual evolution, it is natural to use a rewriting engine and a rewrite theory. Roughly speaking, a rewrite theory is a triple $(\Sigma, E, \mathcal{R})$, where (Σ, E) is an equational theory used for implementing the deterministic computation, therefore (Σ, E) should be terminating and Church-Rosser, and $\mathcal{R}$ is a set of rewrite rules used to implement non-deterministic and/or concurrent computations. Therefore we find rewriting logic suitable for implementing the membrane systems. The Web page of the implementations described here can be found at http://thor.info.uaic.ro/~rewps/index.html.

A P system consists of a maximally parallel application of the evolution rules according to their priorities (if any), the (repeated) steps of internal evolution, communication, and dissolving. This sequence of steps uses a kind of synchronization. The first challenge is to describe the maximally parallel rewriting, because this is not quite natural for rewriting logic. In [2] the reflection property of rewriting logic is used for defining maximally parallel rewriting. A P system was defined at object level, and its semantics at meta-level. The description at meta-level assumes many additional operations, and therefore the checking and the analysis of such a specification was time consuming. Here we use a different approach. The evolution rules are represented as terms at the object level. This allows us to define the operational semantics at the object level and, consequently, the checking and the analysis of the specification is more efficient.

The second challenge is given by the sequence of internal steps: evolution, communication, and dissolving. We decorated the terms denoting membranes with colors, and these colors are used by a rewrite engine to choose the appropriate (sub)set of rewriting rules. The definition of semantics for P systems in rewriting logic reveals an interesting aspect: internal evolution, communication, and dissolving inside a complex membrane may interleave. If only main configurations are observable, then the big-step semantics given by a maximally parallel step and the small steps executed by a single machine to simulate the maximally parallel step are behaviorally equivalent.

16.4.1 Maude Evolution Rules

A P system has a tree like structure with the skin as its root, the composite membranes as its internal nodes, and the elementary membranes as its leaves. The order of the children of a node is not important due to the associativity and commutativity properties of the concatenation operation of membranes.

Since Maude is not able to execute parallel transitions required by a P system, we should use a sequential application of the rules. In this sequential process, in order to prevent the case when the result of one rule is used by another, we use a technique of marking the intermediate configurations with colors. This is achieved by using two operators *blue* and *green*.

The operator blue traverses the tree in a top-down manner, firing the maximally parallel rewriting process in every membrane through the blue operator on Soup terms. The Boolean argument of operation blue is used to show whether a dissolving rule is chosen during the current maximally parallel step in a membrane with dissolving rules. We need this flag because at most one δ symbol is allowed in a membrane.

The multiset of objects is divided into two parts during the maximally parallel rewriting process: *blue* represents the objects available to be "consumed" via evolution rules, while *green* represents the objects resulted from applying evolution rules over the blue objects (therefore the green objects are not available anymore for the current evolution step). When no more rule can be applied, the remaining blue objects (if any) become green.

The operator green traverses the tree in a bottom-up manner as follows:

- green multisets from $(O \cup Msg(O) \cup \{\delta\})_c^*$ merge into one green multiset;
- a leaf becomes green if its multiset is green;
- a set of sibling subtrees (with the roots sibling nodes) becomes green if each subtree is green;
- a subtree becomes entirely green if:
  1. the multiset of the root is green;
  2. the subtrees determined by the children of the root form a green set of sibling subtrees;
  3. there are no messages to be exchanged between the root node and its children.

In the communication stage of a green subtree, a node can send a message only to its parent or to one of its children. The rules for each direction of communication (*out* and *in*) vary on the structure of the destination membrane.

In a P system the dissolving process occurs after the end of the communication process. To fulfil this condition we allow dissolving *only if* there are no messages to be sent. By dissolving the membrane of a node, all of its objects are transferred to the membrane of its parent, the rules are lost, and if it is an internal node, all of its children become children of its parent. The skin membrane is not allowed to be dissolved. The resulting term corresponds to a configuration of P system reachable in one transition step of a P systems starting from a given configuration.

More details on the rules implementing the bottom-up traversal are presented in [5]. In the same paper the dynamics of P systems and their implementations using Maude are related. Such a relationship between operational semantics for P systems and the Maude rewriting relation is given by operational correspondence results.

Let $\Pi = (O, \mu, w_1, \ldots, w_n, (R_1, \rho_1), \ldots, (R_n, \rho_n), i_0)$ be a P system having the initial configuration $\langle L_1 \mid w_1 ; M_{i_1}, \ldots, M_{i_n} \rangle$, $\{i_1, \ldots, i_n\} \subseteq \{2, \ldots, n\}$, with rules$(L_j) = R_j$, priority$(L_j) = \rho_j$ for all membrane labels L_j, and let $\Rightarrow$ be the transition relation between two configurations. We associate to Π a rewriting theory $R(\Pi) = (\Sigma, E, R)$; Σ is the equational signature defining sorts and operation symbols, E is the set of Σ-equations which also includes the appropriate axioms for the associativity, commutativity, and identity attributes of the operators, and R is the set of rewriting rules. Considering $\longrightarrow_{R(\Pi)}$ the rewriting relation, we denote by $\longrightarrow^+_{R(\Pi)}$ the transitive closure of $\longrightarrow_{R(\Pi)}$, and by $\longrightarrow^*_{R(\Pi)}$ its reflexive and transitive closure.

An encoding function $\mathcal{I}_m : \mathcal{M}(\Pi) \to (T_{\Sigma,E})_{\texttt{Membrane}}$ from the set of membranes $\mathcal{M}(\Pi)$ to the ground terms of sort Membrane from the associated rewriting theory is defined by

- if $M = (\!\!(\, L \mid w \,)\!\!)$, then $\mathcal{I}_m(M) = \texttt{< L | w >}$,
- if $M = (\!\!(\, L \mid w\,;\, M_1, \ldots, M_n \,)\!\!)$,
  then $\mathcal{I}_m(M) = \texttt{< L | w ;}\ \mathcal{I}_m(M_1)\,,\ldots,\mathcal{I}_m(M_n)\ \texttt{>}$,

where L is a constant of sort Label, and w is a term of sort Soup. We extend the encoding function $\mathcal{I}_m$ over non-empty sets of sibling membranes by

$$\mathcal{I}_m(M_1, \ldots, M_k) = \mathcal{I}_m(M_1), \ldots, \mathcal{I}_m(M_k),\ k \geq 2.$$

We also define an encoding function $\mathcal{I} : \mathcal{C}(\Pi) \to (T_{\Sigma,E})_{\texttt{Configuration}}$ from the set of configurations to the ground terms of sort Configuration such that $\mathcal{I}(C) = \{\mathcal{I}_m(C)\}$, for every $C \in \mathcal{C}(\Pi)$.

Proposition 16.5 a) *If $blue(L, S) \to green(S')$, then $S \stackrel{mpr}{\Longrightarrow}_L S'$.*
b) *Conversely, if $S \stackrel{mpr}{\Longrightarrow}_L S'$, then there is a rewrite $blue(L, S) \to green(S')$.*

These results are proved by induction on the length of the rewriting (a), and by induction on the number of rules applied in parallel (b). Using similar results which are presented in [5], we finally get the following theorem:

Theorem 16.1 (Operational Correspondence) *The rewriting relation given by the implementing rules represents a correct and complete implementation of $\Rightarrow$ defined by the operational semantics.*

The implementation of the sequential composition using a general rewrite engine like Maude requires some auxiliary operations and verification of conditions. It is possible to define an alternative to the sequential composition, and so having various granularity (between "big" and "small") for the operational semantics for P systems.

Replacing some rules, it is possible to relax the strict separation between the internal steps given by the evolution of membranes, communication, and dissolving. If two parent-child membranes finish their internal evolution, then they can communicate without waiting for the other membranes of the system to finish their evolution step. Similarly, if two parent-child membranes finish the communication, then the child may dissolve whenever it has a δ object. In this way, we reveal two forms of correctness of an implementation with respect to the given operational semantics. In a stronger form, we say that an implementation $\mathcal{I}$ is *faithful* if and only if it is defined by three relations $\leadsto_{mpr}$, $\leadsto_{tar}$, and $\leadsto_{diss}$ such that $\mathcal{I}(C) \leadsto_{mpr} \mathcal{I}(C_1)$ whenever $C \overset{mpr}{\Longrightarrow} C_1$, $\mathcal{I}(C_1) \leadsto_{tar} \mathcal{I}(C_2)$ whenever $C_1 \overset{tar}{\Longrightarrow} C_2$, and $\mathcal{I}(C_2) \leadsto_{diss} \mathcal{I}(C')$ whenever $C_2 \overset{\delta}{\Longrightarrow} C'$, for all configurations C, C' and intermediate configurations C_1, C_2. In a weaker form, we say that an implementation $\mathcal{I}$ is *accurate* if and only if it is defined by a relation $\leadsto$ such that $\mathcal{I}(C) \leadsto \mathcal{I}(C')$ whenever $C \Rightarrow C'$, for all configurations C, C'.

In an accurate implementation it is possible to execute parallel transitions of different phases. This fact can increase the potential parallelism of the rewriting implementations of the P systems. On the other hand, a faithful implementation generates a smaller state space. We can exemplify these aspects by using a simple P system $\langle L_1 \mid aa; \langle L_2 \mid yy; \langle L_3 \mid vv \rangle \rangle \rangle$, where $rules(L_1) = a \rightarrow (b, in(L_2))$, $rules(L_2) = y \rightarrow (x, out)(z, in(L_3))$, and $rules(L_3) = v \rightarrow (u, out)$. We use the Maude command search to generate the whole state space. For the faithful implementation we get 292 states:

```
Maude> search init =>+ C:Configuration .
search in EX : init =>+ C:Configuration .

Solution 1 (state 262)
states: 263  rewrites: 3059 in 41ms cpu (42ms real)
C:Configuration --> {< L1 | x x ; < L2 | b b u u ; < L3 | z z > > >}

No more solutions.
states: 292  rewrites: 3420 in 45ms cpu (46ms real)
```

For the accurate implementation we get 752 states:

```
Maude> search init =>+ C:Configuration .
search in EX : init =>+ C:Configuration .

Solution 1 (state 731)
states: 732  rewrites: 11001 in 90ms cpu (90ms real)
C:Configuration --> {< L1 | x x ; < L2 | b b u u ; < L3 | z z > > >}

No more solutions.
states: 752  rewrites: 11371 in 92ms cpu (92ms real)
```

Since we have a higher level of parallelism in the accurate implementation, there are more paths from the initial configuration to the unique final one. The additional states are given by configurations where different membranes are in different phases, e.g. L_1 evolves, while L_2 and L_3 communicate. These forms of implementation exhibit different levels of parallelism which can be exploited in analyzing P systems. An accurate implementation can be used to analyze more aspects inspired by biology. Such an implementation is also appropriate when we are interested in speeding up the execution on a parallel machine. On the other hand, if we are interested in investigating only the configurations (states), then it is better to use a faithful implementation.

16.5 REGISTER MEMBRANES FOR RULES WITH PROMOTERS AND INHIBITORS

In this section we present two operational semantics of P systems which differ only in the way the maximally parallel application of rules is described. These two operational semantics reflect the fact that resource allocation to rules can be done either statically or dynamically. For P systems with promoters and inhibitors dynamical allocation requires the addition of a register to each of the system's membranes. We define an operational semantics of P systems by means of three sets of inference rules corresponding to maximally parallel rewriting, sending messages, and dissolving. A minimal set of inference rules is defined, and their behavior is detailed along with the presentation of the rewriting logic implementation. We use a uniform representation of rather complex P systems with promoters and inhibitors, and get a flexible interpreter for them in Maude. The main results provide the correspondence between the operational semantics of dynamic allocation and the rewriting theory.

We can associate *promoters* and *inhibitors* with a rule $u \to v$, in the form $(u \to v)|_{w_{prom}, \neg w_{inhib}}$, with w_{prom}, w_{inhib} non-empty multisets of objects. However we no longer consider priorities for rules. A rule $(u \to v)|_{w_{prom}, \neg w_{inhib}}$ associated with a membrane i is applied only if w_{prom} is present and w_{inhib} is absent from the region of the membrane i. The promoters and inhibitors of membrane systems formalize the reaction enhancing and reaction prohibiting roles of various substances present in cells. Membrane systems with promoters or with inhibitors provide characterizations of recursively enumerable sets (of vectors of natural numbers) [7].

For a rule $(u \to v)|_{w_{prom}, \neg w_{inhib}}$ we consider the following notations:

$$lhs(r) = u, \ rhs(r) = v, \ promoter(r) = w_{prom}, \ inhibitor(r) = w_{inhib}.$$

These multisets must satisfy some conditions: u, w_{prom}, w_{inhib} contain only objects, u contains at least one object and v contains only messages. If a rule r has no associated promoter, we set $promoter(r) = empty$; if the rule has no associated inhibitor, we also set $inhibitor(r) = empty$ (this convention is used for the sake of uniformity such that we do not have to differentiate between rules with and without inhibitors).

In this section we consider multisets as functions, since some conditions are easier to express in this manner.

16.5.1 Dynamic Allocation Semantics

In order to give an operational semantics for P systems (operational semantics which can easily be transcribed in a rewriting logic implementation), we present a semantics based on applying rules one by one in a non-deterministic manner until there is no applicable rule left. We call this a *dynamic allocation semantics*.

In a maximally parallel evolution step, a rule's applicability with respect to promoters and inhibitors only depends on the initial content of the membrane. For this reason, when each transition consists of applying only a rule, we need to store somewhere the objects consumed by previous applications of rules. Thus, we consider registers consisting of a multiset of objects which have been consumed previously (in the same maximally parallel rewriting step), to keep track of rule application. In the same manner as in the definition in Section 16.2, we construct the set $\mathcal{M}_h(\Pi)$ of *register membranes*:

- if $M \in \mathcal{M}(\Pi)$ is an elementary membrane and $u : O \to \mathbf{N}$ then the pair $(M, u) \in \mathcal{M}_h(\Pi)$ and is called an *elementary register membrane*;
- if $M = \langle i|w; M_1 \ldots M_n \rangle$ is a composite membrane and $u, u_1, \ldots, u_n : O \to \mathbf{N}$ then $(\langle i|w; (M_1, u_1) \ldots (M_n, u_n)\rangle, u) \in \mathcal{M}_h(\Pi)$ is called a *composite register membrane*.

Registers are used to ensure that dynamic allocation is correct; they are only used during the rule application stage.

We can see a membrane of $\mathcal{M}(\Pi)$ as an equivalence class of register membranes, i.e. obtained by ignoring registers. Namely, we define inductively a relation $\equiv$ on $\mathcal{M}(\Pi)$ as follows: $(M, u) \equiv (M, v)$, $\forall u, v : O \to \mathbf{N}$ and if $H_1 \equiv H_1', \ldots, H_n \equiv H_n'$ then $(\langle i|w; H_1, \ldots H_n\rangle, u) \equiv (\langle i|w; H_1', \ldots H_n'\rangle, v)$. Clearly, $\equiv$ is an equivalence relation.

Proposition 16.6 *The set $\mathcal{M}_h(\Pi)/\equiv$ of equivalence classes is isomorphic with the set $\mathcal{M}(\Pi)$ of membranes of the P system Π.*

The bijection is $\hat{\phi} : \mathcal{M}_h(\Pi)/\equiv \to \mathcal{M}(\Pi)$, induced by $\phi : \mathcal{M}_h(\Pi) \to \mathcal{M}(\Pi)$. which is obtained by considering $\phi(\langle i|w\rangle, u) = \langle i|w\rangle$ and $\phi(\langle i|w; H_1, \ldots H_n\rangle, u) = \langle i|w; \phi(H_1), \ldots \phi(H_n)\rangle$.

The states of the following transition system are register membranes. The labels are taken from the set $Rules \cup \{\tau\}$, where $Rules = \cup_{i \in [n]} Rules(i)$, and τ denotes a silent action—namely the evolution of a membrane in which all rewrites take place in the inner membranes while the content of the top membrane stays the same.

The following definition gives a mathematical description of what it means for a rule r to be applicable in a membrane M with register u.

Definition 16.5 *Consider $H \in \mathcal{M}_h(\Pi)$, $H = (\langle i|w; H_*\rangle, u)$, and $r \in Rules(i)$; H_* is a (possibly empty) set of register membranes. We say that the pair (H, r) is valid when*

- $lhs(r) \leq w$, $promoter(r) \leq w + u$;
- if $inhibitor(r) =$ empty then $\exists a \in O$ such that $(w + u)(a) < inhibitor(r)(a)$;
- if H_* is empty then $rhs(r)(a, in_j) = 0$, $\forall j \in [m]$;
- if $H_* = \{H_1, \ldots, H_n\}$ and $j_1, \ldots, j_n$ are the labels of $H_1, \ldots, H_n$ respectively, then $rhs(r)(a, in_j) = 0$, $\forall j \notin \{j_1, \ldots, j_n\}$.

A register membrane H is *mpr-irreducible* when there is no rule which can be applied in it or in any of the membranes it contains. We define inductively a transition relation $T_{mpr} \subseteq \mathcal{M}_h(\Pi) \times Rules \cup \{\tau\} \times \mathcal{M}_h(\Pi)$ as follows:

- if $H = (\langle i|w\rangle, u)$ is an elementary register membrane and (H, r) is valid, then

$$(\text{seq--elem}) \; \frac{}{(\langle i|w\rangle, u) \xrightarrow{r} (\langle i|w - lhs(r) + rhs(r)\rangle, u + lhs(r))}$$

- if $H = (\langle i|w; H_1, \ldots, H_n\rangle, u)$ is a composite register membrane and $\exists j \in [n]$ such that H_j is not mpr-irreducible, then

$$(\text{silent}) \; \frac{H_j \xrightarrow{l} H'_j}{(\langle i|w; H_1, \ldots, H_n\rangle, u) \xrightarrow{\tau} (\langle i|w; H'_1, \ldots, H'_n\rangle, u)}$$

where $H'_k = H_k$, $\forall k \neq j$ and $l \in Rules \cup \{\tau\}$;
- if $H = (\langle i|w; H_1, \ldots, H_n\rangle, u)$ is a composite register membrane such that H_j are mpr-irreducible $\forall j \in [n]$ and (H, r) is valid, then

$$(\text{rewrite}) \; \frac{}{(\langle i|w; H_1, \ldots, H_n\rangle, u) \xrightarrow{r} (\langle i|w - lhs(r) + rhs(r); H_1, \ldots, H_n\rangle, u + lhs(r))}$$

Note that the rules *seq-elem* and *rewrite* ensure that rules are first applied in elementary membranes until they become irreducible, then in their parents, and so on.

We now present two other transition relations $T_{msg} \subseteq \mathcal{M}(\Pi) \times \{msg\} \times \mathcal{M}(\Pi)$ and $T_{diss} \subseteq \mathcal{M}(\Pi) \times \{diss\} \times \mathcal{M}(\Pi)$ which express the message passing and the dissolving steps in the evolution of a membrane system. We use the isomorphism from Proposition 16.6 to glue together T_{mpr}, T_{msg}, and T_{diss}. We denote by Ω all the multisets of objects together with all the multisets of messages which are in a membrane. We use $promoter(r) = 0_\Omega$ to express that a rule r has no promoter, similarly for no inhibitor or dissolving symbol. A membrane M evolves to a membrane N when

- $(H_M, H) \in T^*_{mpr}$, where H_M is the register membrane with its register and all the registers of its children equal to 0_Ω, $\phi(H_M) = M$, H is a mpr-irreducible register membrane and T^*_{mpr} is the transitive closure of T_{mpr} or H_M is mpr-irreducible and $H := H_M$;

- if $\phi(H)$ is msg-irreducible then $M' := \phi(H)$; otherwise, if $\phi(H)$ is not msg-irreducible, then there is a unique membrane M' such that $(\phi(H), M') \in T_{msg}$;
- if M' is diss-irreducible then $N := M'$; otherwise, if M' is not diss-irreducible, then N is the unique membrane for which $(M', N) \in T_{diss}$.

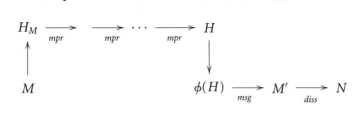

In what follows, let $w(M)$ denote the multiset contained in the membrane M, and $L(M)$ denote the label of M.

We define the following functions over multisets: the *cleanup* function modifies the multiset by "erasing" objects with messages of the form *in_child* and by "transforming" objects of form $(a, here)$ into objects of form a. The *out* function collects the objects a from the objects with messages of form (a, out); the function in_j is similarly defined. The *eraseOut* function removes objects with messages of form (a, out) from a multiset; similarly, *eraseDelta* removes special symbols δ.

The transition relation T_{msg} is given by the following inference rules:

- if $w \neq cleanup(w)$, or $i = 1$ and $w' \neq eraseOut(cleanup(w))$ then

$$(\text{msg1}) \; \frac{}{\langle i | w \rangle \xrightarrow{msg} \langle i | w' \rangle}$$

where $w' = cleanup(w)$ if $i \neq 1$, and $w' = eraseOut(cleanup(w))$ if $i = 1$;
- if M_j are msg-irreducible, $\forall j \in J \subseteq [n]$, then

$$(\text{msg2}) \; \frac{M_k \xrightarrow{msg} M'_k, \forall k \in [n]\setminus J}{\langle i|w; M_1, \ldots, M_n \rangle \xrightarrow{msg} \langle i|w'; M''_1, \ldots, M''_n \rangle}$$

where M''_l have the same structure as M_l, $w(M''_j) = eraseOut(w(M_j)) + in_{L(M_j)}(w)), \forall j \in J$, and $w(M''_k) = eraseOut(w(M'_k) + in_{L(M_k)}(w))$, $\forall k \notin J$, and either $w' = cleanup(w) + \sum_{l \in [n]} out(w(M_l))$ whenever $i \neq 1$, or $w' = eraseOut(cleanup(w)) + \sum_{l \in [n]} out(w(M_l))$ if $i = 1$.

In order to define the transition relation T_{diss}, we first define the notion of diss-irreducibility.

Definition 16.6 *Any elementary membrane is* diss-irreducible. *A composite membrane* $\langle i|w, M_1, \ldots, M_n \rangle$ *is* diss-irreducible *if* $w(M_j)(\delta) = 0$ *and* M_j *is diss-irreducible for all* $i \in [n]$.

In what follows, M_* and N_* range over (possibly empty) sets of membranes.

The transition relation T_{diss} is given by the following inference rules:

- if $M_s = \langle i_s | w_s; M_{*s} \rangle$ are diss-irreducible $\forall s \in [n]$, and there exists a non-empty $J \subset [n]$ such that $w_j(\delta) = 1$, $j \in J$ and $w_k(\delta) = 0$ for $k \notin J$, then

$$(\text{diss1}) \; \frac{}{\langle i|w; M_1, \ldots, M_n \rangle \xrightarrow{diss} \langle i|w'; M_* \rangle}$$

where $w' = w + \sum_{j \in J} eraseDelta(w_j)$ and $M_* = (\bigcup_{k \notin J} \{M_k\}) \cup (\bigcup_{j \in J} M_{*j})$;

- if M_j are diss-irreducible for all $j \in J$ where $J \subseteq [n]$, then

$$(\text{diss2}) \; \frac{M_k \xrightarrow{diss} M'_k, \forall k \notin J}{\langle i|w; M_1, \ldots, M_n \rangle \xrightarrow{diss} \langle i|w; M'_1, \ldots, M'_n \rangle}$$

where $M'_j = M_j$ for all $j \in J$.

16.5.2 Static Allocation Semantics

Let R be a multiset over $Rules(i)$ for some label i. We denote by $lhs(R)$ the multiset over Ω given by $lhs(R)(a) = \sum_{r \in Rules(L)} R(r) \cdot lhs(r)(a)$. The multiset $rhs(R)$ is defined in the same way. We also define $promoter(R)(a) = \max_{r \in supp(R)} promoter(r)(a)$.

Definition 16.7 *For a membrane* $M = \langle i|w; M_* \rangle$ *and for* R *a multiset of rules over* $Rules(i)$ *we say that the pair* (M, R) *is* valid *when*

- $lhs(R) \leq w$, $promoter(R) \leq w$;
- *for all* $r \in supp(R)$, *either* $inhibitor(r) = 0_\Omega$ *or there exists* $a_r \in \Omega$ *such that* $w(a_r) < inhibitor(r)(a_r)$;
- *if* $rhs(R)(a, in_j) > 0$ *then there exists a membrane with label* j *in* M_*.

We say that the pair (M, R) is *maximally valid* if it is valid, and for any multiset R' over $Rules(i)$ such that $R \leq R'$ and (M, R') is valid it follows that $R = R'$.

Note that the multiset R is not required to be non-empty, therefore the pair $(M, 0_{Rules(i)})$ is valid, and can even be maximally valid (when no rule from $Rules(i)$ can be applied).

We define a transition system over the set of membranes by the following rules:

- if $(\langle i|w \rangle, R)$ is maximally valid, then

$$(\text{mpr1}) \; \frac{}{\langle i|w \rangle \xrightarrow{R} \langle i|w - lhs(R) + rhs(R) \rangle}$$

- if $(\langle i|w; M_1, M_2, \ldots, M_k \rangle, R)$ is maximally valid, then

$$\text{(mpr2)} \quad \frac{M_1 \xrightarrow{R_1} N_1, \ldots, M_k \xrightarrow{R_1} N_k}{\langle i|w; M_1, M_2, \ldots, M_k \rangle \xrightarrow{R} \langle i|w - lhs(R) + rhs(R); N_1, N_2, \ldots, N_k \rangle}$$

We say that M is mpr-irreducible if $M \xrightarrow{0_{Rules}} M$, and all its children are also mpr-irreducible.

Note that even if $M \xrightarrow{R} N$, it does not mean that N is mpr-irreducible: since inhibitors can be consumed by the multiset of rules R, it may be the case that there is a non-empty multiset S of rules such that (N, S) is maximally valid.

We prove now the equivalence between static and dynamic allocation semantics. In what follows, let $w(H)$ denote the content of a register membrane H, and $reg(H)$ its register, i.e. $H = (\langle i|w(H); H_* \rangle, reg(H))$.

Proposition 16.7 *Consider a membrane M which is mpr-reducible. Let $H_M \in \phi^{-1}(M)$ denote the register membrane corresponding to M, which has its register and all the registers of inner membranes equal to 0_ϱ. If $M \xrightarrow{R} N$, then there exist $l_1, \ldots, l_k \in Rules \cup \{\tau\}$ such that $H_M \xrightarrow{l_1} \ldots \xrightarrow{l_k} H$, H is mpr-irreducible and $card\{i \in [k]/l_i = r\} = R(r)$. Moreover, $N = \phi(H)$.*

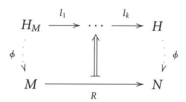

Proposition 16.8 *If H_0 is a register membrane with its register and all registers of inner membranes equal to 0_ϱ and $H_0 \xrightarrow{l_1} \ldots \xrightarrow{l_k} H_k$ in T_{mpr} such that H_k is mpr-irreducible, then there exists a multiset R over the rules in $\phi(H_0)$ such that $\phi(H_0) \xrightarrow{R} \phi(H_k)$. Moreover, $R(r) = card\{i \in [k]/l_i = r\}$.*

These two propositions show that the static allocation can be considered to be *big-step* semantics [13] for the rewriting stage of the evolution of a membrane, while dynamic allocation provides equivalent *small-step* semantics [21]. We use static allocation semantics together with the previously defined T_{msg} and T_{diss} to describe the evolution of a membrane system analogously to the manner in which we used together T_{mpr}, T_{msg}, and T_{diss}.

A Maude implementation of dynamical allocation semantics is presented in [1].

16.6 BISIMULATION

Operational semantics provides a formal way to find out which transitions are possible for the current configurations of a P system. It provides the basis for defining certain equivalences and congruences between P systems. Moreover, operational semantics allows a formal analysis of membrane computing, permitting the study of relations between systems. Important relations include simulation preorders and bisimulation. These are especially useful with respect to P systems, allowing to compare two P systems.

A simulation preorder is a relation between two transition systems associated to P systems expressing that the second can match the transitions of the first. We present a simulation as a relation over the states in a single transition system rather than between the configurations of two systems. Often a transition system consists intuitively of two or more distinct systems, but we also need our notion of simulation over the same transition system. Therefore our definitions relate configurations within one transition system, and this is easily adapted to relate two separate transition systems by building a single transition system consisting of their disjoint union.

Definition 16.8 *Let Π be a P system.*

1. A simulation *relation is a binary relation R over $\mathcal{C}(\Pi)$ such that for every pair of configurations $C_1, C_2 \in \mathcal{C}(\Pi)$, if $(C_1, C_2) \in R$, then for all $C_1' \in \mathcal{C}(\Pi)$, $C_1 \Rightarrow C_1'$ implies that there is a $C_2' \in \mathcal{C}(\Pi)$ such that $C_2 \Rightarrow C_2'$ and $(C_1', C_2') \in R$.*
2. *Given two configurations $C, C' \in \mathcal{C}(\Pi)$, C simulates C', written $C' \leq C$, iff there is a simulation R such that $(C, C') \in R$. In this case, C and C' are said to be similar, and $\leq$ is called the* similarity relation.

The similarity relation is a preorder. Furthermore, it is the largest simulation relation over a given transition system.

A bisimulation is an equivalence relation between transition systems such that one system simulates the other and vice-versa. Intuitively two systems are bisimilar if they match each other's transitions, and their evolutions cannot be distinguished.

Definition 16.9 *Let Π be a P system.*

1. *A bisimulation relation is a binary relation R over $\mathcal{C}(\Pi)$ such that both R and R^{-1} are simulation preorders.*
2. *Given two configurations $C, C' \in \mathcal{C}(\Pi)$, C is bisimilar to C', written $C \sim C'$, iff there is a bisimulation R such that $(C, C') \in R$. In this case, C and C' are said to be bisimilar, and $\sim$ is called the bisimilarity relation.*

The bisimilarity relation $\sim$ is an equivalence relation. Furthermore, it is the largest bisimulation relation over a given transition system.

16.7 RELATED WORK

Structural operational semantics is an approach originally introduced by Plotkin [21] in which the operational semantics of a programming language or a computational model is specified in a logical way, independent of machine architecture or implementation details, by means of rules that provide an inductive definition based on the elementary structures of the language or model. Within "structural operational semantics", two main approaches coexist:

- *Big-step semantics* is also called *natural semantics* in [13, 18], and *evaluation semantics* in [12]. In this approach, the main inductive predicate describes the overall result or value of executing a computation, ignoring the intermediate steps.
- *Small-step semantics* is also called *structural operational semantics* in [21, 18], and *computational semantics* in [12]. In this approach, the main inductive predicate describes in more detail the execution of individual steps in a computation, with the overall computation roughly corresponding to the transitive closure of such small steps.

In general, the small-step style tends to require a greater number of rules than the big-step style, but this is outweighed by the fact that the small-step rules also tend to be simpler. The small-step style facilitates the description of interleaving [17]. The inference rules of P systems provide big-step operational semantics due to the parallel nature of the model. The big-step operational semantics of P systems can be implemented by using a rewriting engine (Maude), and so we get a small-step operational description. The advantages of the implementations in Maude is given by the solid theoretical aspects of the rewriting logic, and by the complex tools available in Maude. By using an efficient implementation of rewriting logic as Maude [10], we can verify various properties of these systems by means of a `search` command (a semi-decision procedure for finding failures of safety properties), and a Linear Temporal Logic model checker. These achievements are presented in [3].

In [8, 9], the nature of parallelism and non-determinism of the membrane systems is expressed in terms of event structures [22], a known formal model using both causality and conflict relations. In event-based models, a system is represented by a set of events (action occurrences) together with some structure on this set, determining the causality relations between the events. The causality between actions is expressed by a partial order, and the non-determinism is expressed by a conflict relation on actions. The behavior of an event structure is formalized by associating to it a family of configurations representing sets of events which occur during the executions of the system. A parallel step is simultaneously executing several rules, each of them producing events which end up in the resulting event configuration. These steps are presumably cooperating to achieve a goal, and so they are not totally independent. They synchronize at certain points, and this is reflected in the events produced.

In [9] the authors determine the event structure given by a membrane system. The paper presents a modular approach of causality in membrane systems, using both string and multiset rewriting. In order to deal with membrane systems, the event structures are extended with notions like maximal concurrent transitions and saturated states with respect to concurrency. The event structure of a membrane system is defined in two steps: first the event structure of a maximally parallel step in membranes is defined, and then it is combined with a communication step. The main result of the paper proves that an event structure of a membrane corresponds to its operational semantics. Event structures for communicating membranes are also defined.

In [8], the author describes the causal dependencies occurring between the reactions of a P system, investigating the basic properties which are satisfied by such semantics.

The paper [6] defines semantics of P systems by means of a process algebra. The terms of the algebra are objects, rules, or membranes; an equivalence of membranes is defined with respect to the objects which enter/exit the membrane. The semantics is compositional with respect to the inclusion of a membrane in another membrane. This is obtained by considering each object and rule separately, with evolution given by possible contexts in which it may find itself embedded.

The paper [14] describes a class of Petri nets suitable for the study of behavioral aspects of membrane systems. Localities are used as an extension of Petri nets in order to describe the compartments defined in P systems, and so leading to locally maximal concurrency semantics for Petri nets. Causality is also considered, using information obtained from the Petri net representation of a P system.

In [11], the authors reason at the abstract level of networks of cells (including tissue P systems) with static structure. They adapt an implementation point of view, and give a formal definition of the derivation step, the halting condition, and the procedure for obtaining the result of a computation. For (tissue) P systems, parameters for rules are employed in order to describe the specific features of the rules.

ACKNOWLEDGEMENTS

Several results mentioned in this chapter were obtained together with Oana Agrigoroaiei, Oana Andrei, and Dorel Lucanu. Many thanks to all of them for their contribution and collaboration.

REFERENCES

[1] O. AGRIGOROAIEI, G. CIOBANU: Rewriting logic specification of membrane systems with promoters and inhibitors. *Proceedings of WRLA 2008*, 1–16. *Electronic Notes in Theoretical Computer Sci.*, 238 (2009), 5–22.

[2] O. ANDREI, G. CIOBANU, D. LUCANU: Executable specifications of the P Systems. *Lecture Notes in Computer Sci.*, 3365 (2005), 127–146.

[3] O. ANDREI, G. CIOBANU, D. LUCANU: A structural operational semantics of the P systems. *Lecture Notes in Computer Sci.*, 3850 (2006), 32–49.

[4] O. ANDREI, G. CIOBANU, D. LUCANU: Operational semantics and rewriting logic in membrane computing. *Electronic Notes of Theoretical Computer Sci.*, 156 (2006), 57–78.

[5] O. ANDREI, G. CIOBANU, D. LUCANU: A rewriting logic framework for operational semantics of membrane systems. *Theoretical Computer Sci.*, 373 (2007), 163–181.

[6] R. BARBUTI, A. MAGGIOLO-SCHETTINI, P. MILAZZO: Compositional semantics and behavioral equivalences for P systems. *Theoretical Computer Sci.*, 395 (2008), 77–100.

[7] P. BOTTONI, C. MARTÍN-VIDE, GH. PĂUN, G. ROZENBERG: Membrane systems with promoters/inhibitors. *Acta Informatica*, 38 (2002), 695–720.

[8] N. BUSI: Causality in membrane systems. *Lecture Notes in Computer Sci.*, 4860 (2007), 160–171.

[9] G. CIOBANU, D. LUCANU: Events, causality, and concurrency in membrane systems. *Lecture Notes in Computer Sci.*, 4860 (2007), 209–227.

[10] M. CLAVEL, F. DURÁN, S. EKER, P. LINCOLN, N. MARTÍ-OLIET, J. MESEGUER, J.F. QUESADA: Maude: Specification and programming in rewriting logic. *Theoretical Computer Sci.*, 285 (2002), 187–243.

[11] R. FREUND, S. VERLAN: A formal framework for P systems. *Lecture Notes in Computer Sci.*, 4860 (2007), 271–284.

[12] M. HENNESSY: *The Semantics of Programming Languages: An Elementary Introduction Using Structural Operational Semantics*. Wiley, 1990.

[13] G. KAHN: Natural semantics. *Lecture Notes in Computer Sci.*, 247 (1987), 22–37.
[14] J. KLEIJN, M. KOUTNY, R. ROZENBERG: Towards a Petri net semantics for membrane systems. *Lecture Notes in Computer Sci.*, 3850 (2005), 292–309.
[15] N. MARTÍ-OLIET, J. MESEGUER: Rewriting logic as a logical and semantical framework. *Handbook of Philosophical Logic*, Kluwer, 2002, 1–87.
[16] R. MILNER: Operational and algebraic semantics of concurrent processes. *Handbook of Theoretical Computer Sci.*, Elsevier, 1990, vol. B, 1201–1242.
[17] P. MOSSES: Modular structural operational semantics. *BRICS RS*, 05-7, 2005.
[18] H.R. NIELSON, F. NIELSON: *Semantics with Applications: A Formal Introduction*. Wiley, 1992.
[19] M. NIELSEN, G. ROZENBERG, P.S. THIAGARAJAN: Transition systems, event structures, and unfoldings. *Information and Computation*, 118 (1995), 191–207.
[20] GH. PĂUN: *Membrane Computing. An Introduction*. Springer, 2002.
[21] G. PLOTKIN: Structural operational semantics. *Journal of Logic and Algebraic Programming*, 60 (2004), 17–140. Initially "A Structural Approach to Operational Semantics", Technical Report DAIMI FN-19, Aarhus University, 1981.
[22] G. WINSKEL: Event structures. *Lecture Notes in Computer Sci.*, 255 (1987), 325–392.

CHAPTER 17

SOFTWARE FOR P SYSTEMS

DANIEL DÍAZ-PERNIL
CARMEN GRACIANI
MIGUEL A. GUTIÉRREZ-NARANJO
IGNACIO PÉREZ-HURTADO
MARIO J. PÉREZ-JIMÉNEZ

17.1 INTRODUCTION

THE *prediction* of the behavior of a computational device through time is usually a hard task. In membrane computing, although P systems are a machine-oriented model of computation, it is even more complex to predict or to guess how a P system will behave.

Thus, the idea of automating the evolution of a P system was one of the first issues in the membrane computing community. The key drawback is the lack of physical support for such a bio-inspired model. In contrast to DNA computing, the existing technology is not able to provide implementations in laboratories (neither *in vitro* nor *in vivo* nor in any electronic medium) for P systems. The only way to get devices for a mechanical application of the rules of a P system is to develop software tools that are able to run on current computers and are capable of simulating computations of P systems.

Over the last ten years, a wide range of such software simulators have been reported. The growth in the number of such software tools has gone in parallel with the development of the theory. Nowadays there exists a large variety of models and submodels of P systems. According to the membrane structure there exist two big families: *cell-like* P systems and *tissue-like* P systems. Spiking neural P systems deserve a separate mention (they have a tissue-like structure, but the information is encoded *in time*, and not by means of the multisets placed in the regions delimited by membranes). These large families can also be classified according to the specific kind of rules which can be applied and on their semantics.

In many cases, ad-hoc software tools have been developed for checking specific models. In this chapter we present several software tools which have been presented over time and we show the perspectives for the immediate future. Since membrane computing is an extremely dynamic area, we strongly encourage the reader to check the P systems web page [61] for updated information about software tools presented after the publication of this book.

The chapter is organized as follows. In the next section, some general features about the processes of the design and development of P systems simulators are given. Section 17.3 is devoted to the first generation of simulators. They usually have a large verbosity and pedagogical purposes. Section 17.4 deals with the second wave of simulators, where the main target is real-life applications and Section 17.5 gives an overview of other software not included in the previous sections. The chapter finishes by showing some projects for the future (Section 17.6) and some conclusions (Section 17.7).

17.2 GENERAL FEATURES

If one considers the large number of P system simulators developed to date, the first impression is that it is difficult to find common features among them. Over ten years, many authors, many P systems models, and many programming languages have been involved in the development of software for P systems.

Nonetheless, as pointed out in [31], the design and development processes for a P system simulator can be structured as follows:

Formal definition of the model: First of all, one has to choose which model is going to be simulated, stating precisely the syntax and semantics to avoid ambiguous interpretations.

Choice of a programming language: A large number of different languages, such as Haskell, Prolog, Java, C, Lisp, Visual C++, CLIPS or MzScheme have been chosen.

A good way to represent the knowledge: This decision is related to the programming language that is used, as specific techniques related to it have to be applied. A good representation allows a quick transition between configurations and, therefore, speeds up the simulation.

Design of an inference engine to carry out the computation: There are two basic difficulties intrinsic to the simulation of a P system in a sequential conventional computer that has a bounded number of processors: the parallelism and the non-determinism.

We can distinguish two waves or generations of these simulators. First, the P systems community needed simulators for a better understanding of the model. These first simulators are useful tools for teachers and researchers. On the one hand, one of the main utilities of this software is its pedagogical use. Such simulators provide exhaustive information to the user of the different changes in the configurations. On the other hand, they have been used for the design and formal verification of complex P systems which solve problems, saving the researchers heavy hand-done calculations. These simulators require a large amount of resources in order to store and show information of the process itself and they can only deal with small instances of the problems. Following this wave, a new generation of simulators has arisen. They share the same diversity. The distinguishing feature is that their purpose is to deal with realistic problems. It should be mentioned that currently the main application areas are real-world problems arisen from biology.

17.3 THE FIRST GENERATION

As pointed out above, the simulators of this first wave have a pedagogical purpose in common. From a technical point of view, the simulated models can be classified into two categories: models where the number of membranes are bounded by the number of membranes in the initial configuration and the models where the number of membranes can increase along the computation.

17.3.1 Simulators of Transition P Systems

In a natural way, the easiest model, the transition P system [40], was the chosen model for the first simulators.

Malița Simulator (2000)

The first P system simulator was presented by Malița [34] in the *Workshop on Multiset Processing* held in Curtea de Argeș, Romania, in 2000. It is a program written in LPA-Prolog for simulating transition P systems. As a good representative of the first generation, its main feature is the balance between understandability and efficiency.

A configuration is represented as a list of labeled nested lists where objects are represented together with their multiplicities. There are flags to distinguish between objects that can be processed or not. The rules are represented by expressions explicitly mentioning four fields: the membrane where it can be applied, the identifier of the rule in its membrane, the initial multiset and, finally, a multiset of products with target indicators or, eventually, the flag *dissolve*.

This simulator applies a restricted parallelism: in each step, *for each* membrane, it selects *only one* rule that is applied as many times as possible.

The simulator receives as input the configuration of a system together with a set of rules, and also a parameter specifying the desired number of evolution steps. The output shows the configurations of a branch of the computation tree until the desired number of evolution steps is reached.

Suzuki and Tanaka Simulator (2000)

In the same year, Suzuki and Tanaka presented in [56] a program written in Lisp for simulating transition P systems without membrane division. This limitation is a feature also shared by Maliţa's simulator. They only considered a class of P systems, which they call Artificial Cell Systems, consisting of a membrane structure, multisets of symbols placed in its regions, and a set of rewriting rules acting in all the regions. In order to control the amount of required resources, they impose an important constraint: the size of the multiset used is bounded.

In spite of the restrictions, this program was developed with the aim of using it in addressing real-life problems. It has been successfully used to simulate the Brusselator model or in modeling and analyzing ecological systems (details can be found in [57]).

Balbontín et al. Simulator (2002)

Two years later, Balbontín, Pérez, and Sancho presented during the *Workshop on Membrane Computing 2002* a simulator [7] for transition P systems written in MzScheme.

This simulator, as Maliţa's one, receives as input the initial configuration of a system including the set of rules and a parameter specifying the desired number of evolution steps, but it outputs the computation tree of the P system, step by step, until the desired number of evolution steps is reached. The expansion of the computation tree is made in a progressive way, level by level (*breadth expansion*). If the branching rate of the computation tree is large, the simulator can only simulate very few steps.

The inference engine that implements the evolution steps follows the formalization from [47]. That is, first of all it checks which are the applicable rules, according to the priority relations, then it calculates the applicability vectors for each membrane and finally it combines such vectors to get the applicability matrices for the system.

The simulator also includes a parser for analyzing the input. If the input is syntactically correct, a compiler rewrites it into an internal grammar.

Baranda Simulator (2002)

In the first years of the development of the theory of P systems, some members of the Natural Computing Group of the Technical University of Madrid [60] proposed frameworks and data structures suitable for P systems, but in an abstract rather than practical context (see, for example, [3, 4, 6]).

In [5], based on that previous theoretical formalization, a simulator for transition P systems implemented by A. Baranda was presented. Two *layers* are considered: there is a static structure, composed by the membranes and objects of the system, and there is a dynamic structure, which refers to the set of rules of the system. They present several specific modules (*Abstract Data Types*) to transfer to the software the concepts of multiset, rule, region, membrane, etc.

The simulator has been written in Haskell and the chosen interpreter has been Hugs98 for Microsoft Windows. The source code can be downloaded from the P system web page [61].

The simulator receives as input a file encoding a system (configuration and rules in each region) and produces another file encoding a system obtained by the application of *one* step of the computation (via a maximal multiset of rules *randomly* selected).

17.3.2 Increasing the Number of Membranes

As pointed out above, one of the first difficulties of the first simulators was to deal with P system models where the number of membranes increases during the computation. Increasing the number of membranes via membrane division or membrane creation is the key point to design polynomial-time solutions to computational hard problems. In such solutions an exponential amount of membranes are created and the parallelism is used to evolve all these membranes simultaneously. Due to the obvious limitations of computational resources, the P systems which can be simulated are of small size.

Ciobanu and Paraschiv Simulator (2002)

In [15] Ciobanu and Paraschiv presented a software application that provides a simulation for the initial version of catalytic hierarchical cell systems and for P systems with active membranes (see [39]). Its main functions are: interactive definition of a membrane system, visualization of a defined membrane system, a graphical representation of the computation and final result, and save and (re)load of a defined membrane system.

The application was implemented in Microsoft Visual C++ using MFC classes. The system is presented to the user with a graphical interface where the main screen is divided into two windows. The left hand window gives a tree representation of the membrane system including objects and membranes. The right hand window provides a graphical representation of the membrane system given by Venn-like diagrams.

Sevilla Team Simulators (2004)

Sevilla Team [63] presented two different simulators in 2004 which were thought as assistant tools for the design and formal verification of cellular solutions to **NP**-complete problems, like Bin Packing [44] and the Common Algorithmic Problem (CAP) [46], via recognizer P systems [43, 49]. In these cases, as only confluent P systems are considered, it suffices to follow one branch of the computation tree.

One of the simulators for dealing with P systems with active membranes is written in CLIPS and it was presented in [43]. The design is based on representing P systems through the *production systems* programming paradigm. In order to carry out the computation, the simulator performs first an initialization stage where the rules are translated into CLIPS rules and then the application of the rules is simulated.

The simulator receives as input the initial configuration of a system and a set of rules. Several options are provided to choose the degree of verbosity of the output: either show all the configurations of the evolution, or show only a concrete instant, or run and show only the final answer. Besides, the user can also decide if the applied rules are also displayed for each step or not.

The second simulator was written in Prolog [19]. It is pretty different from Maliţa's simulator in implementation and it works with P systems with active membranes, instead of transition P systems. It has been successfully used as an assistant in the design and formal verification of P systems to solve **NP**-complete problems (see [18, 19, 29, 49]).

In a similar line to other simulators, this one stores and handles the information related to the P system and tries to show the process to the user in a friendly way. Only one branch of the computation tree is simulated, and therefore the result of the simulation is faithful only in the cases of *confluent* P systems.

This simulator has been also used in [30] as a tool to study the descriptive complexity of P systems. That complexity can be described by a table showing the number of times that the rules of the system are applied at each step. Such tables are known as *Sevilla carpets*. For further details we refer the readers to [30] or [49].

17.3.3 Parallel and Distributed Simulators

One of the main difficulties for the simulation of P systems in current computers is that the main computational power of P systems lies on their intrinsic massive parallelism. Several authors have implemented the first versions of simulators based on parallel and distributed architectures, which is close to the membrane computing paradigm.

Ciobanu and Wenyuan presented in [17] a parallel implementation of transition P systems that was designed for a cluster of computers. It is written in C++ and it makes use of MPI as its communication mechanism.

The program is implemented and tested on a Linux cluster of 64 dual processor nodes at the National University of Singapore. The implementation is object-oriented and the rules are implemented as threads. At the initialization phase, one thread is created for each rule. Rule applications are performed in terms of rounds. To synchronize each thread (rule) within the system, two barriers implemented as mutexes[1] are associated with it. First the barrier that the rule thread is waiting on is released by the primary controlling thread. After the rule application the thread waits for the second barrier, and the primary thread locks the first barrier. Each rule is modeled as a separate thread so it should have the ability to decide its own applicability. When more than one rule can be applied in the same conditions, the simulator picks randomly one among them. For every membrane, the main communication is to send and receive messages to and from its father and children at the end of every round. When there is no rule in any membrane that is applicable, each membrane must inform others about its inactivity, the designated output membrane prints out the result and the system halts.

Also in 2003, Syropoulos, Mamatas, Allilomes, and Sotiriades presented in [58] a purely distributive simulation of P systems. It is implemented using Java's *Remote Methods Invocation* to connect a number of computers that interchange data. The idea of designing a distributed simulator for a network of computers, instead of doing so for a cluster architecture, avoids the problem of a limited hardware compatibility. The class of P systems that the simulator can handle restricts the number of membranes to two, allows cooperation, and the symbol *tar* indicates that the communication rules use target indicators of the type in_j.

Initially, a copy of the simulator is installed on different computers. Randomly, we choose a computer and assign to it the role of the external compartment, while the others play the role of the internal ones. On each computer a membrane-object is ready to participate to the network. Threads are an essential aspect of the implementation. Each membrane class runs in its own thread which operates on a different machine.

---

[1] A mutex object is a synchronization object whose state is set to signalled when it is not owned by any thread, and non-signaled when it is owned.

The simulator has been designed in such way that all membrane-objects send multicast UDP packets to a well known multicast address. Each packet contains the IP address of each sender, and are received by every object participating in the network. Thus, each computer knows which computers are *alive* at any time. In this way, when the simulator has successfully parsed the P system's specification, the main object decides whether there are enough resources or not to start the computation. A universal clock is owned by the object that has the role of the external compartment. This object signals each clock tick by the time the previous macrostep is completed (i.e. when, for a given macrostep, all remote objects have finished their computation).

17.4 THE SECOND GENERATION

After the first generation, where the main interest was the understanding of the processes and the verification of hand-made designs of solutions, a second generation of software has arisen. It shares with the first generation a large heterogeneity as the main feature. In general, the researcher implements a software tool for studying a specific problem or at most, a specific kind of problems. Nonetheless, the main difference with respect to the first generation is that the goal is not the understanding of the computational process. The P system itself is no longer the object of study, but starts to be a tool for the study of real-life problems. In this generation, it is not so important to inform the user which rules are applied, the important thing is to obtain the result as efficiently as possible. Another important feature is that the main application field is the simulation of biological processes.

Ardelean and Cavaliere Simulator (2003)

A very interesting tool for modeling biological processes was presented in [2]. It can be thought of as a transition P system simulator because the number of membranes does not change during the computation. More precisely, the software deals with a special variant of P systems: the allowed rules are rewriting and symport/antiport. This variant was proposed in [12] and its motivations were inspired by the idea of separating the evolutionary mechanism of the cell from the communicative mechanism.

The authors try to bridge the mathematical model with the biological reality, indicating how one can use the P system framework to model very important processes that happen in cells. The simulator takes as input the rules of a system, its membrane structure (which can be any graph, not only a tree), and the multisets of objects associated with the regions. The software assigns to each rule two kinds of probabilities: *probability to be available* and *probability to win a conflict*. At each step, the simulator decides which are the available rules using the above mentioned probability. Then, the available rules are applied in the maximally parallel mode by using the *weak priority* approach. Several important biological processes have been simulated illustrating the usefulness of this software (see [13]).

Verona Team Simulator (2006)

The Group for Models of Natural Computing (MNC group [62]) in Verona developed Psim [9]. It was introduced in order to simulate metabolic P systems, a special class of P systems for expressing the dynamics of some biological systems. Psim's structure is inspired from the application developed by Maliţa [34], the simulator written by Suzuki and Tanaka [56], and the membrane simulator due to Ciobanu and Paraschiv [15].

The program implements the metabolic algorithm in its deterministic version, which transforms populations of objects according to a mass partition principle in a substantially different way with respect to the non-deterministic and maximally parallel paradigm.

The main features are a flexible definition of the membrane structure and rule set via an XML file, a user friendly interface and the possibility of saving intermediate results that can be reloaded. The simulator output consists of a series of graphs representing the objects multiplicity along time. It is written in Java so it is cross-platform.

Each membrane has three different regions: an internal, an external, and an intermediate region located between the outside and the inside regions and which can be seen from both. Each region can contain objects and also further membranes.

In [8] Psim (version 2.4) extends some of its concepts and enhances the simulation environment with many features. Its implementation has always followed some flexibility and extensibility principles so it can be easily extended and integrated with others tools. Its plugging architecture allows one to devise and implement interactions with the simulator engine.

The last version is freely available for download at [66]. In this the input specification is made by means of a transposition of the concept of MP graphs, a formalism for a graphical representation of metabolic systems, into a point and click graphical interface.

Romero-Campero et al. Simulator (2006)

In 2006, the Sevilla and Sheffield Teams presented two software tools for the simulation of biological processes with P systems. In both of them, the authors implement the multi-compartmental Gillespie algorithm, in Scilab and C, respectively. The tools simulate the evolution of multi-compartmental Gillespie algorithm on a hierarchy of compartment structures. The kernel for the system has the following features which aim to provide flexibility in simulation and modularization in modeling along with a model checking strategy[2].

The multi-compartmental algorithm is used to select a compartment and a rule to apply. It is a variation of the stochastic Gillespie algorithm. Rule selection is made by applying the rule which has the shortest time, assigned by the Gillespie algorithm, depending on the availability of its reactants and a kinetic constant, or using that constant as a probability for the rule to be applied. Once a rule is selected, it is applied the maximum possible number of times for the number of reactants available. Each compartment has a set of rules whose scope is the aforementioned compartment and its outside or inside compartment. The set of rules are defined and they can be replicated by assigning the same set to different compartments. A compartment or compartment hierarchy can be modulated in its specification and then replicated many times in the model.

The simulator has been successfully used in addressing several real-world problems as the simulation of pathways associated to the Epidermal Growth Factor (EFR) [45], simulation of FAS-induced apoptosis [14], modeling gene expression control [51], or a first computational model of Quorum Sensing [52] in *Vibrio Fischeri*.

Sedwards and Mazza Simulator (2007)

In [55], a new software for simulating biological processes, called Cyto-Sim, was presented. It is an stochastic simulator of membrane-enclosed hierarchies of biochemical processes, where the membranes comprise an inner, outer, and integral layer. The underlying model allows a formal analysis in addition to simulation. The simulator provides arbitrary levels of abstraction. The paradigm is flexible and extensible, allowing to adapt it to other types of simulation, analysis, and integration within standard platforms. Cyto-Sim supports models described as Petri nets, it can import all versions of SBML and it can export SBML an MATLAB m-files.

Authors have extended the basic membrane systems model to include peripheral and integral membrane proteins and incorporate a Markov chain Monte Carlo algorithm to simulate the time evolution of the systems. The software [65] can simulate micro and macroscopic biological processes using arbitrary kinetic laws best suited to each defined interaction.

Cazzaniga and Pescini Simulators (2006)

New examples of the second generation are the simulators written by Cazzaniga and Pescini. They are available at the P systems web page [61] and have been developed in the C language with the GNU Scientific Library.

One of them simulates the gene regulation system (related to quorum sensing issues) of the bacterium *Vibrio Fischeri* using the multi-compartmental Gillespie algorithm. Bacteria are generally considered to be independent unicellular

---

[2] For details, see [59].

organisms. However it has been observed that certain bacteria, like the marine bacterium *Vibrio Fischeri*, exhibit coordinated behavior which allows an entire population of bacteria to regulate the expression of specific genes in a coordinated way that depends on the size of the population. This cell density dependent gene regulation system is referred to as quorum sensing.

The *Vibrio Fischeri* simulator uses the MPI library of message passing routines that allows a user to write a program in a familiar language, such as C, and carry out a computation in parallel on an arbitrary number of cooperating computers. The source code is available in the software section of the P systems web page [61].

Nishida Simulator (2006)

In [38], Nishida presented a new approximate algorithm for **NP**-complete optimization problems that he called the "membrane algorithm" as it uses P system ingredients. It consists of a number of regions (separated by nested membranes), several subalgorithms placed in regions, a few tentative solutions for the optimization problem (different for each region), and a mechanism to transport tentative solutions between adjacent regions. At each step and for every region solutions are updated by the corresponding subalgorithm. After that, the best solution is sent to the inner region and the worst to the outer. This process is repeated until a termination condition is satisfied. The best solution in the innermost region is the output.

An implementation of this algorithm designed to solve the Travelling Salesman problem was made in Java programming language. There are several variants as different subalgorithms are considered. A version that incorporates a dynamic structure of membranes is also considered. The code is available at [61].

17.5 OTHER SOFTWARE

The list of P systems simulators mentioned above is not exhaustive, as many researchers have developed their own software. In this section we try to describe some of them, but it is impossible to note them all.

17.5.1 Simulators for Non Cell-like Models

All the simulators presented in the previous sections deal with cell-like P system models, but recently, with the development of the research, other simulators have been reported following other membrane structures.

Tissue-like P systems

In [10], a simulator for recognizing tissue P systems with cell division was presented. It is a visual tool called *Tissue Simulator*. The tool allows the user to write in an easy way the rules and the elements of a given system, run the execution of the system, and shows graphically a trace of the simulation with the rules applied in each computation step.

In this way, this simulator corresponds to the first generation. Its new feature is the P system model used: tissue P system with cell division. In this model, membranes are arranged in a general graph instead of a tree-like structure. These systems also have the ability to produce new membranes via membrane division and hence, they can solve NP-complete problems in polynomial time by trading space against time (see, for example, [21]).

Spiking Neural P systems

Spiking neural P systems were introduced in [33]. Only a few months later, a first simulator for spiking neural P systems was presented [32]. It has all the features that belong to the first generation. Its main aim it to give exhaustive information to the user about the computational process. In this way, it can be seen as an assistant for the formal verification of such systems. The tool outputs the transition diagram of a given system in a step-by-step mode. The code is modular and flexible enough to be adapted for further research discoveries.

Conformon P systems

Conformons are defined in molecular biology as sequence-specific mechanical strains embedded in biopolymers, such as DNA supercoils and protein conformational deformations, that provide both the free energy and information needed for biopolymers to drive molecular processes essential for life.

Based on the notion of conformon, Frisco and Ji presented, in 2002, a new model of membrane computing devices, the *conformon P system* [25]. Only three years later, the first simulator for conformon P systems was presented [24]. It has been applied in the study of several biological processes, chemical reactions, and the dynamics of HIV infection [20].

17.5.2 Graphics and P Systems

In [26, 27], a first membrane-based device for computer graphics was presented. It was a hybrid model between L-systems and membrane computing and it used concepts very close to the L-systems model. It is available from the P systems web page [61].

Later, in [53], a new approach was presented for representing the development of higher plants with P systems. It was based on a type of P system with membrane creation. The basic idea was to consider the growing of the structure of the membranes in a P system by means of membrane creation. In [54] the study started in [53] was extended by adding stochastic rules to the P system. In this case, the non-deterministic choice of different rules produces different configurations of the P system and, hence, different graphical representations.

In [50], the authors present a software, JPLANT, which computes the first configurations of a computation and draws the corresponding graphical representation.

17.5.3 Even More Software

Although it is not exactly a simulator, we would like to note the work that Nicolau Jr, Solana, Fulga, and Nicolau published in [37]. They presented an ANSI C library developed to facilitate the implementation and simulation of P systems. Using the library proposed in this paper a user can specify an initial configuration (membrane structure and its contents) and then perform actions on the objects or on the membranes. This library represents an intermediate step towards a practical implementation of P systems *in silico*.

Following the pedagogical purpose of this first generation of simulators, in [36] we can find the description of a software application, *SimCM*, written in Java. We would also like to note the work presented by Acampora and Loia [1] where transition P systems were simulated by a parallel and distributed application, based on multi-agent system technology.

Although it is beyond of the scope of this chapter, we consider Petreska and Teuscher's approach interesting [48]. Instead of developing software they presented a hardware-based parallel implementation.

Finally, the project Xholon [64] also deserves to be cited. Xholon is an open source general-purpose modeling, transformation, and simulation tool, based on XML and Java, that supports the Unified Modeling Language (UML 2.1), systems biology modeling including SBML, other types of modeling, and many of the features needed to support P systems.

17.6 NEW FRONTIERS

We finish our survey of the existing software with two of the most promising projects for the immediate future of P systems.

P-lingua [23] is a whole programming language and it has been created with the aim of becoming the standard for the representation and implementation of future software. The programs in P-lingua define families of P systems in a parametric and modular way. After assigning values to the initial parameters, the compilation tool generates an XML document associated with the corresponding P system from the family and, furthermore, it checks possible programming errors (both lexical/syntactical and semantical). Such documents can be integrated into other applications, thus guaranteeing interoperability. More precisely, in the simulators framework, the XML specification of a P system can be translated into an executable representation.

The second big project [28] that could have a significant impact on the membrane computing area is being carried out in the Chemical Faculty of Technion Institute, Haifa, Israel. It will be the first *in vitro* experiment, using test tubes as membranes and DNA molecules as objects, evolving under the control of enzymes. Moving from software simulations to implementations in vitro leads to new difficulties which should be faced. The first is to determine the P system model which it is possible to implement. In this way, local-loop-free P systems seem to be the appropriate model. The basic issue is to have no loops in the evolution of objects present in a membrane because this would lead to non-desirable cycles. Other technical difficulties need to be solved as well, such as a way of *counting* objects from a multiset or questions related to synchronization and parallelism. There are many questions related to this experiment that need to be successfully answered but, undoubtedly, it can open new research vistas in membrane computing.

17.7 Conclusions

In this paper we have presented some of the P system simulators realized in the first years of research in membrane computing. The evolution of these simulators can be seen as an expression of the evolution of the research itself.

In the first wave, the simulators explored different P system models and were used for checking the correctness of the hand constructed systems. To this end, they communicate to the user every change in the membrane structure in the multisets or give information about which and how many times the rules are applied. These simulators are very good assistants for the design of membrane solutions to hard problems, since hand-made simulations may be wrong if the number of membranes and objects are large.

The main drawback of this first wave of simulators is related to the efficiency and explicitness of the code. If the computer wastes a large amount of resource storing and making explicit the information, efficiency is lost. In this way, the size of the instances of the P systems handled with these simulators is small and they are not appropriate for real-life problems.

After the first simulators, a new generation has arisen. In this new wave, P systems are not the object of study, they have became the tool for studying various processes, especially from biology.

With respect to the future, membrane computing is such a dynamic area that it is impossible to guess the evolution of the simulators. Many researchers are currently interested in different applications of membrane computing, so in the near future new simulators may appear designed for specific application areas.

Acknowledgement

The authors acknowledge the support of the project TIN2006-13425 of the Ministerio de Educación y Ciencia of Spain, cofinanced by FEDER funds, and the support of the project of excellence TIC-581 of the Junta de Andalucía.

References

[1] G. Acampora, V. Loia: A proposal of multi-agent simulation system for membrane computing devices. *2007 IEEE Congress on Evolutionary Computation (CEC 2007)*, 2007, 4100–4107.

[2] I.I. Ardelean, M. Cavaliere: Modelling biological processes by using a probabilistic P system software. *Natural Computing*, 2 (2003), 173–197.

[3] F. Arroyo, A.V. Baranda, J. Castellanos, C. Luengo, L.F. de Mingo: A recursive algorithm for describing evolution in transition P systems. *Pre-Proc. Workshop on Membrane Computing* (C. Martín-Vide, Gh. Păun, eds.), Curtea de Argeş, Romania, 2001. Technical Report GRLMC 17/01, Rovira i Virgili University, Tarragona, Spain, 2001, 19–30.

[4] F. Arroyo, A.V. Baranda, J. Castellanos, C. Luengo, L.F de Mingo: Structures and bio-language to simulate transition P systems on digital computers. In [11], 1–16.

[5] F. Arroyo, C. Luengo, A.V. Baranda, L.F. de Mingo: A software simulation of transition P systems in Haskell. In [42], 19–32.

[6] A. Baranda, J. Castellanos, R. Gonzalo, F. Arroyo, L.F. de Mingo: Data structures for implementing transition P systems in silico. *Romanian J. Information Science and Technology*, 4 (2001), 21–32.

[7] D. Balbontín-Noval, M.J. Pérez-Jiménez, F. Sancho-Caparrini: A MzScheme implementation of transition P systems. In [42], 58–73.

[8] L. BIANCO, A. CASTELLINI: Psim: A computational platform for metabolic P systems. *Lecture Notes in Computer Sci.*, 4860 (2007), 1–20.

[9] L. BIANCO, F. FONTANA, G. FRANCO, V. MANCA: P systems for biological dynamics. In [16], 83–128.

[10] R. BORREGO-ROPERO, D. DÍAZ-PERNIL, M.J. PÉREZ-JIMÉNEZ: Tissue simulator: A graphical tool for tissue P systems. *Proc. Intern. Workshop Automata for Cellular and Molecular Computing* (Gy. Vaszil, ed.) MTA SZTAKI, Budapest, Hungary, 2007, 23–34.

[11] C.S. CALUDE, GH. PĂUN, G. ROZENBERG, A. SALOMAA, eds.: *Multiset Processing. Mathematical, Computer Science, and Molecular Computing Points of View.* LNCS 2235, Springer, 2001.

[12] M. CAVALIERE: Evolution-communication P systems. In [42], 134–145.

[13] M. CAVALIERE, I.I. ARDELEAN: Modelling respiration in bacteria and respiration/photosynthesis interaction in cyanobacteria by using a P system simulator. In [16], 129–158.

[14] S. CHERUKU, A. PĂUN, A, F.J. ROMERO-CAMPERO, M.J. PÉREZ-JIMÉNEZ, O.H. IBARRA: Simulating FAS-induced apoptosis by using P systems. *Progress in Natural Science*, 17 (2007), 424–431.

[15] G. CIOBANU, D. PARASCHIV: P system software simulator. *Fundamenta Informaticae*, 49 (2002), 61–66.

[16] G. CIOBANU, GH. PĂUN, M.J. PÉREZ-JIMÉNEZ, eds.: *Applications of Membrane Computing.* Springer, 2006.

[17] G. CIOBANU, G. WENYUAN: P systems running on a cluster of computers. In [35], 123–139.

[18] A. CORDÓN-FRANCO, M.A. GUTIÉRREZ NARANJO, M.J. PÉREZ JIMÉNEZ, A. RISCOS NÚÑEZ, F. SANCHO-CAPARRINI: Implementing in Prolog an effective cellular solution to the knapsack problem. In [35], 140–152.

[19] A. CORDÓN-FRANCO, M.A. GUTIÉRREZ-NARANJO, M.J. PÉREZ-JIMÉNEZ, F. SANCHO-CAPARRINI: A Prolog simulator for deterministic P systems with active membranes. *New Generation Computing*, 22 (2004), 349–364.

[20] D.W. CORNE, P. FRISCO: Corne, D.W., Dynamics of HIV infection studied with cellular automata and conformon-P systems. *BioSystems*, 91 (2008), 531–544.

[21] D. DÍAZ-PERNIL, M.A. GUTIÉRREZ-NARANJO, M.J. PÉREZ-JIMÉNEZ, A. RISCOS-NÚÑEZ: A uniform family of tissue P system with cell division solving 3-COL in a linear time. *Theoretical Computer Sci.*, 2008, DOI::10.1016/j.tcs.2008.04.005

[22] D. DÍAZ-PERNIL, C. GRACIANI, M.A. GUTIÉRREZ-NARANJO, GH. PĂUN, I. PÉREZ-HURTADO, A. RISCOS-NÚÑEZ, eds.: *Sixth Brainstorming Week on Membrane Computing*, Fénix Editora, Sevilla, 2008.

[23] D. DÍAZ-PERNIL, I. PÉREZ-HURTADO, M.J. PÉREZ-JIMÉNEZ, A. RISCOS-NÚÑEZ, P-LINGUA: A programming language for membrane computing. In [22], 135–155.

[24] P. FRISCO, R.T. GIBSON: A simulator and an evolution program for conformon-P systems. *TAPS, Workshop on Theory and Applications of P Systems*, Timişoara, Romania, IEEE Computer Press, 2005, 427–430.

[25] P. FRISCO, S. JI: Conformons-P systems. *Lecture Notes in Computer Sci.*, 2568 (2003), 291–301.

[26] A. GEORGIOU, M. GHEORGHE: Generative devices used in graphics. *Pre-proc. Workshop on Membrane Computing* (A. Alhazov et al., eds.) Technical Report 28/03, Universitat Rovira i Virgili, Tarragona, 2003, 266–272.

[27] A. GEORGIOU, M. GHEORGHE, F. BERNARDINI: Membrane-based devices used in computer graphics. In [16], 253–282.
[28] R. GERSHONI, E. KEINAN, GH. PĂUN, R. PIRAN, T. RATNER, S. SHOSHANI: Research topics arising from the (planned) P systems implementation experiment in Technion. In [22], 183–192.
[29] M.A. GUTIÉRREZ-NARANJO, M.J. PÉREZ-JIMÉNEZ, A. RISCOS-NÚÑEZ: A fast P system for finding balanced 2-partition. *Soft Computing*, 9 (2005), 673–678.
[30] M.A. GUTIÉRREZ-NARANJO, M.J. PÉREZ-JIMÉNEZ, A. RISCOS-NÚÑEZ: On descriptive complexity of P systems. *Lecture Notes in Computer Sci.*, 3365 (2005), 320–330.
[31] M.A. GUTIÉRREZ-NARANJO, M.J. PÉREZ-JIMÉNEZ, A. RISCOS-NÚÑEZ: Available membrane computing software. In [16], 411–436.
[32] M.A. GUTIÉRREZ-NARANJO, M.J. PÉREZ-JIMÉNEZ, D. RAMÍREZ-MARTÍNEZ: A software tool for verification of spiking neural P systems. *Natural Computing*, 7 (2008), 485–497.
[33] M. IONESCU, GH. PĂUN, T. YOKOMORI: Spiking neural P systems. *Fundamenta Informaticae*, 71 (2006), 279–308.
[34] M. MALIŢA: Membrane computing in Prolog. *Pre-proceedings of the Workshop on Multiset Processing* (C.S. Calude et al., eds.) Curtea de Argeş, Romania, CDMTCS TR 140, Univ. of Auckland, 2000, 159–175.
[35] C. MARTÍN-VIDE, GH. PĂUN, G. ROZENBERG, A. SALOMAA, eds.: *Workshop on Membrane Computing 2003*. LNCS 2933, Springer, 2004.
[36] I.A. NEPOMUCENO-CHAMORRO: A Java simulator for basic transition P systems. In [41], 309–315.
[37] D.V. NICOLAU JR., G. SOLANA, F. FULGA, D.V. NICOLAU: A C library for simulating P systems. *Fundamenta Informaticae*, 49 (2002), 241–248.
[38] T.Y. NISHIDA: Membrane algorithms. *Lecture Notes in Computer Sci.*, 3850 (2006), 55–66.
[39] GH. PĂUN: P systems with active membranes: Attacking NP-complete problems. *J. Automata, Languages and Combinatorics*, 6 (2001), 75–90.
[40] GH. PĂUN: *Membrane Computing. An Introduction*. Springer, 2002.
[41] GH. PĂUN, A. RISCOS, A. ROMERO, F. SANCHO, eds.: *Second Brainstorming Week on Membrane Computing*. Report RGNC 01/04, Sevilla, 2004.
[42] GH. PĂUN, G. ROZENBERG, A. SALOMAA, C. ZANDRON, eds.: *Membrane Computing, WMC-CdeA 2002*. LNCS 2597, Springer, 2003.
[43] M.J. PÉREZ-JIMÉNEZ, F.J. ROMERO-CAMPERO: A CLIPS simulator for recognizer P systems with active membranes. In [41], 387–413.
[44] M.J. PÉREZ-JIMÉNEZ, F.J. ROMERO-CAMPERO: An efficient family of P systems for packing items into bins. *J. Univ. Computer Sci.*, 10 (2004), 650–670.
[45] M.J. PÉREZ-JIMÉNEZ, F.J. ROMERO-CAMPERO: A study of the robustness of the EGFR signalling cascade using continuous membrane systems. *Lecture Notes in Computer Sci.*, 3561 (2005), 268–278.
[46] M.J. PÉREZ-JIMÉNEZ, F.J. ROMERO-CAMPERO: Attacking the Common Algorithmic Problem by recognizer P systems. *Lecture Notes in Computer Sci.*, 3354 (2005), 304–315.
[47] M.J. PÉREZ-JIMÉNEZ, F. SANCHO-CAPARRINI: A formalization of transition P systems. *Fundamenta Informaticae*, 49 (2002), 261–272.
[48] B. PETRESKA, C. TEUSCHER: A reconfigurable hardware membrane system. In [35], 269–285.

[49] A. RISCOS-NÚÑEZ: *Cellular Programming: Efficient Resolution of Numerical NP-complete Problems*. PhD Thesis, University of Seville, 2004.

[50] E. RIVERO-GIL, M.A. GUTIÉRREZ-NARANJO, M.J. PÉREZ-JIMÉNEZ: Graphics and P systems: Experiments with JPLANT. In [22], 241–254.

[51] F.J. ROMERO-CAMPERO, M.J. PÉREZ-JIMÉNEZ: Modelling gene expression control using P systems: The Lac operon, a case study. *Biosystems*, 91 (2008), 438–457.

[52] F.J. ROMERO-CAMPERO, M.J. PÉREZ-JIMÉNEZ: A model of the quorum sensing system in Vibrio fischeri using P systems. *Artificial Life*, 14 (2008), 95–109.

[53] A. ROMERO-JIMÉNEZ, M.A. GUTIÉRREZ-NARANJO, M.J. PÉREZ-JIMÉNEZ: The growth of branching structures with P systems. *Fourth Brainstorming Week on Membrane Computing, Vol. II* (C. Graciani et al., eds.), Fénix Editora, Sevilla, 2006, 253–265.

[54] A. ROMERO-JIMÉNEZ, M.A. GUTIÉRREZ-NARANJO, M.J. PÉREZ-JIMÉNEZ: Graphical modelling of higher plants using P systems. *Lecture Notes in Computer Sci.*, 4361 (2006), 496–506.

[55] S. SEDWARDS, T. MAZZA: Cyto–Sym: A formal language model and stochastic simulator of membrane-enclosed biochemical processes. *Bioinformatics*, 23 (2007), 2800–2802.

[56] Y. SUZUKI, H. TANAKA: On a LISP implementation of a class of P systems. *Romanian J. Information Science and Technology*, 3 (2000), 173–186.

[57] Y. SUZUKI, Y. FUJIWARA, H. TANAKA, J. TAKABAYASHI: Artificial life applications of a class of P systems: Abstract rewriting systems on multisets. In [11], 299–346.

[58] A. SYROPOULOS, E.G. MAMATAS, P.C. ALLILOMES, K.T. SOTIRIADES: A distributed simulation of transition P systems. In [35], 357–368.

[59] P System Modelling Framework at the University of Sheffield: http://www.dcs.shef.ac.uk/marian/PSimulatorWeb/PSystemMF.htm

[60] Natural Computing Group of the Technical University of Madrid: http://www.lpsi.eui.upm.es/nncg

[61] P Systems Web Page: http://ppage.psystems.eu

[62] Models of Natural Computing – University of Verona: http://mnc.sci.univr.it

[63] Research Group on Natural Computing – University of Seville: http://www.gcn.us.es

[64] The Xholon Project: http://www.primordion.com/Xholon

[65] CoSBi's Prototypes – Cyto-Sim: http://www.cosbi.eu/Rpty_Soft_CytoSim.php

[66] The Center for Biomedical Computing - Verona: http://www.cbmc.it

CHAPTER 18

PROBABILISTIC/ STOCHASTIC MODELS

PAOLO CAZZANIGA

MARIAN GHEORGHE

NATALIO KRASNOGOR

GIANCARLO MAURI

DARIO PESCINI

FRANCISCO J. ROMERO-CAMPERO

18.1 INTRODUCTION

MEMBRANE computing introduces a new computational paradigm inspired by cellular biology concepts and principles, with the aim of defining a family of robust, efficient, and easy to use models, called P systems, that, on the one hand, mimic fundamental cellular processes, and, on the other hand, possess the rigor and soundness of mathematical models, and the simplicity and effectiveness of computational approaches. An intensive study of the computational power of these mechanisms together with complexity and decidability aspects has been undertaken

[21]. On the other hand, the use of membrane computing for various application areas has been also developed [10].

Of particular interest are those applications in biology, as this model has emerged from this context. It is worth mentioning that many computational paradigms have been used to model various biological phenomena and processes (different calculi and process algebras, Petri nets, cellular automata, and Boolean networks are among the most successful models in this respect [12]). A plethora of (non)deterministic and stochastic variants have been suggested for various problems and systems to be modeled. Membrane computing models occur in various variants and with different syntax and semantic forms. These models have different structures (hierarchical or cell-like, network or tissue-like, dynamical or population-like), objects involved (multisets of symbols or strings), and rules (evolution, communication, cell division, membrane dissolution, etc.). Although most of the theoretical studies considered so far have imposed an evolution strategy whereby in each evolution step the rules are applied in a non-deterministic and maximally parallel manner, various applications, especially those making use of membrane computing models in biology, have used different execution strategies, namely deterministic or probabilistic/stochastic. These models, by introducing a certain control over the strategy of applying the rules, allow one to perform qualitative and approximate or exact quantitative studies.

In this chapter the modeling approach presented is designated for systems with multiple compartments and exhibiting stochastic behavior. A fundamental result of theoretical statistical physics states that the level of noise in a system is inversely proportional to the square root of the number of interacting particles. Therefore, cellular processes arising from the interactions of a low number of molecules exhibit high levels of noise as it has been reported experimentally [15]. Therefore, in order to accurately model such systems it is necessary to make use of mesoscopic, discrete, and stochastic approaches.

18.2 Basic Definitions and Examples

The probabilistic and stochastic models presented here rely on a basic cell-like P system structure defined below.

Definition 18.1 *A basic P system structure is a construct*

$$\Pi_B = (V, \mu, M_1, \ldots, M_n, R_1, \ldots, R_n),$$

where: *V* is the finite alphabet *of the system;* μ is the membrane structure *consisting of $n \geq 1$ membranes defining compartments identified in a one to one manner with values from* $\{1, \ldots, n\}$; M_i, *for each* $1 \leq i \leq n$, *is a multiset over V defining the* initial state *of the compartment i;* R_i, *for each* $1 \leq i \leq n$, *is the finite set of rules associated to compartment i.*

A dynamical probabilistic P system (DPP, for short) is a P system variant, introduced in [22]. An extensive and detailed description of DPPs and examples are presented in [22, 24, 4, 5].

Definition 18.2 *A dynamical probabilistic P system is a construct*

$$\Pi_D = (V, O, \mu, M_1, \ldots, M_n, R_1, \ldots, R_n, E, I),$$

where V, μ, $M_1, \ldots, M_n, R_1, \ldots, R_n$ have the meaning defined by Definition 18.1 for Π_B; $O \subseteq V$ is the set of analyzed symbols; $E = \{V_E, M_E, R_E\}$ is the environment of the system and consists of an alphabet $V_E \subseteq V$, a feeding multiset M_E, over V_E, and a finite set of feeding rules R_E; $I \subseteq \{0, \ldots, n\}$ is the set of labels of the analyzed regions (the label 0 corresponds to the environment and 1 labels the skin membrane of μ).

The set R_i, $1 \leq i \leq n$, consists of evolution rules of the form $r : u \xrightarrow{c} v$, where *u* is a multiset over *V*, *v* a multiset over $V \times (\{here, out\} \cup \{in_j \mid 2 \leq j \leq n\})$, and *c* a positive real value, called *kinetic constant*. When such a rule is applied, the multiset *u* is replaced by the symbols occurring in the multiset *v*; the symbols marked with *here* will remain in the compartment *i*, those marked with *out* will go to the parent compartment or to the environment, if $i = 1$; the symbols having in_j will be moved to compartment *j*. The set R_E contains rules of the type $u \to (v, in_1)$, for *u, v* multisets over V_E. When such a rule is applied the multiset *v* that replaces *u* is sent to compartment 1.

The alphabet *O* and the set *I* specify the symbols and compartments, respectively, that are considered relevant for the evolution of the analyzed system.

A DPP starts from the initial configuration of the system given by the multisets $M_1, \ldots, M_n$, and M_E. The system then evolves iteratively. The first stage of each iteration consists of evaluating in every compartment *i* the probabilities of all the rules, by using the kinetic constants associated to them and counting all possible distinct combinations of the objects occurring on the left-hand side of these rules with respect to the current multiset in *i*. Obviously the values of these probabilities depend on the multisets present in each compartment, which change every time the rules are applied; consequently the values of these probabilities change accordingly. The rules are selected to be associated to the objects present in each compartment according to their probabilities. The rules are then applied in a maximally parallel manner in each compartment. Exploiting the dynamical way of computing the rule probabilities, the DPPs introduce a sort of control over the use of the rules. The rules from the set R_E follow a different strategy and this is defined according to specific circumstances.

A variant of the above model is represented by the τ-DPP paradigm, firstly presented in [9]; τ-DPP extends the tau-leaping algorithm conceived for a single volume [8], in order to accommodate the requirements of multi-volume systems. The approach based on the τ-DPP paradigm relies on some new features (a new method to compute the probabilities associated to rules and a different strategy to select the rules in a computation step), but does no longer use the environment and its associated elements. In this respect Definition 18.2 will be used without considering the environment E. The system will be denoted by Π_τ.

A τ-DPP starts from the initial configuration given by the n multisets $M_1, \ldots, M_n$ and evolves in steps. In the first stage of every step, in each compartment i, according to tau-leaping method [14, 8], the rule probabilities are computed according to the kinetic constants associated to them and a function depending on the number of objects occurring on the left-hand side of the rules and then the rules to be applied are selected and the time necessary to execute them (called τ_i) is calculated. The minimum τ_i across the entire system Π_τ is determined and all the rules selected in each compartment are applied. So doing, the evolution of the system is synchronized through the use of the same time increment. We refer the reader to [9, 6] for a thorough description of τ-DPP and of the algorithm devised in this respect.

DPP and τ-DPP provide qualitative and approximate quantitative descriptions, respectively, of bio-chemical systems modeled within this probabilistic/stochastic framework. An exact quantitative approach is achieved using a method based on Gillespie's theory of stochastic kinetics [13] applied to a specific type of P system, called stochastic P systems, where multi-compartments are used [3, 29, 30, 28].

Definition 18.3 *A stochastic P system is a construct*

$$\Pi_S = ((V_{obj}, V_{str}), \mu, M_1, \ldots, M_n, (R_1^{obj}, R_1^{str}), \ldots, (R_n^{obj}, R_n^{str})),$$

where V_{obj} is the finite alphabet of simple objects and V_{str} is the finite alphabet of string objects; μ is the membrane structure; $M_i = (obj_i, Str_i)$, $1 \leq i \leq n$, is the initial state of the compartment i, consisting of the multisets obj_i, of objects from V_{obj}, and Str_i, of strings over V_{str}; R_i^{obj} and R_i^{str}, $1 \leq i \leq n$, are the finite sets of evolution (rewriting) rules on multisets of objects and strings, respectively.

The simple objects, V_{obj}, denote simple molecular species, whereas the strings over V_{str} describe more complex molecules, like DNAs, RNAs, proteins. A multiset from M_i, ms, is represented as $ms = e_1 + \cdots + e_p$ with e_i either a simple object from V_{obj} or a string, str, represented as $str = \langle s_1 \cdots s_q \rangle$, where $s_1, \ldots, s_q \in V_{str}$. Each evolution rule on multisets of objects from R_i^{obj} has the following form:

$$r^{obj}: obj_1\,[\,obj_2\,]_i \xrightarrow{c} obj_1'\,[\,obj_2'\,]_i$$

with obj_1, obj_2, obj'_1, obj'_2 some finite multisets of objects from V_{obj}. These multiset rewriting rules operate on both sides of the membrane defining the compartment i, i.e. a multiset obj_1 outside the membrane i and a multiset obj_2 from the compartment i are simultaneously replaced by the multisets obj'_1 and obj'_2, respectively. Each evolution rule on multisets of strings from R_i^{str} is of the following form

$$r^{str} : [\, obj + str \,]_i \xrightarrow{c'} [\, obj' + str'; str'_1 + \cdots + str'_s \,]_i$$

with obj, obj' multisets of objects over V_{obj} and str, str', $str'_1, \ldots, str'_s$ strings over V_{str}. These rules operate on both multisets of objects and strings. Objects obj are replaced by objects obj' and the substring str is replaced by str', whereas the strings $str'_1, \ldots, str'_s$ are released in the compartment. Constants c and c' associated with these rules, called *kinetic stochastic constants*, are used together with the number of objects occurring on their left-hand side to dynamically compute the probabilities associated to them.

A computation in a stochastic P system starts from the initial multisets by setting the simulation time to 0. In each iteration and each compartment, Gillespie's algorithm [13] is used to compute a rule that will be next executed and the time needed to apply it. The rule with the smallest time value is selected to be executed, the simulation time and the waiting time of the rules selected from the other compartments are updated. In the next step, only the compartment(s) affected by the application of the last rule is (are) involved in calculating the rule(s) and associated execution time(s) according to Gillespie method. The rule which is now selected to be executed is based on the smallest time(s) considered out of the new computed time(s) and the updated values. Obviously this algorithm, although very precise, is slow due to a sequential execution of the rules across the entire system.

Two examples will illustrate the modeling capabilities of the P systems introduced above. The first example will present a prey-predator case study using DPPs and the second one will describe a simple regulatory network defined using stochastic P systems.

Example 18.1 *Predator-prey model expresses the population dynamic with respect to a number of interactions between different individuals. This is an example which illustrates an oscillatory system where the interacting elements evolve in different directions, one goes up and the other comes down and vice-versa. This model, known in its continuous variant as Lotka-Volterra model, has a simple representation based on a DPP with one compartment [23]. The two species involved, prey and predator, are designed by symbols X and Y, respectively. Formally the system is defined by*

$$\Pi_{LV} = (V, O, [\,]_1, X^{p_1} Y^{p_2}, R_1, E, I),$$

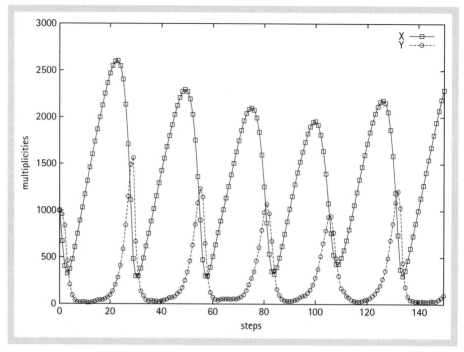

Fig. 18.1 Lotka-Volterra dynamics.

where $V = \{X, Y, A\}$, $O = \{X, Y\}$, $E = (\{A\}, A^s, \{A \to (A, in_1)\})$, $I = \{1\}$, and R_1 consists of

$$r_1 : AX \xrightarrow{c_1} (XX, here), \quad r_2 : XY \xrightarrow{c_2} (YY, here), \quad r_3 : Y \xrightarrow{c} (\lambda, here).$$

The parameters occurring within this specification have the following values: $p_1 = p_2 = 1000$, $s = 200$, $c_1 = 1$, $c_2 = 10^{-2}$, $c_3 = 10$. *The oscillatory behavior of the two species, for the above parameters, is illustrated in Fig. 18.1[23].*

Example 18.2 *A gene regulatory network consisting of a positive expression of a gene is modeled by using a stochastic P system with two compartments. The first compartment contains signalling molecules that diffuse into the second, where a complex molecule is formed which in turn binds a site of the gene. The system is given by the following construction*

$$\Pi_{Reg} = (\{geneR, rnaR, R, S, RS, RSgeneR\}, [\ [\ ]_2\ ]_1, M_1, M_2, R_1, R_2),$$

where $M_1 = \overset{100}{\overbrace{S + \cdots + S}}$ and $M_2 = geneR$, R_1 consists of the following rule:
$r_1 : S[\ ]_2 \xrightarrow{c_1} [\ S\ ]_2 \quad c_1 = 0.3015\, min^{-1}$

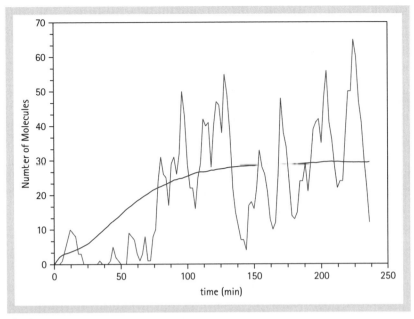

Fig. 18.2 Gene positive expression.

and R_2 contains:

$r_2 : [\, geneR\,]_2 \xrightarrow{c_2} [\, geneR + rnaR\,]_2 \qquad c_2 = 0.12\ min^{-1}$
$r_3 : [\, rnaR\,]_2 \xrightarrow{c_3} [\ ]_2 \qquad c_3 = 0.347\ min^{-1}$
$r_4 : [\, rnaR\,]_2 \xrightarrow{c_4} [\, rnaR + R\,]_2 \qquad c_4 = 3\ min^{-1}$
$r_5 : [\, R\,]_2 \xrightarrow{c_5} [\ ]_2 \qquad c_5 = 0.347\ min^{-1}$
$r_6 : [\, R + S\,]_2 \xrightarrow{c_6} [\, RS\,]_2 \qquad c_6 = 6 \times 10^{-4}\ min^{-1}$
$r_7 : [\, RS + geneR\,]_2 \xrightarrow{c_7} [\, RSgeneR\,]_2 \qquad c_7 = 5 \times 10^{-3}\ min^{-1}$
$r_8 : [\, RSgeneR\,]_2 \xrightarrow{c_8} [\, RS + geneR\,]_2 \qquad c_8 = 10^{-3}\ min^{-1}$
$r_9 : [\, RSgeneR\,]_2 \xrightarrow{c_9} [\, RSgeneR + rnaR\,]_2 \qquad c_9 = 1.2\ min^{-1}$

Figure 18.2 shows how the concentration of the protein molecules evolves in time (one simulation and average behavior over 10,000 runs).

18.3 Modeling Principles

Biological systems are highly complex entities with many compartments and numerous molecules interacting in various ways. The compartments are, in general, tightly inter-related and the entire topology may change over time. In order to properly reflect various aspects of the processes under investigation, it is necessary to work out a set of precise principles of expressing the components of the model and their inter-relationships. Currently there are various computational models used in approaching from different perspectives the systems under investigation [12]. In any modeling framework robust methods and techniques are looked for in order to automate the process of generating the model, especially for highly complex systems with many components and interactions between them, and for devising sound ways to transform it from one formalism to another and to verify certain properties. In this section, we present some principles for rigorously and systematically specifying various transformations and interactions within cellular systems, by using stochastic P systems. More specifically, we will focus on how to express cellular regions and compartments, molecules, molecular and inter-cellular interactions [27].

Compartments specification Membranes play a key role in the hierarchical organization of living cells, delimiting all the compartments contained by them and separating the cells from other cells or environment, and actively taking part in various intra- and inter-cellular processes. Our computational framework based on the P systems model explicitly deals with membranes and compartments. Different processes where membranes are involved in various interactions have been specified and analyzed using suitable variants of P systems, for instance, selective uptake of molecules from the environment [29], signalling at the cell surface [20], pathway signalling networks [7], metapopulations [4, 5], colonies of interacting bacteria which communicate by sending and receiving diffusing signals [2, 30].

Molecular species specification Each molecular species is either represented by a simple object or a string of objects. When the molecules are simple or their internal structure is not relevant for the model, then simple objects are the right representation. When more complex molecules are considered with sites playing distinct roles in the model—proteins with domains in various states, genes containing operons, initiation and termination sites—then strings with objects arranged in a specific order represent the suitable codification.

Molecular interactions specification P systems through their evolution and communication rules model in a natural and direct way a broad area of chemical and biological interactions at the molecular level. Some of these chemical reactions will be illustrated below by defining the rules in the format provided by stochastic P systems with respect to a given compartment i. A thorough description of every type of reaction together with its propensity function is available in [27].

The **transformation** of a molecule, represented by an object a, into another molecule, represented by an object b, and the **degradation** of a molecule, represented by a, are described by the evolution (rewriting) rules r_1 and r_2, respectively:

$$r_1 : [\, a \,]_i \xrightarrow{c_1} [\, b \,]_i, \; r_2 : [\, a \,]_i \xrightarrow{c_2} [\;]_i.$$

The process of **complex formation** consisting of producing a complex molecule, c, from simpler molecules, a and b, and the **dissociation** of the complex c into its initial components is described by the following rules:

$$r_3 : [\, a + b \,]_i \xrightarrow{c_3} [\, c \,]_i, r_4 : [\, c \,]_i \xrightarrow{c_4} [\, a + b \,]_i.$$

The simple passive **diffusion** of a small molecule, specified by the object a, inside and outside a compartment is represented by the communication rules r_5 and r_6, respectively:

$$r_5 : a \,[\,\,]_i \xrightarrow{c_5} [\, a \,]_i, r_6 : [\, a \,]_i \xrightarrow{c_6} a \,[\,\,]_i.$$

In signal transduction processes, molecules hanging on membranes travel through and bind to various receptors; this **binding** mechanism engaging two molecules, represented by an object a outside of the compartment i and another one, b, inside, as well as its reverse operation, called **debinding**, are represented by the evolution-communication rules r_7 and r_8, respectively. These rules are similar to complex formation and dissociation, but the objects are now located in different compartments:

$$r_7 : a \,[\, b \,]_i \xrightarrow{c_7} [\, c \,]_i, r_8 : [\, c \,]_i \xrightarrow{c_8} a \,[\, b \,]_i.$$

The processes of **recruitment** and **releasing** are similar to binding and debinding, but the receptor is located outside the compartment where the complex will be formed. These are represented by rules r_9 and r_{10}, respectively:

$$r_9 : a \,[\, b \,]_i \xrightarrow{c_9} c \,[\,\,]_i, r_{10} : c \,[\,\,]_i \xrightarrow{c_{10}} a \,[\, b \,]_i.$$

In many cellular interactions the expression of various proteins codified in specific genes plays a significant role. The specification of the processes involved in gene expression control can be represented in P systems as rewriting rules on multisets of objects or strings depending on the structural organization of the genes involved in the system and on the level of abstraction of the model. In what follows we present these two alternatives for the transcription factor binding and debinding and the transcription process. When the genes are represented by individual objects the **binding** and **debinding** of **transcription factors** are represented by similar rules to those used for complex formation and dissociation, namely r_{11} and r_{12}, and when the genes are represented as strings where specific sites for transcription factor binding are identified, then rewriting rules on multisets of strings and objects, r_{13}

and r_{14}, are designed:

$$r_{11} : [\ Tf + gene\ ]_i \xrightarrow{c_{11}} [\ \mathit{Tf\text{-}gene}\ ]_i,$$

$$r_{12} : [\ \mathit{Tf\text{-}gene}\ ]_i \xrightarrow{c_{12}} [\ Tf + gene\ ]_i,$$

$$r_{13} : [\ Tf + \langle site \rangle\ ]_i \xrightarrow{c_{13}} [\ \langle site' \rangle\ ]_i,$$

$$r_{14} : [\ \langle site' \rangle\ ]_i \xrightarrow{c_{14}} [\ Tf + \langle site \rangle\ ]_i,$$

where *Tf*, *gene*, *Tf-gene* are objects and $\langle site \rangle$, $\langle site' \rangle$ are strings.

The **transcription** process, the production of mRNAs from genes, can be described as an object-based evolution process using one rule and involving the objects *gene* and *rna*, denoting the gene and the mRNA, respectively:

$$r_{15} : [\ gene\ ]_i \xrightarrow{c_{15}} [\ gene + rna\ ]_i.$$

When the internal structure of the genes entering this process is relevant, i.e. certain operons, operators, or transcription starting points occur in various reactions then they are represented as strings. In this case the complex process of transcription can be represented in detail using rewriting rules on multisets of strings and objects. Here we distinguish three different stages in the gene transcription process. Firstly, the transcription initiation stage consisting of the reversible binding process of the RNA polymerase is described by the rules r_{16} and r_{17} which act upon the object *RNAP* and the gene promoter, represented by the substring $\langle prom \rangle$,

$$r_{16} : [\ RNAP + \langle prom \rangle\ ]_i \xrightarrow{c_{16}} [\ \langle prom \cdot RNAP \rangle\ ]_i,$$

$$r_{17} : [\ \langle prom \cdot RNAP \rangle\ ]_i \xrightarrow{c_{17}} [\ RNAP + \langle prom \rangle\ ]_i.$$

During the second stage of the transcription process, the RNA polymerase moves along different sites of the gene producing complementary RNA sites (although, the growing mRNA hangs out from the RNA polymerase and is not part of the gene, in the current specification, the substring representing the growing mRNA is part of the string representing the gene, but codified with different symbols). The rewriting rule r_{18} describes this stage, where the substring $\langle \overline{site_{ini} \cdot w} \cdot RNAP \cdot site_{mid} \rangle$ represents the current situation when the *RNAP* molecule with a partially formed mRNA chain, $\langle \overline{site_{ini} \cdot w} \rangle$, is ready to transcribe the next site in the gene, $\langle site_{mid} \rangle$; the addition of newly transcribed nucleotides to the current mRNA chain is achieved by adding the substring $\langle \overline{site_{mid}} \rangle$ to the substring $\langle \overline{site_{ini} \cdot w} \rangle$, where w is an arbitrary sequence of nucleotides; then *RNAP* is moved to the next site:

$$r_{18} : [\ \langle \overline{site_{ini} \cdot w} \cdot RNAP \cdot site_{mid} \rangle\ ]_i \xrightarrow{c_{18}}$$

$$[\ \langle site_{mid} \cdot \overline{site_{ini} \cdot w} \cdot site_{mid} RNAP \rangle\ ]_i.$$

In the last stage of the transcription, the RNA polymerase, denoted by the object *RNAP*, after reaching and transcribing the terminal site, dissociates from the gene and the fully transcribed RNA, represented by the string $\langle \overline{site}_{ini} \cdot w \cdot \overline{site}_{ter} \rangle$, is released:

$$r_{19} : [\,\langle \overline{site}_{ini} \cdot w \cdot RNAP \cdot site_{ter} \rangle\,]_i \xrightarrow{c_{19}} [\,RNAP + \langle site_{ter} \rangle;\ \langle \overline{site}_{ini} \cdot w \cdot \overline{site}_{ter} \rangle\,]_i.$$

Multicellular systems can also be represented in this context by using a tissue-like paradigm, whereby the hierarchical structure of compartments is replaced by a network of elements. The approach has been used to model colonies of bacteria [2] and quorum sensing systems in the marine bacterium *Vibrio fischeri* from an artificial life perspective [30].

18.4 FORMAL VERIFICATION AND ANALYSIS

The use of formal computational models together with a systematic methodology to define them, leads not only to robust, executable specifications, but also constitutes the basic and essential starting point for deploying various methods and techniques to improve, evolve, optimize, formally verify, or analyze these systems. In this section we show how formal verification and analysis enrich our knowledge about the models and systems investigated. The type of formal verification investigated is reported for stochastic P systems whereas the formal analysis of the system's dynamics is developed for DPPs.

Once a model has been specified using an adequate formalism, certain techniques can be applied to check for different properties of the model. These properties are formulated as queries in a given formal language, usually a temporal logic framework, and they require the model to be in a specific format and its components, variables, processes, to satisfy certain conditions. One of the most popular methods to verify such properties relies on the use of adequate model checkers. For systems with a stochastic behavior an attractive model checking framework is provided by PRISM (abbreviation for Probabilistic and Symbolic Model Checker), which offers both a specification language, allowing the representation of the processes involved, as transitions of a Markov chain model, and a query language based on temporal logic, helping the process of verifying various properties of the model [17].

Using modeling principles exposed in the previous section, the main parts of a P system specification, components, and objects, are mapped into suitable PRISM elements, modules, and variables, respectively, and the evolution and communication rules, depending on their behavior, are transformed into corresponding commands. A detailed description of these transformations is presented in [3, 26].

A PRISM specification of the model allows it to simulate and get different qualitative information regarding the concentration level of various molecular species with respect to different initial conditions. Moreover, an average behavior can also be provided. Using the temporal logic query language certain properties of the model can be also investigated. More specifically, it supports studying the likelihood that certain molecules satisfy some properties, for instance, reaching a certain concentration level. It also allows one to study some long run, equilibrium and steady state, probabilities (for examples and relevant results see [3, 26]).

In general, the properties a certain model exhibits depend on specific conditions of this model. In this respect it is necessary to have a mechanism that learns either from simulations or experimental data certain properties that hold for those particular instances. Such a method has been reported in [3, 26] using Daikon, a tool for developing reverse-engineering specifications from software systems in terms of invariants or rules that hold true at particular points of a program [11]. Daikon has been used to discover relationships between various molecular species concentrations by analyzing several simulations [3] and generates a series of invariants. These invariants are then translated into properties expressed in temporal logic and analyzed formally using PRISM such as to prove they remain true for the entire model. In [3], by using Daikon, it has been discovered that $0 \leq rna \leq 24$ and $0 \leq protein \leq 205$, which then has been formally confirmed by using PRISM.

A formal model also allows one to analyze its behavior from a different perspective that sheds some light on the dynamic evolution of the system. We briefly present some basic notions, thoroughly explained in [23, 22], that help in describing and analyzing the evolution of some biological systems modeled using the DPP approach.

The evolution of the system can be described as the variation over time of the multisets of objects involved. To simplify the presentation we will refer in the sequel to P systems with one compartment and l molecular species. In order to keep track of the evolution of a system, the multiset at time t, called t-*multiset*, and denoted $M_t = \{a_1^{\alpha_1}, a_2^{\alpha_2}, \ldots, a_l^{\alpha_l}, t\}$, is associated with a vector defining the *position* of M_t, namely $\overrightarrow{M_t} = (\alpha_1, \ldots, \alpha_l, t)$ in the space N^{l+1}, where the coordinates $\alpha_1, \ldots, \alpha_l$ correspond to the multiplicities of the symbols $a_1, \ldots, a_l$ and t is the time value. Based on this codification the distance between two t-multisets $\overrightarrow{M_i}$ and $\overrightarrow{M_j}$, called *displacement*, is introduced. In [22] it is shown that the global behavior of the entire system is influenced by the dispersal of individuals among patches.

Furthermore, the temporal information captured by this codification can be exploited in order to express how fast the DPP has evolved between the two states. This is done by defining the *velocity* of the system which, together with the

t-multiset, produce the *phase point*, $\vec{\varphi_t} = (a_1, \ldots, a_l, v_1, \ldots, v_l) \in \mathbf{N}^l \times \mathbf{R}^l$. These points generate the *phase space*, the set of all possible states of the system during its evolution. In [24] a methodology and a tool, called *vector field over the phase space*, are presented to show certain global properties of the system by looking at the local dynamics, thus avoiding the need to perform long and tedious simulations to identify the regions of homologous behavior. This method allows one to predict, with a certain degree of accuracy, the system behavior. In fact, it is possible to identify attraction basins and periodic or quasi-periodic orbits, or to identify stochastic or equilibrium points.

Another approach aimed at the formal analysis of systems has been developed in [19] to identify distinct areas of the phase space with homologous dynamical behavior, avoiding the use of simulations.

Starting from the observation that most of the information about the structure of a chemical reacting system is stored in its associated stoichiometric matrix, it is possible to algebraically derive notions about the topology of the phase space. Moreover, other detailed information about the system can be obtained exploiting the Markov chains theory.

Having this knowledge about the system, it is then possible to generate its configurations space and check for the mutual reachability of the states. If the "communication" between two states happens in both directions, then these two states are part of a cycle and they are said to be communicating. The importance of the communication property is that, being an equivalence relation, it allows one to partition the state space into communicating classes which correspond to sets of points that belongs to the same dynamics.

Note that the system can be observed in two distinct regimes, one close to the origin of the state space and the other one far from it (called free regime). Near the origin, the small number of molecules involved in the dynamics poses the problem that for a given set of rules (i.e. a cycle), not all possible rules application orders are allowed. This issue raises two main questions: which are the minimal configurations that (1) allow the application of a given sequence of rules, and that (2) allow to apply a set of unordered rules? The answers reside in the definition of the defects notion, which is the formal description of the set of minimal multiset that enables the given sequence of rules. This problem does not affect the free regime where, by definition, all the rules are enabled.

18.4.1 Case Studies

In this section some of the case studies developed around probabilistic and stochastic P systems are presented together with some of their outcomes, proving the benefits and suitability of these models for cellular systems biology.

Metapopulations, or multi-patch systems, are used to analyze the dynamics of ecological systems, where the populations behavior is influenced by the fragmented habitat. The metapopulation consists of some local populations of individuals living in different patches. In the description of metapopulation systems, two distinct dynamics are identified: local interactions inside a patch, following a Lotka-Volterra evolution strategy of preys and predators and a global behavior given by intercommunications between patches, and activities taking place inside. The DPP framework extended concepts to deal with costs related to moves between patches and the process of feeding patches with suitable resources. A system consisting of 4 patches with different areas and either allowing or preventing the individuals to move between them. A detailed description of the metapopulation model and several simulation results are provided in [4, 5]. One of the reported results shows that incrementing the size of patch areas leads to increasing the probability of the population's extinction. Other results show the relationships between the number of individuals in the current populations and the amount of feeding resources provided to patches.

The gene regulation system of the *lac operon* in *Escherichia coli (E. coli)* has been modeled using stochastic P systems [29]. The model focuses on the specification of the mechanisms that allow *E. coli* to sense the sugar source, glucose or lactose, in the environment and to express accordingly specific proteins involved in the uptake and the metabolism of the available sugar. The cell surface including the sensing and transportation mechanisms were explicitly described as well as the sequential arrangements of genes, operators, and promoters in the *lac operon*. The model consists of three different membranes describing the environment, cell surface, and cytoplasm and uses 55 multiset rewriting rules acting on both simple objects and strings. The behavior of the system was studied under different environmental conditions representing the presence and/or absence of glucose and/or lactose, showing that only in the absence of glucose and presence of lactose will *E. coli* fully express the genes involved in the uptake and metabolism of lactose. A simplified version of this system using 36 rules was used to compute in Fig. 18.3 the average number of LacZ proteins over 5000 simulations. The frequency of the different number of LacZ proteins in 5000 simulations at 240 minutes is shown in Fig. 18.4. This experiment shows a bimodal distribution characteristic of autocatalytic systems where even in the presence of the inducer a fraction of the cells remain uninduced. These results prove that computing averages is not enough to characterize a stochastic system.

A variant of stochastic P systems is employed to study signalling cascade pathways. One study investigates the signal transduction system in the Epidermal Growth Factor Receptor cascade looking at the robustness of the system with regard to the number of molecule signals, Epidermal Growth Factor (EGF) in the environment and the number of receptors, Epidermal Growth Factor Receptor (EGFR), on the cell surface [20]. The cell surface was explicitly modeled together with the mechanisms involved in the receptor activation through signal binding, activation

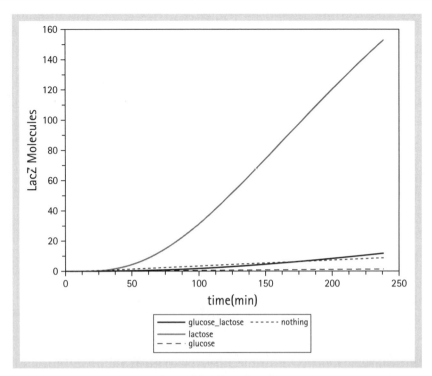

Fig. 18.3 The average number of LacZ molecules.

of specific cytoplasmic kinases through recruitment by active receptor complexes on the cell surface, and cytoplasmic signalling cascades leading to transcription factor activation. The model contains three membranes describing the environment, cell surface, and cytoplasm and uses 160 object rewriting rules. The results obtained show that the EGFR signalling cascade is robust with regard to the number of EGF signals, but it is sensitive to the number of receptors EGFR on the cell surface. Another P system model has been used to study the FAS induced apoptosis signal transduction system [20]. This model analyzed the different possible signalling cascades leading to activation of one of the effectors of apoptosis in the system, caspase3. Similarly to the case of EGFR signal transduction system, the cell surface is explicitly represented as well as the molecular interactions involved in the signal transduction. In addition to these components of the model, an extra membrane was used to represent the mitochondria which play a key role in one of the activation signalling pathways of caspase3. The molecular interactions were represented by 97 rewriting rules acting on multisets of simple objects. The results of the model prove that the interactions between a mitochondria-associated protein, Bcl2, and the proteins Bid and Bax play a key role in caspase3 activation.

Bacterial quorum sensing has been firstly modeled with basic P systems [16]. A stochastic P system has been used to provide an artificial life system that behaves

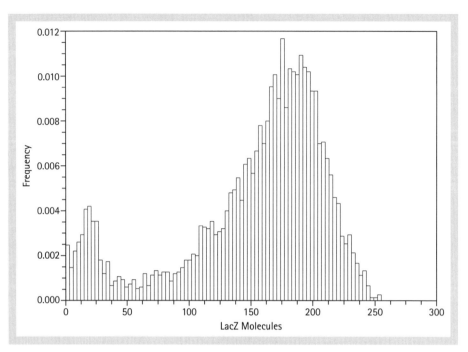

Fig. 18.4 Frequency of the number of LacZ molecules at 240 minutes.

very close to a quorum sensing system as exhibited by the marine bacterium *Vibrio fischeri* [30] relying on the paradigm introduced in [2]. The aim was to show how a colony of bacteria can use a simple communication mechanism consisting of the synthesis, diffusion, accumulation, and sensing of specific signals to achieve coordinated behavior. Large environments are represented as individual membranes connected to each other, forming a so-called *multienvironment* and each bacterium is defined by an individual compartment. The entire colony of bacteria is distributed across the multienvironment system. The processes of synthesis, diffusion, and sensing of molecules are described using rewriting rules on multisets of objects. The simulations performed show that only in the case of high density populations the bacteria are able to communicate and execute coordinated behavior.

A much more complex example deals with the Ras/cAMP/PKA pathway in Yeast *S. cerevisiae*. This pathway is involved in the control of the yeast cell metabolism, stress resistance, and proliferation, in relation to the quantity of available nutrients and is modeled with a τ-DPP approach [7]. The model consists of 30 molecular species interacting through 34 chemical reactions and is structured in four main modules, representing the four main stages of the pathway: the switch cycle of the Ras protein, the synthesis of the signaling molecule cAMP, the activation of the PKA, and the feedback mechanisms. The results prove that the model is able to

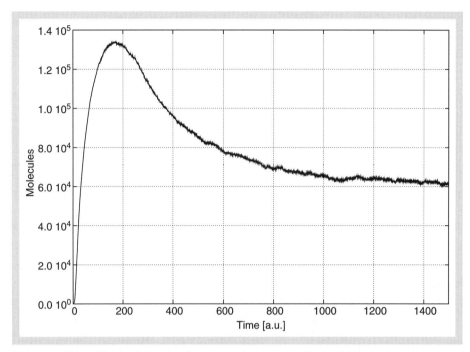

Fig. 18.5 The effect of the feedback mechanism on the cAMP synthesis.

properly simulate the dynamics of the four modules described above that confirm the experimental data about the signalling pathway and provide relevant information about the key regulatory elements of the signalling pathway. The effect of the feedback mechanisms on the cAMP accumulation is shown in Fig. 18.5 [7]. The modeling approach presented is important not only for the study of the effect and the role of the stochastic constants associated to the chemical reactions, but through its modularity it allows to investigate the outcomes produced by some modules and their interactions.

A repressilator system, an oscillator implemented in *Escherichia coli* cells through specific plasmids, coupled with a quorum sensing mechanism, has been developed and simulated by a multi-volume τ-DPP model [6]. The model consists of $n + 1$ volumes, where n of them correspond to the cells placed inside the same environment. In each cell a regulatory network consisting of positive and negative regulations which enhance or inhibit the binding between the RNA polymerase and the promoter site of a gene is present. This mechanism is implemented by using 34 evolution rules and one inside the environment, which deals with the communication of the quorum sensing signal. The detailed description of the model and the results obtained from the simulation can be found in [6]. The quorum sensing signaling mechanism has been studied for a specific case of two cells, one being unable to produce the signaling molecule, whereas the other is fully functional. In this case the communication mechanism is illustrated by showing how the signaling molecules produced by the last cell diffuse into the environment and from there into the former cell.

18.5 Final Remarks

Probabilistic and stochastic P systems contribute consistently to building a mature research area with successful modeling paradigms, strong links with other computational methods, like formal verification and analysis, and a coherent portfolio of case studies.

Apart from the research mentioned in the previous sections, there are other developments regarding different probabilistic approaches [1], theoretical investigations [2], [19] or a new line of research that looks very promising, regarding parameter estimation, structure optimization [25], modularity in system design [31]. Additionally, new research is being initiated that attempts at modeling the microstructure of P systems membranes through dissipative particle dynamics simulations [32], [18]. This work seeks to build a multiscale modeling methodology that captures the benefits that P systems provide as shown in this chapter while extending them in new directions.

Acknowledgements

FRC and NK acknowledge EPSRC grant EP/E017215/1 and BBSRC grant BB/F01855X/1.

References

[1] I. Ardelean, M. Cavaliere: Playing with a probabilistic P simulator: mathematical and biological problems. *Proc. Brainstorming Week on Membrane Computing, 2003* (M. Cavaliere et al., eds.), Tarragona, Spain, 37–45.

[2] F. Bernardini, M. Gheorghe, N. Krasnogor: Quorum sensing P systems. *Theoretical Computer Sci.*, 371 (2007), 20–33.

[3] F. BERNARDINI, M. GHEORGHE, F.J. ROMERO-CAMPERO, N. WALKINSHAW: A hybrid approach to modelling biological systems. *Lecture Notes in Computer Sci.*, 4860 (2007), 138–159.

[4] D. BESOZZI, P. CAZZANIGA, D. PESCINI, G. MAURI: Seasonal variance in P system models for metapopulations. *Progress in Natural Science*, 17 (2007), 392–400.

[5] D. BESOZZI, P. CAZZANIGA, D. PESCINI, G. MAURI: Modelling metapopulations with stochastic membrane systems. *BioSystems*, 91 (2008), 499–514.

[6] D. BESOZZI, P. CAZZANIGA, D. PESCINI, G. MAURI: A multivolume approach to stochastic modelling with membrane systems. *Algorithmic Bioprocesses* (A. Condon et al., eds.), Springer, 2009.

[7] D. BESOZZI, P. CAZZANIGA, D. PESCINI, G. MAURI, S. COLOMBO, E. MARTEGANI: Modeling and stochastic simulation of the Ras/cAMP/PKA pathway in the yeast Saccharomyces cerevisiae evidences a key regulatory function for intracellular guanine nucleotides pools. *Journal of Biotechnology*, 133 (2008), 377–385.

[8] Y. CAO, D.T. GILLESPIE, L.R. PETZOLD: Efficient step size selection for the tau-leaping simulation method. *Journal of Chemical Physics*, 124 (2006), 44–109.

[9] P. CAZZANIGA, D. PESCINI, D. BESOZZI, G. MAURI: Tau leaping stochastic simulation method in P systems. *Lecture Notes in Computer Sci.*, 4361 (2006), 298–313.

[10] G. CIOBANU, M.J. PÉREZ-JIMÉNEZ, GH. PĂUN, eds.: *Applications of Membrane Computing*, Springer, 2006.

[11] M. ERNST, J. COCKRELL, W. GRISWOLD, D. NOTKIN: Dynamically discovering likely program invariants to support program evolution. *IEEE Transactions on Software Engineering*, 27 (2001), 99–123.

[12] J. FISHER, T. HENZINGER: Executable cell biology. *Nature Biotechnology*, 25 (2007), 1239–1249.

[13] D.T. GILLESPIE: Stochastic simulation of chemical kinetics. *Annual Review of Physical Chemistry*, 58 (2007), 35–55.

[14] D.T. GILLESPIE, L.R. PETZOLD: Approximate accelerated stochastic simulation of chemically reacting systems. *Journal of Chemical Physics*, 115 (2001), 1716–1733.

[15] M. KAERN, T.C. ELSTON, W.J. BLAKE, J.J. COLLINS: Stochasticity in gene expression: from theories to phenotypes. *Nature Reviews Genetics*, 6 (2005), 451–464.

[16] N. KRASNOGOR, M. GHEORGHE, G. TERRAZAS, S. DIGGLE, P. WILLIAMS, M. CAMARA: An appealing computational mechanism drawn from bacterial quorum sensing. *Bulletin of EATCS*, 85 (February 2005), 135–148.

[17] M. KWIATKOWSKA, G. NORMAN, D. PARKER: Probabilistic model checking in practice: Case studies with PRISM. *ACM SIGMETRICS Performance Evaluation Review*, 32 (2005), 16–21.

[18] L. LI, P. SIEPMANN, J. SMALDON, G. TERRAZAS, N. KRASNOGOR: Automated self-assembling programming. *Systems Self-assembly: Multidisciplinary Snapshots* (N. Krasnogor et al., eds.), Elsevier, Studies in multidisciplinarity, 2008, 281–308.

[19] M. MUSKULUS, D. BESOZZI, R. BRIJDER, P. CAZZANIGA, S. HOUWELING, D. PESCINI, G. ROZENBERG: Cycles and communicating classes in membrane systems and molecular dynamics. *Theoretical Computer Sci.*, 372 (2007), 242–266.

[20] A. PĂUN, M.J. PÉREZ-JIMÉNEZ, F.J. ROMERO-CAMPERO: Modelling signal transduction using P systems. *Lecture Notes in Computer Sci.*, 4361 (2006), 100–122.

[21] GH. PĂUN: *Membrane Computing. An Introduction.* Springer, 2002.

[22] D. Pescini, D. Besozzi, G. Mauri, C. Zandron: Dynamical probabilistic P systems. *Intern. J. Found. Computer Sci.*, 17 (2006), 183–204.

[23] D. Pescini, D. Besozzi, G. Mauri, C. Zandron: Analysis and simulation of dynamics in probabilistic P systems. *Lecture Notes in Computer Sci.*, 3892 (2006), 236–247.

[24] D. Pescini, D. Besozzi, G. Mauri: Investigating local evolutions in dynamical probabilistic P systems. *Proc. Seventh Intern. Symp. on Symbolic and Numeric Algorithms for Scientific Computing, SYNASC'05, Timişoara, 2005, IEEE Computer Press*, 2005, 440–447.

[25] F.J. Romero-Campero, H. Cao, M. Cámara, N. Krasnogor: Structure and parameter estimation for cell systems biology models. *The Genetic and Evolutionary Computation Conference, GECCO'08, Atlanta*, 2008, 331–338.

[26] F.J. Romero-Campero, M. Gheorghe, G. Ciobanu, L. Bianco, D. Pescini, M.J. Pérez-Jiménez, R. Ceterchi: Towards probabilistic model checking on P systems using PRISM. *Lecture Notes in Computer Sci.*, 4361 (2006), 477–495.

[27] F.J. Romero-Campero, M. Gheorghe, G. Ciobanu, J.M. Auld, M.J. Pérez-Jiménez: Cellular modelling using P systems and process algebra. *Progress in Natural Science*, 17 (2007), 375–383.

[28] F.J. Romero-Campero, M.J. Pérez-Jiménez: P systems, a new computational modelling tool for systems biology. *Lecture Notes in Bio-Informatics*, 4220 (2006), 176–197.

[29] F.J. Romero-Campero, M.J. Pérez-Jiménez: Modelling gene expression control using P systems: the Lac Operon, a case study. *BioSystems*, 91 (2008), 438–457.

[30] F.J. Romero-Campero, M.J. Pérez-Jiménez: A model of the quorum sensing system in Vibrio fischeri using P systems. *Artificial Life*, 14 (2008), 1–15.

[31] F.J. Romero-Campero, J. Twycross, M. Bennett, M. Cámara, N. Krasnogor: Modular assembly of cell systems biology models using P systems. *Intern. J. Found. Computer Sci.*, 20 (2009), 427–442.

[32] J. Smaldon, J. Blakes, N. Krasnogor, D. Lancet: A multi-scaled approach to artificial life simulation with P systems and dissipative particle dynamics. *The Genetic and Evolutionary Computation Conference, GECCO'08, Atlanta*, 2008, 249–256.

CHAPTER 19

FUNDAMENTALS OF METABOLIC P SYSTEMS

VINCENZO MANCA

19.1 INTRODUCTION

P systems were introduced in [35] as a computation model inspired from biology [36]. Applications for modeling biological phenomena, in a discrete mathematical setting, were developed in [2, 4, 10, 11, 18, 20, 37, 38, 40, 41, 42] etc. However, these models are mainly of a qualitative nature, and do not provide criteria for predicting quantitative aspects of biological processes. The main framework analysis for the most part of biological dynamics remains the theory of ordinary differential equations (ODE). However, ODE present some intrinsic limitations in the evaluation of the kinetic reaction rates, which usually refer to a microscopic level, hardly accessible to reliable measurements. In fact, in living organisms these measurements dramatically alter the context of the investigated processes. For overcoming these limitations, Metabolic P systems, MP systems for short, were introduced in [31] and then widely developed, along different directions, in [6, 7, 8, 14, 15, 16, 17, 25, 26, 27, 28, 29, 30]. MP systems intend to model *metabolic systems*, that is, structures where matter of different types is put in a reactor where it is subject to reactions, or transformations of various types (introduction and/or expulsion of matter are also possible).

MP systems introduce a new possibility in modeling real complex phenomena, which is related to the *log-gain theory*, intrinsically based on to the structure of MP systems. In these systems, dynamics is computed by suitable recurrent equations, based on *flux regulation maps*, and the log-gain theory provides a method for deducing the adequate flux maps of an MP model, by means of suitable algebraic manipulations of data coming from macroscopic observation of the system to be modeled. Moreover, a strong connection can be stated between MP systems and ODE [17], that is, from a differential model, an *equivalent* MP system can be deduced, and conversely, any MP system can be transformed into an equivalent differential model too.

According to the *mass action law* of Ordinary Differential Equations (ODE), in any metabolic system, the infinitesimal substance variation produced by a reaction depends on the instantaneous quantities of substances involved in the reaction. Any differential equation of an ODE system expresses dynamical mechanisms at an infinitesimal scale, considered independently from the other equations. The systemic effect results from the combination of the instantaneous effects of all equations. In this perspective, kinetic rate constants of single equations have to be evaluated according to a deep understanding of single molecular events.

Metabolic P systems adopt an opposite viewpoint, where no single instantaneous kinetic is addressed, but rather, the variation of the whole system under investigation is considered, at discrete time instants separated by a specified macroscopic interval. The evolution law of the system consists in the knowledge of the contribution of each reaction in the passage between any two consecutive instants. Therefore, dynamics is given along a sequence of steps, and at each step, it is ruled by a partition of matter among the reactions transforming it. The log-gain theory of MP systems is aimed at reconstructing a kind of "grammar" underlying a phenomenon, by providing the *flux regulation maps* of reactions. Such maps (one for each reaction), in dependence on the state of the system (or only on some substance quantities), specify the amount of matter they consume/produce for each occurrence of their reactants/products.

In this chapter, an outline will be given of fundamental concepts about MP systems. The presentation is mainly focused on the motivations and on the research frontiers of the subject. Many results will be presented by avoiding complete formal explanations, which can be found in the more comprehensive recent papers [17, 28, 29, 30]. However, in many ways the theory of MP systems is based on principles, originating in chemical intuitions, which are mathematically elaborated, but which have their ultimate fundamentals in the experimental adequacy of the theory.

19.2 THE MASS PARTITION PRINCIPLE

In metabolic systems, matter (molecules) is inside a (biological) reactor partitioned in a certain number of (biochemical) substances subject to some reactions.

If we consider the system, along a discrete number of steps, at some specific time interval, we realize that the reactions transform the substance types of matter, but also that matter (of specified types) is introduced from the external environment, or expelled outside. Therefore, abstractly, reactions are agents performing matter transformations. Avogadro's principle, which is fundamental in chemistry, rules the behavior of any reaction r, at any step. In fact r "moves" (consumes/produces) multiples of the same number of objects (molecules), which we call reaction unit or flux of r, according to the specific "stoichiometry" of r establishing its "reactants" (consumed substances) and its "products" (produced substances). This unit can be expressed with respect to a fixed population unit. In this case, a reaction can be identified by a vector of integers, having a dimension equal to the number of substances. For example, if the population unit is of 1000 objects, and there are four kinds of substances, say a, b, c, d, then a reaction can be represented by the vector $(0, 1, 0, -2)$, meaning that it consumes 2000 objects of type d and replaces them by 1000 objects of type b. This is what is usually written $dd \to b$. We call this kind of transformation mechanism *molar multiset rewriting*. It differs from the usual multiset rewriting of P systems. In fact, in the classical case, a rule $ddc \to b$ replaces two d and one c into only one b (the order does not matter), and the application of such a rule is *individual*, because, when it is applied again, other two b and one c become one d. On the contrary, in a molar multiset rewriting perspective, an occurrence of d, c, and b in a rule means a population of d, c, and b respectively, and the size of this population is just the reaction unit of the reaction (a value depending on the state of the system). The place of substances, with respect to the arrow $\to$ of the rule, means increment or decrement (production or consumption). This perspective implies that the time of the system is not the microscopic time of reaction kinetics, but the macroscopic time of the observer, who is most likely able to know the substance variations between two consecutive observation instants. The additivity of the effects of all reactions corresponds to another chemical principle, referred to as Dalton's principle. According to it, if we compute the reaction unit of each reaction, then by adding the effects of all reactions we can compute the next state of the system. This strategy, we call *matter partition principle*, is the essence of MP dynamics. In simple words, the dynamics computation of an MP is obtainable in the following way: *i) compute the reaction units of all reactions; ii) apply all the reactions according to their reaction units; iii) collect their products, by removing the matter they consume.* Another principle, which is implicitly assumed, establishes the matter conservation. It is based on the molecular weights of substances, referred to Lavoisier. According to it, the reactant weights, multiplied by the moles consumed by any reaction, has to be equal to the products' weight, multiplied by the number of produced moles.

In the following, Greek letters $\alpha, \beta, \ldots$ (with possible subscripts) denote finite multisets (strings where symbol order is not pertinent) over an alphabet X of substances. A reaction r is also indicated by $\alpha_r \to \beta_r$, where α_r, β_r are the reactants and products of r. For a multiset α, we denote by $|\alpha|_x$ the multiplicity of x in α

and by $|a|$ the sum $\sum_{x \in a} |a|_x$, where $x \in a$ means that the multiplicity of x in a is different from zero.

Definition 19.1 *The* **stoichiometric matrix** $\mathbb{A}$ *of a set R of reactions over a set X of substances is $\mathbb{A} = (\mathbb{A}_{x,r} \mid x \in X, r \in R)$ where $\mathbb{A}_{x,r} = |\beta_r|_x - |\alpha_r|_x$. The set of reactions having the substance x as a reactant is $R_a(x) = \{r \in R \mid |\alpha_r|_x > 0\}$ and the set of rules consuming or producing x is $R(x) = \{r \in R \mid \mathbb{A}_{x,r} \neq 0\}$. Two reactions r_1, r_2* **compete** *for some substance $x \in X$ if $r_1, r_2 \in R_a(x)$.*

19.3 Metabolic P Systems

A discrete dynamical system is specified by a set of states and by a discrete dynamics on them, that is, by a function from the set $\mathbb{N}$ of natural numbers to the states of the system [23]. In this context, the natural numbers which are the argument of dynamics are called *instants* or *steps*. This general notion of dynamical system is the common basis of both the two main kinds of MP systems we will define. The following definition introduces a class of MP systems, which are called MP systems with flux maps, or MPF systems (F may be omitted) for short.

Definition 19.2 (MPF System) *An MP system with flux regulation maps, shortly an MPF system, is a discrete dynamical system specified by a construct*

$$M = (X, R, V, Q, \Phi, \nu, \mu, \tau, q_0, \delta)$$

where X, R, V are finite disjoint sets, and the following conditions hold, with $n, m, k \in \mathbb{N}$:

- $X = \{x_1, x_2, \ldots, x_n\}$ *is the set of* **substances** *(the types of molecules);*
- $R = \{r_1, r_2, \ldots, r_m\}$ *is the set of* **reactions** *over X, that is, pairs (in arrow notation) of type $\alpha \to \beta$ with α, β strings over the alphabet X (sometimes concatenation is denoted by $+$ for stressing the commutativity implicit in the string notation of multisets);*
- $V = \{v_1, v_2, \ldots, v_k\}$ *is the set of* **parameters** *(such as pressure, temperature, volume, pH, ...) equipped by a set $\{h_v : \mathbb{N} \to \mathbb{R} \mid v \in V\}$ of* **parameter evolution functions**, *the elements of $X \cup V$ are called* **magnitudes**;
- Q *is the set of* **states**, *that is, of the functions $q : X \cup V \to \mathbb{R}$, from magnitudes to real numbers.*
- $\Phi = \{\varphi_r \mid r \in R\}$ *is a set of* **flux (regulation) maps**, *where the function $\varphi_r : Q \to \mathbb{R}$ states the amount (moles) which is consumed/produced, in the state q, for every*

occurrence of a reactant/product of r. We set by $U(q) = (\varphi_r(q)|r \in R)$ *the* **flux vector** *at state q;*

- ν *is a natural number which specifies the number of molecules of a (conventional) mole of M, as its* **population unit**;
- μ *is a function which assigns, to each* $x \in X$, *the* **mass** $\mu(x)$ *of a mole of x (with respect to some measure unit);*
- τ *is the* **temporal interval** *between two consecutive observation steps;*
- $q_0 \in Q$ *is the* **initial state** $(q_0(v) = h_v(0)$ *for all* $v \in V$);
- $\delta : \mathbb{N} \to Q$ *is the dynamics of the system. At the initial instant it provides the initial state, that is,* $\delta(0) = q_0$. *Moreover, for any parameter* $v \in V$ *and for any* $i \geq 0$, $(\delta(i))(v) = h_v(i)$, *while the function* δ *provides the evolution of substance quantities by means of stoichiometric matrix* $\mathbb{A}$ *and vectors U. In fact, let us set, for any magnitude* $w \in X \cup V$ *and* $i \geq 0$

$$(\delta(i))(w) = w[i]$$

and

$$X[i] = (x[i] \mid x \in X)$$

then the dynamics of substances is given by the following recurrent vector equation also called **Equational Metabolic Algorithm (EMA)**, *where* $\times$ *is the usual matrix multiplication, and the sum between vectors is the usual component-wise sum (analogously vector difference and division will be intended):*

$$X[i+1] = \mathbb{A} \times U(\delta(i)) + X[i] \tag{19.1}$$

where $\mathbb{A}$ *is the stoichiometric matrix of R over X.*

If $EMA[i]$ *is the system at step i, given the vectors of* $U(\delta(i))$ *and* $X[i]$, *we can obtain the vector* $X[i+1]$ *by evaluating the right member of the equation above.*

All the components of an MP system, apart the set Q of states (deducible from the other components), and the dynamics, constitute an **MP graph**, easily representable in graphical form, as we will show in the next section. When in a MP graph the elements τ, ν, μ are omitted, then we call it an **MP grammar**, that is, a multiset rewriting grammar where rules are regulated by functions (usually rational algebraic expressions). Such a grammar is completely specified by: i) reactions, ii) flux maps (substances are the elements occurring in the reactions, and parameters are the arguments of flux maps different from substances), iii) parameter evolution maps, and iv) initial values of magnitudes. Parameter evolution maps or initial values may be omitted when only the MP grammar structure is specified. A kind of equivalence between MPF systems and hybrid Petri nets was proved in [14], where it is shown that MP formulation provides logical and computational advantages. In Table 19.1 an MP grammar is given with three substances. Reactions with no

Table 19.1 An MP grammar without parameters (λ is the empty multiset, + denotes the multiset sum on the left, while the arithmetic sum on the right).

$r_1 : \lambda \to a$	$a = 50$	$2 * a/(0.2 * c + 0.4 * b + 100)$
$r_2 : a + a \to b$	$b = 50$	$0.2 * c * a/(0.5 * c + b + 25)$
$r_3 : b \to \lambda$	$c = 50$	$b/110$
$r_4 : a \to c$		$b * a/(0.2 * c + b)$
$r_5 : c \to \lambda$		$0.5 * c/100$

reactants are *input rules* and have symbol λ before the arrow, while rules without product are *output rules* and have symbol λ after the arrow. Numerical values in the middle are the initial values of substances. Expressions on the right side are the flux maps (no parameter is present).

19.4 THE METAPLAB PROJECT

A graphical representation of MP systems has been introduced in [30], where a self-explaining structure of graphs is used for representing all the components of an MP systems: ovals are substances, squares are parameters, circles are reactions, rounded corner squares are flux regulating maps, triangles input and output channels with the external environments and continuous arcs indicate the reactant/product roles in reactions, while discontinuous arcs denote the functional dependencies in flux regulation maps. Fig. 19.1 is an example of MP graph where the different kinds of nodes are put in evidence (in [30] parameter nodes were not explicitly considered). Substance and parameter nodes model specific magnitudes of a system: substances are kinds of matter (with specific weights), while parameters are physical variables which regulate the dynamics, but do not are transformed by reactions. Flux nodes are labeled by functions (usually algebraic expressions) having as arguments magnitudes (substances and parameters) which regulate reactions. This means that parameters are not involved in matter transformations, but can play an important role in the regulation of reaction fluxes (the flux of a reaction is the amount of matter it transforms in a given step).

This graphical representation of [30] evolved toward a software system extending a previous prototypal program of MP systems dynamics simulation, called Psim, developed by Luca Bianco in 2005–2006. With the support of the Computer Science Department and CBMC (Computational BioMedicine Center) at the

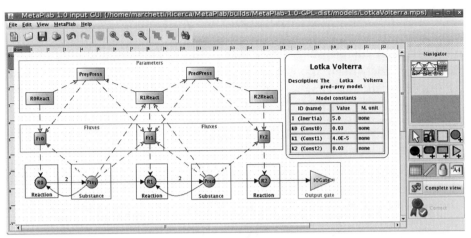

Fig. 19.1 An MP graph visualized by a graphical user interface of MetaPlab. Frame labels point out MP system elements in the MP graph representation. Substances, reactions, and parameters describe the system stoichiometry, while fluxes express the system regulation.

University of Verona, a group of people (Federica Agosta, Alberto Castellini, Giuditta Franco, Luca Marchetti, Roberto Pagliarini, and Michele Petterlini) led by the author, are working on this project, aiming at developing a plugin-based architecture called *MetaPlab*, available at the site mplab.sci.univr.it (an extended version of Psim, called MPsim, is also available at www.cbmc.it). This software intends to assist biologists to understand the internal mechanisms of biological systems and to reproduce and analyze, in silico, biological phenomena: responses to external stimuli, environmental condition alterations, and structural changes. The Java implementation of MetaPlab ensures the cross-platform portability of the software, which will be released under the GPL open-source license. MetaPlab is similar to other well known resources, already available on-line, such as: *COPASI* [22], based on on ODE (for the simulation of biochemical networks), *Cell Illustrator^{TM}* [33], based on Petri nets (for the representation of biological pathways and the simulation of their evolution [13]), or PRISM, based on probabilistic P systems [39] (for model checking analysis). A specific aspect of MetaPlab is the construction of MP models, starting from time series of observation, by implementing the log gain theory for MP systems which will be outlined in the following sections. It is based on an extensible set of plugins, namely Java tools for solving specific tasks relevant in the framework of MP systems, such as parameter estimation for regulative mechanisms of biological networks, simulation, visualization, graphical and statistical curve analysis, importation of biological networks from on-line databases, and possibly other aspects which would be relevant for further investigations.

MetaPlab software is organized to four levels: i) **MP graphs**, ii) Java data structures, called **MP store**, iii) data processing realized by specific **Plug-ins**, organized in an open and cumulative structure (*via* a *Plugin Manager*), and iv) a level of presentation and visualization tools, called **MP vistas**, which provide different analysis of specific modeling aspects, such as, substances and parameters curves, phase diagrams, or statistical indexes. In Fig. 19.2 the schema of this structure is illustrated.

19.5 REACTION MAPS AND DIFFERENTIAL MODELS

MP systems with reaction maps, MPR systems for short, are another class of MP systems, where flux units are not given by flux maps, but are calculated by means

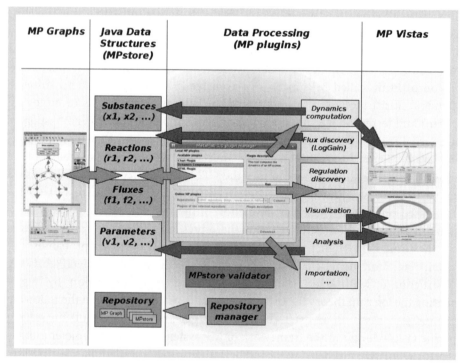

Fig. 19.2 The MetaPlab plugin structure. The user can choose a plug-in from a list, and could download new plugins (lower side) from forthcoming on-line repositories (Figure from: [12]).

of maps $f_r : Q \to \mathbf{R}$, where $r \in R$, called **reaction maps**, and by means of **inertia functions** $\psi_x : Q \to \mathbf{R}$, where $x \in X$. The reaction map f_r provides a value $f_r(q)$, called the **reactivity** of the rule r in the state q, while $\psi_x(q)$ determines the (molar) quantity of substance x that is not consumed in the state q. In MPR systems regulation is obtained in a less direct way, which however turns out to be useful in the comparison between MP models and ODE models. Reaction maps were introduced in [26]. In MPR systems, at any step, the amount of each substance x is distributed among all the rules competing for it, proportionally to the reactivities of reactions in that step, but including, for any substance x, also the reaction $x \to x$ having as reactivity the inertia of x.

Definition 19.3 (MPR System) *An MP system with reaction maps, an MPR system for short, is a discrete dynamical system specified by a construct*

$$M = (X, R, V, Q, F, \psi, \nu, \mu, \tau, q_0, \delta),$$

where, for any $x \in X$, $\psi_x : Q \to \mathbf{R}$ specifies the inertia of substance x in a given state, and all the components different from F, δ and ψ are as in Definition 19.1. The set $F = \{f_r : Q \to \mathbf{R} \mid r \in R\}$ is constituted by functions, called reaction maps, and dynamics δ is the same of the MPF system $M = (X, R, V, Q, \Phi, \nu, \mu, \tau, q_0, \delta)$, where the flux functions are given by means of the following equations:

$$w_{r,x}(q) = \frac{f_r(q)}{\psi_x(q) + \sum_{r' \in R_a(x)} f_{r'}(q)} \tag{19.2}$$

$$\varphi_r(q) = \begin{cases} f_r(q) & \text{if } \alpha_r = \lambda; \\ \min\{\frac{w_{r,y}(q) \cdot q(y)}{|\alpha_r|_x} \mid y \in \alpha_r\} & \text{otherwise.} \end{cases} \tag{19.3}$$

Equation (19.2) provides the partition of x which r is allowed to consume, while Equation (19.3) computes the reaction unit of r, as the minimum of substance quantities available to all reactants of r.

In Fig. 19.3 the MP graph of *Veronator* is depicted, which is an MPR version of the famous Brusselator, based on the Belousov-Zabotinsky reaction exhibiting the same dynamics of the Brusselator.

Definition 19.4 *Two MP systems are substantially equivalent when they have the same sets X (substances), the same ν, μ, $q_{0|X}$, τ (of Definition 19.1), and the same dynamics, restricted to substances.*

From Definition 19.3 it follows that for any MPR an equivalent MPF exists. Also the converse relation holds, that is, for any MPF an equivalent MPR exists [29].

The relationship between ODE and MPR systems was investigated in [17], where the following definition is introduced of ODE-Transform of an MPR system.

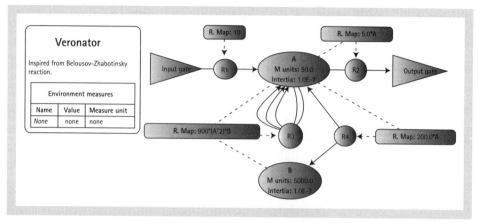

Fig. 19.3 The MP system Veronator, represented by an MP graph.

Definition 19.5 (MPR-ODE Transform) *Let $M = (X, R, \ldots F, \ldots)$ be an MPR systems (F the set of reaction maps). For every $x \in X$, the following is the ODE-transformed of M, where $\Pi(a_r)$ is the product of all the quantities of reactants of reaction r:*

$$x' = \sum_{r \in R} \mathbb{A}_{x,r} f_r(q) \Pi(a_r)$$

Under suitable hypotheses on the structure of an MPR system and of its ODE-Tranform, the dynamics of an MPR system M, computed by its metabolic algorithm, and the dynamics of its ODE-Transform $ODE(M)$, computed by a method of numerical integration, are equivalent, within some given range of approximation [17]. Moreover, according to the equivalence mentioned between MPR and MPF systems, the equivalence between ODE and MPR systems can be easily extended to MPF systems.

19.6 MP Models

MP systems were proven to effectively model the dynamics of several biological processes. Table 19.2 refers to some MP models of interesting chemical and biological dynamics, well known in literature, based on ODE representations. The first MP models were derived from ODE systems and their dynamics was generated by means of a tool of dynamics computation available in Psim. The dynamics of these were MP systems coincident with the dynamics of the corresponding differential models. Table 19.2 summarizes some MP models. The last two were obtained by a systematic application of the procedure outlined in the previous section, for transforming an ODE model in an MP model. In Table 19.3 the MP grammar of

Table 19.2 Some MP Models.

MP Models	References
Belousov-Zhabotinsky Reaction (Brusselator)	[6]
Lotka-Volterra Dynamics	[6]
Susceptible-Infected-Recovered Epidemic	[6]
Protein Kinase C Activation	[8]
Drosophila Circadian Rhythms	[15]
Early Amphibian Mitotic Cycle	[30, 28]
Elementary Metabolic Oscillators	[28]

Table 19.3 Prey-Predator Dynamics.

Reactions	Reaction Maps
Pred-Inertia: $Pd \to Pd$	5.0
Prey-Inertia: $Py \to Py$	5.0
R1: $Py \to Py\,Py$	$3 * 10^{-2}$
R2: $Py\,Pd \to Pd\,Pd$	$(4 * 10^{-5}) * if(Pd < Py, Py, Pd)$
R3: $Pd \to \lambda$	$3 * 10^{-2}$
Initial Pred = 900 moles	Initial Prey = 900 moles

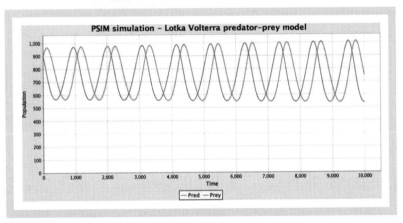

Fig. 19.4 Prey-Predator dynamics generated by the MP system of Table 19.3.

Lotka and Volterra's Predator-Prey dynamics is given, and in Fig. 19.4 the dynamics of this MP system are shown, which coincides with the classical solution via differential equations. The behavior of the mitotic cycle in amphibian embryos [21] was completely reconstructed [28] in terms of MP systems, as illustrated in Fig. 19.6. Its MP dynamics, reported in Fig. 19.7, perfectly coincide with the curve reported in [21].

486 FUNDAMENTALS OF METABOLIC P SYSTEMS

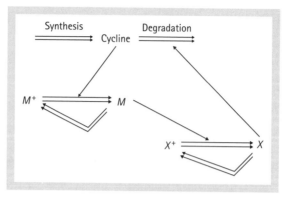

Fig. 19.5 Schema of the mitotic cycle in early amphibian embryos. Double arrows denote transformations, simple arrows denote enhancing influences on the pointed transformations.

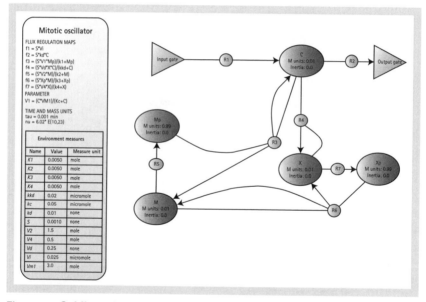

Fig. 19.6 Goldbeter's mitotic cycle represented by an MP graph.

19.7 Log-gain Principles

In order to discover the reaction fluxes at each step, we introduce some *log-gain principles*. The relative variation of a substance x is defined as the ratio $\Delta(x)/x$. In differential notation, this ratio is related to $\frac{dx}{dt}/x$ (t the time variable), and from

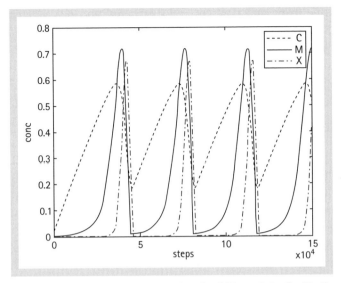

Fig. 19.7 The curves provided by the MP model of mitotic cycle.

elementary calculus we know that it is the same as $\frac{d(\lg x)}{dt}$. This equation explains the term "log-gain" for expressing relative variations [43]. We use a discrete notion of log-gain for stating our principles. In the following, given dynamics δ of an MP system, we use a simplified notations, for $i \in \mathbb{N}, r \in R$, and $w \in X \cup V$ (notation of Definition 19.2):

$$u_r[i] = \varphi_r(\delta(i))$$

$$U[i] = (u_r[i] \mid r \in R)$$

With this notation, Equation 19.1 giving the substance evolutions in MPF systems becomes the following equation system, which we call $ADA[i]$ (**Avogadro and Dalton Aggregation**):

$$X[i+1] = \mathbb{A} \times U[i] + X[i] \tag{19.4}$$

assuming knowing $X[i]$ and $X[i+1]$, it expresses a system of n equations (substances, which are components of $X[i]$) and m variables (fluxes, which are components of $U[i]$). Formally $ADA[i]$ is the same system $EMA[i]$ given in Definition 19.2. However, the two systems have different, or better dual, interpretations. In $EMA[i]$, vector $U[i] + X[i]$ is known, and vector $X[i+1]$ is computed by means of it, while in $ADA[i]$, the vector $X[i+1] - X[i]$ is known and $U[i]$ is computed by solving it.

In the passage from step i to step $i+1$ each reaction has a determined flux, this mean that the system has to be univocally solvable, therefore if $m \leq n$ some redundant equations can be eliminated, by getting a univocally solvable square system which provides as solutions the fluxes between the two states. If $m > n$, as happens very often, then fluxes cannot be univocally deduced by the knowledge of substance values in two consecutive steps. The log-gain principle will allow us to add more equations to the above system in order to get a univocally solvable system which could provide the flux vector.

The log-gain principle extends a very important rule, well known in theoretical biology as the **allometric principle**. According to it, a specific ratio holds between the relative variations of two related biological parameters (e.g. mass of an organism and its superficial area). This principle seems a general property of living organisms which allows them to keep the basic equilibria underlying their internal organization. As is reported in [3], many empirical laws on metabolism are instances of *allometry* and the abundance of power laws in biological systems is also related to this principle. Therefore, it is not surprising that the log-gain mechanism is the basis for identifying the metabolic algorithm of MP systems which fits the observations of given metabolic processes. By using our notation, this principle can be stated in the following way.

Principle 19.1 (Log Gain) *For $i \in \mathbb{N}, r \in R$, and $w \in X \cup V$, let us call*

$$Lg(u_r[i]) = (u_r[i+1] - u_r[i])/u_r[i]$$

the log-gain of the flux unit u_r at the step i, and analogously,

$$Lg(w[i]) = (w[i+1] - w[i])/w[i]$$

the log-gain of the magnitude w at step i. There exists a subset $T_r \subseteq X \cup V$ of elements called (log-gain) tuners of r such that: $Lg(u_r[i])$ is a linear combination, in a unique way, of the tuners of r:

$$Lg(u_r[i]) = \sum_{w \in T_r} p_{r,w} Lg(w[i]) \tag{19.5}$$

In the previous formulation of log-gain principle, some unknown coefficients $p_{r,w}$ with $w \in X \cup V, r \in R$ occur. Therefore, even if we get m new equations (m is the number of reactions), at the same time we introduce new unknown values. This suggests to us the following new formulation of the principle, based on the notion of *log-gain offset values*. Log-gain offsets are seen as the errors introduced in passing from the original formulation of log-gain principle to a formulation where the coefficients of linear log-gain combination are put equal to one.

Principle 19.2 (Offset Log Gain) *For each rule $r \in R$, there exists a subset T_r of magnitudes ($T_r \subseteq X \cup V$), called (log-gain) tuners of r, and a value p_r, called offset log-gain, such that:*

$$Lg(u_r[i]) = \sum_{w \in T_r} Lg(w[i]) + p_r[i] \qquad (19.6)$$

Even if now the number of unknown values was reduced, there is no gain because we have m equations and m variables (the offset values). However, in the system of equations (19.6), we call $OLG[i]$ (**Offset Log-Gain** at step i), some offset values can be avoided, by keeping only n offset values, that is, the same number as the number of substances. This reduced log-gain system, added to (19.4), the system $ADA[i+1]$ (i is replaced by $i+1$, as it will be apparent from the formulation of Principle 19.3), will provide a univocally solvable square system of linear equations [29]. The possibility for such a log-gain offset reduction is due to the fact that all the reactions competing for the same substance are constrained to algebraically equate the variation of this substance. This means that only one offset can be chosen for any set of competing reactions, that is, one offset for each substance. The following property, introduced in [29], formalizes this concept.

Definition 19.6 (Offset Log Gain Covering Property) *Given an MP system M of substances $X = \{x_1, \ldots, x_n\}$, then a set R_0 of n reactions has the **Covering Offset Log Gain Property** if for any $x \in X$ there is an $r \in R_0$ such that $r \in R(x)$.*

We call $COLG[i]$ system (**Covering Offset Log-Gain**) any $OLG[i]$ system having only n among the m offset log-gain values, which correspond to a subset of rules R_0 satisfying the Covering Offset Log Gain Property. It can be shown that a system $COLG[i]$ can be found for any observation step i [29]. We put $P[i] = (p_r[i] \mid r \in R, \ p_r[i] = 0 \text{ if } r \notin R_0)$. In this way, we can write down a system of equations for computing, at each step, the reaction fluxes of a system where states are deduced by observation.

Let

$$W[i] = (w[i] \mid w \in X \cup V)$$

and let $\mathbb{B}$ be a Boolean matrix or m rows and n columns,

$$\mathbb{B} = (\mathbb{B}_{r,w} \mid r \in R, w \in X \cup V)$$

With these positions we can formulate the following log-gain principle, more formally motivated in [29], which allows us to find the fluxes of an MP system at each step.

Principle 19.3 (Offset Log Gain Adaptation) *Let M be an MP system with a set of reactions $R = \{r_1, r_2, \ldots, r_m\}$ and a set of substances $X = \{x_1, x_2, \ldots, x_n\}$.*

The following system, called OLGA[i], is the combination of the system ADA[i + 1] with COLG[i]. When W[i], W[i+1], W[i+2], and U[i] are known values, the unknown values are vectors U[i + 1] and P[i + 1], that is, in OLGA[i] there are $m + n$ variables and the same number of equations:

$$\begin{aligned} X[i+2] - X[i+1] &= \mathbb{A} \times U[i+1] \\ (U[i+1] - U[i])/U[i] &= \mathbb{B} \times ((W[i+1] - W[i])/W[i]) + P[i+1] \end{aligned} \quad (19.7)$$

Starting from a value U[0], the solutions of OLGA[i], for $i \in \mathbb{N}, i > 0$, provide the time series of flux vectors U[i + 1], for $i > 0$.

Let us assume the knowledge of the initial flux vector U[0]. Actually, there are several possibilities for computing this vector, as is explained in [28]. When U[0] is determined, then by means of OLGA[0], the value U[1] is computed, and in general, the vector U[i + 1] can be obtained by means of OLGA[i], for $i = 1, \ldots, t$, by using the observed values $\delta(1), \delta(2) \ldots, \delta(t)$.

Now, let us suppose that reaction fluxes depend on magnitudes with some algebraic rational expression combining some basic rational monomials, then we can use standard interpolation tools for finding the functional dependence of the vector U[i] with respect to the magnitudes. The resulting functions approximate the regulation functions Φ we are searching for, and our task is completed, because the MP metabolic system is completely discovered from them (and the corresponding EMA is well defined).

19.8 MP Cosmogony

Due to the fast development of MP systems, their terminology has not been completely uniform up to now. We will try to make it more stable by means of a dramatization of the subject. We will shortly perform a play, which is also useful for giving a synthesis of the topics of the previous sections. The first notion of a MP system was based on *Metabolic P Algorithm* (MPA), a procedure calculating the state of a MP system along a sequence of steps. The systems based on this algorithm were more specifically called MPR systems, because MPA was defined by means of *Reaction maps*. Then, a wider notion of MP system was defined, called the MPF system, based on flux maps (MP by default). In this extension, the metabolic algorithm reaches a more simple and general formulation by means of a system of recurrent equations we call EMA (*Equational Metabolic Algorithm*). EMA, a goddess, is a sort of *alma mater* (Eva of MP cosmos). She is the first actress of our play. Her generation is the final aim of our cosmogony and corresponds to the solution of an *inverse dynamical problem*, that is, the determination of an MP system representing a natural dynamic which we observe along a temporal sequence of states. Another beautiful goddess, ADA, (*Avogadro–Dalton Action*) corresponds to a sort of mirror image of EMA (known values of EMA (flux units) are unknown in ADA and vice-versa unknown values in ADA (states) are the known values in EMA). ADA married the god COLG (*Covering Offset Log Gain*), son of OLG. From ADA and COLG another beautiful goddess was born, named OLGA (*Offset Log Gain Adaptation*), who marries the god UOZ (*Units Of Zero time*) and generates soldiers U_1, U_2, U_3 .... All of them became slaves of REGCOREMA (*REGression and CORrElation Methods of Approximation*), a queen who generates FLUXMA (the vector of *FLUX MAps*) who eventually generates EMA.

General methods are under investigation which could systematically search for basic rational monomials (depending on substance and parameter values) which also fit with the flux units determined by OLGA modules solutions. In our numerical experiments [34] flux functions were found with a good approximation to the observed dynamics. Moreover, in many cases, we found that, independently from the chosen value of $U[0]$, after a small number of steps, say 3 steps, our procedure will generate, with a great approximation, the same sequence of flux vectors. This means that the data collected in the observation steps are sufficient to determine the functions which, on the basis of substance quantities, regulate the dynamics of the system. This method was tested in all the systems reported in Table 19.2. Numerical elaborations were performed by standard MATLAB® operators (*backslash* operator for square matrix left division or in the least squares sense solution). Specific observation strategies were adopted by using hundreds of steps. Therefore, in the case of natural systems, from suitable observations, we could discover, with good approximation, the underlying regulation maps and consequently reliable computational models of their dynamics can be found.

19.9 Non Photochemical Quenching

A model entirely based on the log-gain theory of MP systems concerns the Non Photochemical Quenching (NPQ) phenomenon, which is a crucial mechanism in plant photosynthesis [1, 32]. The substances involved in NPQ are given in Table 19.4. The modeling details of NPQ can be found in [32], the MP grammar of the NP model is given in Table 19.5. The dynamics generated by the MP model of NPQ is in complete accordance with the observed behavior of the system. No previous differential or probabilistic model of this phenomenon was available, therefore it provides the first example of a model deduced by means of the log-gain theory from the experimental data collected by observing the state of the system along

Table 19.4 Magnitudes involved in the NPQ phenomenon.

Substances	Symbols
Light	l
Cumulative fluorescence	f^+
Cumulative heat	q^+
Light absorbance	r
Hydrogen ions	h
NADPH	p
Closed Photosystem	c
Open Photosystem	o
Violaxanthin	v
Zeaxanthin	z
Active violaxanthin de-epoxidase (VDE$^+$)	x
Inactive violaxanthin de-epoxidase (VDE$^-$)	y

Table 19.5 The MP grammar structure of NPQ, where $L_1, \ldots L_8$ are linear combinations of rational algebraic expressions indicated within parentheses (the values of coefficients of these combinations are given in http://profs.sci.univr.it/~manca/draft/npq-constants.eps).

Reactions	Flux Maps
$r_1 : c \to o + 12h + p$	$L_1(ol, cl, rl, hpl, vl/z, 1)$
$r_2 : c \to c + q^+$	$L_2(c, r, z, l, h, 1)$
$r_3 : c \to c + f^+$	$L_3(c, v, l/r, 1)$
$r_4 : o \to c$	$L_4(ol, cl, rl, hpl, vl/z, 1)$
$r_5 : h \to \lambda$	$L_5(ol, cll, rl, hpl, vl/z, 1)$
$r_6 : p \to \lambda$	$L_6(ol, cl, rl, hpl, vl/z, 1)$
$r_7 : x + 100v \to x + 100z$	$L_7(v, x, 1)$
$r_8 : y + h \to x$	$L_8(y, h, 1)$

a number of steps (about 800 steps). At present this model is too oversimplified to provide useful biological analysis. However, it shows a very important possibility of real applications of MP systems, overcoming the stage of toy models.

An important remark is very appropriate in this regard. A model is never a final result, rather it is only a starting point, and becomes a useful scientific object only when it provides new knowledge about the modeled reality. Therefore, the goal of any model is the possibility of predictions, analyzes, evaluations, or reproductions of similar phenomena, and in general, some kind of comprehension of its application field. This was the case of motion equations of Newtonian physics, and this has to be the aim of new discrete models trying to overcome the limitation of differential equations in the complex phenomena of living organisms. An advantage of the log-gain procedure of deducing MP models, concerns the possibility of applying the method of transforming MP grammars into ODE systems, according to the theory developed in [17]. In this manner the theory of metabolic P systems can be seen as a new tool for constructing classical differential models, where the difficulty of kinetic rate constants evaluation is solved by the log-gain procedure, avoiding analysis at microscopic level.

19.10 Conclusions

MP systems proved to be relevant in the analysis of dynamics of metabolic processes. Their structure clearly distinguishes a reaction level and a regulation level. We showed that an essential component of the regulation level can be deduced by applying the log-gain theory to data that can be collected from observations of the system. The search for efficient and systematic methods for defining MP systems from experimental data is of crucial importance for systematic applications of MP systems to complex dynamics. The log-gain method can deduce, in a given metabolic system, a time series of (approximate) flux unit vectors $U[i]$ (i ranging in time instants), from a time series of observed states. This method is based on the solution of one linear system (of $n + m$ equations and $n + m$ variables) for each value of i. Two crucial tasks remain to be performed for a complete discovery of the underlying MP dynamics which explains an observed dynamic: i) a systematic reliable determination of the initial vector $U[0]$ which is the basis of our iterative method (results in this direction are reported in [28]), and ii) a systematic way for deducing tuners and flux regulation maps from the time series of flux vectors.

MP systems could be applied to the analysis of dynamical systems occurring not only in biology, but also in economic or social systems. Of course, some specific aspects should be adapted to these fields, but in principle, their schema seems to be so general that goods, economical parameters, and transactions should be easily represented in terms of substances, parameters, and reactions. An important aspect of log-gain theory which makes it susceptible to such an extension is that it is based on Avogadro and Dalton principles, which require discreteness and additivity

Table 19.6 Sinus, a MP grammar for the sine function.

SUBSTANCES, RULES, and FLUX MAPS ($K = 10^{-4}$, i *computation step*)
  Tmp   initial value = 1
  S   initial value = 0
  $R1 : Tmp \to S$ Map $F1(\delta(i)) = (Clk[i] == 0) \cdot K \cdot \sqrt{1 - S[i]^2}$
  $R2 : S \to Tmp$ Map $F2(\delta(i)) = (Clk[i] == 1) \cdot K \cdot \sqrt{1 - S[i]^2}$
PARAMETER EVOLUTIONS
  Clk initial value = 0
  $Clk[i + 1] = if(S[i] \leq K, 0, if(S[i] \geq 1 - K, 1, Clk[i]))$
  Sign initial value = 1
  $Sign[i + 1] = if(Tmp[i] \geq 1 - K, 1 - Sign[i], Sign[i])$
  Sinus initial value = 0
  $Sinus[i + 1] = (Sign[i + 1] == 1) \cdot S[i] + (Sign[i + 1] == 0) \cdot (-S[i])$

of transformations (integer transformation ratios and cumulativeness of transformation effects). However, no conservation principle is required by the OLGA system. This is of course an important aspect in real metabolic systems, related to classical Lavoisier's principle (matter is only transformed, but cannot be created or destroyed). But in many cases substance conservation would not be a realistic requirement. Even in some biological phenomena it needs to be relaxed. Signal transduction processes are a remarkable example of this necessity, because protein activation reactions do not correspond to transformations conserving matter. This means that log-gain theory could, in principle, be applied even to this important class of biological phenomena.

Another application of MP systems, along different lines, is the reconstruction of mathematical functions by means of suitable MP systems where the evolution of some magnitude corresponds to the curve of the function. For example, a simple MP system can be defined such that the evolution of a magnitude corresponds to the sine function. Namely, an MP system with a reaction $r : \lambda \to y$, and a corresponding flux map $\varphi_r : \tau\sqrt{1 - y^2}$, with $\tau = 10^{-10}$, provides a dynamic of y which corresponds to the sine function (with an approximation of order 10^{-10}). In fact, $y[n] = \sin n\tau$, for $0 \leq n\tau < \pi/2$ (a tabulation of sine, for the first quadrant, was generated in this way, by using MATLAB®, its extension to the other quadrants is obtained by suitable symmetries. Table 19.6 and Figs. 19.8, 19.9 provide MP representations of the sine function.

This possibility discloses the perspective of searching for MP grammars of real functions. This result shows evidence of the versatility of MP grammars, in fact circular functions are transcendent functions specified by analytic definitions. Our future research will focus on developing the theory of metabolic P systems along three main research lines: i) extending the theoretical aspects, ii) performing suitable biological experiments, and iii) developing computational tools for modeling biological phenomena.

FUNDAMENTALS OF METABOLIC P SYSTEMS 495

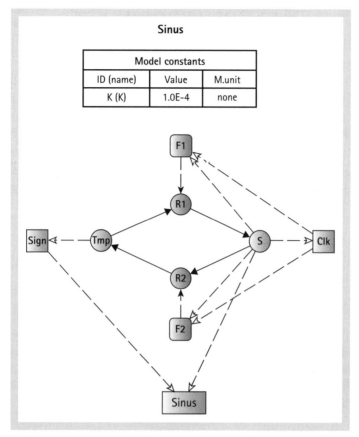

Fig. 19.8 The MP graph of Sinus.

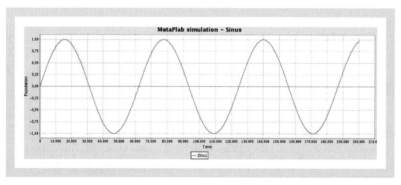

Fig. 19.9 The dynamics of Sinus.

Acknowledgements

I want to express my gratitude to Luca Bianco, Alberto Castellini, Luca Marchetti, and Roberto Pagliarini for their computational experiments which helped me to develop the theory of MP systems and their log-gain analysis. In 2005, Luca implemented the first version of a computational platform, called Psim, which was later incorporated into the more comprehensive system MPsim, which recently evolved into MetaPlab (available at mplab.sci.univr.it). Many bachelor and master students in Computer Science, at the University of Verona, collaborated at several levels in testing and improving some software components of the mentioned systems. I am grateful to Giuditta Franco for helping me to coordinate this activity.

References

[1] T.K. AHN, T.J. AVENSON, M. BALLOTTARI, Y.C. CHENG, K.K. NIYOGI, R. BASSI, G.R. FLEMING: Architecture of a charge-transfer state regulating light harvesting in a plant antenna protein. *Science*, 9 (2008), 794–797.

[2] F. BERNARDINI, M. GHEORGHE, N. KRASNOGOR: Quorum sensing P systems. *Theoretical Computer Sci.*, 371 (2007), 20–33.

[3] L. VON BERTALANFFY: *General Systems Theory: Foundations, Developments, Applications*. George Braziller Inc., New York, 1967.

[4] D. BESOZZI, G. CIOBANU: A P system description of the sodium-potassium pump. *Lecture Notes in Computer Sci.*, 3365 (2005), 210–223.

[5] L. BIANCO: *Membrane Models of Biological Systems*. PhD Thesis, University of Verona, 2007.

[6] L. BIANCO, F. FONTANA, G. FRANCO, V. MANCA: P systems for biological dynamics. In [11], 81–126.

[7] L. BIANCO, F. FONTANA, V. MANCA: Reaction-driven membrane systems. *Lecture Notes in Computer Sci.*, 3611 (2005), 1155–1158.

[8] L. BIANCO, F. FONTANA, V. MANCA: P systems with reaction maps. *Intern. J. Found. Computer Sci.*, 17 (2006), 27–48.

[9] L. BIANCO, V. MANCA, L. MARCHETTI, M. PETTERLINI: Psim: a simulator for biochemical dynamics based on P systems. *2007 IEEE Congress on Evolutionary Computation*, Singapore, September 2007, 883–887.

[10] L. BIANCO, D. PESCINI, P. SIEPMANN, N. KRASNOGOR, F.J. ROMERO-CAMPERO, M. GHEORGHE: Towards a P systems Pseudomonas quorum sensing model. *Lecture Notes in Computer Sci.*, 4361 (2007), 197–214.

[11] G. CIOBANU, G. PĂUN, M.J. PÉREZ-JIMÉNEZ, eds.: *Applications of Membrane Computing*. Springer, 2006.

[12] A. CASTELLINI, V. MANCA: METAPLAB: A computational framework for metabolic P systems. *Pre-proc. WMC 2008*, Edinburgh, UK, July 2008.

[13] A. CASTELLINI, G. FRANCO, V. MANCA: Hybrid functional Petri nets as MP systems. *Natural Computing*, to appear.
[14] A. CASTELLINI, V. MANCA, L. MARCHETTI: MP systems and hybrid Petri nets. *Studies in Computational Intelligence*, 129 (2008), 53–62.
[15] F. FONTANA, L. BIANCO, V. MANCA: P systems and the modeling of biochemical oscillations. *Lecture Notes in Computer Sci.*, 3850 (2005), 199–208.
[16] F. FONTANA, V. MANCA: Predator-prey dynamics in P systems ruled by metabolic algorithm. *BioSystems*, 91 (2008), 545–557.
[17] F. FONTANA, V. MANCA: Discrete solutions to differential equations by metabolic P systems. *Theoretical Computer Sci.*, 372 (2007), 165–182.
[18] G. FRANCO, V. MANCA: A membrane system for the leukocyte selective recruitment. *Lecture Notes in Computer Sci.*, 2933 (2004), 180–189.
[19] G. FRANCO, P.H. GUZZI, T. MAZZA, V. MANCA: Mitotic oscillators as MP graphs. *Lecture Notes in Computer Sci.*, 4361 (2007), 382–394.
[20] G. FRANCO, N. JONOSKA, B. OSBORN, A. PLAAS: Knee joint injury and repair modeled by membrane systems. *BioSystems*, 91 (2008), 473–488.
[21] A. GOLDBETER: A minimal cascade model for the mitotic oscillator involving cyclin and cdc2 kinase. *PNAS*, 88 (1991), 9107–9111.
[22] S. HOOPS, S. SAHLE, R. GAUGES, C. LEE, J. PAHLE, N. SIMUS, M. SINGHAL, L. XU, P. MENDES, U. KUMMER: COPASI – a COmplex PAthway SImulator. *Bioinformatics*, 22 (2006), 3067–3074.
[23] P. KURKA: *Topological and Symbolic Dynamics*. Société Mathématique de France, 2003.
[24] V. MANCA: String rewriting and metabolism: A logical perspective. *Computing with Bio-Molecules* (Gh. Păun, ed.), Springer, 1998, 36–60.
[25] V. MANCA: MP systems approaches to biochemical dynamics: Biological rhythms and oscillations. *Lecture Notes in Computer Sci.*, 4361 (2006), 86–99.
[26] V. MANCA: Metabolic P systems for biochemical dynamics. *Progress in Natural Sciences*, 17 (2007), 384–391.
[27] V. MANCA: Discrete simulations of biochemical dynamics. *Lecture Notes in Computer Sci.*, 4848 (2008), 231–235.
[28] V. MANCA: The metabolic algorithm for P systems: Principles and applications. *Theoretical Computer Sci.*, 404, 1-2 (2008), 142–157.
[29] V. MANCA: Log-gain principles for metabolic P systems. *Algorithmic Bioprocesses* (A. Condon et al., eds.), Springer, 2009, chapter 28.
[30] V. MANCA, L. BIANCO: Biological networks in metabolic P systems. *BioSystems*, 91 (2008), 489–498.
[31] V. MANCA, L. BIANCO, F. FONTANA: Evolutions and oscillations of P systems: Applications to biological phenomena. *Lecture Notes in Computer Sci.*, 3365 (2005), 63–84.
[32] V. MANCA, R. PAGLIARINI, S. ZORZAN: A photosynthetic process modelled by a metabolic P system. *Natural Computing*, DOI 10.1007/s11047-008-9104-x.
[33] M. NAGASAKI, A. DOI, H. MATSUNO, S. MIYANO: Genomic object net: I. A platform for modelling and simulating biopathways. *Applied Bioinformatics*, 2 (2004), 181–184.
[34] R. PAGLIARINI: *Esperimenti per la determinazione computazionale dei parametri regolativi nei P sistemi metabolici*. Master Thesis, Univ. of Verona, 2007.
[35] GH. PĂUN: Computing with membranes. *J. Comput. System Sci.*, 61 (2000), 108–143.

[36] GH. PĂUN: *Membrane Computing. An Introduction*. Springer, 2002.
[37] M.J. PÉREZ-JIMÉNEZ, F.J. ROMERO-CAMPERO: A study of the robustness of the EGFR signalling cascade using continuous membrane systems. *Mechanisms, Symbols, and Models Underlying Cognition, Natural and Artificial Computation, IWINAC 2005*, LNCS 3561, Springer, 2005, 268–278.
[38] D. PESCINI, D. BESOZZI, G. MAURI, C. ZANDRON: Dynamical probabilistic P systems. *Intern. J. Found. Computer Sci.*, 17 (2006), 183–204.
[39] F.J. ROMERO-CAMPERO, M. GHEORGHE, G. CIOBANU, L. BIANCO, D. PESCINI, M.J. PÉREZ-JIMÉNEZ, R. CETERCHI: Towards probabilistic model checking on P systems using PRISM. *Lecture Notes in Computer Sci.*, 4361 (2006), 477–495.
[40] Y. SUZUKI, Y. FUJIWARA, H. TANAKA, J. TAKABAYASHI: Artificial life applications of a class of P systems: Abstract rewriting systems on multisets. *Lecture Notes in Computer Sci.*, 2235 (2001), 299–346.
[41] Y. SUZUKI, H. TANAKA: A symbolic chemical system based on an abstract rewriting system and its behavior pattern. *J. of Artificial Life and Robotics*, 6 (2002), 129–132.
[42] Y. SUZUKI, H. TANAKA: Modelling p53 signaling pathways by using multiset processing. In [11], 203–214.
[43] E.O. VOIT: *Computational Analysis of Biochemical Systems*. Cambridge University Press, 2000.

CHAPTER 20

METABOLIC P DYNAMICS

VINCENZO MANCA

20.1 INTRODUCTION

As the philosopher says [28], "Panta rei" ($\pi\grave{\alpha}\nu\tau\alpha$ $\rho\grave{\epsilon}\iota$), that is, "everything is changing" or also "existence is change". But when something changes, according to a rule, something does not change. In fact, in order to localize and describe a process as an entity, something has to remain stable during it. Therefore, existence is a mysterious mixing of variation and invariance underlying objects and events, at each level of reality. A **dynamical system** is a structure hosting a dynamic. The term "dynamics" points to the process while the term "system" points to the structure. A river, a city, a living organism, considered at different instants of time, are always different in many important aspects, nevertheless, their individualities persist unchanged during all the instants of their life.

Since the epochal discovery of laws of planetary motions, dynamical systems have been mathematically studied by means of differential equations. Recently the study of complex systems arising from life sciences, economy, meteorology, and many other fields, requires novel ideas and different approaches for the analysis and modeling of a wider class of dynamical systems, and for a great variety of aspects concerning the dichotomy structure/behavior. In particular, discrete dynamical systems seem to open new modeling possibilities and pose new problems where the algorithmic aspects replace the geometrical and differential perspective of classical dynamics.

In 1684 the astronomer Edmund Halley traveled to Cambridge to ask Newton about planetary motions. It was known that planets move around the Sun in ellipses. The question Halley posed to Newton was whether and how this elliptic shape was related to the inverse-square law. Sir Isaac replied immediately that it would necessarily be an ellipsis. Halley, struck with joy and amazement, asked him how he knew it. "Why," he said, "I have calculated it". The general rigorous proof of this phenomenon would be definitively given by Euler in 1749. Nowadays, it is a classical topic in physics.

Halley's question is a first example of an *inverse dynamical problem*, that is, given an observed behavior, is there a mathematical system which could explain this behavior? This is a crucial issue, in fact, when we can answer to this question, then we "dominate" the system because we can predict or alter its behavior.

Any system changing in time can be identified with some variables. As they define a system, they must verify some relations which are the *invariant* of the behavior. Newton's solution to Halley's question was obtained firstly by determining some invariants of the planetary orbits, specified in terms of differential equations, then, by solving these equations, the mathematical form of these orbits was deduced. In general, in order to define a mathematical system related to a process, we need to determine its *variables* and its *invariants*. This schema of analysis for dynamical systems is very general and continues to hold in different formal frameworks of dynamical representations. Just to suggest the wide range of this paradigm, let us consider the situation of a discrete phenomenon. Suppose we observe a device generating as outputs words over a particular alphabet. A very natural question is: which is the grammar of the language generated by this device? And, in which manner does its internal structure realize the generation of these words? If we assume that any event can be discretely represented by a suitable word, then the search of rules underlying processes can, in principle, correspond to the search of a grammar. In the case of metabolic processes, we show that what is usually expressed in terms of differential equations can be formulated in terms of a special class of grammars, and very often these grammars are directly related to the biochemical mechanisms of phenomena.

In this paper, we present some general aspects concerning discrete dynamical systems, by focusing the attention on metabolic P systems (MP systems), a special class of dynamical systems, based on P systems [24], which are related to metabolic processes. This paper continues a line of investigation initiated about ten years ago. Discrete mathematical notions of metabolism have been investigated since 1998 [15, 23], but their analysis, in the context of P systems, has been investigated only since 2004 [21] (see Chapter 19 on metabolic P systems in this volume, and [16, 17, 18, 19, 20]). A dynamical perspective in P systems was introduced in [2]. The relationship between dynamical systems and computation systems was discussed in [6]. A qualitative analysis of state transition dynamics, based on quasi dynamics, was investigated in [22], while a relational perspective of this approach was proposed in [25, 26]. A symbolic analysis of oscillations was developed in [4].

In the following, we assume as known the very basic notions in formal language theory and in topology (see appendix A in [13]). Fundamentals of membrane computing and metabolic P systems will be useful for a better understanding of the methodological perspective of the following discussion.

20.2 Discrete Dynamical Systems

Let us consider some mathematical formalizations of dynamical systems. Inspite of their elementary character, there are some subtle points which are very relevant with respect to some basic intuitions related to the intrinsic nature of time. In the following N, Z, Q, R will denote the set of natural, integer, rational, and real numbers respectively.

Given a function $\delta : S \to S$ and $i \in N$, the i-iterated of δ is inductively defined by the following equations, for every $s \in S$:

$$\delta^0(s) = s,$$
$$\delta^{i+1}(s) = \delta(\delta^i(s)).$$

Definition 20.1 *An* **autonomous dynamical system** *is a pair (S, δ) where S is called the* **phase space** *and $\delta : S \to S$.*

In an autonomous dynamical system the dynamics δ is a *next-state-function* from the phase space S to itself. Therefore, time is a consequence of the internal dynamics of the system. In the following definition an external notion of time is assumed.

Definition 20.2 *A* **timed dynamical system** *D is a triple $D = (S, T, \delta)$ given by a set S of states, called* **phase space**, *$T \subseteq R$ is a set of* **time instants**, *and $\delta : T \to S$ is a* **dynamics**, *which assigns to any time instant the state which the system assumes in that instant.*

Any sequence over a set S, or any function $f : R \to R$ determines a dynamical system, in a trivial way. Motions, developments, growing processes, and morphogenesis are representable as dynamical systems. In planetary motions phase space is R^6 because the position and velocity of the planet with respect to three cartesian axes identify the state of the system, and the dynamics is the motion law giving the planet positions and velocities at every instant (a system of two planets needs R^{12} as phase space). A dynamical system $D = (S, T, \delta)$ is **discrete** if T is a subset of Z.

Proposition 20.1 *An* **autonomous** *dynamical system* (S, δ) *determines a family of discrete timed dynamical systems* $D_s = \{(S, \mathbf{N}, \delta_s) \mid s \in S\}$, *where dynamics* δ_s *is called the* **timing** *of* δ *based on the* **initial state** $s \in S$, *and is defined by* $\delta_s(i) = \delta^i(s)$.

The following definition of *flow* generalizes the concept of autonomous system, by using a next-state-function with a real number as further parameter.

Definition 20.3 *A* **flow** *D is a triple* $D = (S, T, \delta)$ *given by i) a set S of states, called* **phase space**, *ii) a set* $T \subseteq \mathbf{R}$ *of* **time intervals** *such that* $0 \in T$ *and when* $t, t' \in T$, *then also* $t + t' \in T$, *and iii) a* **dynamics** δ, *which is a function* $\delta : S \times T \to S$ *providing the state* $\delta(s, t)$ *which the system assumes after time t, starting from s. It is required that* $\delta(s, 0) = s$, *and for any* $s \in S$ *and* $t, t' \in T$ *we have* $\delta(\delta(s, t), t') = \delta(s, t + t')$.

When T is a subset of $\mathbf{Z}$, then D is a **discrete** flow.

Transition graphs are very simple and visual instances of discrete flows over finite phase spaces, where nodes are states and arrows denote the dynamics (in the future along the arrow verse, and in the past along the opposite verse). Instants are integer numbers (positive in the future, and negative in the past), and transitions along edges last a unitary time (see Fig. 20.1).

Corollary 20.1 *Any autonomous dynamical system is a (discrete) flow* (S, T, δ) *where* $T = \mathbf{N}$.

A flow can be also called a *continuous, autonomous dynamical system*. Definitions (20.2) and (20.3) could seem to be apparently equivalent, but they are very different.

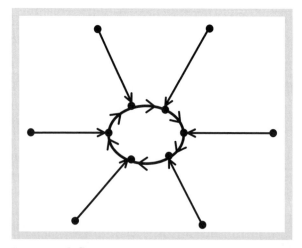

Fig. 20.1 A flow represented by a transition graph.

In a flow, a dynamic is essentially *memoryless*, that is, the behavior following a given state depends exclusively on the state. Conversely, in a timed dynamical system the state at a given instant could depend on the states occurring in all the previous instants. Moreover, in a timed dynamical system, only one behavior is described, while in a flow many behaviors can be obtained in correspondence to different initial states associated to time 0 (see Fig. 20.1). In fact, in correspondence to any specified initial state, a corresponding timed dynamical system is completely determined.

An observation of a dynamical system is realized by recording the values of all variables of the system along a time series (planet motions were investigated starting from this kind of observation). This process can be seen as a coupling of the (observer) clock with the observed system. For this reason, an observed system can be seen as a (discrete) timed system. The discovered flow or autonomous system exhibiting the observed dynamics provides a solution of an inverse dynamical problem which "explains" the observed behavior as a consequence of a mechanism "internal" to the system in question.

In the following, without explicit mention, we consider only discrete flows and discrete timed dynamical systems. A dynamical system (flow) is **metric** or **topological** if its phase space S is a metric or a topological space.

A dynamical system (S, T, δ) is **totally discrete** when both sets S and T are finite or enumerable sets. Even if it sounds a little strange, a real number can be seen as a totally discrete dynamical system, whenever we identify it with the sequence providing its decimal digits. In fact in this case $S = \{0, 1, 2, \ldots 9\}$ and $T = \mathbf{N}$. Infinite words, trees, or graphs are totally discrete dynamical systems on suitable discrete phase spaces, where dynamics is the function giving the structure at some step of its generation. Examples of totally discrete dynamical systems are constituted by deductive and rewriting systems, Chomsky grammars, automata, L systems, cellular automata [31], and Petri nets [27].

Example 20.1 (Chomsky grammar) *A Chomsky grammar can be seen as a (totally) discrete dynamical system where the phase space is the set of languages over an alphabet and dynamics assigns to any natural number n the (finite) language of all the strings generated in at most n applications of grammar rules (at time 0 the language is constituted by $\{S\}$, including only the initial symbol of the grammar).*

Example 20.2 (Fibonacci grammar) *Consider the alphabet of two symbols $\{A, N\}$ and start with the symbol N as the initial word at step 0. Then apply, in a parallel way, at any step $i > 0$ the rewriting rules $N \to A$ and $A \to AN$ to all the symbols of the word of step i, generating in this way the word at step $i + 1$. This system can be easily seen as a dynamical system where the phase space contains all the words over the alphabet $\{A, N\}$, and the dynamic assigns to each number i the word generated at step i. It is easy to show that the length of the word at step i corresponds to the i-th*

Fibonacci number. These numbers can be generated by Binet's formula

$$F_n = \frac{(1+\sqrt{5})^{n+1} - (1-\sqrt{5})^{n+1}}{2^{n+1}\sqrt{5}}.$$

This dynamical system is of fundamental importance in many phenomena of natural development, for example in phyllotaxis (leaf positions in plant developments). Moreover, the generation of the golden ratio $(\sqrt{5}-1)/2$, on which Fibonacci numbers are based, individuates the continuous fraction associated to the equation $\alpha = \frac{1}{1+\alpha}$, which can be seen as a special kind of dynamical system (replace iteratively α by $\frac{1}{1+\alpha}$ in $\frac{1}{1+\alpha}$):

$$\cfrac{1}{1+\cfrac{1}{1+\cfrac{1}{1+\cfrac{1}{1+\cdots}}}}$$

Example 20.3 (**Deterministic P system**) *A deterministic P system [24] is a discrete dynamical system where: i) states are the functions which assign to any region of the system the multiset of objects inside it, ii) time instants are the natural numbers, and iii) dynamics assign to any $n \in \mathbf{N}$ the state which is derived from the initial state after n consecutive evolution steps (according to the specific deterministic evolution strategy of the system). The Fibonacci grammar can be seen a simple case of deterministic P system with multisets over $\{A, N\}$.*

Any computation is a kind of dynamic. However, the two concepts are based on different perspectives. In a computation an input is provided to the system (which reaches an initial state) and an output is expected at the end of the process (extracted from a final state) if the system reaches a state which, according to some termination criterion, is considered as final. This means that the dynamics underlying the passage of a computation from an initial to a final state is functional to the desired relation between input and output, and the sooner a final state is obtained, the better computation is realized. On the contrary, in a dynamical system the focus of the process is in the dynamics itself. Input and output can even be avoided, but when they are present, they are used in order to guarantee that dynamics could proceed by satisfying some behavioral requirements (e.g. acquiring substances or expelling debris). A major example of this dynamical perspective are the living organisms. They are open systems, that is, they need inputs and outputs, to ensure that a dynamic, with specific features, could be maintained in the time, as long as possible. In this perspective, it is very important to individuate dynamical properties which are relevant to life phenomena and to discover principles according to which some artificial dynamics exhibit behaviors with these properties.

The basis of living organisms is the cell, and P systems are the most direct discrete mathematical model of cell. For this reason a dynamical study of P systems, extended and adapted according to specific modeling criteria, could have a very significant role in the recent trend of synthetic biology [12].

20.3 Autonomous Dynamical Systems

Time and space are concepts which are primitive in the notion of dynamical system. However, a more subtle analysis shows that many theoretical possibilities and puzzles arise when Occam's razor is put in action on our intuitions about these concepts. A deep analysis of these issues is matter of philosophical investigation since the beginning of Greek speculations about physics and metaphysics, and therefore could be outside the scope of our discussion. However, some specific points are relevant for our investigation. Here we present some important examples of autonomous systems and will shortly discuss their relationship with the notion of the clock.

Example 20.4 (Quadratic dynamical systems) *A quadratic dynamical system is an autonomous system $Q_r = ([0, 1], \delta_r)$ where $[0, 1]$ is the close interval of reals between 0 and 1, $r \in \mathbb{R}$, and for every $x \in [0, 1]$ we have $\delta_r(x) = rx(1 - x)$. Despite their simplicity these systems exhibit very interesting behaviors and have been intensively studied [14]. The parameter r plays an essential role in determining the properties of these systems. As r increases from 3.4 to 4 we obtain a sequence of values $(r_i \mid i > 0)$ such that $\delta_{r_i}^{2^i}(x) = x$ (period-doubling phenomenon). The sequence $(r_i \mid i > 0)$ has a limit r_∞, and for any $k \in \mathbb{N}$, it happens that $\delta_{r_\infty}^k(x) \neq x$ (infinite period). The system Q_{r_∞} is called a Feigenbaum system, for its importance in the determination of Feigenbaum constant [9].*

Any discrete timed dynamical system $D = (S, \mathbb{N}, \delta)$ can be represented within a particular autonomous system $D_\sigma = (Q, \sigma)$, called the **shift system** of D, where the phase space Q is constituted by the functions q from $\mathbb{N}$ to S, and the dynamic is the shift function $\sigma: Q \to Q$ such that $\sigma(q) = q'$ and $q'(i) = q(i + 1)$ for any $i \in \mathbb{N}$. In fact, for any $i \in \mathbb{N}$ we have $\delta(i) = (\sigma^i(\delta))(0)$. However, in the passage from D to D_σ the phase space changes. Therefore, a natural question arises: for any discrete timed dynamical system D, can we determine an autonomous dynamical system D' having the same phase space of D and such that the timing of D' based on some state of D' could coincide with the dynamics D? The answer is, in general, negative. In fact, in the dynamic of an autonomous discrete dynamical system, the new state of the system depends completely on the previous state of the system, but of course, this could not be true in general when dynamics at some instant time depend on the states of some previous instants.

Let S a set of infinite binary sequences. *Cantor distance* d between two sequence α, β of S is such that $d(\alpha, \beta) = 0$ if $\alpha = \beta$, while $d(\alpha, \beta) = 2^{-m}$, with $m = min\{i \mid a_i \neq \beta_i\}$, if $\alpha \neq \beta$.

Example 20.5 (Symbolic dynamics) *A symbolic dynamic is a shift system associated to a discrete timed dynamical system $(A, \mathbb{N}, \delta)$ where A is a finite alphabet, and Cantor distance is defined over the (infinite) sequences $(\delta(i) \mid i \geq k \in \mathbb{N})$. Symbolic* *dynamics are very important because it has been shown that any autonomous dynamical system with a continuous dynamic over a compact phase space can be "immersed" in a suitable symbolic dynamic* [13].

The definition of autonomous systems implicitly assumes that the application of δ can be considered uniform, that is, the same temporal interval is assumed for any application of δ. In about 1602, Galileo discovered the pendulum, as a physical phenomenon for which this uniformity feature can be assumed. This discovery is the beginning of the experimental science which is intrinsically based on the measure of time. Clocks were available even before Galileo's discovery, and in the middle ages a lot of mechanical clocks were constructed in many European towns (firstly put on public buildings). However, the precision necessary for scientific measurements was essentially based on pendulum isochronism.

Is time before any dynamics, or vice-versa is time an abstraction originated by dynamical phenomena? In fact, time apart dynamics seems to be a metaphysical abstraction. From an epistemological point of view, we have only clocks, and any physical notion of time is always related to a **clock**, that is, a system (S, δ, s_0) where (S, δ) is an autonomous dynamical system, and $s_0 \in S$ is an *initial state*, conventionally associated to the time 0. In a clock it is intrinsically assumed that the application of its next-state-function can be considered a uniform process, in such a way that the passage from one state to the next happens according to the same law, with no appreciable difference in the quantity of time (whatever this could mean) necessary to perform this passage. In other words, clocks are physical entities where we can observe *regular* events of passage from one state to another state.

Assume that we can define a **synchronization** relation between two events. Then the clock $C_1 = (S_1, \delta_1, s_1)$ has a relative ratio $n/m \in \mathbf{Q}$ with the clock $C_2 = (S_2, \delta_2, s_2)$, if the passage from $\delta_1^i(s_1)$ to $\delta_1^{i+n}(s_1))$, for any $i \in \mathbf{N}$, is synchronous with the passage from $\delta_2^i(s_2)$ to $\delta_2^{i+m}(s_2))$. Consider the period of the clock of the *International Bureau of Weights and Measures* as a time unit. If there is a bound k such that no physical clock is available at a ratio equal or inferior to 10^{-k}, with respect to this standard clock, what reason can be used for assuming the existence of time at smaller temporal scales? This argument explains that if, very often, discrete dynamical systems are only approximations of real dynamical systems, on the other side, continuous dynamical systems are always idealizations of real systems. In this sense, in any dynamical system the set T of time instants should be replaced by the states of a clock. Therefore, discrete dynamical systems have a strict physical correspondence with real systems. Of course, this does not diminish the interest of continuous systems, as mathematical objects (mathematical existence could be even more "real" than the physical one). Another aspect related to clocks is that not only is continuously dividable time a mathematical idealization, but even the existence of a set T of time instances which is infinite should be reconsidered. In fact, if T is identified with the states of a clock, then they cannot be an infinite set. Therefore, the notion of an infinite set of instants should be more properly replaced by the notion of an unbound set of instants. In fact, if we combine a periodic clock with an unbound *memory space*, then the completion of any period can be memorized in such a way that we get a system, say a *clock memory*, able to register an unbound amount of events. According to this perspective, if the universe space is essentially finite, then time is essentially finite too.

Example 20.6 (Autonomous Differential Equations) *Differential equations are mathematical conditions identifying flows. A system of autonomous differential equations in a phase space of dimension three where $\delta(t) = (x(t), y(t), z(t))$ has the following form (the attribute autonomous means that time variable t does not occur on the right members of equations):*

$$\frac{dx(t)}{dt} = f(x, y, z)$$

$$\frac{dy(t)}{dt} = g(x, y, z)$$

$$\frac{dz(t)}{dt} = h(x, y, z)$$

Under suitable hypotheses on the nature of the functions in the right members, given an initial state $(x(0), y(0), z(0))$ of the system, it admits a unique solution [11]. Therefore, solving the system, means to get the flow starting from $(x(0), y(0), z(0))$ which satisfies the differential conditions.

Terminological remark. In the following sections, when we say "dynamical system", with no other explicit mention, we mean an autonomous dynamical system.

20.4 TRAJECTORIES, ORBITS, AND PERIODICITY

In this section we will define some basic dynamical concepts. The definitions will be given for discrete autonomous dynamical systems, however their generalizations to timed dynamical systems or to flows are straightforward.

Let us fix a dynamical system $D = (S, \delta)$. A **trajectory** of D, with origin $s \in S$, is a sequence of states $(\delta^i(s) \mid i \in I)$ such that $I = \{i \mid i \leq k \in \mathbb{N}\}$ or $I = \mathbb{N}$. The set of states $\{\delta^i(s) \mid i \in I\}$ which are images of a trajectory constitute an **orbit**. Of course, different trajectories can provide the same orbit. A state $s \in S$ is **periodic** if there exists a number $k \in \mathbb{N}$, such that $\delta^k(s) = s$. The minimum value k such that $\delta^k(s) = s$ is called *period* of s. A state $s \in S$ is a **fixpoint** if $\delta(s) = s$. In this case the orbit of a trajectory with origin s coincides with $\{s\}$. A fixpoint is a periodic point of period 1, while a periodic point of period k is a fixpoint for δ^k. A **cycle** is the orbit of a periodic state.

The state s is **eventually periodic** if a number m exists, called the **transient** of D, such that $\delta^m(s)$ is periodic.

A **non-deterministic** dynamical system is a system $D = (Q, \delta)$ where any element of Q is a set of states belonging to a phase space S. Therefore, the states of phase space Q are subset of S, that is, elements of $P(S)$, the power set of S. A (non-deterministic) state corresponds to a set of *possible* (deterministic) states. Of course a non-deterministic system where (non-deterministic) states are constituted by only one element corresponds to the basic notion of dynamical system which is, by default, **deterministic**. In the case of non-deterministic systems, a trajectory corresponds to a set of possible trajectories. For example, a trajectory $(q_0, q_1, q_2, \ldots)$ describes a system that starts from a state belonging to q_0, then passes to a state belonging to q_1, and then to a state of q_2, and so on. At each step we do not know the state of the system, but only a set of possible states to which the current state belongs.

A motion similar to non-determinism is **quasi determinism** introduced in [22]. Formally it is very close to non-determinism, but it expresses a different perspective.

A **quasi deterministic** dynamical system is a triple (S, δ, μ) where (S, δ) is a (deterministic) dynamical system, and μ is a function defined on $P(S)$ and taking values on positive real numbers. In this context, a subset of S is also called a **quasi state**.

If S is a metric space, the function μ may coincide with the diameter of the subsets of S (the maximum of distance between any two points of a subset), or even μ could be based on more general measures on the space S, but at this level, we can avoid being more explicit in this regard.

Quasi determinism is a powerful concept and in [22] a qualitative analysis of quasi determinism was developed, by showing that even without an explicit use of the function μ many important notions can be investigated and some interesting dynamical properties can be established (in [25, 26] a logical relational perspective of quasi dynamics was developed).

A state s of a quasi deterministic dynamical system is ϵ-**quasi periodic** if there exists a quasi state q with $s \in q$, and $\mu(q) \leq \epsilon$, and a natural number $k > 0$ such that, for any $s' \in q$, the two states s' and $\delta^k(s')$ belong to q.

A state s of a quasi deterministic dynamical system is ϵ-**quasi recurrent** if there exists a quasi state q with with $s \in q$, and $\mu(q) \leq \epsilon$, and for any $s' \in q$ there exists a natural number $k > 0$ such that the two states s' and $\delta^k(s')$ belong to q.

A state s of a quasi deterministic dynamical system is ϵ-**almost periodic** if there exists a quasi state q with with $s \in q$, and $\mu(q) \leq \epsilon$, and a *recurrence bound* $b > 0$ such that, for any $s' \in q$, there exists a natural number $k \leq b$ such that the two states s' and $\delta^k(s')$ belong to q (Fig. 20.2).

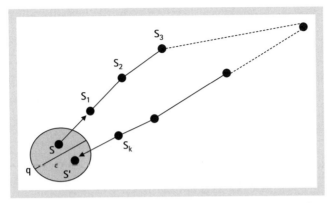

Fig. 20.2 Let $F(s)$ be the condition depicted in the figure. Three main forms of recurrence can be distinguished. *Quasi Periodicity*: $\exists k > 0 \; \forall p' \in q \; F(p)$. *Quasi Recurrence*: $\forall p' \in q \; \exists k > 0 \; F(p)$. *Almost Periodicity*: $\exists b > 0 \; \forall p' \in q \; \exists k \leq b \; F(p)$.

Periodicity has a surprising richness of dynamical possibilities. For example, if some periodic behavior is relative only to a subspace of phase space, then it can be modulated in an extremely interesting series of variants (think in $\mathbf{R}^3$ to the many kinds of spirals and try to generalize similar shapes in general dynamical systems).

A state $s \in S$ is ϵ-**transitive** if the orbit with origin s is dense in the space S, that is, if, for any state $s' \in S$, there exists a natural i such that the states $\delta^i(s)$ and s' belong to some $q \subseteq S$ with $\mu(q) \leq \epsilon$. A dynamical system is ϵ-transitive if some state of the system is ϵ-transitive, and it is transitive if it is ϵ-transitive for any ϵ.

The set of natural numbers, with the successor function over them, is a dynamical system which is not recurrent, but transitive, where 0 is the only transitive state.

Example 20.7 (Champernowne system) *The Champernowne system is the shift system associated to* $(\{0, 1\}, \delta_C)$ *where* δ_C *is the infinite binary sequence of all binary strings concatenated in the lexicographic ordering*:

$$01001010110000100\ldots$$

It is shown in [13] that this dynamical system, equipped with Cantor distance, is transitive and all the states of this space are quasi recurrent.

The following proposition is an easy consequence of the previous definitions.

Proposition 20.2 *In quasi deterministic dynamical systems, periodicity implies quasi periodicity, which implies almost periodicity, which implies quasi recurrence.*

We conclude this section by mentioning the following result, owed to Birkhoff, which is related to the famous property of *eternal return* considered by Poincaré, who initiated a general topological perspective in the investigation of dynamical concepts.

Proposition 20.3 *Any metric autonomous dynamical system with a continuous dynamics has an almost periodic state.*

20.5 Attractors, Basins, and Fluctuations

One of the most important concepts in a dynamical system is that of the **attractor**. There are many ways to formalize this notion. Intuitively an attractor is a quasi state (a set of states) such that when a trajectory reaches it (a state in it), then this trajectory remains inside it. Moreover, this quasi state is included in a bigger quasi state, called its **basin**, such that any trajectory passing through the basin, after a while (or according to a limit process), tends to "fall" in the attractor. A basin is a case of **dynamically invariant** set B, that is in a dynamical system (S, δ), if $x \in B$, then $\delta(x) \in B$. An attractor of basin B is a sort of minimal or limit subset of B where dynamics could remain eternally confined (see Fig. 20.3). This means that attractors can be viewed as "dynamical states", or even, as *second level states*. For example, a living organism in a stable situation, performing life functions, moves along a trajectory which macroscopically seems just a state but corresponds essentially to an attractor. In mammals there are around 200 cell types. The stable states of these cells can be seen as different attractors where cell dynamics can fall, in order to satisfy some specific conditions corresponding to their biological role. We will not go into the details of mathematical definitions of attractors. However, discovering the possible attractors of a system, and their specific properties, is a

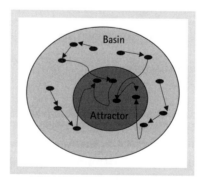

Fig. 20.3 An attractor inside its basin.

crucial aspect of any dynamical investigation. We limit ourselves to outlining some dynamical concepts based on attractors in an informal (geometric) way. A formal treatment of them can be developed in the framework of topological dynamics [13]. In the dynamical system of Fig. 20.1 the set of nodes around the central circle is an attractor, while all the nodes constitute a basin for that attractor.

Example 20.8 *The Collatz dynamical system is the autonomous system on positive natural numbers where dynamics are given, for any $x \in \mathbf{N}$, by $3x + 1$ if x is odd, and $x/2$ if x is even* [30]. *Collatz's conjecture claims that $\{1, 2\}$ is the attractor of this system (starting from any number you fall into 1, and then into 1 and 2 again). It was proved to be true for numbers until around 2^{50} (a web computational project on this topic is active for testing Collatz's conjecture, see http://www.ericr.nl/wondrous/index.html).*

A simple notion of attractor can be defined in the following way.

Definition 20.4 *Given a dynamical system of dynamics δ, an attractor A of basin B, with $A \subseteq B$, is a quasi state such that, for any $x \in A$, $\delta(x) \in A$, and for any $x \in B$ there exists a natural number n such that $\delta^n(x) \in A$.*

An attractor A of basin B is called **minimal** if $A \subseteq C$ for any attractor C of basin B. Usually, attractors are intended as minimal attractors. The following lemma can be easily proved.

Lemma 20.1 *A minimal attractor is constituted by only one state or by a cycle.*

If we consider the notion of *fluctuation* many concepts and properties can be developed in a qualitative manner, avoiding the topological and limit processes on which they are usually based. A **fluctuation** is a shift which displaces the state assigned by the dynamics within a quasi state depending on the intensity of its *perturbing parameters*. This means that, in a dynamical system subjected to fluctuations, the dynamic δ takes as arguments state and fluctuation parameters, represented by a real vector of some finite dimensionality, corresponding to the number of perturbing factors. The notion of fluctuation joined to the notion of quasi determinism provides an interesting framework for expressing dynamical concepts with a strong intuition of aspects typical of living organisms. In mathematical terms, a dynamical system with quasi states and fluctuations is given by:

$$(S, P, \mu, \| \ \|, \delta)$$

where: i) S is a phase space, ii) $P \subseteq \mathbf{R}^m$, with $m \in \mathbf{N}$, iii) μ a measure $\mu : P(S) \to \mathbf{R}$, iv) δ a fluctuating dynamics $\delta : S \times P \to S$, having a vector of perturbing parameters, as an additional argument, which alters the usual deterministic dynamics $\delta : S \to S$, and v) $\| \ \|$ a norm in $\mathbf{R}^m$ (in the usual sense of vector spaces) such that, for any $p \in P$, $\|p\|$ corresponds to the amplitude of the perturbing vector. Further requirements could be considered for adequately expressing the perturbing action of parameters (e.g. continuity and monotony) and for relating the measure μ

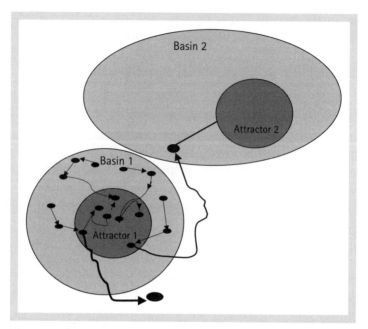

Fig. 20.4 Attractor 1 is a soft attractor. The two big points outside Basin 1 can be reached by fluctuations (bolder lines) from Attractor 1. One of these two points is included in Basin 2. Therefore Attractor 2 can be reached from Attractor 1 as a consequence of a fluctuation.

to the norm $\| \ \|$. Fluctuations are a peculiarity and a necessity of complex systems which interact with an environment. The notions of *rigidity, softness, and plasticity* can be formalized in terms of attractors, basins, and fluctuations. The set P of perturbing parameters can be also analyzed in a more detailed way for distinguishing different types of influences coming from external stimuli, or internal noise, or even from other possible sources of interferences and dependence factors. Given a quasi state q, its fluctuation extension is the set $\{\delta(s, p) \mid s \in q, p \in P\}$. An attractor is **rigid** when its fluctuation extension is strictly included in the basin of the attractor, while an attractor is **soft** when its fluctuation is not included in the basin of the attractor (Fig. 20.4).

In a rigid attractor, there is no possibility of leaving the attractor because the system cannot change its dynamical state. Conversely, if the attractor is soft, a great fluctuation can allow the system to escape from the attractor.

In [12] some informal notions of softness and rigidity are used for analyzing cell differentiation, adaptation, and irreversibility. Rigidity corresponds to the quasi state of a cell which cannot change its functionality. On the contrary, totipotent cells correspond to soft attractors which can differentiate in many possible functionalities.

Let $\epsilon \in R$. A quasi state is ϵ-**stable** when $x \in A, ||p|| \leq \epsilon \implies \delta(x, p) \in A$. It is ϵ-**unstable** when there is a value $p \in P$, with $||p|| \leq \epsilon$ for which $\delta(x, p) \notin A$. Stability and instability with no mention of perturbation amplitude are intended with respect to some unspecified ϵ.

A special case of attractor is the notion of **saddle**, a quasi state, which is stable with respect to some fluctuation parameters, but unstable with respect to other fluctuation parameters.

A concept which is symmetric to that of an attractor is the **repellor**, that is, a quasi state such that when a system reaches it, it after a while surely escapes from it.

Bistability is present when two basins (including two attractors) share a common portion which is ϵ-**unstable**, but the union of the two basins is ϵ-**stable**. In this case when the system, coming from an attractor, after a perturbation reaches the instability region, it may after a while fall into the stable part of the other basin, passing to the other attractor (see Fig. 20.5). More generally, **bifurcation** occurs when two different types of trajectories pass through an unstable region.

Stability and instability can also be referred to for trajectories. Let $x \in S$, then

$$\delta_\epsilon(x) = \{\delta(x, p) \mid p \in P, ||p|| \leq \epsilon\}.$$

Given a trajectory $(\delta^i(s)|i \in I)$ of origin $s \in S$, it is said to be ϵ-stable if, for every $i \in I$, and for every $x \in \delta_\epsilon(\delta^i(s))$, the following inclusion holds:

$$\delta_\epsilon(x) \subseteq \delta_\epsilon(\delta^{i+1}(s)).$$

The trajectory is ϵ-unstable if for some $i \in I$ the inclusion above does not hold. An important notion in theoretical biology is that of **creode** [29], which can be formalized as an ϵ-stable trajectory, for some fluctuation amplitude ϵ (see Fig. 20.6).

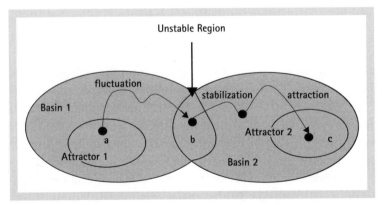

Fig. 20.5 A bistability phenomenon, where a system pass from a point *a* to a point *c* of another attractor, through an unstable region common to the two basins.

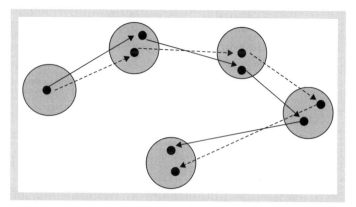

Fig. 20.6 A creode: an ϵ-stable trajectory. The itinerary along intermittent arrows is without perturbation. The itinerary along continuous arrows is fluctuating, but, at any step, it is inside ϵ-quasi states including the states of the unperturbed trajectory.

Finally, two dynamical notions with a strong biological significance are **liveness** and **reachability**. Liveness occurs when a system does not reach a fixpoint. Reachability occurs when from a given (quasi) state another specified (quasi) state can be reached. These concepts have been widely investigated in the context of artificial dynamical systems (especially, Petri nets). However, in the context of biological systems, it is not important to provide general procedures for deciding such properties, but rather, to determine specific cases and properties which can ensure their presence.

20.6 Metabolism

A wide class of systems which are very important from the biological point of view are **metabolic systems**. They are discrete autonomous dynamical systems where the phase space is a vector of real numbers which express the quantities of some substances. In a broad sense substances are different kinds of matter. At each step the substance quantities vary by means of some **reactions**, or more generally **transformations** (a transformation can correspond to a long and complex chain of many reactions). Moreover, some substance quantities can be introduced from outside and/or can be expelled outside. For internal reactions, which do not introduce or expel matter, the matter they consume is equal to the matter they produce (*conservation principle*). At any step, the fluxes of matter (consumed/produced) by all the rules produce the change of state. The matter of a substance which a rule moves, at any step, is an integer multiple (positive if produced and negative if consumed) of some quantities, called moles, specific of each substance. Fluxes depend on the state of the system. The change of each substance, at any step is the sum of contributions of all the rules which consume or produce that substance. This framework proves to be general enough to cover a great variety of dynamical systems, from chemical or biochemical systems, to cellular dynamics, signaling systems seen as protein activations networks, but also population dynamics, and even economic systems seen as goods transformation systems. Indeed, in all these cases some quantities (of different types) change according to rules consuming/producing and inserting/expelling. The interesting fact about these phenomena is their common discrete mathematical representation: multisets of objects and rewriting rules over them. These are the essential ingredients of P systems, which when enriched by membrane structures determine an extraordinary mathematical and applicative potentiality.

A cell is a membrane structure containing molecules. The state of a cell at each point of its life is given by some physical parameters which are essential for sustaining the internal reactions, and by the kinds of molecules inside it, the numbers of them for each type, and where they are localized in the cell structure [12]. The passage from a cell state to another cell state is given by the reactions acting in the cell, that is, the molecules which are consumed and produced for each type and for each compartment, plus the molecules coming from outside into the cell and the molecules going outside from the cell (passages from a compartment to another one can be represented as special reactions).

The following example suggests the enormous relevance that P systems have in modeling cells, or even artificial prototypes of cells. It reports the molecular structure of *Escherichia coli*, a prokaryote which is 500 times smaller than a typical animal or vegetal cell. It has approximately 10^{11} molecules of about 6000 different types (an eukaryote cell has about 10^{15} molecules of about 10, 000 different types).

Example 20.9 (Proto-cells as P systems) *Escherichia coli is a prokaryote which is one of the most investigated primitive organisms. It has an elongated shape of approximately of $2\mu m$ length and a diameter of $1\mu m$, its weight is about one nano gram. Among its 6000 different kinds of molecules about 1000/2000 are proteins. Tables 20.1 and 20.2, from [7], give some quantitative information about the composition of its cell.*

Table 20.1 Grams of macromolecules in a cell of *Escherichia coli*.

DNA	RNA	Proteins	Total Weight
0.017×10^{-12}	0.1×10^{-12}	0.2×10^{-12}	10^{-12}

Table 20.2 Small molecules in a cell of *Escherichia coli*.

Molecule Type	Cell Weight %	Aver. Mol. W.	Number
H_2O	70	18	4×10^{10}
Inorganic Ions	1	40	2.5×10^8
lipids and precursors	2	750	2.5×10^7
Carbohydrates and prec.	3	150	2×10^8
Amino acids and prec.	0.4	120	3×10^7
Nucleotides and prec.	0.4	300	1.2×10^7
Other small molecules	0.2	150	1.5×10^7

The original notion of a P system is not adequate to model cell dynamics, in fact P systems rewriting is performed in a non-deterministic way, according to many possible strategies, and among them is the originally adopted maximal parallel: rewrite all the objects you can, according to a maximal set of rules applicable in a given configuration. This strategy is very useful from the mathematical point of view. In fact, one membrane is enough for simulating any register machine in terms of non-deterministic maximally parallel multiset rewriting[1]. Of course, if our membrane system is a model of a proto-cell, then maximal parallelism is too reactive for modeling in a realistic way the biomolecular dynamics. This was the main motivation for extending P systems to metabolic P systems, MP systems for short. They are essentially multiset grammars where rules are regulated by functions. This means that, if the state of a system is the multiset of objects inside it, for each reaction, a state dependent function provides a value telling the **reaction unit** that has to be moved by the reaction in that state. This reading of reactions is a generalization of the Avogadro principle in chemistry. In fact, if a unit u is assigned to a reaction $aab \to c$ in a given state, then $2u$ objects of type a and u objects of type b are consumed and u objects of type c are produced by the reaction in the passage to the next state. In other words, an MP system can be identified with a *multiset grammar regulated by maps*. For this reason, we adopt two different ways of representing MP systems: MP graphs (see Fig. 20.7) and a set of rules (see Table 20.3) where each rule is equipped with an (algebraic) formula expressing the regulation map, also called the **flux map**, giving, with respect to some conventional population unit, the number of objects transformed by the rule. A theory of these systems has been developed [18] (see also Chapter 19 on metabolic P systems in this volume). They are related to differential equations [8], but they keep

---

[1] Considering Minsky register machines, encode the register r_j containing number n with n copies of the symbol r_j, and encode with the symbol p_i the fact that i is the current instruction. Then, the instruction at i **do** $r_j := r_j + 1$ **and go to** h is represented by the multiset rewriting rule $p_i \to r_j p_h$, while the instruction at i **if** $r_j > 0$ **then do** $r_j := r_j - 1$ **and go to** h **else go to** k is represented by the set of rules $\{p_i \to p'_i \#, \ p'_i r_j \to q_i, \ \# \to \$, \ p'_i \$ \to p_k, \ q_i \$ \to p_h\}$ (symbols $p'_i, \#, \$, q_i$ are auxiliary).

Table 20.3 The regulated multiset grammar of the MP system in Fig. 20.7 (λ is the empty multiset, constant values are given in Fig. 20.7).

$R_1: \lambda \to A$ $2 * K_1 * A/(K_2 * C + K_4 * B + K_1 + K_a)$
$R_2: A \to B$ $K_2 * C * A/(K_1 + K_2 * C + K_4 * B + K_a)$
$R_3: B \to \lambda$ $K_3 * B/(K_3 + K_b)$
$R_4: A \to C$ $K_2 * B * A/(K_1 + K_2 * C + K_4 * B + K_a)$
$R_5: C \to \lambda$ $K_5 * C/(K_5 + K_c)$

an intrinsically discrete algebraic nature, that is, their dynamics can be calculated by means of recurrent equations based on vectors and matrices related to the flux maps and to the stoichiometry of reactions. However, the most important aspect of MP dynamics, from the point of view of complex systems, is its natural potentiality to solve inverse dynamical problems, that is, inferring the flux maps of a given system, which drive the dynamics by means of algebraic data manipulation from some observed trajectories of the system [19].

The picture in Fig. 20.7 is a diagram, called a MP graph, [20] which describes the logic of a metabolic dynamic. Here we distinguish four kinds of nodes: ovals, circles, rectangles, and triangles. Moreover arcs are continuous and discontinuous. Ovals are substances, circles are reactions transforming substances, or introducing or expelling them, and triangles express input or output gates from and to the external environment.

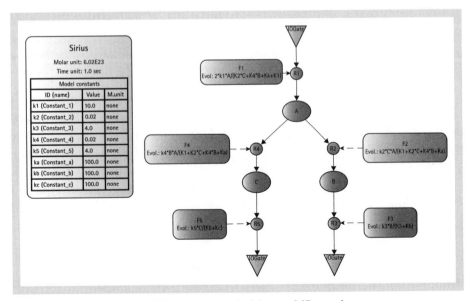

Fig. 20.7 The MP system Sirius represented by an MP graph.

518   METABOLIC P DYNAMICS

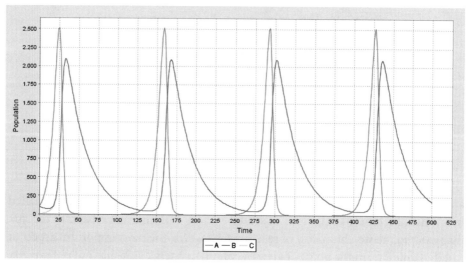

Fig. 20.8 The dynamics of Sirius.

Life phenomena are based, at many different levels, on oscillations. In order to appreciate the relevance of this concept, we quote the words of the Nobel laureate Ilia Prigogine in his foreword to Goldbeter's book [10]:

I remember my astonishment when I was shown for the first time a chemical oscillatory reaction, more than 20 years ago. I still think that this astonishment was justified, as the existence of chemical oscillations illustrates a quite unexpected behavior. We are used to thinking of molecules as traveling in a disordered way through space and colliding with each other according to the laws of chance. It is clear, however, that this molecular disorder may not give rise by itself to supramolecular coherent phenomena in which millions and millions of molecules are correlated over macroscopic dimensions.... It is this synchronization that breaks temporal symmetry.

Here we summarize the main features of the systems given in the following figures. For an easier identification, we call these systems names of well known stars. *Sirius* is depicted in Fig. 20.7. Its oscillating dynamic[2] is given in Fig. 20.8, while the regulated multiset grammar driving the dynamics of *Sirius* is given in Table 20.3. *Sirius* was the first oscillator figured by means of MP systems, its differential formulation is given in [18][3] . However, an oscillatory pattern can be obtained even with only two substances. In fact *Sirius geminus*, in Fig. 20.9, is such a kind

---

[2] In all the examples of MP systems, dynamics are calculated by means of Psim, a computational package [5] developed at the Computer Science Department of the University of Verona, by the group of *Models of Natural Computation* led by the author. The related software can be download from www.cbmc.it. An extended package, under development, called MetaPlab, is available at mplab.sci.univr.it.

[3] See [4] for a general analysis of oscillatory patterns.

METABOLIC P DYNAMICS 519

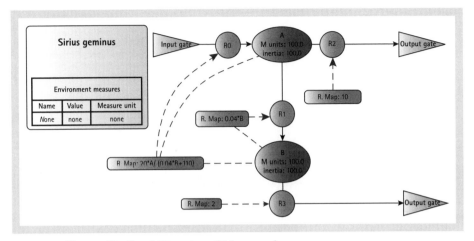

Fig. 20.9 The oscillating MP system Sirius geminus.

of MP system. Its trajectory is given in Fig. 20.10 and its orbit in phase space (phase diagram) is given in Fig. 20.11. The MP system of *Sirius geminus*, and of other MP systems we consider in the following, use *reaction maps* instead of flux maps, which are usually described by simpler formulae but use a more complex mechanism for the determination of reaction units. However, the two kinds of maps are completely equivalent as it is shown in [19]. An even simpler oscillating MP system, *Sirius unarius*, can be found, having only one substance (Fig. 20.12). However, it is interesting that its oscillation (Fig. 20.13) is lost even with very small differences in one regulation map (see Fig. 20.14). Figs. 20.15, 20.16, and 20.17 show

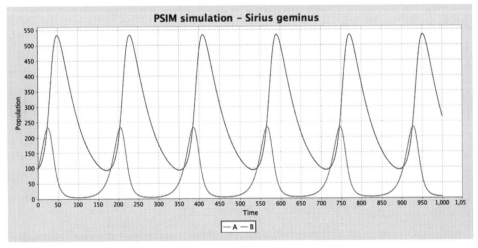

Fig. 20.10 A trajectory of Sirius geminus.

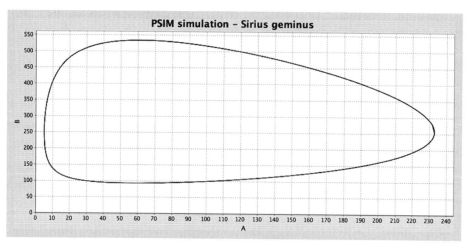

Fig. 20.11 The phase diagram orbit of Sirius geminus.

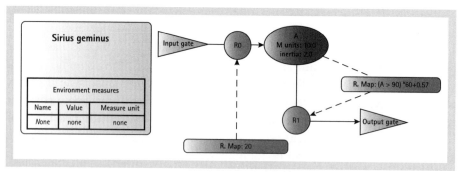

Fig. 20.12 The MP system Sirius unarius, with only one substance.

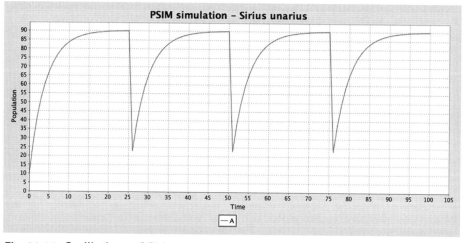

Fig. 20.13 Oscillations of Sirius unarius.

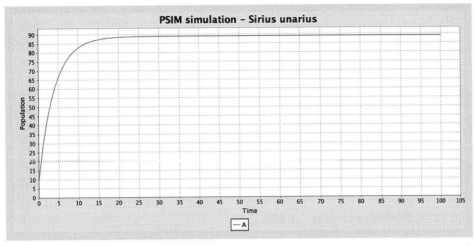

Fig. 20.14 Dynamics of Sirius unarius, where a regulation function was changed by an additive factor of 0.001.

a case of an oscillatory system, *Vega*, which seems to be irregular in its initial phase, but which, after that, becomes very regular. On the contrary, Figs. 20.18, 20.19, and 20.20 report an oscillatory system, *Mizar*, which in the initial steps seems to be regular, but which on a long run, presents punctual anomalies where no evident pattern results to be exactly repeated in time.

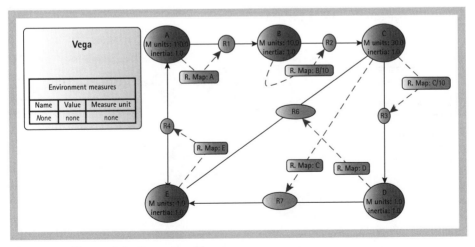

Fig. 20.15 The metabolic system Vega.

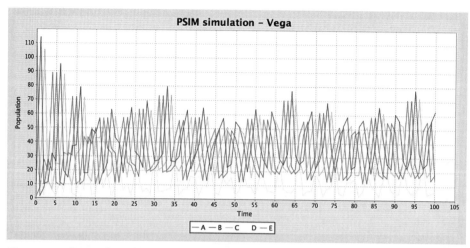

Fig. 20.16 Dynamics of Vega for 100 steps.

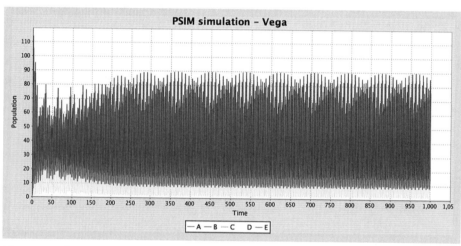

Fig. 20.17 Dynamics of Vega for 1000 steps.

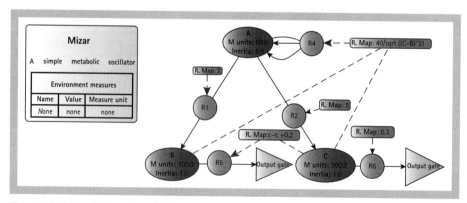

Fig. 20.18 The MP system Mizar.

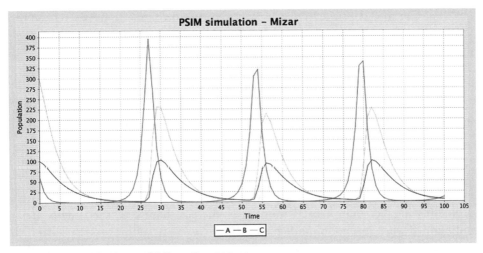

Fig. 20.19 Oscillations of Mizar for 100 steps.

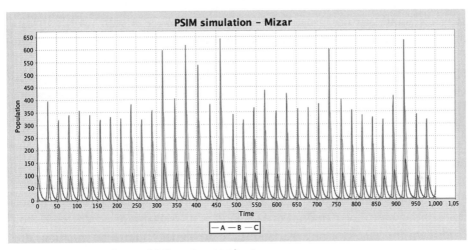

Fig. 20.20 Oscillations of Mizar for 1000 steps.

20.7 Anabolism and Catabolism

Metabolism is a crucial phenomenon of living systems, but a deeper analysis of metabolic phenomena distinguishes two different types of metabolism: **anabolism** and **catabolism**. This distinction is related to the energetic aspect of biochemical reactions. Anabolism is the part of metabolism related to the synthesis of substances which are energetically rich or play a structural role in the life processes, while catabolism is the part of metabolism of degradation of substances into their chemical components, when they are transformed into less energetically rich molecules, or they degrade. In this section we introduce some concepts, related to anabolism and catabolism, that are crucial in the analysis of MP systems modeling life processes.

An MP system is **assimilative** when there is at least one input gate in the corresponding MP graph. It is **dispersive** if there is at least one output gate in the corresponding MP graph. It is **dissipative** if it is simultaneously **assimilative** and **dispersive**, and it is **open** if it is either **assimilative** or **dispersive**, and is **closed** if it is not open.

An MP system is **conservative** when its reactions do not create or destroy matter. It is **creative** when it is not conservative.

An MP system is **bounded** if the total amount of mass of substances in it is a quantity smaller than a fixed number b, the size bound of the system. It is **unbounded** when it is not bounded.

An MP system is **continuative** along a sequence of steps if all the reactions of the system are applied in all the steps.

A simple way for providing oscillations is a closed MP system with only two substances and two reactions: $r_1 : A \to B$ and $r_2 : B \to A$ with two flux maps, the first one giving the flux of r_1 linearly depending on the amount of A, and the second one giving the flux of r_1 linearly depending on the amount of B. However, such a system does not exist in any chemical or biochemical reality. This impossibility is connected to a crucial aspect of chemical transformations which are intrinsically oriented and have to obey a principle based on Gibbs' variation of free energy. In simpler terms, the natural orientation of a chemical transformation is from reactants which are globally *energetically richer* to the correspondent products.

In [18] the principles inspiring MP systems were related to some generalization of chemical laws of Avogadro (*molarity principle*: any reaction involves integer multiples of the same molar amount of each kind of its reactants or products), Dalton (*additivity principle*: substance variations are the sums of substance variations due to all reactions), and Lavoisier (*conservation principle*: in any reaction the mass consumed equates the mass produced) which is generalized by the conservativeness enunciated above. Now we introduce a notion of **dissipation**, which allows us to formulate a sort of generalization of Gibbs' principle, for dealing with the difference between anabolism and catabolism (see [1]).

Given an MP system, a **reaction graph** is a graph where nodes are substances and edges are reactions connecting reactants to products. A **reaction cycle** is a cycle in the reaction graph of an MP system. A reaction cycle is **assimilative** if some of its reactions are connected with reactant substance nodes which are external to the cycle. Analogously, a cycle is **dispersive** if some of its reactions are connected with external product substance nodes. A cycle is **dissipative** if it is assimilative and dispersive. The following is a general principle which introduces in MP systems an energy constraint fundamental for real metabolic processes.

Principle 20.1 (Cycle dissipation) *In a bounded and conservative MP system any reaction cycle must be dissipative.*

The following results show, in a general framework, the importance of open systems and the relationship between dissipation and oscillation in dynamical regimes.

Lemma 20.2 *For any closed, conservative and bounded MP system, there exists a number of steps, such that the system cannot be continuative for a longer sequence of steps.*

Proof. If the system is closed, then from the Cycle dissipation principle no reaction cycles are present in the system, therefore after a number of steps reactants disappear, because the system is conservative and its internal matter cannot increase or decrease. Therefore the system cannot continue to apply its reactions.

Lemma 20.3 *Any assimilative, conservative, bounded, but non-dispersive, MP system, cannot be continuative more than a bounded number of steps.*

Lemma 20.4 *Any dispersive, conservative, bounded, but non-assimilative, MP system, cannot be continuative more than a bounded number of steps.*

Proposition 20.4 *Given a conservative and bounded MP system, there exists a number of steps, such that, if the system is continuative for a longer number of steps, then it has to be dissipative and oscillating along these steps.*

Proof. (outline) From the two previous lemmas the system is open and moreover dissipative. But substance amounts cannot increase beyond a superior level because the system is bounded, and, since it is continuative, they cannot decrease beyond an inferior level too. Therefore, if the system continues to apply its reactions, surely substance quantities vary within some intervals, therefore the system is oscillating.

20.8 Conclusions

In this chapter we have shown that, in the framework of discrete dynamical systems, crucial aspects of living systems can be expressed, especially if classical dynamical notions are formulated by using quasi determinism and fluctuations. P systems are the natural discrete framework for describe cell dynamics. In fact, in a cell there are several chemical components: DNA and RNA molecules, proteins, membranes, and so forth. The state and the function of the cell is directly related to its chemical composition. Through biochemical reactions this composition changes in time. Of course, a cell is a much more complex system, but this standpoint is the basis of any theoretical analysis of its dynamics. In this study we focused on metabolism, which is the basis of life. Metabolic P systems are a starting point for cell dynamics investigations. We aim to extend metabolic P systems in two different directions: i) building specific models of cell processes, and ii) using MP systems, or their suitable extensions, for generating dynamical patterns of interest in the emergence of cell growth, adaptation development, evolution, and reproduction. In this sense, P systems could provide a discrete framework for the kind of investigations which are arising in synthetic biology. In [12] many principles have been individuated, by means of specific mathematical elaborations and biological experiments, such as *minority control, isologous diversification*, or *community effect*, which explain basic cell functions: *cell recursive growth, differentiation, speciation*, and so forth. In order to discover general laws, we need a general type of dynamical system where they can be investigated, and the classical framework of systems biology [3] could be enriched by discrete and algorithmic tools. We claim that these systems are special kinds of P systems. Of course, mathematical theories, computational tools, and biological experiments have to be deeply integrated to achieve good results in this direction. A technical point will soon be addressed. In fact, in metabolic P systems, membrane structure is not very relevant, however, in many of the phenomena we mentioned, it is essential to consider the interactions of many similar systems interacting in a common environment. At this level, membrane structures and related concepts become a crucial aspect. This will be one of the next steps in the application and extension of MP systems.

ACKNOWLEDGEMENTS

I am grateful to many people in the community of P systems who collaborated with me, at several levels and in different ways. I want to express a particular gratitude to Gheorghe Păun for his constant and contagious enthusiasm for developing mathematical research in frontier fields with a free and friendly spirit. I want also thank Erzsébet Csuhaj-Varjú for her warm encouragement to write this chapter, and Giuditta Franco, for her very careful and precise reading of a preliminary version of this chapter, which helped me to correct and improve the work.

REFERENCES

[1] B. ALBERTS, M. RAFF: *Essential Cell Biology. An Introduction to the Molecular Biology of the Cell.* Garland Science, New York, 1997.

[2] F. BERNARDINI, V. MANCA: Dynamical aspects of P systems. *BioSystems*, 70 (2003), 85–93.
[3] L. VON BERTALANFFY: *General Systems Theory: Foundations, Developments, Applications*. George Braziller Inc., New York, 1967.
[4] L. BIANCO, V. MANCA: Symbolic generation and representation of complex oscillations. *Intern. J. Computer Math.*, 83 (2006), 549–568.
[5] L. BIANCO, V. MANCA, L. MARCHETTI, M. PETTERLINI: Psim: a simulator for biochemical dynamics based on P systems. *2007 IEEE Congress on Evolutionary Computation*, Singapore, September 2007.
[6] C. BONANNO, V. MANCA: Discrete dynamics in biological models *Romanian J. Information Science and Technology*, 5 (2002), 45–67.
[7] J. DARNELL, H. LODISH: *Molecular Cell Biology*. Scientific American Books, 1986.
[8] F. FONTANA, V. MANCA: Discrete solutions to differential equations by metabolic P systems. *Theoretical Computer Sci.*, 372 (2007), 165–182.
[9] R.C. HILBORN: *Chaos and Nonlinear Dynamics*. Oxford University Press, 2000.
[10] A. GOLDBETER: *Biochemical Oscillations and Cellular Rhythms*. Cambridge University Press, New York, 2004.
[11] J. JOST: *Dynamical Systems*. Springer, 2005.
[12] K. KANEKO: *Life: An Introduction to Complex Systems Biology*. Springer, 2006.
[13] P. KURKA: *Topological and Symbolic Dynamics*. Société Mathématique de France, 2003.
[14] R.B. MAY: Simple mathematical models with very complicated dynamics. *Nature*, 261 (1976), 459–467.
[15] V. MANCA: String Rewriting and metabolism: A logical perspective. *Computing with Bio-Molecules* (Gh. Păun, ed.), Springer, 1998, 36–60.
[16] V. MANCA: Metabolic P systems for biochemical dynamics. *Progress in Natural Sciences*, 17 (2007), 384–391.
[17] V. MANCA: Discrete simulations of biochemical dynamics. *Lecture Notes in Computer Sci.*, 4848 (2008), 231–235.
[18] V. MANCA: The metabolic algorithm for P systems: Principles and applications. *Theoretical Computer Sci.*, 404, 1–2 (2008), 142–157.
[19] V. MANCA: Log-gain principles for metabolic P systems. *Algorithmic Bioprocesses* (A. Condon et al., eds.), Springer, 2009, Chapter 28.
[20] V. MANCA, L. BIANCO: Biological networks in metabolic P systems. *BioSystems*, 91 (2008), 489–498.
[21] V. MANCA, L. BIANCO, F. FONTANA: Evolutions and oscillations of P systems: Applications to biological phenomena. *Lecture Notes in Computer Sci.*, 3365 (2005), 63–84.
[22] V. MANCA, G. FRANCO, G. SCOLLO: State transition dynamics: basic concepts and molecular computing perspectives. *Molecular Computational Models: Unconventional Approachers* (M. Gheorghe, ed.), Idea Group Inc. UK, 2005, chapter 2, 32–55.
[23] V. MANCA, D.M. MARTINO. From string rewriting to logical metabolic systems. *Grammatical Models of Multi-Agent Systems* (G. Păun, A. Salomaa, eds.), Gordon and Breach Science Publishers, London, 1999.
[24] GH. PĂUN: *Membrane Computing. An Introduction*. Springer, 2002.
[25] G. SCOLLO, G. FRANCO, V. MANCA: A relational view of recurrence and attractors in state transition dynamics. *Lecture Notes in Computer Sci.*, 4136 (2006), 358–372.

[26] G. Scollo, G. Franco, V. Manca: Relational state transition dynamics. *J. Logic and Algebraic Programming*, 76 (2008), 130–144.
[27] W. Reisig: *Petri Nets, An Introduction*. Springer, 1985.
[28] T.M. Robinson: *Heraclitus: Fragments*. University of Toronto Press, Toronto 1987.
[29] R. Thom: *Stabilité Structurelle et Morphogénèse*. Benjamin, Inc. 1972.
[30] G.J. Wirsching: *The Dynamical System Generated by the 3n + 1 Function*. Lecture Notes in Mathematics, 1681, Springer, 1998.
[31] S. Wolfram: *Theory and Application of Cellular Automata*. Addison-Wesley, 1986.

CHAPTER 21

MEMBRANE ALGORITHMS

TAISHIN Y. NISHIDA

TATSUYA SHIOTANI

YOSHIYUKI TAKAHASHI

21.1 INTRODUCTION

P systems offer a theoretical tool to devise an exponentially growing structure in order to solve **NP**-complete problems in deterministic (parallel) polynomial time. When technologies of molecular or biological computing will catch up theory, this computing mechanism will solve such a problem effectively. But, now we have just a serial computer or a parallel system with a small number (compared to the "exponential growth") of processors. Hence we cannot use P systems to obtain an answer of a large instance of an **NP**-complete problem, such as satisfiability of a Boolean formula with hundreds of variables.

However, P systems provide other mechanisms to solve computationally hard problems: membrane separated regions, a multiset in a region, evolution rules in a region, rules of communication across a membrane, parallel iteration of rules, dynamic restructuring of membranes, and so forth. T.Y. Nishida has noticed that a combination of such features and approximate algorithms investigated so far suggests a new class of distributed evolutionary algorithms and he has proposed membrane algorithms to this aim [7, 8, 9, 10]. In this chapter, we describe membrane algorithms as a framework for approximate computations. Then applications of membrane algorithms to some problems are mentioned including results of computer experiments.

21.2 WHAT IS A MEMBRANE ALGORITHM

Since many variants of P system are computationally universal, every algorithm can be, in principle, implemented in these systems. If an instance of a problem is appropriately coded into a P system and the system is realized in vitro, in vivo, or in silico, a number is obtained in the output membrane as the multiplicity of objects in the region. Then an appropriate decoding from this number to the format of the problem gives an answer.

Let us consider an "appropriate coding and decoding" to solve an existing problem, for example, the Traveling Salesman Problem (TSP). Distances between pairs of nodes could be expressed by numbers of objects. What form should a solution have? A solution of TSP is usually represented in a sequence of nodes. Let $(v_1, v_2, \ldots, v_n)$ be a solution of TSP with n nodes where $v_i \in \{1, \ldots, n\}$ is a node for every $1 \leq i \leq n$. The Gödel number $2^{v_1} 3^{v_2} \cdots p_n^{v_n}$ where p_i is the i-th prime number is a possible coding of the solution. A membrane structure $[_1[_2\cdots[_n]_n\cdots]_2]_1$ and a multiset w_i in the region i such that $|w_i| = v_i$ for every $1 \leq i \leq n$ gives another possibility. However, no one thinks that a P system with such coding can effectively solve TSP. In other words, these possibilities just point out theoretical consistency.

In order to solve existing problems effectively, it is important that the structure of a problem is embedded in a P system, rather than they are coded into a P system. For example, a P system which solves TSP should have sequences of nodes as its objects. The evolution rules should look like local search, greedy search, branch and bound, genetic algorithm, and so on.

Now a new P system, called a *membrane algorithm*, is proposed as a "computing platform" in which approximate algorithms are effectively embedded. In order to see features of a membrane algorithm, we summarize a common framework of approximate algorithms. Let $\mathcal{A}$ be the set of all possible solutions or the search space of a problem. There are two operators

$$r : \mathcal{A}^k \to \mathcal{A}^l,$$

and

$$s : \mathcal{A}^l \to \mathcal{A}^k,$$

with $1 \leq k < l$. Operator r changes solutions, into better ones if it can do, and s selects solutions. An algorithm successively computes groups of solutions $A_i = s(r(A_{i-1}))$ for $i \geq 1$, where A_0 is a group of initial solutions. For example, the local search algorithm has $r : \mathcal{A}^1 \to \mathcal{A}^2$ which maps from a solution a to a set $\{a, a'\}$ with a' being a neighbor of a in the search space and $s : \mathcal{A}^2 \to \mathcal{A}$ which simply selects a better solution in $\{a, a'\}$ with respect to the optimizing criterion. As for the genetic algorithm, the mapping $r : \mathcal{A}^n \to \mathcal{A}^{n+2}$ maps a set of solutions A to $A \cup \{b_1, b_2\}$ where b_1 and b_2 are produced from a pair of randomly selected solutions a_1, a_2 in A by recombination and mutation. Operation $s : \mathcal{A}^{n+2} \to \mathcal{A}^n$ selects the best n solutions in $A \cup \{b_1, b_2\}$. If we consider these operations from the biological point of view, especially from the evolution of the virus, then the operation r corresponds to the reproduction of DNA or RNA in a cell and the operation s corresponds to a selection mechanism by which objects go to the outer region across membranes.

The above observations bring to membrane algorithms the following components:

1. A number of regions which are separated by membranes.
2. For every region:
   (a) A few tentative solutions of the optimization problem to be solved.
   (b) A mechanism which modifies the solutions.
   (c) A solution transporting mechanism between adjacent regions.

The operation r is realized by the modification mechanism. The transporting mechanism, which corresponds to the communication rules of P system, selects solutions, i.e. it realizes the operation s. The transporting mechanism plays another important role. It moves solutions to adjacent regions in which solutions are modified by another mechanism.

The two mechanisms of modification and transportation are called the *subalgorithms* of a region. Subalgorithms can be different in different regions. For example, the mutation probability of the genetic algorithm may be varied. Moreover, different regions may have subalgorithms based upon different principles, e.g. local search, genetic algorithm, etc. It should be stressed that in this respect membrane algorithms are different from the heuristic optimization algorithms investigated so far (e.g. [13, 14]).

21.3 Details of a Membrane Algorithm

Membrane algorithms were originally proposed in [7, 8]. In this section, we introduce the details of membrane algorithms according to the references [10, 11].

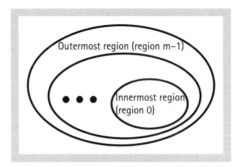

Fig. 21.1 Membrane structure of a membrane algorithm.

21.3.1 Components of a Membrane Algorithm

The original membrane algorithms used a nested membrane structure (Fig. 21.1). The innermost region, surrounded by the only elementary membrane, is the output region. This corresponds to the custom of P systems area that the output region is surrounded by an elementary membrane.

Every region has a subalgorithm as the evolution and communication rules of the region and a number of solutions as the objects. The number of solutions is the smallest one for the subalgorithm to modify solutions. That is, if the subalgorithm is a kind of local search, then the number of solutions is one, and if the subalgorithm is a genetic algorithm, then the number is two. After modification, solutions are sent to the neighboring regions by the communication rule of the subalgorithm. Usually, the best solution is sent to the inner region and the worst solution is sent to the outer region. Every region modifies and sends solutions simultaneously. This corresponds to the parallelism of P systems.

By iterating subalgorithms in all regions, a good approximate solution will appear in the innermost region. The iterations are repeated until a termination condition is satisfied.

The following principle is advised for designing a subalgorithm: "In the inner regions, solutions should be changed slightly; in the outer regions, solutions should be changed largely." Under this principle, it can be expected that a subalgorithm escapes from a local minima in the outer regions and that a subalgorithm makes solutions better or the optimum solution in the inner regions.

21.3.2 Membrane Algorithms for the Traveling Salesman Problem

Membrane algorithms were first applied to the traveling salesman problem in [7, 8]. This subsection describes subalgorithms and parameters of the membrane algorithms solving TSP.

Let m be the number of membranes, let region o be the innermost region, and region $m-1$ be the outermost region.

An instance of TSP with n nodes consists of n pairs of real numbers (x_i, y_i) $(i = 0, 1, \ldots, n-1)$ which correspond to points in the two dimensional space. The distance $d(v_i, v_j)$ between two nodes $v_i = (x_i, y_i)$ and $v_j = (x_j, y_j)$ is the geometrical distance $d(v_i, v_j) = \sqrt{(x_i - x_j)^2 + (y_i - y_j)^2}$. A solution of the instance is a list of nodes $(v_0, v_1, \ldots, v_{n-1})$ in which $v_i \neq v_j$ for every $i \neq j$. The *value* of a solution $v = (v_0, v_1, \ldots, v_{n-1})$ denoted by $W(v)$ is given by

$$W(v) = \sum_{i=0}^{n-2} d(v_i, v_{i+1}) + d(v_{n-1}, v_0).$$

For two solutions u and v, v is better than u if $W(v) < W(u)$. The solution which has the minimum value in all possible solutions is said to be the *optimum solution* of the instance. A solution which has a value close to the optimum solution is a good approximate solution.

Two types of subalgorithms, GA and Brownian, were introduced in [11]. The GA subalgorithm makes two new solutions from two solutions by the following procedure:

1. If the two solutions have the same value, then a part of one solution (which is selected probabilistically) is reversed.
2. The subalgorithm makes two new solutions by the recombination of the two solutions.
3. The subalgorithm modifies the two new solutions by point mutations.
4. The best solution of the four (two old and two new) solutions is sent to the inner region and the worst is sent to the outer region at the communication stage.
5. The best two in the four (the staying and received) solutions are selected as the solutions of the next generation.

The edge exchange recombination method [3] is used in step 2. If the two solutions in a region are identical, recombination makes no new solutions. Step 1 avoids this case and introduces a new solution using reverse operation, which is a kind of mutation.

The Brownian subalgorithm is a kind of local search inspired by the simulated annealing algorithm [2]. The algorithm has a temperature parameter T. The algorithm changes solution u to u' by replacing nodes. Let g be the gain $W(u') - W(u)$. If g is negative, then the new solution is a better solution and it is accepted. Otherwise, it is accepted under the probability $\exp(-g/T)$. That is, if the temperature T is high, then a bad solution may be accepted and it might escape from a local minima. An accepted solution is sent to the adjacent regions.

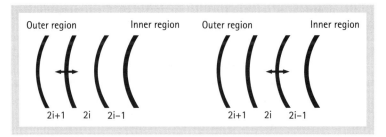

Fig. 21.2 Alternate communications among regions with Brownian subalgorithms. Communications of the left figure and the right figure alternate.

The Brownian subalgorithm puts only one solution to the communication stage. If both adjacent regions have a Brownian subalgorithm, then only one region communicates to the region in order to prevent one solution going to the next region of the neighboring region. A region communicates with the outer region in alternate iterations. A region communicates with the inner region when it does not communicate with the outer region (see Fig. 21.2).

The overall algorithm proceeds as follows:

1. Reads an instance of TSP.
2. Randomly makes appropriate number of tentative solutions for every region.
3. Repeats 3.1 and 3.2 for d times (d is given as a parameter) using the subalgorithms in every region.
   3.1 Modifies solutions.
   3.2 Communicates with the neighboring regions.
4. Outputs the best solution in region 0 as the output of the algorithm.

21.3.3 Computer Experiments

In this subsection, results of computer experiments of the membrane algorithm are mentioned. Three types of membrane algorithm are examined: the first consist of GA subalgorithms only, the second consists of Brownian subalgorithms only, and the third consists of GA and Brownian subalgorithms. The instances of TSP are taken from a collection of benchmark problems[1]. Table 21.1 shows the number of nodes and the optimum values of the instances.

Simulation programs are written in Java. For every pair of an instance and an algorithm, 20 trials have been done and a trial has finished in 500,000 iterations unless otherwise stated.

---

[1] TSPLIB URL: http://www.iwr.uni-heidelberg.de/groups/comopt/software/ TSPLIB95/.

Table 21.1 Optimum values of benchmark problems.

problems	eil51	eil76	eil101	kroA100	kroA150	kroA200
nodes	51	76	101	100	150	200
values	426	538	629	21282	26524	29368

21.3.3.1 Membrane algorithm with GA subalgorithm only

Table 21.2 shows the results of the membrane algorithms with GA subalgorithm only and with 1, 5, 10, or 30 membranes. In this experiment, the termination condition is 200,000 iterations because computation of a GA subalgorithm takes a long time. In region i, a mutation occurs with probability i/m where m stands for the number of membranes. Table 21.2 shows percentages of the errors (the average values—the optimum values) and the standard deviations (in the parentheses) to the optimum values. It can be seen that membrane algorithm obtains good solutions as the number of membranes increases.

21.3.3.2 Membrane algorithm with Brownian subalgorithm only

Table 21.3 shows the results of the membrane algorithms with Brownian subalgorithm only and with 1, 3, 10, or 30 membranes. The temperature T_x at region x is given by

$$T_x = \begin{cases} 0 & \text{if } x = 0, \\ T_{max}\theta^{m-x-1} & \text{otherwise,} \end{cases}$$

Table 21.2 Results of the membrane algorithms with GA subalgorithm only. The first row shows the number of membranes which an algorithm has. The entries of the table stand for the percentages of the errors (average values—the optimum values) and the standard deviations (in the parentheses) to the optimum values. The "*" means that the optimum solution is obtained.

	1	5	10	30
eil51	6.36(2.82)	4.05(1.87)	2.16(1.34)	0.83(0.97)*
eil76	8.61(2.13)	6.96(1.43)	4.16(0.82)	1.98(1.23)*
eil101	8.47(1.55)	7.55(1.28)	5.2(1.42)	3.71(0.87)
kroA100	9.51(3.54)	6.5(3.2)	4.19(1.87)	1.99(1.38)*
kroA150	10.51(2.84)	9.49(2.48)	8.01(1.88)	6.55(1.99)
kroA200	10.76(2.36)	9.9(2.23)	9.01(2.5)	8.36(1.88)

Table 21.3 Results of the membrane algorithms with Brownian subalgorithm only. The meanings of the entries are the same as Table 21.2. The 0(0)* means that all trials get the optimum solution.

	1	3	10	30
eil51	3.45(1.1)	0.04(0.08)*	0(0)*	0.02(0.03)*
eil76	6.16(1.65)	0(0)*	0(0)*	0(0)*
eil101	6(1.27)	0(0)*	0(0)*	0(0)*
kroA100	5.86(2.09)	0.18(0.15)*	0(0)*	0(0)*
kroA150	6.03(2.26)	0.47(0.47)*	0.24(0.11)	0.12(0.09)*
kroA200	6.66(1.49)	0.9(0.57)	0.61(0.19)	0.38(0.13)

where m is the number of regions and T_{max} and θ are defined by

$$T_{max} = \frac{0.5}{\log 2} \max_{v_i/v_{\overline{j}}} d(v_i, v_j),$$

$$T_{min} = \frac{0.25}{\log 20n} \min_{v_i/v_{\overline{j}}} d(v_i, v_j), \text{ and}$$

$$\theta = \sqrt[m-1]{\frac{T_{min}}{T_{max}}},$$

in which $v_0, \ldots, v_{n-1}$ are nodes.

We can conclude that Brownian subalgorithms can obtain quite good solutions[2].

21.3.3.3 Membrane algorithms with Brownian and GA subalgorithms

In this subsection a new possibility of mixing Brownian[3] and GA subalgorithms is described[4]. The "select and move" column in Table 21.4 shows results of a membrane algorithm with 3 membranes, Brownian, GA, and Brownian subalgorithms in regions 0, 1, and 2, respectively. The algorithm exhibits similar behavior to [11]. The results in the "swap" column in Table 21.4 are obtained by an algorithm which has 3 membranes, Brownian subalgorithms with temperature 0 in regions 0 and 2, and a GA subalgorithm with swap communication in region 1.

A GA subalgorithm with swap communication sends the better solution obtained by the recombination to the inner region and the worse solution obtained by the recombination to the outer region. If the solution sent to the inner region

---

[2] Bad values which were reported in [11] were caused by a bug in the program. The results in this section have been computed by a program where the bug has been corrected.

[3] Indeed it is nothing but a local search since the temperature parameter is always 0.

[4] A number of experiments on a mixture of Brownian and GA subalgorithms have been reported in [11].

Table 21.4 Results of the membrane algorithms with Brownian and GA subalgorithms. The algorithms have 3 membranes. Regions 0, 1, and 2 have Brownian, GA, and Brownian subalgorithms, respectively. Temperatures of the Brownian subalgorithms are 0. The GA subalgorithm does not use the mutation. "Select and move" means that the GA subalgorithm communicates with the method in Section 21.3.1. "Swap" means that GA subalgorithm communicates by the swap method which is described in the body text.

	select and move	swap
eil51	4.05(1.61)*	0(0)*
eil76	5.29(2.11)	0(0)*
eil101	6.56(1.89)	0.06(0.08)*
kroA100	4.79(2.69)	0(0)*
kroA150	5.7(1.87)	0.12(0.09)*
kroA200	6.05(1.84)	0.48(0.1)

is better than the solution from the inner region, then they are exchanged. Otherwise, the solution sent to the inner region stays in the original region (and so does the solution from the inner region). If the solution sent to the outer region is worse than the solution from the outer region, then they are exchanged. Otherwise, the solutions do not move. After exchange, the best two solutions of the two original solutions and the exchanged solutions (if any) are selected as solutions of the next generation. At the neighboring regions, a solution from the GA region is accepted as a solution of the next generation only if an exchange occurs.

The "swap" communication method exhibits very good behavior. With only 3 membranes, it is as good as the algorithm with Brownian only and 30 membranes. This means that the method can escape local minima (premature convergence) very well. The following reasoning explains this phenomenon.

Let A_0 be the solution in region 0, A_2 be the solution in region 2, A_1 and B_1 be the original solutions, and C_1 and D_1 be the solutions obtained by recombination in region 1. We assume that A_0 and A_2 fall in local minima, that is, local search cannot improve them. We also assume that recombination can only make worse solutions but some of them have a possibility to become good solutions if they are improved by local search. The following orders of solutions

$$W(A_0) < W(A_1) \le W(B_1) < W(C_1) < W(D_1)$$

and

$$W(B_1) < W(A_2) < W(D_1)$$

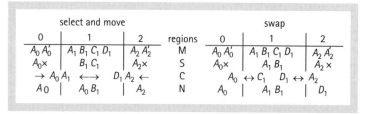

Fig. 21.3 "Select and move" communication (left) and "swap" communication (right). The rows represent solutions after modification (M), solutions after selection (S), solutions sent at communication stage (C), and solutions for the next generation (N), from up to down.

can be assumed since region 2 has a worse solution than region 1. The "select and move" communication method sends A_1 and D_1 to regions 0 and 2, respectively; and receives A_0 and A_2 from regions 0 and 2, respectively. At the next generation, region 0 has A_0, region 1 has A_0 and B_1, and region 2 has A_2 because of the orders. Then no subalgorithms can improve the solutions (Fig. 21.3 left).

On the other hand, the "swap" communication sends C_1 and D_1 to regions 0 and 2, respectively. In this case D_1 and A_2 are exchanged. The solution A_2 is discarded because it is the third better solution at the region. Thus at the next generation, region 0 has A_0, region 1 has A_1 and B_1, and region 2 has D_1. The solution D_1 is expected to be improved by local search (Fig. 21.3 right). Now, if an improved solution of D_1 becomes better than B_1, then it goes to region 1.

21.4 Membrane Algorithms for Various Problems

This and the next section present membrane algorithms which use a different structure and/or solve other problems. For each problem we present the components of the algorithm and computer experiments with instances taken from benchmark data bases.

21.4.1 Job-shop Scheduling Problem

We start with a membrane algorithm which solves Job-Shop Scheduling Problems (JSSP), [12].

21.4.1.1 *The problem*

A JSSP consists of a number of jobs and a number of machines. In what follows we consider a JSSP with m machines $(1, \ldots, m)$ and n jobs $(1, \ldots, n)$. Each job i must be processed by all machines and each machine can process only one job at a time. A machine cannot be interrupted when it begins to process a job. Machine j finishes job i in a process time p_{ij}. For every job, an order of machines on which the job is processed is given by an instance of JSSP. The order is called a *technical order*. For every pair of job i and machine j, the process time p_{ij} is also given.

A schedule consists of assignments of beginning and ending times to all pairs of jobs and machines. The time to complete all jobs is the interval between the earliest beginning time and the latest ending time, which is called a *makespan*. Given a technical order and process times p_{ij}, JSSP asks to find a schedule of shortest makespan.

Example 21.1 *Table 21.5 shows a technical order and process times of JSSP with 3 jobs and 3 machines.*

A schedule is visualized by a Gantt chart. The left chart of Fig. 21.4 shows a schedule in which each machine processes jobs in the order 1, 2, 3. The makespan of the left chart is 27. The right chart shows a schedule which gives the shortest makespan 15. There is more than one schedule which gives the shortest makespan.

21.4.1.2 *Components of the algorithm*

The same membrane structure as in the membrane algorithm solving TSP is used.

We use an $n \times m$ matrix S to denote a solution of JSSP with n jobs and m machines. The i, j element s_{ij} of S denotes that job s_{ij} is processed on machine i at the j-th process of the machine. The beginning and ending times and the makespan are automatically computed from S and the process times. For example, the next

Table 21.5 A technical order (left) and process times (right) of 3×3 JSSP. Each entry of the left table shows the machine on which the job is to be processed at the order. The right table shows the process times p_{ij}, $1 \le i \le 3$ and $1 \le j \le 3$.

	order				machines		
jobs	1st	2nd	3rd	jobs	1	2	3
1	1	2	3	1	3	3	2
2	1	3	2	2	2	3	5
3	2	3	1	3	1	2	6

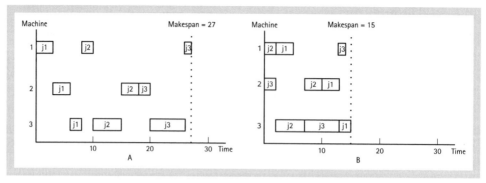

Fig. 21.4 Gantt charts of a schedule (A) and one of the optimal schedule (B).

matrices S_1 and S_2 represent schedules A and B in Example 21.1

$$S_1 = \begin{pmatrix} 1 & 2 & 3 \\ 1 & 2 & 3 \\ 1 & 2 & 3 \end{pmatrix} \quad \text{and} \quad S_2 = \begin{pmatrix} 2 & 1 & 3 \\ 3 & 2 & 1 \\ 2 & 3 & 1 \end{pmatrix},$$

respectively.

Using the above notation, subalgorithms for a membrane algorithm solving JSSP are easily obtained from algorithms commonly used for other optimization problems, since solutions of many optimization problems are expressed as a vector or a matrix.

We have considered two subalgorithms, Brownian and GA subalgorithms, which are using local search and recombination, respectively. A Brownian subalgorithm has one schedule matrix S. It produces a new matrix S' by the local search on one randomly selected row of S. If the makespan τ' of S' is smaller than the makespan τ of S, then S' is accepted as the solution that is sent to the adjacent regions. Otherwise, S' is accepted in probability $\exp(\tau - \tau')/T$ where T is a "temperature" parameter. If S' is not accepted, S becomes the solution to be sent. At the communication stage, if the solution in the outer region is better than the solution in the inner region, then the two solutions are exchanged. In order for a solution not to go through all regions at one iteration of subalgorithms, exchanges between two Brownian regions are done alternatively (see Section 21.3.2 and Fig. 21.2).

A GA subalgorithm has two schedule matrices. The algorithm crosses every pair of corresponding rows of the matrices with the order crossover [3] and makes two new matrices. The best solution of the four (old and new) solutions is sent to the inner region and the worst solution is sent to the outer region.

An execution of the algorithm is terminated after a pre-determined number of iterations of subalgorithms.

We must be careful with deadlocks which are hidden in the matrix notation. In Example 21.1, the next matrix S_d cannot be processed

$$S_d = \begin{pmatrix} 3 & 1 & 2 \\ 2 & 3 & 1 \\ 1 & 2 & 3 \end{pmatrix}.$$

We denote $(M_x, J_x) <_T (M_y, J_y)$ (resp. $(M_x, J_x) <_S (M_y, J_y)$) if the combination of machine M_x and job J_x must be processed earlier than the combination of M_y and J_y by the technical order (resp. solution S). Then we have the next relations:

$$(M_1, J_3) <_{S_d} (M_1, J_2), \quad (M_3, J_2) <_{S_d} (M_3, J_3),$$
$$(M_1, J_2) <_T (M_3, J_2), \quad (M_3, J_3) <_T (M_1, J_3),$$

which make a loop $(M_1, J_3) < (M_1, J_3)$ or a deadlock. Although algorithms specific to JSSP use other data structures, e.g. a graph, we use matrices and a deadlock removing algorithm, which can be found in [12].

21.4.1.3 Computer Experiments

Two types of membrane algorithms have been examined. One has 20 membranes and a Brownian subalgorithm in each region. The temperature of the innermost region is 0. The temperature of region r ($0 < r < 20$) is given by $T_{max}\theta^{19-r}$ where

$$\theta = \sqrt[20]{\frac{T_{min}}{T_{max}}},$$

$T_{max} = 0.5mt_{max}$, and $T_{min} = 0.5t_{min}/m$ in which m is the number of machines and t_{max} and t_{min} are the maximum and minimum process times, respectively. This type is called "all B".

The other type has 20 membranes and one subalgorithm of genetic type and 19 Brownian subalgorithms. The innermost region has a Brownian subalgorithm with temperature 0, region 1 has a genetic subalgorithm, and regions 2 to 19 have Brownian subalgorithms with temperature varying from 0 to the maximum. The temperature T of region r ($2 \leq r < 20$) is given by

$$T = \begin{cases} 0 & \text{if } r = 2 \\ T_{max}\theta^{19-r} & \text{otherwise} \end{cases}$$

where

$$\theta = \sqrt[18]{\frac{T_{min}}{T_{max}}}$$

and T_{max} and T_{min} are the same as in the "all B" algorithm. The genetic subalgorithm in region 1 does the order crossover once in every 16 iterations. It does nothing in

Table 21.6 Best values, averages, and standard deviations (in the parentheses) for "10 tough problems" obtained by membrane algorithms. The size is expressed by (number of jobs)×(number of machines). The number of trials is 20 for each algorithm-problem combination. MA (all B) stands for the membrane algorithm with Brownian subalgorithm only and MA (BG) stands for the membrane algorithm with Brownian and genetic subalgorithms.

Problems (optimum, size)	MA (all B) best	av.(std.)	MA (BG) best	av.(std.)
abz7 (656, 20 × 15)	706	715.4 (4.76)	701	716 (7.67)
abz8 (645, 20 × 15)	718	730.6 (5.92)	716	730.9 (7.73)
abz9 (661, 20 × 15)	727	745.6 (8.51)	728	743.0 (9.46)
la21 (1046, 15 × 10)	1059	1072.9 (7.71)	1058	1074.6 (8.83)
la24 (935, 15 × 10)	951	962.1 (5.03)	946	962.3 (7.89)
la25 (977, 15 × 10)	995	1007.5 (7.90)	988	1008.4 (9.72)
la27 (1235, 20 × 10)	1268	1279.3 (8.35)	1267	1276.0 (5.38)
la29 (1152, 20 × 10)	1202	1220.9 (11.3)	1196	1223.4 (13.4)
la38 (1196, 15 × 15)	1237	1265.0 (14.5)	1221	1251.7 (14.5)
la40 (1222, 15 × 15)	1262	1270.0 (6.09)	1249	1270.5 (12.3)

every other 15 iterations. Since a crossover changes solutions largely, crossover in every iteration is ill-balanced with local search. This algorithm is called "BG".

Table 21.6 shows the results for complex instances which are called the "10 tough problems" [1]. Every experiment consists of 20 trials. One trial is terminated after 200,000 iterations of subalgorithms at every region.

Like the case of TSP, the combination of Brownian and genetic subalgorithms gets slightly better results than the algorithm with Brownian only. However, the results are not as good as those given by algorithms which are especially tuned for JSSP.

21.4.2 Min Storage Problem

This subsection introduces an application of membrane algorithms to the Min Storage Problem [6].

21.4.2.1 Problem

Let $\mathcal{E} = (e_1, e_2, \ldots, e_k)$ be a sequence of integers. Given a positive integer C, the sequence $\mathcal{E}$ is said to be C-feasible if for every $1 \leq i \leq k$ the sum $\sum_{j=1}^{i} e_j$ belongs to the closed interval $[0, C]$.

The Min Storage Problem is an optimization version of the decision problem conscomp which is defined by:

An instance of conscomp consists of a set $\mathcal{E} = \{e_1, e_2, \ldots, e_k\}$ of integers such that $e_1 + e_2 + \cdots + e_k = 0$ and a positive integer C. The question is whether there is a permutation π over $(1, 2, \ldots, k)$ such that the sequence $(e_{\pi(1)}, e_{\pi(2)}, \ldots, e_{\pi(k)})$ is C-feasible or not.

Thus the min storage problem is to minimize the constant C of the conscomp. The formal definition is as follows.

An instance of min storage consists of a set $\mathcal{E} = \{e_1, e_2, \ldots, e_k\}$ of integers such that $e_1 + e_2 + \cdots + e_k = 0$ and a positive integer C (the same instance as conscomp). The question is to find a permutation π such that $\sum_{j=1}^{i} e_{\pi(j)} \geq 0$ for every $1 \leq i \leq k$ and that for every permutation π' which satisfies the constraint $\sum_{j=1}^{i} e_{\pi'(j)} \geq 0$, the inequality

$$\max_{1 \leq i \leq k} \sum_{j=1}^{i} e_{\pi(j)} \leq \max_{1 \leq i \leq k} \sum_{j=1}^{i} e_{\pi'(j)}$$

holds.

It has been proved that conscomp is NP-complete and that Min Storage is NP-hard.

21.4.2.2 Components of the algorithm

The membrane structure is the same as that for the algorithm solving TSP.

Because of the constraint $\sum_{j=1}^{i} e_{\pi(j)} \geq 0$, not all permutations of $(1, 2, \ldots, k)$ are feasible solutions. In order to make the set of all permutations the search space, we consider the function

$$F(\pi) = \begin{cases} \max_{1 \leq i \leq k} \sum_{j=1}^{i} e_{\pi(j)} & \text{if } \sum_{j=1}^{i} e_{\pi(j)} \geq 0 \text{ for all } i \in \{1, \ldots, k\} \\ \sum_{i=1}^{k} |e_i| - \text{NumVPS}_\pi & \text{otherwise} \end{cases}$$

where NumVPS_π is the number of non-negative (that is, valid) prefix sums determined by π. In this way, all feasible solutions get a lower measure with respect to non-feasible solutions. Moreover, every permutation can be measured, and we can also choose which among two non-feasible solutions to prefer: the one which has the lowest number of negative prefix sums.

Two types of membrane algorithms are examined: one uses only the genetic algorithm with PMX recombination [3] and the other uses local search only.

Table 21.7 Results of computer experiments. The column "k" shows the size of the problem. The "GA" and "local search" stand for the membrane algorithms with GA subalgorithm only and local search subalgorithm only, respectively. The entries show the average and the variance (in the parentheses) of MA4MS on 10,000 tests.

k	GA	local search
10	1.0875901(0.0139737)	1.0032719(0.0002333)
20	1.5258556(0.0582978)	1.0039292(0.0003654)
50	2.6124590(0.2444580)	1.0600094(0.0045419)
100	3.9430665(0.5004649)	1.1978124(0.0162296)

After pre-determined iterations, an algorithm terminates and outputs the best solution in the innermost region.

Since the Min Storage Problem is a new problem, instances of the problem must be generated. Interesting considerations about this issue can be found in [6].

21.4.2.3 Computer experiments

Results of computer experiments are evaluated by the index

$$opp_{MA4MA} = \frac{1}{N} \sum_{i=1}^{N} \frac{F_i(\pi)}{opt_i},$$

where N is the number of trials and opt_i has been put equal to the optimal solution of the i-th instance in those experiments for which the size of the instance allowed it to compute. In the experiments for which the size of the instances did not allow it to compute the optimal solution with the brute force approach, we have substituted it with the theoretical lower bound $\max_{1 \leq i \leq k} |e_i|$.

The results are shown in Table 21.7. The number of membranes is 30. The termination condition is 150 iterations. The table shows the average and the variance (in the parentheses) of 10,000 trials. One can see that the membrane algorithm with local search gets good solutions. It should be noted that the local search subalgorithm only does local search at the modification stage. In other words, the good results are the consequences of the membrane structure and the communication rules.

21.5 MEMBRANE ALGORITHMS USING VARIOUS MEMBRANE STRUCTURES

In this section, membrane algorithms using membrane structures different from the nested one are introduced.

21.5.1 Membrane Algorithm with a Star Topology

We start by presenting an algorithm which is proposed by D. Zaharie and G. Ciobanu [16] and solves function optimization problems.

21.5.1.1 Problem

Given a function $f : \mathbf{R}^n \mapsto \mathbf{R}$ and a domain $D \subseteq \mathbf{R}^n$, a function optimization problem is to find a point $x \in D$ such that $f(x)$ is minimum in the range $f(D)$. Hence, a solution of a function optimization problem is an n-dimensional vector $x \in D$.

Three functions are tested in [16]:

Sphere $\quad f(x) = \sum_{i=1}^{n} x_i^2 \quad$ (domain : $[-100, 100]^n$)

Ackley $\quad f(x) = -20 \exp\left(-0.2\sqrt{\frac{\sum_{i=1}^{n} x_i^2}{n}}\right) - \exp\left(\frac{1}{n}\sum_{i=1}^{n} \cos(2\pi x_i)\right)$
$\quad + 20 + e \quad$ (domain : $[-32, 32]^n$)

Griewank $\quad f(x) = \frac{1}{4000}\sum_{i=1}^{n} x_i^2 - \prod_{i=1}^{n} \cos\frac{x_i}{\sqrt{i}} + 1 \quad$ (domain : $[-600, 600]^n$).

All the above functions have the minimum value $f(x) = 0$ at $x = (0, \ldots, 0) \in \mathbf{R}^n$.

21.5.1.2 Components of the algorithm

Membrane structure. The membrane structure represents a star topology of a distributed evolutionary algorithm (Fig. 21.5). That is, the degree is 2 and the skin membrane has a number of elementary membranes. The skin membrane and every elementary membrane have a number of solutions. Solutions are exchanged between the skin membrane and an elementary membrane. Therefore, the skin membrane corresponds to the centre unit of the star topology. The skin membrane surrounds region S_0. Let s be the number of elementary membranes. For every $1 \leq i \leq s$, S_i is a region of an elementary membrane.

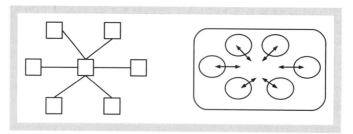

Fig. 21.5 A star topology of a distributed evolutionary algorithm (left) and its realization by a membrane structure (right).

Subalgorithms. Regions $S_0, S_1, \ldots, S_s$ have a number of solutions. Initially the number of solutions is $m(0)$ in every region, but the size can vary during computation. In regions $S_1, \ldots, S_s$, the solutions are modified by an evolutionary algorithm (described below). In the skin membrane, a local search rule is applied.

The evolutionary algorithm constructs a new solution z_i from five solutions $x_i, x_*, x_{r_1}, x_{r_2}, x_{r_3}$, where x_i is the "mother" of z_i, x_* is the best solution in the region, $x_{r_1}, x_{r_2}, x_{r_3}$ are randomly selected solutions. For every $1 \leq j \leq n$, the j-th component z_i^j of z_i is set to

$$z_i^j = \begin{cases} \gamma x_*^j + (1-\gamma)(x_{r_1}^j - x_*^j) + F(x_{r_2}^j - x_{r_3}^j)N(0,1), \\ \qquad\qquad\qquad\qquad \text{with probability } p, \\ x_i^j, \qquad\qquad\qquad\quad\; \text{with probability } 1 - p, \end{cases}$$

where γ, F, and p are parameters and $N(0, 1)$ stands for the normal distribution with mean 0 and standard deviation 1. In the generation mode (variant 1), z_i is build for every $i \in \{1, \ldots, s\}$. In the asynchronous mode (variant 2), a randomly selected x_i is modified. The evolutionary modification is repeated τ steps in an iteration of a subalgorithm.

After an evolutionary modification (and local search in S_0), communications take place. In S_0 all but the best solution are deleted. Every region S_i, $1 \leq i \leq s$, sends a copy of the best solution to the region S_0. The skin membrane adds random solutions in order to make the size of solutions more than $s + 1$. Finally, in every region S_i $1 \leq i \leq s$, the worst solution in the region is replaced with the copy of a randomly selected solution from S_0.

Termination condition and the output. When the best solution in S_0 satisfies the pre-determined accuracy predicate f_*, the algorithm stops. The best solution in the skin membrane is the output.

Table 21.8 Results of the membrane algorithm with the star topology. Each column shows the number of success trials (x/30) and the number of objective function evaluations (lower).

Test function	Generational and random migration	Membrane (variant 1)	Membrane (variant 2)
Sphere	30/30	30/30	30/30
	62002±5123	59049±527	3771±722
Ackley	1/30	30/30	30/30
	84970	240675±55217	3173±880
Griewank	20/30	12/30	26/30
	62902±4272	126304±89874	48724±5044

21.5.1.3 Computer experiments

Computer experiments are done with the following parameters:

$n = 30$ (dimension of the domain of functions)
$m(0) = 10$ (number of initial solutions at every region)
$s = 5$ (number of elementary membranes)
$\tau = 100$ (number of iterations of the evolutionary algorithm)
$f_* = 10^{-5}$ (accuracy for termination)
$F = p = 0.5$ (parameters of recombination)
$\gamma = 1$ (parameter of recombination).

Table 21.8 shows results of successful computations[5] (the simulation program terminates after a solution satisfies the accuracy) in the 30 trials. The generational and random migration shows results by an algorithm which does not use membrane structure for comparison. It can be seen that membrane algorithms obtain good results.

21.5.2 Membrane Algorithm with Global Communications

The next algorithm, which is proposed by L. Huang [4], also solves function optimization problems. Communications are not restricted between adjacent regions in the algorithm.

---

[5] The simulation program should be stopped as a failure computation if the run does not succeed after a long computation; in [16] the failure condition is not mentioned.

21.5.2.1 Problem

In addition to Ackley and Griewank functions, Rosenbrock's function and Rastringin's function are considered in [4]. They are defined by:

$$\text{Rosenbrock } f(x) = \sum_{i=1}^{n-1}(100(x_{i+1} - x_i^2)^2 + (1 - x_i)^2)$$
$$(\text{domain} : [-2.047, 2.048]^n)$$

$$\text{Rastrigin } f(x) = \sum_{i=1}^{n}(x_i^2 - 10\cos(2\pi x_i)) + 10n$$
$$(\text{domain} : [-5.12, 5.12]^n).$$

The two functions have the minimum value $f(x) = 0$ at $x = (0, \ldots, 0) \in \mathbf{R}^n$.

21.5.2.2 Components of the algorithm

Membrane structure. The skin membrane has two nested membranes (Fig. 21.6). The region in the skin membrane is R_0. The nested regions are called $R_{11}, \ldots, R_{1s}$ and $R_{21}, \ldots, R_{2s}$ from the inner region to the outer region. Each region has a number of solutions. In the skin membrane, solutions are not modified.

Subalgorithms. The group of regions $R_{11}, \ldots, R_{1s}$ and the group of regions $R_{21}, \ldots, R_{2s}$ use slightly different subalgorithms. In the first regions, two solutions

$$s = (x_1, \ldots, x_i, \ldots, x_j, \ldots, x_n) \quad \text{and} \quad t = (y_1, \ldots, y_i, \ldots, y_j, \ldots, y_n)$$

produce the next solution by recombination

$$s' = (x_1, \ldots, x_{i-1}, y_i, \ldots, y_j, x_{j+1}, \ldots, x_n),$$

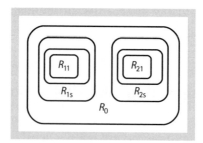

Fig. 21.6 A membrane structure for solving function optimization problems using global communications.

where $1 \leq i < j \leq n$ are randomly selected. Before recombination, a solution $(x_1, \ldots, x_n)$ is modified to $(x_1 + \eta_1, \ldots, x_n + \eta_n)$ by a mutation where

$$\eta_i = \begin{cases} 0, & \text{with probability } 1 - p, \\ \frac{|U_i - L_i|}{v_p} 10^{\frac{i}{s}} u(-0.5, 0.5), & \text{with probability } p, \end{cases}$$

where j is the region number ($1 \leq j \leq s$), U_i and L_i are the upper and lower limits of the i-th coordinate of the domain, v_p is a parameter, and $u(-0.5, 0.5)$ is a uniform random variable between -0.5 and 0.5. A result of mutation is not accepted if it makes a worse solution.

In the second region, two solutions

$$s = (x_1, \ldots, x_i, \ldots, x_j, \ldots, x_n) \quad \text{and} \quad t = (y_1, \ldots, y_i, \ldots, y_j, \ldots, y_n)$$

produce the next solution by recombination

$$s' = (x_1, \ldots, x_{i-1}, w_i, \ldots, w_j, x_{j+1}, \ldots, x_n)$$

where $1 \leq i < j \leq n$ are randomly selected and w_k is given by

$$w_k = \alpha x_k + (1 - \alpha) y_k$$

with a parameter $0 < \alpha <$ for $i \leq k \leq j$. Before recombination, one coordinate x_i of a solution $(x_1, \ldots, x_n)$ is modified to $x_i + \eta_i$ by a mutation in which η_i is given by

$$\eta_i = \frac{|U_i - L_i|}{v_p} e^{\frac{i}{s}} N(0, 1).$$

After modification, each region R_{ki} for $k = 1, 2$ and $1 < i \leq s$ sends copies of the best three solutions to region $R_{k,i-1}$. Every region R_{ki} for $k = 1, 2$ and $1 \leq i < s$ deletes worst three solutions and adds the three solutions from region $R_{k,i+1}$ to the set of solutions for the next generation.

The skin region R_0 gets the best solutions from regions R_{11} and R_{21}. The communication skips the intermediate membranes and goes to the outermost region directory from the innermost region. Region R_0 sends randomly selected solutions to regions R_{1s} and R_{2s}. If the two solutions from the two elementary membranes are identical or if the subalgorithm in region R_0 enters some premature behavior, then the subalgorithm at region R_0 restarts subalgorithms in regions $R_{11}, R_{21}, \ldots, R_{1s}, R_{2s}$. The best solution in R_0 is preserved, other solutions in the region are deleted. In the other regions, randomly generated solutions are chosen.

Termination and output. The whole algorithm terminates when subalgorithms are iterated for a pre-determined number of times. The best solution in the skin membrane is the output of the algorithm.

The two global communications, transport of solutions from regions R_{11}, R_{21} to region R_0 and the restart action, are realized by communications between adjacent

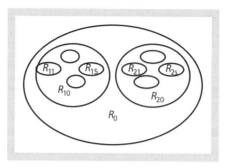

Fig. 21.7 Membrane structure which makes all communications local.

regions if membrane structure shown in Fig. 21.7 is used. There are two new regions R_{10} and R_{20}, which are included in region R_0, R_{10} includes $R_{11}, \ldots, R_{1s}$ and R_{20} includes $R_{21}, \ldots, R_{2s}$. The new regions realizes all communications, i.e. communications from R_{ij} to $R_{i,j-1}$ for $i = 1, 2$ and $2 \leq j \leq s$, from R_{i1} to R_0 for $i = 1, 2$, and from R_0 to R_{ij} for $i = 1, 2$ and $1 \leq j \leq s$.

21.5.2.3 Computer experiments

Computer experiments are done with the following parameters:

$n = 50$ (dimension of the domain of functions)
$m = 10$ (number of initial solutions in every region)
$s = 5$ (number of nested membranes)
$v_p = 0.001$ (parameter of mutation).

Table 21.9 shows the values of tested functions which are obtained by computer experiments. The table also shows results of fuzzy adaptive differential equation (FADE) for comparison. One can see that the membrane algorithm gets better results than FADE. Zaharie and Ciobanu's algorithm optimizes Ackley's function

Table 21.9 Results of membrane algorithm and fuzzy adaptive differential equation (FADE). The iteration column shows the termination conditions for the membrane algorithm.

functions	iterations	FADE	membrane
Rosenbrock	7000	41.57	16.11
Ackley	5000	0.059	1.17×10^{-4}
Rastrigin	10000	258.49	1.38×10^{-6}
Griewank	5000	0.573	1.69×10^{-9}

better than Huang's, while Huang's algorithm seems to optimize Griewank's function better. However, since conditions of experiments are different, these results do not give us evidence to decide which algorithm is better.

21.6 CONCLUSION

We can conclude that a membrane algorithm, which is a P system with solutions of an existing problem as its objects and with modification and transportation mechanisms as its rules, can approximately solve some NP-hard problems very well. The following paragraphs briefly describes in what way each property of the P system contributes to the membrane algorithm.

The membrane structure and the regions which are separated by membranes play the central role. One might think that the structure is similar to that of parallel meta-heuristics (e.g. [13, 14]). However, as it has been pointed out in Section 21.2, the membrane algorithm can combine approximate algorithms which are based upon different principles. Considering the No Free Lunch Theorem for optimization [5, 15], this feature is very important.

Multisets of objects correspond to copies of solutions, which are useful in modifications and communications. Evolution rules at a region and communication rules between regions become subalgorithms. When the nested membrane structure is used, the principle "In the inner regions, solutions should be changed slightly; in the outer regions, solutions should be changed largely" has been established as a guideline for designing a subalgorithm.

Both symport communication (sending a multiset of best n solutions for some $n > 1$) and antiport communication (swapping solutions) are used. The question what combination of communication and modification rules is suitable is an issue for future investigations. Dynamic restructuring of membranes is not introduced here. But a possibility of dynamic restructuring has been proposed in [10].

Although all computer experiments examined here are done by means of sequential processing, a membrane algorithm is essentially parallel. Implementation of membrane algorithms in a parallel processing system is quite interesting future work.

Finally, we mention theoretical studies on membrane algorithms. The result that local search gets good solutions when it is combined with the nested membrane structure (Section 21.4.2.3) suggests a theoretical study on optimization mechanisms involved in the nested membrane structure and communications. The elementary considerations in Section 21.3.3.3 also suggest a theory of modifications and communications, which would become a theory of membrane algorithms.

References

[1] D. APPLEGATE, W. COOK: A computational study of the job-shop scheduling problem. *ORSA J. on Comput.*, 3 (1991), 14–156.

[2] R. AZENCOTT, ed.: *Simulated Annealing*. John Wily & Sons, New York, 1992.

[3] D.E. GOLDBERG: *Genetic Algorithms*. Addison-Wesley, Reading, 1989.

[4] L. HUANG: A variant of P systems for optimization. *Neurocomputing*, to appear. (Paper available as: L. Huang, N. Wang: An optimization algorithms inspired by membrane computing. *Lecture Notes in Computer Sci.*, 4222 (2006), 49–55.)

[5] C. IGEL, M. TOUSSAINT: On classes of functions for which No Free Lunch results hold. *Information Processing Letters*, 86 (2003), 317–321.

[6] A. LEPORATI, D. PAGANI: A membrane algorithm for the min storage problem. *Lecture Notes in Computer Sci.*, 4361 (2006), 443–462.

[7] T.Y. NISHIDA: An application of P-systems: A new algorithm for NP-complete optimization problems. *Proc. 8th World Multi-Conference on Systems, Cybernetics and Informatics* (N. Callaos et al., eds.), 2004, vol. V, 109–112.

[8] T.Y. NISHIDA: An approximate algorithm for NP-complete optimization problems exploiting P-systems. *Proc. Brainstorming Workshop on Uncertainty in Membrane Computing*, Palma de Majorca, 2004, 185–192.

[9] T.Y. NISHIDA: An approximate algorithm for NP-complete optimization problems exploiting P-systems. *Application of Membrane Computing* (G. Ciobanu et al., eds.), Springer, 2006, 301–312.

[10] T.Y. NISHIDA: Membrane algorithms. *Lecture Notes in Computer Sci.*, 3850 (2006), 55–66.

[11] T.Y. NISHIDA: Membrane algorithm with Brownian subalgorithm and genetic subalgorithm. *Intern. J. Found. Computer Sci.*, 18 (2007), 1353–1360.

[12] T.Y. NISHIDA, T. SHIOTANI, Y. TAKAHASHI: Membrane algorithm solving job-shop scheduling problems. *Pre-Proceedings of WMC9* (P. Frisco et al., eds.), Edinburgh, UK, 2008, 363–370.

[13] M. TOMASSINI: Parallelism and evolutionary algorithms. *IEEE Transactions on Evolutionary Computation*, 6 (2002), 443–462.

[14] M. TOMASSINI: *Spatially Structured Evolutionary Algorithms*. Springer, 2005.

[15] D.H. WOLPERT, W.G. MACREADY: No Free Lunch Theorem for optimization. *IEEE Transactions on Evolutionary Computation*, 1 (1997), 67–82.

[16] D. ZAHARIE, G. CIOBANU: Distributed evolutionary algorithms inspired by membranes in solving continuous optimization problems. *Lecture Notes in Computer Sci.*, 4361 (2006), 536–553.

CHAPTER 22

MEMBRANE COMPUTING AND COMPUTER SCIENCE

RODICA CETERCHI
DRAGOŞ SBURLAN

22.1 INTRODUCTION

P systems represent one of the emerging computability models related to the area of molecular computing and it is based upon the notion of the membrane structure of living cells. The model is inspired by the fact that all the cellular-level processes involving different chemical reactions which have precise goals, can be viewed as computing processes. In this respect, any non-trivial biological system is a construct in which different components execute "computations" in a parallel/distributed fashion, sharing the results if needed, in order to accomplish a common goal. Each component present in such a system is well delimited through various types of membranes that keep compartments separated and act as semi-permeable barriers, ensuring that certain substances always stay in the compartment while other substances stay out of it. Membranes can be created, divided, or dissolved if a certain specific request triggers such an action.

All these facts indicate that cells are, in many respects, information processing devices, hence a natural goal is to formalize their behavior and take advantages from their massive parallel molecular features. This brings us to the challenging task one is faced with in designing such systems to perform a specific behavior or to solve a particular problem.

The chapter is intended to survey several practical computer science problems studied in the framework of P systems and to recall some of the fundamental techniques employed for solving these problems algorithmically. We will deal with several areas of applications as sorting, graphics, cryptography, Boolean circuits, and parallel architectures. The covered topics provide an informal introduction to the problem solving, the focus being on the main ideas and results.

22.2 STATIC SORTING P SYSTEMS

Ordering a list of k elements is one of the best known and studied problems in computer science. In general, the most used types of orders are numerical order and lexicographical order. For such numerical sorting problems many sequential and parallel algorithms were proposed and studied. The output of the sorting of a list of numbers has to be a permutation, or reordering, of the input and according to the desired total order. A typical common characteristic was to consider a constant time while comparing two numbers and compute their time complexity with respect to the number of components composing the vector to be sorted. The best known time complexity is $O(k \log k)$ for the sequential case and $O(\log^2 k)$ for the parallel case.

Membrane computing constitutes a challenging framework for implementing sorting algorithms and this is due to its main characteristics: the massive rewriting parallelism and the usage of unstructured data (the multisets) as support for the computations.

22.2.1 Sorting Definitions

The sorting problem can be easily formulated as follows: given a sequence of numbers (usually integers) $\langle x_1, x_2, \ldots, x_k \rangle$, a sorting algorithm has to compute a permutation $\langle x'_1, x'_2, \ldots, x'_k \rangle$ such that $x'_1 \leq x'_2 \leq \ldots \leq x'_k$. Sorting algorithms can be *comparison-based* (in this case the fundamental operation is compare-exchange) or *non-comparison-based* (that uses certain known properties of the elements, e.g. binary representation or distribution). On sequential algorithms, the lower bound on comparison-based sort of k numbers is $O(k \log k)$.

Sorting problems can also be tackled in a parallel framework: sorting networks were introduced in order to execute simultaneously more comparison operations, thus improving sorting performance significantly. Sorting networks are composed of comparators (that have as input pairs of numbers of type (X, Y) and compute $(min(X, Y), max(X, Y))$), wires (that link together the comparators and pass the information among them), input wires, and output wires. The lower bound time complexity on parallel comparison-based sort of k numbers is $O(\log^2 k)$.

For the present work, some notions have to be defined. Let $V = \{\underline{i} \mid 1 \leq i \leq k\}$ be an alphabet such that the symbols are denoted by underlined numbers. By using such an alphabet one can inherently have the implicit order associated with natural numbers—a useful property while defining various types of sorting.

A word over V is denoted by $w = \prod_{j=1}^{m} a_j = a_1 a_2 \ldots a_m$, $m \in \mathbb{N}$, and $a_j \in V$ for each $1 \leq j \leq m$; the product symbol $\prod$ represents the concatenation of symbols.

Let $ord : V \to \{1, \ldots, k\}$ be a bijective function such that $i = ord(\underline{i})$, $1 \leq i \leq k$. Then $i_j = ord(a_j)$ is an ordinal number of the j-th letter of w and $a_j = \underline{i_j}$.

The *alphabet word* represents the string obtained by concatenating elements of the alphabet V in the natural order, i.e. $v = \prod_{j=1}^{k} \underline{j}$.

If $w = \prod_{j=1}^{m} a_j$ is a string, then the set of all strings that can be obtained from w by permuting its symbols is denoted by $Perm(w)$. The set of *scattered subwords* of w is the set $SSub(w)$ of all strings that can be obtained from w by deleting some (0 or more, possibly all) of its symbols, and concatenating the remaining ones, preserving the order.

Example 22.1 *Let $V = \{\underline{1}, \underline{2}, \underline{3}\}$. Then a word over V can be $w = \underline{3}\,\underline{1}\,\underline{2}\,\underline{1}$. The alphabet word is the string $v = \underline{1}\,\underline{2}\,\underline{3}$. The set of permutations of v is given by $Perm(v) = \{\underline{1}\,\underline{2}\,\underline{3},\ \underline{1}\,\underline{3}\,\underline{2},\ \underline{2}\,\underline{1}\,\underline{3},\ \underline{2}\,\underline{3}\,\underline{1},\ \underline{3}\,\underline{1}\,\underline{2},\ \underline{3}\,\underline{2}\,\underline{1}\}$. The set of scattered subwords of w is*

$$SSub(w) = \{\underline{3}\,\underline{1}\,\underline{2}\,\underline{1}\} \cup \{\underline{3}\,\underline{1}\,\underline{2},\ \underline{3}\,\underline{1}\,\underline{1},\ \underline{3}\,\underline{2}\,\underline{1},\ \underline{1}\,\underline{2}\,\underline{1}\}$$
$$\cup \{\underline{3}\,\underline{1},\ \underline{3}\,\underline{2},\ \underline{1}\,\underline{2},\ \underline{2}\,\underline{1},\ \underline{1}\,\underline{1}\} \cup \{\underline{1},\ \underline{2},\ \underline{3},\ \lambda\}.$$

In the framework of P systems, multisets of objects are used to represent the computational support. Hence, while implementing a sorting algorithm, a straightforward idea is to represent the numbers to be sorted as the multiplicity of certain distinct objects. The following cases can be distinguished:

- one can sort only the numbers represented by the multiplicities of the objects but not considering the corresponding objects (therefore, only the "properties" are sorted);
- one can sort both the numbers represented by the multiplicities of the objects and the objects themselves.

Consequently, we can define the following.

Definition 22.1 *Let* $v = \prod_{j=1}^{k} \underline{j}$ *be the alphabet word. The word*

$$w = \prod_{j=1}^{k} a_j \in Perm(v), \ k = card(V) \in \mathbf{N}^+,$$

where $a_j \in V$ *such that* $M(a_j) \leq M(a_{j+1})$, *for each* $1 \leq j \leq k-1$, *is called the* ranking string *of the multiset M.*

Definition 22.2 *The word* $w = \prod_{j=1}^{k} \underline{j}^{M(a_j)}$ *is called* weak sorting string *of the multiset M if* $\prod_{j=1}^{k} a_j$ *is the ranking string of M. Also,* $M' : V \to \mathbf{N}$ *defined as* $M'(\underline{j}) = M(a_j)$ *is the* weak sorting multiset *of M.*

The symbols from the initial multiset are considered to be in a complete relation order and the result of a computation consisting of these objects sorted according to the relation order and having multiplicities sorted.

Definition 22.3 *The word* $w = \prod_{j=1}^{k} a_j^{M(a_j)}$ *is called* strong sorting string *of M if* $\prod_{j=1}^{k} a_j$ *is the ranking string of M.*

Example 22.2 *For the alphabet* $V = \{\underline{1}, \underline{2}, \underline{3}\}$ *and the multiset* $M = \{(\underline{1}, 20), (\underline{2}, 10), (\underline{3}, 30)\}$, *we have: the ranking string is* $\underline{2}\ \underline{1}\ \underline{3}$, *the weak sorting string is* $\underline{1}^{10}\ \underline{2}^{20}\ \underline{3}^{30}$, *and the strong sorting string is* $\underline{2}^{10}\ \underline{1}^{20}\ \underline{3}^{30}$.

In general, the starting configuration of the sorting P system is considered to depend only on the number $k = Card(V)$ of components. Usually, the input for a sorting P system is the multiset $\{(\underline{j}, n_j) \mid 1 \leq j \leq k\}$ over $V \subset O$, placed in a specific region, where O is the system's alphabet.

22.2.2 Static Sorting Using P Systems

Common time complexity lies on the maximum number of steps a Turing machine uses for any input of a given length. Similarly, the time complexity in the P systems framework is based on the number of maximally parallel steps performed by the membranes (assuming that one maximally parallel step is inseparable and the execution time of each such step is always the same, being governed by the universal clock). Based on this assumption we proceed to the investigation of the static sorting algorithms using P systems.

22.2.3 Bead-Sort

One of the first attempts in using P systems framework to solve the sorting problem was done in [3] by implementing the Bead-Sort algorithm.

The general idea was to represent the positive integers to be sorted by a set of beads that are allowed to freely slide along the rods (in a similar manner to

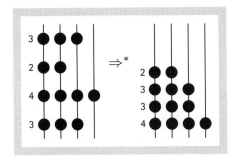

Fig. 22.1 Sorting $< 3, 2, 4, 3 >$.

the functioning of the Abacus) to their appropriate places; consequently, smaller "numbers" always emerge above the larger ones as exemplified in Fig. 22.1.

This simple idea has been implemented with a tissue P system that uses only symport/antiport rules. In this design, the goal was to sort k positive integers, the biggest among them being m. The membrane structure used is of degree $m \cdot k + k$ where $m \cdot k$ membranes are used to represent the m rods with k levels, while the additional k membranes serve as counters. The beads were represented by objects x, while the absence of the beads by symbol $-$. In this way, objects x and $-$ reflect the bead position in the initial state of the frame (see Fig. 22.2). In this bi-dimensional array of membranes, by using antiport rules, the objects x and $-$ are exchanged in a simulation of the "beads falling down" action. The output is read when the system reaches a sorted state; then the objects from the rod membranes will be expelled into the environment in order and such that the objects from the same level will arrive there simultaneously. The following result holds true.

Theorem 22.1 *The time complexity for the bead-sort algorithm using tissue P systems and symport/antiport rules is $O(k)$ where k is the number of elements to be sorted.*

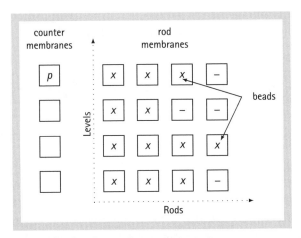

Fig. 22.2 The frame for $< 3, 2, 4, 3 >$.

However, the descriptional complexity for the algorithm is rather high: $m \cdot k + k$ membranes and the usage of antiport rules of unbounded weight.

22.2.4 Communicative Sorting

Communicative sorting represents another sorting algorithm using standard P systems and antiport rules with priorities introduced in [6]. In this case, the integers to be sorted were codified as the multiplicities of a certain object (say, symbol a) and, in order to be distinguished, they were disposed in a nested structure of membranes (each region contained at most one number because only one symbol was used to represent the set of integers). An important detail regards the design of a comparator whose functioning is presented in Fig. 22.3. There, because of the priority relation, the rule $(a, out; a, in)$ will be first performed $\min(i, j)$ times, and only afterwards (and only in the case $j > i$) the rule (a, in) will start its execution. Consequently, in one computational step, the membranes of the reached stable configuration will hold objects a with multiplicities in ascending order from the outermost to the innermost membrane.

The system uses a promoter that travels between the regions of the system and triggers the rules of the comparators (a modified version of the one schematically presented in Fig. 22.3) such that an odd-even transposition network is simulated. The total order among the membranes was used to have a codification of the sorting result and this is obtained by adult halting. Consequently, the subsequent result is true.

Theorem 22.2 *The time complexity for the communicative sorting algorithm using P systems with antiport promoted rules and priorities is $O(k)$, where k is the number of elements to be sorted.*

The same idea of simulating the parallel odd-even transposition network using weak priorities and cooperative object rewriting rules (with at most two objects on

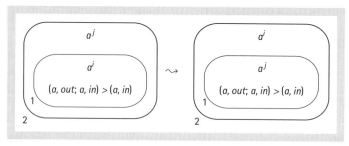

Fig. 22.3 The comparator if $j > i$. The P system in the right-hand side is the stable configuration obtained after the application of the rules.

the left-hand side of any used rule) has been addressed in [6]. There, the numbers to be sorted were represented as multiplicities of distinct objects (say, $A, B, \ldots$) and the rules of type $AB \to A'B' > \{A \to B', B \to B'\}$ perform in just one step the compare-swap-if-needed operator:

$$A^{k_1} B^{k_2} \Longrightarrow A'^{\min(k_1,k_2)} B'^{\max(k_1,k_2)}.$$

As a consequence, the following result holds true.

Theorem 22.3 *The time complexity for the weak sorting algorithm using P systems with cooperative rules with weight at most 2 and weak priorities is $2k + 1$, where k is the number of elements to be sorted.*

In [2] yet another comparator was proposed; however in this case the used features were very "weak": three mobile catalysts and non-cooperative rules. A codification scheme for the numbers to be sorted was used as above in a design that involves four nested membranes. The main idea used is that if a catalyst changes the region during the computation, when it comes back to its original region it will find a different context, and so it will react in a different manner. More precisely, two mobile catalysts were used to synchronously decrease the multiplicities of the objects representing the integers to be sorted; the third catalyst was required for some "cleaning" tasks (in case of the auxiliary used objects) – an important feature while considering the comparator reusability. For two integers i and j this comparator produces the output after $7 \cdot (|i - j|) + 1$ steps.

22.2.5 Sorting by Carving

Another fruitful idea regards the *sorting by carving* and it was addressed in [1] and [2]. There, the numbers to be sorted were represented by the multiplicities of different species of objects. Such pairs of type (*object, multiplicity*) were called components. The general idea of the algorithm was to simultaneously rewrite objects from all components (one symbol from each of the components at once) and to "detect" when one (or more, in case in the sequence of numbers to be sorted there exist two or more numbers with the same values) component is "exhausted" (i.e. was completely rewritten). When this happens, a copy of the exhausted component can be sent out by the sorting P system (as representing the smallest number(s) in the sequence) in order to form the strong/weak sorting string. The above procedure is repeated until all the components are exhausted. Based on this strategy several results involving different P system models were obtained.

Theorem 22.4 *The time complexity for the strong sorting algorithm using P systems with cooperative rules and promoters is $2k + 1$, where k is the number of elements to be sorted. The time complexity is constant with respect to the values of the elements and the number of different objects used while implementing the algorithm is exponential with respect to k.*

A similar result also holds for P systems using inhibitors.

Theorem 22.5 *The time complexity for the strong sorting algorithm using P systems with cooperative rules and inhibitors is 2k, where k is the number of elements to be sorted.*

Strong priorities and stable catalysts can be employed for solving the strong sorting problem.

Theorem 22.6 *The time complexity for the strong sorting algorithm with P systems with strong priorities and s-stable ($s = k + 1$) catalysts is $\max_{1 \leq i \leq k}(x_k) + \sum_{i=1}^{k} x_k$, i.e. the sum of the elements to be sorted plus their maximum.*

Using membrane dissolution and cooperative rules for solving first the ranking problem and afterwards to release properly the output into the environment leads to the following result.

Theorem 22.7 *The time complexity for the weak sorting algorithm with P systems with cooperative object rewriting rules and membrane dissolution is $2k^2 + 3k + 4$ where k is the number of elements to be sorted. However, the number of different objects used is exponential with respect to k.*

Although the above result has a poor time/space complexity, it shows the membrane dissolution potential to perform elaborate computations.

22.2.6 Ranking

Mobile catalysts and non-cooperative rules were used with unexpectedly good time complexity for the ranking problem (see [2]). For solving this problem three nested membranes were used and $k + 1$ mobile catalysts, where the first k are associated with the objects whose multiplicities are sorted, plus one which was used to "clean" unneeded symbols. The ranking algorithm proceeds as in the sorting by the carving algorithm presented in Section 22.2.5, i.e. by simultaneously rewriting an object from each component until one component is exhausted. In this case, this task is accomplished by the first mentioned k mobile catalysts. The procedure is repeated and, at each moment a component is exhausted, a symbol representing the exhausted component is sent outside. In this way the ranking string is formed in a time complexity proportional with the maximum number in the sequence of the numbers that are ranked.

The presented ranking construction can be furthermore used to implement a strong sorting algorithm (see [39]) whose time complexity is linear with respect to the maximum number of elements to be sorted; it also depends on the number of components to be sorted and on the sum of the elements.

22.2.7 Other Types of Sorting P Systems

In [5] two types of dynamic P systems are proposed for sorting. Integers are represented as multiplicities of a symbol in a structure of nested membranes. A new membrane is added to the system, containing a new integer, and by activating the comparator rules the new integer "sinks" to its appropriate place. The dynamic aspect, the appearance of a new membrane, is governed by an external mechanism, contextual grammars acting on string descriptions of membrane structures.

In [7] two other models for sorting with P systems are presented, based on the powerful sorting algorithm known as bitonic sort. The first model uses a membrane for each integer, membranes are placed in the nodes of a 2D-mesh structure, and the P system is with *dynamic communication graph*, i.e. it has a virtual communication graph like tissue P systems, but at each time moment only some of the edges are active. The evolution of the P system is described by a sequence of sets of pairs [*graph, rules*], with the rules in the set *rules* either rewriting rules if *graph* is a subset of the identity graph, or communication rules if *graph* is a subset of proper edges. (In [8] a simulation with the same formalism is presented, but in which several integers are represented and several comparisons take place in each membrane.) The second model realizes a simulation of the bitonic sort in one single membrane, with distinct integers codified over an alphabet of distinct symbols, using only rewriting rules.

In [9] and [22] sorting is implemented with spiking neural P systems.

22.3 Graphics

Computer graphics represent another topic studied within the P systems framework. This section emphasizes the flexibility of the P systems to produce various types of structures and computer-aided visualizations of them.

22.3.1 Turtle Graphics with P Systems

Lindenmayer systems were among the first mathematical formalisms conceived for modeling plant development by expressing the neighborhood relations between cells (or larger plant modules) in terms of division, growth, and death of individual cells. In the abstract framework, these organisms were considered as being composed of discrete units (symbols from an alphabet) which evolve in time due to some rewriting rules. These rules are applied in parallel and they were intended to describe the cell division in multicellular organisms. The parallel application of the rules (which is unlike the sequential rewriting in Chomsky grammars) induce an important effect over the formal properties of rewriting systems, and this constitutes the basis for the geometric aspect of the underlying obtained branching topology (see [33],[37]). The common graphical interpretation of L systems relies on the following: a special designed L system starts from an initial string (the axiom) and evolves iteratively according to the rewriting rules; any resulting string obtained can be the subject of the turtle interpreter (a cursor which can move, rotate, and draw in two or three dimensions as it encounters a command symbol in the generated string) which produces the corresponding graphics. In this way, astonishing graphics representing high order plants were produced by L systems. However, in order to represent branching structures from strings some memory pointers were needed; they were used to remember the location and orientation in which the branches were developed, and in particular for developing several branches from the same point.

On the other hand, P systems inherit in a certain sense the rewriting parallelism of L systems. Moreover, the topology of P systems is inherently a branching structure based on the inclusion relationship among membranes. These features were used as main arguments while defining sub-LP systems (see [20]) in an attempt to define a more flexible and expressive model for plant development and graphical representation. The result was a hybrid model between L systems and membrane computing. In this way, one can make use of the representational compactness of L system-like grammatical rules, but one can also use the membranes in order to localize the computation (an important aspect that can be used while modeling events that trigger some sudden change in the growth of a plant). More precisely, sub-LP systems make use of strings as support for computation, rewriting conditional or non-conditional rules (including dissolution of membranes), numerical real variables, and arithmetical rules acting on the numerical variables. The arithmetical rules allow for the contents of a numerical variable to be set to the result of an arithmetical expression. The strings are composed of symbols which can have associated several parameters specified using arithmetical expressions. A simple sub-LP system given using the syntax specification as defined in [20] is presented in Example 22.1

Example 22.3 *Let the sub-LP system:*

$$\{ \text{ region } 0;$$
$$\{ \text{ region } 1;$$
$$A[1, 2] A[2, 3] A[3, 2];$$
$$\text{define } y = 8;$$
$$A[x, 2] : x > 2 \to B[2, x + 10]C;$$
$$\} \}$$

Fig. 22.4 A tree generated by a sub-LP system.

Here two nested regions are specified, the inner one containing a string composed of three parameterized symbols A[1; 2], A[2; 3], and A[3; 2] (each symbol has two parameters), a numerical initialized variable y = 8 and one conditional rule acting on the parameterized symbols.

The usage of such a specification language allows the construction of modules that are more easily understandable and that can be intuitively assembled together. In this way, graphs like the one presented in Fig. 22.4 are obtained by a sub-LP system specified in a compact manner.

The advantages of using LP systems reside in the usage of membranes as subsystems, hence on the localized control mechanisms (specified for instance by the arithmetical rules and numerical variables). In addition, the natural hierarchical structure of membranes in P systems (represented, for instance, in a Venn diagram) improve readability as opposed to the difficulty of finding out the branching structure in a string generated by an L system.

Another effort to represent higher plants using P systems has been made in [36]. The approach was to use evolution rules to expand the membrane structure in a similar manner as mature parts in higher plants (e.g. trunks and branches) maintain their morphology during their growth while new branches evolve from already existing points. In particular, in order to generate tree-like structures and represent them using a turtle interpreter, a restricted P systems model with membrane creation was employed. This model has all the components defined as for the standard P systems, the only constraints being the usage of a restricted alphabet O (composed by some symbols representing target indications for the turtle cursor) and of only two kinds of rules: (a) $a \to v$, where $a \in O$ and v is a multiset over O (this rule, acting in a given region, replaces object a by the multiset v), (b) $a \to [v]$, where $a \in O$ and v is a multiset over O (this rule, acting in a given region, replaces an object a by a new membrane with the same label and containing the multiset v).

The functioning of such a system (say Π) is the same as in the case of a standard P system. In a computation of Π, each reachable configuration has associated a rooted tree representing the membrane structure (where the root represents the skin membrane and the leaves are the elementary membranes). This rooted tree will set up the topology of the modeled biological item, while the objects which reside inside the membranes will control the turtle movements in the bi-dimensional space. More precisely, let $O = \{L, E, W, F, +, -, B_L, B_R, B_{S_1}, B_{S_2}\}$; for a given configuration of Π, a depth-first search of the rooted tree is performed and the graphic is generated according to the objects found within membranes as follows. If a membrane contains the multiset F^m, then a segment of length $m \times l$, where l is a specified fixed length is drawn. The segment is drawn rotated with respect to the segment corresponding to the parent membrane with an angle of $n \times d$, where n is the multiplicity of objects "+" minus the multiplicity of objects "-" in the membrane, and δ is a specified fixed angle. The width of the segment is specified by the multiplicity of the object W. The objects B_{S_1} and B_{S_2} represent straight branches to be created, whereas the objects B_L and B_R represent branches to be created rotated to the left and to the right, respectively. The objects L and E do not have a graphical interpretation; they can be considered as seeds for growing the branch in length and width. In [36] the following simple example illustrates the above concepts.

Example 22.4 *Let $\Pi = (O, \mu = [\,], R, w_1 = LEWFB_L B_{S_1})$ and where the set of rules is*

$$R = \{B_{S_1} \to [LEWFB_{S_2} B_R], \quad B_L \to [+LEWFB_L B_{S_1}],$$
$$B_{S_2} \to [LEWFB_L B_{S_1}], \quad B_R \to [-LEWFB_L B_{S_1}],$$
$$L \to LF, \quad E \to EW\}$$

The third configuration in the computation performed by Π as well as the corresponding graphics are presented in Fig. 22.5.

A similar approach using P systems with membrane creation was used in [38] for generating polygons, spirals, and friezes. For instance, the construction of the Archimedean spiral is given in Example 22.3.

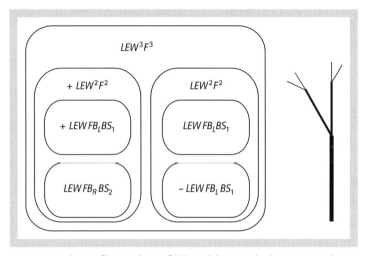

Fig. 22.5 A configuration of Π and its turtle interpretation.

Example 22.5 *Let the P system*

$$\Pi = (O, \mu = [\,], w_1 = F^n WHL, R)$$

with the set of rules $R = \{H \to [-F^n WLH], L \to LF\}$. *In Fig. 22.6 an Archimedes spiral is represented for* $n = 5$, *length of* $F = 0.01$, *width* $W = 1.0$, *angle* $\delta = 15$, *and step* 120.

It is also worth mentioning that several software tools for representing graphics obtained using P systems have been developed (for more details we indicate [19],[38]).

Fig. 22.6 Archimedean spiral.

22.4 TILING RECTANGULAR PICTURES WITH P SYSTEMS

For a given alphabet V, a *picture* p of size $m \times n$ (or a $(m \times n)$ picture) over V is an array of the form $p = (a_{ij})_{1 \le i \le m, 1 \le j \le n}$, with $a_{ij} \in V$ for $1 \le i \le m$, $1 \le j \le n$. We denote by V^{**} the set of all pictures over V (including the empty picture denoted by λ). A *picture language* (or a two-dimensional language) over V is a subset of V^{**}. A *subpicture* of a picture p is a subarray of p. A (2×2) subpicture of p is called a *tile* of p. The set of all tiles of p is denoted by $B_{2,2}(p)$. In the sequel, we will identify the boundaries of a picture by surrounding it with the marker #. For a picture p, we denote with $\hat{p}$ the picture that corresponds to p whose boundaries have been identified with a marker #.

Definition 22.4 *Let V be an alphabet and let $L \subseteq V^{**}$ be a picture language over V. The language L is called a* local *picture language if there exists a set Θ of tiles over $V \cup \{\#\}$ such that $L = \{p \mid p \in V^{**} \text{ and } B_{2,2}(\hat{p}) \subseteq \Theta\}$. We write $L = L(\Theta)$.*

In [11] and [12] special kinds of tissue P systems were proposed as generative devices for picture languages, with each membrane capable of holding one pixel of the picture, and with emphasis on local and recognizable picture languages.

22.4.1 Tiling Picture Languages

We consider now another class of languages of rectangular pictures, a class which arises from a tiling of the integer plane $Z^2 = (Z \times Z, +, (0, 0))$.

We consider $\mathcal{P} = \{P_1, P_2, \ldots, P_k\}$ a finite collection of finite connected subsets of Z^2, called *prototiles*. The prototiles can be normalized such that the lexicographically least point is the origin, $(0, 0) \in Z^2$.

A *translate* of a prototile $P \in \mathcal{P}$ by a $t \in Z^2$ is the subset $t + P$ of Z^2 (where $+$ is the addition in Z^2) and is called a *tile*.

A *tiling of the integer plane* Z^2 by the prototiles $\mathcal{P}$ is an expression of Z^2 as a disjoint union of tiles, $Z^2 = \sqcup (t_j + P_{k_j})$, where $(t_i + P_{k_i}) \cap (t_j + P_{k_j}) = \emptyset$ if $i \ne j$ (we say that the set of prototiles $\mathcal{P}$ *tiles* the integer plane).

Consider $A = \{1, 2, \ldots, k\}$ the finite set of labels of prototiles of $\mathcal{P}$ and think of it as a finite alphabet of shapes.

Consider A^{Z^2} the set of functions defined on Z^2 and with values in A. A tiling of the plane $Z^2 = \sqcup(t_j + P_{k_j})$ will be associated with an element $x \in A^{Z^2}$ in the following way: for every $v \in Z^2$, $x(v) = r$ iff the point v lies in a tile that is a translate of P_r, i.e., iff $v \in t_j + P_{k_j}$ and $k_j = r$. For $x \in A^{Z^2}$ and $v \in Z^2$, we will also use the notation x_v for $x(v)$.

A *2-dimensional shift* is an application $\sigma : \mathbb{Z}^2 \to Homeo(A^{\mathbb{Z}^2})$ such that for any $v \in \mathbb{Z}^2$, $\sigma_v : A^{\mathbb{Z}^2} \to A^{\mathbb{Z}^2}$ is the translation of the plane by the vector v, i.e. $(\sigma_v(x))_w = x_{w+v}$, for any $x \in A^{\mathbb{Z}^2}$ and any $w \in \mathbb{Z}^2$.

A subset $X \subseteq A^{\mathbb{Z}^2}$ is called σ-*invariant* iff $\sigma_v(X) \subseteq X$, for any $v \in \mathbb{Z}^2$. We call (X, σ) a *subshift* of $(A^{\mathbb{Z}^2}, \sigma)$ iff X is a topologically closed and σ-invariant subset of $A^{\mathbb{Z}^2}$. $(A^{\mathbb{Z}^2}, \sigma)$ is called the *full-shift*.

Consider now the set $T(\mathcal{P})$ of all $x \in A^{\mathbb{Z}^2}$ which correspond to tiling of $\mathbb{Z}^2$ by $\mathcal{P}$. This set is a σ-invariant closed subset of $A^{\mathbb{Z}^2}$, and thus $(T(\mathcal{P}), \sigma)$ is a subshift of $(A^{\mathbb{Z}^2}, \sigma)$.

We extend the set of prototiles $\mathcal{P}$ with one more prototile $P_\# = \{(0, 0)\}$. We will think about the tiles obtained by translating this prototile as 1×1 blocks (pixels) filled with #. Then $\bar{\mathcal{P}} = \{P_1, \ldots, P_k, P_\#\}$ is the extended set of tiles, $\bar{A} = \{1, \ldots, k, \#\}$ is the extended alphabet of shapes. (Note that if there is a singleton in $\mathcal{P}$, then it will be considered as distinct from $P_\#$.)

Consider now the tiling $\mathbb{Z}^2 = \sqcup(t_j + P_{k_j})$ of the integer plane by $\bar{\mathcal{P}}$, which have the property that the set of all $\{t_j \mid k_j \neq \#\}$ lie inside an $m \times n$ rectangular region of $\mathbb{Z}^2$ and $\{t_j \mid k_j = \#\}$ is precisely the whole $\mathbb{Z}^2$ out of which we have "cut" that rectangular region. Such tiling will be called *compatible with a rectangular tiling*.

Formally, there exist $c \in \mathbb{Z}^2$ and $m, n \in \mathbb{N}$, such that, if we denote $I = \{v \in \mathbb{Z}^2 \mid v = c + (i, j), 1 \leq i \leq m, 1 \leq j \leq n\}$ and $E = \mathbb{Z}^2 - I$, then $\bar{x} \in \bar{A}^{\mathbb{Z}^2}$, which identifies the above tiling, is such that:

$$\bar{x}(v) = \begin{cases} r \in A, & \text{iff } v \in I, \\ \#, & \text{iff } v \in E. \end{cases}$$

Note that $\mathbb{Z}^2 = I \cup E$ is itself a tiling of the plane, which is called *rectangular tiling*. A tiling is compatible with a rectangular tiling iff it is decomposable as $\mathbb{Z}^2 = \sqcup(t_j + P_{k_j}) = (\sqcup_{k_j \in A}(t_j + P_{k_j})) \sqcup (\sqcup_{k_i = \#}(t_i + P_{k_i}))$, and this last decomposition is a rectangular tiling.

It is easy to see that both rectangular tiling and tiling compatible with rectangular tiling are shift invariant. Due to this property we can speak about (finite) rectangular pictures obtained by keeping from the integer plane only the $m \times n$ region which is covered with tiles from $\mathcal{P}$. In other words, we consider the equivalence classes of tiling compatible with rectangular tiling that can be obtained one from the other by translations, and we associate a rectangular picture with any such equivalence class.

Up to now, we defined rectangular shapes coverable with tiles of given shapes. Further, we can introduce an alphabet of symbols V, such that any pixel of the picture will hold such a symbol.

Formally, if $\bar{\mathcal{P}} = \{P_1, \ldots, P_k, P_\#\}$ is a set of prototiles, V is an alphabet and # is a special symbol, denoting the blank, then for any $i = 1, \ldots, k$, we can define the application $f_i : P_i \to V$, which associates a symbol from V with any pixel of P_i. For $P_\#$, we take $f_\# : P_\# \to \{\#\}$. Then, for any tiling of the integer plane $Z^2 = \sqcup(t_j + P_{k_j})$, we can define an application $f : Z^2 \to V \cup \{\#\}$, such that $f(t_j + w) = f_{k_j}(w)$, for any j and any $w \in P_{k_j}$. If the tiling is compatible with a rectangular tiling of size $m \times n$, we can extract from it a rectangular picture, $p = (a_{ij})_{1 \le i \le m, 1 \le j \le n}$, with $a_{ij} \in V$, in other words, a word $p \in V^{**}$.

A triple $(\mathcal{P}, V, \mathcal{F})$, consisting of a set of prototiles $\mathcal{P} = \{P_1, \ldots, P_k\}$, an alphabet V, and a family of applications $\mathcal{F} = \{f_1 : P_1 \to V, \ldots, f_k : P_k \to V\}$, will be called a *rectangular cover system* iff there exists at least one tiling of the plane which uses the prototiles $\mathcal{P} = \{P_1, \ldots, P_k\}$ and is compatible with a rectangular tiling.

A word $p \in V^{**}$ is called *tileable by the system* $(\mathcal{P}, V, \mathcal{F})$ if it can be extracted by the procedure described above from a tiling of the plane. Alternatively, p (which is a rectangle filled with symbols from V) can be disjointly covered by tiles in $\mathcal{P}$, filled with symbols in V by the functions in $\mathcal{F}$.

We denote by $L(\mathcal{P}, V, \mathcal{F})$ the two-dimensional language of all rectangular words over V, $p \in V^{**}$, which are tileable by the system $(\mathcal{P}, V, \mathcal{F})$.

For example, let the set of prototiles $\mathcal{P} = \{P_1 = \{(0, 0), (0, 1)\}$, $P_2 = \{(0, 0), (1, 0)\}, P_3 = \{(0, 0)\}\}$, the alphabet $V = \{0, 1\}$ and the applications $\mathcal{F} = \{f_1 : P_1 \to V, f_2 : P_2 \to V, f_3 : P_3 \to V\}$, with $f_i(x) = 1$, for any $x \in P_i$, $i = \{1, 2\}$ and $f_3(0, 0) = 0$. The language $L_3 = L(\mathcal{P}, V, \mathcal{F})$ contains all rectangular pictures that can be covered with the tiles in Fig. 22.7.

In Fig. 22.8, two rectangular pictures are given, both of size 3×4, one that belongs and the other one that does not belong to the language L_3.

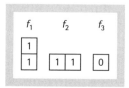

Fig. 22.7 Example of prototiles filled-in with symbols from {0, 1}.

Fig. 22.8 Example of pictures that belong (the left picture), respectively do not belong (the right picture) to the language L_3.

Lemma 22.1 *[13] For a fixed given rectangular cover system $(\mathcal{P}, V, \mathcal{F})$, the language $L(\mathcal{P}, V, \mathcal{F})$ is closed to both column and row catenation.*

Theorem 22.8 *[13] The class of local two-dimensional languages is incomparable with the class of tileable picture languages.*

22.4.2 Tiling P Systems

In [11] and in [12] P systems were introduced which generate picture languages. The emphasis was on local and recognizable languages. In [13], on which the presentation here is mainly based, the purpose is the generation of tileable languages. The P systems in question combine features of tissue-like P systems and P systems with active membranes. Each membrane is capable of holding one pixel of the picture. Starting with the upper-left corner, a picture is generated pixel by pixel from adjacent membranes (pixels).

A *tiling P system* is formally a construct

$$A = (O, V, M, (cont_m)_{m \in M}, R),$$

where: $O = V \cup \{\#\} \cup \{p_t \mid t \in T\} \cup Q \cup \{s, k\}$ is the alphabet of symbols. V is the *output alphabet* over which the pictures will be; # is the special marker for the picture boundaries; $T = \{ij \mid i, j \in \{0, 1, 2\}\}$ and a symbol p_{ij} codifies a position out of 9 possible ones inside a rectangular picture (p_{00} stands for the upper-left corner, p_{01} for the upper edge, p_{11} for the interior of the picture, etc.); s is the *output start symbol*—its presence in a membrane determines application of rules which fill that membrane with symbols from $V \cup \{\#\}$ and the generation of neighboring membranes; k is the *killer symbol*, which destroys a membrane if either it is of two distinct types, or if different types of tiles are in the same membrane, or if the type of tile does not match the type of membrane; $M \subset \{d, r\}^*$ are the labels of the membranes, where d stands for *down* and r for *right*, each membrane being labeled according to a path from the upper-left corner and its present position.

The rules from R are divided into 3 groups: the *creation rules*, *contamination rules*, and *destruction rules*. The contamination rules produce the killer symbol in one membrane in any of the 3 cases mentioned above, and the destruction rules are responsible for the dissolution of a membrane which contains it, not before the killer symbol is propagated to neighboring membranes. The creation rules are responsible for generating a grid of membranes, for checking the integrity of the grid (the rectangular shape of the picture), and generating the tiles on the grid. The functions $cont_m$ describe the contents of the membrane occupying the position m, among other symbols, needed to continue the generation process, they contain triples (α, β, γ) with the significance that the respective membrane holds pixel (β, γ) of the tile obtained by shifting the prototile P_α.

The generation of tileable pictures is accomplished by an evolving sequence of tiling P systems. An *initial* P system over V will be one with $M = \{\lambda\}$ and $cont_\lambda = \{p_{00}s\}$. This means that the picture generation starts with the upper-left corner. It will spread downwards and to the right. Each application of the evolution rules leads us from a tiling P system to another one, said to *immediately evolve* from the first. A P system becomes *stable* if it immediately evolves only into itself. It is called *dead* if $M = \emptyset$, otherwise it is called *alive*. The *stable universe* of a tiling P system A is the set of all stable alive P systems into which A can evolve.

With any stable alive P system A we associate a *picture* over V in the following way: first, we define the natural numbers s and t by: $s = max\{i \mid \exists j, d^i r^j \in M\} - 1$ and $t = max\{j \mid \exists i, d^i r^j \in M\} - 1$. If s and t are greater than 1, then we consider the picture $(a_{ij})_{1 \leq i \leq s, 1 \leq j \leq t}$ with $a_{ij} = h$ iff $h \in cont_{d^i r^j} \cap V$, otherwise we consider the empty picture.

We consider the 2D-language defined in the example, $L_3 = L(P, V, \mathcal{F})$. We define an initial tiling P system A with the set of headed prototiles labels $Q = \{(1, 0, 0), (1, 0, 1), (2, 0, 0), (2, 1, 0), (3, 0, 0)\}$, and the subset of starting prototiles labels $Q_0 = \{(1, 0, 1), (2, 0, 0), (3, 0, 0)\}$. Then, the language of rectangular pictures associated with all tiling P systems from the stable universe of A is exactly L_3. Further details can be found in the papers mentioned above.

22.5 CRYPTOGRAPHY USING P SYSTEMS

One of the first attempts at using P systems in cryptography was made in [4]. There, two algorithms for message authentication were proposed (with and without confirmation from the sender). The employed P system model in both cases was the one with active membranes. Here, we briefly comment the authentication of messages without confirmation algorithm.

Consider two parties, Alice and Bob, that have two identification keys a_{Alice} and a_{Bob}, respectively. Also, assume that both Alice and Bob are aware of each other's key.

Let Π be a special designed P system having the property that there are two multisets w_i and w_j placed in two distinct membranes (labeled i and j, respectively) which codify a message m and a sender authentication data ID, respectively. In addition, we also presume that both the message m and the ID can be read when they are outside the skin membrane, in the environment.

Suppose now that Alice receives Π and wants to read the message signed by Bob m_{Bob}. Alice introduces in Π the object a_{Alice} which will generate the following actions: a path to the node j is found and a copy of the ID_{Bob} is expelled into the environment; Π is transformed using membrane handling rules into Π' (in this way, Alice can also detect whether or not Π was previously attacked by an intruder). Alice identifies Bob (by using ID_{Bob}) and then she selects and introduces a_{Bob} into Π'. This will generate the following actions: if a_{Bob} is the valid one, then a path to the node i is found and the message m_{Bob} is expelled into the environment.

The design of Π also includes rules that destroy the whole information in Π or Π' if any illegal attempt is performed.

In [28] the focus on using the P system formalisms in the area of communication protocol and security has a different perspective. There, instead of specifying new protocols, a new verification tool for an existing one is developed. More precisely, the specification of the Needham-Schroeder public-key protocol (a communication protocol over an insecure network that provides mutual authentication between two parties exchanging messages) and the implementation of an attack-search procedure by state exploration using P systems are considered. The starting point was the logical analysis of the protocol and the goal was to find an interleaving of elementary actions (sending and answering messages) that allow an intruder to obtain confidential information.

The declarative style supported by the P system framework has represented a convincing reason for studying the intruder-centric model (which is usually used when applying formal methods to cryptographic protocol verification). Using such an approach, a security hole of the protocol (an intruder who interferes in the communication between two parties, impersonates one of them and sets up a false session with other one by executing two simultaneous runs of the protocol) is easily discovered by the state exploration procedure. However, the employed method was brute force and consists in a systematic search of attacks.

In reference [23] the study was on the class of P systems with active membranes breaking one of the most known cryptosystems—DES. This cryptosystem encrypts 64-bit messages using a 56-bit key, and by breaking DES it is commonly understood that given one (plain-text, cipher-text) pair, one can find a key mapping the plain-text to the cipher-text. The method used for breaking DES with P systems was to check through all 2^{56} keys to find out which key maps the plain text to the cipher-text. The division of membranes produces exponential space by giving rise to more and more membranes involving more and more rules, and in this way all the 2^{56} keys could be simultaneously checked. Consequently, the DES key was recovered in linear time with respect to the length of the key.

Many cryptographic algorithms are based on some **NP-Complete** problems like the Integer Factorization Problem, the Knapsack Problem, or the SUBSET SUM Problem. These problems were addressed in the P system framework and this is why we briefly recall them here.

In [29] a polynomial solution to the Integer Factorization Problem is given by using deterministic P systems with active membranes and cooperative rules. The method employed was to use the standard algorithm for computing the remainder of the division of two binary presented natural numbers. Liberal loop programs (see [27]) were used to describe algorithms of elementary number theory in terms of bit operation. These programs were implemented in a very intuitive manner into P systems with active membrane framework. However, the number of involved membranes grows exponentially.

Uniform polynomial solutions to the Knapsack and SUBSET-SUM problems have been proposed—Chapter 12. As above, the model used was P systems with active membranes and the general "strategy" for solving the mentioned problems was an exhaustive search of the solution: first, the entire candidate's space was simultaneously generated, then the appropriate conditions (which are particular for each problem) were checked in parallel, and finally the solution was provided.

22.6 THE SIMULATION OF PARALLEL ARCHITECTURES

In [14], [15], and [16], the simulations of two particular parallel models of computation—the shuffle-exchange and the 2D-mesh one—with P systems with dynamic communication graphs are proposed.

In [17] the focus is on a *class* of parallel architectures, known as SIMD (Single Instruction Multiple Data) models of computation, with fixed number of processors, and communication of data among processors according to fixed network patterns. We proposed a coherent manner of simulating any SIMD-X architecture, and furthermore, any particular algorithm Y in a given SIMD-X architecture, with P systems with dynamic communication graphs. The current presentation is based mainly on [17] because of its generality.

The notion of *P systems with dynamic communication graphs* has been developed in order to achieve the simulation of any particular parallel architecture (the notion could be used for other purposes a well). They resemble tissue-like P systems, but with dynamic features, the membranes being connected by graph structures which change in time.

22.6.1 P Systems with Dynamic Communication Graphs

Let *Graphs* denote the set of all possible graphs having n nodes labeled $P_1, \ldots, P_n$. Having fixed the nodes, each element of *Graphs* will be uniquely identified by the specific set of edges.

A distinguished element of *Graphs* is the *identity* graph, denoted in the sequel Id: (the set of nodes is fixed as mentioned above) the set of edges is defined as $Id = \{(i, i) \mid 1 \leq i \leq n\}$. Another distinguished element of *Graphs* is the *total graph*, denoted G_{total}: the set of edges is defined as $G_{total} = \{(i, j) \mid 1 \leq i, j \leq n\}$.

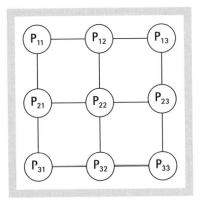

Fig. 22.9 2D-mesh with $n = 3$.

Definition 22.5 *A P system with dynamic communication graphs is a construct* $\Pi = (V, P_1, \ldots, P_n, R_\mu)$, *where* $P_1, \ldots, P_n$ *are elementary membranes, and V is an alphabet of symbols used to codify the contents of the membranes.*

R_μ *is a set of pairs* [graph, rules], *with graph* $\in$ *Graphs and such that:* (i) *if graph* $\subseteq$ *Id then its associated rules are rewriting rules;*

(i) *if graph* $\subseteq G^+_{total} = G_{total} - Id$ *then its associated rules are communication rules.*

Application of rules in a classical P system is replaced in the case of P systems with dynamic communication graphs, with applications of pairs [*graph, rules*], in order to pass from one configuration to another. By applying such a pair [g, r] to a configuration we mean "apply r to the nodes/edges described by g", which makes sense in view of Definition 22.1.

For the purposes of simulating algorithms on parallel architectures we will use such P systems *with finite sequential support*, i.e. with the set R_μ being a finite sequence.

Let X be a fixed certain subset of *Graphs*. We will call X the set of *admissible communication graphs*.

Definition 22.6 *A P system with dynamic communication graphs will be called of X-type iff for any pair* [graph, rules] $\in R_\mu$ *we have graph* $\in X$.

22.6.2 SIMD Networks of Processors

Some networks organizations for processors are briefly described here (see [34]).

In *mesh networks* the processors/nodes form a q-dimensional lattice, and communication is allowed only between neighboring nodes. For the case $q = 2$, we are dealing with 2D(two-dimensional)-mesh networks. The n^2 processors/nodes of such a network will be denoted P_{ij}, with $1 \leq i, j \leq n$.

The allowed communication is described by the *strict total graph*,

$$G^+_{total} = G_h \cup G_v = \bigcup_{i=1}^{n} G_{i*} \cup \bigcup_{j=1}^{n} G_{*j}, \text{ where}$$

$G_{i*} = \{((i, j), (i, j+1)) \mid 1 \leq j \leq n-1\}$, for all $1 \leq i \leq n$,

$G_{*j} = \{((i, j), (i+1, j)) \mid 1 \leq i \leq n-1\}$, for all $1 \leq j \leq n$.

A *shuffle–exchange network* consists of $k = 2^n$ nodes, representing processors, labeled with the elements of the set $L = \{0, 1, \ldots, k-1\}$, and two kinds of connections.

Exchange connections link pairs of nodes whose labels, written in binary form, differ in the least significant bit. The following characterizes exchange connections.

Let $a_{n-1}a_{n-2}\ldots a_1 a_0$ be the label of a node in a shuffle–exchange network, expressed in binary. Then this node is exchange connected with the node labeled by $a_{n-1}a_{n-2}\ldots a_1 a_0^*$, where $a_0^* = 1 - a_0$.

The set $\{(j, r) \mid j \in L\}$ of exchange connections will be considered the set of edges of a graph, with set of nodes L (or, equivalently, the set of processors labeled by L). We call this graph the *exchange graph*, and denote it G_e. Note that the exchange graph is a non-oriented one.

Shuffle connections link a node labeled by i ($i \neq k-1$) with a node labeled by j (in the direction from i to j!) if and only if $j \equiv 2i$ modulo $k-1$ (excepting the node labeled by $k-1$ that is connected to itself); that is,

$$j = \begin{cases} 2i & \text{for } 0 \leq i \leq k/2 - 1 \\ 2i + 1 - k & \text{for } k/2 \leq i \leq k-1 \end{cases}$$

We have the following characterization of the shuffle connection.

Lemma 22.2 *Let* $a_{n-1}a_{n-2}\ldots a_1 a_0$ *be the label of a node in a shuffle–exchange network, expressed in binary. Then this node is shuffle connected with the node labeled by* $a_{n-2}a_{n-3}\ldots a_0 a_{n-1}$.

The set $\{(i, j) \mid i \in L\}$ of shuffle connections can be considered the set of edges of a graph, with set of nodes L (or, equivalently, the set of processors labeled by L). We call this graph the *shuffle graph*, and denote it G_s. Note that the shuffle graph is an oriented graph.

Shuffle/exchange connections are used by the processors to perform *shuffle/exchange operations*, which consist of transferring values of internal variables "along" a given connection.

The shuffle–exchange network corresponding to eight nodes is depicted in Fig. 22.10. The arrows (oriented edges) depict the *shuffle* connections, while the double-arrows (the non-oriented edges) depict the *exchange* connections.

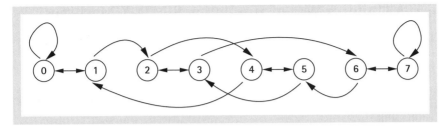

Fig. 22.10 Shuffle–exchange network.

A parallel machine consists of a large number of processors, each one having an arithmetic logic unit with registers and a private memory, able to solve problems in a cooperative way.

According to Flynn's classification of computers (see [18]), a form of synchronous parallelism is called **SIMD** (Single–Instruction–Multiple–Data). A **SIMD** machine consists of a set of identical processors capable of simultaneously performing the same instruction (issued by a central control unit), on different sets of data, and in a synchronous manner: each processor executing an instruction in parallel must be allowed to finish before the execution of the next instruction starts.

Alternatively, one can speak of **SIMD** models of parallel computation. Here, only the SIMD models with a *fixed number of processors* are considered. In addition, SIMD models differ in the ways communication of data is achieved between processors. On the one hand, we have communication of data achieved via *shared memory*, and, on the other hand, we have communication of data achieved via a predefined *network organization*. Here only the SIMD models with predefined *network organization* are considered. The focus will be on the basic types described previously.

The following notations are used:

- MC^q for a q-dimensional mesh network;
- S–E(n) a shuffle–exchange network with $k = 2^n$ nodes;
- P^p a pyramid network of size p;
- B(k) a butterfly network with $k + 1$ ranks;
- CC^k a cube-connected network of dimension k.

The descriptions of pyramid, butterfly, and cube-connected networks, which we do not present here, can be found for instance in [34].

If $X \in \{ MC^q, S\text{-}E(n), P^p, B(k), CC^k \}$ is one of the above network types, by **SIMD-X** we will refer to a SIMD model which communicates data according to the X network structure. The term **SIMD-X** machine is also used.

Graphs(X) denotes the active graphs associated to the particular network organization (or architecture) X. These will be the sets of edges used for actual communication of data between processors in a **SIMD-X** machine.

22.6.3 A Uniform Way of Simulating SIMD-X Machines

We achieve a uniform simulation of any **SIMD-X** machine by applying the following rules (guidelines):

1. Simulation of internal computations:
   - We associate to each processor an elementary membrane, and use the same labeling.
   - We codify the contents of each processor's variables in a suitable manner over an alphabet V.
   - We simulate the internal computations performed by the processors (recall that in the case of the **SIMD** architecture, we have the same internal computations inside each processor), by the action of symbol or object rewriting rules, at work simultaneously inside each individual corresponding membrane.

     In order to have a unitary formalism, we will associate such rules to the identity graph, Id, on the corresponding set of nodes, or to subsets of Id, depending on the parallel model simulated, or even on the particular algorithm simulated.

2. Simulation of data communication:
   - We will simulate the exchange of data performed by the processors with communication rules (symport/antiport rules) between membranes.
   - The communication rules will be associated to edges of graphs belonging to the admissible set $Graphs(X)$ which characterizes the **SIMD-X** model.

As we have seen, the graphs are of different types for each particular **SIMD-X** architecture. Moreover, even for a given architecture, they can be different, or used in different manners, depending on the specifics of algorithms implementations on that architecture.

Theorem 22.9 *For any **SIMD-X** model, if there exists an alphabet of symbols, V, able to codify the contents of the processors, and a (finite) set R_μ of pairs [graph, rules], with graph $\in Graphs(X)$, and such that:*

(i) *if graph $\subseteq Id$, then its associated rules are rewriting rules, able to simulate the internal processing operations;*

(ii) *if graph $\subseteq G^+_{total}$, then its associated rules are communication rules, able to simulate communication of data between processors;*

*then we can construct a corresponding P system, with dynamic communication graphs, $\Pi(X) = (V, P_1, \ldots, P_n, R_\mu)$, where $P_1, \ldots, P_n$ are elementary membranes in bijection with the processors via the same labels, and it will be a simulation of the **SIMD-X** model.*

For $X = $ S–E (the perfect–shuffle) we have (see [14], [15]): $Graphs($S–E$) = \{G_s, G_e, Id\}$, and the conditions for pairs are:

(i) if *graph* = *Id*, then its rules are rewriting rules;
(ii) if *graph* ∈ {G_s, G_e}, then its rules are communication rules.

For X = MC² (the 2D-mesh) we have (see [16]): *Graphs*(MC²) = $\mathcal{P}(G_{total}^+) \cup \mathcal{P}(Id)$, and the conditions for pairs are:

(i) if *graph* ⊂ *Id*, then its rules are rewriting rules;
(ii) if *graph* ⊂ G_{total}^+, then its rules are communication rules.

22.6.4 A Uniform Way of Simulating Algorithms on SIMD-X Machines

Any particular deterministic algorithm Y, implemented on a SIMD-X model, can be simulated, in the same conditions as in Theorem 22.1, by a P system with dynamic communication graphs

$$\Pi(X, Y) = (V, P_1, \ldots, P_n, R_\mu(Y)),$$

with a finite sequential support, obtained from $\Pi(X)$, which simulates the architecture, by particularizing the sequence $R_\mu(Y)$ in such a way that its application simulates the action of algorithm Y.

We give next as an example the solving of the reduction problem on the S–E architecture. Let ∗ be a binary associative and commutative operation over a set, and let $a_0, \ldots, a_{k-1}$ be elements of this set. The *reduction* (see [34], Section 2.3.1) is the process of computing $a_0 \ast \cdots \ast a_{k-1}$.

For X = S–E, Y is the following algorithm for solving the reduction problem:

```
procedure reductionPS(a0, a1, ..., ak−1)
begin
    for i ← 1 to n do
        for all j where 0 ≤ j ≤ k − 1 do
            shuffle (sj)
            tj ← sj
            exchange (tj)
            sj ← sj ∗ tj
        endfor
    endfor
end
```

Since ∗ is binary (and the network organizations in question are fit for parallelizing the computation of the final result, $S = a_0 \ast \cdots \ast a_{k-1}$ by computing as many $a_i \ast a_j$ as possible in parallel) each processor has two variables, s, and an auxiliary t. Variable s is used to store the entry data, as well as intermediate and final results, t is used for transfer of data between adjacent processors.

Correspondingly, an alphabet V_s can be used to codify contents of s and a disjoint alphabet V_t can be used to codify contents of t. V_s and V_t are in bijective correspondence; if $a \in V_s$ then $a' \in V_t$ is its image through the bijection. If the values to be represented are positive integers, the codifications need only one symbol. For $V_s = \{a\}$, integer x as value of variable s is represented as a^x. For $V_t = \{a'\}$, integer y as value of variable t is represented as a'^y. Computing $x * y$ and putting the result in s needs at least the rewriting of a''s to a's (in the case $* = +$ on positive integers, only such a rewriting).

Let $(A, *, 0)$ be a commutative monoid and consider the following set of elements of A: $\{a_0, \ldots, a_{k-1}\}$. Assume that every element of A can be codified over an alphabet V_s (and over the alphabet V_t) and assume that there exists a set of symbol rewriting rules r_* over $V_s \cup V_t$, such that its action inside each membrane simulates the instruction $s = s * t$ (computing the value of $s * t$ and putting the result in s). Consider the P system with sequential periodic dynamic communication graphs, of S–E type

$$\Pi(\text{S-E}, Y) = (V_s \cup V_t, P_0, \ldots, P_{k-1}, R_\mu(Y)),$$

with period

$$R_\mu = \{[\{(a, out) \mid a \in V_s\}, G_s], [\{a \to aa' \mid a \in V_s\}, Id],$$
$$[\{(a', out) \mid a' \in V_t\}, G_e], [r_*, Id]\}),$$

such that $R_\mu(Y) = R_\mu^{\log_2(k)}$. Then, the system $\Pi(\text{S-E}, Y)$ computes $\{a_0 * \cdots * a_{k-1}\}$, starting from an initial configuration in which each P_j contains a codification of a_j over V_s, for all $0 \leq j \leq k-1$, and applying the sequence $R_\mu(Y)$ (or, equivalently, applying $n = \log_2(k)$ times the sequence R_μ). The result $S = a_0 * \cdots * a_{k-1}$ is obtained in each membrane, codified over V_s.

22.7 Boolean Circuits

Boolean circuits are well-known classical computing devices, which encapsulate the parallelism features at the resolution of a single bit. Their main components are the *logical gates* able to compute the elementary logical functions ($\neg$ NOT, $\wedge$ AND, $\vee$ OR) and the bit-carrying *wires*. Boolean circuits are finite oriented acyclic directed graphs such that the edges represent the wires and the nodes represent the gates (*input gates* if the indegree is 0, *output gates* if the outdegree is 0, or logical gates— the nodes labeled $\wedge$, $\vee$ have indegree 2 and outdegree 1, while the nodes labeled $\neg$ have indegree and outdegree 1). Circuits do not have memory nor states and

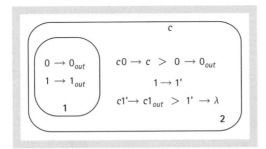

Fig. 22.11 The P system Π_{OR} simulating the OR gate.

they compute functions from the set $B_k = \{f \mid f : \{0, 1\}^k \to \{0, 1\}\}$—the set of all k-ary Boolean functions.

In [10] Boolean circuits were simulated using P systems. The simulation is performed in two steps, by first reproducing the functioning of the logical gates in terms of P systems and creating modules, and next by assembling these modules into actual circuits. Several P systems models were used to carry out these simulations, the goal being to show the model flexibility while solving such problems.

For example, in Fig. 22.11 the OR gate is simulated by using a P system with non-cooperative, catalytic rules, and weak priorities. In this simulation the input to the gate is considered to arrive in the innermost membrane and the result of the computation is sent out into the environment. Based on this the following result holds:

Lemma 22.3 *The P system Π_{OR}, acting on pairs of input values $x_1, x_2 \in \{0, 1\}$, is deterministic and produces into the environment in three computational steps the value $x_1 \vee x_2$.*

Similar results also stand for the simulation of the NOT and AND gates. In this way, any Boolean circuit whose underlying graph structure is a binary tree structure can be simulated with P systems using such modules. However, apart from the modules representing the logical gates, the simulation of a circuit needs the *SYNC* modules—such a module synchronizes two output values when they become input values for a gate. A simple example is given in Fig. 22.12.

The following result holds true:

Theorem 22.10 *Every Boolean circuit a, whose underlying graph structure is a rooted binary tree, can be simulated by a P system, Π_a. Π_a is constructed from standard P systems of type Π_{AND}, Π_{OR}, and Π_{NOT}, by reproducing in the architecture of the membrane structure, the structure of the tree associated to the circuit. The time complexity for computing a Boolean function depends linearly on the depth of the underlying tree structure of the P system.*

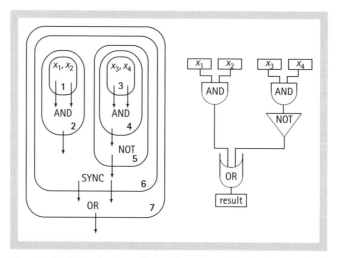

Fig. 22.12 The design of the P system simulating a circuit.

A similar approach, but involving different models of P systems, has been done in [21], [32], and [22].

Boolean circuit related problems have been addressed in [24], [25], and [26], from the conservative logic point of view. A reversible Fredkin gate is a three-input/three-output Boolean gate whose input/output map $FG : \{0, 1\}^3 \rightarrow \{0, 1\}^3$ associates the input (x_1, x_2, x_3) to its corresponding output (y_1, y_2, y_3) such that $y_1 = x_1$, $y_2 = (\neg x_1 \wedge x_2) \vee (x_1 \wedge x_3)$, and $y_3 = (x_1 \wedge x_2) \vee (\neg x_1 \wedge x_3)$. In [24], the Fredkin gate is simulated using an *energy-based* P system (a P system model in which the amount of energy manipulated and/or consumed during computations is taken into account). The constructed P system turns out to be itself reversible and conservative as the Fredkin gate. Moreover, in [25] and [26] it is shown how any reversible circuit composed by Fredkin gates can be simulated by a corresponding reversible and conservative energy-based P system. In this context, the simulating P system can be made self-reversible, i.e. the same system can also perform backward computations. In addition, the number of membranes used by the constructed P system is independent of the number of gates occurring in the simulated circuit.

22.8 Conclusions

We have surveyed the formalisms regarding practical computer science issues in the P systems framework. In addition, we reviewed some of the most known results and employed techniques concerning algorithms for such problems. In this regard, several P systems models were discussed and their ability to perform specific tasks was commented. We were also interested in the comparison between the classical sequential/parallel algorithms and those implemented in the framework of P systems. The presented results showed that, even if P systems are using unstructured data to perform computations (the multisets), the model is robust enough to have good complexity outcomes while solving practical problems.

Acknowledgement

D. Sburlan gratefully acknowledges the support of the Romanian CNCSIS IDEI-PCE 551/2009 research grant.

References

[1] A. Alhazov, D. Sburlan: Static sorting algorithms for P systems. *Pre-Proc. 4thWMC* (A. Alhazov et al., eds.), GRLMC Rep. 28/03, Tarragona, 2003, 17–40.

[2] A. Alhazov, D. Sburlan: Static sorting P systems. *Applications of Membrane Computing* (G. Ciobanu et al., eds.), Springer, 2006, 215–252.

[3] J.J. Arulanandham: Implementing bead-sort with P systems. *Lecture Notes in Computer Sci.*, 2509 (2002), 115–125.

[4] A. Atanasiu: Authentication of messages using P systems. *Lecture Notes in Computer Sci.*, 2597 (2002), 33–42.

[5] R. Ceterchi, C. Martín-Vide: Dynamic P systems. *Lecture Notes in Computer Sci.*, 2597 (2003), 146–186.

[6] R. Ceterchi, C. Martín-Vide: P systems with communication for static sorting. *Proc. Brainstorming Week on Membrane Computing*, GRLMC Rep. 26/03 (M. Cavaliere et al., eds.), Tarragona, 2003, 101–117.

[7] R. Ceterchi, M.J. Pérez-Jiménez, A.I. Tomescu: Simulating the bitonic sort using P systems. *Lecture Notes in Computer Sci.*, 4860 (2007), 172–192.

[8] R. Ceterchi, M.J. Pérez-Jiménez, A.I. Tomescu: Sorting omega networks simulated with P systems: Optimal data layouts. *Proc. 6th Brainstorming Week on Membrane Computing* (D. Diaz-Pernil et al., eds.), Fenix Editora, Sevilla, 2008, 79–92.

[9] R. Ceterchi, A.I. Tomescu: Implementing sorting networks with spiking neural P systems. *Fundamenta Informaticae*, 87 (2008), in press.

[10] R. Ceterchi, D. Sburlan: Simulating Boolean circuits with P systems. *Lecture Notes in Computer Sci.*, 2933 (2003), 104–122.

[11] R. Ceterchi, R. Gramatovici, N. Jonoska, K.G. Subramanian: Generating picture languages with P systems. *Proc. Brainstorming Week on Membrane Computing*, GRLMC Rep. 26/03 (M. Cavaliere et al., eds.), Tarragona, 2003, 85–100.

[12] R. CETERCHI, R. GRAMATOVICI, N. JONOSKA, K.G. SUBRAMANIAN: Tissue-like P systems for picture generation. *Fundamenta Informaticae*, 56 (2003), 311–328.

[13] R. CETERCHI, R. GRAMATOVICI, N. JONOSKA: Tiling rectangular pictures with P systems. *Lecture Notes in Computer Sci.*, 2933 (2003), 88–103.

[14] R. CETERCHI, M.J. PÉREZ JIMÉNEZ: Simulating shuffle–exchange networks with P systems. *Proc. Second Brainstorming Week on Membrane Computing* (Gh. Păun et al., eds.), Rep. RGNC 01/04, Seville, 117–129.

[15] R. CETERCHI, M.J. PÉREZ JIMÉNEZ: A perfect shuffle algorithm for reduction processes and its simulation with P systems. *Proc. ICCC 2004* (I. Dzitac et al., eds.), Ed. Univ. Oradea, 2004, 92–97.

[16] R. CETERCHI, M.J. PÉREZ JIMÉNEZ: On two-dimensional mesh networks and their simulation with P systems. *Lecture Notes in Computer Sci.*, 3365 (2005), 259–277.

[17] R. CETERCHI, M.J. PÉREZ JIMÉNEZ: On simulating a class of parallel architectures. *Intern. J. Found. of Computer Sc.*, 17 (2006), 91–110.

[18] M.J. FLYNN: Very high-speed computing systems. *Proc. of the IEEE*, 54 (1966), 1901–1909.

[19] A. GEORGIOU: SubLP-Studio software. Available from the P systems web page at http://ppage.psystems.eu/software.html, 2003.

[20] A. GEORGIOU, M. GHEORGHE, F. BERNARDINI: Membrane-based devices used in computer graphics. *Applications of Membrane Computing* (G. Ciobanu et al., eds.), Springer, 253–282.

[21] M. IONESCU, I. TSEREN-ONOLT: Boolean circuits and a DNA algorithm in membrane computing. *Lecture Notes in Computer Sci.*, 3850 (2006), 272–291.

[22] M. IONESCU, D. SBURLAN: Some applications of spiking neural P systems. *Proc. 8th WMC* (G. Eleftherakis et al., eds.), 2007, 283–394.

[23] S.N. KRISHNA, R. RAMA: Breaking DES using P systems. *Theoretical Computer Sci.*, 299 (2003), 495–508.

[24] A. LEPORATI, C. ZANDRON, G. MAURI: Simulating the Fredkin gate with energy-based P systems. *J. Univ. Computer Sci.*, 10 (2004), 600–619.

[25] A. LEPORATI, C. ZANDRON, G. MAURI: Universal families of reversible P systems. *Lecture Notes in Computer Sci.*, 3354 (2005), 257–268.

[26] A. LEPORATI, C. ZANDRON, G. MAURI: Reversible P systems to simulate Fredkin circuits. *Fundamenta Informaticae*, 74 (2006), 529–548.

[27] A.R. MEYER, D.M. RITCHIE: The complexity of loop programs. *Proc. ACM National Meeting, ACM Pub.* P-67 (1967), 465–469.

[28] O. MICHEL, F. JACQUEMARD: An analysis of a public key protocol with membranes. *Applications of Membrane Computing* (G. Ciobanu et al., eds.), Springer, 2006, 283–302.

[29] A. OBTULOWICZ: On P systems with active membranes: Solving the integer factorization problem in a polynomial time. *Lecture Notes in Computer Sci.*, 2235 (2001), 267–285.

[30] GH. PĂUN: (DNA) Computing by carving. *Soft Computing*, 3 (1999), 30–36.

[31] M.J. PÉREZ JIMÉNEZ, A. ROMERO-JIMÉNEZ, F. SANCHO-CAPARRINI: Computationally hard problems addressed through P systems. *Applications of Membrane Computing* (G. Ciobanu et al., eds.), Springer, 2006, 315–346.

[32] Y.J. PRAKHSH, K. KRITHIVASAN: Simulating Boolean circuits with tissue P systems. *Pre-Proc. 5th WMC*, Milano, Italy, 2004, 343–359.

[33] P. PRUSINKIEWICZ, A. LINDENMAYER: *The Algorithmic Beauty of Plants*. Springer, 1990.
[34] M.J. QUINN: *Parallel Computing. Theory and Practice*. McGraw-Hill Series in Computer Science, 1994.
[35] A. ROMERO-JIMÉNEZ, M.A. GUTIÉRREZ-NARANJO, M.J. PÉREZ JIMÉNEZ: The growth of branching structures with P systems. *Proc. Fourth Brainstorming Week on Membrane Computing* (C. Graciani-Diaz et al., eds.), Vol. II. Fenix Editora, Sevilla, 2006, 253–265.
[36] A. ROMERO-JIMÉNEZ, M.A. GUTIÉRREZ-NARANJO, M.J. PÉREZ-JIMÉNEZ: Graphical modelling of higher plants using P systems. *Lecture Notes in Computer Sci.*, 4361 (2006), 496–506.
[37] G. ROZENBERG, A. SALOMAA: *The Mathematical Theory of L Systems*. Academic Press, 1980.
[38] E. RIVERO-GIL, M.A. GUTIÉRREZ-NARANJO, A. ROMERO-JIMÉNEZ, A. RISCOS-NÚÑEZ: A software tool for generating graphics by means of P systems. Submitted, 2007.
[39] D. SBURLAN: A static sorting algorithm for P systems with mobile catalysts. *An. Şt. Univ. Ovidius Constanţa*, 11 (2003), 195–205.

23. OTHER DEVELOPMENTS

CHAPTER 23.1

P COLONIES

ALICA KELEMENOVÁ

23.1.1. INTRODUCTION

A P colony is a membrane computing model, which reflects motivation from colonies of grammar systems, i.e. the idea to use agents, which are as simple as possible and placed in a common environment, composing a system which produces non-trivial emergent behavior, using the environment as the communication medium. For *colonies* of simple formal grammars, which were introduced in [14], see also [16, 18].

Agents of the P colonies are single cells with a small number of objects inside them, processed by simple programs. In comparison with P systems, P colonies consist of single cells "floating" in a common environment. Cells in P colonies are not connected into a graph structure like the cells of the P systems, but, on the other hand, the unstructured sets of rules of cells in P systems are structured to simple programs in P colonies. A tuple of rules in a program have to act in parallel in order to change all objects placed in the cell, in one derivation step.

Cells in P colonies process symbolic objects appearing in their environment. All objects are atomic, identified with symbols. These cells are basic computing *agents* of P colonies.

We restrict the complexity and the capabilities of agents as much as we can. Each agent is associated with a small number of *objects* present inside it, and with a set of *rules*, forming *programs* to process these objects. The rules available to an agent are either of the form $a \to b$, specifying that an internal object a is transformed into an internal object b, or of the form $c \leftrightarrow d$, specifying the fact that an internal object c is sent out of the agent, to the environment, in exchange of the object d, which was present in the environment and is now brought inside the agent. We consider here the object-to-object exchanges and one-object internal evolutions as the simplest forms of rules. We also add priority rules r_1/r_2 to agents (originally called checking rules), with the following meaning: if the rule r_1 can be applied then it must be applied; if not, then rule r_2 must be performed. The program of an agent allows it to change simultaneously and deterministically all objects in the cell by different rules, so the number of objects in an agent is identical with the number of rules in each of its programs.

The environment is elementary in the following sense: at the beginning of the computation, all objects from the environment are identical. An arbitrarily large number of copies of a generic object e is available in it, and during the computation, although arbitrarily many copies of e can be changed in other objects (after being introduced in agents), still arbitrarily many copies of e remain available. The objects initially present inside each agent are also objects e.

Thus, a P colony consists of a finite number of agents (identified by the sets of programs) placed in a common environment, where arbitrarily many copies of e are present.

We discuss parallel and sequential P colonies depending on the number of agents acting in one derivation step. In the first case, each agent which can apply any of its programs has to non-deterministically choose one, and apply it; in the sequential case one agent, which is non-deterministically chosen, is allowed to act. The computation halts when no agent can apply any of its programs to the existing objects. We associate a result with a halting computation, in the form of the number of copies of a distinguished object present in the environment.

P colonies are computationally complete, i.e. all the number sets computable by Turing machines are computable also by P colonies. As we summarize in Subsection 23.1.3 of this section, this computing power can emerge from agents with a rather small number of programs. Also a trade-off result holds: if each agent can contain arbitrarily many programs, then one agent can compute at the level of Turing machines. More objects in cells reach universality with less programs or less cells. Priority rules are not necessary to achieve universality, but they can simplify the size of P colonies, namely they can decrease the number of agents and/or the number of their programs.

Universality can also be obtained in the case where both the number of programs and agents are bounded as we present in Subsection 23.1.4.

These results allow the following interpretation: the environment is essential as a medium for communication and for storing information during the computation, even with no structure and no information in the environment at the beginning of the computation. The power of cooperating agents of a very restricted form can be dramatically different from the power of individual agents.

23.1.2. DEFINITIONS

A *P colony* of capacity $k \geq 1$ is a construct $\Pi = (A, e, f, C_1, \ldots, C_n)$, where A is an alphabet (its elements are called *objects*), e and f are two distinguished objects of A (environmental object and final object), and $C_1, \ldots, C_n$ are *cells (agents)*; each cell C_i is a pair $C_i = (O_i, P_i)$, where O_i is a multiset of k symbols over A (the initial state of the cell), and P_i is a finite set $\{p_{i,1}, \ldots, p_{i,k_i}\}$ of *programs*; each program $p_{i,j}$ consists of k rules of the forms $a \to b$ (internal point mutation), $c \leftrightarrow d$ (one object exchange with the environment), or r_1/r_2 (priority rule, where r_1 and r_2 are arbitrary combination of point mutation and/or exchange rules).

In this section, we consider P colonies having capacity one, two, or three, i.e. P colonies with one, two, or three objects in each cell, all of them being copies of e at the beginning of a computation, and, correspondingly, one, two, or three rules in programs, respectively, associated with each cell.

No restriction is imposed on the rules in programs. In the case of colonies with capacity 2 we call *restricted* the case when all programs are of the forms $\langle a \to b; c \leftrightarrow d \rangle$ and $\langle a \to b; c \leftrightarrow d/c' \leftrightarrow d' \rangle$ (always, one point mutation and one exchange rule are applied, the second one possibly a checking rule). We call a program *homogeneous* if all its rules are of the same type. A P colony is homogeneous if all its programs are homogeneous.

At the beginning of the *computation* performed by a given P colony, the *environment* contains arbitrarily many copies of the symbol e (and nothing else); moreover, each cell contains k copies of e.

At each step of computation, the contents of the environment and of the agents change in the following manner. In the *parallel derivation mode*, each agent which can apply any of its programs has to non-deterministically choose one and apply it simultaneously with all the other agents; in *the sequential mode*, one agent uses one of its programs at a time. When using a program, each of its rules must be applied to different objects from the cell. Using the programs, the P colony passes from a configuration (contents of the cells and of the environment) to another configuration. A sequence of transitions starting with the initial states of the agents and the environment is a computation. A computation halts, if it reaches a configuration where no cell can use any program. We associate a result with a halting computation, in the form of the number of copies of the object f present in the environment in the halting configuration. With a P colony Π we associate a set of numbers, denoted by $N(\Pi)$, computed by all possible halting computations of Π.

The number of cells in a given P colony Π is called the *degree* of Π; the maximal number of programs of the cells of Π is called the *height* of Π.

The family of all sets of numbers $N(\Pi)$ computed in x-mode derivation for $x \in \{par, seq\}$ by P colonies of capacity k, degree at most $n \geq 1$ and height at most $h \geq 1$, without using priority rules in their programs, is denoted by $NPCol_x(k, n, h)$. If one of the parameters n, h is not bounded, then we replace it with $*$. In the case of colonies with capacity 2 and acting with restricted programs, we write $NPCol_{par} R$ instead of $NPCol_{par}$. If checking rules are allowed, then we write $NPCol_x K$ instead of $NPCol_x$.

In following parts of the section we determine the values of parameters k, h, n which make the families $NPCol_x(k, n, h)$, $NPCol_x R(k, n, h)$, $NPCol_x K(k, n, h)$, and $NPCol_x K R(k, n, h)$ equal to NRE, the family of sets of numbers computed by Turing machines.

23.1.3. UNIVERSALITY RESULTS

Following the idea of the simplicity of cells we study P colonies with the capacity one, two, and three, respectively. This allows us to obtain the universal generative power of P colonies for each of the mentioned capacity and to determine degree n and height h, as small as possible, which guarantee the universal generative power of P colonies of a given capacity.

The typical proof technique which is used to achieve universality results for P colonies is to simulate the computation of the register machines by P colonies. Let us consider a register machine $M = (m, H, l_0, l_h, I)$ with m registers, set of labels H, starting instruction l_0, halting instruction l_h, and set of the labeled program instructions I. A P colony which generates the same set of integers as the register machine M can be constructed in the following manner. All the labels from H will be objects of the P colony. The content of the ith register, for $1 \leq i \leq m$, will be represented by the number of copies of a specific object a_i placed in the environment. We assume that at the beginning of the simulation of the instruction labeled by l_j, the object l_j is in the environment (or exceptionally in a cell, in the case of P colonies with exactly one cell).

The behavior of the register machine is simulated by the computation of the P colony. We proceed in following steps:

1. *Initialization:* We create the configuration corresponding to the initial state of the register machine, i.e. the configuration with the object l_0 in the environment. (Register machines start with all the registers empty, so no a_i occurs in the environment.)
2. *Simulation of an instruction l_i:* (ADD$(r), l_j, l_k$) *by the P colony:* We add programs to some agents which identify l_i in the environment, create one new copy of a_r and place it in the environment. Moreover, the object l_i is consumed by some membrane and a new object l_j or l_k is created and placed in the environment.
3. *Simulation of an instruction l_i:* (SUB$(r), l_j, l_k$) *by the P colony:* We add programs to some agents which identify l_i in the environment, consume one copy of a_r from the environment and place there object l_j instead of the object l_i. If no a_r is in the environment, the program just place object l_k instead of the object l_i in the environment.
4. *Simulation of the halting instruction l_h:* HALT *by the P colony:* We add programs to some membrane(s), which identify l_h in the environment. Programs of cells have to guarantee that no membrane will be active in this case, after eventually consuming l_h.

The results presented in this section illustrate a trade-off among the capacity, degree, and height of the P colonies. We demonstrate influences of various program limitations to these values. We will also present an example of typical proof for the case of restricted P colonies. For proofs of all other presented results the reader is referred to the references.

P colonies of degree one contain only one agent, so the parallel and sequential rewriting coincide in this case, and we write simply *NPCol* $(k, 1, h)$ for the corresponding classes. To obtain computational completeness two objects in agents are sufficient (and no restriction to the number of programs is considered). Universality can be obtained by using priority rules in the restricted as well as in the homogeneous P colonies.

Theorem 23.1.1. [5, 9, 11, 15]: *NPCol KR*$(2, 1, *)$ = *NPCol KH*$(2, 1, *)$ = *NRE*.

Remark: *NPCol R*$(2, 1, *)$ is not computationally complete (see Section 23.1.4). Are *NPCol R*$(3, 1, *)$ and *NPCol*$(2, 1, *)$ computationally complete?

P colonies of capacity one have one object in each agent and just one rule in each program. These systems with priority rules are computationally complete for at least 4 agents (and no limitation on the number of programs) or with no priority rules for at least 6 programs in each agent (and no limitation on the number of agents).

Theorem 23.1.2. [3, 5, 6]: *NPCol$_{par}$ K*$(1, 4, *)$ = *NPCol$_{par}$*$(1, *, 6)$ = *NRE*.

Problem: What can be said about the colonies with no checking rules in the first case and with checking rules in the second case? The sequential model was not discussed for P colonies of capacity one.

P colonies of capacity two. We already know that one agent with checking rules is sufficient. As for the number of programs we start with results for the restricted P colonies, which form the original model.

Theorem 23.1.3. [9, 11, 13, 15]: $NPCol_xKR(2, *, 5) = NRE$, $x \in \{par, seq\}$.

Sketch of proof Let us consider a register machine $M = (m, H, l_0, l_h, I)$. We construct a P colony $\Pi = (A, f, e, C_1, \ldots, C_s)$, with $f = a_1$, set of objects $A = H \cup \{a_i \mid 1 \leq i \leq m\} \cup \{e, d, d'\}$, and with $s = card(I) + 2$ cells containing the following programs:

1. Starting cells C_1, C_2 with sets of programs:
   $P_1 = \{\langle e \to d; e \leftrightarrow e \rangle, \langle e \to l_0; d \leftrightarrow e \rangle, \langle e \to e; l_0 \leftrightarrow d' \rangle\}$,
   $P_2 = \{\langle e \to d'; e \leftrightarrow e \rangle, \langle e \to e; d' \leftrightarrow e \rangle, \langle e \to e; e \leftrightarrow d \rangle\}$.
   The cell C_1 sends to the environment one copy of d and l_0; l_0 is exchanged with d' and C_1 stops. The cell C_2 can produce several copies of d' and it stops with the copy of d in it.

2. For each *ADD* instruction $l_1 : (ADD(r), l_2, l_3)$ from I we consider the cell C_{l_1} with the following five programs
   $P_{l_1} = \{\langle e \to a_r; e \leftrightarrow l_1 \rangle, \langle l_1 \to l_2; a_r \leftrightarrow e \rangle, \langle e \to e; l_2 \leftrightarrow e \rangle,$
   $\langle l_1 \to l_3; a_r \leftrightarrow e \rangle, \langle e \to e; l_3 \leftrightarrow e \rangle\}$.
   This cell brings the object l_1 inside, and at the same time it changes its inner e with a_r; in the next step, this a_r is released into the environment, while l_1 is transformed either to l_2 or to l_3, non-deterministically; finally the labels l_2, l_3 are sent to the environment making possible the simulation of next instruction.

3. For each *SUB* instruction $l_1 : (SUB(r), l_2, l_3)$ from I we consider the cell C_{l_1} with the following five of programs
   $P_{l_1} = \{\langle e \to e; e \leftrightarrow l_1 \rangle, \langle l_1 \to l_2; e \leftrightarrow a_r/e \leftrightarrow e \rangle, \langle a_r \to e; l_2 \leftrightarrow e \rangle,$
   $\langle l_2 \to l_3; e \leftrightarrow e \rangle, \langle e \to e; l_3 \leftrightarrow e \rangle\}$.
   Again, the cell brings l_1 inside, in the next step it checks whether a copy of a_r is present in the environment; in the positive case the cell brings a_r inside (and changes it here into e); in the negative case the inner object e remains unchanged. In the former case, l_2 moves into the environment; in the latter case l_2 is transformed into l_3 and this label is sent to the environment. The simulation of the SUB instruction is correct, the number of copies of a_r from the environment is decreased by one if possible, and the next available label indicates the correct action.

4. No program is added for instruction l_h.

The computation in Π stops at the moment when the label l_h is sent to the environment and at that time the number of copies of a_1 present in the environment is equal to the contents of the first register of M. Consequently, $N(M) = N(\Pi)$. Each cell contains at most five programs, which complete the proof.

Problem: Determine the computational power of P colonies with height of at most 4.

The absence of checking rules does not influence the bound for the number of programs, but it increases the bound for the number of agents by 1.

Theorem 23.1.4. [6, 13]: $NPCol_{par}R(2, *, 5) = NPCol_{par}R(2, 2, *) = NRE$.

Problem: Determine bounds in the sequential rewriting.

In non-restricted P colonies we allow the use of arbitrary combinations of rules in programs including checking rules. Four programs in cells are sufficient to enrich the universality.

Theorem 23.1.5. [9, 13]: $NPCol_{par}K(2, *, 4) = NPCol_{seq}K(2, *, 4) = NRE$.

Homogeneous P colonies have programs with all rules of the same type (evolving, exchanging, and priority programs.) For the number of programs the bounds 4 and 5 were established depending on the presence of checking rules.

Theorem 23.1.6. [5]: $NPCol_{par}KH(2, *, 4) = NPCol_{par}H(2, *, 5) = NRE$.

P colonies of capacity three. Three objects in a cell allow a decrease in the number of programs needed for universality to three. If no checking rules are used, then at most five programs are needed.

Theorem 23.1.7. [9]: $NPCol_{par}K(3, *, 3) = NPCol_{seq}K(3, *, 3) = NPCol_{par}(3, *, 5) = NRE$.

Some classes of P colonies, which do not use priority rules, are known to have the generative power of the partially blind register machines NRM_{pb}:

Theorem 23.1.8. [6, 11, 15]: $NPCol(2, 1, *) = NPCol_{seq}(2, *, *) = NPColR(2, 1, *) = NRM_{pb} \subseteq NPCol_{par}(1, 2, *)$.

23.1.4. P COLONIES WITH BOUNDED CELLS

Universal register machines with a small number of registers and instructions were studied by I. Korec in [19]. The simulation of those small universal register machines by P colonies allows one to eliminate $*$'s from previous results and to obtain P colonies with all parameters bounded. P colonies constructed in [10], which we summarize bellow, use a different initialization of the computation. Besides the infinitely many copies of the environmental object, these P colonies start with finitely many other objects in the environment. Classes of P colonies with initialization will be indicated by adding I in front of other symbols, i.e. we write $IPCol(k, l, h)$, for instance.

Theorem 23.1.9. [10]: $IPCol\ KR(2, 23, 5) = IPCol\ KR(2, 22, 6) = IPCol\ KR(2, 1, 142) = IPCol\ K(2, 22, 5) = IPCol\ R(2, 57, 8) = IPCol\ (2, 35, 8) = IPCol(3, 35, 7) = NRE.$

P colonies with and without the initialization were compared by L. Cienciala and L. Ciencialová ([4]):

A P colony with capacity 1 (resp. 2), n agents, and non-empty starting environment w over m letter alphabet can be simulated by a P colony with capacity 1 (resp. 2) and n + 2m agents with empty starting environment. This fact together with the previous theorem leads to the bounded estimations for the P colonies with empty initializations, too.

23.1.5. MODIFICATIONS AND EXTENSIONS

A program to be used by an agent of a P colony (parallel or sequential) is non-deterministically chosen from all possible applicable programs. A more sophisticated way to apply rules in a derivation step can be found in the case of *P colonies with prescribed teams* [12]. These P colonies also use some new types of rules in their programs in addition to those from the original P colonies. The reader can find all details of the model in [12].

Colonies of synchronizing agents (CSA) introduced in [2] are formed by multisets of agents determined by objects (not necessary of the same size), which act with global rules, i.e. all agents obey the same rules. The only feature that may distinguish the agents are their contents. Rules of CSA are of two types. The evolution rules used here are multiset rewriting rules, instead of the objects to object evolution rules from P colonies. Moreover the agents can influence each other using pairwise synchronizing rules. An application of a synchronizing rule changes the content of two agents simultaneously. In [2] one finds computational power results (computational completeness, power identical with partially blind register machine, ...) and robustness, where the authors consider the ability of colonies to generate core behaviors despite of the removal of agents or rules. Decidable temporal logic is used to specify and investigate dynamical properties of CSAs.

One can deal also with P colonies whose environment is dynamically changing not only by actions of the cells but also by its own rules. The idea, which follows the similar motivation as the eco-grammar systems [8] or eco-colonies [20], leads to *EP colonies* proposed in [7], where also some bio-inspired research topics are suggested.

The *colonies of linguistic P systems*, LP colonies for short, were proposed for modeling the linguistic generation and evolution by P systems in [1]. Each agent of an LP colony is a P system (i.e. not a single cell, but rather a complicated P system), which is understood as a special module for each part of language (e.g. syntax, semantics, etc.) that evolve in parallel. Agents acting on a common tape are composed to the structure of a colony typical for colonies of grammars ([14]). See [1] for details of the model and its motivation.

References

[1] G. Bel Enguix, M.D. Jiménez López: LP colonies for language evolution. A preview. *Proc. 6th Intern. Workshop on Membrane Computing*, Vienna, July 2005, 179–192.

[2] M. Cavaliere, M. Mardare, S. Sedwards: A multiset-based model of synchronizing agents. Computability and robustness. *Theoretical Computer Sci.*, 31 (2008), 216–238.

[3] L. Ciencialová, L. Cienciala: Variation on the theme: P colonies. *Proc. 1st Intern. Workshop on Formal Models* (D. Kolář, A. Meduna, eds.), Ostrava, 2006, 27–34.

[4] L. Cienciala, L. Ciencialová: On the low degree of P colonies. Manuscript.

[5] L. Cienciala, L. Ciencialová, A. Kelemenová: Homogeneous P colonies. *Computing and Informatics*, 27 (2008), 481–496.

[6] L. Cienciala, L. Ciencialová, A. Kelemenová: On the number of agents in P colonies. *Lecture Notes in Computer Sci.*, 4860 (2007), 193–208.

[7] E. Csuhaj-Varjú: EP-colonies: Micro-organisms in a cell-like environmentit *Proc. 3rd Brainstorming Week on Membrane Systems*, Sevilla, 2005, 123–130.

[8] E. Csuhaj-Varjú, A. Kelemenová, J. Kelemen, Gh. Păun: Eco-grammar systems – A grammatical framework for life-like interactions. *Artificial Life*, 3 (1997), 1–28.

[9] E. Csuhaj-Varjú, J. Kelemen, A. Kelemenová, G. Păun, G. Vaszil: Computing with cells in environment: P colonies. *J. Multi-Valued Logic*, 12 (2006), 201–215.

[10] E. Csuhaj-Varjú, M. Margenstern, G. Vaszil: P colonies with a bounded number of cells and programs. *Lecture Notes in Computer Sci.*, 4361 (2007), 352–366.

[11] R. Freund, M. Oswald: P colonies working in the maximally parallel and in the

sequential mode. *Proc. 7th Intern. Symposium on Symbolic and Numeric Algorithms for Scientific Computing, SYNASC 2005*, Timişoara, Romania, 2005, 419–426.

[12] R. FREUND, M. OSWALD: P colonies and prescribed teams. *Intern. J. Computer Math.*, 83 (2006), 569–502.

[13] R. FREUND, M. OSWALD: P colonies working in the maximally parallel and in the sequential mode. *Pre-Proc. 1st Intern. Workshop on Theory and Application of P Systems* (G. Ciobanu, Gh. Păun, eds.), Timişoara, Romania, 2005, 49–56.

[14] J. KELEMEN, A. KELEMENOVÁ: A grammar-theoretic treatment of multiagent systems. *Cybernetics and Systems*, 23 (1992), 621–633.

[15] J. KELEMEN, A. KELEMENOVÁ: On P colonies, a simple biochemically inspired model of computation. *Proc. 6th Intern. Symposium of Hungarian Researchers on Computational Intelligence*, Budapest TECH, Hungary, 2005, 40–56.

[16] J. KELEMEN, A. KELEMENOVÁ, V. MITRANA: Neo-modularity and colonies. *Where Mathematics, Linguistics and Biology Meet* (V. Mitrana, C. Martín-Vide, eds.), Kluwer, 2001, 63–74.

[17] J. KELEMEN, A. KELEMENOVÁ, GH. PĂUN: Preview of P colonies. A biochemically inspired computing model. *Proc. 9th Intern. Conf. Simulation and Synthesis of Living Systems (Alife IX)* (M. Bedau at al., eds.) Boston Mass., 2004, 82–86.

[18] A. KELEMENOVÁ, E. CSUHAJ-VARJÚ: Languages of colonies. *Theoretical Computer Sci.*, 134 (1994), 119–130.

[19] I. KOREC: Small universal register machines. *Theoretical Computer Sci.*, 168 (1996), 267–301.

[20] Š. VAVREČKOVÁ, A. KELEMENOVÁ: Properties of eco-colonies. *Proc. Intern. Workshop Automata for Cellular and Molecular Computing* (G. Vaszil, ed.), MTA SZTAKI, Budapest, 2007, 129–143.

CHAPTER 23.2

TIME IN MEMBRANE COMPUTING

MATTEO CAVALIERE
DRAGOŞ SBURLAN

23.2.1 INTRODUCTION

A standard feature of membrane computing (P systems) is the fact that each rule (of whatever type) is executed in exactly one time-unit (one step); however, this assumption in not justified by the corresponding biological reality. In fact, chemical reactions, or even more complex biological operations, such as cellular division, take a certain time to be completed. In many cases, different biological processes take different times of completion, also depending on environmental conditions, and the processes are synchronized by using appropriate biological signals. For these reasons in [8] a timed version of P systems (timed P systems) was introduced. In timed P systems, to each rule a natural number representing the time of execution of the rule is associated. In [8] also the concept of time-independent P systems was introduced. A time-independent P system is a P system that always produces the same result, independently from the execution times of the rules. The notion of time-independent P systems tries to capture the class of systems that are robust against environment changes that could affect, in an unpredictable manner, the execution time of the rules of the system. Timed and time-independent P systems are reviewed in Subsections 23.2.2 and 23.2.3.

Another direction of research is to use time itself to encode the result of a computation. This can be done in several ways. For instance, the time of occurrence of certain events can be used to compute numbers (e.g. as done in [10] and [14]). There, two general conclusions were obtained. If the specific events (such as the use of certain rules, the entering/exit of certain objects into/from the system) can be freely chosen, then, it is easy to obtain computational completeness results; however, if one takes the length (number of steps) as the result of the computation, non-universal systems can be obtained. A related approach is the one followed in spiking neural P systems, where the interval between two specific events (firing of neurons) is considered (we refer to the corresponding chapter). An overview of models that use time as support for the computation is given in Subsection 23.2.4.

Time is also important in the definition of paradigm of computations whose capabilities go beyond Turing's. In fact, if one supposes the existence of two scales of time (an external, global one of the user) and an internal, local time of the device, then it is possible to implement accelerated computing devices that can have more computational power than Turing machines. This approach has been used in [4] to construct accelerated P systems where acceleration is obtained by either decreasing the size of the reactors or by speeding-up the communication channels. We review this approach in Subsection 23.2.5.

We conclude by providing an overview of other works that consider the use of time in membrane computing.

23.2.2 TIMED AND TIME-FREE P SYSTEMS

Timed and time-independent P systems were originally introduced in [8] in the context of P systems with promoters (used as "signals") and bi-stable catalysts. We recall here a simpler definition, using P systems with bi-stable catalysts (as done in [5]).

Definition 23.2.1 *A P system with (bi-stable) catalysts, of degree $m \geq 1$, is a construct*

$$\Pi = (V, C_b, \mu, w_1, \ldots, w_m, R_1, \ldots, R_m, i_0), \text{ where:}$$

- *V is the alphabet of Π; its elements are called objects;*
- *$C_b \subseteq V$ is the set of bi-stable catalysts, i.e. objects c that have two states, c and $\bar{c}$; if the two states coincide, then c is a catalyst; if only catalysts are used, then their set is denoted by C;*
- *μ is a membrane structure with m membranes labeled $1, 2, \ldots, m$;*

- w_i, $1 \leq i \leq m$, specifies the multiset of objects present in the corresponding region i at the beginning of a computation;
- R_i, $1 \leq i \leq m$, are finite sets of evolution rules over V associated with regions $1, 2, \ldots, m$ of μ; the rules can be non-cooperative or catalytic, of the forms usual in P systems with symbol objects.
- $i_0 \in \{0, 1, \ldots, m\}$ is the output region; if $i_0 = 0$, then it indicates the environment.

A *timed P system* $\Pi(e) = (V, C_b, \mu, w_1, \ldots, w_m, R_1, \ldots, R_m, i_0, e)$ can be constructed by adding to the P system Π with bi-stable catalysts, a computable mapping $e : R_1 \cup \cdots \cup R_m \longrightarrow \mathbf{N}$, that specifies the execution times for the rules. A timed P system $\Pi(e)$ works in the following way. We suppose to have an external clock that marks time-units of equal length, starting from time 0. In each region a finite number of objects (symbol objects or bi-stable catalysts) and a finite number of evolution rules are present. At each step in the regions of the system we have together rules in execution and rules not in execution and all the rules that can be started in each region have to be applied. When a rule $r \in R_i$, $1 \leq i \leq m$, is applied, then all objects that can be processed by the rule have to evolve by this rule. Hence, rules are used in a non-deterministic maximally parallel manner. *When a rule r is started at time j, then its execution terminates at time $j + e(r)$ (the objects produced by the rule can be used starting from the time $j + e(r) + 1$).* Clearly, when a rule r is started, then the occurrences of symbol-objects used by this rule are not anymore available for other rules. A computation halts when no rule can be applied in any region and there are no rules in execution (such configuration is called *halting configuration*).

The output of a halting computation is the vector of numbers representing the multiplicities of objects present in the output region in the halting configuration. Collecting all the vectors obtained, for all possible halting computations, we get the set $Ps(\Pi(e))$ of vectors of natural numbers generated by the system $\Pi(e)$. Notice that the obtained set of vectors depends on the mapping e.

We are interested in a particular class of systems, called *time-free*. A P system Π using bi-stable catalysts is *time-free* if and only if every system in the set

$$\{\Pi(e) \mid e : R_1 \cup R_2 \cup \cdots \cup R_m \longrightarrow \mathbf{N}\}$$

produces the same set of vectors of natural numbers.

The set of vectors generated by a time-free P systems Π is indicated by $Ps(\Pi)$ (there is no ambiguity in this case).

The notation $fPsOP_m(\alpha)$, $\alpha \in \{ncoo\} \cup \{2cat_k, cat_k \mid k \geq 0\}$, denotes the family of sets of vectors of natural numbers generated by time-free P systems with at most m membranes, evolution rules that can be non-cooperative (*ncoo*), or catalytic (*cat_k/$2cat_k$*), using at most k catalysts/k bi-stable catalysts (as usual $*$ is used if the corresponding number of membranes or catalysts/bi-stable catalysts is unknown). The letter f stands for *time-free*.

The notations used in the paper are standard: for a family of languages *FL*, we denote by *NFL*, *PsFL* the family of length sets and of Parikh images, respectively, of languages in *FL*.

23.2.3 COMPUTATIONAL POWER AND DECIDABILITY

From the definition, it is clear that systems using only non-cooperative rules are always time-free (the formal proof can be found in [5]) and from this one can get the following result.

Theorem 23.2.1 [5] $PsCF = fPsOP_m(ncoo)$ for all $m \geq 1$.

However, time-freeness is not guaranteed for systems that use catalytic rules. This is shown in the following example that also presents the functioning of a timed P system and the notion of a time-free P system.

Example 23.2.1 Consider the following P system with one catalyst

$$\Pi = (\{B', B'', c, X, D, b, a\}, \{c\}, [_1[_2\ ]_2]_1, B'B''c, \lambda, R_1, \emptyset, 2), \text{ where:}$$

$$R_1 = \{r_1 : B' \to b_{in_2}X,\ r_2 : cB'' \to c,\ r_3 : X \to D,$$

$$r_4 : cX \to c,\ r_5 : D \to a_{in_2}\}.$$

The system Π is not time-free. In fact it generates two different outputs by using two different time-mappings e' and e''.

First, consider the time-mapping e' defined in the following way:
$e'(r_1) = 1,\ e'(r_2) = 2,$
$e'(r_i) = k'$, for some $k' \in \mathbf{N}$, for $i = 3, 4, 5$.

Consider now the time-mapping e'' defined in the following way:
$e''(r_1) = 2,\ e''(r_2) = 1,$
$e''(r_i) = k''$, for some $k'' \in \mathbf{N}$, for $i = 3, 4, 5$.

If the times of execution of the rules in Π are defined by the mapping e', then is easy to see that $Ps(\Pi(e')) = \{(1, 1)\}$. In fact, the two rules r_1 and r_2 are started in parallel. When r_1 terminates, the objects b and X are produced. Then, at step 2, the rule r_3 is applied and the object D is produced. Notice that in step 2 the rule r_3 is the only one that can be applied because rule r_2 is still in execution (and then c is busy). In step 3 the object a is produced by $D \to a$ and sent to the output region. Therefore, this is the only possible computation and the output of the system is the set $\{(1, 1)\}$.

If the time of execution of the rules in Π are defined by the mapping e'', then $Ps(\Pi(e'')) = \{(1, 0), (1, 1)\}$.

In fact, rules r_1 and r_2 are started in parallel as in the previous case, but because of the time-mapping e'', rule r_1 ends after rule r_2. Then, after two steps, X is produced and b is sent to the output region; in the next step X can be rewritten in two possible ways. In the first way the catalyst c can be used to apply rule r_4 and then the computation halts producing as output the vector $(1, 0)$. In the second case the rule r_3 is used, and then rule r_5, that produces and sends the object a to the output region. Therefore, the output of the system is $\{(1, 1), (1, 0)\}$.

Notice that for any time-mapping e, $Ps(\Pi(e)) \neq \emptyset$.

It is not known whether or not time-free systems using only catalysts are universal (this is particularly interesting since standard catalytic P systems are known to be universal, [17]). However, it is known that time-free P systems using bi-stable catalysts are universal. This was originally shown in [5] and the result (in number of bi-stable catalysts) improved in [1].

Theorem 23.2.2 [1] *$PsRE = fPsOP_1(2cat_4)$.*

Using Theorem 23.2.2 we know that, for any P system using bi-stable catalysts, there exists an equivalent P system with bi-stable catalysts that is time-free (possibly, paying a price in the number of bi-stable catalysts). An interesting question concerns the possibility of checking whether or not an arbitrary P system is time-free. The answer to such question is negative, if one considers P systems with bi-stable catalysts.

Theorem 23.2.3 [5] *Given an arbitrary P system Π with bi-stable catalysts, it is undecidable whether or not Π is time-free.*

It is not known whether or not it is possible to check, in an algorithmic manner, the time-freeness of an arbitrary P system using catalysts.

23.2.4 TIME AS SUPPORT FOR COMPUTATION

There is another way of using time in membrane computing that consists in counting (and considering as produced result) the time between two specific events during the computation. In the simplest case the length of a computation (the number of steps from an initial configuration to a halting one) can be taken as the number computed by the system. This line of research was started in [10] from which we recall definitions and main results. Other developments are reviewed in Section 23.2.6.

This idea can be applied to any standard model of P systems, and, in particular, we show how it can be employed in P systems with symport/antiport rules and P systems with cooperative evolution rules.

It has been shown in [10] that computational completeness can be obtained in a rather straightforward manner for P systems with symport/antiport rules or P systems with catalysts if one takes as result produced by a system the *time elapsed between two freely chosen events*. More complex is the situation where one considers the *length of a computation as the number computed by a system*. For a system Π we denote by $lg(\Pi)$ the set of numbers of steps (lengths) of halting computations in Π. Then, we denote by $N_{lg}OP_m(features)$ the family of sets of numbers $lg(\Pi)$ computed by systems with at most $m \geq 1$ membranes, using the features specified by *features*. The features can be $sym_k, anti_r, ncoo, coo, cat_s, \delta$ where k, r are the maximal weights for symport and antiport rules, and s is the maximal number of catalysts which are allowed. We use αsym and $\beta anti$, with $\alpha, \beta \in \{p, i\}$ to denote that promoters (p) or inhibitors (i) are used.

One can see ([10]) that *the length set of any regular language can be computed in this way, both by symport/antiport systems of weight one and by non-cooperative P systems.*

Therefore, the families $N_{lg}OP_1(sym_1, anti_1)$, $N_{lg}OP_1(ncoo)$ contain all semilinear sets of numbers (it is not known if the converse is true).

One can get more by using catalytic rules (one catalyst) and membrane dissolution, or using cooperative rules.

Theorem 23.2.4 [10] *The families $N_{lg}OP_2(cat_1, \delta)$ and $N_{lg}OP_2(coo)$ contain non-semilinear sets of numbers.*

These families are non-trivial but they do not contain all Turing computable sets of numbers because during a computation of length n one can only construct a working space which is of the order of k^n, for a constant k depending on the system at hand.

Let us first recall from [15] two results from computational complexity.

Given a proper function f, let $SPACE(f)$ denote the deterministic space and $NSPACE(f)$ the non-deterministic space with respect to f.

Theorem 23.2.5 *If $f(n)$ is a proper function, then $SPACE(f(n))$ is a proper subset of $SPACE(f(n)\log f(n))$.*

Theorem 23.2.6 *If $f(n)$ is a proper function with $f(n) \geq \log n$, then $NSPACE(f(n)) \subseteq SPACE(f^2(n))$.*

The following non-universality result holds (we recall from [10] a sketch of the proof because the complexity arguments employed are rather general and can find other applications in this framework).

Theorem 23.2.7 *None of the families $N_{lg}OP_*(asym_*, \beta anti_*, \delta)$, for any $\alpha, \beta \in \{p, i\}$, nor $N_{lg}OP_*(coo, \delta)$ equal NRE.*

Proof. Consider a P system Π with any type of rules (symport/antiport – with or without promoters or inhibitors, cooperative rules, membrane dissolving included) generating as the length of its halting computations the set $lg(\Pi)$. For each halting computation C of length n of Π, $C = C_1 \Longrightarrow C_2 \Longrightarrow \ldots \Longrightarrow C_n$, define

$$W_\Pi(n, C) = \max\{\text{number of objects present in } C_i \mid 1 \leq i \leq n\}.$$

Now define $W_\Pi(n) = \max\{W(n, C) \mid C \text{ is a halting computation of length } n \text{ of } \Pi\}$.

It is clear that $W_\Pi(n)$ is in $O(k^n)$, where k is a constant depending on Π.

We can construct the following non-deterministic Turing machine M accepting $lg(\Pi)$. The machine M uses two tapes. On one tape it writes the input (in binary), and a counter (in binary), separated by a marker, and on the other the machine simulates the computation of the system Π (in a non-deterministic way, if Π is non-deterministic). Each parallel step in Π is simulated by several successive steps in M—not that the duration of the computation in M is relevant here, but the space it uses. For each simulated parallel step of Π, the machine M increases the counter, storing in this way the number of parallel steps simulated. When the simulation of Π reaches an halting configuration, the machine M checks whether the counter has reached exactly the number given as input. If this is the case, then it answers yes, otherwise it answers no.

The machine M uses a space in $O(2^{k^n})$ where n is the size of the input (written in binary; notice that the system Π generates strings in unary, hence we have to input them in binary but work with them in unary).

Therefore, $lg(\Pi) \in NSPACE(f(n))$, for $f(n) \in O(2^{k^n})$, hence, by Theorem 23.2.6 we get $lg(\Pi) \in SPACE(g(n))$, for $g(n) \in O((2^{k^n})^2)$. In this way, we get the following upper bound,

$$N_{lg}OP_*(coo) \subseteq \bigcup_{j \in \mathbb{N}} SPACE(O((2^{j^n})^2)).$$

According to Theorem 23.2.5, given any proper function $h(n)$ such that $h(n) \geq (2^{j^n})^2 \log((2^{j^n})^2)$, for any $j \in \mathbb{N}$, there exists a set in $SPACE(h(n))$ that is not in the family $N_{lg}OP_*(coo)$, hence the theorem follows (note that although the construction of M depends on Π, the space used by M is always bounded in the same way).

It is interesting to note that, as shown in [10], P systems with symport/antiport rules of weight two, as well as catalytic P systems (with at least two catalysts) can simulate the instructions of a register machine in exactly k steps, where k is a constant depending only on the type of the P system, but not on the particular system itself.

From this it follows that there exist P systems with symport/antiport systems of weight two and catalytic P systems with two catalysts that can generate non-semilinear sets of numbers as length sets of computations (hence, not using membrane dissolution as in Theorem 23.2.4). Moreover, it is known that the length sets of all these P systems are recursive ([10]).

23.2.5 Acceleration in Membrane Systems

Hypercomputation seeks to discover and implement computational devices that transcend the boundaries given by the Turing machine model. In this respect, regardless of the physical constraints, a fruitful mathematical principle consists in performing an infinite computation in a finite elapsed time (see, e.g., [18]). Such a speed-up principle can be implemented in several ways. One method starts from some universality results of certain deterministic computing machines, that, when "accelerated", are able to accomplish the mentioned principle ([19]). In [4], [20], the notion of a biological computing agent with super-Turing capabilities is proposed in the framework of membrane computing. There, a class of P systems that allow deterministic characterization of Turing computability is used as support for the acceleration principle. The main idea is to have two scales of time: an external global one (where the results is collected), and an internal, local time of the accelerating device. The problem to be solved is formulated in global time, at some moment t, and introduced into the accelerated device which is able to perform an "inner" infinite computation in a finite number, T, of external time units, when the answer to the problem is collected.

We provide an intuition on how time can be used to construct timed P systems that can solve Turing undecidable problems (detailed proofs can be found in [4] and [20]).

We start from a deterministic universal P system i.e. a P system that deterministically simulates a deterministic universal register machine (several of them are present in literature). We can then consider a timed version of such a universal P system, called Π_{acc}, by admitting a time mapping such that the execution times of the rules always decrease by half at each application (i.e. the first applied rule takes one unit of the global time, the second applied rule takes $\frac{1}{2}$ unit of global time, and so on); in this way, infinitely many computational steps can be performed in at most two global units of time since $\sum_{i=0}^{\infty} \frac{1}{2^i} = 2$. Starting from Π_{acc} one can construct a P system Π_{halt} with two regions that *decides the halting* problem for register machines.

The inner region of Π_{halt} contains the accelerated device Π_{acc} that receives as input a multiset—the code of an arbitrary deterministic register machine M—and an input vector of numbers v and simulates, in a deterministic and accelerated fashion, the computation of M on v, i.e. in at most two time units. The outer region contains rules that can check, after two time units, whether or not the *HALT* instruction of the register machine has been simulated in the inner region. This is equivalent to checking whether or not M has halted on the input v, a well-known undecidable problem for register machines.

23.2.6 OTHER APPLICATIONS OF TIME IN P SYSTEMS

The idea of time-freeness has also been applied to EC P systems, in [5] with universality results obtained in [5] and then improved in [1].

An extension of timed P systems (called frequency P systems), has been introduced and investigated in [12]. In frequency P systems each membrane is clocked independently from the others, and each membrane operates at a certain work frequency that could change during the execution. Dynamics of such systems have been investigated in [12].

Other types of time-free systems, using signal-promoters or symport/antiport rules have been considered in [9].

In [8] time-free P systems using one catalyst and signal-promoters have been shown to produce the Parikh image of the languages generated by Indian parallel grammars, while time-free P systems with priorities over the rules are universal. A generalization of time-free P systems, called clock-free P systems, has also been considered in [8]. In this case, different applications, even of the same rule, may take different times to be executed. Clock-free P systems are independent from the time associated with the applications of the rules. Universality of clock-free P systems with one catalyst and promoters has been proved in [8]. In [8] it is also considered the case of "partial" time-free systems, obtained by adding restrictions and conditions on the execution times of the rules.

In [7] the time of execution of rules is stochastically determined. Experiments on the reliability of the computations have been considered and links with the idea of time-free systems are also discussed.

Time can also be used to "control" the computation, for instance by opportune changes in the execution times of the rules during a computation, and this possibility has been considered in [11]. Moreover, timed P automata have been proposed and investigated in [3] where ideas from timed automata have been incorporated into timed P systems.

In [14], [13], and [16] one considers the time as the result of the computation (as in Section 23.2.4) but considering special "observable" configurations taken in regular sets (with the time elapsed between such configurations considered as output). In particular, in [14] and [13], one considers P systems with symport and antiport rules showing that universality results of standard P systems with symport/antiport can be improved in such a model. In [16] one applies the idea to P systems with proteins embedded on the membranes. Considering observable configurations is very much in the spirit of having an external observer, able to watch part of the system (see, e.g. [6]). Finding non-trivial and realistic observers remains a research topic.

ACKNOWLEDGEMENT

D. Sburlan gratefully acknowledges the support of the Romanian CNCSIS IDEI-PCE 551/2009 research grant.

REFERENCES

[1] A. ALHAZOV: Number of protons/bi-stable catalysts and membranes in P systems. Time-freeness. *Lecture Notes in Computer Sci.*, 3850 (2006), 79–95.

[2] A. ALHAZOV, M. CAVALIERE: Evolution-communication P systems: Time-freeness. *Proc. Third Brainstorming Week on Membrane Computing* (M.A. Gutiérrez-Naranjo et al., eds.), Sevilla University, 2005, 11–18.

[3] R. BARBUTI, A. MAGGIOLO-SCHETTINI, P. MILAZZO, L. TESEI: Timed P automata. *Proc. MECBIC 2008*, ENTCS, to appear.

[4] C.S. CALUDE, GH. PĂUN: Bio-steps beyond Turing. *BioSystems*, 77 (2004), 175–194.

[5] M. CAVALIERE, V. DEUFEMIA: Further results on time-free P systems. *Intern. J. Found. Computer Sci.*, 17 (2006), 69–89.

[6] M. CAVALIERE, P. LEUPOLD: Evolution and observation. A new way to look at P systems. *Lecture Notes in Computer Sci.*, 2933 (2004), 70–87.

[7] M. CAVALIERE, I. MURA: Experiments on the reliability of stochastic spiking neural P systems. *Natural Computing*, 43 (2008).

[8] M. CAVALIERE, D. SBURLAN: Time-independent P systems. *Lecture Notes in Computer Sci.*, 3365 (2005), 239–258.

[9] M. CAVALIERE, D. SBURLAN: Time and synchronization in membrane systems. *Fundamentae Informaticae*, 64 (2005), 65–77.

[10] M. CAVALIERE, R. FREUND, A. LEITSCH, GH. PĂUN: Event-related outputs of computations in P systems. *J. Automata, Languages and Combinatorics*, 11 (2006), 263–278.

[11] M. CAVALIERE, C. ZANDRON: Time-driven computations in P systems. *Proc. Fourth Brainstorming Week on Membrane Computing* (M.A. Gutiérrez-Naranjo et al., eds.), 2006, 133–143.

[12] D. MOLTENI, C. FERRETTI, G. MAURI: Frequency membrane systems. *Pre-Proc. 8th Workshop on Membrane Computing* (G. Eleftherakis et al., eds.), 2007, 445–455.

[13] H. NAGDA, A. PĂUN, A. RODRÍGUEZ-PATÓN: P systems with symport/antiport and time. *Lecture Notes in Computer Sci.*, 4361 (2006), 463–476.

[14] O.H. IBARRA, A. PĂUN: Computing time in computing with cells. *Lecture Notes in Computer Sci.*, 3892 (2006), 112–128.

[15] C.H. PAPADIMITRIOU: *Computational Complexity*. Addison-Wesley, 1984.

[16] A. PĂUN, A. RODRÍGUEZ-PATÓN: On flip-flop membrane systems with proteins. *Lecture Notes in Computer Sci.*, 4860 (2007), 414–427.

[17] R. FREUND, L. KARI, M. OSWALD, P. SOSIK: Computationally universal P systems without priorities: Two catalysts are sufficient. *Theoretical Computer Sci.*, 330, (2005), 251–266.

[18] M. HOGARTH: Does general relativity allow an observer to view an eternity in a finite time? *Foundations of Physics Letters*, 5 (1992), 73–81.

[19] P.H. POTGIETER: Zeno machines and hypercomputation. *Theoretical Computer Sci.*, 358 (2006), 23–33.

[20] D. SBURLAN: *Promoting and Inhibiting Contexts in Membrane Computing*. PhD Thesis, University of Seville, Spain, 2006.

CHAPTER 23.3

MEMBRANE COMPUTING AND SELF-ASSEMBLY

MARIAN GHEORGHE
NATALIO KRASNOGOR

23.3.1 INTRODUCTION

SELF-ASSEMBLY is a process in which simple components or parts of a pre-existing system form complex structures or spontaneously create hierarchical and ordered aggregates. The components that are self-assembled have a very limited behavioral repertoire performing simple actions under a reduced set of well defined conditions and obeying simple, but formally defined, rules. Self-assembly covers a broad range of phenomena and embraces many disciplines: physics, chemistry, biology, engineering, computing [6]. In this section we focus on specific computational aspects of self-assembly systems. In this respect, a particular interest will be given to natural computing paradigms which in many cases are intuitive and efficient in solving complex problems.

This presentation points to nature inspired paradigms, like sticker systems, P systems, and evolutionary approaches, at specific ways of adequately identifying the most suitable variants that model self-assembly systems, as well as to adapting them by tuning specific parameters to a certain behavior.

23.3.2 Computing by Self-Assembly

Various models of self-assembly have been considered so far, tackling different aspects of this complex phenomenon. We refer in this section to a theoretical computational model based on formal language theory concepts and addressing problems regarding computability and complexity aspects for three types of self-assembly paradigms, namely *DNA self-assembly* of sticker systems, *computing with 2D tiles of arbitrary shapes* formed with pixel-like elements and *self-assembling membrane systems*. This approach, although similar to those searching for optimal solutions of highly complex combinatorial problems [1], focuses on assessing the computational power of the self-assembly systems studied. In order to show what and how these three self-assembly phenomena compute, a generic theorem called *self-assembly universality* (SAU) lemma [5] is used, which is based on a well-known result in language theory stating that each recursively enumerable language can be written as the projection of the intersection of the equality set of two morphisms with a regular set [4].

In DNA self-assembly one starts from a "soup" of DNA double strand molecules, where one is the seed and most of them are either single strands or incomplete double strand molecules with single strands at one or both ends, called sticky ends. Based on the Watson–Crick complementarity principle these DNA fragments self-assemble forming more complex double stranded structures. The computation ends when complete DNA molecules are obtained and the result is read from the upper strand of each complete DNA molecule by discarding the right end which occurs after a given marker. The model employed in this case is called a sticker system.

In the case of computing with 2D tiles corresponding to arbitrary shapes aggregating pixel-like elements, a formal model, called shape grammar, is used to build up the self-assembly system. One considers again a "soup", but now of tiles (also known as polyominoes), each one in an arbitrary number of copies and with a distinguished element, called the seed. Starting from the seed, in each step, a new polyomino is added to the current assembly in such a way that some edges of the current construction and the new tile match up; in the newly obtained assembly there is no overlapping and holes. In the end the shape of this construction needs to completely cover a given area, generally a rectangle, and the result is the sequence of labels of the polyominoes utilized. As this model is not powerful enough to achieve universality, some extra control mechanisms are added, one of which is the existence of a next function. This provides for each polyomino the set of polyominoes that may be utilized in the next step of self-assembling the system.

In these two examples the key elements of the self-assembly process have been the building units of the entire process. For this reason this type of process is called *mechanical self-assembly* [5]. In the next self-assembly paradigm the building blocks, the bricks, represent the basic layer of the process, let us call it the "hardware", whereas the set of interactions that take place within the self-assembled system form the software component which is executed on this machinery, and for this reason it is called *behavioral self-assembly* [5]. In this case the computational model employed is a specific variant of population P systems [2], namely a self-assembly P system [3]. Such a P system consists of cells, each with a number of binding sites that will be used to create bonds with other cells. In addition to these binding sites, each cell contains some objects and rules either transforming these objects or sending them to a neighboring cell after the assembly is formed. Certain types of trees and graphs can be self-assembled with this model, where cells and bonds account for nodes and edges of these constructions, respectively. A particular form of this model, called restricted self-assembly P systems, is introduced in [5]. In this case, each cell contains two binding sites from the set {*in, out*} and only communicating (antiport) rules. The self-assembly process starts with a seed cell, an arbitrary number of cells plus a finite number of some distinguished cells. It consists of two stages: first, the self-assembled construction is wired from the seed using the existing cells such that no further bonds can be created between the current construction and the cells available; secondly, the objects residing in various cells are moved around by the antiport rules. The result of a halting computation is the string of symbols from a terminal set which are sent into the environment. It is proved that this model is universal. Three complexity measures are considered in this case and $LOP(types_n, obj_m, anti_q)$ denotes the family of languages computed by restricted self-assembly P systems with at most n types of cells (including both the cells available in an infinite number and those occurring in a bounded number), at most m objects, and using antiport rules with the size of each component bounded by q; when one of these parameters is unbounded, it is replaced by $*$. The following result holds [5]:

Theorem 23.3.1 $RE = LOP(types_3, obj_1, anti_4)$.

If the symbols sent into the environment are counted rather than being aggregated into strings then sets of number of objects are obtained. The corresponding family is denoted by $NOP(types_n, obj_m, anti_q)$ and the following universality holds [5]:

Theorem 23.3.2 $NRE = NOP(types_*, obj_2, anti_2)$

23.3.3 Evolving Self-Assembled Systems

The first two computational paradigms presented in the above section are deeply related to tile self-assembly. They both have direct connections to a widely known formal system, called Wang tiles model. The tile self assembly model has been investigated, looking both for a systematic search of counters [8] and proposing evolutionary algorithms for the structural properties of the discrete self-assembling tile systems [12], [13].

A Wang tile system consists of a set of rectangles each of which having associated to each edge a color or a glue type. For each pair of colors the strength of their binding is provided. Two colliding sides will bind if the strength of the bond is greater than or equal to a given global parameter, called temperature.

In [8] one demonstrates a hybrid human-computer methodology for enumerating correct "counters" implemented for the Tile Assembly Model [14]. Through a sophisticated domain-dependent enumerative process the author was able to find a seven tiles family that can self-assemble into an $N \times \lceil log_2 N \rceil$ rectangle in optimal $\Theta(N)$.

Evolutionary algorithms are employed to evolve the design of a set of tiles into a predefined shape through self-assembly, by starting from a set of tiles with their specific interactions on their edges and a given temperature level. The entire process consists of two stages: the tile simulation and structure comparison. In the first part, four different models of self-assembly are considered by either using a random or probabilistic driven walk for each tile, or by permitting or inhibiting the rotation of the tiles. In the second stage, the resulting self-assembled blocks are compared to the target shape using the Minkowski functionals [6] and the best individuals are selected for the next iteration. The most interesting and challenging part of this investigation was related to the study of emergent, more complex structures obtained from two, three, four, or five tiles that persisted during the evolution process, by assessing all possible self-assembling conformations. These specific configurations mentioned above, called generalized secondary structures, play an important role in the entire self-assembly process with respect to the model used. The general conclusion of this investigation is that for the simplest models, the two-tile interactions are the most common blocks of the self-assembly process leading to uncontrolled structures, whereas in the more complex case of probabilistic motion, aggregations occur as a result of interactions between three-, four- and five-tile blocks, so intermediate self-assembled structures are more often used.

In self-assembling P systems, components, called cells, with a given structure of inputs and outputs are aggregated in more complex units by creating bonds between inputs and outputs of various cells. In a similar manner, the study of self-assembling program units with one input and a number of outputs is developed by using an evolutionary approach [7] and considering new features, like the spatial distribution of the units and their proximity to the units they collide with. These program units are kept in a virtual compartment of a fixed size and are allowed to walk randomly and self-assemble given they collide and have compatible inputs and outputs [7]. The model is studied in different conditions, with one or more compartments (in the latter case program units are allowed to move between compartments) and when the equilibrium principle (maintaining the same number of program units throughout the entire evolution) is abolished. It is shown the diversity and complexity of the programs obtained increase when far from equilibrium systems are considered.

With the aim of producing a multi-scale design and simulation tool for artificial life, a methodology based on a fully formal, but biologically intuitive, model, namely stochastic P systems (see Chapter 16 of this handbook) and dissipative particle dynamics (DPD), a coarse grained molecular simulation platform, has been proposed [11]. The P system model captures the three main functional elements of a minimal cell, namely compartments, metabolism, and information polymers, whereas DPD is able to simulate bottom-up self-assembly of membranes and the evolution of such a system by representing rules and symbols with a chemistry extension of DPD. The P system compartments are modeled as DPD vesicles composed of amphiphiles and incorporating pores to facilitate the exchange of molecules between neighboring compartments.

In order to prove the functionality of this methodology, a simple diffusive system consisting of two compartments, one containing the other, has been studied with respect to poration by tracking the movement of solvent particles from the system to the external environment. In the first stage, porated vesicles were self-assembled from a number of amphiphiles and pore polymers randomly distributed in a confined space. In the next step, the P system diffusion model has been executed and the results compared to the DPD simulation. Although some results were promising, there were also some inadequacies between some of the values produced. In order to remedy this problem, the P system specification has been refined by introducing the volume of each compartment into the model and by including this parameter in the propensity function of each rule. As a consequence highly similar behaviors of the DPD and P system have been obtained. This methodology is very promising as it opens a number of important research perspectives in artificial life, by providing a fully formal model and a simulation engine that match up very well and allow one to perform both theoretical investigations on the model as well as to get insights into the behavior of the system through robust and faithful simulations.

23.3.4 CONCLUSIONS

In this section we have discussed links between various computing paradigms (sticker systems, P systems, evolutionary computing) and self-assembly. The main emphasis of this work is on summarizing new research avenues regarding the interplay between the above mentioned areas, the usefulness of nature inspired and intuitive computational models, and the need to develop complex methodologies that fine-tune the model to specific requirements which might involve not only adjusting parameters, like in [11], but also more complex structural optimizations of the model [9].

Acknowledgements

NK acknowledges EPSRC grant EP/E017215/1 and BBSRC grant BB/F01855X/1.

References

[1] L.M. ADLEMAN, Q. CHENG, A. GOEL, M. HUANG, D. KEMPE, P. MOISSET DE ESPANES, P.W.K. ROTHEMUND: Combinatorial optimization problems in self-assembly. *Proc. Annual ACM Symposium on Theory of Computing (STOC)*, ACM Press, 2002, 23–32.

[2] F. BERNARDINI, M. GHEORGHE: Population P systems. *J. Universal Computer Sci.*, 10 (2004), 509–539.

[3] F. BERNARDINI, M. GHEORGHE, N. KRASNOGOR, J.-L. GIAVITTO: On self-assembly in population P systems. *Lecture Notes in Computer Sci.*, 3699 (2005), 46–57.

[4] K. CULIK II: A purely homomorphic characterization of recursively enumerable sets. *Journal of the ACM*, 26 (1979), 345–350.

[5] M. GHEORGHE, GH. PĂUN: Computing by self-assembly: DNA molecules, polyominoes, cells. In [6], 49–78.

[6] N. KRASNOGOR, S. GUSTAFSON, D.A. PELTA, J.L. VERDEGAY, eds.: *Systems Self-Assembly: Multidisciplinary Snapshots*. Elsevier, 2008.

[7] L. LI, P. SIEPMANN, J. SMALDON, G. TERRAZAS, N. KRASNOGOR: Automated self-assembling programming. In [6], 281–308.

[8] P. MOISSET DE ESPANES: Computer aided search for optimal self-sssembly systems. In [6], 225–243.

[9] F.J. ROMERO-CAMPERO, H. CAO, M. CÁMARA, N. KRASNOGOR: Structure and parameter estimation for cell systems biology models. *Proc. Genetic and Evolutionary Computation Conf., GECCO'08*, Atlanta, 2008, 331–338.

[10] F.J. ROMERO-CAMPERO, J. TWYCROSS, M. BENNETT, M. CÁMARA, N. KRASNOGOR: Modular assembly of cell systems biology models using P systems. *Intern. J. Found. Computer Sci.*, 20 (2009), 427–442.

[11] J. SMALDON, J. BLAKES, N. KRASNOGOR, D. LANCET: A multi-scaled approach to artificial life simulation with P systems and dissipative particle dynamics. *Proc. Genetic and Evolutionary Computation Conference, GECCO'08*, Atlanta, 2008, 249–256.

[12] G. TERRAZAS, N. KRASNOGOR, G. KENDALL, M. GHEORGHE: Automated tile design for self-assembly conformations. *Proc. 2005 IEEE Congress on Evolutionary Computation*, Edinburgh, UK, vol. 2, 2005, 1808–1814.

[13] G. TERRAZAS, M. GHEORGHE, G. KENDALL, N. KRASNOGOR: Evolving tiles for automated self-assembly design. *Proc. 2007 IEEE Congress on Evolutionary Computation*, 2007, 2001–2008.

[14] P. ROTHEMUND, E. WINFREE: The program-size complexity of self-assembled squares. *Proc. Thirty Second Annual ACM Symposium on Theory of Computing*, ACM press, 2000, 459–468.

CHAPTER 23.4

MEMBRANE COMPUTING AND X-MACHINES

PETROS KEFALAS
IOANNA STAMATOPOULOU
MARIAN GHEORGHE
GEORGE ELEFTHERAKIS

23.4.1 INTRODUCTION TO X-MACHINES

X-MACHINES, a state-based formal model introduced by Eilenberg [5], are considered suitable for the formal specification of a system and its components. More particularly, the *stream X-machine (sXM)* model was found to be well-suited for specifying reactive systems. Since then, valuable findings, using X-machines as a formal notation for specification, verification, testing and simulation purposes, have been reported [13, 6, 8]. A sXM model consists of a number of states and a memory, which can accommodate various complex data structures. The transitions between states are labeled by functions operating on inputs and memory values and producing outputs and new memory values.

Definition 23.4.1 *A deterministic sXM [8] is an 8-tuple*

$$\mathcal{X} = (\Sigma, \Gamma, Q, M, \Phi, F, q_0, m_0),$$

where Σ and Γ are the input and output alphabets, respectively; Q is the finite set of states; M is the (possibly) infinite set called memory; Φ is a set of partial functions φ; each such function maps an input and a memory value to an output and a, possibly, different memory value, $\varphi : \Sigma \times M \to \Gamma \times M$; F is the next state partial function, $F : Q \times \Phi \to Q$, which given a state and a function from Φ determines the next state (F is often described as a state transition diagram); q_0 and m_0 are the initial state and initial memory, respectively.

Although the sXM paradigm provides an intuitive method for modeling, the task of specifying a large-scale system such as a single sXM can prove to be rather cumbersome. The process of decomposing a complex system into smaller parts naturally leads to less fault-prone models and, in consequence, more reliable systems. To this end, a *communicating X-machines system (CsXM)* consists of several sXMs that are able to exchange messages. By "message" it is implied that the output of a X-machine can be accepted as input by another one. Broadly speaking, a CsXM consists of several communicating X-machine components and a relation defining the communication among them. Several attempts to define a communicating X-machine have been reported, but the most significant for this presentation are those introduced in [2, 12]. CsXM defined in [2] is using a communication matrix and links between components, whereas the model presented in [12] is relying on communication channels and relationships between functions occurring inside components. The last model (Fig. 23.4.1) preserves more faithfully the legacy for complete testing and verification of individual components.

The computation in any of these models is performed in an asynchronous manner, whereby a function is executed for as long as it takes and the synchronization between components is obtained through specific mechanisms involving

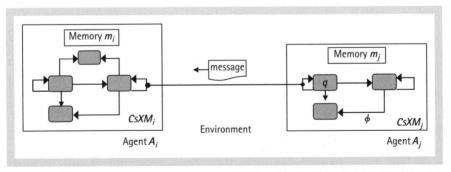

Fig. 23.4.1 An abstract example of two CsXM components, $CsXM_i$, $CsXM_j$; the • symbol denotes that a function receives its input from a machine component ($CsXM_j$) and the ♦ symbol means that a function φ sends its output to another machine component ($CsXM_i$).

local functions and global communication elements. The number of components running in parallel is also arbitrarily defined.

In what follows, we bridge X-machines with membrane computing, considering both P systems with symbol objects and string objects.

P systems have been investigated for their computational power and complexity aspects [20], or used to solve hard problems, specify applications in graphics, linguistics or model various biological systems, but they have been rarely used for modeling complex systems, such as *multi-agent systems (MAS)*. Attempts in this respect have been reported in [22], demonstrating the power of P systems for modeling the MAS dynamic structure. However, the main problem which appears in such modeling activity is that the resulting model and the object interaction within an agent is not always as intuitive to develop as it is when CsXM modeling is used [15]. A brief comparison between various classes of P systems and CsXMs [26, 27] shows that these paradigms have quite complementary features. Some P systems are very suitable for specifying systems that dynamically change their topology [3] or express transformations occurring inside various compartments [21]. CsXMs are very appropriate for modeling complex data structures, communication and synchronization aspects, and sophisticated computations [26, 27].

23.4.2 Transformations Between P Systems and CsXMs

Most of the variants of P systems and (communicating) X-machines are Turing complete and have, as mentioned in the previous section, quite complementary features. One aspect that interested the interactions between these two computational paradigms was related to finding ways of transforming one mechanism into another.

23.4.2.1 Transforming P Systems to CsXMs

Transforming various classes of P systems to CsXMs appeared a natural way of linking these mechanisms, because CsXM model offers a number of attractive and generic features related to specifying local computations in a very abstract and formal manner through functions from Φ and linking them by different communication procedures.

In [1] the process of transforming P systems with replicated rewriting into stream X-machines and CsXMs using communication matrices [2] has been studied. A P system with replicated rewriting uses strings and the rules have the format $X \to (v_1, tar_1)|| \ldots ||(v_n, tar_n)$, $n \geq 1$. Such a rule rewrites a string $x_1 X x_2$ by n strings $x_1 v_i x_2$, $1 \leq i \leq n$, which are sent to the regions indicated by targets tar_i. In this case a specific CsXM system is built consisting of so called PX components [1], with functions processing strings that simulate the above mentioned rewriting rules. Complexity aspects regarding the cost of the parallel computation and communication have been established, together with estimates to implement standard operations in distributed environments, like routing, broadcast, and convergecast.

Another study involved the transformation of a class of P systems using simple objects with evolution-communication and dissolution rules, all obeying a partial order relationship [11]. More particularly, in this case, a set of principles was derived that, when applied to a P system, leads to an CsXM with communication channels [12] that models the same behavior. The idea of this transformation process is that each region of the P system is modeled by a sXM able to communicate with another sXM that corresponds to one of the inner membranes or to its outer membrane. There are two important aspects regarding the transformation of such a P system to the corresponding CsXM. The first is that a CsXM, being asynchronous in its nature, needs to model the maximally parallel computation of a P system. This was resolved through the use of an additional *clock sXM* in the system, responsible for synchronizing all the sXM components and designating the beginning and end of the P system's macro-cycle. The second aspect is related to a specific feature of the class of P systems involved in the transformation process, namely its dynamic structure, implied by the use of the membrane dissolution rule, that does not have a direct corresponding operation within the CsXM framework. This has been managed by introducing an additional state, the *dissolved* state that designates that the dissolved membrane that the sXM represents no longer exists. However, as in the resulting CsXM system, the communication channels between the machine components are static, any machine in the dissolved state still performs one operation, that of passing along objects from its outer to its inner membranes and vice-versa. The time complexity of simulating a derivation step of a P system has been estimated.

The above presented results have not only a theoretical and conceptual importance, but they also allow, the latter directly and the former with some adjustments, to specify P systems in a language such as X-Machine Description Language (XMDL) [9] and then animate the sXM model to observe the behavior of the P system, through its corresponding CsXM system. Additionally, such an approach can be used in order to verify certain properties of the P system model, by expressing these properties in $\mathcal{X}m$CTL and using the sXM model checking technique [7, 6] or to test it by automatically generating suitable test sets [15].

23.4.2.2 Transforming CsXMs to P Systems

Transforming a CsXM into an adequate P system raises a number of non-trivial problems: how to codify the functions occurring in the components of a CsXM; how to match up the stream-oriented process defined by each CsXM component with the multiset-based process specific to most of the P systems variants. In order to address these issues a number of assumptions are made for the CsXM considered: the memory set associated to every component is finite, and consequently the number of distinct computations defined by each function is finite too; each function will process a symbol from a multiset at any given moment in time, rather than an input from a stream. As a CsXM is defined by a network of components related through communication channels, rather than a cell-like P system, a tissue P system [20] will be considered instead. This model has its regions arranged in a network structure rather than a tree. The rationale behind such transformations is that the CsXM models may rely on components thoroughly verified and tested and the obtained tissue P systems may be further extended with other important features like cell differentiation, death, birth, and bond making rules, and transformed into a more suitable model for MAS specification. We will briefly describe this process, while a comprehensive discussion can be found in [18, 19].

Every sXM component of a CsXM is associated with a tissue P system region with objects, transformation, and communication rules. The objects of the tissue P system are obtained from the states, memory values, input and output symbols. For each computation of each function that links a state, a memory value, and an input symbol with another state, a new memory value and an output symbol, a suitable evolution, or communication rule is considered.

The produced model may be enhanced with features that deal with a potential dynamic structure of a MAS, by enriching the model with features such as cell death, bond making rules, and cell division, specific to population P systems [18, 19]. Appropriate tools have been constructed to automatically transform CsXMs to tissue P systems, enhance them, and animate the computation [16]. The basic notation used in these transformations is X-Machine Description Language *(XMDL)* [9] and Population P System Description Language *(PPSDL)* [23].

23.4.3 Combining P Systems and X-Machines

The combination of these paradigms is meant to lead to a new model which is able to solve some hard problems or to provide a better modeling framework.

A specific combination of these paradigms is studied in [4] with the aim of producing more efficient models. Such a model, called PX system, consists of an X-machine structure, where the memory is organized into layers corresponding to regions of a P system and each function is a tuple of sets of rules which are applied in one step in the regions defined by the memory. In this way the rules in each region are divided into subsets and distributed alongside the edges of the X-machine. A parallel variant is also defined which is able to efficiently solve some NP-complete problems [4].

In order to produce better models for multi-agent systems the best features of each of these paradigms, namely the memory and the graph structure of a CsXM and the dynamic structure of the population P systems, have been considered [17]. This attempt to combine the two models has led to the definition of a generic framework, namely $OPERAS$, that allows to define a MAS with a dynamic structure [25, 24].

The basic principle behind the $OPERAS$ framework is that each agent can be defined in terms of two separate characteristics, one for its behavior (modeling its knowledge, actions, and control over its internal state) and one responsible for the reconfiguration of the system structure (adding and removing agents and communication links between them). Various instantiations of the framework have been defined, using CsXMs and population P systems: in $OPERAS_{XX}$ CsXMs are used for both characteristics of an agent and, on the contrary, $OPERAS_{CC}$ uses only population P systems [27]. $OPERAS_{XC}$, combining the two formalisms, uses a CsXM for the definition of the behavioral part of an agent, and a population P system model for the structure reconfiguration part [26, 25]. For the latter, bond-making and cell division and death rules, operating on objects representing the computation state of the underlying CsXM, are responsible for dynamically changing the structure of the MAS.

A number of MAS models have been developed under the $OPERAS$ framework using both formalisms, such as NASA's Autonomous Nano-Technology Swarm (ANTS) [26, 25] and other biology-inspired systems, e.g. the behavior of social insects, like ants and bees [19, 28].

23.4.4 Tools

The XMDL notation was developed to assist with the modeling and animation of models [9, 10]. The idea behind XMDL was to use a simple, yet powerful, declarative notation which would be close to the mathematical, yet practical, notation for sXM and CsXM. XMDL possesses constructs with which one can define an input and an output alphabet set, a memory structure including an initial memory, a set of states including an initial state, transitions between states, and functions. Functions get an input and a memory and give an output and a new memory, if certain conditions hold on input or memory values. The modeler can define any kind of different types of values by combining built-in types, such as natural numbers, with user-defined types, such as sets, sequences, tuples, etc. A tool, called X-System, has also been implemented [14]. It includes a parser, a compiler of XMDL to Prolog, and an animator. Models written in XMDL are compiled and animated, that is, the synchronous computation of the CsXM model is imitated through inputs provided by the user.

Similarly to XMDL, PPSDL has been designed so as to allow the experimentation with some population P system models [23]. We decided to keep the concept and, occasionally, the look of XMDL to some extent, and came up with a simple declarative notation, as close as possible to the formal definition of a population P system. PPSDL possesses constructs that allow one to define types of cells, cells as instances of those types, objects in cells and in the environment, as well as all types of rules. What makes PPSDL practical for modeling is the ability to associate objects with types, by combining built-in types with user-defined types. Therefore, objects in cells are characterized by a type identifier. PPSDL is the core of PPS-System, which includes a compiler of PPSDL to Prolog and an animator, similar to X-System. The animator simulates the computation of a model. PPS-System allows the user to input objects directly to cells during computation, if needed, thus allowing more flexibility in the animation.

23.4.5 SUMMARY

State-based models, such as stream X-machines, and P systems have complementary features that can be used together either to develop more powerful computational paradigms or to model complex dynamic systems, such as multi-agent systems. It appears that the internal behavior of an agent can be easily modeled as a state machine, whereas the dynamics of the structure of the system can be suitably modeled by P systems reconfiguration rules. For communication between agents any of the two could be utilized. On the other hand, there have been investigations into transforming communicating X-machines to P systems variants and vice-versa so as to use the features specific to them as well as various existing methods to check certain properties or tools to verify, test, or animate.

Future developments in this area are expected to focus more on relating the models, methodologies, and tools already developed with other computational paradigms, model checking techniques, and simulation frameworks so as to increase their computational and modeling capabilities.

References

[1] J. Aguado, T. Bălănescu, T. Cowling, M. Gheorghe, M. Holcombe, F. Ipate: P systems with replicated rewriting and stream X-machines (Eilenberg machines). *Fundamenta Informaticae*, 49 (2001), 1–17.

[2] T. Bălănescu, A.J. Cowling, H. Georgescu, M. Gheorghe, M. Holcombe, C. Vertan: Communicating stream X-machine systems are no more than X-machines. *J. Univ. Computer Sci.*, 5 (1999), 494–507.

[3] F. Bernardini, M. Gheorghe: Population P systems. *J. Univ. Computer Sci.*, 10 (2004), 509–539.

[4] F. Bernardini, M. Gheorghe, M. Holcombe: PX systems = P systems + X-machines. *Natural Computing*, 2 (2003), 201–213.

[5] S. Eilenberg: *Automata, Languages and Machines*. Academic Press, 1974.

[6] G. Eleftherakis: *Formal Verification of X-Machine Models: Towards Formal Development of Computer-Based Systems*. Ph.D. thesis, Department of Computer Science, University of Sheffield, 2003.

[7] G. Eleftherakis, P. Kefalas, A. Sotiriadou: XmCTL: Extending temporal logic to facilitate formal verification of X-machine models. *Ann. Univ. Bucureşti. Mathematics-Informatics*, 50 (2001), 79–95.

[8] M. Holcombe, F. Ipate: *Correct Systems: Building a Business Process Solution*. Springer, 1998.

[9] E. Kapeti, P. Kefalas: A design language and tool for X-machines specification. *Advances in Informatics* (D.I. Fotiadis, S.D. Spyropoulos, eds.), World Scientific, 2000, 134–145.

[10] P. Kefalas: *XMDL User Manual*. CITY College, Thessaloniki, Greece, 2000.

[11] P. Kefalas, G. Eleftherakis, M. Holcombe, M. Gheorghe: Simulation and verification of P systems through communicating X-machines. *BioSystems*, 70 (2003), 135–148.

[12] P. Kefalas, G. Eleftherakis, E. Kehris: Modular modelling of large-scale systems using communicating X-machines. *Proc. 8th Panhellenic Conference in Informatics* (Y. Manolopoulos, S. Evripidou, eds.), Livanis Publishing Company, 2001, 20–29.

[13] P. Kefalas, G. Eleftherakis, E. Kehris: Communicating X-machines: A practical approach for formal and modular specification of large systems. *J. Information and Software Technology*, 45 (2003), 269–280.

[14] P. Kefalas, G. Eleftherakis, A. Sotiriadou: Developing tools for formal methods. *Proc. 9th Panhellenic Conference in Informatics*, 2003, 625–639.

[15] P. Kefalas, M. Holcombe, G. Eleftherakis, M. Gheorghe: A formal method for the development of agent-based systems. *Intelligent Agent Software Engineering* (V. Plekhanova, ed.), Idea Publishing Group Co., 2003, 68–98.

[16] P. Kefalas, I. Stamatopoulou, G. Eleftherakis, M. Gheorghe: Transforming state-based models to P systems models in practice. *Pre-Proc. 9th Workshop on Membrane Computing (WMC9)* (P. Frisco et al., eds.), 2008, 247–261, and *Lecture Notes in Computer Sci.*, 5391 (2009), 260–273.

[17] P. Kefalas, I. Stamatopoulou, M. Gheorghe: A formal modelling framework for developing multi-agent systems with dynamic structure and behaviour. *Lecture Notes in Artificial Intelligence.*, 3690 (2005), 122–131.

[18] P. KEFALAS, I. STAMATOPOULOU, M. GHEORGHE: Principles of transforming communicating X-machines to population P systems. *Proc. Intern. Workshop on Automata for Cellular and Molecular Computing (ACMC'07)* (G. Vaszil, ed.), 2007, 76–89.

[19] P. KEFALAS, I. STAMATOPOULOU, I. SAKELLARIOU, G. ELEFTHERAKIS: Transforming communicating X-machines into P systems. *Natural Computing*, to appear.

[20] GH. PĂUN: *Membrane Computing. An Introduction.* Springer, 2002.

[21] F.J. ROMERO-CAMPERO, M. GHEORGHE, G. CIOBANU, J.M. AULD, M.J. PÉREZ-JIMÉNEZ: Cellular modelling using P systems and process algebra. *Progress in Natural Science*, 17 (2007), 375–383.

[22] I. STAMATOPOULOU, M. GHEORGHE, P. KEFALAS: Modelling dynamic configuration of biology-inspired multi-agent systems with communicating X-machines and population P systems. *Lecture Notes in Computer Sci.*, 3365 (2005), 389–401.

[23] I. STAMATOPOULOU, P. KEFALAS, G. ELEFTHERAKIS, M. GHEORGHE: A modelling language and tool for P systems. *Proc. 10th Panhellenic Conference in Informatics (PCI'05)* (P. Bozanis, E. Houstis, eds.), 2005, 142–152.

[24] I. STAMATOPOULOU, P. KEFALAS, M. GHEORGHE: Modelling the dynamic structure of biological state-based systems. *BioSystems*, 87 (2007), 142–149.

[25] I. STAMATOPOULOU, P. KEFALAS, M. GHEORGHE: OPERAS: a formal framework for multi-agent systems and its application to swarm-based systems. *Proc. 8th Intern. Workshop on Engineering Societies in the Agents World (ESAW'07)* (A. Artikis et al., eds.), 2007, 208–223.

[26] I. STAMATOPOULOU, P. KEFALAS, M. GHEORGHE: OPERAS for space: Formal modelling of autonomous spacecrafts. *Proc. 11th Panhellenic Conference in Informatics (PCI'07)* (T. Papatheodorou et al., eds.), Current Trends in Informatics, 2007, 69–78.

[27] I. STAMATOPOULOU, P. KEFALAS, M. GHEORGHE: OPERAS$_{CC}$: An instance of a formal framework for MAS modelling based on population P systems. *Lecture Notes in Computer Sci.*, 4860 (2007), 551–566.

[28] I. STAMATOPOULOU, I. SAKELLARIOU, P. KEFALAS, G. ELEFTHERAKIS: *OPERAS* for social insects: Formal modelling and prototype simulation. *Romanian J. Information Sci. and Technology*, 11 (2008), 267–280.

CHAPTER 23.5

Q-UREM P SYSTEMS

ALBERTO LEPORATI

23.5.1 INTRODUCTION

THE quest for a quantum version of P systems started at the beginning of 2004. As a result, two first ideas were proposed in [6]: either to follow the steps usually performed in Quantum Computing to define the quantum version of a given computation device, or to propose a completely new computation device based on the most elementary operation which can be conceived in physics: the exchange of a quantum of energy among two quantum systems. In the former case we would have obtained yet another quantum computation device whose computation steps are defined as the action of unitary operators, whose computations are logically reversible, and in which there are severe constraints on the amount of information which can be extracted from the system by measuring its state. In the latter case, instead, we felt that a new and interesting computational model could be introduced.

Hence, after a careful investigation, we decided to adopt creation and annihilation operators as the most elementary operations which can be performed by our computation devices. Our first proposal of quantum inspired P systems, explored in [6], was based on *energy-based P systems*, in which a given amount of energy is associated to each object of the system. Further, special symbols which denote "free" energy float into the system; when a given amount of free energy attaches to an object, this latter is transformed to another object, whose type is uniquely determined by the amount of resulting energy. However this approach presented several problems, the most important being the difficulty to control to which objects the free energy symbols have to attach in order to successfully complete an intended (deterministic) computation.

Looking for some alternatives, we considered the model of *UREM P systems* introduced in [1], in which a non-negative integer value is assigned to each membrane. Such a value can be conveniently interpreted as the *energy* of the membrane. In these P systems rules are assigned to the *membranes* of the system, rather than to the regions, and act as filters. When an object crosses a membrane, it may be transformed to another object, and possibly change the energy value associated to the membrane. This model of computation has thus been chosen as the base to define in [5] a *quantum inspired* version of UREM P systems (here referred to as Q-UREM P systems, for short). In Q-UREM P systems, the rules are realized through (not necessarily unitary) linear operators, which can be expressed as an appropriate composition of a truncated version of creation and annihilation operators. The operators which correspond to the rules have the form $|\beta\rangle \langle\alpha| \otimes O$, where O is a linear operator which modifies the energy associated with the membrane (implemented as the state of a quantum harmonic oscillator).

In the rest of this section we overview the basic notions of quantum mechanics which have led to the definition of Q-UREM P systems, and the results obtained so far about their computational power.

23.5.2 QUANTUM COMPUTING

From an abstract point of view, a quantum computer can be considered as made up of interacting parts. The elementary units (memory cells) that compose these parts are two-level quantum systems called *qubits*. A qubit is typically implemented using the energy levels of a two-level atom, or the two spin states of a spin-$\frac{1}{2}$ atomic nucleus, or a polarization photon. The mathematical description, independent of the practical realization, of a single qubit is based on the two-dimensional complex Hilbert space $\mathbb{C}^2$. The Boolean truth values 0 and 1 are represented by the unit vectors of the canonical orthonormal basis, called the *computational basis* of $\mathbb{C}^2$:

$$|0\rangle = \begin{bmatrix} 1 \\ 0 \end{bmatrix} \qquad |1\rangle = \begin{bmatrix} 0 \\ 1 \end{bmatrix}$$

While bits can only take two different values, 0 and 1, qubits are not confined to their two basis (pure) states, $|0\rangle$ and $|1\rangle$, but can also exist in states which are coherent superpositions such as $\psi = c_0 |0\rangle + c_1 |1\rangle$, where c_0 and c_1 are complex numbers satisfying the condition $|c_0|^2 + |c_1|^2 = 1$. Performing a *measurement* of the above state ψ will return 0 with probability $|c_0|^2$ and 1 with probability $|c_1|^2$; the state of the qubit after the measurement (*post–measurement state*) will be $|0\rangle$ or $|1\rangle$, depending on the outcome.

A *quantum register* of size n (also called an *n-register*) is mathematically described by the Hilbert space $\otimes^n \mathbb{C}^2 = \underbrace{\mathbb{C}^2 \otimes \ldots \otimes \mathbb{C}^2}_{n \text{ times}}$, representing a set of n qubits labeled by the index $i \in \{1, \ldots, n\}$. An *n-configuration* (also *pattern*) is a vector $|x_1\rangle \otimes \ldots \otimes |x_n\rangle \in \otimes^n \mathbb{C}^2$, usually written as $|x_1, \ldots, x_n\rangle$, considered as a quantum realization of the Boolean tuple $(x_1, \ldots, x_n)$. Let us recall that the dimension of $\otimes^n \mathbb{C}^2$ is 2^n and that $\{|x_1, \ldots, x_n\rangle \mid x_i \in \{0, 1\}\}$ is an orthonormal basis of this space called the *n-register computational basis*.

Computations are performed as follows. Each qubit of a given n-register is prepared in some particular pure state ($|0\rangle$ or $|1\rangle$) in order to realize the required n-configuration $|x_1, \ldots, x_n\rangle$, quantum realization of an input Boolean tuple of length n. Then, a linear operator $G : \otimes^n \mathbb{C}^2 \to \otimes^n \mathbb{C}^2$ is applied to the n-register. The application of G has the effect of transforming the n-configuration $|x_1, \ldots, x_n\rangle$ into a new n-configuration $G(|x_1, \ldots, x_n\rangle) = |y_1, \ldots, y_n\rangle$, which in the quantum realization of the output tuple of the computer. We interpret such modification as a computation step performed by the quantum computer. The action of the operator G on a superposition $\Phi = \sum c^{i_1 \ldots i_n} |x_{i_1}, \ldots, x_{i_n}\rangle$, expressed as a linear combination of the elements of the n-register basis, is obtained by linearity: $G(\Phi) = \sum c^{i_1 \ldots i_n} G(|x_{i_1}, \ldots, x_{i_n}\rangle)$. We recall that linear operators which act on n-registers can be represented as order 2^n square matrices of complex entries. Usually (but not in this section) such operators, as well as the corresponding matrices, are required to be unitary. In particular, this implies that the implemented operations are logically reversible (an operation is *logically reversible* if its inputs can always be deduced from its outputs).

All these notions can be easily extended to quantum systems which have $d > 2$ pure states. In this setting, the d-valued versions of qubits are usually called *qudits* [3]. As it happens with qubits, a qudit is typically implemented using the energy levels of an atom or a nuclear spin. The mathematical description—independent of the practical realization—of a single qudit is based on the d-dimensional complex Hilbert space $\mathbb{C}^d$. In particular, the pure states $|0\rangle, \left|\frac{1}{d-1}\right\rangle, \left|\frac{2}{d-1}\right\rangle, \ldots, \left|\frac{d-2}{d-1}\right\rangle, |1\rangle$ are represented by the unit vectors of the canonical orthonormal basis, called the *computational basis* of $\mathbb{C}^d$:

$$|0\rangle = \begin{bmatrix} 1 \\ 0 \\ \vdots \\ 0 \\ 0 \end{bmatrix}, \quad \left|\frac{1}{d-1}\right\rangle = \begin{bmatrix} 0 \\ 1 \\ \vdots \\ 0 \\ 0 \end{bmatrix}, \quad \ldots, \quad \left|\frac{d-2}{d-1}\right\rangle = \begin{bmatrix} 0 \\ 0 \\ \vdots \\ 1 \\ 0 \end{bmatrix}, \quad |1\rangle = \begin{bmatrix} 0 \\ 0 \\ \vdots \\ 0 \\ 1 \end{bmatrix}$$

As before, a *quantum register* of size n can be defined as a collection of n qudits. It is mathematically described by the Hilbert space $\otimes^n \mathbf{C}^d$. An *n-configuration* is now a vector $|x_1\rangle \otimes \ldots \otimes |x_n\rangle \in \otimes^n \mathbf{C}^d$, simply written as $|x_1, \ldots, x_n\rangle$, for x_i running on $L_d = \left\{0, \frac{1}{d-1}, \frac{2}{d-1}, \ldots, \frac{d-2}{d-1}, 1\right\}$. An n-configuration can be viewed as the quantum realization of the "classical" tuple $(x_1, \ldots, x_n) \in L_d^n$. The dimension of $\otimes^n \mathbf{C}^d$ is d^n and the set $\{|x_1, \ldots, x_n\rangle \mid x_i \in L_d\}$ of all n-configurations is an orthonormal basis of this space, called the *n–register computational basis*.

Let us now consider the set $\mathcal{E}_d = \left\{\varepsilon_0, \varepsilon_{\frac{1}{d-1}}, \varepsilon_{\frac{2}{d-1}}, \ldots, \varepsilon_{\frac{d-2}{d-1}}, \varepsilon_1\right\} \subseteq \mathbf{R}$ of real values; we can think of such quantities as energy values. To each element $v \in L_d$ we associate the energy level ε_v; moreover, let us assume that the values of $\mathcal{E}_d$ are all positive, equispaced, and ordered according to the corresponding objects: $0 < \varepsilon_0 < \varepsilon_{\frac{1}{d-1}} < \cdots < \varepsilon_{\frac{d-2}{d-1}} < \varepsilon_1$. As explained in [6, 4], the values ε_k can be thought of as the energy eigenvalues of the infinite dimensional quantum harmonic oscillator truncated at the $(d-1)$-th excited level, whose Hamiltonian on $\mathbf{C}^d$ is:

$$H = \begin{bmatrix} \varepsilon_0 & 0 & \cdots & 0 \\ 0 & \varepsilon_0 + \Delta\varepsilon & \cdots & 0 \\ \vdots & \vdots & \ddots & \vdots \\ 0 & 0 & \cdots & \varepsilon_0 + (d-1)\Delta\varepsilon \end{bmatrix} \qquad (23.5.1)$$

The unit vector $|H = \varepsilon_k\rangle = \left|\frac{k}{d-1}\right\rangle$, for $k \in \{0, 1, \ldots, d-1\}$, is the eigenvector of the state of energy $\varepsilon_0 + k\Delta\varepsilon$. To modify the state of a qudit we can use creation and annihilation operators on the Hilbert space $\mathbf{C}^d$, which are defined respectively as:

$$a^\dagger = \begin{bmatrix} 0 & 0 & \cdots & 0 & 0 \\ 1 & 0 & \cdots & 0 & 0 \\ 0 & \sqrt{2} & \cdots & 0 & 0 \\ \vdots & \vdots & \ddots & \vdots & \vdots \\ 0 & 0 & \cdots & \sqrt{d-1} & 0 \end{bmatrix} \qquad a = \begin{bmatrix} 0 & 1 & 0 & \cdots & 0 \\ 0 & 0 & \sqrt{2} & \cdots & 0 \\ \vdots & \vdots & \vdots & \ddots & \vdots \\ 0 & 0 & 0 & \cdots & \sqrt{d-1} \\ 0 & 0 & 0 & \cdots & 0 \end{bmatrix}$$

It is easily verified that the action of $a^\dagger$ on the vectors of the canonical orthonormal basis of $\mathbf{C}^d$ is the following:

$$a^\dagger \left|\frac{k}{d-1}\right\rangle = \sqrt{k+1} \left|\frac{k+1}{d-1}\right\rangle \qquad \text{for } k \in \{0, 1, \ldots, d-2\}$$

$$a^\dagger |1\rangle = 0$$

whereas the action of a is:

$$a \left| \begin{matrix} k \\ d-1 \end{matrix} \right\rangle = \sqrt{k} \left| \begin{matrix} k-1 \\ d-1 \end{matrix} \right\rangle \quad \text{for } k \in \{1, 2, \ldots, d-1\}$$

$$a |0\rangle = 0$$

Using $a^\dagger$ and a we can also introduce the following operators:

$$N = a^\dagger a = \begin{bmatrix} 0 & 0 & 0 & \cdots & 0 \\ 0 & 1 & 0 & \cdots & 0 \\ 0 & 0 & 2 & \cdots & 0 \\ \vdots & \vdots & \vdots & \ddots & \vdots \\ 0 & 0 & 0 & \cdots & d-1 \end{bmatrix} \qquad aa^\dagger = \begin{bmatrix} 1 & 0 & \cdots & 0 & 0 \\ 0 & 2 & \cdots & 0 & 0 \\ \vdots & \vdots & \ddots & \vdots & \vdots \\ 0 & 0 & \cdots & d-1 & 0 \\ 0 & 0 & \cdots & 0 & 0 \end{bmatrix}$$

The eigenvalues of the self–adjoint operator N are $0, 1, 2, \ldots, d-1$, and the eigenvector corresponding to the generic eigenvalue k is $|N = k\rangle = \left| \begin{matrix} k \\ d-1 \end{matrix} \right\rangle$.

One possible physical interpretation of N is that it describes the *number of particles* of physical systems consisting of a maximum number of $d-1$ particles. In order to add a particle to the k particles state $|N = k\rangle$ (thus making it switch to the "next" state $|N = k + 1\rangle$) we apply the creation operator $a^\dagger$, while to remove a particle from this system (thus making it switch to the "previous" state $|N = k - 1\rangle$) we apply the annihilation operator a. Since the maximum number of particles that can be simultaneously in the system is $d-1$, the application of the creation operator to a full $d-1$ particles system does not have any effect on the system, and returns as a result the null vector. Analogously, the application of the annihilation operator to an empty particle system does not affect the system and returns the null vector as a result.

Another physical interpretation of operators $a^\dagger$ and a, by operator N, follows from the possibility of expressing the Hamiltonian (23.5.1) as $H = \varepsilon_0 \mathbb{I} + \Delta\varepsilon N = \varepsilon_0 \mathbb{I} + \Delta\varepsilon a^\dagger a$. In this case $a^\dagger$ (resp., a) realizes the transition from the eigenstate of energy $\varepsilon_k = \varepsilon_0 + k \Delta\varepsilon$ to the "next" (resp., "previous") eigenstate of energy $\varepsilon_{k+1} = \varepsilon_0 + (k+1) \Delta\varepsilon$ (resp., $\varepsilon_{k-1} = \varepsilon_0 + (k-1) \Delta\varepsilon$) for any $0 \le k < d-1$ (resp., $0 < k \le d-1$), while it collapses the last excited (resp., ground) state of energy $\varepsilon_0 + (d-1)\Delta\varepsilon$ (resp., ε_0) to the null vector.

The collection of all linear operators on $\mathbb{C}^d$ is a d^2–dimensional linear space whose canonical basis is $\{E_{x,y} = |y\rangle \langle x| \mid x, y \in L_d\}$. Since $E_{x,y} |x\rangle = |y\rangle$ and $E_{x,y} |z\rangle = 0$ for every $z \in L_d$ such that $z \ne x$, this operator transforms the unit vector $|x\rangle$ into the unit vector $|y\rangle$, collapsing all the other vectors of the canonical orthonormal basis of $\mathbb{C}^d$ to the null vector. Each of the operators $E_{x,y}$ can be expressed, using the whole algebraic structure of the associative algebra of operators, as a suitable composition of creation and annihilation operators, as explained in [6, 4].

23.5.3 CLASSICAL AND QUANTUM-LIKE UREM P SYSTEMS

As stated in section 23.5.1, Q-UREM P systems have been introduced in [5] as a quantum inspired version of UREM P systems. Hence, let us start by recalling the definition of the classical model of computation.

A UREM P system [1] of degree $d + 1$ is a construct Π of the form:

$$\Pi = (A, \mu, e_0, \ldots, e_d, w_0, \ldots, w_d, R),$$

where: A is an alphabet of *objects*; μ is a *membrane structure*, with the membranes labeled by numbers $0, \ldots, d$ in a one-to-one manner; $e_0, \ldots, e_d$ are the initial energy values (non-negative integers) assigned to the membranes $0, \ldots, d$; $w_0, \ldots, w_d$ are multisets over A associated with the regions $0, \ldots, d$ of μ; R is a finite set of *unit rules* of the form $(\alpha_i : a, \Delta e, b)$, where $\alpha \in \{in, out\}$, $0 \leq i \leq d$, $a, b \in A$, and $|\Delta e|$ is the amount of energy that, for $\Delta e \geq 0$, is added to or, for $\Delta e < 0$, is subtracted from e_i (the energy assigned to membrane i) by the application of the rule. The set of all rules $(\alpha_i : a, \Delta e, b)$ associated with membrane i is denoted by R_i, $0 \leq i \leq d$.

The *initial configuration* of Π consists of $e_0, \ldots, e_d$ and $w_0, \ldots, w_d$. The transition from a configuration to another one is performed by non-deterministically choosing one rule from R and applying it (observe that here we consider a *sequential* model of applying the rules instead of choosing rules in a maximally parallel way, as it is often required in P systems). Applying $(in_i : a, \Delta e, b)$ means that an object a (being in the membrane immediately outside of i) is changed into b while entering membrane i, thereby changing the energy value e_i of membrane i by Δe. On the other hand, the application of a rule $(out_i : a, \Delta e, b)$ changes object a into b while leaving membrane i, and changes the energy value e_i by Δe. The rules can be applied only if the amount e_i of energy assigned to membrane i fulfills the requirement $e_i + \Delta e \geq 0$. Moreover, we use some sort of local priorities: if there are two or more applicable rules in membrane i, then one of the rules with $\max |\Delta e|$ has to be used. Since the rules transform one copy of an object to (one copy of) another object, in [1] they are referred to as *unit* rules. Hence, for conciseness, this model of P systems with unit rules and energy assigned to membranes has been abbreviated as *UREM P systems*.

A sequence of transitions is called a *computation*; it is *successful* if and only if it halts. The *result* of a successful computation is considered to be the distribution of energies among the membranes (a non-halting computation does not produce a result). If we consider the energy distribution of the membrane structure as the input to be analyzed, we obtain a model for accepting sets of (vectors of) non-negative integers. The following result, proved in [1], establishes the computational completeness for these systems.

Proposition 23.5.1 *Every partial recursive function $f : \mathbf{N}^\alpha \to \mathbf{N}^\beta$ can be computed by a UREM P system with (at most) $\max\{\alpha, \beta\} + 3$ membranes.*

Note that local priorities are necessary to obtain computational completeness. In fact, in [1] it is also proved that UREM P systems without priorities and with an arbitrary number of membranes do not reach the power of Turing machines, since they characterize the family $Ps\,MAT^\lambda$.

In Q-UREM P systems, all the elements of the model (multisets, the membrane hierarchy, configurations, and computations) are defined just like the corresponding elements of the classical P systems, but for objects and rules. The objects of A are represented as pure states of a quantum system. If the alphabet contains $d \geq 2$ elements then, recalling the notation introduced in Subsection 23.5.2, without loss of generality we can put $A = \{|0\rangle, |\frac{1}{d-1}\rangle, |\frac{2}{d-1}\rangle, \ldots, |\frac{d-2}{d-1}\rangle, |1\rangle\}$, that is, $A = \{|a\rangle \mid a \in L_d\}$. As stated above, the quantum system will also be able to assume as a state any superposition of the kind $c_0 |0\rangle + c_{\frac{1}{d-1}} |\frac{1}{d-1}\rangle + \ldots + c_{\frac{d-2}{d-1}} |\frac{d-2}{d-1}\rangle + c_1 |1\rangle$ with $c_0, c_{\frac{1}{d-1}}, \ldots, c_{\frac{d-2}{d-1}}, c_1 \in \mathbf{C}$ such that $\sum_{i=0}^{d-1} |c_{\frac{i}{d-1}}|^2 = 1$. A multiset is simply a collection of quantum systems, each in its own state.

To represent the energy values assigned to membranes we must use quantum systems which can exist in an infinite (countable) number of states. Hence we assume that every membrane of a Q-UREM P system has an associated infinite dimensional quantum harmonic oscillator whose state represents the energy value assigned to the membrane. To modify the state of such harmonic oscillator we can use the infinite dimensional version of creation ($a^\dagger$) and annihilation (a) operators described in Subsection 23.5.2. The actions of $a^\dagger$ and a on the state of an infinite dimensional harmonic oscillator are analogous to the actions on the states of truncated harmonic oscillators; the only difference is that in the former case there is no state with maximum energy, and hence the creation operator never produces the null vector. Also in this case it is possible to express the operators $E_{x,y} = |y\rangle \langle x|$ as appropriate compositions of $a^\dagger$ and a.

As in the classical case, rules are associated to the membranes rather than to the regions of the system. Each rule in R_i is an operator of the form

$$|y\rangle \langle x| \otimes O, \qquad \text{with } x, y \in L_d \qquad (23.5.2)$$

where O is a linear operator which can be expressed by an appropriate composition of operators $a^\dagger$ and a. The part $|y\rangle \langle x|$ is the *guard* of the rule: it makes the rule "active" (that is, the rule produces an effect) if and only if a quantum system in the basis state $|x\rangle$ is present. The semantics of rule (23.5.2) is the following: If an object in state $|x\rangle$ is present in the region immediately outside membrane i, then the state of the object is changed to $|y\rangle$ and the operator O is applied to the state of the harmonic oscillator associated with the membrane. Notice that the application of O can result in the null vector, so that the rule has no effect even if its guard is satisfied; this fact is equivalent to the condition $e_i + \Delta e \geq 0$ on the energy of membrane i required in the classical case. Conversely to the classical case, no local priorities are assigned to the rules. If two or more rules are associated to membrane i, then they are summed. This means that, indeed, we can think of each membrane as having only one rule with many guards. When an object is present, the inactive parts of the rule (those for which the guard is not satisfied) produce the null vector as a result. If the region in which the object occurs contains two or more membranes, then all their rules are applied to the object. Observe that the object which activates the rules never crosses the membranes. This means that the objects specified in the initial configuration can change their state but never move to a different region. Notwithstanding, transmission of information between different membranes is possible, since different objects may modify in different ways the energy state of the harmonic oscillators associated with the membranes.

The application of one or more rules determines a *transition* between two configurations. A *halting configuration* is a configuration in which no rule can be applied. A sequence of transitions is a *computation*. A computation is *successful* if and only if it reaches a halting configuration. The *result* of a successful computation is considered to be the distribution of energies among the membranes in the halting configuration. A non-halting computation does not produce a result. Just like in the classical case, if we consider the energy distribution of the membrane structure as the input to be analyzed, we obtain a model for accepting sets of (vectors of) non-negative integers.

23.5.4 COMPUTATIONAL POWER

In [5] the following theorem has been proved, establishing computational completeness for Q-UREM P systems.

Theorem 23.5.1 *Every partial recursive function $f : \mathbf{N}^\alpha \to \mathbf{N}^\beta$ can be computed by a Q-UREM P system with (at most) $\max\{\alpha, \beta\} + 3$ membranes.*

The proof is obtained by simulating register machines by means of Q-UREM P systems whose membrane structure is *flat*, that is, composed of a skin membrane that encloses one elementary membrane for each register of the simulated machine M. The input values $x_1, \ldots, x_\alpha$ are expected to be in the first α registers of M, and thus are encoded as the energies of the first α elementary membranes. Similarly, the output values are encoded as the energies of the first β elementary membranes at the end of a successful computation. The region enclosed by the skin contains only one object, which mimics the program counter of M. Precisely, if the program counter of M has the value $k \in \{1, 2, \ldots, m\}$ then the object present in region 0 is $|p_k\rangle = \left|\frac{k-1}{m-1}\right\rangle$. The sets R_i of rules depend upon the instructions of the simulated machine. Each increment instruction $j : (INC(i), k)$ is simulated by a guarded rule of the kind $|p_k\rangle \langle p_j| \otimes a^\dagger \in R_i$, whereas each decrement instruction $j : (DEC(i), k, l)$ is simulated by a guarded rule of the kind $|p_l\rangle \langle p_j| \otimes |\varepsilon_0\rangle \langle \varepsilon_0| + |p_k\rangle \langle p_j| \otimes a \in R_i$. In the former case, if the object $|p_j\rangle$ is present in region 0, then the rule transforms it to $|p_k\rangle$ and increments the energy level of the harmonic oscillator associated with membrane i. Similarly, in the latter case the object that represents the program counter of M, and the state of the harmonic oscillator, are modified in one of two possible ways, according to the current energy level of the harmonic oscillator. The set R_i of rules is obtained by summing all the operators which involve register i. When the object $|p_m\rangle$ appears in region 0, a projection operator can be applied to the states of the harmonic oscillators of the output membranes to retrieve the result of the computation.

Notice that objects $|p_j\rangle$ never cross any membrane. This fact avoids one of the problems raised in [6]: the existence of a "magic" quantum transportation mechanism which should be able to move objects between the regions of the system, according to the targets contained into the rules. Further, Q-UREM P systems do not need to exploit the membrane hierarchy to reach computational completeness: a flat membrane structure suffices.

23.5.4.1 Solving 3-SAT with Q-UREM P Systems

Q-UREM P systems can also be very efficient computation devices. Indeed, in [4] it has been shown that, under the assumption that an external observer is able to discriminate a null vector from a non-null vector, any instance of the NP-complete problem 3-SAT [2] can be solved in a polynomial time by using Q-UREM P systems. This solution is presented in the so-called *semi-uniform* setting, which means that for every instance of 3-SAT a specific Q-UREM P system that solves it is built.

Here we just give an abstract view of the solution; for the details, we refer the reader to [4]. An instance $\phi_{n,m}$ of 3-SAT is a set $C = \{C_1, C_2, \ldots, C_m\}$ of 3-clauses, built on a finite set $\{x_1, x_2, \ldots, x_n\}$ of Boolean variables. Every 3-clause is a disjunction of three literals, each of which is either a Boolean variable or a negated Boolean variable. The instance $\phi_{n,m}$ is *positive* if there exists an assignment to the variables $x_1, x_2, \ldots, x_n$ that satisfies (that is, make *true*) all the clauses in C. Note that the number of all possible assignments to n Boolean variables is 2^n, and that the number m of possible 3-clauses is polynomially bounded with respect to n: in fact, since each clause contains exactly three literals, we can have at most $(2n)^3 = 8n^3$ clauses.

Given an instance $\phi_{n,m}$ of 3-SAT, composed of m 3-clauses, built on n Boolean variables, it is not difficult to design a register machine $M_{\phi_{n,m}}$ that computes the value of $\phi_{n,m}$ for any fixed assignment to $x_1, x_2, \ldots, x_n$. Indeed, such a register machine can be obtained from $\phi_{n,m}$ in a straightforward (mechanical) way. Then, it is possible to build a Q-UREM P system $\Pi_{\phi_{n,m}}$ that evaluates $\phi_{n,m}$ by simulating the machine $M_{\phi_{n,m}}$, as described in the proof of Theorem 23.5.1. Such a simulation computes the value of ϕ_n for a *single* assignment to its variables; however, if we initialize the harmonic oscillators of the n input membranes with a uniform superposition of all possible classical assignments to $x_1, x_2, \ldots, x_n$, then at the end of the computation the harmonic oscillator of the output membrane will be in one of the following states:

- $|0\rangle$, if $\phi_{n,m}$ is not satisfiable;
- a superposition $\alpha_0 |0\rangle + \alpha_1 |1\rangle$, with $\alpha_1 \neq 0$, if $\phi_{n,m}$ is satisfiable.

By adding the rule $|p_{\text{end}}\rangle \langle p_{\text{end}}| \otimes 2^n |1\rangle \langle 1|$ (where p_{end} denotes the label of the halting instruction of $M_{\phi_{n,m}}$) to the output membrane, we can extract the result of the computation: in fact, if $\phi_{n,m}$ is not satisfiable then by applying this rule we obtain the null vector, whereas if $\phi_{n,m}$ is satisfiable we obtain a non-null vector. We can thus conclude that if an external observer is able to discriminate between a null vector and a non-null vector, then we have a semi-uniform family of Q-UREM P systems that solve any instance of 3-SAT in a polynomial time.

23.5.5 FINAL REMARKS

In this section we have briefly overviewed the state of the art concerning Q-UREM P systems. We believe that the study of their computational properties is of interest; for example, does the computational power or the efficiency of the system augment when assuming the presence of pairs (or tuples) of entangled object? Are Q-UREM P systems, with or without entangled objects, able to solve harder than NP-complete problems?

References

[1] R. Freund, A. Leporati, M. Oswald, C. Zandron: Sequential P systems with unit rules and energy assigned to membranes. *Lecture Notes in Computer Sci.*, 3354 (2005), 200–210.
[2] M.R. Garey, D.S. Johnson: *Computers and Intractability. A Guide to the Theory on NP-Completeness*. W.H. Freeman and Company, 1979.
[3] D. Gottesman: Fault-tolerant quantum computation with higher-dimensional systems. *Chaos, Solitons, and Fractals*, 10 (1999), 1749–1758.

[4] A. LEPORATI, S. FELLONI. Three "quantum" algorithms to solve 3-SAT. *Theoretical Computer Sci.*, 372 (2007), 218–241.
[5] A. LEPORATI, G. MAURI, C. ZANDRON: Quantum sequential P systems with unit rules and energy assigned to membranes. *Lecture Notes in Computer Sci.*, 3850 (2006), 310–325.
[6] A. LEPORATI, D. PESCINI, C. ZANDRON: Quantum energy-based P systems. *First Brainstorming Workshop on Uncertainty in Membrane Computing*, Palma de Mallorca, Spain, November 2004, 145–167.
[7] M.A. NIELSEN, I.L. CHUANG: *Quantum Computation and Quantum Information*. Cambridge Univ. Press, 2000.

CHAPTER 23.6

MEMBRANE COMPUTING AND ECONOMICS

GHEORGHE PĂUN

RADU A. PĂUN

23.6.1 INTRODUCTION

TRADITIONALLY, (continuous or discrete) mathematical models of economic processes have mainly been of an analytic or operational-research type. However, in the last decade, computer simulations (e.g. agent-based computational economics models) were investigated more and more, and increasingly complex economic systems could be modeled. The book [4] is a good introduction to this research area.

Modeling economic processes in terms of membrane computing is a rather promising approach, complementary but related to the multi-agent techniques. The reasons are obvious: the compartmental structure (in the form of cell-like or tissue-like arrangements of membranes) can describe the organization of economic units, the "chemical reactions" can describe "production rules", etc. (some details will be uncovered in the next section). In addition, the experience gained with modeling biological processes can be very useful when modeling economic processes. However, an essential observation needs to be made: the economic "reactions" do not develop only under the control of stoichiometric parameters, the psycho-social determination is also important (e.g. the trust among transacting partners), and this imposes changes in models. Then, the general features which make P systems suitable models for many biological processes can also be invoked for economic modeling: modularity, scalability, ease to understand and program, non-linearity, etc.

The possibility of using membrane computing in modeling economic processes has already been proposed by various researchers. This section's bibliography indicates several papers by Polish authors (see, e.g. [1], [7]) who mainly use category theory to build membrane computing models of economic processes (e.g. accounting, management, etc.). For space reasons, we do not recall details from this direction of research. We also refrain from detailing the work of the Vienna group (see, e.g. [5], [8]), although these papers are closer to the general membrane computing approach, as used in biology. We also ignore here, as the topic is briefly mentioned in Section 22.7, the so-called numerical P systems introduced in [12], which are directly inspired by economic modeling (but not used to model real processes yet).

The next subsections recall a few ideas from [9] and some simulation results from [10].

23.6.2 Economic Counterparts of Some Membrane Computing Ingredients

Many ingredients of membrane computing can easily be interpreted in economic terms, which makes the biological paradigm/metaphor directly relevant for modeling economic processes. Our goal is not to produce a "membrane computing economics dictionary", but rather to illustrate the parallel between the notions. The case study presented in the next subsection will elaborate on this topic.

An *object* can correspond to any unitary item which is produced, transferred, consumed, ordered, planned in an economic system, such as commodities of any kind, parts of commodities (if they explicitly appear in the production process), monetary units, labor, the need for any one of these, or the order placed to obtain them. Then, as mentioned in Subsection 23.6.1, a *membrane* can delimit any economic entity handling objects, at any level of aggregation (individuals, working places within the enterprise, the firm, or the whole economy). In turn, a *membrane structure* can model the organization of an economic entity, indicating its subsystems and higher entities, as well as the communication among them (either in the tree or graph form).

Several differences from biology are already apparent: in economies we can have membranes with a non-empty intersection and, mainly, the possibility of communicating (objects) across several membranes at once, not just across a single membrane.

We mentioned above that *multiset rewriting rules* can model production operations (assembly/de-assembly operations). Then, *symport rules* can correspond to taking inputs (freely) or sending objects out (e.g. waste materials), while *antiport rules* can model economic exchanges (trading materials, money, orders, etc. among compartments). In this setup, the environment can be seen as both a source of raw material, and a market for delivering products, waste, etc. Like in biology, the rules (production or exchange) can be applied under various types of *controls*, such as promoters and inhibitors (e.g. bank restrictions or facilities), priority relations, etc.

The operations with membranes are also of interest. For instance, membrane creation can be seen as the founding of a new economic agent, whereas membrane dissolution can represent the ceasing of activity of such an agent; as in the case of dissolving a membrane, when a company section stops functioning, its activity is terminated but its employees and inventory become a direct part of the company, which can redistribute them among the remaining sections. Then, phagocytosis directly corresponds to the economic process of (vertical) integration, where previously independent companies start functioning under unified ownership and control.

Many other membrane computing ingredients can be interpreted in terms of economic models. However, several economic aspects are not covered by membrane computing (as it is developed so far). We mentioned above the possibility of overlapping membranes and communicating at long distances, not only across a single membrane.

More important from an economic point of view can be the lack of performance criteria, almost absent in membrane computing (one exception is the case of P systems with energy associated with rules, where one can impose that the rules chosen to apply consume a minimal/maximal quantity of energy), but central in economics. In addition, "reaction rules" change continuously in economics, for instance, because of technological advance and market functioning; the only membrane computing aspect resembling this feature is the possibility of creating rules (of given types) during computation, in so-called P systems with rule creation.

In economics, one also frequently observes correlation between remote compartments, but this can be easily modeled by means of *bi-rules*, of the following form:

$$([u_1 \to v_1]_i, [u_2 \to v_2]_j),$$

where i, j are membranes and u_1, v_1, u_2, v_2 are non-empty multisets of objects. The idea is that rule $u_1 \to v_1$ is applied in membrane i simultaneously with rule $u_2 \to v_2$ being applied in membrane j. No restriction is imposed on labels i, j.

Hence, in particular, they may be equal, or one or both may represent the environment (thus relating the system's inner evolution to the evolution of the environment, in particular, allowing the evolution of objects placed in the environment).

23.6.3 MEMBRANE COMPUTING REPRESENTATIONS OF THE PRODUCERS–RETAILERS MODEL

We pass now to illustrating the usefulness of the membrane computing formalism as a framework for economic modeling, starting with a simplified case and gradually adding new features to it.

23.6.3.1 The Model

Let us consider k *producers*, $P_1, P_2, \ldots, P_k$, with *production capacities* $n_1, n_2, \ldots, n_k$, respectively. This means that each P_i can produce n_i copies of a good d in each time unit, $1 \leq i \leq k$. The time unit is not specified. We denote by b_i a unit of production capacity at producer P_i, $1 \leq i \leq k$.

We also consider l *retailers*, $R_1, R_2, \ldots, R_l$, with *retailing capacities* $m_1, m_2, \ldots, m_l$, respectively; a unit of capacity at R_j is denoted by c_j, $1 \leq j \leq l$.

We denote $N = \sum_{i=1}^{k} n_i$, $M = \sum_{j=1}^{l} m_j$, and assume that $N = M$, that is, the total production and retailing capacities are equal.

The production is based on some raw material provided by a *source S*; one unit of raw material is denoted by a. As before, we assume only one input is needed to produce d, and, further, that a producer needs exactly one a for one d (these assumptions can easily be modified/relaxed).

Retailers sell objects d (bought from producers) in order to satisfy the need for d, as induced by an aggregate consumer denoted by C. The need for an unit (we also say copy) of good d is denoted by $\bar{d}$.

The system's evolution is presented next. We start with no copy of a and no copy of $\bar{d}$ in the system. In the first step, the source S provides "around" N copies of a, with various probabilities for each number of copies. For instance, we can consider the following rules for producing raw material:

$$S \rightarrow Sa^{N+g}[.01], \quad S \rightarrow Sa^{N}[.95], \quad S \rightarrow Sa^{N-g}[.03], \quad S \rightarrow Sa^{N-2g}[.01],$$

where the probability of using each rule is mentioned in square brackets.

In turn, at the same time, the aggregate consumer C introduces "around" M copies of $\bar{d}$ in the system, that is, "around" M needs for good d. The respective rules are similar to those associated with S:

$$C \to C\bar{d}^{M+g'}[.03], \quad C \to C\bar{d}^{M}[.90], \quad C \to C\bar{d}^{M-g'}[.04], \quad C \to C\bar{d}^{M-2g'}[.03],$$

where again we associate probabilities with rules, in order to regulate their frequency of application.

The source S of raw material and the source C of consumption needs use a rule at the beginning of each day. During the day, producers and retailers perform several steps. Specifically, each producer P_i takes as many copies of a (if available) as many b_i's it has available, and transforms them into d_i, $1 \leq i \leq k$. The significance is that one "item" of production capacity of P_i is used/active and one copy of d is produced in P_i. The respective operation can be written in the form

$$b_i a \to d_i.$$

Note that this way both a copy of b_i and one of a are "consumed". Initially, there are n_i copies of b_i for each P_i in the system, $1 \leq i \leq k$.

The above rule has no probability associated, meaning that it is necessarily used as soon as b_i and a are available.

Simultaneously, retailers take the needs for d, depending on their capacities, by means of rules of the form

$$c_j \bar{d} \to \bar{d}_j, \quad 1 \leq j \leq l,$$

with the interpretation that the need for one d is transformed by R_j into an "order" for d, indicated by $\bar{d}_j$.

At this point, the purchasing operation takes place, written in the form

$$d_i \bar{d}_j \to b_i c_j \; [Rscore_{i,j}]. \tag{23.6.1}$$

The interpretation is that one item of d is purchased by R_j from P_i, thus satisfying the order $\bar{d}_j$ and setting free one unit of capacity for both P_i and R_j. As above, we have indicated in square brackets the probability that R_j purchases a d from P_i. In the current version of the model, this probability depends directly on the trust retailer R_j places on producer P_i, and is inversely related to the price P_i charges R_j. This coefficient/probability can be computed once (at the beginning of the system evolution) or can be computed at each step, depending on the previous transactions between P_i and R_j. In the latter case we can write $Rscore_{i,j}(t)$, making the time explicit in the form of the rules.

This ends a cycle (a day), therefore we can apply rules for S and C again, and repeat the whole process.

As standard in membrane computing, in each time unit, all objects which can evolve according to the previous scenario should do it—in the sense of maximally parallel use of rules.

If we use dynamically computed probabilities for producer–retailer interaction rules (rules of type (23.6.1)), then after each step we have to recompute the probabilities of these rules (by updating the trust coefficients and considering the new prices charged by each producer).

The previous process can be enriched by also considering prices of the handled items (raw material and good d).

Let us denote by u a monetary unit. When u belongs to a producer i or a retailer j, we write u_i, v_j, respectively. Similarly, we use u_S, u_C to denote a monetary unit belonging to S or C, respectively. Then, we consider *prices* for a and d, with the following cases: $p_S(a)$ is the price of a when S sells one item of a to any producer; $p_C(d)$ is the price of one d when C "sells" one d—hence this is the amount of u associated with $\bar{d}$ by C; $ps_j(d)$ is the price of d when sold by R_j to C; finally, $pp_i(d)$ is the price of d set by P_i.

Thus, the rules considered above can be rewritten as follows (we also provide explanations to each rule's meaning):

1. $S \to S a^{N \pm g} [prob(g)]$
   S produces $N \pm g$ copies of a;
2. $b_i u_i^{p_S(a)} a \to d_i u_S^{p_S(a)}$
   for each copy of a passing from S to P_i, the producer pays $p_S(a)$ monetary unit, hence $p_S(a)$ copies of u_i are transformed into u_S;
3. $C \to C \bar{d}^{M \pm g'} u_C^{p_C(t)(M \pm g')} [prob(g')]$
   the aggregate consumer C introduces on the market $M \pm g'$ copies of the need for d, along with the corresponding amount of money ($p_C(d)$ is the price of d as estimated by C);
4. $c_j \bar{d} u_C^{ps_j(d)} \to \bar{d}_j v_j^{ps_j(d)}$
   the need for one $\bar{d}$ goes from C to R_j, along with the price of d when sold by R_j;
5. $d_i \bar{d}_j v_j^{pp_i(d)} \to b_i c_j u_i^{pp_i(d)} [Rscore_{i,j}(d)]$
   one copy of d is purchased by R_j from P_i, at the price $pp_i(d)$ set by P_i.

Clearly, by considering money, the evolution of the system is much more complex. Prices can influence the probabilities of using the rules, available d_i, and orders $\bar{d}_j$. In addition, the amount of money present in the system and in each of its components can evolve in time, entailing possible shortages of money, hence the impossibility of using some rules, etc.

In an attempt to bring the model closer to reality, we can add further details. One of the easiest (and most interesting) additions is to consider investments—the possibility for producers and retailers to increase their capacity. For instance, we can use rules of the form

$$v_j^x \to c_j,$$

with the meaning that x monetary units can be spent for increasing the capacity of R_j by one unit. Because investments are usually made following positive financial results, we can impose that such a rule is only used if no rule of the form $d_i \bar{d}_j v_j^{pp_i(t)} \to b_i c_j u_i^{pp_i(t)} [Rscore_{i,j}(t)]$ can. Stated otherwise, these latter rules have priority over the "investment rules" $v_j^x \to c_j$.

Furthermore, we can associate probabilities with investment rules, or an upper bound on the number of times they are used, thus modeling a more cautious investing behavior.

23.6.3.2 Computer Simulations

A computer program was written to simulate the previous model, starting with its simple version and then adding more features to it. Specifically, we consider two producers with production capacities 65 and 35 units, and three retailers with selling capacities 50, 30, and 20 units, which satisfy our initial (simplifying) condition $M = N$. So, our initial multiset will contain b_1^{65}, b_2^{35}, c_1^{50}, c_2^{30}, and c_3^{20}.

The raw material is generated by S and we only introduce 60 units of a so that to have competition among the producers:

$$S \to Sa^{60}.$$

The aggregate consumer C generates the need for good d, labeled $\bar{d}$, also in a way that leads to competition among retailers (but is balanced with the available amount of a):

$$C \to C\bar{d}^{60}.$$

The purchasing rules are set as follows. For R_1 we have:

$$d_1 \bar{d}_1 \to b_1 c_1 \; [1], \quad d_2 \bar{d}_1 \to b_2 c_1 \; [0],$$

meaning that R_1 only deals with the first producer. This way we model possible geographical barriers prohibiting trade with other producers, or an overwhelming trust in producer P_1. For R_2 we have:

$$d_1 \bar{d}_2 \to b_1 c_2 \; [0.5], \quad d_2 \bar{d}_2 \to b_2 c_2 \; [0.5],$$

meaning that R_2 randomly chooses between P_1 and P_2. For R_3 we have:

$$d_1 \bar{d}_3 \to b_1 c_3 \; [0.15], \quad d_2 \bar{d}_3 \to b_2 c_3 \; [0.85],$$

meaning that this retailer places different trust in each producer.

In our initial model the time unit was one day, and the idea was to follow model's evolution over time, for example, for a period of one year. In this framework, each day consists of three time units (three clock steps): one when the source S generates the raw material simultaneously with the aggregate consumer C generating the need for d, one when production capacities are used and the finite good d is produced, while retailing capacities are used and orders for d are produced, and, finally, one when producers and retailers interact and capacities are re-activated.

Several simulations were run with these data, following the evolution of our system over time. We do not present these simulations here, and instead proceed to the more complex case when monetary units are introduced. The prices we consider are the following: $p_S(a) = 11$, $pp_1(d) = 12$, $pp_2(d) = 13$, $ps_1(d) = 13$, $ps_2(d) = 14$, $ps_3(d) = 14$, and $p_C(d) = 14$.

The aggregate consumer C generates the need for good d (denoted $\bar{d}$) along with the money needed to buy that quantity of d (note that C generates the money according to the estimated price $p_C(d) = 14$). The rule is:

$$C \to C\bar{d}^{60} u_C^{840}.$$

Each producer P_i makes one unit of good d from a copy of a and one of b_i, paying 11 monetary units to source S. Hence, we have the rules:

$$b_1 a u_1^{11} \to d_1 u_S^{11}, \quad b_2 a u_2^{11} \to d_2 u_S^{11}.$$

At the beginning of the simulation, producers P_1 and P_2 need an initial capital to purchase a and transform it into d, and we set producers' initial endowment to 750 and 400 monetary units. This means that our initial multiset will be appended with u_1^{750}, and u_2^{400}.

Once the aggregate consumer has issued the need for good d and the accompanying amount of money, retailers transform this need into orders for good d, and receive a certain payment for this, in the following way:

$$c_1 \bar{d} u_C^{13} \to \bar{d}_1 v_1^{13}, \quad c_2 \bar{d} u_C^{14} \to \bar{d}_2 v_2^{14}, \quad c_3 \bar{d} u_C^{15} \to \bar{d}_3 v_3^{15}.$$

We do not present further rules of the model, but we pass to simulation results. We are interested in the financial performance of producers and retailers; for producers we follow the number of copies of u_1 and u_2, while for retailers we follow the number of copies of v_1, v_2. The simulation was run for 100 steps and the results are given in Fig. 23.6.1.

As expected, all our variables increase, and some do so quicker than others—for example, u_2 seems to have a steeper slope than u_1, even though u_2 represents the money held by the smaller producer. This is not counter-intuitive since the number of copies of a released on the market is 60 at each step. These units are distributed almost evenly among producers, and each of them gets about 30 copies of a (less than their individual capacities). In addition, the second producer is the one charging the higher price—13, rather than 12.

Now, the idea arises to use the extra money accumulated by producers and retailers for investments. We focus on producer P_2 and retailer R_2, to exemplify

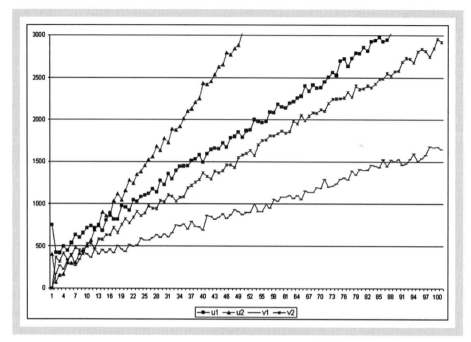

Fig. 23.6.1 Evolution of u_1, u_2, v_1, and v_2 in the case of no investments.

the new "investment rules" we add to the model:

$$u_2^{20} \to b_2, \quad v_2^{20} \to c_2.$$

Such rules should be used with lower priority when compared with the "production rules" (buying raw material), and we implement this by assigning a low probability to these investment rules.

We add another feature to the model: not all money should be invested, some should be saved. We therefore also consider "saving rules" of the form

$$u_2 \to u_2, \quad v_2 \to v_2,$$

with probabilities equal to those of the investment rules (by adjusting these probabilities we can model a more risky or a more cautious behavior).

We simulate the system's evolution over 200 time steps and illustrate this for several variables. The amounts of money producers P_1 and P_2 possess at each step are shown in Fig. 23.6.2. Unlike before, the money held by the two producers no longer constantly increases, but fluctuates around the 450 value. Fig. 23.6.3 shows the evolution of retailer investments (linearly increasing over time).

We extend our exercise further and place an upper bound on the number of investments an agent can make (a bound on the capacity units newly created).

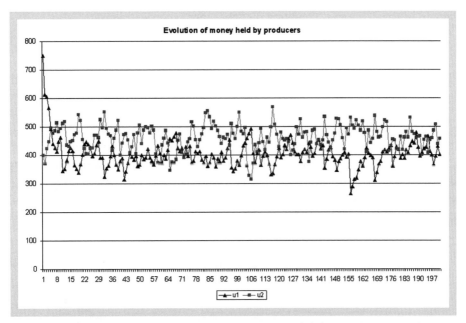

Fig. 23.6.2 Evolution of u_1 and u_2 in the unbounded investment case.

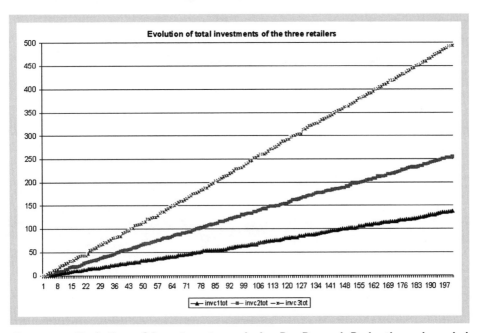

Fig. 23.6.3 Evolution of investments made by R_1, R_2, and R_3 in the unbounded investment case.

This can be easily done by changing the investment rules as follows:

$$f_2 u_2^{20} \to b_2, \quad g_2 v_2^{20} \to c_2,$$

where f_2, g_2 are new objects, denoting the possible new capacity units of P_2 and R_2, respectively. Initially, we introduce into the system 200 copies of each f_2 and g_2 (hence only 200 times the investment rules can be used, until exhausting the possible "space" for development). The simulation for producers is depicted in Fig. 23.6.4 (total investments allowed = 200). The second part of the figure "zooms in" on the interesting section of the graph, by restricting attention to monetary values below 800 units, and hence makes this figure comparable to Fig. 23.6.2. We see that, as before, the monetary units held by producers fluctuate around the 450 value. Then, once the investing limit has been reached, monetary units

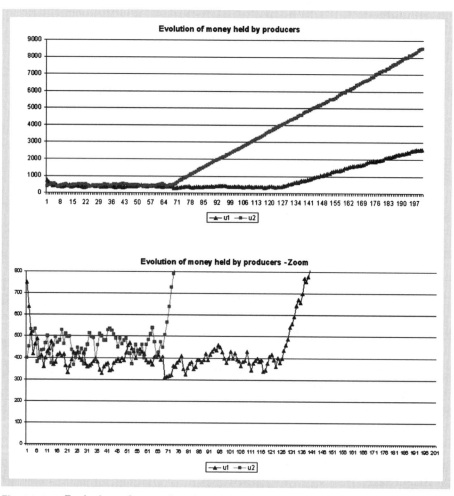

Fig. 23.6.4 Evolution of u_1 and u_2 in the bounded investment case.

accumulate linearly (as in the case with no investment). It is worth noting that P_2 reaches the investing limit first, around step 70, while it takes P_1 approximately 60 more steps to reach the limit.

What is also worth pointing is that we have considered producers to be equally interested in saving and investing (we have used equal probabilities for these rules). Attitude towards savings/investments can be adjusted by properly choosing these probabilities.

Further features can be added to the model: we can increase the demand generated by C, along with the money made available by C, we can model the accumulation of trust, etc. We omit the details.

23.6.4 FINAL REMARKS

The usefulness of the membrane computing framework as an intermediate step between the verbal (linguistic) model and computer simulations should be obvious—given the ambiguity/imprecision of natural languages, it is not at all a trivial task to go directly from the linguistic model to a computer program. Moreover, writing a model in terms of membrane computing is equivalent to writing a program in the "programming language" of P systems, a precise, algorithmic framework, which then can be easily converted into a program in a standard language. We expect significant progress to be made in this area, comparable with that made in using P systems as models of biological processes.

REFERENCES

[1] J. BARTOSIK: Paun's systems in modeling of human resource management. *Proc. Second Conf. Tools and Methods of Data Transformation*, WSU Kielce, 2004.

[2] J. BARTOSIK, W. KORCZYNSKI: Systemy membranowe jako modele hierarchicznych struktur zarzadzania. *Mat. Pokonferencyjne Ekonomia, Informatyka, Zarzadzanie. Teoria i Praktyka*, Wydzial Zarzadzania AGH, Tom II, AGH 2002.

[3] F. BERNARDINI, M. GHEORGHE, M. MARGENSTERN, S. VERLAN: Producer/consumer in membrane systems and Petri nets. *Lecture Notes in Computer Sci.*, 4497 (2007), 43–52.

[4] J.M. EPSTEIN, R. AXTELL: *Growing Artificial Societies—Social Science from the Bottom Up*. Brookings Institution Press and The MIT Press, 1996.

[5] R. FREUND, M. OSWALD, T. SCHIRK: How a membrane agent buys goods in a membrane store. *Progress in Natural Science*, 17 (2007), 442–448.

[6] W. KORCZYNSKI: On a model of economic systems. *Second Conf. Tools and Methods of Data Transformation*, WSU Kielce, 2004.
[7] W. KORCZYNSKI: Păun's systems and accounting. *Pre-proc. Sixth Workshop on Membrane Computing*, Vienna, Austria, July 2005, 461–464.
[8] M. OSWALD: Independent agents in a globalized world modelled by tissue P systems. *Artificial Life and Robotics*, 11 (2007), 171–174.
[9] GH. PĂUN, R. PĂUN: Membrane computing as a framework for modeling economic processes. In *Proc. SYNASC 05*, Timişoara, Romania, IEEE Press, 2005, 11–18.
[10] GH. PĂUN, R. PĂUN: A membrane computing approach to economic modeling: The producer-retailer interactions and investments. *Economic Analysis and Forecasting*, Part I: 3 (2006), 30–37, Part II: 4 (2006), 47–54.
[11] GH. PĂUN, R. PĂUN: Membrane computing models for economics. An invitation-survey. *Economic Studies and Research*, 40 (2006), 5–19.
[12] GH. PĂUN, R. PĂUN: Membrane computing and economics: Numerical P systems, *Fundamenta Informaticae*, 73 (2006), 213–227.

CHAPTER 23.7

MOBILE MEMBRANES AND MOBILE AMBIENTS

BOGDAN AMAN
GABRIEL CIOBANU

23.7.1 INTRODUCTION

MOBILE ambients [8] and mobile membranes [10] have similar structures and work with common concepts. Both have a hierarchical structure representing locations, and are used to model various aspects of distributed systems. The mobile ambients are suitable to represent the movement of ambients through ambients by consuming capabilities and the communication which takes place inside the boundaries of ambients. Mobile membranes are suitable to represent the evolution of objects and the movement of membranes through membranes. The mobility of membranes is expressed using the biological operations of *exocytosis* and *endocytosis*.

In Subsection 23.7.2 we introduce the safe ambients, in order to realize an encoding of the safe ambients into mobile membranes in Subsection 23.7.3 (for more details see [2, 4]). We continue by providing an operational correspondence between the safe ambients and their encodings, as well as various related properties of P systems ([3, 4]).

To illustrate how these formalisms work and can be used to model biological phenomena, in [4] the sodium-potassium exchange pump is described using the safe ambients, and then this description is translated into mobile membranes. The obtained description is compared with a direct description of the pump given first by using P systems.

In Subsection 23.7.4 we investigate the enhanced mobile membranes. The contextual evolution rules describe how an object from a membrane can evolve only in some contexts. The other rules describe the objective endocytosis and exocytosis. We use the class of enhanced mobile membrane system to model some evolutions in the immune system. Finally we study the computational power of the enhanced mobile membranes. In particular, we focus on the power of mobility given by the operations *endo*, *exo*, *fendo*, and *fexo*. The computational universality is obtained with 12 membranes, while systems with 8 membranes subsume *EToL*, and those with 3 membranes are contained in *MAT*.

After the bridge between these two formalisms is established, we investigate in Subsection 23.7.5 the problem of reaching a configuration from another configuration in mobile membranes, and prove that the reachability can be decided by reducing it to the reachability problem of the pure and public ambient calculus without the capability open.

23.7.2 Safe Ambients

Safe ambients are a variant of mobile ambients in which any movement of an ambient takes place only if both participants agree. The mobility is provided by the consumption of certain pairs of capabilities. The safe ambients differ from mobile ambients by the addition of co-actions: if in mobile ambients a movement is initiated only by the moving ambient and the target ambient has no control over it, in safe ambients both participants must agree by using matching action and co-action.

We give a short description of pure safe ambients (SA); more information can be found in [11]. Given an infinite set of names $\mathcal{N}$ (ranged over by $m, n, \ldots$), we define the set $\mathcal{A}$ of SA-processes (denoted by $A, A', B, B', \ldots$) together with their capabilities (denoted by $C, C', \ldots$) as follows:

$$C ::= \text{in } n \mid \overline{\text{in}} \, n \mid \text{out } n \mid \overline{\text{out}} \, n \mid \text{open } n \mid \overline{\text{open}} \, n$$
$$A ::= 0 \mid A \mid B \mid C.A \mid n[\, A\,] \mid (\nu n) A$$

Process 0 is an inactive mobile ambient. A movement $C.A$ is provided by the capability C, followed by the execution of A. An ambient $n[A]$ represents a bounded place labelled by n in which an SA-process A is executed. $A \mid B$ is a parallel composition of mobile ambients A and B. $(\nu n) A$ creates a new unique name n within the scope of A.

The operational semantics of pure ambient safe calculus is defined in terms of a reduction relation $\Rightarrow_{amb}$ by the following axioms and rules.

Axioms:
(In) $n[\ in\ m.A\ |\ A'\ ]\ |\ m[\ \overline{in}\ m.B\ |\ B'\ ] \Rightarrow_{amb} m[\ n[\ A\ |\ A'\ ]\ |\ B\ |\ B'\ ]$,
(Out) $m[\ n[\ out\ m.A\ |\ A'\ ]\ |\ \overline{out}\ m.B\ |\ B'\ ] \Rightarrow_{amb} n[\ A\ |\ A'\ ]\ |\ m[\ B\ |\ B'\ ]$,
(Open) $open\ n.A\ |\ n[\ \overline{open}\ n.B\ |\ B'\ ] \Rightarrow_{amb} A\ |\ B\ |\ B'$.

Rules:
(Res) $\dfrac{A \Rightarrow_{amb} A'}{(\nu n)A \rightarrow_{umb} (\nu n)A'}$, (Comp) $\dfrac{A \Rightarrow_{amb} A'}{A\ |\ B \Rightarrow_{amb} A'\ |\ B}$,

(Amb) $\dfrac{A \Rightarrow_{amb} A'}{n[\ A\ ] \Rightarrow_{amb} n[\ A'\ ]}$, (Struc) $\dfrac{A \equiv A',\ A' \Rightarrow_{amb} B',\ B' \equiv B}{A \Rightarrow_{amb} B}$.

$\Rightarrow^*_{amb}$ denotes a reflexive and transitive closure of the binary relation $\Rightarrow_{amb}$.

23.7.3 Translating Safe Ambients into Mobile Membranes

A translation from the set $\mathcal{A}$ of safe ambients to the set $\mathcal{M}$ of mobile membranes is given formally as follows:

Definition 23.7.1 *A translation* $\mathcal{T} : \mathcal{A} \to \mathcal{M}$ *is given by* $\mathcal{T}(A) = dlock\ \mathcal{T}_1(A)$, *where* $\mathcal{T}_1 : \mathcal{A} \to \mathcal{M}$ *is*

$$\mathcal{T}_1(A) = \begin{cases} cap\ n[\ ]_{cap\ n} & \text{if } A = cap\ n \\ cap\ n[\ \mathcal{T}_1(A_1)\ ]_{cap\ n} & \text{if } A = cap\ n.\ A_1 \\ [\ \mathcal{T}_1(A_1)\ ]_n & \text{if } A = n[\ A_1\ ] \\ [\ ]_n & \text{if } A = n[\ ] \\ (\nu n)\mathcal{T}_1(A_1) & \text{if } A = (\nu n)A_1 \\ \mathcal{T}_1(A_1), \mathcal{T}_1(A_2) & \text{if } A = A_1\ |\ A_2 \end{cases}$$

An object $dlock$ is placed near the membrane structure to prevent the consumption of capability objects in a membrane system which corresponds to a mobile ambient which cannot evolve further.

Using the structural congruence relations $\equiv_{amb}$ and $\equiv_{mem}$ defined in [4] for safe ambients and mobile membranes, the next result holds:

Proposition 23.7.1 *Structurally congruent ambients are translated into structurally congruent membrane systems; moreover, structurally congruent translated membrane systems correspond to structurally congruent ambients:*

$$A \equiv_{amb} B \text{ iff } \mathcal{T}(A) \equiv_{mem} \mathcal{T}(B).$$

Denoting by r an instance of the rules from the particular set of developmental rules used in [4], we use $M \xrightarrow{r} N$ to denote the transformation of a membrane system M into a membrane system N by applying a rule r. Even when some rules can be applied in parallel, we prefer to write them as a sequence of rules. It is worth noting that the application order of the rules is not important because we get the same result. Similarly to [6] where a structural operational semantic for a particular class of P systems was defined, we can define the corresponding relation $\Rightarrow_{mem}$. Considering two membrane systems M and N with only one object $dlock$, we say that $M \Rightarrow_{mem} N$ if there is a sequence of rules $r_1, \ldots, r_i$ such that $M \xrightarrow{r_1} \ldots \xrightarrow{r_i} N$. The *operational semantic* of the membrane systems is defined in terms of the transformation relation $\xrightarrow{r}$ by the following rules:

$$(DRule)\ M \xrightarrow{r} N \text{ for each developmental rule } a), \ldots, k)$$

$$(Res)\ \frac{M \xrightarrow{r} M'}{(\nu n)M \xrightarrow{r} (\nu n)M'}\ ;\ (Comp)\ \frac{M \xrightarrow{r} M'}{M, N \xrightarrow{r} M', N};$$

$$(Amb)\ \frac{M \xrightarrow{r} M'}{[\,M\,]_n \xrightarrow{r} [\,M'\,]_n}\ ;\ (Struc)\ \frac{M \equiv_{mem} M',\ M' \xrightarrow{r} N',\ N' \equiv_{mem} N}{M \xrightarrow{r} N}.$$

Using the reduction relations defined previously for safe ambients and mobile membranes, we provide an operational correspondence between the safe ambients and their encoding.

Proposition 23.7.2 *If A and B are two ambients and M is a membrane system such that $A \Rightarrow_{amb} B$ and $M = \mathcal{T}(A)$, then there exists a chain of transitions $M \xrightarrow{r_1} \ldots \xrightarrow{r_k} N$ such that $r_1, \ldots, r_k$ are developmental rules, and $N = \mathcal{T}(B)$.*

Proposition 23.7.3 *Let M and N be two membrane systems with only one $dlock$ object, and an ambient A such that $M = \mathcal{T}(A)$. If there is a sequence of transitions $M \xrightarrow{r_1} \ldots \xrightarrow{r_k} N$, then there exists an ambient B with $A \Rightarrow^*_{amb} B$ and $N = \mathcal{T}(B)$. The number of pairs of non-star objects consumed in membrane systems is equal to the number of pairs of capabilities consumed in ambients.*

Theorem 23.7.1 (*operational correspondence*)

1. *If $A \Rightarrow_{amb} B$, then $\mathcal{T}(A) \Rightarrow_{mem} \mathcal{T}(B)$.*
2. *If $\mathcal{T}(A) \Rightarrow_{mem} M$, then exists B such that $A \Rightarrow_{amb} B$ and $M = \mathcal{T}(B)$.*

23.7.4 Enhanced Mobile Membranes

The movement in mobile membranes is given mainly by two operations: exocytosis and endocytosis, each of them working either in a "subjective" or an "objective" manner. According to [8], an "objective" movement is expressing that the moving membranes are controlled by objects placed inside the membranes that are passed through, and a "subjective" movement is expressing that a membrane control its own moving.

Starting from this observation, a new class of P systems is defined in [5], namely the enhanced mobile membrane systems. The distinction is made by three new rules which are inspired by some evolution of the immune system. The contextual evolution rule states that a multiset from a membrane can evolve only in a certain context. The other two rules describe the objective endocytosis and exocytosis; one is called "forced endocytosis", while the other is called "forced exocytosis".

Definition 23.7.2 *An enhanced mobile membrane system is a construct*

$$\Pi = (V, H, \mu, w_1, \ldots, w_n, R), \text{ where:}$$

1. $n \geq 1$ *(the initial degree of the system);*
2. *V is an alphabet (its elements are called objects);*
3. *H is a finite set of labels for membranes;*
4. *μ is a hierarchical membrane structure, consisting of n membranes, labeled with elements of H (two different membranes can have the same label).*
5. *$w_1, w_2, \ldots, w_n$ are strings over V describing the multisets of objects placed in the n regions of μ;*
6. *R is a finite set of developmental rules of the following forms:*
   (a) $[\,[u \to v]_m]_k$ *for $k, m \in H$, $u \in V^+$, $v \in V^*$;* local evolution
   These rules are called local because the evolution of a multiset of objects u of membrane m is possible only when membrane m is inside membrane k. If the restriction of nested membranes is not imposed, that is, the evolution of the multiset of objects u in membrane m is allowed irrespective of where membrane m is placed, then we say that we have a global evolution rule, and write it simply as $[u \to v]_m$.
   (b) $[[w]_m [u]_h]_k \to [[w]_m [v]_h]_k$ *for $h, m \in H$, $u \in V^+$, $v, w \in V^*$;* contextual evolution
   These rules are called contextual because the evolution of a multiset of objects u of membrane h is possible only when membrane h is sibling with membrane m containing the multiset of objects w and both membranes h and m are placed inside the same membrane k. If the multiset of objects w is not specified, then the evolution is allowed only in the context of a sibling membrane m placed in the same membrane k.
   (c) $[u]_h [\ ]_m \to [[v]_h]_m$ *for $h, m \in H$, $u \in V^+$, $v \in V^*$;* endocytosis
   An elementary membrane labeled h enters the adjacent membrane labeled m, under the control of the multiset of objects u. The labels h and m remain unchanged during this process; however the multiset of objects u may be

modified to the multiset of objects v during the operation. Membrane m is not necessarily an elementary membrane.

(d) $[[u]_h]_m \to [v]_h[\ ]_m$, for $h, m \in H, u \in V^+, v \in V^*$; exocytosis

An elementary membrane labeled h is sent out of a membrane labeled m, under the control of the multiset of objects u. The labels of the two membranes remain unchanged, but the multiset of objects u from membrane h may be modified during this operation. Membrane m is not necessarily elementary.

(e) $[\ ]_h[u]_m \to [\ [\ ]_h v]_m$ for $h, m \in H, u \in V^+, v \in V^*$; forced endocytosis

An elementary membrane labeled h is engulfed into the adjacent membrane labeled m, under the control of the multiset of objects u. The labels h and m remain unchanged during the process; however the multiset of objects u may be transformed into the multiset of objects v during the operation. Membrane m is not necessarily elementary. The effect of this rule is similar to the effect of rule (c). The main difference from rule (c) is that the movement is not controlled by a multiset of objects placed inside the moving membrane h, but by a multiset of objects u placed inside the membrane m which engulfs membrane h. Namely, the membrane which initiates the move is the passive membrane m, and not the active membrane h as in rule (c).

(f) $[u[\ ]_h]_m \to [\ ]_h[v]_m$ for $h, m \in H, u \in V^+, v \in V^*$; forced exocytosis

An elementary membrane labeled h is sent out of a membrane labeled m under the control of the multiset of objects u. The labels of the two membranes remain unchanged; however, the multiset of objects u may be transformed into the multiset v during the operation. Membrane m is not necessarily elementary. The effect of this rule is similar to the one of rule (d), the main difference being that that the movement is not controlled by a multiset of objects placed inside the moving membrane h, but by a multiset of objects u placed inside the membrane m which engulfs membrane h. Namely, the membrane which initiates the move is the passive membrane m, and not the active membrane h as in rule (d).

(g) $[u]_h \to [v]_h[w]_h$, $h \in H, u, v, w \in V^*$; elementary division

In reaction with a multiset of objects u, the elementary membrane labeled h is divided into two membranes labeled h, with the multiset of objects u replaced in the two new membranes by possibly new multisets of objects v and w.

The rules of type (c) and (d) are used to simulate the movement of membranes in the membrane systems controlled by objects placed inside the moving membranes ("subjective" movement). The rules of type (e) and (f) are used to simulate the movement of membranes in the membrane system that are controlled by objects placed inside the membranes which are passed through ("objective" movement).

The contextual evolution rules, namely the rules of type (b), express the fact that a multiset of objects from a membrane can evolve in the presence of a multiset of objects placed in a sibling membrane, where the two sibling membranes are placed into the same membrane.

In order to dissolve some membranes we use the special object $\delta \in V$ which once created by a rule of type (a) or (b), dissolves the surrounding membrane.

The motivation for these new rules can be found in [5] where some aspects of the immune system are modeled: cells "eat" other cells (so the membrane movement is controlled by the eater, not by the eaten membrane).

Example 23.7.1 *Dendritic cells can engulf bacteria, viruses, and other cells (forced endocytosis). Once a dendritic cell engulfs a bacterium, it dissolves this bacterium and places portions of bacterium proteins on its surface. These surface markers serve as an alarm to other immune cells, namely helper T cells, which then infer the form of the invader. This mechanism makes the T cells sensitive to recognize the antigens or other foreign agents which trigger a specific reaction of the immune system (contextual evolution). Antigens are often found on the surface of bacterium and viruses.*

The computational power of the P system with enhanced mobile membranes has been studied in [9]. The operations governing the mobility of the enhanced mobile membranes are endocytosis (*endo*), exocytosis (*exo*), forced endocytosis (*fendo*) and forced exocytosis (*fexo*). The interplay between these four operations is quite powerful, and the computational power of a Turing machine is obtained by using 12 membranes. The exact computational power of the four operations against a subset of the operations (using more than one operation) is not clear. Moreover, the optimality of the results presented below is an open problem.

The result of a halting computation consists of the vector describing the multiplicity of objects from a membrane designated as the output membrane; a non-halting computation provides no output. The set of vectors of natural numbers produced in this way by a system Π is denoted by $Ps(\Pi)$. The family of all sets $Ps(\Pi)$ generated by systems of degree at most n using rules $\alpha \subseteq \{exo, endo, fendo, fexo, cevol\}$, is denoted by $PsEM_n(\alpha)$. Here *cevol* represents contextual evolution. The main results are the following.

Theorem 23.7.2 *(i) $PsEM_{12}(endo, exo, fendo, fexo) = PsEM_3(cevol) = PsRE$.*
(ii) $PsEM_3(endo, exo) = PsEM_3(fendo, fexo)$.

An interesting problem to look at is whether systems using all the four rules can be simulated by systems using only a pair of rules from $\{endo, exo, fendo, fexo\}$, namely either pair (*endo, exo*) or pair (*fendo, fexo*).

23.7.5 REACHABILITY PROBLEM

In [1] the reachability problem for a special class of mobile membranes is studied. It is proven that reachability in mobile membranes can be decided by reducing it to the reachability problem of the pure and public ambient calculus from which the open capability has been removed. It is known [7] that the reachability for this fragment of ambient calculus is decidable by reducing it to marking reachability for Petri nets.

For two mobile membrane systems M and N we say that M reduces to N if there is a sequence of rules applicable in the membrane system M in order to obtain the membrane system N.

Theorem 23.7.3 *For two arbitrary mobile membranes M_1 and M_2, it is decidable whether M_1 reduces to M_2.*

The main steps of the proof are as follows:

1. mobile membrane systems are reduced to pure and public mobile ambients without the capability *open*;
2. the reachability problem for two arbitrary mobile membranes can be expressed as the reachability problem for the corresponding mobile ambients;
3. the reachability problem is decidable for the pure and public mobile ambients without the capability *open*.

For more details see [1].

REFERENCES

[1] B. AMAN, G. CIOBANU: Reachability problem in mobile membranes. *Lecture Notes in Computer Sci.*, 4860 (2007), 113–123.

[2] B. AMAN, G. CIOBANU: Translating mobile ambients into P systems. *Electronic Notes in Theoretical Computer Sci.*, 171 (2007), 11–23.

[3] B. AMAN, G. CIOBANU: Structural properties and observability in membrane systems. *Proceedings SYNASC*, IEEE Computer Society, 2007, 74–84.

[4] B. AMAN, G. CIOBANU: On the relationship between membranes and ambients. *BioSystems*, 91 (2008), 515–530.

[5] B. AMAN, G. CIOBANU: Describing the immune system using enhanced mobile membranes. *Electronic Notes in Theoretical Computer Sci.*, 194 (2008), 5–18.

[6] O. ANDREI, G. CIOBANU, D. LUCANU: Structural operational semantics of P systems. *Lecture Notes in Computer Sci.*, 3850 (2006), 32–49.

[7] I. BONEVA, J.-M. TALBOT: When ambients cannot be opened. *Lecture Notes in Computer Sci.*, 2620 (2003), 169–184.

[8] L. CARDELLI, A. GORDON: Mobile ambients. *Lecture Notes in Computer Sci.*, 1378 (1998), 140–155.

[9] S.N. KRISHNA, G. CIOBANU: On the computational power of enhanced mobile membranes. *Lecture Notes in Computer Sci.*, 5028 (2008), 326–335.

[10] S.N. KRISHNA, GH. PĂUN: P systems with mobile membranes. *Natural Computing*, 4 (2005), 255–274.

[11] F. LEVI, D. SANGIORGI: Mobile safe ambients. *ACM TOPLAS*, 25 (2003), 1–69.

[12] A. REGEV, E.M. PANINA, B. SILVERMAN, L. CARDELLI, E.Y. SHAPIRO: Bioambients: an abstraction for biological compartments. *Theoretical Computer Sci.*, 325 (2004), 141–167.

CHAPTER 23.8

OTHER TOPICS

GHEORGHE PĂUN
GRZEGORZ ROZENBERG

23.8.1 NUMERICAL P SYSTEMS

WE start with a rather "exotic" class of P systems, motivated by issues related to economics, where one uses numerical variables rather than symbol or string objects placed in the regions of a membrane structure. These variables can evolve in time, starting from initial values, and their values can be distributed to other variables, somewhat similar to communication between regions using objects with target indications *here, out, in*. The evolution of variables is directed by "production functions" (below we consider polynomials) and the distribution is done according to a "repartition protocol". By a synchronized use of production functions followed by repartition of variable values, we get transitions between system configurations. The values assumed by a distinguished variable during a computation form the set of numbers computed by the system.

This general idea can be implemented in a number of ways—here we describe one of them considered in [21].

We consider usual cell-like membrane structures (with the standard $1, 2, \ldots, m$ labeling of membranes). Variables in region i are written in the form $x_{j,i}$, $j \geq 1$. The value of $x_{j,i}$ at time $t \in \mathbf{N}$ is denoted by $x_{j,i}(t)$. These values can be of various types—in what follows we consider integers as values of variables (although in many applications one would most probably use real numbers).

In order to evolve the values of variables, we use *programs*, composed of two components, a *production function* and a *repartition protocol*. The former can be any function with variables from a given region—here we consider only polynomials with integer coefficients. Using such a function (chosen non-deterministically if there are several programs in a given region), we compute a *production value* of the region at a given step. This value is distributed to variables from the region where the program resides, and to variables in its upper and lower neighbors according to the repartition protocol associated with the used production function. For a given region i, let $v_1, \ldots, v_{n_i}$ be all these variables. Here we consider as repartition protocols expressions of the form

$$c_1|v_1 + c_2|v_2 + \cdots + c_{n_i}|v_{n_i},$$

where $c_1, \ldots, c_{n_i}$ are natural numbers. The idea is that the coefficients $c_1, \ldots, c_{n_i}$ specify the proportion of the current production distributed to each variable $v_1, \ldots, v_{n_i}$.

This is precisely defined as follows. Consider a program

$$(F_{l,i}(x_{1,i}, \ldots, x_{k_i,i}), c_{l,1}|v_1 + c_{l,2}|v_2 + \cdots + c_{l,n_i}|v_{n_i})$$

and let

$$C_{l,i} = \sum_{s=1}^{n_i} c_{l,s}.$$

At a time instant $t \geq 0$ we compute $F_{l,i}(x_{1,i}(t), \ldots, x_{k_i,i}(t))$. The value $q = F_{l,i}(x_{1,i}(t), \ldots, x_{k_i,i}(t))/C_{l,i}$ represents the "unitary portion" to be distributed according to the repartition expression to variables $v_1, \ldots, v_{n_i}$. Thus, $v_{l,s}$ will receive $q \cdot c_{l,s}, 1 \leq s \leq n_i$.

A production function may use only some of the variables from a region. Those variables "consume" their values when the production function is used (they become zero)—the other variables retain their values. To these values—zero in the case of variables contributing to the region production—one adds all "contributions" received from the neighboring regions.

Thus, a *numerical P system* is a construct of the form

$$\Pi = (\mu, (Var_1, Pr_1, Var_1(0)), \ldots, (Var_m, Pr_m, Var_m(0)), x_{j_0,i_0}),$$

where μ is a membrane structure with m membranes labeled injectively by $1, 2, \ldots, m$, Var_i is the set of variables from region i, Pr_i is the set of programs from region i (all sets Var_i, Pr_i are finite), $Var_i(0)$ is the set of initial values for the variables in region i, and x_{j_0,i_0} is a distinguished variable (from a distinguished region i_0), which provides the result of a computation.

Each program is of the form specified above: $pr_{l,i} = (F_{l,i}(x_{1,i}, \ldots, x_{k_i,i}), c_{l,1}|v_1 + c_{l,2}|v_2 + \cdots + c_{l,n_i}|v_{n_i})$ denotes the lth program from region i, where $Var_i = \{x_{1,i}, \ldots, x_{k_i,i}\}$.

Such a system evolves in the way informally described before. Initially, the variables have the values specified by $Var_i(0), 1 \leq i \leq m$. A transition from a

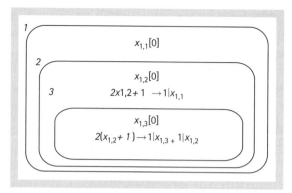

Fig. 23.8.1 The system Π_1.

configuration at time instant t to a configuration at time instant $t+1$ is made by (i) choosing non-deterministically one program from each region, (ii) computing the value of the respective production function for the values of local variables at time t, and then (iii) computing the values of variables at time $t+1$ as directed by repartition protocols. A sequence of such transitions forms a computation, with which we associate a set of numbers, viz., the numbers which occur as values of the variable x_{j_0, i_0}; this set of numbers is denoted by $N(\Pi)$.

As an example we consider the numerical system Π_1 given in Fig. 23.8.1 with the distinguished variable $x_{1,1}$. One can easily see that variable $x_{1,3}$ increases by 1 at each step, also transmitting its value to $x_{1,2}$. In turn, region 2 transmits the value $2x_{1,2} + 1$ to $x_{1,1}$, which is never consumed, hence its value increases continuously. In the initial configuration all variables are set to 0. Thus, $x_{1,1}$ starts from 0 and continuously receives $2i + 1$, for $i = 0, 1, 2, 3, \ldots$, which implies that in n steps the value of $x_{1,1}$ becomes $\sum_{i=0}^{n-1}(2i+1) = n^2$, and consequently $N(\Pi_1) = \{n^2 \mid n \geq 0\}$.

The system Π_1 was deterministic; let us consider also a non-deterministic system:

$$\Pi_2 = ([\ ]_1, (\{x_{1,1}\}, \{(2x_{1,1}, 1|x_{1,1}), (3x_{1,1}, 1|x_{1,1})\}, 1), x_{1,1}).$$

The production is assigned to the unique variable, but in each step we can choose either the first program or the second one; in the former case $x_{1,1}$ is multiplied by 2, and in the latter case it is multiplied by 3. Thus, the values of $x_{1,1}$ will be of the form $2^i 3^j$, with $i \geq 0$, $j \geq 0$. Actually, *all numbers of this form are values of $x_{1,1}$*, where the value $2^i 3^j$ is obtained in step $i + j$.

In these two examples we have chosen the programs in such a way that the production value is divisible by the total sum of coefficients c_j from each region (let us denote this case with *div*). When a current production is not divisible by the given total value of coefficients, then we can take the following decisions: (i) the remainder is lost (the production which is not immediately distributed is lost), (ii) the remainder is added to the production obtained in the next step (the non-distributed production is carried over to the next step), (iii) the system simply stops and aborts, no result is associated with that computation. We denote these three cases by *lost, carry, stop*, respectively.

Thus, we can distinguish many types of systems, depending on the programs and their use. The family of sets of numbers $N(\Pi)$ computed by numerical P systems with at most m membranes, production functions which are polynomials of degree at most n, with at most r variables in each polynomial, with non-negative coefficients, and the distribution of type α is denoted by $NNP_m(poly^n(r), nneg, \alpha)$, $m \geq 1, n \geq 0, r > 0, \alpha \in \{div, lost, carry, stop\}$. The restriction to deterministic systems is indicated by adding the letter D in front of NNP. If arbitrary coefficients are allowed, then the indication "nneg" is removed. If one of the parameters m, n, r is not bounded, then it is replaced by $*$. The set of positive numbers occurring as values of the output variable is denoted by $N^+(\Pi)$, and NN gets the superscript + when considering the family of such sets.

Here are some results concerning these families—mode details can be found in [21].

Theorem 23.8.1 (i) $DNN^+ P_1(poly^1(1), nneg, div) - SLIN_1^+ / \emptyset$.
(ii) $SLIN_1^+ \subset DNN^+ P_*(poly^1(1), nneg, div)$.

The main result of [21] shows that, surprisingly enough, numerical P systems of a rather restricted type are Turing complete, even when using small numbers of membranes and polynomials of low degrees with a small number of variables:

Theorem 23.8.2 $N^+RE = NN^+ P_8(pol^5(5), div) = NN^+ P_7(poly^5(6), div)$.

The proof is based on the characterization of recursively enumerable sets of numbers as positive values of polynomials with integer values, [19]. It is not known whether a similar result holds for deterministic numerical P systems.

Many research topics are open for numerical P systems; among others, we mention: a thorough investigation of all classes of systems mentioned above, considering also vectors of numbers, looking for non-universal classes (and decidability results for those classes), hierarchies, and normal forms.

23.8.2 Around a Bio-Lab Implementation

At this moment there does not exists an implementation of a P system in a biolab. Recently, plans for an explicit experiment for implementing a P system in biological media were made at E. Keinan laboratory in Technion, Haifa, Israel. The experiment is still in progress, but the project has already inspired some theoretical research issues.

First of all, one needs to ask what does it mean to have "a P system implementation". The basic features of membrane computing are: (1) *compartmentalization* by means of membranes, (2) *multisets* (which means *counting* the objects present in regions), (3) *evolution rules*, (4) *synchronization* of evolution of regions, e.g. by using evolution rules in a maximally parallel manner, (5) *communication* between regions, and (6) defining the result of a computation (mainly) for *halting* computations.

In order to implement a P system in a laboratory, all or most of these features should be implemented. Regions (compartments) can be created by using standard test tubes or similar labware, and multisets are usual in biochemistry, although without precise counting—still, by defining carefully "moles" of substances, one can count in terms of such moles. Anyway, full synchronization and parallelism cannot be guaranteed by biochemical reactions, hence a certain degree of non-determinism/approximation should be allowed in the experiment. In particular, a good degree of synchronization can be obtained by "waiting long enough" so that all reactions that can take place in a test tube actually take place. This raises an important issue: these reactions should not cycle, the process should be finite in each region of the system. Note that counting is also needed when reading the result of a computation.

In the experiment planned at Technion, the above concerns are addressed as follows: (1) test tubes implement regions, (2) multisets are implemented as predefined "moles" of DNA molecules, (3) evolution rules are implemented as enzyme driven operations on DNA molecules, and care is taken that there are no cycles of substances in any region, (4) synchronization is achieved by waiting long enough for reactions to take place, (5) communication is implemented by moving all relevant objects to the next tube in a mechanical way, and (6) the result is read of by spectrophotometry (certain molecules are marked and their number is estimated).

These laboratory solutions require finding a suitable problem or a class of problems for which no cycle of substances is possible in any region, and the solution allows a degree of approximation. In this experiment one does not need to implement the notion of halting computations, because one computes Fibonacci sequences of numbers and just produces sequences of consecutive outputs (measurements) at specific time moments.

From a theoretical point of view, the central issue is to find a non-trivial class of P systems such that the reactions from each region are completed in a finite (small) number of steps. Therefore, no region can "produce" a cycle of objects which can run forever.

This intuitive goal can be reached in various formal ways. For instance, we can request that no local transition graph contains a cycle (the catalysts are ignored).

Such local transition graphs are defined as follows. For each region i of a P system with the set of objects O and set of rules R_i in region i, the transition graph $\gamma_i = (O, E)$ of region i has the set of edges defined as follows: for each $a, b \in O$,

$(a, b) \in E$ iff there is $u \to v \in R_i$ such that $|u|_a \geq 1, |v|_b \geq 1$.

A stronger condition is to require that no object produced in a region can evolve in the same region. In the case of non-cooperative systems, this means that the local transition graph contains no paths of length longer than one. For cooperative systems this statement is not true: having the rules $a \to b$, $bc \to cc$, the local transition graph contains the path (a, b, c), but still it is not possible to have two reactions in a row, because without c, the product b of the first reaction cannot evolve.

Thus, formally, we can formulate several properties which ensure the local loop-freeness. Defining such properties and investigating P systems satisfying them is one of the *research topics* we present in what follows, recalling some details from [12].

In what follows, we briefly discuss the P systems which are local-loop-free (LL-free for short) in the sense of the previous definition: no cycle exists in any local transition graph.

We start by considering a simple P system, illustrated in Fig. 23.8.2, which generates the Fibonacci sequence:

$$N(\Pi) = \{1, 2, 3, 5, 8, 13, \dots\}.$$

This system can also be represented in a more intuitive way (as far as the reactions taking place in regions, and communicated objects are concerned) as a tissue-like P system with immediate communication—see Fig. 23.8.3. The labels of the edges/arrows indicate the objects which are communicated between the respective nodes/cells.

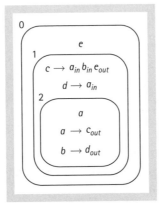

Fig. 23.8.2 A P system computing Fibonacci numbers.

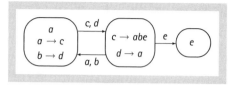

Fig. 23.8.3 The P system in Fig. 23.8.2 as a tissue-like system.

For both cell-like and tissue-like systems, the result is the number of copies of a special object e present within a designated "output" membrane which has no other role in the system.

Many problems can now be formulated—we state a few examples here:

- Prove universality for LL-free P systems.
- Are LL-free systems with only two membranes for computing and one additional membrane for collecting the result of a computation (hence of the form of the systems illustrated in Figs 23.8.2 and 23.8.3) universal?
- Is the framework of LL-free systems suitable for considering complexity issues?
- What about sorting, ranking, or other computer science applications of P systems (as those discussed in Chapter 22) based on LL-free P systems?

We recall now some results from [12] concerning the first of the above problems.

We denote by $tNOP_m^{llf}(coo)$ the family of sets of numbers $N(\Pi)$ generated by LL-free tissue-like P systems with cooperative rules having at most $m \geq 1$ membranes, with immediate communication and with the result collected in a special output membrane which has only this role (no objects evolve in this membrane, it has no evolution rule inside).

In this framework, it is easy to prove the following result:

Theorem 23.8.3 $tNOP_m^{llf}(coo) = NRE$ for all $m \geq 4$.

A similar result, for conformon-P systems, is given in [11].

23.8.3 FURTHER RESEARCH DIRECTIONS

The handbook presents most of the well developed research directions in membrane computing. Still, there are a number of research topics/issues which were not discussed. Here is a list of some of them:

1. The relationships between membrane computing and λ calculus, as established in [9], [14] (among others) using specific operations on membranes and communication rules of *broadcasting* type, e.g. objects are sent to all lower level membranes.

2. Logical approaches, e.g. working with incompletely specified P systems, inferring the behavior of a system from an incomplete specification of it—see [7], [18].
3. "Computing beyond Turing" was addressed not only through the use of acceleration, as discussed in Chapter 11 and Section 23.2, but also by investigating *non-uniform lineages of P systems*, in the sense of [25].
4. Solving various mathematical problems which are not necessarily over-polynomial, but are known to be hard, such as, e.g. classifying the states of a finite Markov chain ([6]) or computing the permanent of a binary matrix ([1]).
5. Inferring membrane structures, evolution rules, or even complete P systems, using various learning techniques. This direction of research was explored in, e.g. [23] and [24]—see also the website of membrane computing [26].
6. Considering P systems based on "non-crisp" mathematics, especially fuzzy P systems; a workshop was devoted to this topic—see [22].
7. Process algebra models of membranes modeled by circular strings—see, e.g. [2].
8. Several new ideas were launched during the ninth edition of the Workshop on Membrane Computing; we only mention some titles of papers from the workshop pre-proceedings volume, [10], which is also accessible through [26]: Dual P systems, Fast synchronization in P systems, On testing P systems, Event-driven metamorphoses of P systems, How redundant is your universal computation device?, Applications of page ranking in P systems, Towards a wet implementation of τ-DPP, Defining and executing P systems with structured data in K.
9. Applications to linguistics, at various levels: modeling sociolinguistic issues ([3]), conversation, natural language semantics ([4]), and parsing with P automata ([13]).
10. General links to other cell processes or cell-oriented modeling approaches, as proposed and preliminarily explored in [16] and [15].
11. General investigations of multisets, especially in computer science inspired frameworks (multiset automata, operations with multisets and AFL-like issues formulated in this context, fuzzy multisets, etc.). We mention here [5] as a basic source of information; more references can be found at the website [26].

References

[1] A. ALHAZOV, L. BURTSEVA, S. COJOCARU, Y. ROGOZHIN: Computing solutions of NP-complete problems by P systems with active membranes. In [10], 59–69.
[2] R. BARBUTI, A. MAGGIOLO-SCHETTINI, P. MILAZZO, A. TROINA: The calculus of looping sequences for modeling biological membranes. *Lecture Notes in Computer Sci.*, 4860 (2007), 54–76.

[3] G. BEL-ENGUIX: Unstable P systems. Applications to linguistics. *Lecture Notes in Computer Sci.*, 3365 (2004), 190–209.

[4] G. BEL-ENGUIX, M.D. JIMÉNEZ-LÓPEZ: Linguistic membrane systems and applications. In [8], 347–388.

[5] C. CALUDE, GH. PĂUN, G. ROZENBERG, A. SALOMAA, eds.: *Multiset Processing. Mathematical, Computer Science, and Molecular Computing Points of View*. LNCS 2235, Springer, 2001.

[6] M. CARDONA, M.A. COLOMER, M.J. PÉREZ-JIMÉNEZ, A. ZARAGOZA: Classifying states of a finite Markov chain with membrane computing. *Lecture Notes in Computer Sci.*, 4361 (2006), 266–278.

[7] M. CAVALIERE, R. MARDARE: Playing with partial knowledge in membrane systems: A logical approach. *Lecture Notes in Computer Sci.*, 4361 (2006), 279–297.

[8] G. CIOBANU, GH. PĂUN, M.J. PÉREZ-JIMÉNEZ, eds.: *Applications of Membrane Computing*. Springer, Berlin, 2006.

[9] L. COLSON, N. JONOSKA, M. MARGENSTERN: λP systems and typed λ-calculus. *Lecture Notes in Computer Sci.*, 3365 (2004), 1–18.

[10] P. FRISCO, D.W. CORNE, GH. PĂUN, eds.: *Proceedings of the 9th Workshop on Membrane Computing*, Edinburgh, UK, 2008.

[11] P. FRISCO, GH. PĂUN: No cycles in compartments. Starting from conformon-P systems. *Proc. Sixth. Brainstorming Week on Membrane Computing* (D. Diaz-Pernil et al., eds.), Fenix Editora, Sevilla, 2008, 157–170.

[12] R. GERSHONI, E. KEINAN, GH. PĂUN, R. PIRAN, T. RATNER, S. SHOSHANI: Research topics arising from the (planned) P systems implementation experiment in Technion. *Proc. Sixth. Brainstorming Week on Membrane Computing* (D. Diaz-Pernil et al., eds.), Fenix Editora, Sevilla, 2008, 183–192.

[13] R. GRAMATOVICI, G. BEL-ENGUIX: Parsing with P automata. In [8], 389–410.

[14] N. JONOSKA, M. MARGENSTERN: Tree operations in P systems and λ-calculus. *Fundamenta Informaticae*, 59 (2004), 67–90.

[15] S. MARCUS: Membranes versus DNA. *Fundamenta Informaticae*, 49 (2002), 223–227.

[16] S. MARCUS: Bridging P systems and genomics. A preliminary approach. *Lecture Notes in Computer Sci.*, 2597 (2003), 371–376.

[17] S. MARCUS: A typology of imprecision. In [22], 169–184.

[18] R. MARDARE, C. PRIAMI: Logical analysis of biological systems. *Fundamenta Informaticae*, 64 (2005), 275–289.

[19] Y. MATIJASEVITCH: *Hilbert's Tenth Problem*. MIT Press, Cambridge, London, 1993.

[20] A. OBTULOWICZ: Fuzzy P systems and fuzzy rule-based decision-making. In [22], 193–195.

[21] GH. PĂUN, R. PĂUN: Membrane computing and economics: Numerical P systems. *Fundamenta Informaticae*, 73 (2006), 213–227.

[22] F. ROSSELLÓ, ed.: *Brainstorming Workshop on Uncertainty in Membrane Computing*. Palma de Mallorca, November 2004.

[23] J.M. SEMPERE: P systems with external input and learning strategies. *Lecture Notes in Computer Sci.*, 2933 (2003), 341–356.

[24] J.M. SEMPERE, D. LOPEZ: Identifying P rules from membrane structures with an error-correcting approach. *Lecture Notes in Computer Sci.*, 4361 (2006), 507–520.
[25] P. SOSÍK, O. VALÍK: On evolutionary lineages of membrane systems. *Lecture Notes in Computer Sci.*, 3850 (2005), 67–78.
[26] The Membrane Computing Website: http://ppage.psystems.eu.
[27] Z. PAWLAK: Rough sets. *Intern. J. Computing and Information Sci.*, 11 (1982), 341–356.
[28] Z. PAWLAK: *Rough Sets. Theoretical Aspects of Reasoning About Data.* Kluwer, 1991.

Selective Bibliography
(Books, Collective Volumes, PhD Theses, Special Issues of Journals)

[1] A. ALHAZOV: *Communication in Membrane Systems with Symbol-Objects*. PhD Thesis, Rovira i Virgili University, Tarragona, Spain, Spain, 2006.

[2] A. ALHAZOV, C. MARTÍN-VIDE, GH. PĂUN, eds.: *Pre-proceedings of Workshop on Membrane Computing*. WMC 2003, Tarragona, Spain, July 2003, Technical Report 28/03, Rovira i Virgili University, Tarragona, 2003.

[3] F. ARROYO-MONTORO: *Estructuras y biolenguaje para simular computacion con membranas*. PhD Thesis, Polytechnical Univ. of Madrid, Spain, 2004.

[4] A.V. BARANDA GARCIA-SOTOCA: *Simulating P Systems on the Electronic Computer*. PhD Thesis, Polytechnical Univ. of Madrid, Spain, 2003.

[5] F. BERNARDINI: *Membrane Systems for Molecular Computing and Biological Modelling*. PhD Thesis, Univ. of Sheffield, UK, 2006.

[6] D. BESOZZI: *Computational and Modelling Power of P Systems*. PhD Thesis, Univ. degli Studi di Milano, Italy, 2004.

[7] L. BIANCO: *Biological Modelling Using P Systems: the MP Approach*. PhD Thesis, Verona University, Italy, 2007.

[8] R. BRIJDER: *Models of Natural Computation: Gene Assembly and Membrane Systems*. PhD Thesis, Leiden Univ., The Netherlands, 2008.

[9] C. CALUDE, M.J. DINNEEN, GH. PĂUN, eds.: *Pre-proceedings of Workshop on Multiset Processing*. Curtea de Argeș, Romania, August 2000, TR 140, CDMTCS, Univ. Auckland, New Zealand, 2000.

[10] C.S. CALUDE, GH. PĂUN, G. ROZENBERG, A. SALOMAA, eds.: *Multiset Processing. Mathematical, Computer Science, Molecular Computing Points of View*. LNCS 2235, Springer, Berlin, 2001.

[11] M. CAVALIERE: *Evolution, Communication, Observation: From Biology to Membrane Computing and Back*. PhD Thesis, Univ. Sevilla, Spain, 2006.

[12] M. CAVALIERE, C. MARTÍN-VIDE, GH. PĂUN, eds.: *Proceedings of the Brainstorming Week on Membrane Computing; Tarragona, February 2003*. Technical Report 26/03, Rovira i Virgili University, Tarragona, 2003.

[13] L. CIENCIALA: *P Automata*. PhD Thesis, Univ. Opava, Czech Republic, 2005.

[14] G. CIOBANU, GH. PĂUN, eds.: *Pre-proceedings of Workshop on Theory and Applications of P Systems, TAPS'05*. Timişoara, Romania, September 26–27, 2005.

[15] G. CIOBANU, GH. PĂUN, M.J. PÉREZ-JIMÉNEZ, eds.: *Applications of Membrane Computing*. Springer, Berlin, 2006.

[16] D. CORNE, P. FRISCO, GH. PĂUN, eds.: *Pre-proceedings of Ninth Workshop on Membrane Computing, WMC9.* Edinburgh, UK, July 2008.

[17] C. DÍAZ-PERNIL, C. GRACIANI, M.A. GUTIÉRREZ-NARANJO, GH. PĂUN, I. PÉREZ-HURTADO, A. RISCOS-NÚÑEZ, eds.: *Proceedings of the Sixth Brainstorming Week on Membrane Computing.* Sevilla, 2008, Fenix Editora, Sevilla, 2008.

[18] G. ELEFTERAKIS, P. KEFALAS, GH. PĂUN, eds.: *Proceedings of Eighth Workshop on Membrane Computing.* Thessaloniki, Greece, June 2007.

[19] G. ELEFTHERAKIS, P. KEFALAS, GH. PĂUN, G. ROZENBERG, A. SALOMAA, eds.: *Membrane Computing, International Workshop, WMC8, Thessaloniki, Greece, 2007, Selected and Invited Papers.* LNCS 4860, Springer, Berlin, 2007.

[20] R. FREUND, G. LOJKA, M. OSWALD, GH. PĂUN, eds.: *Proceedings of Sixth International Workshop on Membrane Computing, WMC6.* Vienna, Austria, July 18–21, 2005.

[21] R. FREUND, GH. PĂUN, G. ROZENBERG, A. SALOMAA, eds.: *Membrane Computing, International Workshop, WMC6, Vienna, Austria, 2005, Selected and Invited Papers.* LNCS 3850, Springer, Berlin, 2006.

[22] P. FRISCO: *Theory of Molecular Computing. Splicing and Membrane Systems.* PhD Thesis, Leiden University, The Netherlands, 2004.

[23] P. FRISCO: *Computing with Cells. Advances in Membrane Computing.* Oxford Univ. Press, 2009.

[24] M. GHEORGHE, ed.: *Membrane Computing.* Special issue of *Romanian Journal of Information Science and Technology*, vol. 11, nr. 3 (2008).

[25] M. GHEORGHE, N. KRASNOGOR, M. CAMARA, eds.: *P Systems Applications to Systems Biology.* Special issue of *BioSystems*, vol. 91, nr. 3 (2008).

[26] C. GRACIANI, GH. PĂUN, A. ROMERO-JIMÉNEZ, F. SANCHO-CAPARRINI, eds.: *Proceedings of the Fourth Brainstorming Week on Membrane Computing.* Sevilla, 2006, vol. II, Fenix Editora, Sevilla, 2006.

[27] M.A. GUTIÉRREZ-NARANJO, GH. PĂUN, M.J. PÉREZ-JIMÉNEZ, eds.: *Cellular Computing. Complexity Aspects.* Fenix Editora, Sevilla, 2005.

[28] M.A. GUTIÉRREZ-NARANJO, GH. PĂUN, A. RISCOS-NÚÑEZ, F.J. ROMERO-CAMPERO, eds.: *Proceedings of the Fourth Brainstorming Week on Membrane Computing.* Sevilla, 2006, vol. I, Fenix Editora, Sevilla, 2006.

[29] M.A. GUTIÉRREZ-NARANJO, GH. PĂUN, A. ROMERO-JIMÉNEZ, A. RISCOS-NÚÑEZ, eds.: *Proceedings of the Fifth Brainstorming Week on Membrane Computing.* Sevilla, 2007, Fenix Editora, Sevilla, 2007.

[30] S. HEMALATHA: *A Study on Rewriting P Systems, Splicing Grammar Systems and Picture Array Languages.* PhD Thesis, Anna Univ., Chennai, India, 2007.

[31] H.J. HOOGEBOOM, GH. PĂUN, G. ROZENBERG, eds.: *Workshop on Membrane Computing, WMC7.* Leiden, The Netherlands, July 17–21, 2006.

[32] H.J. HOOGEBOOM, GH. PĂUN, G. ROZENBERG, A. SALOMAA, eds.: *Membrane Computing, International Workshop, WMC7, Leiden, The Netherlands, 2006, Selected and Invited Papers.* LNCS 4361, Springer, Berlin, 2007.

[33] HUANG LIANG: *Research on Membrane Computing. Optimization Methods.* PhD Thesis, Institute of Advanced Process Control, Zhejiang University, China, 2007.

[34] A.-M. IONESCU: *Membrane Computing. Traces, Neural Inspired Models, Controls.* PhD Thesis, URV Tarragona, Spain, 2008.

[35] M. IONESCU, GH. PĂUN, T. YOKOMORI, eds.: *Spiking Neural P Systems.* Special issue of *Natural Computing*, vol. 7, nr. 4 (2008).

[36] T.-O. ISHDORJ: *Membrane Computing, Neural Inspirations, Gene Assembly in Ciliates*. PhD Thesis, Sevilla Univ., Spain, 2007.

[37] N. JONOSKA, GH. PĂUN, eds.: *Membrane Computing*. Special issue of *New Generation Computing*, vol. 22, nr. 4 (2004).

[38] J. KELEMEN, GH. PĂUN, eds.: *Membrane Computing*. Special issue of *Computers and Informatics*, vol. 27 (2008).

[39] S.N. KRISHNA: *Languages of P Systems. Computability and Complexity*. PhD Thesis, Univ. Madras, India, 2001.

[40] L. LAKSHMANAN: *On the Crossroads of P Systems and Contextual Grammars: Variants, Computability, Complexity and Efficiency*. PhD Thesis, Dept. of Mathematics, Indian Institute of Technology, Madras, India, 2003.

[41] C. MARTÍN-VIDE, G. MAURI, GH. PĂUN, G. ROZENBERG, A. SALOMAA, eds.: *Membrane Computing, International Workshop, WMC 2003, Tarragona, July 2003, Selected Papers*. LNCS 2933, Springer, Berlin, 2004.

[42] C. MARTÍN-VIDE, GH. PĂUN, eds.: *Pre-proceedings of Workshop on Membrane Computing*. Curtea de Argeş, Romania, August 2001, TR 16/01, Univ. Rovira i Virgili, Tarragona, Spain, 2001.

[43] C. MARTÍN-VIDE, GH. PĂUN, eds.: *Membrane Computing*. Special issue of *Fundamenta Informaticae*, vol. 49, nr. 1–3 (2002).

[44] C. MARTÍN-VIDE, GH. PĂUN, eds.: *Membrane Computing*. Special issue of *Natural Computing*, vol. 2, nr. 3 (2003).

[45] G. MAURI, GH. PĂUN, M.J. PÉREZ-JIMÉNEZ, G. ROZENBERG, A. SALOMAA, eds.: *Membrane Computing, International Workshop, WMC5, Milano, Italy, 2004, Selected Papers*. LNCS 3365, Springer, Berlin, 2005.

[46] G. MAURI, GH. PĂUN, C. ZANDRON, eds.: *Pre-proceedings of Fifth Workshop on Membrane Computing, WMC5*. Milano, Italy, 2004.

[47] M. MUTYAM: *Studies of P Systems as a Model of Cellular Computing*. PhD Thesis, Indian Institute of Technology Madras, India, 2003.

[48] M. OSWALD: *P Automata*. PhD Thesis, Technical Univ., Vienna, Austria, 2003.

[49] L. PAN, GH. PĂUN, eds.: *Pre-proceedings of the International Conference Bio-Inspired Computing – Theory and Applications, BIC-TA 2006, Volume of Membrane Computing Section*. Wuhan, China, September 18–22, 2006.

[50] L. PAN, GH. PĂUN, eds.: *Membrane Computing at BIC-TA 2006*. Special issue of *Progress in Natural Sciences*, vol. 17, nr. 4 (2007).

[51] A. PĂUN: *DNA and Membrane Computing*. PhD Thesis, Univ. London-Ontario, Canada, 2003.

[52] GH. PĂUN, ed.: *Membrane Computing*. Special issue of *Romanian Journal of Information Science and Technology*, vol. 4, nr. 1–2 (2001).

[53] GH. PĂUN: *Membrane Computing. An Introduction*. Springer, Berlin, 2002.

[54] GH. PĂUN, M.J. PÉREZ-JIMÉNEZ, eds.: *Membrane Computing*. Special issue of *Journal of Universal Computer Science*, vol. 10, nr. 5 (2004).

[55] GH. PĂUN, M.J. PÉREZ-JIMÉNEZ, eds.: *Membrane Computing*. Special issue of *Soft Computing*, vol. 9, nr. 9 (2005).

[56] GH. PĂUN, M.J. PÉREZ-JIMÉNEZ, eds.: *Membrane Computing*. Special issue of *International Journal of Foundations of Computer Science*, vol. 17, nr. 1 (2006).

[57] GH. PĂUN, M.J. PÉREZ-JIMÉNEZ, eds.: *Membrane Computing*. Special issue of *Journal of Automata, Languages and Combinatorics*, vol. 11, nr. 3 (2006).

[58] GH. PĂUN, M.J. PÉREZ-JIMÉNEZ, eds.: *Membrane Computing*. Special issue of *Theoretical Computer Science*, 372, 2–3 (2007).

[59] GH. PĂUN, M.J. PÉREZ-JIMÉNEZ, eds.: *Membrane Computing*. Special issue of *Fundamenta Informaticae*, vol. 87, nr. 1 (2008).

[60] GH. PĂUN, A. RISCOS-NÚÑEZ, A. ROMERO-JIMÉNEZ, F. SANCHO-CAPARRINI, eds.: *Proceedings of the Second Brainstorming Week on Membrane Computing, Sevilla, February 2004*. Technical Report 01/04 of Research Group on Natural Computing, Sevilla University, Spain, 2004.

[61] GH. PĂUN, G. ROZENBERG, A. SALOMAA, C. ZANDRON, eds.: *Membrane Computing. International Workshop, WMC 2002, Curtea de Argeș, Romania, August 2002. Revised Papers*. LNCS 2597, Springer, Berlin, 2003.

[62] GH. PĂUN, C. ZANDRON, eds.: *Pre-proceedings of Workshop on Membrane Computing*. Curtea de Argeș, Romania, August 2002, MolCoNet Publication No 1, 2002.

[63] I. PÉREZ-HURTADO: *P-Lingua: A Programming Language for Membrane Computing*. PhD Thesis, Sevilla Univ., Spain, in preparation.

[64] M.J. PÉREZ-JIMÉNEZ, A. RISCOS-NÚÑEZ, eds.: *Modelos de computacíon molecular, celular y cuántica*. Fenix Editora, Sevilla, 2007.

[65] M.J. PÉREZ-JIMENEZ, A. ROMERO-JIMÉNEZ, F. SANCHO-CAPARRINI, *Teoría de la complejidad en modelos de computacíon celular con membranas*. Kronos, Sevilla, 2002.

[66] M.J. PÉREZ-JIMENEZ, A. ROMERO-JIMÉNEZ, F. SANCHO-CAPARINNI, eds.: *Recent Results in Natural Computing*. Fenix Editora, Sevilla, 2004.

[67] M.J. PÉREZ-JIMENEZ, F. SANCHO-CAPARRINI, *Computacíon celular con membranas: Un modelo no convencional*. Kronos, Sevilla, 2002.

[68] B. POPA: *Membrane Systems with Bounded Parallelism*. PhD Thesis, Louisiana Tech Univ., Ruston, LA, USA, 2006.

[69] A. RISCOS-NÚÑEZ: *Cellular Programming. Efficient Resolution of Numerical NP-complete Problems*. PhD Thesis, Univ. Sevilla, Spain, 2004.

[70] F.J. ROMERO-CAMPERO: *P Systems, a Computational Modelling Framework for Systems Biology*. PhD Thesis, Sevilla Univ., Spain, 2008.

[71] A. ROMERO-JIMÉNEZ: *Computational Complexity and P Systems*. PhD Thesis, Univ. of Sevilla, Spain, 2003.

[72] F. ROSSELLÓ, ed.: *Brainstorming Workshop on Uncertainty in Membrane Computing*. Palma de Mallorca, Spain, November 2004.

[73] F. SANCHO-CAPARRINI: *Verification of Programs in Unconventional Computing Models*. PhD Thesis, Sevilla Univ., Spain, 2002.

[74] D. SBURLAN: *Promoting and Inhibiting Contexts in Membrane Computing*. PhD Thesis, Univ. Sevilla, Spain, 2006.

[75] CL. ZANDRON: *Computing with Membranes. P Systems*. PhD Thesis, Univ. Milano, Italy, 2002.

Index of Notions

action potential 34, 38
activator arc 402
active P automaton 164
aerobic oxidation 43
allometric principle 488
alphabet 59
antibody 41
antigen 41
antiport 119
antiporter 39
apoptosis 469
asynchronous 252
asynchronous SN P system 345
ATP 43
   hydrolysis 43
ATPase 39
Avogadro principle 477
axon 34

barb-event 400
bead sort 556
behavioral self-assembly 607
bi-stable catalyst 595
big-step semantics 433
binary normal form 71
binding 463
bio-lab implementation 658
biofilm 54
bisimulation 432
bistable catalyst 101
Boolean circuit 578
boundary rules 137
boundedness 401
Brusselator 483

carrier 38
catalytic P system 83
   purely 84
causality semantics 399
cell-like P system 438
channel 38
channel state 229

chemotaxis 48
Chomsky grammar 62
Chomsky hierarchy 63
Chomsky–Schützenberger theorem 66
cilia 41
CLIPS 442
coding 61
colonies 584
compartment 393
complex formation 463
complexity class 79
computation 395
   free-, min-, max-, lmax- 395
conditional uniport 137
configuration 393
   initial 393
confluent P system 307, 442
conformon P system 253, 448
   conformon 252
   grid 277
      cell 277
   interaction rule 252
   module 257
      decreaser 262
      increaser 262
      separator 259
      splitter 258
      strict interaction 263
   operation 253
   simple 277
   total value 264
   with negative values 271
cooperating distributed grammar system 74
corresponding
   marking 398
   PTL-net 397
   step 398
coverability tree 401
cryptography 570
Cyto-Sim 446
cytoplasm 31
cytoskeleton 33

deadlock 187
debinding 463
dissociation 463
DNA 30
DNA self-assembly 606
Dyck language 66
dynamic allocation semantics 427

electrochemical potential 37
enabled step 396
 free-, min-, max-, lmax- 396
enabled vector multi-rule 394
 free-, min-, max-, lmax- 394
endocytosis 47, 645
 receptor-mediated 47
enhanced mobile membrane system 649
epidermal growth factor receptor 468
ET0L system 67
eukaryotic cell 30, 32
evolution rule 393
evolutionary algorithm 546
exocytosis 47, 645
extended H system 200
extended SN P system 344

Fibonacci sequence 659
finite automaton 67
flagellum 42
fluid mosaic model 36
flux regulation map 478
forbidding condition 178
forgetting rule 339
formal verification 465
Fredkin gate 580
free-computation 395
free-enabled
 step 396
 vector multi-rule 394
free-step sequence 396
function optimization problem 545

G protein 40
gap junction 52
Geffert normal forms 65
gemmation 240
gene expression 44
gene regulatory network 460
genetic algorithm 531
Gillespie algorithm 242, 446, 458

H system 77, 200
 axiom 200
 finite 200

language generated 200
time-varying distributed, see TVDH
 system 200
halting
 adult 102, 110
 partial 102, 107, 121
 signal 102, 110
 total 102
 unconditional 102, 110
 with states 102, 110
Head splicing system 200
hypercomputation 601

inhibitor 426
inhibitor arc 402
inhibitors 402
initial marking 396
insertion-deletion 77
invariant analysis 398
inverse morphism 61
ion channel 39
iterated splicing 199

Java 443
job-shop scheduling problem 538
JPLANT 449

k-determinism 100
Kleene closure 61
Kuroda normal form 65

L-uniformity 321
Lac operon 468
lambda-calculus 660
language 60
Lindenmayer system 66
linear set 60
Lisp 440
lmax-computation 395
lmax-enabled
 step 396
 vector multi-rule 394
lmax-step sequence 396
local search 531
locality 396
locality mapping 396
log-gain principle 486
loop-free P system 659
looping sequences 661
Lotka-Volterra model 459, 484

marking 396
Markov chain 661
mass partition principle 477

matrix grammar 70
Maude 421
max-computation 395
max-enabled
   step 396
   vector multi-rule 394
max-step sequence 396
Mealy membrane automaton 164
Mealy multiset automaton 164
mechanical self-assembly 606
membrane 35, 393
   dissolving 404
   thickening 404
membrane algorithm 447, 530
membrane algorithm
   Brownian subalgorithm 533, 540
   GA subalgorithm 533, 540, 543
   global communication 547
   local search subalgorithm 543
   membrane structure
      star topology 545
   subalgorithm 532
membrane creation 326
membrane dissolution 171, 420
membrane division 310, 650
membrane protein 35
   integral 36
   peripheral 36
membrane structure 393
membrane system 393
   basic 393
   reachability graph 400
   with promoters and inhibitors 402
membrane thickness 170
metabolic P system 445, 478
metabolism 43
   anabolism 43
   catabolism 43
MetaPlab 481
microvilli 41
Milano theorem 311
min storage problem 542
min-computation 395
min-enabled
   step 396
   vector multi-rule 394
min-step sequence 396
minimal communication 132
minimal parallelism 104, 323
mitotic cycle 485
mobile catalyst 101
model checking 465
morphism 61

MP grammar 479
MP graph 479
multiset 3, 392
MzScheme 440

natural semantics 433
Needham-Schroeder protocol 571
neural P system 235
neuron 33
non photochemical quenching 491
normal form 180, 343
NP-complete problem 80
numerical P system 654

occurrence net 399
operational semantics 415
operon 45
ordered grammar 73

P automaton 146
   ω-P automaton 162
   analyzing P system 145, 156
   evolution communication P automaton 163
   extended P automaton 155
   one-way P automaton 145, 148
   over infinite alphabets 159
   sequential mode 146
   with conditional rules 163
   with marked membranes 163
   with membrane channels 163
   with priorities 162
   with states 163
P colony 586
   of linguistic P systems 592
   with prescribed teams 591
P system
   energy-based 621
   evolution-communication 137
   generalized communicating 245
   polarizationless 316
   with active membranes 310
   with input 305
P transducer 164
P-lingua 449
Păun's conjecture 317, 324
parallel architecture 572
parallel communicating grammar system 74
parallel implementation 443
parallel rewriting 184
Parikh image 60
Parikh theorem 66
Penttonen normal form 65
permitting condition 178

Petri net 396
  activator arc 402
  boundedness 401
  causality semantics 399
  coverability tree 401
  inhibitor arc 402
  initial marking 396
  invariant analysis 398
  marking 396
  occurrence net 399
  place 396
  PTL-net 396
  PTRL-net 402
  range arc 402
  reachability graph 400
  reachability problem 401
  step 396
  transition 396
  weight function 396
  with localities 396
  with range arcs 402
phagocytosis 47
photosynthesis 43
pinocytosis 47
place 396
plasma membrane 31
polarity 51
population P system 241
Population P system description language (PPSDL) 616
post-translational modification 44
pre-computed resources 359
predator-prey model 459, 484
priority 171
PRISM 465
probabilistic P system 457
programmed grammar 72
prokaryotic cell 30, 31
Prolog 439, 442
promoter 426
promoters 402
protein 30
Psim 445, 481
PSPACE 315
PTL-net 396
PTRL-net 402

quantum computing 621
quorum sensing 53, 242, 447, 469

random context grammar 73
range arc 402
ranking 560

reachability graph 400
reachability problem 401
receptor 40
recognizer P system 305
recruitment 463
rection map 483
register machine
  restricted 263
    instruction 264
register machine 69
regular expression 61
releasing 463
rewriting P system 170
ribosome 44
RNA 30
  polymerase 44
rotate-and-simulate 205

safe ambient 646
SAT problem 311, 323, 327, 357, 629
second messenger 46
self-assembly P systems 606
self-assembly universality
    lemma 606
semi-uniform solution 308, 359
semilinear set 60
sequential SN P system 348
Sevilla carpet 442
Sevilla theorem 309
shuffle-exchange network 574
signal transduction 46
SIMD model 573
simple P machine 164
small-step semantics 433
sorting by carving 559
spike 339
spiking neural P system/SN P system 339, 448
spiking rule 339
splicing 44, 198
splicing operation 76
splicing P system 201
  configuration 202
    initial 202
  definition
    variants 202
  generating 201
  immediate communication 217
  non-extended 202, 218
  one compartment
    decidability 213
    universality 214
  transition 202
  type 203

splicing P system *(cont.)*
   universality 206
splicing rule 199
   diameter 199
splicing tissue P system 204
   one-way communication 215
   simple 204
   universal
      small 218
1-splicing 199
2-splicing 199
static allocation semantics 430
static sorting 554
step 396
step sequence 396
   free-, min-, max-, lmax- 396
stochastic P system 458
stoichiometric matrix 478
structural operational semantics 433
subset-sum problem 358
symport 119
symport/antiport 228
symporter 39
synapse 34

T cell 41
tag system 218
τ-DPP paradigm 458
tau-leaping algorithm 458
tiling 566
tiling P system 569
time-free P system 596
timed P system 596
tissue 49
tissue P system 127, 229
   with terminal communication 238
   with receptors 237
tissue simulator 448

tissue-like P system 438
traces 139
transcription 44, 464
transition 396
transition graph 659
transition mode
   k-restricted minimally parallel 107
   asynchronous 103, 121
   maximally parallel 120
   minimally parallel 120
   sequential 103, 121
translation 44
transporter 38
   ABC (ATP-Binding Cassette), 40
traveling salesman problem 530, 532
   optimum solution 533
Turing machine 68
turtle graphics 561
TVDH system 200
   component 200

uniform solution 308
uniporter 39

vector multi-rule 394
Vibrio Fischeri 446

Wang tile system 608
weight function 396
workspace theorem 65

X-machine 612
   communicating 613
   stream 612
X-machine description language (XMDL) 616

Z-binary normal form 72